NOLAMP- Nordic Laser Materials Processing Conference (19TH-NOLAMP-2023)

IOP Conference Series: Materials Science and Engineering
Volume 1296

Turku, Finland
22 - 24 August 2023

Editors:

Antti Salminen
Ashish Ganvir
Heidi Piili
Mohsen Amraei
Andrey Mityakov

ISBN: 978-1-7138-8733-1
ISSN: 1757-8981

Printed from e-media with permission by:

Curran Associates, Inc.
57 Morehouse Lane
Red Hook, NY 12571

Some format issues inherent in the e-media version may also appear in this print version.

Institute of Physics
Dirac House, Temple Back
Bristol BS1 6BE UK

Phone: 44 1 17 929 7481
Fax: 44 1 17 920 0979

techtracking@iop.org

Additional copies of this publication are available from:

Curran Associates, Inc.
57 Morehouse Lane
Red Hook, NY 12571 USA
Phone: 845-758-0400
Fax: 845-758-2633
Email: curran@proceedings.com
Web: www.proceedings.com

TABLE OF CONTENTS

IOP Conf. Series: Materials Science and Engineering 1296 (2023) 011001 doi:10.1088/1757-899X/1296/1/011001

About the conference:

Laser material processing is growing in its importance of manufacturing. In Nordic countries the tradition is carried out with NOLAMP Conference.

The first conference on **No**rdic **La**ser **M**aterial **P**rocessing, NOLAMP was held in Oslo in 1987. Then the laser material processing was a rather new and revolutionary way to manufacture products with high quality and high efficiency. The laser research in the Nordic countries was in its pioneering age and the enthusiasm among researchers for the new possibilities was unrivalled. The intention for creating this conference, which was originally suggested by Dr. Bernt Thorstensen at SI in Oslo, Norway, was to establish a fruitful cooperation among Nordic laser researchers. Also, NOLAMP intended to provide a forum for young researchers to present on-going works and to establish contacts in the scientific community as well as in industry.

Ever since then, the conference has been reaching its goals and acted as discussion forum between academia and industry in the field. Currently the conference gets growing attention all over Europe also. The conference is now organized 5th time in Finland and this time it is organized in Turku at the south-west coast of Finland. The conference agenda includes various topics in field of laser materials processing and as current trend is, there is quite a lot of papers on the newcomer of laser processing the additive manufacturing.

Conference date and location

August 22nd - 24th 2023, Turku, Finland

Conference topics:

Laser materials processing on:

- Additive manufacturing
- Welding
- Surface treatment
- Laser systems and monitoring
- Artificial intelligence
- Other processing

Keynote speakers:

- **Eduard Hryha**, Chalmers University of Technology, Sweden: *"Powder for Metal Additive Manufacturing: Manufacturing, Properties, Reuse and Degradation"*
- **Klaus Löffler**, Precitec GmbH & Co. KG, Germany: *"Will the European Industry suffer from Chinas business model for Lasers and what might be a way out?"*
- **Teemu Tolonen**, Pemamek Ltd., Finland: *"Utilization of Laser-hybrid Process in Heavy Welding Automation -Application examples for different industries"*

- **Niklas Strömberg**, Örebro University, Sweden: *"Topology Optimization, TMPS-based Lattice Structures, and Implicit-based Geometry for 3D Printing"*
- **Olli Nyrhilä**, EOS Metal Materials, Finland: *"Metal Materials in Additive Manufacturing"*
- **Chunlin Xu**, Åbo Akademi University, Finland: *"3D printing in extending the value of wood biopolymers"*
- **Edward W. (Ted) Reutzel**, Pennsylvania State University, USA: *"Progress Towards Sensing and Mitigating Flaws in Additive Manufacturing"* (online lecture)
- **Dhruv Bhatt**, Valmet Automotive Oy, Finland: *"Laser Material Processing in the EV industry"*
- **Mikko Vänskä**, Meyer Turku Oy, Finland: **"Why Laser Hybrid Welding for Shipbuilding"**

Sponsors:

- Gold Sponsor: Linde Gas
- Silver Sponsor: Hamamatsu Photonics
- Bronze Sponsors: Cavitar ltd, EOS, ADLOptica
- Sponsors: Photonics Finland, PREIN Flagship, Finnish Welding Society, 4D photonics, Machine technology centre Turku ltd, Turku University of Applied sciences

Supporters:

- City of Turku
- Visit Turku

Scientific Committee members:

- Klas Boevie, Sintef, Norway
- Ivan Bunaziv, Sintef, Norway
- Inigo Flores, Tampere University, Finland
- Ashish Ganvir, University of Turku, Finland
- Eduard Hryha, Chalmers University of Technology, Sweden
- Vahid Javaheri, University of Oulu, Finland
- Paul Kah, University West, Sweden
- Alexander Kaplan, Lulea University of Technology, Sweden
- Morten Kristiansen, Aalborg University, Denmark
- Ville Laitinen, LUT university, Finland
- Sini Metsä-Kortelainen, VTT, Finland
- Olli Nyrhilä, EOS Metal Materials, Finland
- Jouni Partanen, Aalto University, Finland
- Heidi Piili, University of Turku, Finland

IOP Conf. Series: Materials Science and Engineering 1296 (2023) 011001 doi:10.1088/1757-899X/1296/1/011001

- Xianbo Ren, Sintef, Norway
- Antti Salminen, University of Turku, Finland
- Mikko Vänskä, Meyer Finland Oy, Finland

Editors:

- Main editor: Antti Salminen, University of Turku
- Ashish Ganvir, University of Turku
- Heidi Piili, University of Turku
- Mohsen Amraei, University of Turku
- Andrey Mityakov, University of Turku

Reviewers:

- Mohsen Amraei, University of Turku
- Ivan Bunaziv, SINTEFF
- Inigo Flores, Tampere University
- Ashish Ganvir, University of Turku
- Eduard Hryha, Chalmers University of Technology
- Vahid Javaheri, University of Oulu
- Alexander Kaplan, Lulea University of Technology
- Veli Kujanpää, LUT university
- Ville Laitinen, LUT university
- Nikhil Kamboj, University of Turku
- Sini Metsä-Kortelainen, VTT
- Andrey Mityakov, University of Turku
- Chinmayee Nayak, University of Turku
- Patricia Nyamekye, LUT university
- Jouni Partanen, Aalto University
- Joonas Pekkarinen, Turku University of Applied Sciences
- Heidi Piili, University of Turku
- Asim Rashid, University of Turku
- Kandice S. B. Ribeiro, University of Turku
- Antti Salminen, University of Turku
- Jari Tuominen, Tampere University
- Tero Tuovinen, Jyväskylä University of Applied Sciences
- Anna Unt, LUT university
- Joerg Volpp, Lulea University of Technology
- Mikko Vänskä, Meyer Finland Oy

National Organizing Committee

- Antti Salminen, Professor, Faculty of Technology, University of Turku
- Heidi Piili, Adjunct Professor, Faculty of Technology, University of Turku
- Mohsen Amraei, University Teacher, Faculty of Technology, University of Turku
- Kandice Barros Ribeiro, Postdoctoral Researcher, Faculty of Technology, University of Turku
- Ashish Ganvir, Assistant Professor, Faculty of Technology, University of Turku
- Anna Huusko, Project Coordinator, Faculty of Technology, University of Turku
- Nikhil Kamboj, Postdoctoral Researcher, Faculty of Technology, University of Turku
- Andrey Mityakov, Docent, University Lecturer, Faculty of Technology, University of Turku
- Katri Motuste, Financial Secretary, Faculty of Technology, University of Turku
- Chinmayee Nayak, Postdoctoral Researcher, Faculty of Technology, University of Turku
- Asim Rashid, University Lecturer, Faculty of Technology, University of Turku

Antti Salminen, Conference host

Ashish Ganvir, Conference Vice Host

Heidi Piili, Chair of National Organising Committee

Mohsen Amaraei, Vice Chair of National Organising Committee

IOP Conf. Series: Materials Science and Engineering 1296 (2023) 011002 doi:10.1088/1757-899X/1296/1/011002

Peer Review Statement

All papers published in this volume have been reviewed through processes administered by the Editors. Reviews were conducted by expert referees to the professional and scientific standards expected of a proceedings journal published by IOP Publishing.

- **Type of peer review:** Double Anonymous
- **Conference submission management system:** Morressier
- **Number of submissions received:** 57
- **Number of submissions sent for review:** 47
- **Number of submissions accepted:** 46
- **Acceptance Rate (Submissions Accepted / Submissions Received × 100):** 80.7
- **Average number of reviews per paper:** 2
- **Total number of reviewers involved:** 25
- **Contact person for queries:**
  Name: Antti Salminen
  Email: antti.salminen@utu.fi
  Affiliation: University of Turku

Enhancing industry 5.0 goals through laser based additively Manufactured High-Performance Metals

P Nyamekye, S Westman and V Tepponen

patricia.nyamekye@lut.fi, vesa.tepponen@lut.fi, sami.westman@lut.fi

Corresponding author's patricia.nyamekye@lut.fi

Abstract. In the current industry 4.0 (I4.0) and emerging Industry 5.0 (I5.0) eras, the integration of digitalization and sustainable goals is driving nature-centric and human-centric products. The production of these products are also creating digital supply chain that boosts agility and resiliency in operations, management, and supply chain interlinks. Additive manufacturing (AM), an element of I4.0, for instance, offers an integrated connection of all process steps and value chains using computer-based designing, data-driven simulations, cloud-based processing, storing, and managing software along seamless digital threads to create agile and resilient supply chain. The integration of AM and other I4.0 technologies enables greater flexibility offering intrinsic sustainable, human-centric and resilience advantages. Laser based additive manufacturing (LBAM), one of the subcategories of AM, offers opportunities to manufacture new intricate and conventionally impractical metal product designs in an ecological and economic competitive stance. High performance metals (HPMs) suited for high-stress and corrosive demanding applications are tough to machine and prone to thermal cracking in case of welding. Conventionally difficult and lengthy to manufacture yet unavoidable HPMs such as nickel alloys and titanium alloys are effortlessly possible via LBAM. Different industrial sectors that utilize these grades of metal alloys continue to adopt AM for the offered design flexibility for achieving goals such as customization, lightweight, on-demand manufacturing, raw material efficiency and cost saving. This study uses literature review and manufacturing case studies to demonstrate the flexibility and digital nature of LBAM towards I5.0 goals. The study objectively highlights the promising responsiveness of AM in the eventuality of supply disruptions that may be caused by sudden changes. The novelty of the study lies in the pragmatic emphasis on the potentials of LBAM and paired I4.0 technologies in revolutionizing the industry towards industry 5.0 goals. The study shows how I4.0 elements can be paired to enable operational efficiencies, lower carbon emissions, and foster sustainability in promoting I5.0 transformation. This study offers a fundamental understanding of the role of LBAM in the advancement of sustainability, human-centricity, and resilience.

Keywords: Additive manufacturing, agility, high performance metals, industry 5.0, laser-based AM, sustainability, human-centric, resilient.

1. Introduction

In a span of two decades; 2001-2021 [1–3], industry 4.0 (I4.0) and industry 5.0 (I5.0) have emerged within the EU giving humans the capability to unlimited creativity, decentralization, customization coupled with high degree of flexibility, control, and ease of work within virtual and physical environs.

Laser based manufacturing leverages precise, resource-efficient, less waste, clean energy, and negligible emission production routes [4–6]. The digital nature of additive manufacturing (AM) facilitates coupling with other I4.0 technologies to overcome production complexities [7], and supports digital inventory, decentralized production, mass customization. These helps maintain agility, improve resource efficiency, reduce waste and emissions, being exemplary decisive indicators for sustainability and value creation [8]. This study presents a state-of-the-art review of the prospects and contributions of emerging technologies (mainly lasers and AM) to accelerate I5.0. In affirming the contributions of lasers and AM to I5.0 case examples of additively manufactured high-performance metals (HPMs) are presented. HPMs are essential for the manufacturing industry, particularly for high-stress applications, where standard metals are not suitable. HPMs such as titanium alloys, nickel alloys, and aluminum alloys have been widely used in critical engine and structural transport applications for their excellent properties, including high strength-to-weight ratio, corrosion resistance, and thermal stability [9–11]. However, the conventional manufacturing (CM) of these metals involves lengthy, expensive, and complicated processes [8,12–14] which may limit their adoption. The suitability of AM for HPMs has supported the adoption for traditional and new materials [6,13,15]. AM has the potential to overcome manufacturing limitations and unlock new possibilities for scenarios requiring flexibility, small-lot production, mass-customization, sustainable and functional end-use parts for various industrial sectors [15,16]. Among the categories of AM, laser-based additive manufacturing (LBAM) has gained significant attention owing to its ability to produce high precision and complex metal components. LBAM of HPM is becoming increasingly popular in the aerospace, automotive, energy and biomedical industries. The unique properties of AM specific HPMs feedstock enable highly efficient, accurate, and reliable functional products not amenable to CM counterparts. AM can be used solely or in a hybrid manufacturing route to offset CM constraints [17]. Manufacturers can leverage the unique capabilities of LBAM through paired I4.0 technologies to produce parts with intricate designs and optimized shapes that were traditionally impossible or too expensive to produce. The use of LBAM for HPMs through complementary I4.0 technologies are propelling to achieving I5.0 goals. I4.0 refers to the integration of digital technologies in manufacturing processes, while I5.0 represents the integration of human and machine intelligence to create more personalized products and services [18].

1.1 Research questions, objectives and expected results

In advancing I5.0 goals, the exploration of the integration of AM with other complementary I4.0 technologies is being investigated across multiple industries [8,19,20]. This study aims to examine how AM prospectively promote I5.0 in different industries. The anticipated outcome of this study is to identify pragmatic means to create smarter AM production. A seamless integration of human intelligence and I4.0 capabilities to develop microsystems that incorporate sensory and assistive technologies will be proposed. These microsystems are expected to facilitate an autonomous production of highly customized products and services in support of I5.0 goals. The underlying research questions, objectives, and expected results are as follows:

Research Questions:

- What are the industrial benefits and challenges of AM especially LBAM?
- What added values are attainable with LBAM in industries?
- How does AM enhance industry 5.0 achievement?

Objectives:

- To identify the opportunities and limitations of industrial AM.
- To examine the potential of using LBAM to create value with HPMs.
- To explore the use of AM with complementary I4.0 technologies towards I5.0.

Expected Results:

- Identification of the key benefits of using LBAM, including improved product functionality, quality, reduced material waste, and faster production times.
- Identification of the value trends with the use of LBAM for HPMs production.
- Recommendations for successful integration of human intelligence and I 4.0 capabilities

towards the creation of self-configurable, and interoperable smart systems capable of boosting self-adjustable parameters, self-optimizable designs, and manufacturing flexibilities.

1.2 Methodology

This article explores I5.0 goals in relation to I4.0 technologies, particularly AM and discusses the characteristics and applications using existing laser based additively manufactured HPMs cases. This study is carried out as a literature review where databases including Scopus, Google Scholar (covers wide and diverse scholarly literature for specific research studies) and the internet were scrutinized to locate reviewed publications and relatable manufacturing cases. Most relatable existing articles and cases used in this study were identified using the search string combinations including “Industry 5,0” “Industry 4.0”, “Additive manufacturing” “3D printing” “High performance metal” and all related key words. Once suitable articles were identified, the cited references were reviewed to assist in locating additional articles, resulting in “snowball sampling” effect. The search string additive manufacturing car engine part high performance metal; additive manufacturing medical surgical implants and devices; Application of laser based directed energy deposition” were used to find suitable cases from the internet. A stepwise generalization of the study, route to finding and selecting inclusive references (also search hits) is illustrated in figure 1.

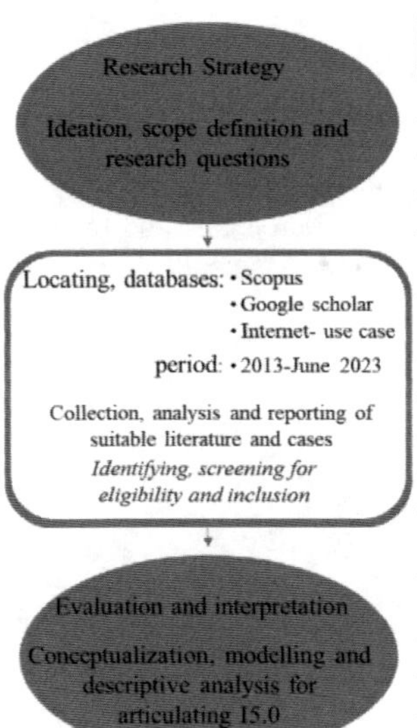

Search string (Boolean operators) (2013 - June 2023)	Records identified
Scopus	
“Industry 5.0” AND “additive manufacturing” OR “3D printing”	24
“Industry 5.0” AND “industry 4.0” AND “additive manufacturing” AND “3d printing”	2
“Industry 5.0” And “Laser” And “Additive Manufacturing” Or “3d Printing”	2
“Industry 5.0” AND “Additive manufacturing” OR “3D printing” AND “human intelligence” OR “artificial intelligence” OR “Cyber physical systems”	10
“Industry 5.0” AND “Additive manufacturing” OR “3D printing” AND “Laser” AND “metal” AND “human intelligence” OR “artificial intelligence” OR “Cyber physical systems”	-
“Additive manufacturing” OR “3D printing” AND “High performance metals”	55
Google scholar	
“Industry 5.0” AND “additive manufacturing” OR “3D printing” AND “metals”	64
“Industry 5.0” And “Laser” And “Additive Manufacturing” Or “3d Printing”	59
“Industry 5.0” AND “Additive manufacturing” OR “3D printing” AND “human intelligence” OR “artificial intelligence” OR “Cyber physical systems”	308
“Industry 5.0” AND “Additive manufacturing” OR “3D printing” AND “Laser” AND “metal” AND “human intelligence” OR “artificial intelligence” OR “Cyber physical systems”	45
“Additive manufacturing” OR “3D printing” AND “High performance metals”	28
Total identified records	*597*
Eligible for inclusion (also includes 6 industrial cases from internet and older references on operational resilient aspects and sustainability). ⇨	**104**

Figure 1: Diagram of overview, review process and literature review outcome in this study

The systematic literature review identified combined 597 scientific published reviewed articles indexed in the selected databases assuming no duplicates, which though may be far from the reality. Firstly, an initial screening of abstract containing preselected keywords and later full texts that expatiated the key themes including industry 5.0 goals, industry 4.0, their sub themes, laser based additive manufacturing, powder bed fusion, directed energy deposition and additive manufactured metals. Major research themes on co-human-artificial intelligence-CPS were identified, and closely evaluated and analyzed to develop a conceptual articulating model of industry 5.0 via LBAM of HPM.

2. Literature review

In LBAM, a high-power laser beam is used to melt and fuse successive layers of metal powder or wire to create a three-dimensional object. The laser beam is controlled by a computer program that follows a digital model of the object being manufactured. As each layer of metal is melted and fused, the platform holding the object is lowered, and the next layer is added on top until the object is complete. LBAM

IOP Conf. Series: Materials Science and Engineering 1296 (2023) 012001 doi:10.1088/1757-899X/1296/1/012001

allows for the precise creation of complex geometries that would be difficult or impossible to produce using conventional manufacturing (CM) methods. LBAM, especially powder bed fusion (PBF) and directed energy deposition (DED) are well-suited to produce HPM, such as titanium and nickel alloys. These metals have excellent strength-to-weight ratios and are widely used in aerospace, medical, and other industries that require materials with exceptional properties.

2.1. Background

Calls to enhance sustainability have propelled some industrial commitments to cooperate social responsibilities. The societal aspects of sustainability often remained unmeasurable in industrial sectors. Evolving into I5.0 gives a clear and concise element by which enabling technologies such as photonics and AM can be translated into realistic societal benefits [3]. Photonics is defined as *the science and technology of light* [21] of which lasers are core components products. AM is defined as a *process of joining materials to make parts from 3D model data, usually layer upon layer, as opposed to subtractive manufacturing and formative manufacturing methodologies* [22].
The European Union (EU) in defining core of I5.0 stated that *Industry 5.0 recognises the power of industry to achieve societal goals beyond jobs and growth to become a resilient provider of prosperity, by making production respect the boundaries of our planet and placing the wellbeing of the industry worker at the centre of the production process* [3]. I5.0 is identified as a concept fostering the provision of prosperity beyond jobs and growth through the capabilities of modern technologies while respecting the production limits of the planet. I5.0 is an emerging concept along I4.0 that prioritizes the well-being of human force, pivotal in industrial activities amidst the facilitating of value creation within constraints of natural resources.

The EU identifies three main pillars of I5.0, *sustainability, human-centricity, and resilience*. According to the EU, these necessary hallmarks are desirable features to remain relevant and maintain futuristic competitiveness for European industries. An illustration of the I5.0 key elements based on the definition of EU is summarized as Figure 2 shows.

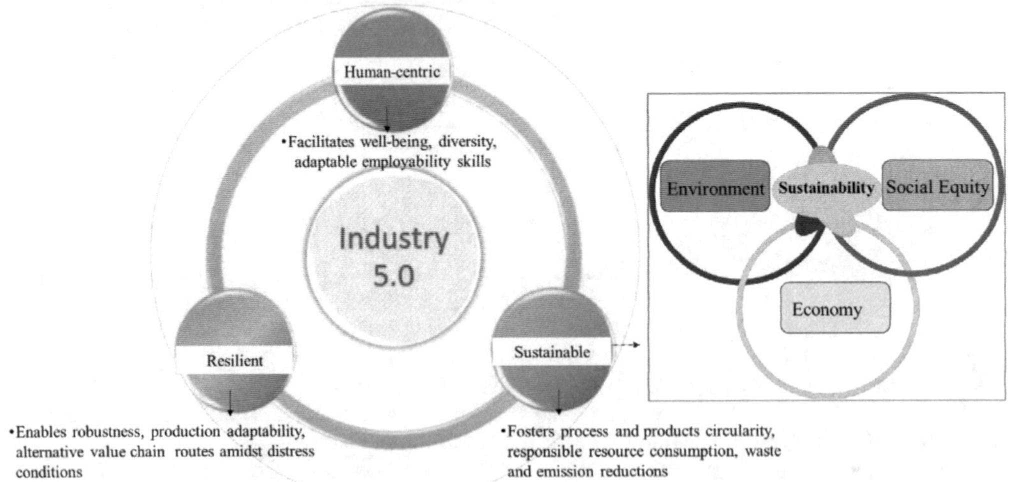

Figure 2. Illustration of Industry 5.0 and its relation to sustainability, adapted from [3,23].

As can be seen from Figure 2, the model describing I5.0 mimics the intersecting representation of sustainability. Sustainability is commonly explained as the ability to *meet the needs of the present generation without compromising the ability of future generations to meet their needs* [24]. In the context of I5.0, sustainability refers to the ability of industries to effectively combine social, economic, and environmental sustainability principles into their operations and processes. Examples of sustainability strategically in I5.0 includes:

- Use of clean and renewable energy sources: I5.0 encourages the use of clean and renewable energy sources in manufacturing processes to reduce carbon emissions and promote environmental sustainability. For instance, transport sectors prefer electrical driven to conventional fuel driven motors, usually fossil based This helps.to reduce fossil fuel and to facilitate emissions reliance and at same time promotes innovation.
- Resource efficiency: Resource efficiency involves minimizing waste, promoting recycling and reusing materials to reduce the environmental impact of production processes. For example, using smart sensors and predictive maintenance to optimize equipment use, minimize downtime, and reduce the need for overproduction.
- Supply chain sustainability: Companies can ensure sustainability in their supply chain by prioritizing suppliers that adopt sustainable practices such as reduced carbon emissions, ethical business practices, and adherence to regulatory compliance.

According to [25] Human-centricity can be defined to mean *focusing on something variously called (self-) sovereignty, self-determination, self-governance, autonomy, agency or the like, in terms of the people involved with the generation of data*. Examples of concepts aiming to maintain and promote human centricity include: (1) General Data Protection Regulation (GDPR) targeted at protection of the personal data within cooperative and governmental institutions, (2) the White Paper on Artificial Intelligence which *sets out the principles for eventual AI regulation safeguards for the users of certain categories of AI technologies* and (3) Digital Education Action Plan (2021-2027) which aims to bridge the knowledge gap, inclusiveness of workforce and capabilities required to effectively function in ongoing digitalization [3]. Pioneering the implementing of such value-based policies in several of key policies makes the EU a frontier to such human-centric and social-centric approaches.
Resilience is defined as the ability to endure all transitions caused by change, or the degree of change tolerated before deterioration in performance occurs without any corrective action. Similarly, resilience is defined as "*the ability to anticipate and withstand disruptions, respond to them, and effectively recover from disruptions*; *the ability to cope flexibly with change* [3]. Adaptability offers industries a good stance to prevail in competitive environments [26]. Been able to adapt products, processes and service with existing capabilities and competencies to respond to changes matching customer needs are prospective to maximizing agile competitiveness through flexibility, time efficiency and cost savings [26,27]. Agility is defined as *the continual readiness of an entity to rapidly or inherently, proactively or reactively, embrace change, through high quality, simplistic, economical components, and relationships with its environment* [28]. Agility fosters quick recognition of supply chain changes to proactively respond to changes through their adaptive redesigning [29]. I5.0 promotes innovation and creates new markets across several industrial sectors [25,30].

Following the studies of [28,31] agility, robustness or resilience and sustainability may be necessitated by change or pressure persuaders. Industrial evolution likewise is propelled by change or pressure, technological changes, constantly changing customer demands, and environmental concerns [28] broadly classify common interdisciplinary change/pressure persuaders into competition, customers, technology, social and overhead. The study further defines the different forms of changes and how a working team may affect operability changes at the diverse levels of the organization. The emergence of sustainability along industrial evolution has guided the way these technologies utilize natural resources to maximize economic growth. For instance, through the offered design flexibility of AM, engineers are becoming more creative and sustainability-centric in product designing amidst market competitiveness. Aspects of sustainability are consciously or unconsciously incorporated into product designing through computer generated designs and intelligent manufacturing systems.

Industry has emerged from manual production methods to digitally driven machines from (1) steam or water powered (first revolution), (2) electricity powered for mass production, (second revolution), (3) computers, electronics, and automation powered production (third revolution), (4) digitalization, connectivity, and cycler physical systems (CPS) driven production (fourth revolution). The different industrial transitions were driven by the need for better productivity and economic benefits. The digital era of I4.0 continues along the emergence of I5.0 unabatedly with some industrial sectors yet to adopt the offered technological advances such as AM, AI, automation photonics and simulations. The shift from I4.0 to I5.0 is characterized with a change from technological-centricity to value-centricity [2]. While I4.0 assisted with digital and interconnectivity to the realization of several of industrial benefits with focus on easing laborious work, increasing flexibility, and maximizing efficiency [32], I5.0 anchors on the delivery of added values for societal benefits through these benefits [2,3].

According to the study of [2,33] I4.0 technologies relief workers from routine tasks through extensively open virtual technological tools to becoming more creative and value-creation oriented. Automation and robots for instance reduce laborious human tasks, increase productivity, and promote swift informed decision-making [2,27,34]. Empowering workers skills with capabilities along the needs of present smart factories equip employees to gaining the required skills and understanding to maintaining job security, authority and to make decisive choices complimentary to the complex, dynamic, and flexible manufacturing systems in pursuit of I5.0.

Several studies highlight emergent technologies including AM, Artificial Intelligence (AI), cyber physical systems (CPS), simulations, digital twin, autonomous machine to mention but a few as exemplary innovations propelling the core of I5.0 including sustainability, human-centricity, and resilience [33,35,36]. AI and autonomous systems for example allow machine systems to sense, perform, control, and even adapt parameters where necessary to complete full scenario in manufacturing [27,33]. Conventionally difficult and lengthy to manufacture yet unavoidable HPMs including nickel alloys, titanium alloys and tungsten carbide composites are effortlessly possible via AM [6,37,38]. Through LBAM high-end critical metal components with small features can be customized and accurately manufactured for applications such as medicine, aerospace, and energy [39]. This is possible through the accuracy and high precision of lasers and flexibility for layerwise manufacturing for a wide range of material. LBAM also omits traditional manufacturing constraints [40,41].

The continuous advancement of new generation of intelligent and functional gradient materials are promising to LBAM applications [42–46]. I4.0 technological capabilities enable sustainable, human-centric and resilience via intelligent machines, optimal scheduling, process optimizing, agile and customized manufacturing, consequently, maximizing efficiency, reducing defects, and omitting laborious human-tasks within manufacturing value chains. Value chain describes *the full range of activities that firms and workers do to bring a product from its conception to its end use and beyond* [47]. Simplifying and reducing value chain interlinks is viable to enhance efficiency, productivity, competitiveness, and effectively respond to change or unforeseeable disruptions. Some enablers required to simplify supply interlinks in modern value chains are connectivity, cyber security, digital inventory, on-demand, and decentralized manufacturing [26,33,48].

2.2. Laser based additive manufacturing

The most used AM for metal materials (in powder, wire, sheet forms, etc.) [49,50] is the laser based, namely, selective laser sintering (SLS) or full melting and bonding (selective laser melting, (SLM)); powder bed fusion, laser metal deposition (LMD); directed energy deposition (DED). These AM machine systems use laser beam to manufacture high precision and low-cost complex high-performance materials. LBAM can are leveraging to new materials (single, multiple, dissimilar, biomaterials, nanomaterials, optical materials) and new product design geometries (topology, biomimetic structures, and lattice or cellular structures) [51–56] which are otherwise impossible via comparable AM methods or CM methods. Different industries use LBAM to sustainably create new parts or to remanufacture existing worn-out parts [5,57].

LBAM can be grouped into three (see Figure 3, based on material feed, material type, mode of fusion or adhesion, namely, (1) powder bed, (2) powder blown, and (3) wire fed. [57,58].

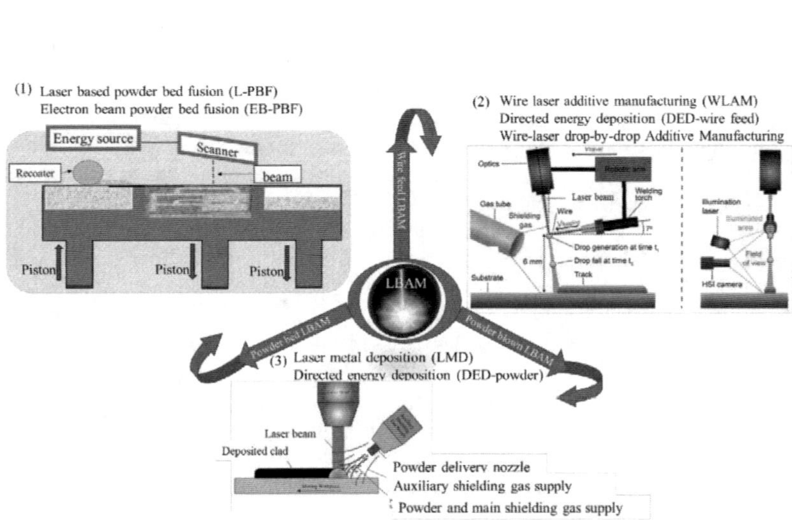

Figure 3. Illustration of the basic layout and difference in LBAM, modified from [23,56,59].

2.2.1. Powder bed fusion: this class of AM, is based on a powder feedstock and a build bed on which the feedstock is laid layer by layer for the successive melting. L-PBF is one of the earliest, most common, and advanced metal AM process methods to date. The laser utilizing variant of the PBF processes for metals is the most developed research- and industry-wise. The process uses the laser beams thermal energy to selectively fuse regions of a powder bed typically either by melting or sintering [60]. The process principle seen in Figure 3 typically includes guided laser, powder dispensing and spreading system, build platform and powder collecting chamber. The process is done in a shielding gas filled chamber. During L-PBF workflow, a thin layer of powder is spread along the build platform, which is melted or sintered according to sliced layer geometry by the laser. The built platform is then lowered according to chosen layer thickness, and a new layer of powder material is spread on top of the powder bed, which now contains the first solidified material layer. Melting/sintering of the second layer is started, and this procedure is repeated until finished part(s) [61]. The main process parameters of L-PBF include laser power, spot size, pulse duration and frequency, scan speed, spacing and pattern, powder particle shape, shape size and distribution, layer thickness, material properties, powder bed temperature, etc. Substantial portion of these parameters are strongly interdependent [62].

Basis for adapting new materials for the process is excellent since all materials that can be melted and resolidified are potential candidates in principle. Material catalogue for PBF process has grown large and versatile in conjunction with process development. PBF can be used to manufacture a variety of materials including polymers, metals, ceramics and their composite variants. Common suitable polymers include e.g., polyamide (PA), polypropylene (PP), polyaryletherketone (PAEK) and polystyrene (PS) -based materials. Commonly found metal materials for PBF include stainless and tool steels, titanium-alloys, nickel-based alloys, aluminum-alloys, cobalt-chrome, and precious metals such as gold and silver. Typical commercially available ceramic materials e.g., aluminum and titanium oxides. A few ceramic biocompatible materials, such as calcium hydroxyapatite, have also been under research and development for specific medical applications [62,63].

All AM process methods have their typical advantages and disadvantages. Few specific benefits for L-PBF include wide material catalogue, great mechanical properties of metal parts, good accuracy and surface finish with fine particle size and competitiveness with geometrically complex parts for low to medium production volumes. Polymer-wise the method enables support-free processing and easy 3D nesting of parts. Typical drawbacks for L-PBF include challenges with metal supports, post-processing needs, powder handling and comparably slow part production times [62].

2.2.2. Directed energy deposition (powder and wire based): Laser based directed energy deposition (DED) is a standardized AM process variant in which focused thermal energy is used to fuse materials by melting at the same time they are deposited to the process. In laser-based process variant the laser beam acts as the thermal energy source and material feedstock can typically be in either powder or wire form. Material binding happens by melting and solidification of subsequent layers [60]. In typical process workflow laser beam is focused on predefined region of substrate, in which deposition material is then fed to either through the energy beam or to the beam focal point. Energy source and material feed (deposition head) are moved along defined processing path forming a track of solidified material. Adjacent material tracks form single layers, while subsequent layers form the three-dimensional geometry. The main process parameters for laser-based DED include laser beam power, spot size, traverse speed, material feed rate and track scan spacing. Material feeding angle and orientation (front, back, side feeding) also have an affect when off-axial material feeding is utilized. Material catalogue for laser-based DED processing includes metal materials when focusing on commercialized processing systems. Metallic materials that show good weldability and melt pool stability can be processed via laser-based DED. Specific limitations for metal materials include materials with high reflectivity and thermal conductivity values, which can cause challenges [62].

Laser-based DED specific advantages include the ability to utilize it for repairing damaged parts and direct feature addition [5]. Existing high-end components might have worn and/or corroded sections, which can be refurbished close to original geometry by this method. Simplistic 3D geometries or coatings can be deposited directly on existing parts to add value. Material related benefits include excellent control of microstructure and the ability to produce multi-material parts e.g., in situ manufactured heterogenous and composite structures. DED processes in general cannot produce complex structures and have challenges producing small-scale features. Build accuracy is typically a trade-off for higher build speed. With low build accuracy, complexity and poor surface roughness, the produced parts typically require extensive post-processing [5,62].

2.3. High-performance metals characteristics

HPM are characterized by their exceptional mechanical, thermal, and chemical properties. These materials exhibit excellent strength, toughness, hardness, and ductility, and are often resistant to corrosion, oxidation, and wear. HPM includes titanium, nickel, cobalt, and some of their alloys [64] which possess unique combination of properties like mechanical strength, corrosion resistance, biocompatibility, and temperature resistance, among others. LBAM through the capabilities of computer-based design and simulation software can maximize performance through the offered potentials of optimized near net-shape, lightweight designs, customized parts, and flexible production to enhance quality, agility, and cost benefits [4,5,65]. Product design used for heat or mass transfer can effortlessly be optimized with increased surface area and tortuosity suitable for performance [66]. This feature can be particularly useful for heat exchangers, turbine blades, and other high-temperature applications. Lasers allows for the creation of varied microstructural gradients and heterogeneous structures, which are intentionally designed to possess distinct features and functionalities [5]. The aerospace industry extensively uses LBAM for HPMs such as titanium alloys, high strength steel and aluminum alloys to produce engine and structural parts [67]. The energy sector uses LBAM to produce parts like heat exchangers, turbine blades, and fuel cells [68]. The automotive sector utilizes LBAM to produce lightweight components with improved mechanical properties, such as aluminum alloys, steel, and titanium [30,65]. In the biomedical sector, LBAM is used to produce complex parts with biocompatible metals, such as titanium and cobalt-chrome-molybdenum alloys. Dental implants, prosthetics, and surgical instruments are some examples of biomedical applications of HPMs with LBAM [69].

Some examples of high-performance metals and their characteristics include -high strength (e.g., nickel alloy, IN718) and intelligent (e.g., magnetic shape memory alloy, Ni-Mn-Ga) materials.

- High strength nickel alloy, IN718: Inconel 718 is a nickel chromium based super alloy that has high strength and oxidation resistance even at high temperature [65]. This makes IN718 optimal in aerospace and power generation fields [66]. IN718 is difficult to manufacture traditionally due to low thermal conduction and rapid work hardening quickly wearing out tools in prolonged machining [63]. Low thermal conduction increases temperature in the tool. Machining passes should be minimized as material hardens after each pass. AM is a preferred manufacturing method as it lowers the need for machining [63].

- Intelligent material (Ni-Mn-Ga): Ni-Mn-Ga, a magnetic memory shape alloy that produces strains when exposed to magnetic fields. This behaviour causes displacement in the part which can be used for fast micro actuators, precision dosing pumps and energy generation from vibrations. Fatigue life of 2x10^9 cycles also support this. Parts from the memory alloy have been generally manufactured using binder jetting but the change is chemical composition due to binder material made the method suboptimal. This is not a problem when using L-PBF process which is able to produce purer material. The fast melting and cooling of the process may affect the properties of the material, however, any changes that may occur could be mitigated with the use of post heat treatment. Process also opens up the optimization of the structure as optimal artificial porosity (net like structure) could increase magnetic induced strains [70,71].

3. Application, benefits, challenges and contributions to industry 5.0

AM substantially accelerates I5.0 evolvement through several offered benefits in overcoming limitations imposed by CM methods. However, the former method is often ignored due to its unpopularity, known limitations and expenses involved. LBAM is recognized to offer overwhelming benefits including flexible, efficient, agile, and customization to niche markets that utilize HPM. AM contributes to the advancement of sustainability, human centricity, and resilience goals of I5.0 which are crucial for the long-term survival of businesses and the environment.

3.1. Application of laser based additive manufacturing

Exemplary industrial applications of additively manufactured HPMs

- AIMEN ADDIFLAP project: The ADDIFLAP project utilizes LB DED to optimize (for complexity and lightweight) flap track mechanism for a jet. This mechanism integrates a flap track support system with a carriage system that is based on a sliding pad (see Figure 4).

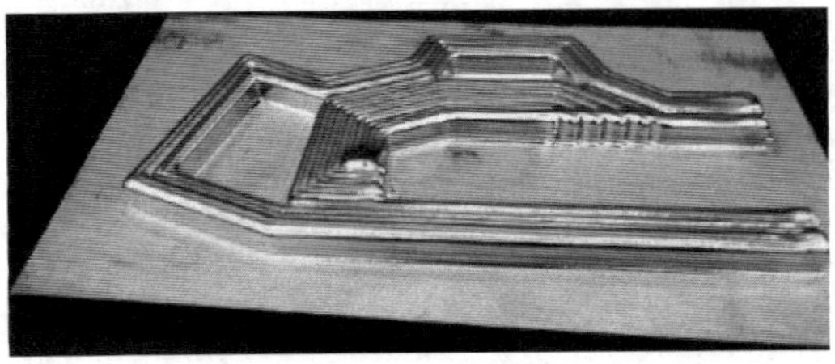

Figure 4. Representation of wire-based LB-DED flap track. Courtesy of Aimen, [72].

Figure 4 is a representation of the flap track support system, the structures on which the hydraulic systems that move the flaps of an aircraft are placed during take-off, approach and landing maneuvers. The use of DED technology allows for precise and complex designs to be manufactured in a shorter amount of time than CM methods would allow. The new additively manufactured part is promising to reduce carbon dioxide emissions by 26%. The result is a more efficient and cost-effective production process for aerospace components with approximately 60% a material waste reduction [73]. This project showcases the potential of AM in the aviation industry and its ability to revolutionize the way aircraft parts are produced.

- Porsche drive housing and pistons: Porsche, an automotive company, utilized the capabilities of simulations and LBAM to design and manufacture innovative high-stress car parts. With LBAM, it is possible to integrate multiple functional parts, which reduces assembly work and improves the overall product quality. Through the potentials of simulations and LBAM, an integration of unique structures (bionic, lattices, ribbing, topology) can be used to create optimized product designs, for instance the housing (See Figure 5 (a)) and pistons (See Figure 5 (b)).

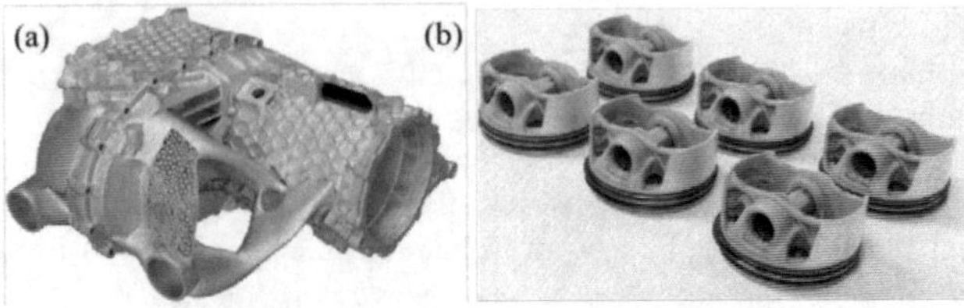

Figure 5. Representation of high-performance metal (a) electric drive housing prototype and (b) sports car pistons. Courtesy of Porsche, [30,74].

LBAM offers numerous benefits for these parts, including functional integration, reduced weight, enhanced performance, lower costs, shorter lead times, and less material waste. The LBAM-manufactured parts satisfactorily passed quality and stress tests and, in some instances, outperformed comparable conventional manufactured counterpart. For instance, each of the additively manufactured pistons weighed 10% lighter than comparable forged counterparts and exhibited enhanced efficiency and performance [65]. Additionally, the optimized housing with the special structures reduced 40% of weight and increased performance, for instance, stiffness in highly stressed regions were increased in the case of the housing [30]. Overall, the housing case demonstrate that LBAM can effectively produce larger and highly stressed components providing numerous advantages [30,65]. This example is a promising step to overcoming the size limitation of earlier age PBF machine systems.

- Stryker surgical devices: Stryker, a medical device company, uses LBAM to produce surgical tools and orthopedic implants made from titanium and cobalt-chromium-molybdenum alloys. LBAM allows for highly precise and intricate designs, leading to better medical products performance (e.g., accurate bone preparation and implantation), (See Figure 6).

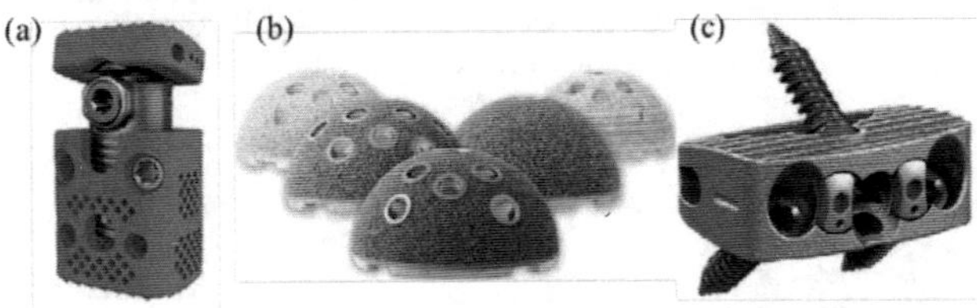

Figure 6: (a) Cervical 3D Expandable Corpectomy Cage, (b) Acetabular and (c) Monterey AL Interbody Systems. Courtesy of Stryker, [69,75,76].

The additively manufactured cages in Figure 6 (a, b) hold the cam locks (the gold elements) and provide a passage for the screws into the vertebral bodies. Through LBAM new alloys (e.g., Tritanium, a highly porous titanium material) and product designs (e.g., lattices or roughened surface) that mimic the complex characteristics of cancellous bone are possible [69]. Tritanium are biocompatible, possesses excellent strength and corrosion resistance properties, making them ideal for use in medical applications These features provide effective medical solutions to specific tissue needs (e.g., protein expression). LBAM also allow the flexible designs suitable for clinical fixation considerations (e.g., stability height, lordotic angulation adjustment). Additionally, LBAM reduces material waste, making the manufacturing process more sustainable and cost-effective for such applications.

3.2. Benefits and challenges of AM for high-performance metals

LBAM enables digital production route with proof-of-concept designs, build planning, and simulation to optimize time, raw material, energy consumption. LBAM can create end-use parts directly from virtual data without the need of intermediate steps, improving development time and efficiency [40,53,77]. LBAM facilitates the integration of digital technologies in manufacturing processes by enabling the creation of complex geometries and parts with high precision. This is achieved using computer-aided design (CAD) software, which allows for the creation of digital 2D layers from 3D models that can be directly translated into physical parts [78]. In addition, interoperability in LBAM enhances the agility and resiliency of manufacturing processes. The digital designing, data sharing, cloud storing and internet sharing of product designs are fostering to localized production and key to reduce supply chains interlinks [34]. This is achieved with on-demand manufacturing, which enables the production of parts as they are needed, eliminating the need for large inventories of finished goods. LBAM produces fewer negative environment impacts, possess minimal health risk and reduces the risk of toxic emissions typically associated with CM processes [79,80]. LBAM allows the combining of multiple materials to suit varying requirements, a possibility cable of maximizing product performance with functionally gradient and nature inspired designs [51,55].

Despite the several LBAM offered benefits, there are also some challenges and limitations associated method [17]. The application of AM example for HPM comes with specific challenges which may deter adoption rate, for example, high machine cost [50], limited materials [43] and slow build time [81]. LBAM especially requires feedstock of specific properties (high purity, spherical, particle size) which may not be readily achievable with all HPMs. Required feedstock size and high pricing, especially metals [14,82] restricts its application to high end industries. Another prominent challenge to adoption is the prohibitive cost of AM machines and materials, which may deter smaller companies from adopting the technology. Again, AM sometimes requires a considerable amount of time during the product design, manufacturing, and post-processing phases. Finding suitable software, build planning, process parameters etc. to meet specific needs can also be lengthy and time consuming [83,84]. Furthermore, the quality and consistency of AM parts can vary depending on the capabilities of the used category. The final surface quality of LBAM often requires post-processing to bring it to the expected requirement. The final quality of the printed parts as well as the intended use case thus influences the type of post processing method [85]. This requires a good consideration of the suitability of the available LBAM system (including the several process parameters) prior manufacturing, a task which often increases cost and time.

Additionally, the LBAM can be slow and the required additional post-processing steps may increase the overall production time and cost. For a holistic evaluation of the merits and demerits of LBAM manufactured parts, a comprehensive study of the different lifecycle stages is necessary to evaluate their full contributions as there can be inefficiencies depending on the used process category, materials or application. L-PBF for instance is often impede due to recordings of high energy consumptions, the need of sacrificial supports which are removed during the mandatory post-processing in manufacturing phase. Limiting the judgement to such factors alone may block novices from utilizing the process without a holistic evaluation of design, use and end-of-life benefits.

3.3. Laser based additive manufacturing contribution to sustainability

Through LBAM resource consumption can be improved as the different methods use materials exactly at regions they are required. Product designs of small features are possible due to lasers ability to precisely melt and fuse small regions [5,84,86]. This capability of LBAM have propelled the development of varying scale (micro-macro) HPM highlighted in studies [44,51,83,87,88] to meet extreme working conditions and increasing demand of sustainable goals of industry. Through the offered benefits of complex product designs, multiple HPM parts can be consolidated, and cost effectively manufactured with simplified process steps and shortened time [89]. LBAM offer exceptional route to achieve precisely and accurately superior end-use HPM void of defects and errors including lack of fusion, grain morphology, poor surface quality using suitable feedstock [86,90], process parameters, orientation [49], in-situ monitoring and control [91,92] and complimentary CM methods [93]. Sustainable manufacturing is often considered to have less negative impact on the ecosystem. This aspect is very much attainable via LBAM of HPM since there is little or no waste and need of process fluid (e.g., water, coolant, filter) in laser powdered machines compared to conventional energy powered machines [93,94]. Studies have shown that using AM can reduce scrap by 10%) [95]. For instance, LBAM creates about 20% of waste in the form of supports scrap, an offsetable amount via recycling (as input for new raw material) [96]. LBAM effectively replaces CM methods such as chrome coating which has significant contribution to carbon dioxide (CO_2) emissions [58]. LBAM allows the reconditioning

IOP Conf. Series: Materials Science and Engineering 1296 (2023) 012001 doi:10.1088/1757-899X/1296/1/012001

of worn-out parts for extended service life without the need of total replacement [86], consequently, reducing raw material consumption, downtime and costs, lead time.

3.4. Laser based additive manufacturing contribution to human-centricity in industry 5.0
Pivotal to the AM categories is the upskill of user digital knowledge and technological application. AM induces a learning custom in its users as conventional manufacturing workflows are not applicable. LBAM in driving human-centricity offers means to create high valued individualized HPM to fit specific needs of application, particularly medical applications [97]. Examples by enabling the production of patient-specific metal implants and prosthetics in the medical industry. A potential to improve medical functionalities by tailoring the products to the individual needs of patients. In contribution to well-being, LBAM offers automatic manufacturing without needing human presence [98], reducing fatigue and possibility for human error during the process. Emerging digital thread through integrated intelligence and CPS AM machines provide virtual monitoring, and controlling allowing secured flexible and remote working [98,99]. LBAM safely support remote and harsh conditions working as most processes occur in a closed chamber promoting decentralization and safety, fostering equal accessibility of the technology [81,100].

3.5. Laser based additive manufacturing contribution to resilience in industry 5.0
LBAM inherently offers a simplified supply chain especially for spare parts through swift product designing, and localized manufacturing consequently reduces downtime and lead times [101–103]. AM allows quick alterations and iterations of product designs to suit required needs and to be manufactured near to the point of application [104]. Due to the ability to consolidate multiple parts, LBAM reduces the value chain interlinks and offers a swift route to manufacture HPMs which otherwise will require multiple conventional steps to manufacture. Via digital thread and digital twin capabilities, physical counterparts of virtual models can be manufactured on-demand to meet stringent functional requirements, budgets, and timelines. This is achievable either with in-house or through business-to-business (B-2-B) or Business-to-consumer (B-2-C) AM machine enabling global digital co-creation. AM potentially promotes innovation by altering business models e.g., creating new markets across several industrial sectors [30]. Founded by proprietary companies, AM machines owners can engage in B-2-B or B-2-C operational models for agile development, flexible or customized production as best fit for peculiar organization.

4. Discussion
Secrecy and lack of reporting for competitive autonomous and monopoly market hindered the fast advancement of AM. Early ages of AM had lots of companies protecting their expertise to maintain monopoly or competitive edge. The new form of industrial revolution has necessitated ecosystem workforce amongst all industrial sectors fostering data exchange yet offer a level of protection among working groups. Such ecosystem has and continue to foster the omission of secrecy in AM e.g., among machine manufacturers, feedstock suppliers, and software developers. The material market has extended beyond machine manufactures which was formerly the norm. Notably to such expansion is the increase in material-related research that continue to offer wide range of high performance materials suitable for demanding applications. There are several ways by which industries are overcoming some of the already mentioned limitations in AM, including:

- High investment cost in AM especially in metal based AM [8]. This limitation can be countered through AM service bureaus and ecosystems. Companies aiming to adopt AM can do so firstly through contract manufacturing service providers without the need to commit to high cost of software, hardware, materials, and overheads.
- Limited materials and required feedstock. Through continuous efforts of existing and new start-ups the amount of material portfolio for AM has increased. Paramountcy to this also are the new feedstock forms like sheets, pellets and slurries which are possible following the layer-by-layer route of AM to create geometrically complex products [42,50,78]. This often requires extensive research into new materials along with the advancement of AM machines. This is partially good,

as it creates new materials of better properties and performance, the workflow may be hectic and worrisome and in some case increase cost and developmental time.

- Metal materials are not necessarily easy to be manufactured using LBAM as several factors including the varying material properties, process parameters, data discrepancies in modelling and simulation software, among others can affect the build rate, part quality, or usage [34,44,45]. The emergence of Automation makeup language eliminated data discrepancies and promotes systems integration for modern machine system. AI-AM research is promising to develop intelligent closed-loop in-situ monitoring and controlling AM machines [7,17,34,91,92]. For example, integrated self-configuration and adaptability AM machine systems capable of correcting in-process fluctuations [4]. Robotic-assisted AM and multiple laser systems offer means increase productivity [34,81].

Overall, the use of AM and other I4.0 technologies is an important aspect of I5.0, which is driving the evolution of manufacturing towards more efficient, agile, and personalized production. The capabilities of I4.0 technologies including simulation, AI, CPS are promising in providing secured timely picture of ongoing work with real-time detection and responding to manufacturing anomalies for increasing effectiveness, quality, and accurate forecasts [97,98].

4.1. Prospect to overcome maximizes additively manufacturing toward industry 5.0
AM workflow can be adjusted to be autonomous, interoperable, and self-driving for strategic changes. Achievable through integrated complemental cloud-based, digital sensing and assisting microsystems AM machines [16,81] and thereby increase flexibility amidst worker fatigue reduction. LBAM promises new routes to maximize productivity, resource efficiency, quality and to minimize waste and emission creation [77]. This may be possible via cognitive thoughts, interactive iterative and sensing for swift adaptable and flexible and production [101]. These potentials are promising to deliver smart designing, robust manufacturing, autonomous systems capable of achieving the goals of I5.0, namely, sustainability, human-centric, resilience. Figure 7 summarizes the key existing and futuristic leverages for I5.0 via LBAM and complimentary I4.0 technologies.

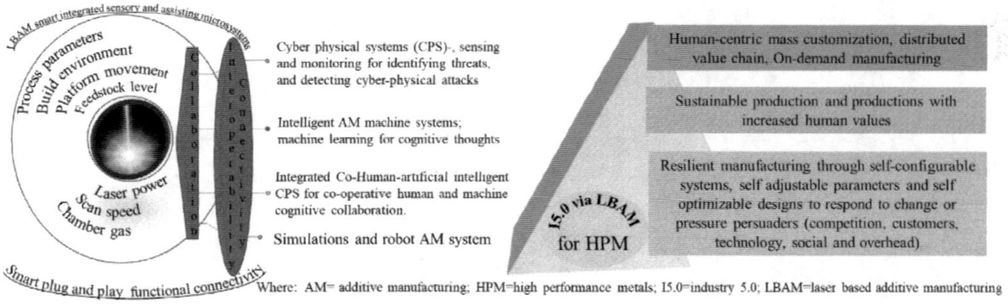

Figure 7. Representation of prospective co-human-artificial intelligence-CPS AM towards industry 5.0.

- Simulations and robot assisting AM system: Through computer-based simulations and programmable robots, AM machines can outperform tasks beyond human cognitive spectrum [4,19]. Simulations can perform interactive iterations of designs based on human-defined functional requirements to generate multiple suitable geometric response [4]. Robots can perform laborious tasks; relieving humans of tediousness and offering the possibility of humans to make judgement and decisions on accuracy and quality.
- CPS; Augmented reality and digital twin: Through augmented reality and assisted reality to remotely visualize and interact for human-human and human-machine interactive diagnostics and decision-making. Through digital representation of physical scenery, operators can monitor and give corrective instruction via remote connectivity where necessary [20]. Such capabilities

enable secured manufacturing process and virtual controlling of ongoing production consequently enhancing resource and cost effectiveness.

- Intelligent flexible AM machine systems; machine learning for cognitive thoughts: Through closed-loop controlled LBAM (currently missing). The best intelligence of LBAM existing machines is to halt manufacturing process at the sense of abnormalities. Emerging AM machines can be equipped with more interoperable intelligent, autonomous systems to achieve total close-loop control manufacturing using existing digital workflow. Such advancement may be limiting as result of lack of data format standardization for the different interoperating systems [34]. Enhancing interoperability in AM will complement the variations and limitations of machine systems to increase machine flexibility.
- Integrated human-artificial intelligence CPS; for co-operative human-sensory-machine cognitive collaboration for autonomous AM to reduce downtime, maximize productivity, and support informed decision-making. Equipping AM machines with digital assisted tools capable of identifying threats, sensing abnormalities, monitoring parameters, and controlling processes can articulate automation and self-adaptable responsiveness to change [27,98]. This will require the upskilling of workers knowhow to suit the requirements smart factories and consequently improve creativeness, confidence and employability.

4.2. Industrial relevance of the study

This study demonstrate how industries can incorporate I4.0 technologies in their workflow to articulate e.g., co-creation, productivity, and competition while decreasing inefficiencies. Optimal utilization of software and AM machine systems can be enhanced by integrated co-human-artificial intelligence-CPS. Such integration is promising to efficient, interactive, connective, interoperable, monitoring, and controlling needed to create added values. The summarized benefits of I4.0 technologies applicable to LBAM, highlighted in this study, are conceptualized in Figure 8.

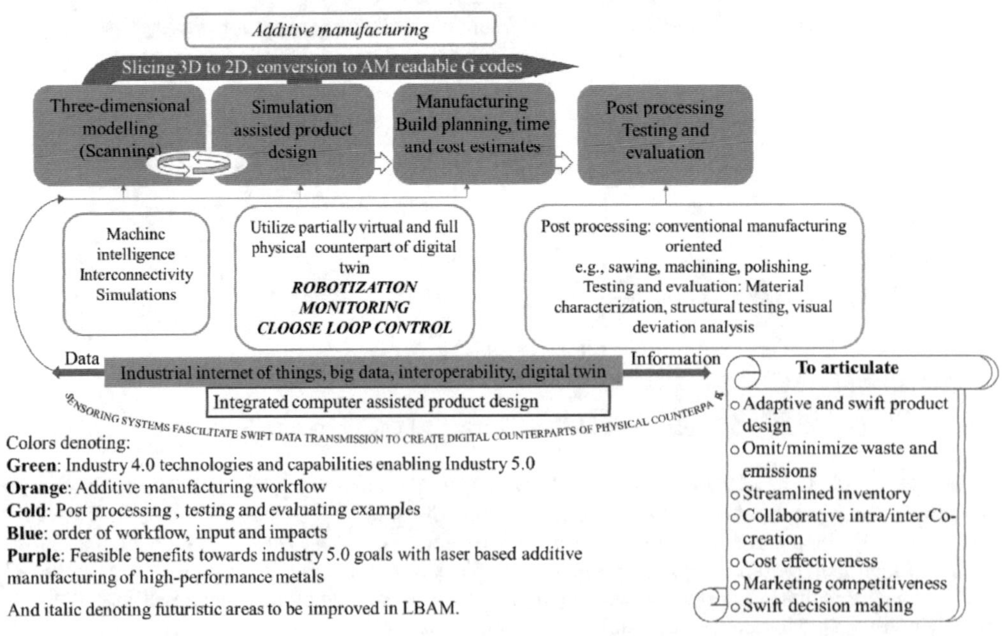

Figure 8. Illustration of potential LBAM working route to articulate industry 5.0.

AI is currently used to automatically analyze input data to predict valuable optimized product designs and insightful interpretation to aid decision-making. Interconnectivity allows the interoperating of different AM systems, e.g., cloud computing allows virtual storing and accessing data from separate locations and unifying data exchange allow robotization in large-scale AM. Continuous development of example AI- data- robot- and simulation-, assisted large scale and/or volume productions increase productiveness, and promote well-being. Modern AM machines are equipped with monitoring capability but mostly record layer by layer building data and store them for viewing and human interpretation. Such data are useful to developing a true self controlling AM system. Nevertheless, the existing monitoring can further be developed to give an interpretation of recorded data for effective human justification or comparison with human interpretations. Closed-loop control currently may be farfetched from AM systems. Such cognitive and self-adaptive systems, however, are anticipated as research continues, especially, as calls for resilience begin to emerge from diverse governmental, industrial and research institutions. The emergence of AI-AM research is likely to revolutionize AM adoption, making it faster, reliable, and more cost-effective. Potential research on incorporating AI capabilities to AM is promising to improve the efficiency and accuracy of product design. AI-AM could effectively reduce errors and minimize waste in the manufacturing process with advanced monitoring systems and self-configuring machines. AI-AM could significantly be used to correct in-process fluctuations, a potential to improve quality of as-build part.

4.3. Novelty of the study

The novelty of this study is in demonstrating how LBAM provides a means of promote efficient production, and flexible manufacturing operations, while maintaining a high degree of resilience for present and emerging industries. The study identifies optimal route to LBAM implementation coupled with other I4.0 technologies. Such integration fosters agility and adaptability consequently, empowering businesses to maximize value, increase competitiveness, reduce their carbon emissions, and make significant strides toward accomplishing a holistic I5.0. The study shows how LBAM can revolutionize the manufacturing industry and aid in creating sustainable, human-centric, and resilient future. There is no doubt that AM is simultaneously disrupting existing processes while driving value for human and society [42]. The evidence is summarized in Figure 9 based on presented literature and cases.

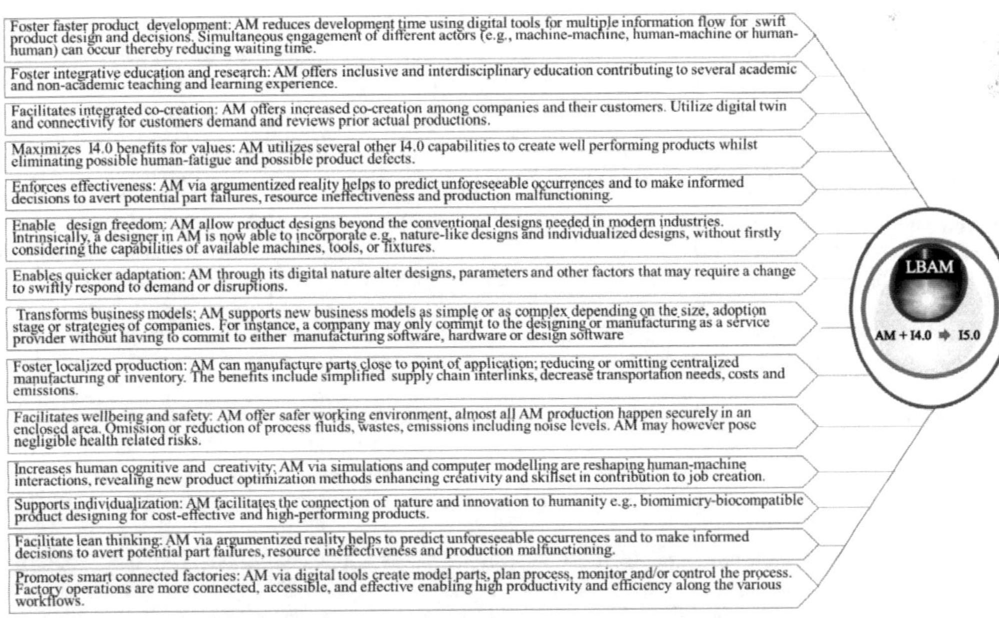

Figure 9. Prospective LBAM enablers of I5.0.

As it can be seen from Figure 9, applying AM with capabilities of other technologies is promising to achieve the identified leverages towards I5.0. A seamless collaboration of value changers, connectivity of systems and their interoperability might offer the required levers needed to accelerate value creation within industries as opined in studies [3,8,16].

5. Conclusions

In conclusion, AM is recognized for its potential in promoting agility through simplified supply chains, reduced lead time, and faster delivery times. This can be particularly useful in situations where swift and adaptable response will be needed for unexpected changes in demand or supply chain disruptions.

Exemplary case studies show complementarity AM and I4.0 technologies in fulfilling I5.0 goals. Integrating the core principles of I5.0 into industrial processes can lead to increased efficiency, safety, cost savings, and reduced carbon emissions. For instance, self-adaptive LBAM machine can self-adjust process parameters based on stored data to continue manufacturing without halting the process or building defected parts to the last layer. The benefits of I5.0 cores can be shared between stakeholders, including industry, consumers, and the environment. Improving human cognitive utilization with simulations minimize human fatigue with robotized AM system. Upskilling workers knowledge in conformity with current industrial trends tends to increase human creativeness, and participation in industrial workflow boosting job security, inclusiveness, confidence, among others. Example, by combining the expertise of designers and software to fine-tune generated intelligent designs to suit specific AM design rules.

Equipping the already integrated LBAM machines with a self-controlling functionality can boost sustainable aspects and resilience to propel a full utilization of the leverages of digital transformation. LBAM leverages I5.0 by expanding production capacity (enhancing accuracy and precision) of HPMs through machine intelligence and assisted robots. The benefits of such digitally integrated workflow reduces physical inventory with virtual inventory. AM through the capabilities of other I4.0 technologies create new career opportunities, business models and boost customer satisfaction via co-creation, customization, and decentralized.

This study has shown that the application of LBAM through the capabilities of other I4.0 technologies are promising to transform manufacturing practices. LBAM is identified to create better performing HPM products, optimize raw material consumption, reduce material waste and emissions; offering the potential to enhance sustainability. LBAM also can create optimized and intricate designs offering manufacturers a significant route to better functional parts. Finally, LBAM feasibly improves resource efficiency, and cost savings.

This study also shows that the use of LBAM and other AM are prone to challenges and limitations during production especially metal parts. This study has highlighted some limitations of LBAM which require continuous research efforts for continued leveraging of sustainable, human centric and resilient manufacturing value chains. Nevertheless, the advantages of AM are substantial and can facilitate the advancement of the manufacturing industry towards a more efficient, agile, and customizable productions. The study offers novices an understanding of AM contribution to I5.0 and gives reference documents for academia and industrial researchers aiming to undertake similar studies.

5.1. Limitations, and scope of future research

The study was limited by the availability of data on the use of LBAM for HPMs especially in relation to the emerging I5.0 as the concept is merely gaining attention. Partly to this data scarcity is the tendency of researchers to use generic term AM for all related studies without specific distinguishing of the distinctive AM categories.

Ongoing AM research could intensify collaborative research that can enhance human cognitive abilities and creativity, leading to the development of novel manufacturing capabilities, raw materials, software, and hardware. Efforts must be made to integrate true self-configurable, self-adjustable and self-optimizable intelligent capabilities for total autonomous workflow. This way, the benefits of design flexibility, and process specific varying conditions (e.g., material properties, process parameters) can be matched to reduce errors (e.g., porosity, increased quality (e.g., accuracy and repeatability)). Again, co-intelligent systems capable of performing manual tasks (e.g., powder sieving and refilling, controlling) are yet to be explored. Using integrated human-machine intelligence for AM can bolster agility and resilience, leading to the creation of more efficient and effective production.

6. References

[1] Wang L 2022 A futuristic perspective on human-centric assembly *J. Manuf. Syst.* **62** 199–201
[2] Xu X, Lu Y, Vogel-Heuser B and Wang L 2021 Industry 4.0 and Industry 5.0—Inception, conception and perception *J. Manuf. Syst.* **61** 530–5
[3] European Commission, Directorate-General for Research and Innovation, Breque, M., De Nul, L., Petridis, A. 2021 *Industry 5.0 : towards a sustainable, human-centric and resilient European industry,*
[4] Stathatos E and Vosniakos G C 2019 Real-time simulation for long paths in laser-based additive manufacturing: a machine learning approach *Int. J. Adv. Manuf. Technol.* **104** 1967–84
[5] Cooper K P 2014 Laser-based additive manufacturing: where it has been, where it needs to go *Laser 3D Manufacturing* vol 8970 p 897003
[6] Zhang Y, Sogn H, Cai M and Santana R 2023 An Additive Manufacturing Process Enables the 3D-Printed Application of Armors for Drill Bits *SPE/IADC Middle East Drilling Technology Conference and Exhibition* (SPE)
[7] Silbernagel C, Aremu A and Ashcroft I 2020 Using machine learning to aid in the parameter optimisation process for metal-based additive manufacturing *Rapid Prototyp. J.* **26** 625–37
[8] José Horst D and De Almeida Vieira R 2018 Additive Manufacturing at Industry 4.0: a Review *Artic. Int. J. Eng. Tech. Res.* 3–8
[9] Gupta M K, Etri H El, Korkmaz M E, Ross N S, Krolczyk G M, Gawlik J, Yaşar N and Pimenov D Y 2022 Tribological and surface morphological characteristics of titanium alloys: a review *Arch. Civ. Mech. Eng.* **22**
[10] Karthick M, Anand P, Meikandan M, Sekar S, Natrayan L and Bobe K 2022 Optimization of Plasma Arc Cutting Parameters on Machining of Inconel 718 Superalloy *J. Nanomater.* **2022**
[11] Nitoi A, Cristea D, Pop M A, Bedo T, Varga B and Munteanu D 2019 Aluminum based metastable alloys for additive manufacturing *IOP Conference Series: Materials Science and Engineering* vol 682
[12] Pramanik A 2014 Problems and solutions in machining of titanium alloys *Int. J. Adv. Manuf. Technol.* **70** 919–28
[13] Kotadia H R, Gibbons G, Das A and Howes P D 2021 A review of Laser Powder Bed Fusion Additive Manufacturing of aluminium alloys: Microstructure and properties *Addit. Manuf.* **46**
[14] Vafadar A, Guzzomi F, Rassau A and Hayward K 2021 Advances in metal additive manufacturing: A review of common processes, industrial applications, and current challenges *Appl. Sci.* **11** 1–33
[15] Niaki M K, Torabi S A and Nonino F 2019 Why manufacturers adopt additive manufacturing technologies: The role of sustainability *J. Clean. Prod.* **222** 381–92
[16] Forkel E, Baum J, Schumann C-A and Mueller E 2018 Smart Interoperable Logistics and Additive Manufacturing - Modern Technologies for Digital Transformation and Industry 4.0
[17] Leirmo T S and Martinsen K 2019 Evolutionary algorithms in additive manufacturing systems: Discussion of future prospects *Procedia CIRP*
[18] Maddikunta P K R, Pham Q-V, B P, Deepa N, Dev K, Gadekallu T R, Ruby R and Liyanage M 2022 Industry 5.0: A survey on enabling technologies and potential applications *J. Ind. Inf. Integr.* **26** 100257
[19] Urhal P, Weightman A, Diver C and Bartolo P 2019 Robot assisted additive manufacturing: A review *Robot. Comput. Integr. Manuf.* **59** 335–45
[20] Chu Y B and Chang C W 2021 A Mobile Augmented Reality Interface on Additive Manufacturing *Lecture Notes in Electrical Engineering* vol 741 LNEE pp 1–12
[21] European Commission 2021 Photonics and our daily life
[22] ISO/ASTM 2021 *ISO/ASTM 52900:2021 Additive manufacturing — General principles — Fundamentals and vocabulary* vol 2
[23] Nyamekye P 2021 *Life cycle cost-driven design for additive manufacturing: the frontier to sustainable manufacturing in laser-based powder bed fusion* (Lappeenranta-Lahti University of Technology, Lappenranta, Finland)
[24] WCED 1987 *Our Common Future: Report of the World Commission on Environment and Development*
[25] World Economic Forum 2021 *On the Importance of Human-Centricity and Data* vol 4075

[26] Trappey A J C, Trappey C V., Govindarajan U H, Sun J J and Chuang A C 2016 A Review of Technology Standards and Patent Portfolios for Enabling Cyber-Physical Systems in Advanced Manufacturing *IEEE Access* **4** 7356–82

[27] Boër C R and Jovane F 2002 Flexible and reconfigurable manufacturing automation for mass customisation *IFAC Proceedings Volumes (IFAC-PapersOnline)* vol 15 (IFAC Secretariat) pp 55–60

[28] Conboy K and Fitzgerald B 2004 Toward a conceptual framework of agile methods 37–44

[29] Sawyerr E and Harrison C 2020 Developing resilient supply chains: lessons from high-reliability organisations *Supply Chain Manag.* **25** 77–100

[30] Porsche AG. 2020 Prototype for small-series production: electric drive housing from a 3D printer

[31] Adebanjo D, Teh P L and Ahmed P K 2016 The impact of external pressure and sustainable management practices on manufacturing performance and environmental outcomes *Int. J. Oper. Prod. Manag.* **36** 995–1013

[32] Turner C and Oyekan J 2023 Personalised Production in the Age of Circular Additive Manufacturing *Appl. Sci.* **13** 4912

[33] Salminen K, Viljakainen A, Virkkunen R, Helaakoski H, Kokkonen A, Vita J and Alahuhta P 2021 *Towards Sustainable Manufacturing: How is autonomy paving the way towards a sustainable industry?* (Espoo)

[34] Babcinschi M, Freire B, Ferreira L, Señaris B, Vidal F, Vaz P and Neto P 2020 An integrated and interoperable AutomationML-based platform for the robotic process of metal additive manufacturing *Procedia Manufacturing* vol 51

[35] Massaro A 2022 *Electronics in Advanced Research Industries: Industry 4.0 to Industry 5.0* (Wiley-IEEE Press)

[36] Rahmani R, Karimi J, Resende P R, Abrantes J C C and Lopes S I 2023 Overview of Selective Laser Melting for Industry 5.0: Toward Customizable, Sustainable, and Human-Centric Technologies *Machines* **11** 522

[37] Alrubaye I D K and Fantoni G 2023 Toward green electrical discharge machining (EDM): state of art and outlook *Mach. Sci. Technol.* **27** 63–105

[38] M. P 2020 Additive Manufacturing of Tungsten Carbide Hardmetal Parts by Selective Laser Melting (SLM), Selective Laser Sintering (SLS) and Binder Jet 3D Printing (BJ3DP) Techniques *Lasers Manuf. Mater. Process.* **7** 338–71

[39] Deng D 2018 *Additively Manufactured Inconel 718 : Microstructures and Mechanical Properties*

[40] Paul C P, Yadav S, Nayak S K, Jinoop A N and Bindra K S 2022 Is Laser Additive Manufacturing Sustainable? pp 29–54

[41] Lee H, Lim C H J, Low M J, Tham N, Murukeshan V M and Kim Y J 2017 Lasers in additive manufacturing: A review *Int. J. Precis. Eng. Manuf. - Green Technol.* **4** 307–22

[42] Gibson I, Rosen D, Stucker B and Khorasani M 2021 Sheet Lamination *Additive Manufacturing Technologies* (Cham: Springer International Publishing) pp 253–83

[43] Chen T and Lin Y C 2017 Feasibility Evaluation and Optimization of a Smart Manufacturing System Based on 3D Printing: A Review *Int. J. Intell. Syst.* **32**

[44] Laitinen V, Sozinov A, Saren A, Chmielus M and Ullakko K 2021 Characterization of as-built and heat-treated Ni-Mn-Ga magnetic shape memory alloy manufactured via laser powder bed fusion *Addit. Manuf.* **39**

[45] Yan L, Chen Y and Liou F 2020 Additive manufacturing of functionally graded metallic materials using laser metal deposition *Addit. Manuf.* **31** 100901

[46] Rafiee M, Farahani R D and Therriault D 2020 Multi-Material 3D and 4D Printing: A Survey *Adv. Sci.* **7**

[47] WBCSD 2014 ' Value Chain ' Definitions and Characteristics *Value Chain* **6** 4–6

[48] Varriale V, Cammarano A, Michelino F and Caputo M 2023 Industry 5.0 and Triple Bottom Line Approach in Supply Chain Management: The State-of-the-Art *Sustainability* **15** 5712

[49] Kok Y, Tan X P, Wang P, Nai M L S, Loh N H, Liu E and Tor S B 2018 Anisotropy and heterogeneity of microstructure and mechanical properties in metal additive manufacturing: A critical review *Mater. Des.* **139** 565–86

[50] Martin V, Witz J F, Gillon F, Najjar D, Quaegebeur P, Benabou A, Hecquet M, Berté E, Lesaffre F, Meersdam M and Auzene D 2022 Low cost 3D printing of metals using filled polymer pellets *HardwareX* **11** e00292

[51] Gu D, Shi X, Poprawe R, Bourell D L, Setchi R and Zhu J 2021 Material-structure-performance integrated laser-metal additive manufacturing *Science (80-.).* **372**

[52] Zhu Y, Li J, Saleh M S, Pham H, Plateau T P, Panat R and Park J 2020 Towards high-performance Li-ion batteries via optimized three-dimensional micro-lattice electrode architectures *J. Power Sources*

[53] Wei C, Zhang Z, Cheng D, Sun Z, Zhu M and Li L 2021 An overview of laser-based multiple metallic material additive manufacturing: From macro: From micro-scales *Int. J. Extrem. Manuf.*

[54] Sola A, Trinchi A and Hill A J 2023 Self-assembly meets additive manufacturing: Bridging the gap between nanoscale arrangement of matter and macroscale fabrication *Smart Mater. Manuf.* **1** 100013

[55] Lawrence Livermore National Laboratory 2023 Fabrication of Functionally Graded Optical Components Using Additive Manufacturing *Lasers Opt. Sci. Technol.*

[56] Khodabakhshi F, Farshidianfar M H, Bakhshivash S, Gerlich A P and Khajepour A 2019 Dissimilar metals deposition by directed energy based on powder-fed laser additive manufacturing *J. Manuf. Process.* **43**

[57] Pinkerton A J 2016 Lasers in additive manufacturing *Opt. Laser Technol.* **78** 25–32

[58] Ur Rahman N, Matthews D T A, de Rooij M, Khorasani A M, Gibson I, Cordova L and Römer G W 2019 An Overview: Laser-Based Additive Manufacturing for High Temperature Tribology *Front. Mech. Eng.* **5**

[59] Da Silva A, Frostevarg J, Volpp J and Kaplan A F H 2021 Additive Manufacturing by laser-assisted drop deposition from a metal wire *Mater. Des.* **209** 109987

[60] Finnish Standards Association 2016 *SFS-EN ISO 17296-2 : 2016 Additive manufacturing . General principles . Part 2 : Overview of process categories and feedstock*

[61] Sun S, Brandt M and Easton M 2017 Powder bed fusion processes: An overview *Laser Additive Manufacturing: Materials, Design, Technologies, and Applications* (Elsevier Inc.) pp 55–77

[62] Gibson I, Rosen D, Stucker B and Khorasani M 2021 *Additive Manufacturing Technologies* (Springer International Publishing)

[63] Wohlers 2022 *Wohlers report 2022: 3D Printing and Additive Manufacturing Global State of the industry*

[64] Gu D 2015 *Laser additive manufacturing of high-performance materials*

[65] Porsche AG. 2023 3D printing technology optimises pistons for the powerful 911 GT2 RS

[66] Sun H, Zhu J, Baumann D, Peng L, Xu Y, Shakir I, Huang Y and Duan X 2019 Hierarchical 3D electrodes for electrochemical energy storage *Nat. Rev. Mater.* **4** 45–60

[67] Flower H M 1995 *High Performance Materials in Aerospace* ed H M Flower (Dordrecht: Springer Netherlands)

[68] Anush Raj B, Winowlin Jappes J T, Adam Khan M, Dillibabu V and Brintha N C 2020 Direct metal laser sintered (DMLS) process to develop Inconel 718 alloy for turbine engine components *Optik (Stuttg).* **202**

[69] Stryker 2023 Capri Cervical 3D Expandable

[70] Milleret A, Laitinen V, Ullakko K, Fenineche N and Attallah M M 2022 Laser powder bed fusion of (14 M) Ni-Mn-Ga magnetic shape memory alloy lattices *Addit. Manuf.* **60**

[71] Laitinen V 2021 *Laser powder bed fusion for the manufacture of Ni-Mn-Ga magnetic shape memory alloy actuators*

[72] ADDIFLAP Project 2021 Deposition of Ti6Al4V by W-DED-LB process

[73] Pressgraph Mediagroup, S L 2023 Aircraft parts using laser-wire additive manufacturing

[74] MAHLE GmbH 2023 MAHLE produces high-performance aluminum pistons using 3D printing for the first time

[75] Stryker 2022 Fixated on science, Monterey AL Interbody System

[76] Stryker 2023 Trident II Acetabular System

[77] Majeed A, Zhang Y, Ren S, Lv J, Peng T, Waqar S and Yin E 2021 A big data-driven framework for sustainable and smart additive manufacturing *Robot. Comput. Integr. Manuf.* **67**

[78] Mühler T, Gomes C M, Heinrich J and Günster J 2015 Slurry-based additive manufacturing of ceramics *Int. J. Appl. Ceram. Technol.* **12** 18–25

[79] Kokare S, Oliveira J P and Godina R 2023 A LCA and LCC analysis of pure subtractive manufacturing, wire arc additive manufacturing, and selective laser melting approaches *J. Manuf. Process.* **101** 67–85

[80] Shah I H, Hadjipantelis N, Walter L, Myers R J and Gardner L 2023 Environmental life cycle assessment of wire arc additively manufactured steel structural components *J. Clean. Prod.* **389**

[81] Khorasani A M, Gibson I, Veetil J K and Ghasemi A H 2020 A review of technological improvements in laser-based powder bed fusion of metal printers *Int. J. Adv. Manuf. Technol.* **108** 191–209

[82] Sing S L, Huang S, Goh G D, Goh G L, Tey C F, Tan J H K and Yeong W Y 2021 Emerging metallic systems for additive manufacturing: In-situ alloying and multi-metal processing in laser powder bed fusion *Prog. Mater. Sci.* **119**

[83] Zhu L, Li N and Childs P R N 2018 Light-weighting in aerospace component and system design *Propuls. Power Res.* **7** 103–19

[84] Oliveira J P, LaLonde A D and Ma J 2020 Processing parameters in laser powder bed fusion metal additive manufacturing *Mater. Des.* **193** 108762

[85] Peng X, Kong L, Fuh J Y H and Wang H 2021 A review of post-processing technologies in additive manufacturing *J. Manuf. Mater. Process.* **5**

[86] Daraban A E O, Negrea C S, Artimon F G P, Angelescu D, Popan G, Gheorghe S I and Gheorghe M 2019 A deep look at metal additive manufacturing recycling and use tools for sustainability performance *Sustain.* **11** 5494
[87] Dursun T and Soutis C 2014 Recent developments in advanced aircraft aluminium alloys *Mater. Des.* **56** 862–71
[88] Uhlmann E, Kersting R, Klein T B, Cruz M F and Borille A V 2015 Additive Manufacturing of Titanium Alloy for Aircraft Components *Procedia CIRP* vol 35 pp 55–60
[89] Bosio F, Aversa A, Lorusso M, Marola S, Gianoglio D, Battezzati L, Fino P, Manfredi D and Lombardi M 2019 A time-saving and cost-effective method to process alloys by Laser Powder Bed Fusion *Mater. Des.* **181**
[90] Karmuhilan M and Kumanan S 2022 A Review on Additive Manufacturing Processes of Inconel 625 *J. Mater. Eng. Perform.* **31** 2583–92
[91] Yuan B, Guss G M, Wilson A C, Hau-Riege S P, DePond P J, McMains S, Matthews M J and Giera B 2018 Machine-Learning-Based Monitoring of Laser Powder Bed Fusion *Adv. Mater. Technol.* **3**
[92] Liu C, Law A C C, Roberson D and Kong Z (James) 2019 Image analysis-based closed loop quality control for additive manufacturing with fused filament fabrication *J. Manuf. Syst.* **51** 75–86
[93] Pragana J P M, Sampaio R F V, Bragança I M F, Silva C M A and Martins P A F 2021 Hybrid metal additive manufacturing: A state–of–the-art review *Adv. Ind. Manuf. Eng.* **2** 100032
[94] Dahmus J B and Gutowski T G 2004 An environmental analysis of machining *American Society of Mechanical Engineers, Manufacturing Engineering Division, MED* vol 15 pp 643–52
[95] Brandt M 2017 The role of lasers in additive manufacturing *Laser Additive Manufacturing: Materials, Design, Technologies, and Applications*
[96] Stieberova B, Broumova M, Matousek M and Zilka M 2022 Life Cycle Assessment of Metal Products Produced by Additive Manufacturing: A Metal Mold Case Study *ACS Sustain. Chem. Eng.* **10** 5163–74
[97] Liu C, Tian W and Kan C 2022 When AI meets additive manufacturing: Challenges and emerging opportunities for human-centered products development *J. Manuf. Syst.* **64** 648–56
[98] Mamun A Al, Liu C, Kan C and Tian W 2022 Securing cyber-physical additive manufacturing systems by in-situ process authentication using streamline video analysis *J. Manuf. Syst.* **62** 429–40
[99] Godina R, Ribeiro I, Matos F, Ferreira B T, Carvalho H and Peças P 2020 Impact assessment of additive manufacturing on sustainable business models in industry 4.0 context *Sustain.* **12**
[100] Bogers M, Hadar R and Bilberg A 2016 Additive manufacturing for consumer-centric business models: Implications for supply chains in consumer goods manufacturing *Technol. Forecast. Soc. Change* **102** 225–39
[101] Öberg C 2019 Additive manufacturing – digitally changing the global business landscape *Eur. J. Manag. Bus. Econ.* **28** 174–88
[102] Blakey-Milner B, Gradl P, Snedden G, Brooks M, Pitot J, Lopez E, Leary M, Berto F and du Plessis A 2021 Metal additive manufacturing in aerospace: A review *Mater. Des.* **209** 110008
[103] Gebler M, Schoot Uiterkamp A J M and Visser C 2014 A global sustainability perspective on 3D printing technologies *Energy Policy*
[104] Narsimlu, K; Pathak G. Arhendu; Mulky G. Avinash; Yavarna C 2017 A market analysis on the impact of additive layer manufacturing technologies on aerospace and defense supply chain *Int. J. Manag.* **8** 171–87

Acknowledgments

The authors would like to acknowledge Analytics-based Management for Business and Manufacturing Industry (AMBI) platform and Laboratory of Laser Materials Processing and Additive Manufacturing at Lappeenranta University of Technology (LUT) for supporting this study. Authors would also like to express their gratitude to all partners and companies Verkostoitumisella voimaa 3D-tulostukseen (VERKOTA) project for supporting this study.

IOP Conf. Series: Materials Science and Engineering 1296 (2023) 012002 doi:10.1088/1757-899X/1296/1/012002

Preliminary study of knowledge transfer aspect in the creation of a framework for education and training of digital manufacturing technologies

H Piili[1], A Huusko[1], A Kurvinen[2]

[1] Department of Mechanical Engineering, Faculty of Technology, University of Turku, Turku, 20520, Finland
[2] Department of Business, International Business, LAB University of Applied Sciences, Lappeenranta, 53850, Finland

anu.kurvinen@lab.fi

Abstract. Digitalization is changing the industry. As this change accelerates its speed, it also requires a transformation process where knowledge transfer between industry and research institutes play a significant role. There is a need to be more fluent, flexible, and efficient in order to get the latest research results into industrial implementation as quickly as possible. The challenge in knowledge transfer is that its speed in the current stage is too slow compared to the speed of development and changes required by digitalization of traditional manufacturing industries. The motivation for this study is the gap in knowledge transfer. One emerging digital transformation is the establishment of modern digital manufacturing technologies, e.g., additive manufacturing (AM). There are different approaches to supporting the industry in this transformation. Knowledge transfer can happen, for example, through education (e.g., master students) and industrial training, but also the fluent transfer of the latest research results from research institutes to companies is needed. University education needs to support the requirements of the manufacturing industry by providing future experts with skills to smooth the transformation process and bring novel technology applications, such as AM, to industrial-scale use. The article discusses how university education can support future competence-building in the industry. Similar needs are also present in industrial training, which universities often provide. Both education and training need to be improved from fundamental approaches to explain how this new knowledge should be created, i.e., how knowledge transfer happens most efficiently. The outcome of this article is the basis of the framework for education and training of digital manufacturing technologies by using modern learning methods and tools. More detailed pedagogical and knowledge transfer models can be developed and applied when this framework is created.

1. Introduction

Additive manufacturing (shortened as AM), or more commonly 3D printing, consists of wide variety of different modern manufacturing technologies. AM is based on direct fabrication of a digital 3D model to a final product which is fabricated adding material layer by layer. This is where term AM has its origin. It is not only material what is added, but it is also value, properties etc. which are added. AM enables production of different and even better products compared to conventional manufacturing technologies. As in many cases, AM is a new method introduced in production, one of the challenges for design processing and engineers, is to identify suitable applications for AM. When successfully applied, AM can help companies to create additional value in e.g. new product development and order fulfillment processes. [1]

An estimation of potential of AM can be gathered by considering the potential of laser cutting, which is one of the most widely used modern manufacturing technologies, as in [2] and [3] With a market value of around four billion euros and an annual growth rate of approximately 10%, laser cutting has been a significant technology for over four decades. However, the emergence of AM and 3D printing promises radical advancements in the production of three-dimensional parts, mirroring the transformative impact of laser cutting on flat sheet products. Currently, laser devices utilized in AM constitute merely 1% of all laser devices employed across various fabrication technologies worldwide. This statistic highlights the growth potential for lasers in AM in the coming years. Even with a conservative estimate, it is highly likely that the technology will experience a minimum of 100% growth, making it remarkable in manufacturing, as e.g. in [4], [5] and [6].

Metal AM is a decades old manufacturing method which has grown to a point where it is a potential way of manufacturing for example for certain aviation and medical applications. AM in general was originally used for prototyping purposes, but nowadays it is mainly used for fabricating of functional parts. AM is utilized in many industries globally, but still lacks for example quality assurance systems ([7] [8] etc.). Utilization level of metal AM is presumable lower than it could be in Finland. Lack of knowledge about possibilities of metal AM might be a reason for that. Finding suitable parts for metal AM also requires lot of knowledge which has not been taught to most engineers during their education. Therefore, there are companies utilizing metal AM on a daily basis whilst some have never tried it. [9]

Education and training play a crucial role in the successful industrial implementation of modern technologies. Education in this study means education in universities, and training is industrial training. When both the new generation entering the workforce and those who have been in it for some time comprehend the potential and limitations of these technologies, it opens opportunities for rethinking product design. AM and 3D printing offer freedom to design new products, new methods of production and generally new ways of doing things. The vast potential of product design serves as a catalyst for the widespread adoption of AM and 3D printing, as a deeper understanding of the technology allows for innovative and optimized designs that leverage the capabilities of these advancements, as e.g. in [10], [11] and [12].

AM and 3D printing present a significant opportunity for Finland and its industrial sector, leveraging the strong industrial expertise and proficiency in information and communication technology (ICT) in country, as concluded in [13] and [14]. Authors have concluded to discuss in this study both AM and 3D printing, as both industrial applications (additive manufacturing) and consumer applications (3D printing) are included to this theme in upper level. To fully harness the potential of this technology, companies must explore the suitability of their products for AM and 3D printing, as well as test product enhancements and consider complete redesigns. Equally important is providing education and training opportunities for the existing workforce, enabling successful industrial implementation. This holds particular significance for Finland, as it strives to maintain its competence in global markets. By equipping both new and experienced professionals with the necessary knowledge and skills, Finnish industry can maximize the benefits of AM and 3D printing, fostering innovation and competitiveness in the global arena (e.g.in [10], [15]).

A number of publications related to topic of this study were surveyed via Scopus with different keyword combinations. Scopus is a comprehensive database covering a wide range of academic disciplines, ensuring it to be trustful database for literature survey.

2. Aim, purpose and methodology of this study

The aim of this study is to address the gap in knowledge transfer between research institutes, universities and industry amidst the rapid digitalization of traditional manufacturing industries, as in [16] and [17]. The focus is on supporting transformation of industry through education, such as master's programs, and industrial training. These education means in this article all AM education provided by universities (also universities of applied sciences, UASs) or other similar institutes, and it is called university education. The study explores how university education can equip future experts with the skills needed for the implementation of modern digital manufacturing technologies, like AM, at an industrial scale. It emphasizes the need for improved education and training approaches that efficiently facilitate knowledge transfer. The study lays the groundwork for a framework to enhance education and training in digital manufacturing technologies, highlighting the importance of further research and development in this area. Discussion on right balance between studying explicit content that will, depending on subject, become easily out-of-date and meta-skills that are useful but at the same time may leave learners an experience of not learning anything specific has been going on for a long time. Skills that are easy to teach and where learning is easily measured, are skills that are also easily automated, as concluded in [18], [19] and [20]. Demand for AM courses has grown dramatically. Therefore, even complete master's degrees are offered by some universities nowadays. Individual AM study courses are offered by all education levels and low-cost 3D printers can be found even in elementary schools. However, industrial training should also be offered in the same scale in order to increase utilization of AM. [21]

This study is scientifically important as it addresses the crucial issue of knowledge transfer between research institutes and the industry in the context of digitalization in traditional manufacturing industries. It provides insights into the role of education in equipping future experts and emphasizes the need for improved approaches to facilitate efficient knowledge transfer. The study establishes a foundation for further research and development in enhancing education and training in digital manufacturing technologies. The industrial importance of this study lies in its potential to bridge the gap in knowledge transfer between research institutes and the industry during the digitalization of traditional manufacturing sectors. By focusing on education and training, the study aims to equip the industry with skilled professionals capable of implementing digital manufacturing technologies effectively.

Background and basis of framework for this study was done as literature review. A Scopus survey was carried out also to see number of research articles related to this study between 2010-2022. In this article survey following searches were done: "additive manufacturing" OR "3D printing", "engineering" AND "additive manufacturing" OR "3D printing", "industry" AND "additive manufacturing" OR "3D printing", "education" AND "additive manufacturing" OR "3D printing", "learning" AND "additive manufacturing" OR "3D printing", "knowledge transfer" AND "additive manufacturing" OR "3D printing". Based on the survey result, the research on additive manufacturing/3D printing and engineering has increased during the period of 2010--2022, which reflects the fact that interest towards industrial use of AM and its different variations has been increasing. At the same time there is a corresponding development and increasing trend in research concerning education, learning and training of AM. When looking for studies handling knowledge transfer in AM, there were only 1-5 research publications annually between 2010-2015. When approaching year 2019 and after that there is a clear increase in research concerning knowledge transfer. This shows that understanding the full potential of a novel manufacturing methodology to a company requires a more in-depth learning and adoption of new knowledge on different levels of organization. In this article different ways of learning and acquiring new knowledge are discussed.

3. Learning in education and training of additive manufacturing

AM and 3D printing are revolutionary technologies poised to shape the future of global industries. These emerging technologies, including additive manufacturing, have already begun to disrupt the manufacturing sector, unlocking potential for previously unimagined applications. Across Europe, several universities are offering master's degree programs focused on AM. These academic programs are essential in cultivating the next generation of experts who will drive innovation and advancement in AM and 3D printing. By equipping students with specialized knowledge and skills, these universities are paving the way for the widespread adoption of these technologies and fostering a competitive edge for industries in the global business arena. As AM continues to evolve, the demand for educational programs in this field is likely to grow, fueling further advancements and unlocking even larger possibilities. These are deduced e.g. in [3], [22], [23] and [24].

Deloitte and The Manufacturing Institute [25] conducted a study among US manufacturing industry, in order to reveal the future of manufacturing work. According to the study, there is a widening gap between the amount of skilled new employees and the jobs that need to be done. [25] University pedagogics and range of available studies are thus in key role in formulating the path towards potential exploitation of the AM technology in industry applications ([26], [27] etc.)

As this study focuses on educating a field of technology that is expected to some extent change and develop the traditional manufacturing industry, there is a need to think pedagogical models that enable learning and development but at the same time encourage creativity and innovativeness. In the modern world, there is lot of information available. The challenge is that, one has to be able to think critically and have skills to synthesize and put the information into action in a wise way. [28] According to Fadel et al. [29], there is also an important dimension needed – meta-learning. Meaning that there are internal processes required for learning to take place, namely reflection and adaptation of our learning. The educators have to rethink the methods of learning and the choice needs to be done between teaching skills vs. teaching knowledge, as in [29] and [30]. A summary of the choices to be made in education is presented in table 1. [30]

Teacher-directed	Learner-centered
Direct instruction	Interactive exchange
Knowledge	Skills
Content	Process
Basic skills	Applied skills
Facts and principles	Questions and problems
Theory	Practice
Curriculum	Projects
Time-slotted	On-demand
One-size-fits-all	Personalized
Competitive	Collaborative
Classroom	Global community
Text-based	Web-based
Summative tests	Formative evaluations
Learning for school	Learning for life

Table 1. Comparison of teacher-directed and learner-centered approaches. [30]

In an ideal case, the process of learning enables the student to test or experiment the theories learnt and reflect the learnings with the lecturer or coach. That way the process of learning and new knowledge creation can take place.

3.1 Lifelong, life-wide and life-deep learning

Lifelong, life-wide and life-deep learning is very often mentioned as crucial part of education, training and generally working life in the future [31]. According to UNESCO, lifelong learning means the integration of learning into various aspects of life, accommodating people of all ages and contexts. It includes diverse learning activities within different settings, such as family, school, community, and the workplace, utilizing both formal and informal educational methods to address a wide range of learning needs and demands. [32]

The ideal learner is someone who is flexible and motivated to continually develop themselves, particularly in response to evolving work demands or shifts in their profession. Lifelong learners view professional development as both a challenge and an essential aspect of modern work life, recognizing its significance in staying relevant and thriving in their careers. [33] Lifelong learning is very often mixed with life-wide learning. Lifelong learning consists of informal/formal learning in context of professional learning, whereas life-wide learning includes whole life with all its aspects. These are discussed e.g. in [34], [35] and [36]. Hietanen and Rubin [37] conclude that education and training should be discussed in context of human lifespan and situations in human life, as whole human life is one kind of learning situation. This idea is shown in figure 1.

The presence of a third dimension in lifelong and life-wide learning, known as life-deep learning, is highlighted by [31], [38] and [39]. This concept is intricately linked to the holistic and multilayered development of individuals, encompassing the formation of their lifespan in its entirety, including the diverse changes, transitions, and challenges they experience along the way. Life-deep learning emphasizes a comprehensive and profound approach to personal growth and development. Lifelong, life-wide and life-deep learning can be shown as three-dimensional illustration as figure 2 shows.

Lifelong, life-wide and life-deep learning are consequences of changes in todays´ working life. Digitalization and modern technologies require that learning should happen in all aspects of human life. If not, it will lead to lowered level of skills needed in work life and in worst case to digital exclusion. [39] It can be concluded that constant changes in work life and skills needed there is permanent situation nowadays. This can be observed from expectations set to education and industrial learning of AM and 3D printing. This education and training should in first hand be able to provide such tools for students and trainees that they can update frequently their know-how proactively and independently. [40]

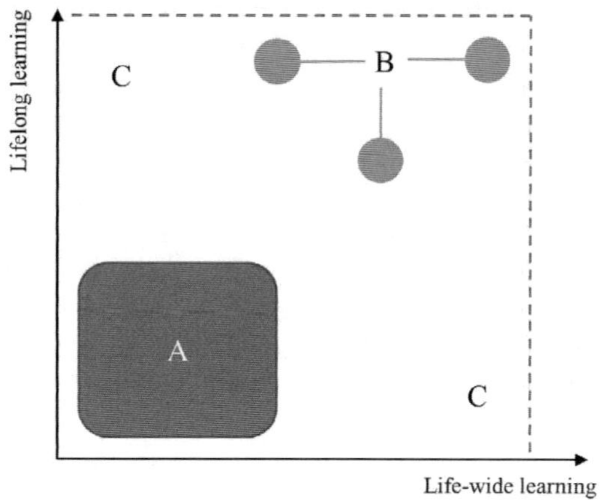

Figure 1. Lifelong and life-wide learning cover whole human life. A = From comprehensive school to university, B = industrial training in work life, C = human lifespan. [37]

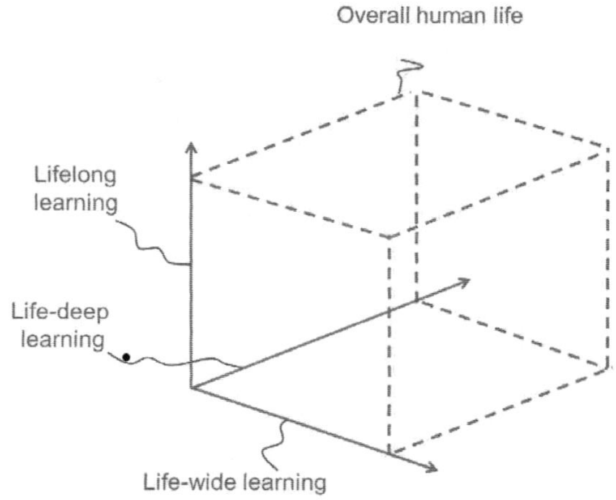

Figure 2. Lifelong, life-wide and life-deep learning is widened to concern lifespan, situations and emotions of human being and this way they form three-dimensional integrity. [36]

3.2 Creative thinking

According to [10] and [12], role of education is important when a new technology is industrially implemented. Development, planning and execution of education for AM and 3D printing is challenging as this area develops very fast. Planning of education for AM and 3D printing requires collection pieces of data from various sources. [12]

Introduction of new technologies or new ways to utilize them requires encouraging innovative thinking and also creative thinking in a company [41]. New technology cannot just be copied between industries and organization, also innovativeness and creation of new type of solutions is important [42].

Education plays a crucial role in the industrial implementation of new technologies. Additive manufacturing and 3D printing offer opportunities for designing new products and revolutionizing production methods. However, the rapid development of this field poses challenges in the development, planning, and execution of education programs, as in [2] and [24]. To effectively plan education for AM and 3D printing, it is necessary to gather data from various sources to stay abreast of the latest advancements. Incorporating new technologies or innovative ways of utilizing them requires fostering innovative and creative thinking within companies. It is not enough to simply copy technologies between industries and organizations; it is essential to encourage innovativeness and the creation of novel solutions. These are further discussed in [43] and [44]. By incorporating these principles into education, students are motivated to engage in lifelong, life-wide, and life-deep learning. They are encouraged to continuously learn and apply their knowledge and skills in various contexts, expanding their learning experiences beyond traditional classroom settings. [45]

Companies are looking for ways to add value to the product development and order fulfillment processes by introducing AM in the production. Before implementing reliable identification and selection processes to find out where AM promises the most added value. A model of value clusters of AM has been studied [1]. The study mapped e.g. design strategies and evaluation matrix for assessing the parts to be produced by AM. The proposed method for value-driven part identification and selection was applied and evaluated in various technology transfer projects between AM experts and experienced engineers in the industry both in minor design change cases and major design changes. Engineer ingenuity to identify suitable parts was combined with AM expertise. In the study, it was proposed that selecting parts manufactured by AM consists of two phases. First, identifying parts that might benefit from a change to additive manufacturing. Secondly, selection phase that evaluates the identified parts and leads to selection of ones that best fit by the properties and features of the application. This study example confirms that modern AM education can not only rely on introducing copying the technology itself but also needs learner-centered methodologies that encourage creativity thinking in the use cases. [1]

4. Discussion

In the era of Industry 4.0, robots and artificial intelligence are transforming the economy. There is something that cannot be replaced by automation, that is creativity. According to the study conducted by IBM, creativity on all levels of actions, is a way to respond to the complexity of the business environment of current times and turn the challenges to advantages. [17] The ongoing development of pedagogical methods in AM and 3D printing provides an opportunity to foster creativity and thinking beyond traditional manufacturing approaches. To enhance creative thinking of students, it is crucial to include a more focused emphasis on innovation and problem-solving in their courses. Practical exercises and concrete steps can help students gain hands-on experience and apply their skills effectively. It is obvious that there is the profound impact of digitalization on industries and it underscores the crucial role of knowledge transfer between industry and research institutes in enabling the transformation process. There is clear need for a more efficient and agile knowledge transfer approach to quickly implement the latest research findings into industrial practices. [16] However, the current speed of knowledge transfer lags behind the rapid pace of development required by digitalization in traditional manufacturing sectors, presenting a significant challenge.

Results of Scopus survey done in this study show that there is a constant, and steady increase in research related to additive manufacturing or 3D printing during past ten years (2010-2022). At the same time, it can be stated that there is an increase in the research of education and learning of AM and 3D printing. Knowledge transfer in AM and 3D printing is relatively new research topic according to this Scopus search, but there is some research already done in the area. This reflects the fact that traditional education and training methods also need to be complemented with new methods and learning environments that support knowledge transfer of research results to the use of manufacturing industry.

There is a gap in knowledge transfer, particularly in the context of emerging digital manufacturing technologies like AM, as this was shown in literature survey done with Scopus. This issue needs further study. It is important to provide new approaches to support the industry during this transformation, including education (such as through master's programs) and industrial training. University education plays a vital role in equipping future experts with the necessary skills to facilitate the smooth integration of novel technologies like AM into industrial-scale applications. [46]

Both education and training need improvement, with a focus on fundamental approaches that explain how knowledge transfer can happen most efficiently. This is why there is a need of framework for education and training in digital manufacturing technologies. The framework proposed in this article is for education and training in digital manufacturing technologies, aiming to improve knowledge transfer efficiency and provide a basis for further research and the development of pedagogical and knowledge transfer models. [22] Engaging students in real-world problem-solving tasks assigned by companies can further stimulate their creativity and motivation. [47] By working on these assignments, students develop a deeper understanding of how their skills can be applied in practical situations and are encouraged to think outside the box. Expanding the scope of study courses to include students from diverse disciplines beyond mechanical engineering can also enhance creativity. By forming multidisciplinary groups, students can benefit from different perspectives, knowledge, and skill sets. Collaborating with peers from various backgrounds fosters creativity and enables the exploration of unconventional solutions that may not have been considered within a single discipline.

In Finland, there is active cooperation between AM industry and research which has been developed during last years to be closer than ever. For example, there are ecosystem of Finnish AM Ecosystem (FAME) between industry and research was established 2020, and it has several companies in field of AM. FAME has several AM projects going on between industry and research institutes, this is further introduced in [48] and [49]. There is a rich variety of companies operating in different parts of AM sector – as AM production companies, various type of equipment providers from AM printers for metal to biotechnological technology as well as programming companies. COVID-19 pandemic challenged global supply chains and enabled creation and implementation of emerging innovative ideas. These are discussed e.g. in [9], [50] and [51].

5. Conclusions

The aim and purpose of this study was to address the gap in knowledge transfer of AM technologies between research institutes and the industry organizations. The study was done as literature review study in Scopus. Number of published research articles handling topics 3D printing, AM, knowledge transfer and training of the technologies was analysed as a Scopus study. The time range the articles were published was 2010-2020. Main results of the study are that the meaning of education in AM and 3D printing, and also sharing the knowledge of different methods has increased heavily during the analysed period. The number of research articles handling knowledge transfer analysis also showed that knowledge transfer as a research topic is not yet as deeply investigated, and assumedly thus meaning of it not as deeply recognized. The article survey shows that development of pedagogical methods is an ongoing process that evolves over time. AM and 3D printing are methodologies where all the application areas are not found yet.

As a result, to the main purpose of the article, "finding solutions for knowledge transfer aspect in the creation of a framework for education and training of digital manufacturing technologies", it can be stated that knowledge transfer process is a complex process. When developing training organized for industry representatives, involving different company internal groups participating the design processes and making the evaluations of customer and business values of applying AM in the production could be one solution enforcing knowledge transfer. This in general gives the requirement to learn creativity and ability to think outside the traditional manufacturing industry schemes. Collaboration between industry, universities and research institutes in AM in Finland continues to be all the time closer, fostering a diverse range of companies across the AM sector. These companies encompass AM production, equipment providers for metal and biotechnological technologies, and programming firms. The COVID-19 pandemic disrupted global supply chains, prompting the birth and adoption of novel, ingenious concepts.

In the future, one possibility to strengthen the knowledge transfer, creativity thinking and innovativeness of the students, would be to include even a clearer step of innovation creation and problem-solving to the study course since the fact that the lecturers are telling about this, is not making the idea concrete enough to all the students. Other ways of increasing creativity among university students could be trying to get some company-given tasks to be the student groups which would in some cases give more cause to real problem-solving assignment. Also increasing the interest of the study course towards student groups outside mechanical engineering and forming more multidisciplinary groups, when possible could be one solution.

6. Author contribution

During the preparation and writing of this work no AI based tools were used. Authors reviewed and edited the content of this work independently and take full responsibility of this contribution. All authors drafted the idea of this publication. Further brainstorming was executed by all authors. The actual drafting of the framework for education and training of digital manufacturing technologies introduced in this article was discussed, commented on, modified and concluded by all the authors. All authors have contributed to the literature review part of the article, Chapter "2. Aim, purpose and methodology of this study" and Chapter "3. Learning in education and training of additive manufacturing" introduction. All authors have contributed to chapter "4. Discussion" and "5 Conclusions". Heidi Piili was responsible for sub-chapter "3.1 Lifelong, life-wide and life-deep learning" and the Scopus research. Anu Kurvinen is responsible for sub-chapter "3.2 Creative thinking". The final draft of the publication was discussed, commented on, edited, modified and concluded by all authors.

7. References

[1] Klahn C, Fontana F, Leutenecker-Twelsiek B, Meboldt M. Mapping value clusters of additive manufacturing on design strategies to support part identification and selection. Rapid Prototyp J 2020;26:1797–807. https://doi.org/10.1108/RPJ-10-2019-0272.

[2] Khorasani M, Loy J, Ghasemi AH, Sharabian E, Leary M, Mirafzal H, et al. A review of Industry 4.0 and additive manufacturing synergy. Rapid Prototyp J 2022;28:1462–75. https://doi.org/10.1108/RPJ-08-2021-0194.

[3] Stavropoulos P, Lianos AK, Bikas H, Mourtzis D. Skills Requirements for the 4th Industrial Revolution: The Additive Manufacturing case. MATEC Web Conf 2020;318:01021. https://doi.org/10.1051/matecconf/202031801021.

[4] Naresh, Khatak P. Laser cutting technique: A literature review. Mater Today Proc 2022;56:2484–9. https://doi.org/10.1016/j.matpr.2021.08.250.

[5] Gonzalez G, Roppolo I, Pirri CF, Chiappone A. Current and emerging trends in polymeric 3D printed microfluidic devices. Addit Manuf 2022;55:102867. https://doi.org/10.1016/j.addma.2022.102867.

[6] Kantaros A, Diegel O, Piromalis D, Tsaramirsis G, Khadidos AO, Khadidos AO, et al. 3D printing: Making an innovative technology widely accessible through makerspaces and outsourced services. Mater Today Proc 2022;49:2712–23. https://doi.org/10.1016/j.matpr.2021.09.074.

[7] Gisario A, Kazarian M, Martina F, Mehrpouya M. Metal additive manufacturing in the commercial aviation industry: A review. J Manuf Syst 2019;53:124–49. https://doi.org/10.1016/J.JMSY.2019.08.005.

[8] Blakey-Milner B, Gradl P, Snedden G, Brooks M, Pitot J, Lopez E, et al. Metal additive manufacturing in aerospace: A review. Mater Des 2021;209:110008. https://doi.org/10.1016/j.matdes.2021.110008.

[9] Kaivo-oja J, Ahlqvist T, Kuusi O, Linturi R, Roth S. New industrial platforms and radical technology foresight: the case of 3D printing in Finland and Europe. Int J Manuf Technol Manag 2018;32:65–84. https://doi.org/10.1504/IJMTM.2018.089464.

[10] Piili H. Näkökulmia 3D-tulostuksen opetukseen ja koulutukseen. Lappeenranta University of Technology; 2015.

[11] Flores Ituarte I, Huotilainen E, Mohite A, Chekurov S, Salmi M, Helle J, et al. 3D printing and applications: academic research through case studies in Finland. 85-2 Proc Nord 2016 Vol 2 Trondheim Nor 10th - 12th August 2016 2016:032–41.

[12] Pikkarainen A, Piili H. Implementing 3D Printing Education Through Technical Pedagogy and Curriculum Development. Int J Eng Pedagogy IJEP 2020;10:95–119. https://doi.org/10.3991/ijep.v10i6.14859.

[13] Ituarte IF, Salmi M, Ballardini RM, Tuomi J, Partanen J. Additive Manufacturing in Finland: Recommendations for a Renewed Innovation Policy. Phys Procedia 2017;89:70–9. https://doi.org/10.1016/j.phpro.2017.08.002.

[14] Lindqvist M, Piili H, Salminen A. Benchmark Study of Industrial Needs for Additive Manufacturing in Finland. Phys Procedia 2016;83:854–63. https://doi.org/10.1016/j.phpro.2016.08.089.
[15] Mäkitie AA, Salmi M, Lindford A, Tuomi J, Lassus P. Three-dimensional printing for restoration of the donor face: A new digital technique tested and used in the first facial allotransplantation patient in Finland. J Plast Reconstr Aesthet Surg 2016;69:1648–52. https://doi.org/10.1016/j.bjps.2016.09.021.
[16] Ma Y, Li B. Effect of digitalization on knowledge transfer from universities to enterprises: Evidence from postdoctoral workstation of Chinese enterprises. Technol Soc 2022;71:102102. https://doi.org/10.1016/j.techsoc.2022.102102.
[17] Vuori V, Helander N, Okkonen J. Digitalization in knowledge work: the dream of enhanced performance. Cogn Technol Work 2019;21:237–52. https://doi.org/10.1007/s10111-018-0501-3.
[18] Schooling Redesigned: Towards Innovative Learning Systems | READ online. Oecd-Ilibraryorg n.d. https://read.oecd-ilibrary.org/education/schooling-redesigned_9789264245914-en (accessed May 18, 2023).
[19] Sony M, Mekoth N. Employee adaptability skills for Industry 4.0 success: a road map. Prod Manuf Res 2022;10:24–41. https://doi.org/10.1080/21693277.2022.2035281.
[20] Cliff A, Walji S, Jancic Mogliacci R, Morris N, Ivancheva M. Unbundling and higher education curriculum: a Cultural-Historical Activity Theory view of process. Teach High Educ 2022;27:217–32. https://doi.org/10.1080/13562517.2019.1711050.
[21] Rof A, Bikfalvi A, Marques P. Digital Transformation in Higher Education: Intelligence in Systems and Business Models. In: Kahraman C, Haktanır E, editors. Intell. Syst. Digit. Transform. Theory Appl., Cham: Springer International Publishing; 2023, p. 429–52. https://doi.org/10.1007/978-3-031-16598-6_18.
[22] Borgianni Y, Maccioni L, Spena PR, Shunmugavel MK. University Education in Additive Manufacturing and the Need to Boost Design Aspects. Proc Des Soc Int Conf Eng Des 2019;1:629–38. https://doi.org/10.1017/dsi.2019.67.
[23] Kianian B, Tavassoli S, Larsson TC, Diegel O. The Adoption of Additive Manufacturing Technology in Sweden. Procedia CIRP 2016;40:7–12. https://doi.org/10.1016/j.procir.2016.01.036.
[24] Kozák Š, Ružický E, Štefanovič J, Schindler F. Research and education for industry 4.0: Present development. 2018 Cybern. Inform. KI, 2018, p. 1–8. https://doi.org/10.1109/CYBERI.2018.8337556.
[25] 2018 Manufacturing Skills Gap Study | Deloitte US n.d. https://www2.deloitte.com/us/en/pages/manufacturing/articles/future-of-manufacturing-skills-gap-study.html (accessed May 18, 2023).
[26] Smart work: The transformation of the labour market due to the fourth industrial revolution (I4.0). Int J Bus Econ Sci Appl Res IJBESAR 2017;10:47–66.
[27] Holmwood J, Marcuello Servós C. Challenges to Public Universities: Digitalisation, Commodification and Precarity. Soc Epistemol 2019;33:309–20. https://doi.org/10.1080/02691728.2019.1638986.
[28] Juvonen P, Kurvinen A. Developing Professional Skills Through Facilitated Dialogue. In: Auer ME, Tsiatsos T, editors. Interact. Mob. Commun. Technol. Learn., Cham: Springer International Publishing; 2018, p. 501–11. https://doi.org/10.1007/978-3-319-75175-7_49.
[29] (3) (PDF) Four-Dimensional Education n.d. https://www.researchgate.net/publication/318430582_Four-Dimensional_Education (accessed May 18, 2023).
[30] Trilling B, Fadel C. 21st century skills: Learning for life in our times. Hoboken, NJ, US: Jossey-Bass/Wiley; 2009.
[31] Banks J, Au K, Ball AF, Bell P, Gordon E, Gutierrez K, et al. Learning in and out of school in diverse environments: Life-Long, Life-Wide, Life-Deep 2007.

[32] Embracing a culture of lifelong learning: contribution to the Futures of Education initiative;report;a transdisciplinary expert consultation - UNESCO Digital Library n.d. https://unesdoc.unesco.org/ark:/48223/pf0000374112 (accessed May 18, 2023).
[33] Häkkinen T. Ajatuksia alumnitoiminnan kehittämisestä 2008. http://www.theseus.fi/handle/10024/20215 (accessed May 18, 2023).
[34] Elinikäisen oppimisen politiikassa määritellään käsitys vaadittavasta kansalaisuudesta (Väitös: KM Heikki Kinnari, 28.3.2020, aikuiskasvatustiede) n.d. https://www.utu.fi/fi/ajankohtaista/vaitos/elinikaisen-oppimisen-politiikassa-maaritellaan-kasitys-vaadittavasta (accessed May 18, 2023).
[35] Desjardins R. Determinants of literacy proficiency: a lifelong-lifewide learning perspective. Int J Educ Res 2003;39:205–45. https://doi.org/10.1016/j.ijer.2004.04.004.
[36] HEIKKINEN HLT, TYNJÄLÄ HJAP. Teacher education and development as lifelong and lifewide learning. Peer-Group Mentor. Teach. Dev., Routledge; 2012.
[37] Hietanen O, Rubin A. Oppimisympäristöjen tulevaisuus. Tutkimuksen ja yhteiskunnan haasteita. n.d.
[38] Sacco K, Falk JH, Bell J. Informal Science Education: Lifelong, Life-Wide, Life-Deep. PLOS Biol 2014;12:e1001986. https://doi.org/10.1371/journal.pbio.1001986.
[39] Smith R. Self-construction and social transformation: lifelong, lifewide and life-deep learning. Int J Lifelong Educ 2018;37:393–5. https://doi.org/10.1080/02601370.2018.1456031.
[40] Chong S, Pan G-T, Chin J, Show PL, Yang TCK, Huang C-M. Integration of 3D Printing and Industry 4.0 into Engineering Teaching. Sustainability 2018;10:3960. https://doi.org/10.3390/su10113960.
[41] Crittenden AB, Crittenden VL, Crittenden WF. The digitalization triumvirate: How incumbents survive. Bus Horiz 2019;62:259–66. https://doi.org/10.1016/j.bushor.2018.11.005.
[42] Almeida F, Duarte Santos J, Augusto Monteiro J. The Challenges and Opportunities in the Digitalization of Companies in a Post-COVID-19 World. IEEE Eng Manag Rev 2020;48:97–103. https://doi.org/10.1109/EMR.2020.3013206.
[43] de Almeida JF, Amaral DC, Coelho RT. Innovative Framework to manage New Product Development (NPD) Integrating Additive Manufacturing (AM) and Agile Management. Procedia CIRP 2021;103:128–33. https://doi.org/10.1016/j.procir.2021.10.020.
[44] Lee C-H, Liu C-L, Trappey AJC, Mo JPT, Desouza KC. Understanding digital transformation in advanced manufacturing and engineering: A bibliometric analysis, topic modeling and research trend discovery. Adv Eng Inform 2021;50:101428. https://doi.org/10.1016/j.aei.2021.101428.
[45] Vijayan KK, Mork OJ. IdeaLab: A Learning Factory Concept for Norwegian Manufacturing SME. Procedia Manuf 2020;45:411–6. https://doi.org/10.1016/j.promfg.2020.04.045.
[46] Argote L, Ingram P. Knowledge Transfer: A Basis for Competitive Advantage in Firms. Organ Behav Hum Decis Process 2000;82:150–69. https://doi.org/10.1006/obhd.2000.2893.
[47] Kirchheim A, Dennig H-J, Zumofen L. Why Education and Training in the Field of Additive Manufacturing is a Necessity. In: Meboldt M, Klahn C, editors. Ind. Addit. Manuf. - Proc. Addit. Manuf. Prod. Appl. - AMPA2017, Cham: Springer International Publishing; 2018, p. 329–36. https://doi.org/10.1007/978-3-319-66866-6_31.
[48] Mörk P. New billion-euro export-oriented growth engines kicking in n.d. https://www.businessfinland.fi/en/whats-new/news/2022/new-billion-euro-export-oriented-growth-engines-kicking-in (accessed June 1, 2023).
[49] Fame3D n.d. https://fame3d.fi/ (accessed June 1, 2023).
[50] 3D-tulostuksen teollinen läpimurto saavuttamassa vihdoin Suomen – voi jopa tuoda tehtaita takaisin maailmalta. Yle Uut 2021. https://yle.fi/a/3-11782958 (accessed June 1, 2023).
[51] Merkittäviä 3D-tulostusinnovaatioita ja onnistumisia maailmalta; 3D-tulostamisen hyödyntämisessä pitää käyttää harkintaa kriisitilanteessakin n.d. https://mfg40.fi/merkittavia-3d-tulostusinnovaatioita-ja-onnistumisia-maailmalta-3d-tulostamisen-hyodyntamisessa-pitaa-kayttaa-harkintaa-kriisitilanteessakin/ (accessed June 1, 2023).

Acknowledgments
This research was done in collaboration with the project Osuva ja vaikuttava 360-vr-materiaali modernien digitaalisten valmistusteknologioiden kouluttamiseen ja perehdyttämiseen eri kohderyhmien (teknologiayritykset, loppuasiakkaat, koulutuslaitokset) tarpeisiin, *Focused and effective 360-VR material for training and familiarizing with modern digital manufacturing technologies for the needs of different target groups (technology companies, end customers, educational institutions)* (shortened as Osuu ja Uppoaa) funded by European Social Fund (S22515). The funding is part of Covid-19 pandemic related actions by the EU. The duration of the project is 1.8.2021-30.11.2023. Project aims to provide extensive additive manufacturing training materials for SMEs utilizing virtual reality to enhance the competitiveness of companies. Project partners are University of Turku and industrial partners. Authors of this article would like to acknowledge this project and its partners for knowledge, assistance and collaboration to be able to conduct this study. Project has enabled also availability of such data of industrial training, user experience and training concept that has been essential for carrying out this study. Authors would like to express their gratitude also to staff of LAB UAS Business Department and Department of Mechanical Engineering in Faculty of Technology at University of Turku for their knowledge and help to be able to execute this study.

IOP Conf. Series: Materials Science and Engineering 1296 (2023) 012003 doi:10.1088/1757-899X/1296/1/012003

The meaning of knowledge transfer in educating engineers for modern manufacturing technologies

H Piili[1], A Huusko[1], A Kurvinen[2], A Pikkarainen[3]

[1] Department of Mechanical Engineering, Faculty of Technology, Research group of Digital Manufacturing and Surface Engineering (DMS), University of Turku, Turku, 20520, Finland
[2] Department of Business, International Business, LAB University of Applied Sciences, Lappeenranta, 53850 Lappeenranta, Finland
[3] Department of Mechanical engineering, New Industry expertise group, Lapland University of Applied Sciences, Kemi 94600, Finland

ari.pikkarainen@lapinamk.fi

Abstract. The development of modern manufacturing technologies such as additive manufacturing and other laser-based manufacturing technologies have increased their usage, especially in engineering education. Education must correspond to different applications of these technologies to ensure the quality of their usage in the manufacturing industry. Laser-based technologies have an impact on society through companies and industry in the manufacturing sector as they are employing graduated students, who are more aware of the possibilities of the technologies. Traditional manufacturing industries are facing the era of digital transformation. There are vast differences between industries and different industry players as well as their ability to keep up with the transformation process. The pace of the process can depend on many different items that are related to the company organization and leadership, technology base and to the abilities to modify these. The pace can also depend on the processes of cooperation and development between other ecosystem players e.g., supply chain partners or development drivers such as customer industries. This study concentrates on one of the key issues, namely learning and adoption of new skills to tackle the change. Universities have a major role in the knowledge transfer from education to working life. To develop this, there is a need to identify the factors in university training or pedagogical choices that can speed up the process of transformation. The aim of this research is to identify the factors for knowledge transfer through arranging efficient education for laser-based manufacturing technologies, especially in engineering education. It was concluded that by educating engineers with sufficient knowledge about the technologies, the manufacturing industry can utilize the knowledge and implement laser-based technologies better in their operations.

1. Introduction

Additive manufacturing (AM), known also as 3D printing, is a rapidly developing and expanding manufacturing process, in which material is added layer by layer. Other conventional manufacturing technologies the contrary rely on e.g., subtract or join materials. This approach of AM makes it a unique technology to create totally new applications by utilizing design freedom, as e.g., in [1] and [2]. Recent studies have also indicated that AM is gaining acceptance in novel material engineering and industrial engineering. Novel industrial applications are developed in the electronics, automobile, aerospace, and petrochemical industries, as discussed in [3] and [4]. For instance, AM has been playing a crucial role

in recent years in the aerospace industry. Boeing is developing a communications satellite with over a 1000 additively manufactured parts made of titanium and aluminium and a high-performance polymer, as in [5]. Perseverance rover of NASA was the first one to carry additively manufactured ceramic parts to Mars. This has increased the reliability of the rover according to NASA, as in [6] and [7]
Additive manufacturing includes wide variety of different manufacturing techniques, each with device-specific variations. In addition, each equipment manufacturer has named its own processes differently. [8] Different technologies are suitable for different products, and additive manufacturing is characterized not only by a wide range of technologies but also by applications. New techniques and variations on existing technologies are also being developed, which further increases the wide range of terminology in the field, as e.g., in [9]. New technologies and upgrades enable, for example, the use of new materials, higher manufacturing speeds and, as construction chambers grow, the manufacture of larger pieces. One of the hot areas of research is the so-called additive manufacturing of multi-material, which is studied in different parts of the world. For this technology, the application can be, for example, process industry products where the outer casing of the tank is a pressure vessel made of steel and the inside is corrosion resistant, e.g., stainless duplex steel. More discussion in [10] and [11]. Software development also opens new opportunities for designing better and optimised structures, which is one of the cornerstones of the utilisation of AM. On the software side, development is fast and significant. Several very successful projects have already been carried out in Finland, where the functions of hydraulic components have been optimised with computational fluid dynamics (CFD) software and manufactured with additive manufacturing. Often the weight of these parts drops significantly and e.g. pressure drop in flow components drop to a fraction compared to a traditional design. In the future, Finland can also see strong development in these areas, as in [13].
The rapidly developing field of additive manufacturing also increases the need for continuous training to keep up with the development of the field. Although additive manufacturing can offer benefits in optimizing and manufacturing many parts, the growth seen now will not increase the share of additive manufacturing in the manufacturing process market very quickly unless the issue is highlighted and promoted in all possible situations. Technological development in different areas of AM is happening rapidly now and will fuel new applications and growth opportunities [14] Education is key to increasing the use of additive manufacturing. Training is needed at all levels of the organization to ensure management support for product development print ideas. Training experts, such as to the area of AM, is the key function where engineering graduates and work-life professionals are trained through engineering programs and short-term courses. This is concluded e.g., in [15] and [16]. These allow them to apply the technologies in companies and support and industrialize the technologies. This enables the development of the technology and though using it in companies, support the usage of the full potential in company operations [17].
Knowledge embedded in the interactions of people, tools, and tasks forms a basis for competitive advantage in firms. [18] Knowledge transfer is needed in the shop floor level in production work, but it is also reflected in product development, and finally company management understanding in the topic and meaning of new knowledge to the business. Universities have a role in knowledge transfer of new technologies to the manufacturing industry. As technological change is very rapid, also more thought should be given to the way technology transfer can be speeded up. Degree education and training offered by universities are both ways of knowledge transfer in traditional perspective. Traditional ways and university pedagogy need to be adjusted to meet the needs of the industry and the pace of change in industry. These facts are revealed by figure 1 which shows number of publications according to Scopus with different keywords used.
It can be concluded from figure 1 that number of publications generally in field of additive manufacturing and 3D printing has started to increase strongly in 2016. The same trend can be noted also with number of publications in field of engineering and simulation. Figure 1 reveals also that number of publications in field of learning and training in significantly less than corresponding number in additive manufacturing. Figure 1 indicates also that number of publications in field of learning has been in 2022 only 12% from corresponding number of additive manufacturing, and number of publications in training has been 6% and number of publications in knowledge transfer 0.3%. It can be

concluded from figure 1 that this study has huge importance, and there are only very few publications available in this field.

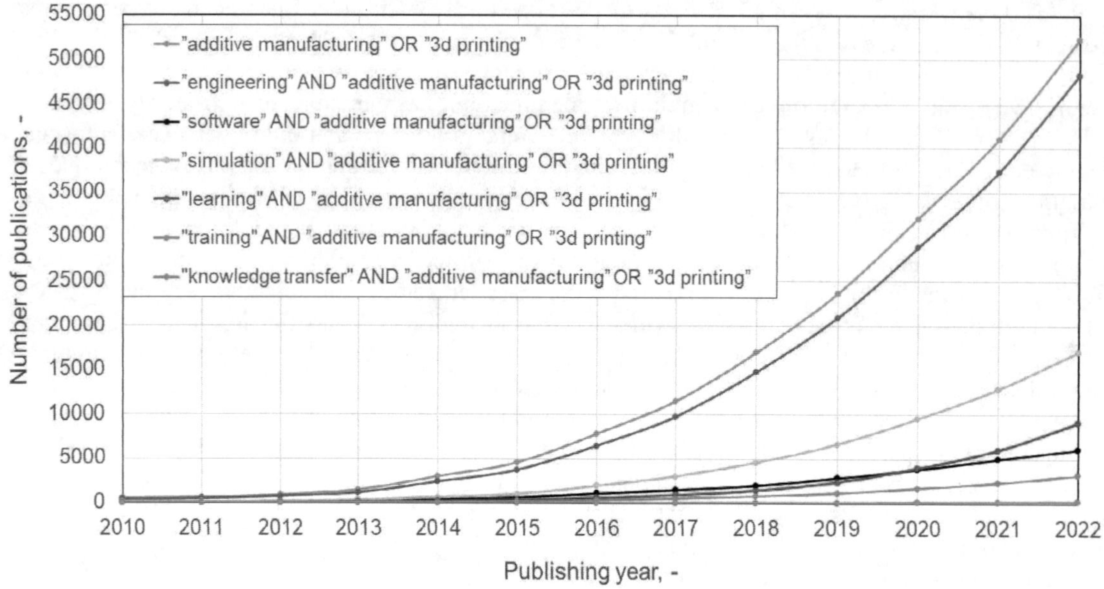

Figure 1. Number of publications according to Scopus when different keywords were used.

This study examines the adoption of additive manufacturing knowledge as a modern manufacturing technology, emphasizing the need to stay updated in the evolving field. In this study, two different cases are discussed, knowledge transfer through a Master level degree programme and AM training offered directly to companies. This study emphasizes the importance of a trained workforce for companies adopting these technologies. The study aims to propose solutions for technology-based knowledge transfer from education to companies, enabling increased technological proficiency and productivity. The scarcity of articles in this field, as identified through a Scopus survey, highlights the significance of this research. Industrial importance lies in manufacturing companies adopting and testing new ideas for business development, while change processes depend on various factors such as organizational structure, leadership, and technology base.

2. Literature review

2.1. Laser-based manufacturing technologies

Additive manufacturing includes large number of applications and subcategories. It is a process that builds up a physical object layer-by-layer, using a digital model as a guide. Additive manufacturing is divided into seven subcategories, including material extrusion (ME), vat photopolymerization (VP), powder bed fusion (PBF), directed energy deposition (DED), binder jetting (BJ), material jetting (MJ), and sheet lamination (SL) [19] Each of these subcategories has its own set of advantages and disadvantages, and they are used in different applications. Under these seven subcategories, there are dozens of different technologies with different material options, different possibility to fabricate various sizes of product, various device types starting from consumer devices to industrial scale large AM systems and different costs of devices. These technologies are constantly evolving, and new ones are being developed all the time, as observed in [20] and [21]. One of the most promising area of additive manufacturing is metal additive manufacturing. It has more than 30 different technologies, including powder bed fusion, binder jetting, and directed energy deposition [22]. These technologies are used to produce parts that are strong, durable, and heat-resistant, making them suitable for industrial applications. The vast number of technologies makes metal AM a versatile tool for a wide range of applications. However, it also means that it can be challenging to choose the right technology for a specific application. It is important to understand the capabilities and limitations of each technology and subcategory to make an informed decision [23].

2.2. Educational point-of-view of the technologies and digital transformation in manufacturing

Global manufacturing and production chains are focusing more and more in embracing factors of digitalization in different areas targeting increased efficiency and flexibility in their operations. The fast transformation speed of modern manufacturing technologies sets new kinds of challenges to the education of such technologies to educate experts who can adopt the technologies in real-life. According to The Global Smart Industry Readiness Index Initiative [24], the manufacturing sector needs to emphasize education and training of the workforce to meet the requirements of digitalization and the development of company strategies. This kind of transformation in the operational structure increases the competitiveness of companies in the digitalization race. Digital transformation covers areas such as information processing, computation and connectivity within company operations. It offers a platform in distributing information in a new way which boosts collaboration possibilities within digital networks. This kind of environment in the manufacturing sector offers new ways to apply modern manufacturing technologies and create a positive atmosphere for innovation and improved efficiency with perspective to competitiveness, as in [25].

The digital transformation related especially to higher education (HE) needs strategic planning to be able to fulfil the needs from modern manufacturing technologies and overcome obstacles in arranging the required workforce training and education. One key issue is for higher education institutes (HEIs) stakeholders to develop digital skills in personal and operational level. This increases the digital competitiveness of the HE organization and enables improved possibility to arrange high quality education. Digital trends such as artificial intelligence (AI), machine learning, cloud computing and big data are factors required to be implemented in arranging the education as the same factors are spreading also in the manufacturing sector in technological level. Strategic digitalization planning in education requires systematic operation such as roadmaps which enables the development in management and HEI business operations. [26]

2.3. Knowledge and knowledge transfer in different forms

Knowledge can be defined as explicit formal knowledge that can be documented or as tacit knowledge where personal experience or organizational experience can be embedded and that is often referred as know-how. Knowledge transfer can take place as series of actions that include identifying the knowledge, capturing the knowledge and utilizing the knowledge. In this respect knowledge transfer also benefits from interaction and sharing of information between different actors ([27], [28] etc.). Work integrated learning (WIL) philosophy, where students learn outside the university while working in company environment [29] on real company cases is one way of bringing new knowledge of AM to the industry. Students have learnt the theoretical background in the university courses, and they learn to apply it in the industry with real industry cases where one needs to be able to find solutions and to argument the choices. This approach supports the knowledge transfer of new methodologies to the manufacturing of industrial applications. WIL is also about creating new knowledge that results from complex, new and real industry need problem solving. The transfer across university and organizational boundaries is primarily transfer of knowledge that originally had been acquired in one specific situation or context to other contexts [30]. The core idea of WIL is that employability of graduates will increase if they are able to integrate theoretical and academic studies with work-experiences in their respective educational fields [29]. Knowledge is acquired by working with partner companies on challenges that require solutions. University can become a partner that is appreciated for problem-solving in its region. This type of co-operation in knowledge transfer requires the creation of trust between company and university as well recognition of expertise of the other. These act as the basis for working co-operation.

2.4. Processes of cooperation to enhance knowledge transfer

According to Luomaranta et al. [31], value-focused approach to adopting AM and innovating means understanding the value that new technology can create to all organizations of supply chain involved. According to [32], adoption of AM in companies in industrial scale use can be incremental innovation in one industry, whereas in another industry it represents radical innovation. Also, the ways in which AM is used in companies can represent radical innovation, incremental innovation or disruptive technology e.g., producing complex parts. In actual company processes, it means adoption of AM on different levels; concept level, process innovation level as well as product innovation level, as e.g., in [33]. All levels of taking advantage of AM also require different ways of knowledge transfer into a company – people management. Businesses are part of value chains or value networks and the whole development path of the industry can be a driver to the development needs or need for innovativeness in a company. [34] For knowledge transfer to be successful within an organization, it is imperative that learning be recognized as a critical component of overall success, as in [35] and [36]. Despite its importance, Gino and Staats [37] have identified several challenges that can impede the learning process. One challenge is the fear of failure, which can trigger negative emotions and cause individuals to avoid mistakes at all costs. As a result, projects may be structured in a manner that does not allow for experimentation, with little time or resources allocated for trial and error. This approach can impede an ability of organisation to develop new capabilities and take appropriate risks. Instead, managers should cultivate a culture that tolerates failure and encourages open discussion about it. [38] Another challenge is the overreliance on past performance. Relying too heavily on previous successes may inhibit innovation and limit the potential for growth. Additionally, a lack of reflection can impede the learning process. In an environment where workers are expected to be "always on," there may be limited opportunities for individuals to reflect on their performance and identify areas for improvement. Encouraging reflection is essential for enhancing learning and development within an organization, as e.g., in [39] and [40].

3. Case studies of knowledge transfer in training

This article analyses AM knowledge transfer in two case-studies in Finland for training industry representatives. One example presents Northern university of applied sciences learning environment and the other introduces a university project from Southwest Finland where also a new technological means of introducing AM to the industry is tested.

3.1. Case-study 1: Lapland University of Applied Sciences new Master´s degree programme for training workforce

Lapland University of Applied Sciences in Finland (Lapland UAS) started to develop a new Master´s degree program in 2022 called "Master of Renewing Industry" based on the needs coming from Northern Finland labour market and companies. The purpose for the degree program was to provide training for workforce during one academic year and transfer knowledge about modern manufacturing technologies to companies. In Finland, traditional engineering Bachelor´s degree length is four years and 240 ECTS in length. Therefore, UASs can provide Master´s degrees in engineering with 60 ECTS during one year [41]. Main aim for the new education was to provide efficient knowledge transfer from education to companies and therefore support companies in their development of technological expertise and competitiveness. The planning work consisted of mapping out future needs for training engineers for the year 2030-2035 [42] which aimed to point out required professional skills in educating future engineers. In addition, Lapland UAS collaborated closely with Northern Finland companies in mapping out the needs of the companies for the education. This led to the creation of the themes for the degree program. Themes were constructed around the concept of twin transition, digitalization and sustainability in company operations. Other important themes were modern manufacturing technologies, strategical leadership in companies, modelling and simulation of industrial processes and identification of technological development possibilities, just a few to mention. This kind of short-term education is a good example how higher education institutes (HEI´s) can support companies in their pursuit of developing their workforce expertise to match the demands coming from the fast development of technology.

3.2. Case study 2: Industrial training of additive manufacturing by using virtual tools in Osuu ja Uppoaa project

In the ongoing Osuu ja Uppoaa project (August, 2021 – July, 2023) funded by the European Social Fund at the University of Turku, the understanding of industrial AM is being brought to small and medium-sized enterprises through VR technology [43]. The operational area is Southwest Finland. The training is structured around a combination of theoretical lectures delivered by researchers from the Department of Mechanical Engineering at the University of Turku and practical demonstrations provided by companies specializing in AM services. These practical examples illustrate the various stages of the AM processes and showcase the final products. [44] The training materials are divided into short micro-learning units, allowing users to navigate between them seamlessly. The content is meticulously designed in accordance with micro-learning principles, as shorter modules delivered through VR goggles offer enhanced user-friendliness. Additionally, incorporating brief, independent learning sessions allows easy integration into the daily work routine. Employing VR goggles as the training platform aligns with the concept of micro-learning, benefiting from its technical simplicity. Using VR goggles simplifies the training experience by eliminating the need for complex login procedures or application downloads, allowing users to seamlessly engage in the training without wasting excessive time.

4. Results and discussion

4.1. Case-study 1: Lapland University of Applied Sciences new Master´s degree programme for training workforce

Lapland UAS has a concept called Smart Lab which integrates three different themes together regarding manufacturing. Smart Lab is concentrated around FESTO Cyber-Physical Factory (CP Factory) which provides digital manufacturing concept to mechanical engineering education. The environment is being used both in Bachelor and Master level of engineering. Smart Lab includes also additive manufacturing and automation laboratory (pneumatics and hydraulics). Main purpose of the environment is to provide education combining different manufacturing themes and actors. One key issue is to enable company participation in the function of the laboratory environment. Companies can participate e.g., by bringing own project work topics or assignments for development. Through workshop actions, companies can also use the environment physically in participation with Lapland UAS mechanical engineering degree. This is enabled through different partnership agreements which ensure close collaboration with actors from different areas. The basic concept can be seen in figure 2.

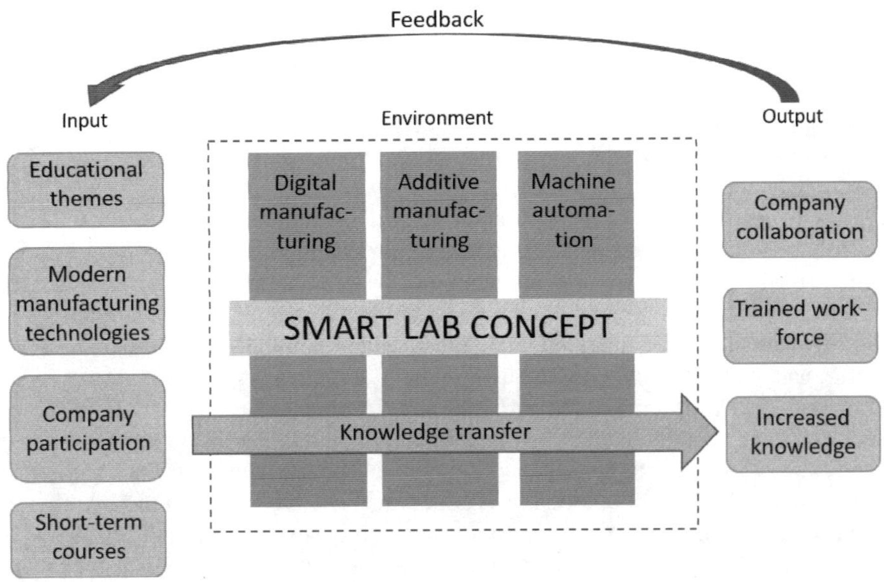

Figure 2. Lapland UAS Smart Lab concept as knowledge transfer enabler.

As seen in figure 2, the Smart Lab concept works through three different elements. In input, traditional engineering education themes from modern manufacturing technologies provide the necessary platform for the training. Company participation (through their own projects and assignments) form the actors functioning in the laboratory environment. The education and training can be tailored e.g., as short-term courses based on company needs and wishes. In the environment, the Smart Lab concept provides all the necessary surroundings for physical laboratory work and education of the actors. This can be built through the three different themes (separately or by combining e.g., digital manufacturing and additive manufacturing). In output, the environment has produced different elements through knowledge transfer. The concept enables results in a form of increased company collaboration, trained workforce (either through graduated mechanical engineering students or trained company personnel) and increased knowledge. This element of knowledge as an output benefits companies through identifying the possibilities of utilising modern manufacturing technologies. This can lead to increased productivity as the possibilities are used in operational and strategical development in companies. One important element of the model is collecting feedback in the output phase. E.g., as companies function in the SmartLab environment (e.g., in a short-term course), it is important to collect feedback from the participants for development of the input phase. This improves the equivalency of what the SmartLab concept offers to companies vs. the needs of the companies.

4.2. Case study 2: Industrial training of additive manufacturing by using virtual tools in Osuu ja Uppoaa project

Information on the free AM training via VR Goggles has been shared through local media, social media (such as the Mechanical Engineering Department of the University of Turku on LinkedIn), participation in various training and business events related to new technologies, and articles written in the industry magazine (Welding Technology Magazine) about the training. In addition to these actions, pilot training material has been presented at the Turku University of Applied Sciences and student organization events. Despite of direct and indirect interactions in project with numerous key personnel in SMEs, minimal interest from the side of companies was shown. Remarkable is that every interested company already has experience in additive manufacturing. According to survey done in this study, students who had some previous contact with additive manufacturing, either through their studies or personal experiences, were more interested in learning about the technology.

5. Conclusions

The aim of this study was to identify the factors for efficient knowledge transfer and education in laser-based manufacturing technologies, particularly in engineering education, to ensure the manufacturing industry can effectively utilize and implement these technologies in their operations, thus contributing to the digital transformation of traditional manufacturing industries. In this study, two learning environments were discussed and the learning ideas of them explained. The Lapland UAS Smart lab concept has now been operational for over two years and experiences have showed that the environments functions well as a supporter of digital manufacturing skills for engineering students. The environment promotes the connection of different themes such as digital manufacturing and storage control, machine vision, automatic process control and the connection of other manufacturing technologies such as AM. AM has been used in the environment for developing products to the manufacturing system. The students have been able to design and manufacture different assemblies where the parts can be processed through the CP factory enabling e.g., modularity in the manufacturing process. This gives new kind of approach methodology to view real-life manufacturing processes in educational environment. Companies have at the same time gained new knowledge related to AM and its possibilities in their own operations and they are willing to continues this type of co-operation with the UAS. As the students graduate to companies, they can adopt the knowledge transfer elements from modern themes and adopt them while working in companies. One important factor in ensuring the quality of knowledge transfer is collecting feedback. As the different operators such as companies and other participants participate to the SmartLab concept, it is imporant to collect their feedback afterwards to further improve the contents of the education and respond better to the needs of the operators.

In this study, it was also found out that learning environments can be integral part of knowledge transfer, learning environments can at the same time be combining industries with university research and education. The learning environments studied and presented in the article are pioneering methods of knowledge transfer in their particular geographical areas in Northern Finnish UAS and in Southwestern Finland. In both cases new knowledge is transformed in the companies via university-company collaboration, trained and skilled workforce as well as increased knowledge. The digital learning environment supports knowledge transfer by utilizing various channels such as local and social media, participation in relevant events, articles in industry magazines, and presentations at educational and student organization events, to share information about additive manufacturing training via VR goggles, as tested in Osuu ja Uppoaa project, with students who had prior exposure to additive manufacturing displaying a higher interest in learning about the technology.

In the future, studies around comparative analysis of learning environments, long-term impact of university-industry collaboration and evaluation of digital learning tools will be carried out. Conducting a comparative analysis of different learning environments, such as Lapland UAS Smart lab and similar setups in other regions, to assess their effectiveness in facilitating knowledge transfer and education in laser-based manufacturing technologies. This study would help identify the strengths and weaknesses of various learning environments and provide insights for further improvement.

6. Author contribution

During the preparation and writing of this work no AI based tools were used. Authors reviewed and edited the content of this work independently and take full responsibility of this contribution.

All authors drafted the idea of this publication. Further brainstorming was executed by all authors. The actual drafting of the concept of knowledge transfer in the education of engineers approach introduced in this article was discussed, commented on, and modified by all the authors.

Heidi Piili participated in the literature review from the areas of additive manufacturing, additive manufacturing education and training, and knowledge transfer of additive manufacturing knowledge. She was responsible for sub-chapter "2.1. Laser-based manufacturing technologies".

Anna Huusko has participated in this publication from areas of learning, education and industrial training and discusses obstacles that industrial training can meet. She was also responsible of sub-chapter "3.2. Case study 2: Industrial training of additive manufacturing by using virtual tools in Osuu ja Uppoaa project" and sub-chapter "4.2 Case study 2: Industrial training of additive manufacturing by using virtual tools in Osuu ja Uppoaa project".

Anu Kurvinen has participated in this article from the areas of learning, education and training and discussing how knowledge transfer is taking place in industry. She was responsible for Literature review's sub-chapters "2.2. Educational point-of-view of the technologies and digital transformation in manufacturing", "2.3. Knowledge and knowledge transfer in different forms" and "2.4. Processes of cooperation to enhance knowledge transfer".

Ari Pikkarainen participated in the literature review from the areas of additive manufacturing and knowledge transfer especially from the educational point-of-view. He was responsible for sub-chapter "3.1. Case-study 1: Lapland University of Applied Sciences new Master´s degree programme for training workforce" and for sub-chapter "4.1. Lapland University of Applied Sciences new Master´s degree programme for training workforce". He also contributed to chapter "5. Conclusions" related to previous chapters. Ari Pikkarainen is also the corresponding author of this publication.

7. References

[1] Dilberoglu UM, Gharehpapagh B, Yaman U, Dolen M. The Role of Additive Manufacturing in the Era of Industry 4.0. Procedia Manufacturing 2017;11:545–54. https://doi.org/10.1016/J.PROMFG.2017.07.148.

[2] Feng SC, Moges T, Park H, Yakout M, Jones AT, Ko H, et al. Functional Requirements of Software Tools for Laser-Based Powder Bed Fusion Additive Manufacturing for Metals. Journal of Computing and Information Science in Engineering 2023;23. https://doi.org/10.1115/1.4054933.

[3] Daminabo SC, Goel S, Grammatikos SA, Nezhad HY, Thakur VK. Fused deposition modeling-based additive manufacturing (3D printing): techniques for polymer material systems. Materials Today Chemistry 2020;16:100248. https://doi.org/10.1016/j.mtchem.2020.100248.

[4] Shrinivas Mahale R, Shamanth V, Hemanth K, Nithin SK, Sharath PC, Shashanka R, et al. Processes and applications of metal additive manufacturing. Materials Today: Proceedings 2022;54:228–33. https://doi.org/10.1016/j.matpr.2021.08.298.

[5] Altıparmak SC, Yardley VA, Shi Z, Lin J. Challenges in additive manufacturing of high-strength aluminium alloys and current developments in hybrid additive manufacturing. International Journal of Lightweight Materials and Manufacture 2021;4:246–61. https://doi.org/10.1016/j.ijlmm.2020.12.004.

[6] Kelly JW, Hamlin E, McCulloch D, Moxey L, Seibold RW, Crispi G, et al. A Decade of Space Technology Maturation through NASA's Flight Opportunities Program. ASCEND 2020, Reston, Virginia: American Institute of Aeronautics and Astronautics; 2020. https://doi.org/10.2514/6.2020-4135.

[7] Bhuvanesh Kumar M, Sathiya P. Methods and materials for additive manufacturing: A critical review on advancements and challenges. Thin-Walled Structures 2021;159:107228. https://doi.org/10.1016/J.TWS.2020.107228.

[8] Khajavi SH, Partanen J, Holmström J. Additive manufacturing in the spare parts supply chain. Computers in Industry 2014;65:50–63. https://doi.org/10.1016/j.compind.2013.07.008.

[9] Javaid M, Haleem A, Singh RP, Suman R, Rab S. Role of additive manufacturing applications towards environmental sustainability. Advanced Industrial and Engineering Polymer Research 2021;4:312–22. https://doi.org/10.1016/J.AIEPR.2021.07.005.

[10] Xu X, Tan YH, Ding J, Guan C. 3D Printing of Next-generation Electrochemical Energy Storage Devices: from Multiscale to Multimaterial. ENERGY & ENVIRONMENTAL MATERIALS 2022;5:427–38. https://doi.org/10.1002/eem2.12175.

[11] Shaeri Karimi MH, Yeganeh M, Alavi Zaree SR, Eskandari M. Corrosion behavior of 316L stainless steel manufactured by laser powder bed fusion (L-PBF) in an alkaline solution. Optics & Laser Technology 2021;138:106918. https://doi.org/10.1016/J.OPTLASTEC.2021.106918.

[12] Alshare AA, Calzone F, Muzzupappa M. Hydraulic manifold design via additive manufacturing optimized with CFD and fluid-structure interaction simulations. Rapid Prototyping Journal 2018;25:1516–24. https://doi.org/10.1108/RPJ-03-2018-0064.

[13] The Pressure Vessel 3D Printed by ANDRITZ Savonlinna Works Oy and FAME Ecosystem Is a European Giant – Fame3D n.d. https://fame3d.fi/the-pressure-vessel-3d-printed-by-andritz-savonlinna-works-oy-and-fame-ecosystem-is-a-giant/ (accessed May 18, 2023).

[14] Amankwah-Amoah J, Khan Z, Wood G, Knight G. COVID-19 and digitalization: The great acceleration. Journal of Business Research 2021;136:602–11. https://doi.org/10.1016/j.jbusres.2021.08.011.

[15] Godina R, Ribeiro I, Matos F, T. Ferreira B, Carvalho H, Peças P. Impact Assessment of Additive Manufacturing on Sustainable Business Models in Industry 4.0 Context. Sustainability 2020;12:7066. https://doi.org/10.3390/su12177066.

[16] Rowan NJ, Galanakis CM. Unlocking challenges and opportunities presented by COVID-19 pandemic for cross-cutting disruption in agri-food and green deal innovations: Quo Vadis? Science of The Total Environment 2020;748:141362. https://doi.org/10.1016/j.scitotenv.2020.141362.

[17] Kirchheim A, Dennig H-J, Zumofen L. Why Education and Training in the Field of Additive Manufacturing is a Necessity. In: Meboldt M, Klahn C, editors. Industrializing Additive

Manufacturing - Proceedings of Additive Manufacturing in Products and Applications - AMPA2017, Cham: Springer International Publishing; 2018, p. 329–36. https://doi.org/10.1007/978-3-319-66866-6_31.

[18] Argote L, Ingram P. Knowledge Transfer: A Basis for Competitive Advantage in Firms. Organizational Behavior and Human Decision Processes 2000;82:150–69. https://doi.org/10.1006/obhd.2000.2893.

[19] Singh T, Kumar S, Sehgal S. 3D printing of engineering materials: A state of the art review. Materials Today: Proceedings 2020;28:1927–31. https://doi.org/10.1016/J.MATPR.2020.05.334.

[20] Lee JY, An J, Chua CK. Fundamentals and applications of 3D printing for novel materials. Applied Materials Today 2017;7:120–33. https://doi.org/10.1016/J.APMT.2017.02.004.

[21] Khosravani MR, Reinicke T. On the environmental impacts of 3D printing technology. Applied Materials Today 2020;20:100689. https://doi.org/10.1016/J.APMT.2020.100689.

[22] Cabanettes F, Joubert A, Chardon G, Dumas V, Rech J, Grosjean C, et al. Topography of as built surfaces generated in metal additive manufacturing: A multi scale analysis from form to roughness. Precision Engineering 2018;52:249–65. https://doi.org/10.1016/j.precisioneng.2018.01.002.

[23] Zaman UK uz, Rivette M, Siadat A, Mousavi SM. Integrated product-process design: Material and manufacturing process selection for additive manufacturing using multi-criteria decision making. Robotics and Computer-Integrated Manufacturing 2018;51:169–80. https://doi.org/10.1016/J.RCIM.2017.12.005.

[24] World Economic Forum. World Economic Forum n.d. https://www.weforum.org/whitepapers/the-global-smart-industry-readiness-index-initiative-manufacturing-transformation-insights-report-2022/ (accessed May 26, 2023).

[25] Vial G. Understanding digital transformation: A review and a research agenda. The Journal of Strategic Information Systems 2019;28:118–44. https://doi.org/10.1016/j.jsis.2019.01.003.

[26] Rof A, Bikfalvi A, Marques P. Digital Transformation in Higher Education: Intelligence in Systems and Business Models. In: Kahraman C, Haktanır E, editors. Intelligent Systems in Digital Transformation: Theory and Applications, Cham: Springer International Publishing; 2023, p. 429–52. https://doi.org/10.1007/978-3-031-16598-6_18.

[27] Fernie S, Green SD, Weller SJ, Newcombe R. Knowledge sharing: context, confusion and controversy. International Journal of Project Management 2003;21:177–87. https://doi.org/10.1016/S0263-7863(02)00092-3.

[28] Foss NJ, Pedersen T. Microfoundations in international management research: The case of knowledge sharing in multinational corporations. J Int Bus Stud 2019;50:1594–621. https://doi.org/10.1057/s41267-019-00270-4.

[29] Yorke M, Knight PT. Embedding Employability n.d.

[30] Kurvinen A, Juvonen P, Svensson L. LEARNING SKILLS TO STAND OUT IN THE COMPLEX WORLD – DISCUSSION OF PEDAGOGICAL REVOLUTION IN HIGHER EDUCATION. ICERI2020 Proceedings 2020:3722–8. https://doi.org/10.21125/iceri.2020.0841.

[31] Luomaranta T, Martinsuo M. Additive manufacturing value chain adoption. Journal of Manufacturing Technology Management 2022;33:40–60. https://doi.org/10.1108/JMTM-07-2021-0250.

[32] Steenhuis H-J, Pretorius L. The additive manufacturing innovation: a range of implications. Journal of Manufacturing Technology Management 2017;28:122–43. https://doi.org/10.1108/JMTM-06-2016-0081.

[33] Ghobadian A, Talavera I, Bhattacharya A, Kumar V, Garza-Reyes JA, O'Regan N. Examining legitimatisation of additive manufacturing in the interplay between innovation, lean manufacturing and sustainability. International Journal of Production Economics 2020;219:457–68. https://doi.org/10.1016/j.ijpe.2018.06.001.

[34] Stornelli A, Ozcan S, Simms C. Advanced manufacturing technology adoption and innovation: A systematic literature review on barriers, enablers, and innovation types. Research Policy 2021;50:104229. https://doi.org/10.1016/j.respol.2021.104229.

[35] Alabi MO, de Beer DJ, Wichers H, Kloppers CP. Framework for effective additive manufacturing education: a case study of South African universities. Rapid Prototyping Journal 2020;26:801–26. https://doi.org/10.1108/RPJ-02-2019-0041.

[36] Kipper LM, Iepsen S, Dal Forno AJ, Frozza R, Furstenau L, Agnes J, et al. Scientific mapping to identify competencies required by industry 4.0. Technology in Society 2021;64:101454. https://doi.org/10.1016/j.techsoc.2020.101454.

[37] Staats BR, Gino F. Specialization and Variety in Repetitive Tasks: Evidence from a Japanese Bank. Management Science 2012;58:1141–59. https://doi.org/10.1287/mnsc.1110.1482.

[38] Clausen TH, Demircioglu MA, Alsos GA. Intensity of innovation in public sector organizations: The role of push and pull factors. Public Administration 2020;98:159–76. https://doi.org/10.1111/padm.12617.

[39] Chaithanapat P, Punnakitikashem P, Khin Khin Oo NC, Rakthin S. Relationships among knowledge-oriented leadership, customer knowledge management, innovation quality and firm performance in SMEs. Journal of Innovation & Knowledge 2022;7:100162. https://doi.org/10.1016/j.jik.2022.100162.

[40] Mergel I, Ganapati S, Whitford AB. Agile: A New Way of Governing. Public Administration Review 2021;81:161–5. https://doi.org/10.1111/puar.13202.

[41] Ylemmät AMK-tutkinnot - Lapin AMK. Lapland University of Applied Sciences n.d. https://www.lapinamk.fi/fi/Hakijalle/YAMK-tutkinnot (accessed May 18, 2023).

[42] Osaamispulssi • Teknologiateollisuus. Osaamispulssi n.d. https://osaamispulssi.fi/ (accessed May 18, 2023).

[43] Osuu ja Uppoaa | Work Informatics 2021. https://workinformatics.utu.fi/hankkeet/osuujauppoaa/ (accessed May 28, 2023).

[44] Osuu ja Uppoaa -hankkeessa kehitetään virtuaalinen koulutusmenetelmä 3D-tulostukseen n.d. https://www.utu.fi/fi/ajankohtaista/mediatiedote/osuu-ja-uppoaa-hankkeessa-kehitetaan-virtuaalinen-koulutusmenetelma-3d (accessed May 23, 2023).

Acknowledgments

This research was done in collaboration with the project Osuva ja vaikuttava 360-vr-materiaali modernien digitaalisten valmistusteknologioiden kouluttamiseen ja perehdyttämiseen eri kohderyhmien (teknologiayritykset, loppuasiakkaat, koulutuslaitokset) tarpeisiin, *Focused and effective 360-VR material for training and familiarizing with modern digital manufacturing technologies for the needs of different target groups (technology companies, end customers, educational institutions)* (shortened as Osuu ja Uppoaa) funded by European Social Fund (S22515). The funding is part of Covid-19 pandemic related actions by the EU. The duration of the project is 1.8.2021-30.11.2023. Project partners are University of Turku and industrial partners. Authors of this article would like to acknowledge this project and its partners for knowledge, assistance and collaboration to be able to conduct this study. Project has enabled also availability of such data of industrial training, user experience and training concept that has been essential for carrying out this study. Authors would like to express their gratitude also to staff of LAB UAS Business Department, Lapland UAS Mechanical Engineering Department and Department of Mechanical Engineering in Faculty of Technology at University of Turku for their knowledge and help to be able to execute this study.

IOP Conf. Series: Materials Science and Engineering 1296 (2023) 012004 doi:10.1088/1757-899X/1296/1/012004

Virtual Reality-Based Training in Additive Manufacturing: Developing a Service Concept that Adapts to Company-Specific Requirements

A Huusko[1], J Susi[2], S Parchegani[1], H Hosseini[1], A Tuomisto[2], A Helin[2], H Piili[1], A Salminen[1]

[1] Department of Mechanical Engineering, Faculty of Technology, University of Turku, Turku, 20520, Finland
[2] Department of Management and Entrepreneurship, Turku School of Economics, University of Turku, Turku, 20024, Finland

anna.huusko@utu.fi

Abstract. Additive manufacturing (AM), commonly known as 3D printing, has vast amount of technological opportunities, but it also necessitates specialized expertise among its specialists, adopters, users etc. As AM continues to demonstrate its potential and has high growth rates in fields of its use, applications, systems etc., the importance of AM education and training becomes increasingly evident. This paper serves as a preliminary study outlining a novel training concept for industrial AM training including microlearning units and conducted in virtual reality, focusing on the accessibility and flexibility of users. The concept has been developed as part of the ongoing industrial training project. Microlearning and virtual tool-assisted training methods offer effective and agile approaches to enhance the continuous learning of new industrial skills, enabling companies to keep pace with evolving technologies. These tools facilitate extensive customization of training, making company-specific training even more crucial for effective skill development and knowledge transfer. The design of this training concept leverages the user story mapping (USM) method, enabling a deeper understanding of company-specific requirements and needs concerning additive manufacturing training. Moreover, through questionnaires and interviews conducted during the USM process helps identify technology readiness levels (TRLs) of companies in relation to their AM adoption. Measuring TRLs provides valuable insights into the technical maturity of companies regarding additive manufacturing, as readiness levels tend to vary significantly among different organizations. The preliminary framework provided serves as a foundation for designing a modern training concept that cater to the unique requirements of different companies, fostering their successful adoption of AM technologies.

1. Introduction

Modern technologies profoundly influence each society, and digitalization impacts daily life. This also includes the interaction of people with one another and extends to the methods they employ for product exploration and procurement. At the forefront of this is the fourth industrial revolution, characterized by a range of innovative technologies; it is predicted to change the manufacturing and distribution of products. Correspondingly, modern, digital manufacturing can now leverage the digital flexibility and efficiency especially offered by additive manufacturing (AM), widely recognized as 3D printing, one of the most interesting technologies. [1] The rapidly developing field also increases the need for continuous training to keep up with the development of AM. New hardware, terminology, software, and materials are constantly being developed, and information needs to be updated. [2] Despite the abundance of information available on additive manufacturing, it remains crucial to establish dependable research-based knowledge to ensure that learning of the technology is built upon accurate information right from the start avoiding any misconceptions. [3] According to Ernst & Young [4], the need for more experts/professionals for AM is the second most significant obstacle to implementing AM, the first being the high cost of the machines. The reports highlight a considerable need for AM training at the EU level to bridge the gap between AM progress and the knowledge of experts and students. The landscape of AM training in the EU has constantly been evolving in various parts of the EU separately. European standardized regulations are required to tackle this need for more knowledge [5].

Some projects have been developed in the past few years in Europe to provide general guidelines and instruction on addressing the gap between AM knowledge and technology development, such as the European harmonized qualification system CLLAIM [6] and Knowledge Alliance for AM between Industry and universities, ADMIRE [7], to define the qualification required for the whole AM chain, beginning from design to the quality assurance of the finished part. These projects were conducted in collaboration with European Federation for Welding, Joining and Cutting (EWF) to integrate AM qualification system in Europe. [8] In another research [9], a similar review was done internationally by reviewing the recent research in this area, proposing a comprehensive training framework for AM that combines the knowledge from research and education in academia with industrial training to promote the innovative aspects of AM in all aspects to increase its competitiveness. [10]

Digitalisation and time-pressure dynamics in workplaces have revolutionized training and learning approaches. Previously, training was constrained by time and physical location, and the methods employed were not customized to meet the specific needs of individuals. [11] It has been demonstrated that effective learning necessitates individualization and concise learning segments to optimize the learning experience. [12] New avenues have emerged for accessing knowledge, particularly in highly specialized and technology-oriented fields of work. The challenge with SMEs is often that company experts are too busy to join any professional training courses or participate in any presentations, listen to lectures, or even visit YouTube to browse any interesting work-related content. [13] While companies recognize the need for training, resources are often limited for extensive training programs targeting business experts. To overcome this challenge, training can be carried out based on microlearning. This principle allows business experts to focus on their work and, in the event of a problem or need for development, to seek only an answer to a question or development needs in the current situation. [14] Industrial training is characterized by the pervasive presence of technology.

Virtual reality (VR) learning has been discovered to be more effective than using technical manuals or multimedia presentations, especially in complex tasks. This is why new training tools have emerged as the primary instruction means. The sense of physical presence and the inclusion of interactive elements enhance trainee engagement and motivation, contributing to positive learning outcomes. [15] The visual aspect of 360° videos can also make the learning subjects more appealing, especially when combined with effective video thumbnails designed to capture attention and interest. [16] This VR-based training content was specifically designed to facilitate the acquisition of new work-related information by company experts. The content will be accessed through VR goggles provided by the project group. [17]

1.1. Background and motivation

This article emerges from project of the Osuu ja uppoaa, which focuses on designing a virtual reality-based (VR) training concept for industrial additive manufacturing (AM) technology. Osuu ja Uppoaa project is funded by European Social Fund (S22515) as a part of COVID-19 pandemic-related actions by the EU. [17] It explores the process and methodology employed to create an effective training experience for AM within an industrial setting. The content curated for the virtual reality-assisted training concept is meticulously selected, drawing upon decades of experience in AM training with diverse target groups. The experts have a rich history of experience and expertise in the field of AM research and training. This paper discusses a pioneering approach to the design of the industrial additive manufacturing (AM) training concept. It combines cutting-edge virtual reality (VR) technology with one of the latest learning concepts microlearning. The target training group comprises product designers, production managers, middle management supervisors, and individuals responsible for strategic decision-making and investments, specifically within small and medium-sized enterprises (SMEs) in the South-west area of Finland. Although additive manufacturing has been selected as the target technology in this study, the solutions produced are applicable to other technologies.

1.2. Methodology

The method employed in designing the training concept in this study is user story mapping (USM). This approach provides a structured framework for capturing user requirements, prioritizing features, and creating a cohesive roadmap for developing the training concept. Using USM, this study ensures a user-centric design that aligns with the needs and preferences that consider users and their prior experiences with AM. By utilizing USM, the training concept fosters meaningful interactions, tailors content to individual requirements, and incorporates the valuable insights gained from past experiences of users in the AM domain. To be able to ensure that this training effectively addressed the diverse needs and skill levels of different companies, this has to be taken into account in USM tool. Authors have long (c. 15 years) experience of different skill levels of companies in South-West Finland, and also a preliminary questionnaires were carried out before pilot training to understand training needs of companies in AM, and also to understand their skill levels. By utilizing the USM tool to identify company-specific requirements, skill level and training needs in AM, an additional benefit is gained: a deeper understanding of the Technology Readiness Level (TRL) of the companies. [18] The technology readiness level (TRL) measures the maturity of technology for practical implementation [19]. In training, this helps to situate companies based on their maturity to use AM and thus enables them to identify which kind of training can be introduced for the company.

1.3. Scopus survey

A thorough survey was conducted using Scopus, a widely recognized scholarly database to gain a comprehensive understanding of the industrial training landscape. The survey was performed using specific keywords (years 2010-2023) related to the focus of this study. This approach allowed extensive exploration of relevant literature and publications, enabling a comprehensive analysis of the current state of industrial training within the targeted field. The number of published articles in AM training is about 6% of the total published articles in AM in 2022. The trend shows a 75% increase in the number of published articles in AM training since 2018, implying the importance of this topic in the whole spectrum of AM-related subjects. The utilization of virtual reality in AM training also shows an upward trend. The corresponding number of articles in which virtual reality is utilized to improve AM training is 1 % of the total number of AM articles in 2022. However, the number has increased by 80% since 2018. Interestingly, the keywords “user story mapping” and “additive manufacturing” resulted in no publications. Therefore, an examination was conducted on the number of publications containing the keywords "user story mapping" and "training," revealing a noticeable scarcity of studies in this field. Despite the broad applicability of the user story mapping (USM) method in different areas of product design [20], the search on this topic only covered a mere 0.08% of the total available articles on AM. This limited data availability highlights the need for further research. It leads to the conclusion that the case of this study holds significant relevance, as there is limited or negligible availability of publications addressing the subject matter discussed in this article.

2. Description of digital training tool approach

Microlearning is an effective training approach for organizational learning. Subject matter experts are crucial in microlearning, providing accurate narration and demonstrations. [21] Some industrial training institutions combine traditional exercises with microlearning tools, enabling focused assignments and reinforcing understanding. [13] Microlearning values active problem-solving as an immediate application of knowledge, and prompt feedback contributes to efficient information absorption. However, it may only be suitable for some skills. Numerous case studies consistently demonstrate the efficacy of microlearning in improving learning outcomes, as e.g. in [22]. Comparisons with control groups using traditional methods reveal higher test scores, self-reported competence, and learner satisfaction among microlearning participants. Retention assessments prove that microlearning groups outperform control groups in long-term knowledge recall. Nevertheless, microlearning has some limitations. These include the potential for information overload, distractions from device features, internet connectivity issues, affordability of hardware, and limited access to immediate teacher guidance, particularly in remote learning scenarios. [23]

Virtual reality has been employed as a training tool in various fields. It allows users to examine and interact with training subjects closely. For instance, virtual simulations can assist in motor skill development, such as practicing basketball throws. [24] The VR-based simulation will also improve operational efficiency when creating a safe, reliable, realistic educational environment. Similarly, VR can be utilized for training in operating and maintaining complex machinery, allowing users to virtually explore the equipment, interact with its components, and receive detailed instructions. [25] This shortens the learning time, lowers any risk perception, and allows training not to affect production without decreasing user and machinery safety. [26] The VR experience can even work better than a real-world experience based on the security of this experience and its costs. [27] 360° videos have been explored as an affordable alternative to accurate virtual reality solutions in industrial training.

While VR requires complex 3D modeling and software development, 360° videos offer a more straightforward approach. [28] They can be filmed and played back using basic systems without extensive customization. The user can choose the target he/she wants to view, and he/she does not have to settle for the selection of video makers. [29] Some proprietary 360° video solutions even allow for the addition of interactive elements, such as displaying icons that trigger playback of related sub-topic videos. [30] The user can choose the target he/she wants to view, and he/she does not have to settle for the selection of video makers. [29] This combination of 360° videos and interactive features aligns well with microlearning principles. This is precisely why these two elements are integrated into the AM training concept. Use of microlearning in virtual learning training aligns well with the characteristics of digital natives and the modern learner. These individuals often prefer quick, on-demand access to information and have shorter attention spans. Microlearning meets their learning preferences by offering concise, interactive, and engaging modules that cater to their learning needs. The microlearning approach in virtual learning training also offers a learner-centred and efficient way to deliver knowledge and optimize the effectiveness of training programs.

3. Virtual reality-based training for small and medium-sized companies

The most challenging aspect of making digital capability visible lies in providing a purposeful description of the work system, of which digital capabilities are an integral part [31]. The technical expertise of the training concept plays a mediating role, aiming to enhance the understanding of experts in their operating environment, thereby improving the successful implementation of desired development steps, agile experiments, and more profound changes. Consequently, the new training concept and knowledge transfer approach encompasses the skills and means necessary for comprehending and advancing the contents and processes of business, process, and work digitalization. When describing the training concept, it was found helpful to incorporate a third interpretation in the training concept representation in addition to the technical-economic and socio-political interpretations. This third interpretation emphasizes key concepts of trust, social capital, and collaboration [50,51]. The social nature of capabilities within complex situations and subjects is a significant factor that should be highlighted in the design of technical content (e.g., fundamentals of AM) and technical implementation (e.g., distance and micro-learning). Otherwise, attempts to enhance capabilities may remain superficial or get lost in the information overflow if the personal and social dimensions are not considered. In this sense, this is also a matter of management and HR issues in every participating organization. [35,36]

Contents of VR training material have been created based on the understanding of the skills of AM experts and knowledge needed for proper AM work and based on questionnaires identifying the needs of company experts for AM-related information. Based on these requirements, theoretical and practical content is included, all in a virtual reality environment. An entire AM theory introduction course is included for the theory part but is divided into presentations lasting only 5-10 minutes according to the microlearning principles. The practical part consists of presentations of providers of AM products and services and also presentations of a few companies who are keen to understand more about the possibilities of AM production. Brief company introductions are included for all participating companies as well as contact details of university experts, the training concept, and a reference to the participating individuals. [37]

4. Developing training product

4.1 User story mapping as a tool for concept design

In the traditional approach to training concept development, designers rely on their ideas and assumptions about how the product or service will be used. However, the actual value of the final product can only be realized if end-users are included in the design phase. One of the main challenges in this process is determining the most effective method for documenting, describing, and transferring the knowledge gathered from users, particularly in the context of services. *[38]* This challenge becomes even more critical in the case of services related to learning, where user feedback and dialogue are crucial components of the design training process. While primarily introduced and employed in software development, the user story mapping (USM) has become a versatile approach for organizing and disseminating information about a product or service *[38]*. As previously mentioned, adopting a learner-centric approach is a crucial element for an effective deep-learning process. To accommodate learners with diverse knowledge backgrounds, the USM method has been selected as the foundation for designing this virtual reality-based training concept, especially when dealing with new technologies such as AM. This technique plays a pivotal role in comprehending user needs and experiences and ultimately in devising practical solutions to real-world problems *[20]*. In the AM training, using the USM method in designing a training concept emphasizes dialogue with Users and time efficiency, which are critical elements in an industrial context.

User story mapping (USM) for domain modeling is a method derived from the software functionality definition domain, which puts the end-user and his perspective in focus. [20] The domain of interest is defined through the collection of user activities, which indirectly gathers all actors, resources, processes, and overall dynamics. [38] The original USM method is divided into two key dimensions (see Figure 1): 1. User and 2. Usage. In this study, the term "user" encompasses a diverse group, including experts such as product designers, production managers, middle management supervisors, and individuals responsible for strategic decision-making and investments within SMEs. Engineering students and other stakeholders who actively participated in piloting the training concept are also considered users in this context. The "Usage" dimension in Figure 1 includes standardized Steps, while the user can select the specific Activities within these Steps. This allows flexibility and personalization, as the training can be adjusted to address specific skill levels, learning styles, and objectives of each user. [20] chosen by users form the backbone of the training concept. In this virtual-based AM training context, four User Steps are identified. The first is Pilot training in Figure 1, that provides a concise overview of the AM theme. It serves as an introduction to familiarize users with the fundamental concepts and principles of additive manufacturing. In the second step, preparation for training, Users receive guidance on effectively using the virtual reality goggles and navigating between different training content. The focus is ensuring users are comfortable with the technology and can access the training materials seamlessly. In the third, Training step, users engage in the core training activities by selecting relevant learning content that supports their specific individual and organizational needs and objectives. The training materials cover various aspects of AM, such as design principles and manufacturing processes. The last step, Evaluation in Figure 1, assesses what is learned and how that can be implemented in a real-life work context. The user experiences obtained from various Activities consist of User Stories. [20] These user stories encapsulate valuable data that guides the development and individualization of the training concept, ensuring that it aligns closely with the needs and preferences of the users. These User stories build the backbone of the USM model.

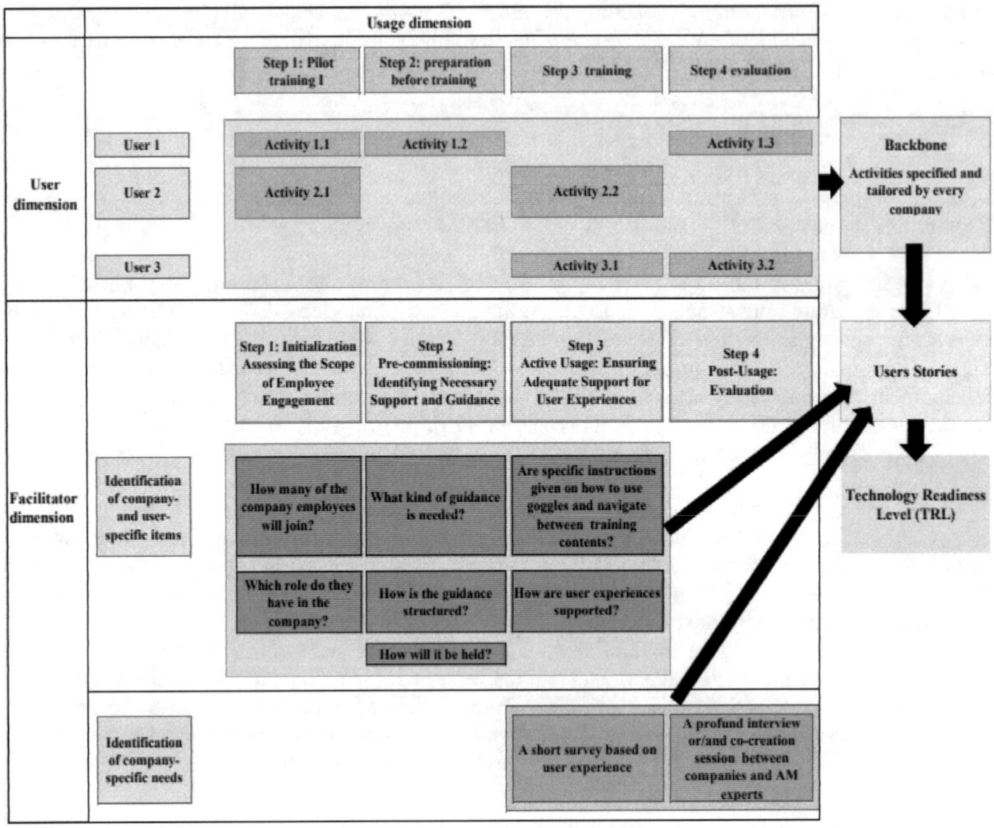

Figure 1. User story mapping (USM) modified from Kiritsis et al.[20]

The USM (user story mapping) model (Figure 1) has been modified to meet the requirements of the Virtual Reality-based Additive Manufacturing training concept. Because of the nature of training design, a new dimension called "Facilitator" has been introduced. In the literature, facilitator has various roles and activities. [39] According to Stetler at al. [40] facilitator is defined as a person who has a supportive and problem-solving function. According to Kolb et al. [39], the facilitation process involves skilfully managing interaction to create a positive and productive experience, where the facilitator plays a crucial role in actively encouraging interaction among participants to achieve the goals effectively (in this case, learning goals). Through this dimension, users gain valuable support relating to technical and educational issues in virtual reality-based AM training. By incorporating the Facilitator dimension (as shown in Figure 1), the model enables to design a more tailored and relevant learning experiences with learning supportive dialogue, aligning with the unique needs of each organization and individual. In this particular model, the utilization of user stories contributes to a comprehensive understanding of various aspects of the training process, enabling the design of a unique and customized training concept. Furthermore, user stories serve as an additional output by providing insights into the Technology Readiness Level (TRL) of the companies involved.

4.2 Technology readiness level

The technology readiness level (TRL) measures the maturity and readiness of the technology for practical implementation, progressing from early stages to deployment on a scale of 1 to 9. NASA initially developed TRL as they needed to evaluate the maturity level of different new technologies. [41] TRLs are used, for example, in aerospace, to assess technology readiness, guide decision-making, and funding, and facilitate integration into real-world applications. [42] TRLs, ranging from 1 to 9, were developed by NASA to assess technology maturity. [43] Lower TRLs (1-3) represent the conceptual phase, while higher TRLs (4-6) involve validation and prototyping. TRLs 7-9 indicate full maturity and commercial availability, as in [44]. The TRLs help assess the technology readiness of companies for adopting AM by evaluating their maturity and preparedness. Aligning these readiness levels ensures effective training that meets specific company needs and enables customization through virtual reality-based tools. This systematic approach serves as a reference for decision-making, resource allocation, and continuous improvement in training technology. Factors such as adaptability, openness to new technologies, and willingness to embrace change impact the successful implementation of training technologies. Additionally, the characteristics of the technology itself, including user-friendliness, reliability, and scalability, play a crucial role in determining its readiness for integration into training practices. Recognizing the AM traits allows for a comprehensive evaluation of technology readiness (as in [44] and [45], facilitating the selection of suitable AM technologies and maximizing their effectiveness in supporting learning and development. [46] and [44] introduce Education Readiness Levels (ERL), which aim to assess the "readiness" of the training module. Ten readiness levels have been defined for each element, based on the Technology Readiness Level (TRL) definitions. Authors will study this topic in the future more closely to be able to construct Training Readiness Level for training purposes.

During the development of the training concept, a significant challenge arose with SMEs: these organizations often operate under constraints that limit the time available for dedicated training or information gathering. Their resources are primarily allocated toward daily work activities and routines. Investing in new machinery or developing new work processes on modern technologies is a big challenge among SMEs because they have invested in their current machinery, and the investments must be worth the money. Considering the ongoing challenges of companies due to post-pandemic time and the complex global logistics and wartime conditions in Europe, small and medium-sized enterprises (SMEs) find themselves compelled to adopt short-term planning strategies and avoid long-term investments, particularly in emerging technologies like additive manufacturing (AM). The limited availability and realistic understanding of AM within their specific fields make it difficult for SMEs to envision its implementation and justify long-term commitments. Recognizing the limitations and challenges faced by SMEs in terms of work environment efficiency, it becomes even more crucial to meticulously customize company-specific training opportunities. User story mapping is a valuable tool for this kind of concept design, as it allows designers to gather user feedback and dialogue, leading to practical solutions. The method focuses on user activities and usage, providing flexibility and personalization. User stories form the backbone of the training concept in this study, guiding its development. The user story mapping model has been modified for virtual reality-based training, introducing the "Facilitator" dimension to provide technical and educational support. Technology readiness level (TRL) measures the maturity of technology for practical implementation. Aligning training technologies with appropriate TRLs ensures testing, validation, and pedagogical suitability.

The study recognizes AM as an emerging manufacturing technology with rapid growth. As a result, the demand for industrial AM training is escalating. The findings underline the significance of developing comprehensive training frameworks encompassing various types of training to cover the diverse needs of different companies. Furthermore, utilizing new, virtual reality-based tools provides opportunities for extensive customization, enhancing the effectiveness and efficiency of training programs. This study concludes by presenting a preliminary framework for creating company-specific training programs in additive manufacturing.

This framework serves as a foundation for designing effective training concept that accommodates the specific requirements of companies with different background knowledge. By establishing a structured approach to training development, organizations can maximize the benefits derived from AM technologies and accelerate their adoption and integration into their operations. In addition to the technology readiness level (TRL), Education Readiness Levels (ERL) have been introduced to assess the "readiness" of training modules in AM training. [44] ERLs align with TRL definitions and define ten readiness levels for each element. [46] This approach allows for a comprehensive evaluation of technology readiness, enabling the selection of suitable technologies for effective learning and development. Further research will aim to construct Training Readiness Levels for training purposes.

Evaluating maturity, adaptability, and technology characteristics ensures effective training and supports the successful integration of AM practices.

5. Conclusions

The rapidly evolving field of AM requires continuous training to keep up e.g. with hardware, software, and material advancements. Establishing research-based knowledge is crucial to avoid misconceptions. Training, especially industrial training, is essential at all levels of the organization to support product development and enhance expertise. This study investigates the feasibility of virtual reality (VR) based on industrial training in additive manufacturing. The aim is to develop an effective training concept for small and medium-sized enterprises (SMEs) in the AM industry. The study utilizes microlearning methods and 360° video materials to create an engaging and tailored training experience. User story mapping (USM) is employed to understand the specific training needs of companies in the AM field and ensure that the training program aligns with their requirements. The study also assesses the technology readiness levels (TRLs) of companies regarding the adoption of AM. The research findings contribute to the development of a VR-based training concept for AM, considering the unique challenges and requirements of SMEs.

Combining these approaches tailored by the user story mapping method can revolutionize industrial AM training, and offer personalized, immersive, and efficient learning experiences. By leveraging modern technologies and methodologies, companies can enhance their competitiveness and bridge the gap between AM progress and expert knowledge. As the field of AM continues to grow, the importance of such training approaches will expand, ensuring that professionals are equipped with the necessary skills and knowledge to leverage the potential of this innovative technology. Topics for further studies were found in exploring the effectiveness of virtual reality-based industrial training in AM, in enhancing user engagement and interaction in virtual reality-based AM training, optimizing user story mapping for tailored additive manufacturing training, and assessing the readiness levels of companies for adopting additive manufacturing.

During the preparation of this work, the authors used Chat GPT and Grammarly in order to reformulate and improve the structures of sentences and as a tool to find synonyms and different expressions to avoid tautology. After using these AI tools, the authors reviewed and edited the content as needed and take full responsibility for the content of the publication.

Piili and Huusko drafted the idea of this publication. Further brainstorming was executed by all authors. The actual drafting of the concept of education approach introduced in this article was discussed, commented on, and modified by all the authors. Susi drafted the first approach to VR/360 training concept and practices for it, and this was used as part of this article. Huusko, Piili, Tuomisto, Helin, and Parchegani carried out a theoretical review and editing of it. Huusko developed a USM method draft for this article, and it was discussed and modified by all authors. Salminen reviewed, supervised, provided Resources and Funding acquisition. The final draft of the publication was discussed, commented on, edited, and modified by all authors.

Acknowledgments

This research was done in collaboration with the project Osuva ja vaikuttava 360-vr-materiaali modernien digitaalisten valmistusteknologioiden kouluttamiseen ja perehdyttämiseen eri kohderyhmien (teknologiayritykset, loppuasiakkaat, koulutuslaitokset) tarpeisiin, *Focused and effective 360-VR material for training and familiarizing with modern digital manufacturing technologies for the needs of different target groups (technology companies, end customers, educational institutions)* (shortened as Osuu ja Uppoaa) funded by European Social Fund (S22515). The funding is part of Covid-19 pandemic related actions by the EU. The duration of the project is 1.8.2021-30.11.2023. Project aims to provide extensive additive manufacturing training materials for SMEs utilizing virtual reality to enhance the competitiveness of companies. Project partners are University of Turku and industrial partners. Authors of this article would like to acknowledge this project and its partners for knowledge, assistance and collaboration to be able to conduct this study. Project has enabled also availability of such data of industrial training, user experience and training concept that has been essential for carrying out this study. Authors would like to thank also to staff of Turku School of Economics and Department of Mechanical Engineering in Faculty of Technology at University of Turku for their knowledge and help to be able to execute this study.

References

[1] Charles A, Bayat M, Elkaseer A, Scholz S. Simulation in Additive Manufacturing and Its Implications for Sustainable Manufacturing in the Era of Industry 4.0. In: Scholz SG, Howlett RJ, Setchi R, editors. Sustain. Des. Manuf., Singapore: Springer Nature; 2023, p. 370–7. https://doi.org/10.1007/978-981-19-9205-6_36.
[2] Borgianni Y, Maccioni L, Spena PR, Shunmugavel MK. University Education in Additive Manufacturing and the Need to Boost Design Aspects. Proc Des Soc Int Conf Eng Des 2019;1:629–38. https://doi.org/10.1017/dsi.2019.67.
[3] Leutenecker-Twelsiek B, Ferchow J, Klahn C, Meboldt M. The Experience Transfer Model for New Technologies - Application on Design for Additive Manufacturing. In: Meboldt M, Klahn C, editors. Ind. Addit. Manuf. - Proc. Addit. Manuf. Prod. Appl. - AMPA2017, Cham: Springer International Publishing; 2018, p. 337–46. https://doi.org/10.1007/978-3-319-66866-6_32.
[4] EY's Global 3D printing Report 2016 2016.
[5] AM-platform.eu - Home n.d. http://www.rm-platform.com/ (accessed May 23, 2023).
[6] CLLAIM Project n.d. https://cllaimprojectam.eu/ (accessed May 23, 2023).
[7] Admire Project n.d. https://admireproject.eu/background.html (accessed May 23, 2023).
[8] European Welding Federation n.d. https://www.ewf.be/qualification/iamqs.aspx (accessed May 23, 2023).
[9] Stavropoulos P, Foteinopoulos P, Stavridis J, Bikas H. Increasing the industrial uptake of additive manufacturing processes: A training framework. Adv Ind Manuf Eng 2023;6:100110. https://doi.org/10.1016/j.aime.2022.100110.
[10] Ma Y, Li B. Effect of digitalization on knowledge transfer from universities to enterprises: Evidence from postdoctoral workstation of Chinese enterprises. Technol Soc 2022;71:102102. https://doi.org/10.1016/j.techsoc.2022.102102.
[11] Bygstad B, Øvrelid E, Ludvigsen S, Dæhlen M. From dual digitalization to digital learning space: Exploring the digital transformation of higher education. Comput Educ 2022;182:104463. https://doi.org/10.1016/j.compedu.2022.104463.
[12] Abduraxmanova SA. INDIVIDUALIZATION OF PROFESSIONAL EDUCATION PROCESS ON THE BASIS OF DIGITAL TECHNOLOGIES. World Bull Soc Sci 2022;8:65–7.
[13] Zhang J, West RE. Designing Microlearning Instruction for ProfessionalDevelopment Through a Competency Based Approach. TechTrends 2020;64:310–8. https://doi.org/10.1007/s11528-019-00449-4.
[14] Busse J, Schumann M. Towards a Pedagogical Pattern Language for Micro Learning in Enterprises. 26th Eur. Conf. Pattern Lang. Programs, New York, NY, USA: Association for Computing Machinery; 2022, p. 1–8. https://doi.org/10.1145/3489449.3489973.
[15] Kraus N, Marchenko O. Innovative-digital entrepreneurship as key link of Industry X.0 formation in the conditions of virtual reality. Balt J Econ Stud 2021;7:47–56.
[16] Paul M, Maglaras L, Ferrag MA, Almomani I. Digitization of healthcare sector: A study on privacy and security concerns. ICT Express 2023. https://doi.org/10.1016/j.icte.2023.02.007.
[17] Osuu ja Uppoaa | Work Informatics 2021. https://workinformatics.utu.fi/hankkeet/osuujauppoaa/ (accessed May 28, 2023).
[18] A bibliometric method for assessing technological maturity: the case of additive manufacturing | SpringerLink n.d. https://link.springer.com/article/10.1007/s11192-018-2941-1 (accessed June 13, 2023).
[19] Huang Y, Leu MC, Mazumder J, Donmez A. Additive Manufacturing: Current State, Future Potential, Gaps and Needs, and Recommendations. J Manuf Sci Eng 2015;137. https://doi.org/10.1115/1.4028725.
[20] Kiritsis D, Milicic A, Perdikakis A. User Story Mapping-Based Method for Domain Semantic Modeling. In: Karagiannis D, Mayr HC, Mylopoulos J, editors. Domain-Specif. Concept. Model., Cham: Springer International Publishing; 2016, p. 439–54. https://doi.org/10.1007/978-3-319-39417-6_20.

[21] Jahnke I, Lee Y-M, Pham M, He H, Austin L. Unpacking the Inherent Design Principles of Mobile Microlearning. Technol Knowl Learn 2020;25:585–619. https://doi.org/10.1007/s10758-019-09413-w.
[22] Gagne JCD, Park HK, Hall K, Woodward A, Yamane S, Kim SS. Microlearning in Health Professions Education: Scoping Review. JMIR Med Educ 2019;5:e13997. https://doi.org/10.2196/13997.
[23] Lin J, Sun G, Cui T, Shen J, Xu D, Beydoun G, et al. From ideal to reality: segmentation, annotation, and recommendation, the vital trajectory of intelligent micro learning. World Wide Web 2020;23:1747–67. https://doi.org/10.1007/s11280-019-00730-9.
[24] Pagé C, Bernier P-M, Trempe M. Using video simulations and virtual reality to improve decision-making skills in basketball. J Sports Sci 2019;37:2403–10. https://doi.org/10.1080/02640414.2019.1638193.
[25] Naranjo JE, Sanchez DG, Robalino-Lopez A, Robalino-Lopez P, Alarcon-Ortiz A, Garcia MV. A Scoping Review on Virtual Reality-Based Industrial Training. Appl Sci 2020;10:8224. https://doi.org/10.3390/app10228224.
[26] Monetti FM, de Giorgio A, Yu H, Maffei A, Romero M. An experimental study of the impact of virtual reality training on manufacturing operators on industrial robotic tasks. Procedia CIRP 2022;106:33–8. https://doi.org/10.1016/j.procir.2022.02.151.
[27] Lee GA, Yang U, Son W, Kim Y, Jo D, Kim K-H, et al. Virtual Reality Content-Based Training for Spray Painting Tasks in the Shipbuilding Industry. ETRI J 2010;32:695–703. https://doi.org/10.4218/etrij.10.1510.0105.
[28] Calvert J, Abadia R. Impact of immersing university and high school students in educational linear narratives using virtual reality technology. Comput Educ 2020;159:104005. https://doi.org/10.1016/j.compedu.2020.104005.
[29] Interactive virtual technologies in engineering education: Why not 360° videos? | SpringerLink n.d. https://link.springer.com/article/10.1007/s12008-019-00553-y (accessed May 23, 2023).
[30] Snelson C, Hsu Y-C. Educational 360-Degree Videos in Virtual Reality: a Scoping Review of the Emerging Research. TechTrends 2020;64:404–12. https://doi.org/10.1007/s11528-019-00474-3.
[31] "Work Informatics—Mon Amour" by Markku I. Nurminen n.d. https://aisel.aisnet.org/sjis/vol29/iss1/3/ (accessed May 29, 2023).
[32] Kumar K, van Dissel HG, Bielli P. The Merchant of Prato-Revisited: Toward a Third Rationality of Information Systems. MIS Q 1998;22:199–226. https://doi.org/10.2307/249395.
[33] misqresearchcurations. Trust. MIS Q 2016. https://www.misqresearchcurations.org/blog/2017/5/10/trust-1 (accessed May 29, 2023).
[34] Ghaffarian V. The new stream of socio-technical approach and main stream information systems research. Procedia Comput Sci 2011;3:1499–511. https://doi.org/10.1016/j.procs.2011.01.039.
[35] Emerging technologies as pedagogical tools for teaching and learning science: A literature review - Oliveira - 2019 - Human Behavior and Emerging Technologies - Wiley Online Library n.d. https://onlinelibrary.wiley.com/doi/full/10.1002/hbe2.141 (accessed May 29, 2023).
[36] Viljoen S. A Relational Theory of Data Governance Feature. Yale Law J 2021;131:573–654.
[37] Osuu ja Uppoaa -hankkeessa kehitetään virtuaalinen koulutusmenetelmä 3D-tulostukseen n.d. https://www.utu.fi/fi/ajankohtaista/mediatiedote/osuu-ja-uppoaa-hankkeessa-kehitetaan-virtuaalinen-koulutusmenetelma-3d (accessed May 23, 2023).
[38] Milicic A, Perdikakis A, El Kadiri S, Kiritsis D, Ivanov P. Towards the Definition of Domain Concepts and Knowledge through the Application of the User Story Mapping Method. Prod. Lifecycle Manag. Knowl.-Rich Enterp., Springer Berlin, Heidelberg; 2012, p. 58–69.
[39] Kolb JA, Jin S, Hoon Song J. A model of small group facilitator competencies. Perform Improv Q 2008;21:119–33. https://doi.org/10.1002/piq.20026.
[40] Stetler CB, Legro MW, Rycroft-Malone J, Bowman C, Curran G, Guihan M, et al. Role of "external facilitation" in implementation of research findings: a qualitative evaluation of facilitation experiences in the Veterans Health Administration. Implement Sci 2006;1:23. https://doi.org/10.1186/1748-5908-1-23.
[41] Technology readiness levels: Shortcomings and improvement opportunities - Olechowski - 2020 - Systems Engineering - Wiley Online Library n.d. https://incose.onlinelibrary.wiley.com/doi/full/10.1002/sys.21533?casa_token=Rk_h98LtpAQAAAAA%3Au_yTDIlvRC7I_0jvz_O_0Q92FMLQACEoWIuUzqMYVySa9PS039Zq7JHE064BayMx-W8d-FnFoXeKO8k (accessed May 26, 2023).

[42] Beims RF, Simonato CL, Wiggers VR. Technology readiness level assessment of pyrolysis of trygliceride biomass to fuels and chemicals. Renew Sustain Energy Rev 2019;112:521–9. https://doi.org/10.1016/j.rser.2019.06.017.
[43] Straub J. In search of technology readiness level (TRL) 10. Aerosp Sci Technol 2015;46:312–20. https://doi.org/10.1016/j.ast.2015.07.007.
[44] Dinda S, Simpson TW, Gluck L. Education Readiness Levels (ERLs): A Scale for Assessing Educational Coursework and Training Modules, American Society of Mechanical Engineers Digital Collection; 2017. https://doi.org/10.1115/DETC2017-68086.
[45] Zutin GC, Barbosa GF, de Barros PC, Tiburtino EB, Kawano FLF, Shiki SB. Readiness levels of Industry 4.0 technologies applied to aircraft manufacturing—a review, challenges and trends. Int J Adv Manuf Technol 2022;120:927–43. https://doi.org/10.1007/s00170-022-08769-1.
[46] Espadinha-Cruz P, Neves A, Matos F, Godina R. Development of a maturity model for additive manufacturing: A conceptual model proposal. Heliyon 2023;9:e16099. https://doi.org/10.1016/j.heliyon.2023.e16099.

IOP Conf. Series: Materials Science and Engineering 1296 (2023) 012005 doi:10.1088/1757-899X/1296/1/012005

Using artificial neural networks to model single bead geometries processed by laser-wire direct energy deposition

Reza Asadi[1], Antoine Queguineur[1,2], Aapo Ylä-Autio[1], Antti Martikkala[1,3] Olli Wiikinkoski[1], Hossein Mokhtarian[1], Iñigo Flores Ituarte[1]

[1]Department of Automation and Mechanical Engineering, Faculty of Engineering and Natural Sciences, Tampere University, Korkeakoulunkatu 6, 33014 Tampere, Finland

[2] Ecole Centrale Nantes, GeM - UMR CNRS 6183, 1 rue de la Noé, 44321 Nantes – France

[3] Department of Mechanical and Industrial Engineering, Norwegian University of Science and Technology, Richard Birkelands Vei 2b, NO-7034, Trondheim, Norway

reza.asadi@tuni.fi

Abstract. Wire-feed laser additive manufacturing processes have gained researchers' attention because of their potential to reduce material waste, guarantee accuracy, increase material quality and density, and produce a wide dimensional range of final products. Nevertheless, printing materials with desired geometrical properties of the beads is still challenging in such processes. This might be attributed to the need for more sufficient experimental data and precise modeling approaches. In this study, an architecture based on Artificial Neural Networks (ANNs) is developed to model the bead geometries (width, height, and area), considering the wire feed rate, laser power, and travel speed as process parameters. A design-of-experiment based on full factorial design is considered for processing single beads with a Fraunhofer coaxial wire-feed laser system. Inconel 625 wire with a diameter of 1.14 mm and stainless steel substrate are utilized as the experimental materials. Geometrical data is obtained using a laser scanner model RA-7525 SE with 0.026mm volumetric accuracy. The beads' geometrical details are provided as the feeding data for the proposed ANN. For each bead, a length of 10 mm is considered to calculate the average geometrical parameters, which increases the accuracy of the dataset in comparison to the values acquired via a macroscopic picture of the cross-section of each weld bead. A variety of hyperparameters are chosen and compared regarding precision criteria, including Mean Square Error (MSE), to increase the model's accuracy. A train-test separation strategy is considered to evaluate the model's accuracy on independent data points. The outcome of this research is an ANN-based geometry prediction model that can be utilized to enhance the development of offline path planners and optimize process parameter selection for a precise geometry toward process control.

Keywords: Artificial neural networks, direct-energy deposition, additive manufacturing, bead geometry modeling, Inconel 625

1. Introduction

Metal Additive Manufacturing (AM) has become an increasingly interesting field of research due to its capacity to produce high-performance products, achieve high-level automation, and reduce materials waste [1]. There is an array of methods based on the deposition strategies, utilized materials and source of energy adopted in AM processes. Each method has its own specifications making them suitable for unique applications [2]. Wire-feed Laser Additive Manufacturing (WLAM) is an economical additive manufacturing method for complex geometries, good surface quality and near net shapes [3]. In comparison to the AM methods using electric arc to melt and deposit the material, in WLAM processes energy coming from the laser can be more focused which results in smaller melt pools, smaller feature size, and more efficient energy consumption [4]. These characteristics of WLAM result in the final product with more controlled residual stresses and phase transformations [5]. However, the deposition rate in such methods is less than in wire arc additive manufacturing methods.

Among the nickel-based alloys, Inconel 625 is a superalloy widely used in the aerospace sector and especially the engines and turbines [6], [7]. This material is strengthened through solid solution hardening that enables the achievement of high mechanical properties. Hence, the addition of alloying elements such as chromium, molybdenum, and niobium to the nickel matrix lead to the formation of carbides and intermetallic [8] [9]. Inconel 625 has excellent characteristics, including high corrosion resistance, strength and good tensile properties at high and cryogenic temperatures [10]. The properties of Inconel 625 are well-suited in WLAM processes thanks to a high-energy-density laser used to fuse the wire and the near-net-shape required in aerospace applications. However, rapid melting and solidification as well as cyclic heating and cooling during the deposition of layers in WLAM impact the characteristics of the deposited materials and especially the risk of solidification cracking [11]. Therefore, a comprehensive study is essential for discovering the properties of final products.

There is a growing interest in laser-wire AM [12]–[15] . However, metal AM widespread acceptance is hindered by lack of knowledge in organizational structures and high costs [1], [16], [17], parts quality (i.e., geometry, defects, microstructure, mechanical properties, etc.) [18], lack of manufacturing robustness, stability, and repeatability in the process [19]. Therefore, modeling the geometrical details of the metal AM processes and specifically WLAM can have significant effect on the process control and stability. The impact of related process parameters on the product geometry [13], microstructure [14], and surface quality [15] have been considered. Ali Gökhan Demir [20] has studied the correlation between geometrical characteristics and process parameters. Naïve Bayes (NB) networks and Gaussian process regression (GPR) model were utilized to assess the quality of the model and predict the printed bead geometries [13].

Based on the literature review, more investigations are required in modeling and prediction of the bead geometrical details. To that end, modeling single bead geometry can be assumed as the basis and steppingstone for such details. Developing an accurate model considering the overlapping factors, weld pool details and geometrical details of layers is a prerequisite for developing an efficient process. Consequently, the adaption of a G-code generator to design and print multilayer complex parts is achievable. In this paper, a combination of different Design of Experiments (DoE) including, Full Factorial (FF) and Central Composite Design (CCD), is adopted to implement the experiments on WLAM of Inconel 625. The experimental data is fed to a Neural Network (NN) to model the geometrical properties, and finally the sensitivity of the process parameters to the geometrical properties and the accuracy of the utilized ANN is evaluated. This research endeavor makes contributions in the following areas: (a) developing ANNs by employing the most efficient hyperparameters, including activation functions. (b) utilizing and selecting the most suitable optimization algorithm for training ANNs specifically for WLAM experiments. (c) incorporating a comprehensive statistical analysis to evaluate the sensitivity of process parameters and their interactions with respect to variations in bead geometrical details.

2. Experimental setup and data acquisition plan

The experiments are conducted on a LW-DED device introduced in Figure 1 with predetermined sets of process parameters. Fraunhofer CoaxWire welding head is implemented on an Aerotech 3-axis CNC machine. The welding head feeds the wire from the central axis, while three laser beams converge at a common focal point to the wire to make the process direction independent. Dinse, FDE 100 L is utilized as the wire feeder and process control system. A 3 KW Coherent Corelase fiber laser is used as the energy resource of the process. A UTP A 6222Mo-3 is selected as alloy 625 wire feedstock material with a diameter of 1.14 mm. Table 1 presents the chemical composition of the selected wire. A 10 mm plate of 316L stainless steel (SS) has been chosen as the substrate. Single beads of Inconel 625 with 35 mm length under various process parameter levels, without in-situ or ex-situ thermal treatment are printed and analyzed in this study. Laser Power (LP), Travel Speed (TS), and Wire Feed Rate (WFR) are used as process parameters, and bead height, width, and area are the response parameters.

Table 1. Chemical composition of the filler metal

Filler metal chemical composition - UTP A 6222Mo-3, wt%													
C	Si	Mn	Ni	Cr	Mo	Cu	Ti	Al	Nb	P	S	Fe	Ta
<0.01	0.12	0.02	64.66	22.30	8.71	0.02	0.22	0.13	3.45	0.004	0.001	0.29	0.004

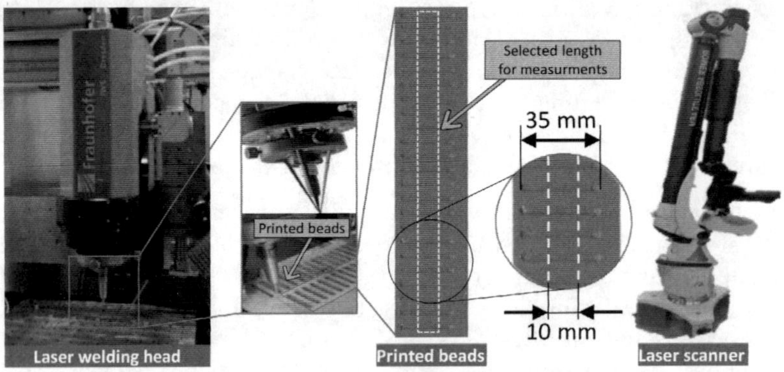

Figure 1. The utilized experimental setup for tests

After the screening tests to determine the ranges of the process parameters, as provided in Table 2, three different Design of Experiments (DoE) were selected. Two different FF DoEs with small and large range of variables and a CCD DoE are chosen. All the experimental results obtained from the mentioned DoEs are collected and fed into the proposed network to predict the geometrical bead values. For each FF DoE and 3 levels for each factor, 27 experiments are required. Furthermore, 20 tests with 6 repetitions in center point for CCD DoE are required. With the reduction of common experimental points between different DoEs, in total there are 66 experiments needed to be conducted.

Table 2. WLAM process parameter ranges

Process Parameters	Range
A: WFR (m/min)	[0.4 – 1.1]
B: TS (mm/s)	[3 – 8]
E: LP (kW)	[1.4 – 2.1]

Bead geometrical details including the printed beads shapes, their height, width, and area are obtained from the laser scanning system model RA-7525 SE with volumetric accuracy of 0.026 mm (Figure 1). The obtained data are evaluated, and the bead geometries are calculated for 10 mm length in the middle of the beads. The overall length of printed beads is set to 35 mm. To validate the scanning measurements, the theoretical WFR is compared with the overall printed volume and if the difference is less than 5%, the values are accepted. Since the main aim of this study is evaluation of machine learning based methods, 20 random tests, which are not similar to the training data but are within the validated ranges of process parameters, are conducted and the results are fed to the developed networks.

3. Modeling approach

In this study, statistical analysis techniques and Deep Learning (DL) methods are utilized to assess the sensitivity of the response parameters to the process parameters and modeling the bead geometry respectively.

IOP Conf. Series: Materials Science and Engineering 1296 (2023) 012005 doi:10.1088/1757-899X/1296/1/012005

3.1. Statistical Analysis

In statistics, to evaluate the correlation between explanatory variables, which are process parameters in manufacturing, and response parameters, Response Surface Methodology (RSM) is the most popular method. RSM, which was introduced by P. Box and K. B. Wilson [21], deploys second-degree polynomial model on the sequence of DoEs data to attain such an aim. In this method, some criteria are provided to consider the sensitivity of the responses to the process parameters.

Coefficient of determination (R^2) is the relationship between variations in one variable and differences in another. R^2 is defined as equation (1):

$$R^2 = 1 - \frac{Sum\ of\ squared\ error\ (SSE)}{Total\ sum\ of\ squares(TSS)} = 1 - \frac{\sum_{i=1}^{m}(y_i - f_i)^2}{\sum_{i=1}^{m}(\bar{y} - y_i)^2} \quad (1)$$

Which y_i is the *i*-th target value, f_i is the *i*-th modeled value and $\bar{y}$ is the mean of target values. The best case is when the model is exact, and SSE equals to 0. In this case, R^2 equals to one. However, R^2 values more than 0.85 is considered as a sign of significant correlation, and values more than 0.80 is acceptable. R^2_{adj} is a robust version of R^2 that is used for analysis and the acceptable value for this criterion is equivalent or smaller than R^2.

Pareto chart is a chart that visually depicts the most significant factors for modeling a specific response parameter. In the present study, all the mentioned criteria are utilized to assess the sensitivity of geometrical attributes including width, height, and bead to the variation of process parameters.

3.2. Artificial Neural Network (ANN)

Manufacturing processes signatures and expected outcomes behave non-linearly and containing many complex factors [22]. Consequently, mathematical modeling of the required phenomena is challenging and usually not accurate. The assumed phenomenon to be modeled is bead geometry (width, height and area) as it is introduced inFigure ***2***-a. ANNs are among the commonly used methods due to the simplicity of understanding their concept, and the ability to improve their performance by feeding them more experimental data. ANNs are bio-inspired methods based on the structure of neurons and neural system. As it is shown in Figure ***2***-b, an artificial neuron consists of input parameters multiplied by their dedicated weights (W_i). To handle the non-linearities, after adding a bias (b), an activation function (Ψ) is applied before getting the neuron's output. The training process helps to adjust the weights, and bias (b) values [23].

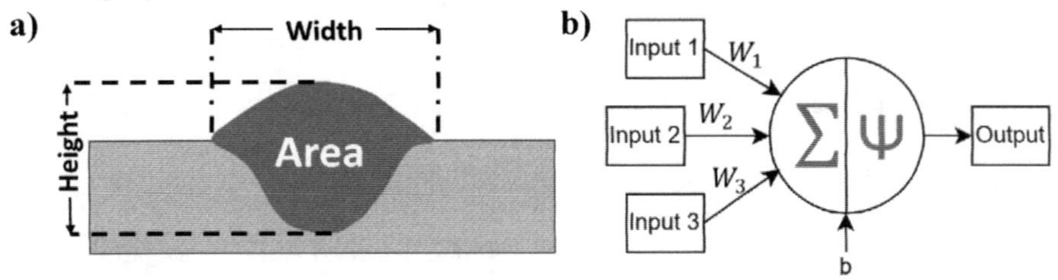

Figure 2. Schematic representation of a) bead geometrical features and b) perceptron

Due to the complexity of AM processes specially WLAM and existence of unknown parameters, modeling the geometrical phenomena cannot be implemented by a single perceptron. Consequently, multi-layer ANNs have been deployed. In this study, an ANN with three inputs and three outputs is utilized for the modeling of bead geometry (Figure 3-a). The specification of the network is as following:
Input layer: The same as all ANNs, input layer is the layer for feeding the process parameters to the network. Here, WFR, TS, and LP are the input nodes.
Hidden layer 1: This layer is the first computational layer. Each neuron gets the inputs to the neurons and apply the activation function on it to create the outputs for the next layer. The output of the *j*-th

neuron in layer k is calculated as it is shown in equation (2). Ψ is the assumed activation function. $w_{i,k}^{j}$ is the weight to be multiplied to the input from the i-th output of the previous layer. Each neuron j has its own bias and added is before applying Ψ.

$$y_k^j = \Psi\,(\sum_{i=1}^{n} w_{i,k}^{j}\, x_i + b_j) \qquad (2)$$

The utilized Ψ is Gaussian Error Linear Unit (GELU) and is calculated based on equation (3).

$$GELU(x) = xP(X \leq x) \triangleq x\phi(x) = x.\frac{1}{2}[1 + \operatorname{erf}(\frac{x}{\sqrt{2}})] \qquad (3)$$

$x\phi(x)$ is defined as GELU, and $\phi(x)$ is the standard Gaussian cumulative distribution function. The term erf(z) is the "error function" and its function defined in equation (4):

$$erf(z) = \frac{2}{\sqrt{\Pi}}\int_0^z e^{-t^2}\, dt \qquad (4)$$

Based on the evaluations, GELU activation function performance proved to be better than others like Rectified Linear Unit (ReLU), tangent hyperbolic (Tanh) or sigmoid. This is attributed to the shape of such an activation function since it takes the advantage of linear function for positive values and opposite to ReLU it is not returning zero values abruptly for negative values. As it can be seen in Figure 3-b, it converges to zero smoothly and this is helpful for training the network for bead geometrical features. The number of neurons in this layer is set to 32, and all the neurons utilizes GELU as activation function.
Hidden layer 2: Details for the second hidden layer including the activation function and the relation between inputs and outputs are the same as the previous layer. The number of neurons in this layer is set to 64 based on the assessment of different values for it.

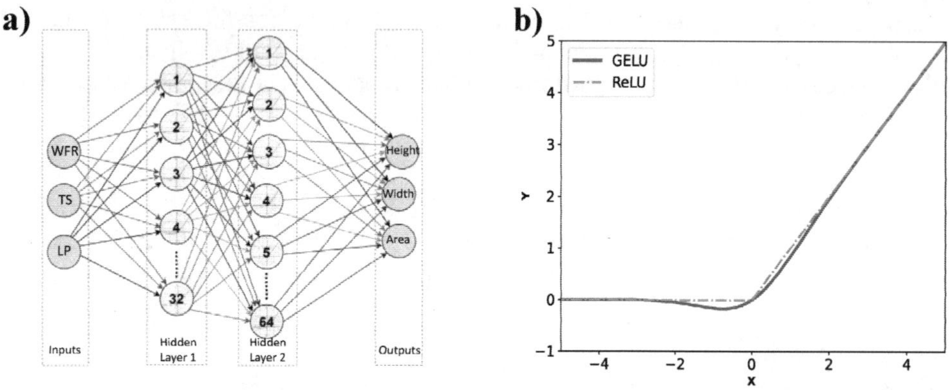

Figure 3. a) Architecture of the proposed ANN, b) comparison between GELU and RELU

Output layer: In this layer as the last layer of the architecture, the assumed response parameters including bead width, height, and area is obtained.
ANN evaluation details: The loss function to calculate the difference between ANN output and the target values is set to Mean Square Error (MSE) loss function (equation (5)).

$$mean\ squared\ error\ (MSE) = \frac{1}{m}\sum_{i=1}^{m}(y_i - f_i)^2 \qquad (5)$$

In DL, optimization method is one of the key parts of the architecture and determines the way by which the network tries to learn and make the model accurate. The optimization method utilized in this study is from Adaptive Moment Estimation (Adam) family. In Adam optimization method, which is an adaptive method for training the network, by combining the RMSprop and weighted average concept of momentum, it is tried to increase the speed and accuracy of training procedure. In the traditional Adam method, while adding the regularization term (as it can be seen in Table 3, step 7 yellow part), the algorithm is adding the regularization term to the cost function. As a result, it can easily track the regularization term and affect the performance of training. To tackle this problem, AdamW method is introduced [24]. Instead of adding the L2 regularization to the gradient g_t, it is added to weight updated part as weight decay parameter (Table 3, step 10 green part). It is proved that this method has better performance than Adam, and weight decay coefficient is set to 0.01.

Table 3. The steps of implementing Adam and AdamW algorithms

Adam and AdamW optimization algorithms
1. Given α_t (learning rate), β_1 (momentum), β_2 (RMSprop factor), ω (weight decay), and ϵ (constant equals to 10^{-8})
2. Initialization Ƞ (learning rate), m and v (first and second momentum vector), time t
Loop:
3. t → t+1
4. Applying the gradient on batch: $\nabla f_t(x_{t-1})$
5. Calculation of g_t: $\nabla f_t(x_{t-1}) + w_t x_{t-1}$
6. Calculation of 1st momentum: $m_t \rightarrow \beta_1 m_{t-1} + (1-\beta_1)\, g_t$
7. Calculation of 2nd momentum: $v_t \rightarrow \beta_2 v_{t-1} + (1-\beta_2)\, g^2{}_t$
8. Bias correction of m_t and v_t: $m_t \rightarrow m_t/(1-\beta^t{}_1)$, $v_t \rightarrow v_t/(1-\beta^t{}_2)$
9. readjust Ƞ if any decay is needed
10. Weight update: $x_t \rightarrow x_{t-1} - \eta(\frac{\alpha_t m_t}{\sqrt{v_t} + \epsilon} + w_t x_{t-1})$
Continue until the defined criterion is satisfied.
Returning the updated weights.

4. Results and discussion

4.1. Statistical analysis of experimental data

In the present study, CCD DoE is selected to evaluate the sensitivity of the response parameters (width, height, and area of a single bead) to a variation of process parameters. Due to the specification of the selected DoE, all the interaction terms and quadratic terms can be considered for statistical analysis. Pareto charts, which introduce the most influencing parameters to the geometrical features, are illustrated in Figure 4. It is worth mentioning that p-values less than 0.05 is considered as the criteria for considering a parameter as sensitive term. Whereas all the response parameters are sensitive to WFR and TS, there are trivial differences in the sensitivity of other terms, including individual LP, quadratic or interaction terms.

For bead height, quadratic terms of TS (BB), and WFR (AA) are among the parameters that are affecting the output. Bead width is the response parameter with the highest number of effective factors. As can be seen in Figure 4-b, along with AA and BB as the quadratic terms of WFR and TS, LP is a sensitive term to the variation of bead width as the output. Bead area is the parameter with TS, WFR, quadratic term of TS (BB), and the interaction term of WFR and TS (AB) as the most sensitive parameters. While all the geometrical features have their own sets of sensitive parameters, WFR, TS, and quadratic term of TS are their common factors as the effective parameters to model them.

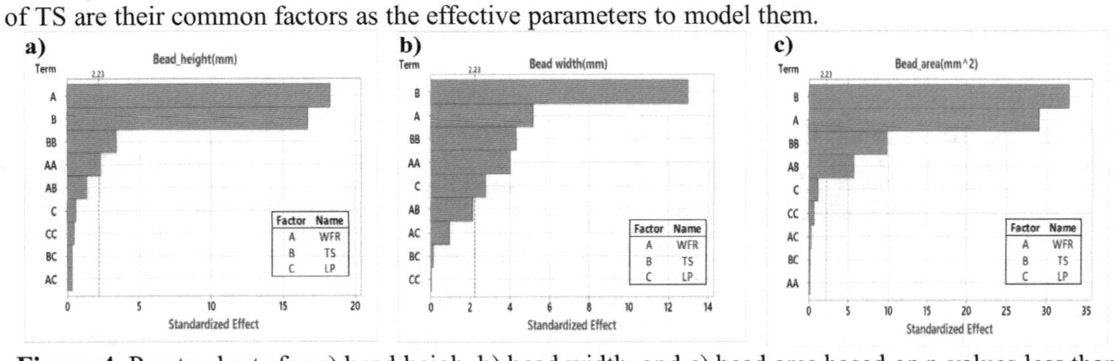

Figure 4. Pareto charts for a) bead heigh, b) bead width, and c) bead area based on p-values less than 0.05 as the sensitivity line

After specifying the most sensitive parameters to the assumed geometrical features, the degree of correlation between all the process parameters and each response parameter is essential for modeling the bead geometry. Table 4 provides the values for the criteria including R^2 and R^2_{adj} for each response parameter. All the values are over 90%, and as a result the inputs-outputs correlation is satisfying. However, the lowest R^2 and R^2_{adj} are attributed to width and should be considered for ANN modeling.

Table 4. R^2 and R^2_{adj} for geometrical features

Response parameter	R^2	R^2_{adj}
Bead height (mm)	98.45	97.05
Bead width (mm)	96.00	92.41
Bead area (mm^2)	99.51	99.08

4.2. ANN-based modeling of geometrical parameters

The introduced ANN is utilized to model bead geometries, including width, height, and area. To ensure the accuracy of the training process and prevent over or underfitting, 15% of training dataset was separated and utilized as the validation data. Whereas the training process is initialized with 2000 epochs, it stops as soon as the validation dataset loss increases continuously. The best model coefficients and specification in terms of validation results are saved and loaded for evaluation of testing data.

The training and validation loss reduction and the stopping epochs are presented in Figure 5. The best epochs for training bead height, width, and area are 790, 280, and 295, respectively. It can be observed that after rapid fluctuations and decrease of MSE as the loss function for the considered response parameters, the improvement rate decreases and stops with less than 1000 epochs.

After training and validating the performance of the model, the testing process is required to be accomplished. The best approach is to plot the fit line between model output as the predicted values and the target values as ground truth data. Figure 6 provides the fit line and the line y=x as the best results with R^2 equals to 1. As it can be observed, the fitted lines for bead height and area are perfectly matching the model outputs with R^2 more than 0.98 and 0.99 respectively. R^2 for bead width is less than other two features and is slightly over 0.82. As it was shown in Table 4, R^2 values for width based on statistical analysis of CCD DoE is less than R^2 for height and area. However, this value for our ANN model is even lower and it is due to two main reasons: a) In general, the bead width is more challenging for modeling based on statistical analysis, and b) in the modeling approach, all the conducted DoEs are utilized to create the model. As a result, while it is more generalized and reliable model, the variations of parameters are higher in comparison to a single CCD DoE and modeling the bead geometry based on the whole data is more sophisticated.

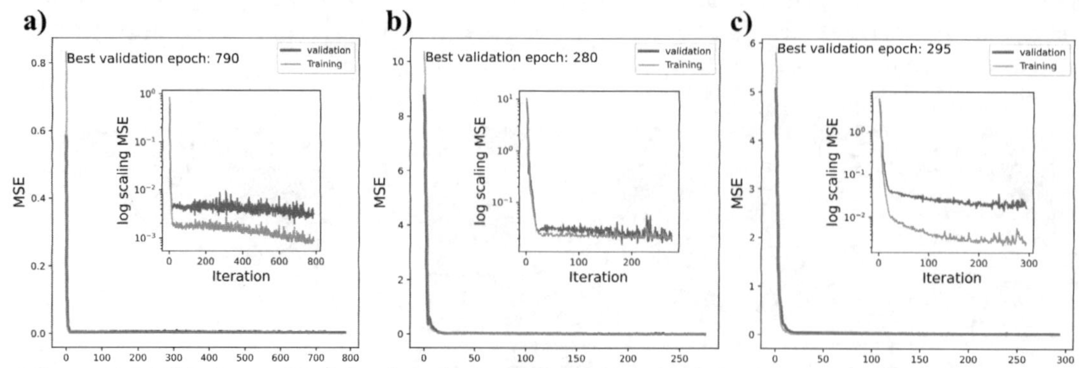

Figure 5. Validation and training loss for modeling a) bead height, b) bead width, and c) bead area, and their best validation epoch

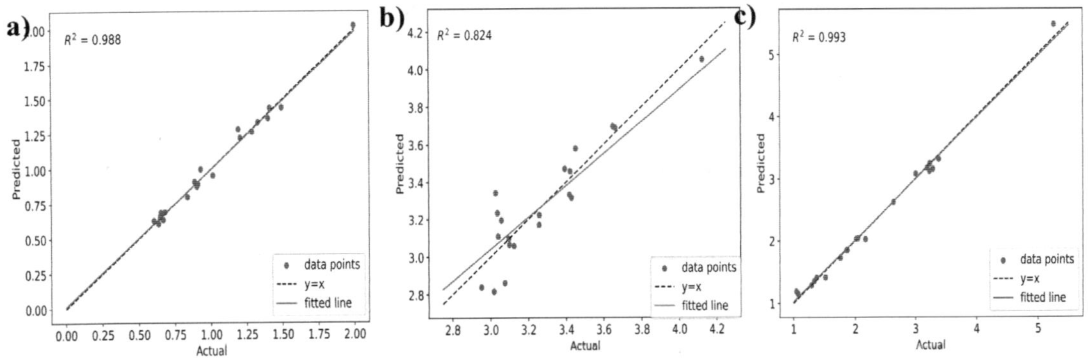

Figure 6. ANN model fit line for a) bead height, b) bead width, and c) bead area, and comparison of them with line y=x

As the last analysis, Table 5 provides the values for MSE and RMSE of training and testing datasets as the criteria for assessing the accuracy of the ANN model. Since the model is built and trained based on training dataset, it is obvious that MSE and RMSE values are higher for testing data. Bead height has the lowest MSE and RMSE for both training and testing datasets, approving that the model for height is the most accurate model. In addition, the values of the MSE and RMSE loss functions for the bead area are quite close to those for height with less than 0.00005 and 0.00123 differences, respectively for training. However, the same criteria for bead width are higher in comparison to values for height and area. MSE and RMSE for testing data in modeling of bead width are 0.01642 and 0.12817, respectively.

Table 5. MSE and RMSE for training and testing datasets in terms of bead geometrical features

Response parameter	Training		Testing	
	MSE	RMSE	MSE	RMSE
Bead height (mm)	0.00037	0.01939	0.00155	0.03939
Bead width (mm)	0.00305	0.05523	0.01642	0.12817
Bead area (mm^2)	0.00042	0.02062	0.00730	0.08546

5. Conclusion

In this study aimed to model bead geometries of WLAM processes on an Inconel 625 wire based on ANN. Three different DoEs and statistical analysis are deployed to consider the correlation between process and response parameters and model the geometrical features. A length of 10 mm in the middle of printed beads is considered for geometrical details. The main inferred conclusions are:

- The best performance is attributed to an ANN with two hidden layers including 32 and 64 neurons, respectively. The assumed activation function is GELU which takes advantage of linear function for positive values and opposite to ReLU it is not returning zero values abruptly for negative values. These specifications are proved to be effective in the training process. 15% of training data was separated and fed to the network as the validation data. It is proved that for all the response parameters, the required epoch numbers were less than 1000.
- Based on statistical analysis, wire feed rate, travel speed and the quadratic term of travel speed are common influencing parameters to all geometrical features. However, quadratic term of wire feed rate is a sensitive factor to height, and laser power and quadratic term of wire feed rate is other sensitive terms for bead width. For bead area, the interaction between wire feed rate and travel speed is an additional sensitive factor.

- While the ANN models R^2 for response parameters are acceptable (0.988, 0.824, and 0.993 for height, width, and area), the values for height and area are more significant. As a result, MSE values for height and area are less than 0.0004 and 0.007 for training and testing data, respectively.
- While MSE and R^2 values for bead width were acceptable, they were lower in comparison to the statistics for other response parameters. There are reasons for that. The statistics obtained from RSM with lower values for beads width approve that modeling the bead width is more challenging than height and area. In addition, some instabilities in the process including the wobbling of wire can make the printed bead weaving and complicated to model.

The future research plan includes improving the stability and accuracy of the WLAM process. Additionally, investigating the printing of walls and thermal management challenges during the process.

Acknowledgements
This research has been supported by the project TANDEM (4056/31/2021) Business Finland under the SMART EUREKA cluster on Advance manufacturing program and partially supported by the project Multi-disciplinary Digital Design and Manufacturing - D2M (346874) Academy of Finland / Academy Research Fellow.

References
[1] J. Foshammer, P. V. Søberg, P. Helo, and I. F. Ituarte, "Identification of aftermarket and legacy parts suitable for additive manufacturing: A knowledge management-based approach," *Int. J. Prod. Econ.*, vol. 253, p. 108573, Nov. 2022, doi: 10.1016/j.ijpe.2022.108573.
[2] I. Flores Ituarte, S. Panicker, H. P. N. Nagarajan, E. Coatanea, and D. W. Rosen, "Optimisation-driven design to explore and exploit the process–structure–property–performance linkages in digital manufacturing," *J. Intell. Manuf.*, vol. 34, no. 1, pp. 219–241, Jan. 2023, doi: 10.1007/s10845-022-02010-2.
[3] W. Zhang, Y. Xu, Y. Shi, G. Su, Y. Gu, and K. Volodymyr, "Intergranular corrosion characteristics of high-efficiency wire laser additive manufactured Inconel 625 alloys," *Corros. Sci.*, vol. 205, p. 110422, Aug. 2022, doi: 10.1016/j.corsci.2022.110422.
[4] A. E. Odermatt, F. Dorn, V. Ventzke, and N. Kashaev, "Coaxial laser directed energy deposition with wire of thin-walled duplex stainless steel parts: Process discontinuities and their impact on the mechanical properties," *CIRP J. Manuf. Sci. Technol.*, vol. 37, pp. 443–453, May 2022, doi: 10.1016/j.cirpj.2022.02.019.
[5] M. Froend, V. Ventzke, F. Dorn, N. Kashaev, B. Klusemann, and J. Enz, "Microstructure by design: An approach of grain refinement and isotropy improvement in multi-layer wire-based laser metal deposition," *Mater. Sci. Eng. A*, vol. 772, p. 138635, Jan. 2020, doi: 10.1016/j.msea.2019.138635.
[6] S. Ranjan, V. Handa, S. Sehgal, and S. Singh, "Development of Inconel 625 joints using modified microwave hybrid heating technology," *Mater. Today Proc.*, p. S2214785323014499, Mar. 2023, doi: 10.1016/j.matpr.2023.03.402.
[7] R. Vijayakumar, N. Srirangarajalu, M. Santhanakumar, N. E. E. Paul, and M. Rajesh, "Investigation of Abrasive Aqua Jet Hole Making (AAJHM) parameters using desirability analysis on Inconel-625 space alloy," *J. Manuf. Process.*, vol. 92, pp. 311–328, Apr. 2023, doi: 10.1016/j.jmapro.2023.03.008.
[8] G. H. S. F. L. Carvalho, G. Venturini, G. Campatelli, and E. Galvanetto, "Development of optimal deposition strategies for cladding of Inconel 625 on carbon steel using wire arc additive manufacturing," *Surf. Coat. Technol.*, vol. 453, p. 129128, Jan. 2023, doi: 10.1016/j.surfcoat.2022.129128.
[9] E. Pavithra and V. S. Senthil Kumar, "Microstructural evolution of hydroformed Inconel 625 bellows," *J. Alloys Compd.*, vol. 669, pp. 199–204, Jun. 2016, doi: 10.1016/j.jallcom.2016.02.011.

[10] P. Ganesh *et al.*, "Fatigue and fracture toughness characteristics of laser rapid manufactured Inconel 625 structures," *Mater. Sci. Eng. A*, vol. 527, no. 29–30, pp. 7490–7497, Nov. 2010, doi: 10.1016/j.msea.2010.08.034.

[11] Y. L. Hu, X. Lin, X. B. Yu, J. J. Xu, M. Lei, and W. D. Huang, "Effect of Ti addition on cracking and microhardness of Inconel 625 during the laser solid forming processing," *J. Alloys Compd.*, vol. 711, pp. 267–277, Jul. 2017, doi: 10.1016/j.jallcom.2017.03.355.

[12] F. Chen, Q. Wang, C. Zhang, Z. Huang, M. Jia, and Q. Shen, "Microstructures and mechanical behaviors of additive manufactured Inconel 625 alloys via selective laser melting and laser engineered net shaping," *J. Alloys Compd.*, vol. 917, p. 165572, Oct. 2022, doi: 10.1016/j.jallcom.2022.165572.

[13] S. Liu, C. Brice, and X. Zhang, "Hierarchical bead materials multi-property design for wire-feed laser additive manufacturing," *J. Manuf. Process.*, vol. 80, pp. 546–557, Aug. 2022, doi: 10.1016/j.jmapro.2022.05.028.

[14] B. J. Hayes *et al.*, "Predicting tensile properties of Ti-6Al-4V produced via directed energy deposition," *Acta Mater.*, vol. 133, pp. 120–133, Jul. 2017, doi: 10.1016/j.actamat.2017.05.025.

[15] H. Gong, K. Rafi, H. Gu, T. Starr, and B. Stucker, "Analysis of defect generation in Ti–6Al–4V parts made using powder bed fusion additive manufacturing processes," *Addit. Manuf.*, vol. 1–4, pp. 87–98, Oct. 2014, doi: 10.1016/j.addma.2014.08.002.

[16] I. Flores, N. Kretzschmar, A. H. Azman, S. Chekurov, D. B. Pedersen, and A. Chaudhuri, "Implications of lattice structures on economics and productivity of metal powder bed fusion," *Addit. Manuf.*, vol. 31, p. 100947, Jan. 2020, doi: 10.1016/j.addma.2019.100947.

[17] R. M. Ballardini, I. Flores Ituarte, and E. Pei, "Printing spare parts through additive manufacturing: legal and digital business challenges," *J. Manuf. Technol. Manag.*, vol. 29, no. 6, pp. 958–982, Jul. 2018, doi: 10.1108/JMTM-12-2017-0270.

[18] I. F. Ituarte, S. H. Khajavi, and J. Partanen, "Challenges to implementing additive manufacturing in globalised production environments," *Int. J. Collab. Enterp.*, vol. 5, no. 3/4, p. 232, 2016, doi: 10.1504/IJCENT.2016.082335.

[19] I. F. Ituarte, E. Coatanea, M. Salmi, J. Tuomi, and J. Partanen, "Additive Manufacturing in Production: A Study Case Applying Technical Requirements," *Phys. Procedia*, vol. 78, pp. 357–366, 2015, doi: 10.1016/j.phpro.2015.11.050.

[20] A. G. Demir, "Micro laser metal wire deposition for additive manufacturing of thin-walled structures," *Opt. Lasers Eng.*, vol. 100, pp. 9–17, Jan. 2018, doi: 10.1016/j.optlaseng.2017.07.003.

[21] G. E. P. Box and K. B. Wilson, "On the Experimental Attainment of Optimum Conditions," *J. R. Stat. Soc. Ser. B Methodol.*, vol. 13, no. 1, pp. 1–38, Jan. 1951, doi: 10.1111/j.2517-6161.1951.tb00067.x.

[22] M. Ostolaza, J. I. Arrizubieta, A. Queguineur, K. Valtonen, A. Lamikiz, and I. Flores Ituarte, "Influence of process parameters on the particle–matrix interaction of WC-Co metal matrix composites produced by laser-directed energy deposition," *Mater. Des.*, vol. 223, p. 111172, Nov. 2022, doi: 10.1016/j.matdes.2022.111172.

[23] R. Asadi, A. Yeganefar, and S. A. Niknam, "Optimization and prediction of surface quality and cutting forces in the milling of aluminum alloys using ANFIS and interval type 2 neuro fuzzy network coupled with population-based meta-heuristic learning methods," *Int. J. Adv. Manuf. Technol.*, vol. 105, no. 5–6, pp. 2271–2287, Dec. 2019, doi: 10.1007/s00170-019-04309-6.

[24] Ilya Loshchilov and Frank Hutter, "Fixing Weight Decay Regularization in Adam," presented at the ICLR 2018 Conference Blind Submission, Oct. 2017.

Industrial IoT system for laser-wire direct energy deposition: data collection and visualization of manufacturing process signals

Antti Martikkala*',, Olli Wiikinkoski*, Reza Asadi*, Antoine Queguineur*'***, Aapo Ylä-Autio*, Iñigo Flores Ituarte***

* Department of Automation and Mechanical Engineering, Faculty of Engineering and Natural Sciences, Tampere University, Korkeakoulunkatu 6, 33014 Tampere, Finland
** Department of Mechanical and Industrial Engineering, Norwegian University of Science and Technology, Richard Birkelands Vei 2b, NO-7034, Trondheim, Norway
*** Ecole Centrale Nantes, GeM - UMR CNRS 6183, 1 rue de la Noé, 44321 Nantes – France

antti.martikkala@tuni.fi

Abstract. Industry 4.0, also known as the Fourth Industrial Revolution, is a term used to describe the current trend of automation and data exchange in manufacturing and other industries. The Internet of Things (IoT) plays a crucial role in Industry 4.0 by connecting devices, machines, and products to the Internet and enabling real-time data exchange. Moreover, additive manufacturing is a key developing manufacturing technology in Industry 4.0. New technologies such as data analysis with Artificial Intelligence and machine vision are widely used in optimization. However, in a lab environment, these technologies depend on the data collected from the process. For such work, the researchers should be able to focus on their core research rather than on the development of infrastructure to collect and analyse the data. This research presents an open software and hardware IoT solution to monitor a laser wire direct energy deposition system installed in a cartesian type 3-axis machine tool. The IoT solution adopts three open-source tools for core issues, such as 1) interoperability, flexibility, and availability; 2) data storage; and 3) data visualization of sensor data and manufacturing process signals. The system architecture is based on one or more edge devices connected to sensors and forwarding their data toward a local API endpoint. The endpoint is created with Node-RED, an open-source visual flow-based development tool for IoT data. Node-RED forwards the data to an open-source InfluxDB database. Finally, the data is visualized with an open-source Grafana application. The system is prototyped, designed, implemented, and tested in a lab environment to monitor a laser-wire direct energy deposition process. The significance of such a flexible IoT data collection system for research and development projects can be integral. Thus, providing savings in time and money can substantially speed up the development of new technologies where the value arises from the sensor data and its analysis.

Keywords: Internet of Things, Additive Manufacturing, Manufacturing process monitoring, open source

1. Introduction

Metal 3D printing, also known as metal additive manufacturing, is one of the most relevant technologies in Industry 4.0. This is because it offers several benefits that are highly relevant to modern manufacturing, such as the ability to produce complex geometries [1], improve part performance, and reduce the cost and lead time [1] associated with traditional manufacturing methods [2]. By fine-tuning the printing parameters, such as laser power, travel speed, and wire feed rate, manufacturers can achieve better control over the properties of the printed parts [3]. This can result in better control of the part geometry and improve mechanical properties, such as strength, hardness, and ductility [4], [5].

A laser wire direct energy deposition system is an additive manufacturing technology that uses a laser to melt a wire feedstock material as it is deposited onto a substrate. The process involves feeding a wire into the melt pool created by the laser and depositing it layer by layer to build up a 3D object. The laser melts the wire and fuses it with the substrate, creating a strong and precise bond. This technology can create complex geometries and repair damaged or worn parts, making it useful in aerospace, automotive, and medical [2].

Optimizing the printing process can also help to reduce defects and improve the quality of the printed parts [6]. For example, by predicting the temperature and stress distributions during the printing process, manufacturers can identify potential defects and adjust the printing parameters to prevent them [7], [8]. Additionally, in-situ monitoring techniques, such as infrared cameras and acoustic sensors, can help detect defects in real time and enable corrective actions to be taken during printing [9]. Through the optimization of the printing process, manufacturers can achieve greater control over the properties of the printed parts and improve the quality of the final product.

Data collection is an integral part of the research, as it serves as the foundation for generating accurate and reliable findings [10]. It enables researchers to gather information and evidence that helps in answering research questions, testing hypotheses, and drawing conclusions [10]. Furthermore, this work presents an affordable and flexible solution for data collection from the laser wire direct energy deposition system used in the lab. Moreover, user-friendly visualization tools make it easier for researchers to get firsthand knowledge from the process in real time. Further, the system is tested, and its performance is evaluated.

Research questions are defined as follows:

1. *How to collect data from various sensors to monitor the 3D printing process.*
2. *How to make the collected data easily available for the researchers.*

IoT-related research work often centers around the challenge of achieving efficient and affordable data collection. Researchers at the University of Iowa have highlighted the need for a data acquisition infrastructure that can enable them to focus on research and scholarship rather than being bogged down with hardware and software development [11] This is also reflected in a survey conducted in 2015, which identified the need for tools to facilitate the development of various IoT applications and services [12].

1.1. Context

Here the aim is to optimize the laser-wire direct energy process for additive manufacturing by monitoring the process using a suite of sensors. However, the system's architecture has yet to be discovered, and new sensors may be added in the future. The current plan is to include the following sensors.

Firstly, the energy consumption of the laser, wire feed, and the movement of the printing head and table are monitored. This information can be used to optimize the energy consumption and reduce the environmental impacts of the additive manufacturing.

A Cavitar camera can be used to capture high-speed footage of the laser-wire interaction, providing detailed insight into the process and allowing researchers to identify areas for improvement or optimization. For example, the camera can capture the formation of the molten pool, the size and shape of the deposited material, and the behavior of the wire feed and laser during the deposition process. [13]

A pyrometer is an infrared thermometer used to measure the temperature of an object from a distance. In a laser-wire direct energy process, pyrometers can be used to measure the temperature of the material being processed, as well as the temperature of the laser itself. This information can be used to control the laser power and maintain the desired processing temperature [14], [15]

Laser distance sensors can be used to monitor the distance between the laser and the material being processed [16]. This information can be used to adjust the focus of the laser and ensure that the material is being processed at the desired depth. In addition, laser distance sensors can be used to measure the height of the printing head and the position of the table, providing additional information for controlling the process.

Acoustic sensors can be used to monitor the printing process and detect any potential defects. These sensors can detect acoustic emissions generated by the process, such as the sound of wire feeding, laser melting, and the movement of the printing head and table. By analyzing the acoustic signals, it is possible to detect anomalies such as wire breakage, improper melting, and movement irregularities [17], [18]

A profile scanner can generate a high-resolution 3D model of the object, which can be used to monitor the quality of the printing process and detect any defects or deviations from the desired shape. The profile scanner can also be used to adjust the printing process in real-time, for example by adjusting the position or power of the laser or the speed of the printing head, to ensure that the object is built to the desired specifications. [19]

In conclusion, the integration of a suite of sensors in a laser-wire direct energy process for additive manufacturing has the potential to optimize the process and improve the quality of the final product. The combination of energy monitoring, high-speed camera footage, temperature measurement, laser distance sensing, acoustic sensing, and profile scanning can provide comprehensive insights into the process and enable researchers to identify areas for improvement. Furthermore, the real-time feedback provided by the sensors can be used to adjust the process parameters and improve the overall quality and efficiency of the additive manufacturing process. I.e., optimize the process parameters such as the laser power, wire feed rate, and the speed of the printing head and table.

As such, the development of such a sensor suite represents an important step forward in the advancement of additive manufacturing and could have significant implications for a range of industries.

1.2. Literature and prior works

This section presents an overview of some related studies and their developed systems in the realm of IoT-based data acquisition, storage, processing, and visualization. The aim is to provide a foundation for understanding the current state of the art in this field and to identify potential areas for future research and development.

The Iowa Quantified (IQ) infrastructure, developed by researchers at the University of Iowa, provides an end-to-end solution for sensor deployment and data collection based on Amazon Web Services (AWS), MQTT, and Jupyter notebooks. The infrastructure focuses on enabling researchers to concentrate on their work by offering wireless sensors, a cloud-based toolkit, and a web-accessible portal for real-time and offline data processing and visualization. [1]

An integrated data collection platform designed for smart grid and microclimate research combines wired energy subsystems and wireless climate and heating subsystems, using LoRaWAN technology. The platform employs custom scripts and MQTT protocol to send data to the IoT platform ThingsBoard. The ThingsBoard platform is used for visualization, enabling the monitoring of sensor data, climate sensor mapping, and displaying energy disaggregation algorithm results. [20]

A performance evaluation and resource monitoring system for heterogeneous edge devices was developed using open-source software like Docker, Kubernetes, Prometheus, Grafana, and Node

Exporter. This system creates a containerized environment and edge computing-based cluster, allowing for streamlined management, deployment, and real-time monitoring and analysis of CPU and memory usage. The system is designed for practical utility in deep learning models, big data analysis, distributed systems, and parallel computing applications. [21]

A study focused on creating a data acquisition system for solar power systems, collecting battery voltage, room temperature, and humidity data through an IoT platform. The system employed an ESP8266 Node Micro-Controller Unit with a DHT11 sensor for data sensing, and a Raspberry Pi for data access, processing, and visualization. The MQTT protocol, Grafana, and InfluxDB were used for messaging and data visualization. Implemented through Docker containers, the system allowed for seamless migration to other platforms, flexibility in mounting locations, and the potential for future adaptation to cloud-based servers for database replication. [22]

While each system serves a distinct purpose, they collectively demonstrate the potential of open-source IoT platforms in addressing various data collection and analysis challenges.

2. Materials and Methods

The ultimate goal of the work with laser-wire printing is to control the printing process in real time using a feedback loop from the sensor data.

"Keep It Simple Stupid" (KISS) principle suggests that developers should choose the simplest system or tool that can meet their needs rather than selecting a more complex tool that may have additional features that are not necessary [23]. Moreover, the definition of "best" may vary depending on the specific application, with some prioritizing reliability, speed of implementation, ease of use, and so on [23]. However, ultimately the best solution is one that is cost-effective and provides value for money [23]. In the context of this work the aim for simplicity is based on lowering the requirements of the system performance and focusing on the interoperability and usability. Which does not actually lead to the selection of simplest tool, but rather on the selection of most flexible and easiest tools.

Subsequently, the development of a fast and optimized real-time control system will be deferred. Moreover, it is important to note that the definition of 'real-time' may vary depending on the context in which it is applied [24]. This approach later allows to narrow down the development of the real-time control system to specific sensors and gives more information to define its minimum requirements. I.e., the real-time control can be developed once enough data and knowledge is gathered to develop models to control the process. After further analysis of the collected data from the process, it could be also possible to better define what real-time means in the context of laser-wire direct energy process.

The system architecture can be divided into separate tasks, adopting the idea of IoT platform classification [25]. Consequently, the architecture should consist of three main parts: A) data flow handling, B) data storage, and C) visualization. The design focuses on extensibility, interoperability, performance, and low cost. The use of open-source tools is prioritized, and the system should not be reliant on any individual application [26]. Edge processing is adopted to handle vast amounts of data and reduce the load on the data collection system [27]–[29]. Edge processing becomes crucial when working with high-frequency data, and helps ensure accurate timestamps [30]. It is also more suitable for developing real-time control with less latency [31].

Consequently, with edge processing, the requirements for the data handling system are lowered. The main requirements for the system are extensibility, interoperability, and low cost. The next three chapters introduce different options for the main parts of the system architecture. The remainder of the paper is organized as follows: Sections 2.1. , 2.2. , and 2.3. present possible tools for each part in the system. Section 3 presents the developed system and its architecture, and other technical choices. Section 4 discusses the implications, proposes future improvements, concludes the paper, and summarizes the results.

During the preparation of this work the authors used ChatGPT versions 3.5 and 4 to refine the narrative, improve readability and language. After using this tool, the authors reviewed and edited the content as needed and take full responsibility for the content of the publication.

IOP Conf. Series: Materials Science and Engineering 1296 (2023) 012006 doi:10.1088/1757-899X/1296/1/012006

2.1. Data flow handling
Node-RED is an open-source visual programming tool built on Node.js, enabling users to create complex workflows and automate tasks using a drag-and-drop interface. Node-RED offers a library of pre-built nodes for interacting with various devices and services [32]. Crosser.io, an edge computing platform, provides a visual programming interface for building and deploying real-time data processing applications, featuring pre-built modules, debugging tools, and deployment support [33]. Apache NiFi, an open-source data integration platform, allows for managing data flows, connecting data sources, and processing data in real-time. It is designed for handling large data volumes and offers a visual interface, data transformation tools, and monitoring, security, and scalability features [34].

2.2. Databases
InfluxDB is a time-series database designed for storing and querying time-stamped data, suitable for IoT data storage due to its ability to handle large volumes and built-in downsampling and retention policies [35]. Prometheus is a monitoring and alerting system with a time-series database, ideal for IoT data collection and storage, offering powerful query and visualization tools and a push-based data model [36]. Elasticsearch is a distributed search and analytics engine with built-in support for time-series data, handling large volumes of data and providing powerful search and visualization capabilities, as well as real-time data ingest [37]. Cassandra is a distributed NoSQL database designed for handling large data volumes, with column-based storage and support for replication and fault tolerance, suitable for IoT environments requiring high availability [38]. PostgreSQL is a scalable relational database suitable for IoT data storage, offering powerful indexing and querying capabilities, as well as data replication and high availability support [39].

2.3. Visualization
Grafana is an open-source platform for collecting, storing, and visualizing data from various sources, including IoT devices. It supports over 50 data sources, powerful visualization tools, and is compatible with time-series databases like InfluxDB, MySQL, and PostgreSQL [40]. Kibana, an open-source analytics and visualization platform, works with Elasticsearch and is suitable for IoT data collection and visualization [41]. ThingsBoard, an open-source IoT platform, offers data collection, processing, and visualization with support for multiple data sources and databases, and features a built-in visualization engine [42]. Prometheus, an open-source monitoring and alerting system, is ideal for IoT data collection and visualization, supporting time-series data and numerous data sources, including custom exporters for IoT devices [36].

3. Results
The Node-RED flow editor is chosen for its open-source nature, ease of use, local execution, and compatibility with various devices and APIs. As a popular visual programming tool for creating IoT applications, Node-RED can integrate with methods like MQTT, CoAP, and HTTP REST [43]. MQTT, a lightweight protocol designed for IoT applications with low overhead [44], is selected as the primary protocol for sending data from edge to Node-RED, while still allowing flexibility for other protocols if needed. CoAP, a lightweight protocol for constrained devices and networks [45], and REST, a widely used architecture for web services and APIs [45], [46], are also considered.

According to the methodology the system is set up in the most profitable manner, i.e., using the easiest tools. Grafana, an open-source option known for its flexibility, scalability, and community support, supports a wide range of data sources and is compatible with various databases and IoT platforms. Its intuitive interface and extensive visualization options make it an attractive choice for users with different technical backgrounds. InfluxDB is often preferred over other databases for use with Grafana due to its specific design for handling time-series data, making it a highly efficient and scalable option for data storage and retrieval [35]. The simplified system architecture in the Figure 1 shows the main parts and selected tools and their connections.

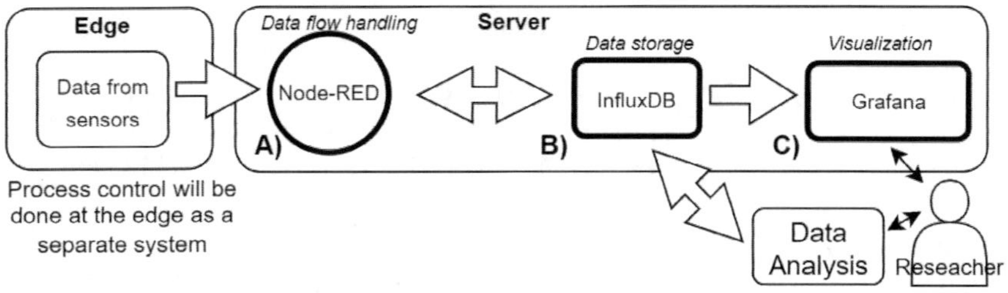

Figure 1. The system architecture.

After the data collection system is set up, the focus is on collecting the data from sensors. The data collection is primarily done with PLCs and computers at the IoT edge, but also microcontrollers as edge devices could be used. Consequently, the system consists of one or more edge devices with sensors connected to them. These edge devices are further connected to the server through Ethernet. Moreover, these edge devices must be programmed for their tasks to collect and transmit their data which sets another challenge for the researchers. In addition, if using multiple computers on the same network, it is important to ensure that the clocks on all the devices are synchronized to the same time source. Timestamps and synchronization are especially critical in IoT edge devices that require time-sensitive data collection such as acoustic sensor data [47]. This can be achieved for example, through the use of software tools such as Chrony or NTP to synchronize the system clock with a common time server [48], [49]. The more detailed architecture in the Figure 2 shows the complete system with various sensors. The system is designed to be expandable for real-time, closed-loop control to optimize the process, and potentially enable data backup and sharing via cloud databases.

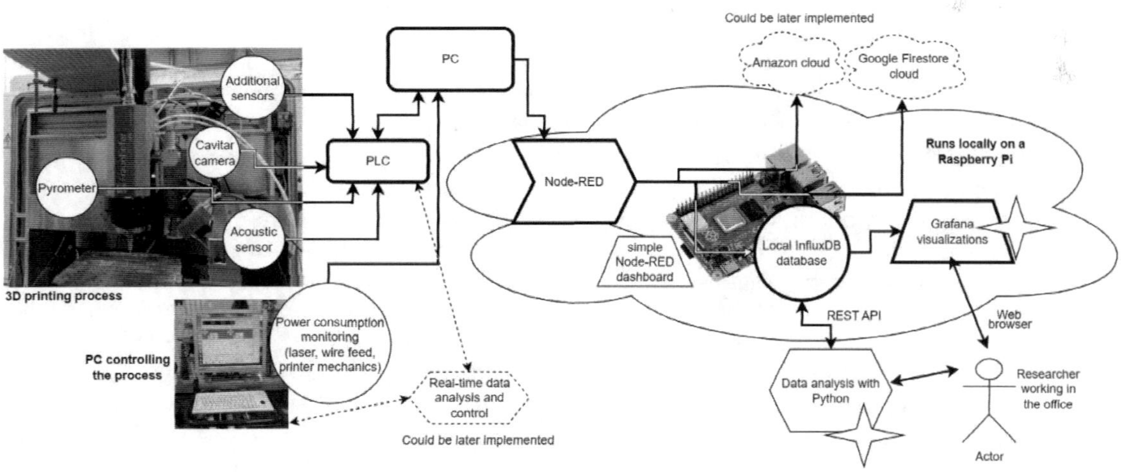

Figure 2. Detailed illustrative architecture of the system in the lab.

The deployment of an outdated Windows XP computer, which controls the laser-wire direct energy deposition and gathers system data sets additional challenges to system architecture. Firstly, it poses a security risk due to its discontinued support and potential vulnerabilities. To mitigate these risks, a segregated lab network is established without internet access. This is facilitated through a dedicated router and switch, connecting all necessary devices. Moreover, older operating systems without built-in support for modern time synchronization protocols may pose further challenges in data collection.

Figure 3 illustrates the wire direct energy deposition system under observation, along with data obtained from various sensors. The system has been performing as anticipated. Existing Grafana plugins facilitate the visualization of images captured by the Cavitar camera as seen in the figure. Real-time visualizations enable operators to oversee the details of the printing process from a safe distance, allowing them to respond as necessary. Creating new dashboards featuring data from a selection of sensors is effortless, and the wide range of plugins, coupled with community support, greatly expands the available possibilities.

The system is straightforward to configure, and, importantly, incorporating a new sensor into the system is relatively simple as the biggest changes are done on the edge. Presently, all sensor data is transmitted to the cloud from a single PC which can be later replaced with a PLC. This setup is effectively addressing the majority of synchronization challenges. However, servo and power consumption data from an older Windows XP machine can only be integrated into the system after the printing process. This limitation arises because the data file is reserved by another process and cannot be accessed until the process is complete. In this instance, synchronization is achieved by identifying the corresponding timestamp from the acoustic sensor data and adjusting the timestamps accordingly.

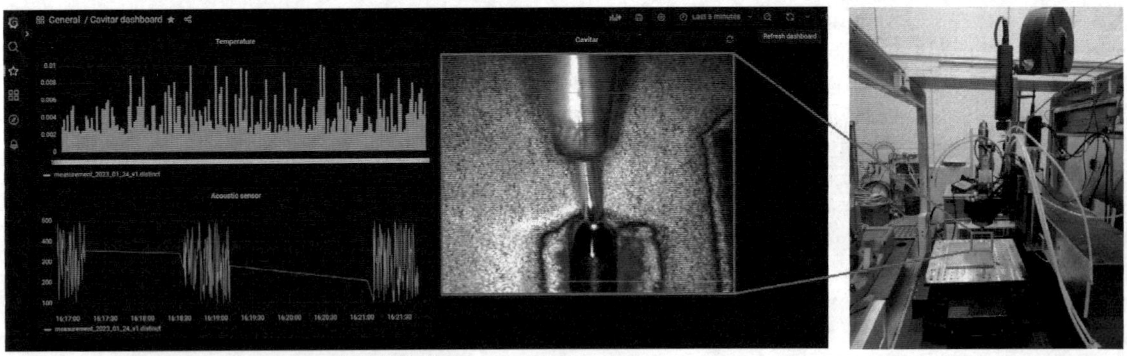

Figure 3. Hardware and data visualization.

From the perspective of researchers conducting data analysis, the system offers numerous advantages. Data can be accessed remotely via the internet, and Grafana's visualization tools enable preliminary analysis before employing more sophisticated techniques. Animations of recorded datasets can be created with tools such as grafanimate [50]. For advanced analysis, researchers primarily utilize Python. Consequently, the InfluxDB-Python library is recommended for extracting data from InfluxDB using Python, as it provides a Python API to interface with InfluxDB. However, the ease of data access presents potential risks to the system's data integrity. Consequently, higher privileges are necessary for altering data stored in the cloud. Finally, data retrieval templates are provided and shared among the researchers. These include scripts for example for extracting and storing requested data in both Pandas DataFrames and CSV files.

4. Conclusions

In conclusion, the presented open-source tools based IoT solution offers an open software and hardware approach for real-time data collection and visualization from a laser-wire direct energy deposition system. It consists of three main components: data flow handling, data storage, and data visualization. Node-RED, InfluxDB, and Grafana are employed due to their flexibility, efficiency, and ease of use. This solution aligns with the Industry 4.0 paradigm and enables researchers to focus on their core research. Real-time monitoring and data collection can facilitate process optimization, improve quality control, and reduce energy consumption.

The proposed system architecture based on edge devices and API endpoints handled with Node-RED makes it easy to add new sensors and expand the system's functionality. The collected data is easily accessible to researchers through the use of Grafana, which offers a user-friendly interface and a variety of visualization options. Additionally, the use of InfluxDB-Python library provides an easy-to-use Python API for data retrieval for analysis. This allows researchers to easily import data into their preferred data analysis tools, such as Pandas and NumPy. Overall, the combination of user-friendly visualization tools and flexible data retrieval options makes it easy for researchers to access and analyze the collected data in a meaningful way.

The developed system offers a flexible, cost-effective solution for efficient data collection and analysis. Future directions involve process optimization on edge and integration of additional sensors or data sources. More comprehensive comparative analysis of different tools for similar systems can also be explored, testing system performance and modifiability.

Acknowledgements

This research has been supported by the project TANDEM (4056/31/2021) Business Finland under the SMART EUREKA cluster on Advance manufacturing program and partially supported by the project Multi-disciplinary Digital Design and Manufacturing - D2M (346874) Academy of Finland / Academy Research Fellow.

Antti Martikkala: Conceptualization, Methodology, Writing - Original Draft and Review, Investigation (system development), Olli Wiikinkoski, Aapo Ylä-Autio, Reza Asadi, Antoine Queguineur: Investigation (involved in conducting experiments), Writing - Review & Editing, Validation (ensured experimental accuracy). Iñigo Flores Ituarte: Conceptualization, Supervision.

References

[1] I. F. Ituarte, N. Boddeti, V. Hassani, M. L. Dunn, and D. W. Rosen, "Design and additive manufacture of functionally graded structures based on digital materials," *Addit. Manuf.*, vol. 30, p. 100839, Dec. 2019, doi: 10.1016/j.addma.2019.100839.

[2] S. A. M. Tofail, E. P. Koumoulos, A. Bandyopadhyay, S. Bose, L. O'Donoghue, and C. Charitidis, "Additive manufacturing: scientific and technological challenges, market uptake and opportunities," *Mater. Today*, vol. 21, no. 1, pp. 22–37, Jan. 2018, doi: 10.1016/j.mattod.2017.07.001.

[3] I. Flores Ituarte, S. Panicker, H. P. N. Nagarajan, E. Coatanea, and D. W. Rosen, "Optimisation-driven design to explore and exploit the process–structure–property–performance linkages in digital manufacturing," *J. Intell. Manuf.*, vol. 34, no. 1, pp. 219–241, Jan. 2023, doi: 10.1007/s10845-022-02010-2.

[4] S. Liu, C. Brice, and X. Zhang, "Hierarchical bead materials multi-property design for wire-feed laser additive manufacturing," *J. Manuf. Process.*, vol. 80, pp. 546–557, Aug. 2022, doi: 10.1016/j.jmapro.2022.05.028.

[5] H. Y. Chia, J. Wu, X. Wang, and W. Yan, "Process parameter optimization of metal additive manufacturing: a review and outlook," *J. Mater. Inform.*, vol. 2, p. 16, Oct. 2022, doi: 10.20517/jmi.2022.18.

[6] I. F. Ituarte, F. Nilsén, V. K. Nadimpalli, M. Salmi, J. Lehtonen, and S.-P. Hannula, "Towards the additive manufacturing of Ni-Mn-Ga complex devices with magnetic field induced strain," *Addit. Manuf.*, vol. 49, p. 102485, Jan. 2022, doi: 10.1016/j.addma.2021.102485.

[7] Z.-C. Fang, Z.-L. Wu, C.-G. Huang, and C.-W. Wu, "Review on residual stress in selective laser melting additive manufacturing of alloy parts," *Opt. Laser Technol.*, vol. 129, p. 106283, Sep. 2020, doi: 10.1016/j.optlastec.2020.106283.

[8] M. Gan, Q. Wu, and L. Long, "Prediction of Residual Deformation and Stress of Laser Powder Bed Fusion Manufactured Ti-6Al-4V Lattice Structures Based on Inherent Strain Method," *Mater. Res.*, vol. 26, p. e20220516, 2023, doi: 10.1590/1980-5373-mr-2022-0516.

[9] R. Jafari-Marandi, M. Khanzadeh, W. Tian, B. Smith, and L. Bian, "From in-situ monitoring toward high-throughput process control: cost-driven decision-making framework for laser-based additive manufacturing," *J. Manuf. Syst.*, vol. 51, pp. 29–41, Apr. 2019, doi: 10.1016/j.jmsy.2019.02.005.

[10] H. Taherdoost, "Data Collection Methods and Tools for Research; A Step-by-Step Guide to Choose Data Collection Technique for Academic and Business Research Projects," Aug. 2021.

[11] K. P. Lee, S. J. Kuhl, H. J. Bockholt, B. P. Rogers, and D. A. Reed, "A cloud-based scientific gateway for internet of things data analytics," *ACM Int. Conf. Proceeding Ser.*, vol. 7, Jul. 2018, doi: 10.1145/3219104.3219123.

[12] M. A. Razzaque, M. Milojevic-Jevric, A. Palade, and S. Clarke, "Middleware for Internet of Things: a Survey," *IEEE INTERNET THINGS J.*, vol. 0, no. 0, 2015.

[13] "Cavitar Welding Camera," *Cavitar Ltd - See What You Have Missed.* https://www.cavitar.com/product/cavitar-welding-camera/ (accessed Mar. 22, 2023).

[14] D. V. Nishar, J. L. Schiano, W. R. Perkins, and R. A. Weber, "Adaptive control of temperature in arc welding," *IEEE Control Syst. Mag.*, vol. 14, no. 4, pp. 4–12, Aug. 1994, doi: 10.1109/37.295964.

[15] S. Tofil, G. Witkowski, and K. Mulczyk, "A temperature control system for laser welding of plastics differing in optical parameters," in *2019 20th International Carpathian Control Conference (ICCC)*, May 2019, pp. 1–5. doi: 10.1109/CarpathianCC.2019.8766037.

[16] G. Masinelli, T. Le-Quang, S. Zanoli, K. Wasmer, and S. A. Shevchik, "Adaptive Laser Welding Control: A Reinforcement Learning Approach," *IEEE Access*, vol. 8, pp. 103803–103814, 2020, doi: 10.1109/ACCESS.2020.2998052.

[17] I. S. genannt Wersborg, T. Bautze, F. Born, and K. Diepold, "A cognitive approach for a robotic welding system that can learn how to weld from acoustic data," in *2009 IEEE International Symposium on Computational Intelligence in Robotics and Automation - (CIRA)*, Dec. 2009, pp. 108–113. doi: 10.1109/CIRA.2009.5423224.

[18] S. Jinachandran, J. Xi, G. Rajan, C. Shen, H. Li, and B. G. Prusty, "Fibre optic acoustic emission sensor system for hydrogen induced cold crack monitoring in welding applications," in *2016 IEEE Sensors Applications Symposium (SAS)*, Apr. 2016, pp. 1–6. doi: 10.1109/SAS.2016.7479896.

[19] M. Chalus and J. Liska, "3D robotic welding with a laser profile scanner," in *2016 International Conference on Control, Automation and Information Sciences (ICCAIS)*, Oct. 2016, pp. 7–12. doi: 10.1109/ICCAIS.2016.7822426.

[20] N. I. Bazenkov *et al.*, "Intensive data collection system for smart grid and smart building research," in *2019 1st International Conference on Control Systems, Mathematical Modelling, Automation and Energy Efficiency (SUMMA)*, Nov. 2019, pp. 411–415. doi: 10.1109/SUMMA48161.2019.8947469.

[21] Y.-W. Chan, H. Fathoni, H.-Y. Yen, and C.-T. Yang, "Implementation of a Cluster-Based Heterogeneous Edge Computing System for Resource Monitoring and Performance Evaluation," *IEEE Access*, vol. 10, pp. 38458–38471, 2022, doi: 10.1109/ACCESS.2022.3166154.

[22] O. Sudjana and H. Septanto, "Data Acquisition and Visualization for Solar Power Battery using IoT Open Source Stack Solution," in *2020 IEEE International Conference on Sustainable Engineering and Creative Computing (ICSECC)*, Dec. 2020, pp. 177–181. doi: 10.1109/ICSECC51444.2020.9557548.

[23] L. Sheppard, T. W. Lumpkin, and T. Kelly, "A KISS Approach to Embedded Health Management," in *2006 IEEE Autotestcon*, Sep. 2006, pp. 174–180. doi: 10.1109/AUTEST.2006.283622.

[24] G. C. Buttazzo, *Hard Real-Time Computing Systems: Predictable Scheduling Algorithms and Applications*, vol. 24. in Real-Time Systems Series, vol. 24. Boston, MA: Springer US, 2011. doi: 10.1007/978-1-4614-0676-1.

[25] A. Martikkala, A. Lobov, M. Lanz, and I. Flores, "Towards the Interoperability of IoT Platforms : A Case Study for Data Collection and Data Storage," *Proc. 17th IFAC Symp. Inf. Control Probl. Manuf. Bp. Hung. June 7-9 2021*, 2021.

[26] L. Bass, P. Clements, and R. Kazman, *Software architecture in practice*, 2nd ed. in SEI series in software engineering. Boston: Addison-Wesley, 2003.

[27] M. Satyanarayanan, "The Emergence of Edge Computing," *Computer*, vol. 50, no. 1, pp. 30–39, Jan. 2017, doi: 10.1109/MC.2017.9.

[28] C. Perera, Y. Qin, J. C. Estrella, S. Reiff-Marganiec, and A. V. Vasilakos, "Fog Computing for Sustainable Smart Cities: A Survey." arXiv, Mar. 21, 2017. doi: 10.48550/arXiv.1703.07079.

[29] Z. Lin and L. Zhang, "Data synchronization algorithm for IoT gateway and platform," in *2016 2nd IEEE International Conference on Computer and Communications (ICCC)*, Oct. 2016, pp. 114–119. doi: 10.1109/CompComm.2016.7924676.

[30] P. Jia, X. Wang, and X. Shen, "Digital-Twin-Enabled Intelligent Distributed Clock Synchronization in Industrial IoT Systems," *IEEE Internet Things J.*, vol. 8, no. 6, pp. 4548–4559, Mar. 2021, doi: 10.1109/JIOT.2020.3029131.

[31] E. Badidi and K. Moumane, "Enhancing the Processing of Healthcare Data Streams using Fog Computing," in *2019 IEEE Symposium on Computers and Communications (ISCC)*, Jun. 2019, pp. 1113–1118. doi: 10.1109/ISCC47284.2019.8969736.

[32] "Node-RED." https://nodered.org/ (accessed Dec. 16, 2020).

[33] "Stream Analytics & Integration," *Crosser*. https://crosser.io/ (accessed Mar. 22, 2023).

[34] "Apache NiFi." https://nifi.apache.org/ (accessed Mar. 22, 2023).

[35] "InfluxDB Time Series Data Platform," *InfluxData*, Jan. 15, 2022. https://www.influxdata.com/home/ (accessed Mar. 22, 2023).

[36] Prometheus, "Prometheus - Monitoring system & time series database." https://prometheus.io/ (accessed Mar. 22, 2023).

[37] "Elastic Observability and Security — built on Elasticsearch." https://www.elastic.co/ (accessed Mar. 22, 2023).

[38] "Apache Cassandra | Apache Cassandra Documentation." https://cassandra.apache.org/_/index.html (accessed Mar. 22, 2023).

[39] P. G. D. Group, "PostgreSQL," *PostgreSQL*, Mar. 22, 2023. https://www.postgresql.org/ (accessed Mar. 22, 2023).

[40] "Grafana: The open observability platform," *Grafana Labs*. https://grafana.com/ (accessed Mar. 22, 2023).

[41] "Kibana: Explore, Visualize, Discover Data," *Elastic*. https://www.elastic.co/kibana (accessed Mar. 22, 2023).

[42] thingsboard, "ThingsBoard - Open-source IoT Platform," *ThingsBoard*. https://thingsboard.io/ (accessed Mar. 22, 2023).

[43] A. Melnikov and I. Fette, "The WebSocket Protocol," Internet Engineering Task Force, Request for Comments RFC 6455, Dec. 2011. doi: 10.17487/RFC6455.

[44] "MQTT - The Standard for IoT Messaging." https://mqtt.org/ (accessed Mar. 22, 2023).

[45] N. Naik, "Choice of effective messaging protocols for IoT systems: MQTT, CoAP, AMQP and HTTP," in *2017 IEEE International Symposium on Systems Engineering, ISSE 2017 - Proceedings*, Institute of Electrical and Electronics Engineers Inc., Oct. 2017. doi: 10.1109/SysEng.2017.8088251.

[46] R. Kumar N.V. and M. Kumar P., "Survey on State of Art IoT Protocols and Applications," in *2020 International Conference on Computational Intelligence for Smart Power System and Sustainable Energy (CISPSSE)*, Jul. 2020, pp. 1–3. doi: 10.1109/CISPSSE49931.2020.9212227.

[47] P. Sommer and R. Wattenhofer, "Gradient clock synchronization in wireless sensor networks," in *2009 International Conference on Information Processing in Sensor Networks*, Apr. 2009, pp. 37–48.

[48] J. Martin, J. Burbank, W. Kasch, and P. D. L. Mills, "Network Time Protocol Version 4: Protocol and Algorithms Specification," Internet Engineering Task Force, Request for Comments RFC 5905, Jun. 2010. doi: 10.17487/RFC5905.

[49] A. E. Dinar, B. Merabet, and S. Ghouali, "NTP Server Clock Adjustment with Chrony," J. K. Mandal, S. Mukhopadhyay, and A. Roy, Eds., in Lecture Notes in Networks and Systems, vol. 137. Singapore: Springer Singapore, 2021, pp. 177–185. doi: 10.1007/978-981-15-6198-6_16.

[50] "panodata/grafanimate." panodata, Mar. 10, 2023. Accessed: Apr. 20, 2023. [Online]. Available: https://github.com/panodata/grafanimate

IOP Conf. Series: Materials Science and Engineering 1296 (2023) 012007 doi:10.1088/1757-899X/1296/1/012007

Artificial intelligence methods for in-process high-speed image analysis in laser beam welding of hairpins

Hake, C.[1], Omlor, M.[1] ✉, Breitbarth, A.[2, 3], Notni, G.[2, 3], Dilger, K.[4]

[1] Dr. Ing. h.c. F. Porsche AG, Stuttgart, Germany
[2] Fraunhofer Institute for Applied Optics and Precision Engineering, Jena, Germany
[3] Technische Universität Ilmenau, Ilmenau, Germany
[4] Technische Universität Braunschweig, Braunschweig, Germany

Hake, C. and Omlor, M. contributed equally to this work and should be considered as joint first authors.

✉ markus.omlor@porsche.de https://orcid.org/0000-0002-6963-9238

Abstract. In the production of modern electric drives for battery electric vehicles, hairpin technology is used to increase the copper fill factor in the stator of a permanently excited synchronous machine. A central process in the production of these stators is the contacting of the hairpin ends by means of laser beam welding. This welding process is characterized by geometric and process-related deviations from previous process steps, which influence the result of the welded joint. It is desirable to find an in-process method for monitoring. As part of the process monitoring of welded joints, high-speed camera images are often used to detect weld spatter. These can be detected by a program based on a static algorithm. For this reason, a feasibility analysis is performed regarding the application of AI for the detection of spatters, in which the methods of semantic segmentation and single-image classification prove to be useful. In a preliminary experiment, three base networks for each of the two methods are evaluated with respect to the best training results. The single-image classification method will then be extended by a subsequent static algorithm, so that a hybrid use of AI and static algorithm will be investigated. The evaluation and final comparison of all evaluation methods is performed using data from a welding experiment. It turns out that the hybrid approach of single-image classification and static algorithm has numerous advantages in the detection of spatter compared to semantic segmentation and the static algorithm.

Keywords: artificial intelligence, hairpin stator, laser beam welding, in-process measurement

Abbreviations

AI – Artificial Intelligence
BF1 – Boundary F1
CCD – Charged Coupled Device
CNN – Convolutional Neural Network
Cu-OF – Oxygen free Copper
FCN – Fully Convolutional Network
GAN – Generative Adversarial Network
IoU – Intersection over Union
LED – Light Emitting Diode
NST – Neural Style Transfer
PEEK – Polyetheretherketone
RAM – Random-Access Memory
ReLU – Rectified Linear Unit
RGB – Red, Green, Blue
SVM – Support Vector Machine
TNN – Thresholding Neural Network
UV – Ultraviolet
VGG – Visual Geometry Group

1 Introduction

1.1 Hairpin technology in automotive industry

The winding of the stator of a permanently excited synchronous machine can nowadays be produced by different winding technologies. The winding process is highly automated and largely dependent on the type of winding. Low process and cycle times and a low reject rate are required in the automotive industry, as well as a high efficiency of the electrical machine. Therefore, hairpin winding technology is increasingly used [1]. Hairpins are solid, rectangular copper flat wires that are inserted into the stator slots of the laminated core after several pre-processing steps. The individual hairpins are formed in a U-shape. In the automotive industry, hairpins thus replace round wire winding processes. The major advantage of hairpin technology is essentially the rectangular cross-section of the flat copper wire and the small number of turns; both are responsible for increasing the copper fill factor from 45% to up to 70% [2].

One of the most important process steps in the manufacturing of hairpin windings is the contacting of two corresponding hairpins by laser beam welding [3]. During the actual welding process, which lasts < 150 milliseconds, a high heat input into the material occurs by absorption [4, 5]. A deep penetration welding is used for welding copper with infrared wavelength. A vapor capillary is formed and surrounded by the molten pool. During the welding process, the laser beam follows a predetermined welding path. The vapor presses against the back of the capillary as the laser moves. This vapor pressure, in combination with the flow of the molten pool, can lead to spatters [3, 6].

The joint generates the same properties as the base material, so that high mechanical strength and current-carrying capacity are achieved. [7]. Depending on the variation of the process parameters and numerous other influences of the workpiece, process media and handling of the equipment, seam defects can occur. Since not all seam defects can be detected by the high-speed camera used in this paper, the directly detected seam defects are limited to spatter and melt ejection. Spatters can be seen as a quality feature during welding.

Large losses of the melt pool due to spatters reduce the melt volume and thus also the weld penetration depth. All spatters can lead to damage of the component and the tool, including the insulation of the copper wire. For this reason, it is useful for a development environment to find an easy way to automatically count spatters in order to optimize the parameters of the hairpin welding process.

1.2 Inline and in-process monitoring technologies with artificial intelligence

Numerous other works have already investigated the visual inline and in-process inspection of laser beam welded copper joints. "In-process" describes the measurement in a welding process itself, whereas inline measurement is the measurement taken for a weld as part of a series production line. Vater et al (2021) [8] examine hairpin ends made of copper material welded by a laser beam welding system. In this work, the copper ends are recorded before and after welding by a 3D camera of the model XR-HT40M of the company Keyence and are classified as correct welding, insufficient welding, welding craters and welding spatter. Vater et al. (2021) [8] use a 10-layer deep CNN, the VGG-10 network, as the method of evaluation. As a result, the trained VGG-10 detects the above features with an accuracy of 99.58%.

Also dedicated to the analysis of hairpin welds is Mayr et al. (2018) [9], who perform and analyze 500 copper hairpin welds in their work. The analysis of previously defined features such as resistance, irregularities, undercut, recessed seam, holes, spatter, and weld bead is performed by varying the welding factors such as laser power and feed rate. Mayr et al. (2018) [9] use a CCD camera for visual detection and analyze its images using a custom-built CNN. Furthermore, Mayr et al. (2018) [9] train an SVM on different machine parameters so that it can predict the quality of the weld pre-processing with a probability of 87%. The CNN, which analyzes the image files, can detect weld defects with a probability of 70%. A combined quality monitoring system consisting of SVM and CNN has an overall

IOP Conf. Series: Materials Science and Engineering 1296 (2023) 012007 doi:10.1088/1757-899X/1296/1/012007

detection rate of 92% according to Mayr et al. (2018) [9]. There is a high number of papers that use high-speed cameras to detect weld spatter. For example, You et al. (2014) [10], Hartung et al. (2021) [11, 12], and Tan et al. (2020) [14] use different high-speed cameras, each equipped with various filters such as UV [10, 11] or bandpass filters [10, 11, 13, 14], to examine laser-beam welded components. Hartung et al. (2021) [11] analyze 18 mm long welds and conclude that the occurrence of spatter as well as the quality of the weld strongly depends on the welding factors. In their experiments, Hartung et al. (2021) [11] vary the laser power between 4 kW and 6 kW, create a gap between the welded components or defocus the laser optics. This affected the process in such a way that spatters, and unstable welds were produced [11]. Here, the spatters are detected by a proprietary CNN based on U-Net with an accuracy of 75%. In Hartung et al. (2021) [12] this approach is followed up and optimized. The images from the camera already mounted on the laser beam welding system, which are taken from three different perspectives during the welding process of hairpin pairs, are superimposed and then evaluated by the modified U-Net. Process light and spatters are considered as the main process deviations. Hartung et al. (2021) [12] achieve a training accuracy of the CNN architecture of 98%. Furthermore, the spatter detection method of Tan et al. (2020) [14] must be mentioned, as it consists of a combination of CNN and TNN. In the first step, the CNN classifies the images into those with and without spatters. In the second step, the image is subjected to examination by the TNN, which segments the image into spatters, melt pool, and irrelevant area. As a result, spatters can be detected with an accuracy of 80.48% in the examined images.

Based on the above-mentioned work, a methodology for the in-process evaluation of high-speed videos of the laser contacting process of electric drives was developed previously by Omlor et al. (2022) [15]. For this purpose, Omlor et al. (2022) [15] determined technical boundary conditions and a suitable recording environment based on predefined good and defect videos. Furthermore, typical image features of the laser beam welding process were determined, based on which specific process features were selected, and possible interfering influences were identified. The developed methodology is implemented using a static algorithm [16] that follows a fixed program flow and can therefore detect spatter. The detected spatter is correlated with spatter manually counted by an expert, resulting in a coefficient of determination of $R^2 = 83.6\%$, thus confirming the basic suitability of the method [15]. Since the method shown in [15] is based on experimentally determined thresholds and is therefore not prepared for extrinsic changes such as changes in brightness, changes in the setup of the welding system, etc., it is advisable to investigate an additional method, in this case the use of artificial networks.

2 Methods and materials

2.1 Experimental laser beam welding setup

The section of the welding system relevant to this work, used for welding the hairpins investigated in the experiment, is shown in figure 1. The main component is a disk laser (TruDisk6001 by Trumpf GmbH & Co. KG, Ditzingen, Germany) with scanner optics (PFO33/2 by Trumpf GmbH & Co. KG, Ditzingen, Germany), which consists of a 6.0 kW laser that welds in the wavelength of 1,060 nm and has a focus diameter of 110.0 µm.

The laser optics can be moved in all three axes x, y and z. The components to be welded, either individual hairpin pairs in a special test rack or an entire stator, are inserted into the system on a workpiece carrier. The workpiece carrier is locked in the welding system, and the stator to be welded, driven by a stepper motor, can rotate on the workpiece carrier at certain rotation angles around the subsequent direction of the shaft rotation. The high-speed videos are recorded by a high-speed camera (FASTCAM Nova S9 type 900K-M-32 GB, Photron, Reutlingen, Germany). A bandpass filter is mounted on the camera, which is transparent in a wavelength range from 600 to 660 nm and provides protection against overexposure. The high-speed camera is mounted on the laser optics at an angle to the outer hairpin edge at a fixed distance of 220 mm and is supported during video recording by an artificial illumination in the form of an LED emitting in a wavelength of 630 nm. Attached to the camera is a 12x zoom lens (Navitar, New York, USA) that does not require an aperture. Mounting the high-speed camera to the laser optics ensures a constant distance to the weld. The camera's trigger is activated automatically by a trigger connected to the welding system. Once the trigger is active, the camera records a high-speed video at a frame rate of 10,000 frames per second. The exposure time of each frame is less than 0.1 ms. The resolution of the video is 896 × 896 pixels, with the color spectrum limited to 12-bit gray values. Photron Fastcam Viewer version 4.0.5.0 operates as the recording program with no filtering operation applied other than the bandpass filter. The hairpin ends are always in the position and arrangement to the laser optics shown in figure 1. The hairpins are composed of Cu-OF material and, except for the area to be welded, are covered with a PEEK plastic protective layer. The welding process is divided into four process phases P_1 to P_4.

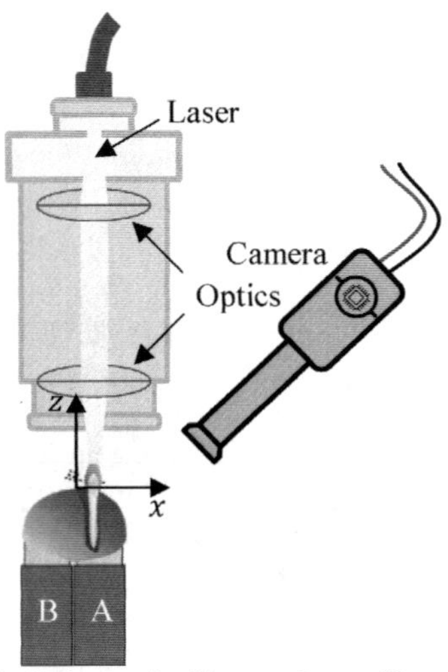

Figure 1. Schematic illustration of laser optics, high-speed camera, and hairpins A, B

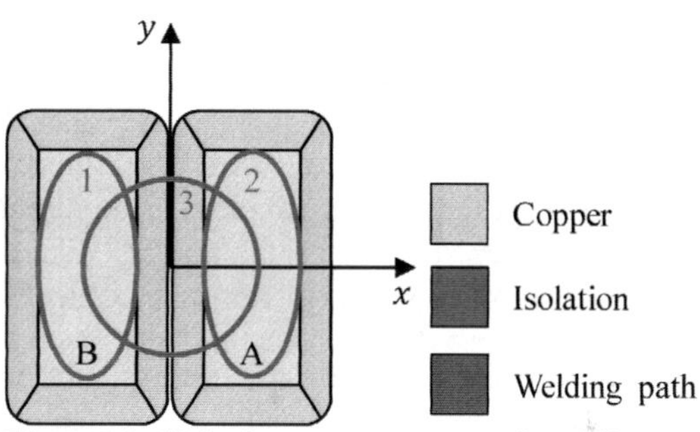

Figure 2. Hairpin arrangement and welding paths [16]

In P_1, the upper surface of the outer hairpin B in arrangement, relative to the camera, is welded for 33.5 ms in the form of an ellipse, as shown in figure 2, followed by the same ellipse on the surface of the inner hairpin A for 28.3 ms in P_2. Due to the different process times, the melting volume is formed uniformly. The melt pool of both hairpins fills the gap that originally existed between them. In P_3, the two melt pools are joined by welding for 47.0 ms in the form of a circle across both melt pools of P_1 and P_2. P_4 consists of the cooling process, which lasts 125.0 ms [15].

2.2 Potential-analysis of artificial intelligence methods

Three principal methods exist to apply CNNs in the field of computer vision. These are semantic segmentation, object identification and classification.

In semantic segmentation, each pixel of an image is assigned to a class trained to the network and then clustered. Semantic segmentation networks essentially consist of an encoder-decoder structure. In the first step, the encoder extracts relevant features of the image. In the second step, there is a decoder that is used to upsample the feature vectors computed by the encoder. The decoder is symmetrical to the encoder so that the segmented image has the same size as the input image. To compute the class probabilities of each pixel, the decoder uses pooling indices computed in the max-pooling step of the corresponding encoder to perform nonlinear upsampling [17]. The final layer is a softmax activation function to compute class probabilities. In addition to this classical segmentation network, there are two other ways to perform semantic segmentation. One is a fully convolutional network and the other is a network that uses dilated convolutions. An FCN consists of a downsampling path, which is used to extract and interpret the image content, and an upsampling path, which is used to localize the features.

The last layer of the FCN, which is usually the fully connecting layer, is transformed into a convolutional layer with multiple 1×1 filters, where the number of filters must match the number of segmentable classes [18]. To increase the resolution afterwards, transposed convolution is applied with a filter size corresponding to the original resolution. The step size is half of the filter size. The last layer is a softmax activation function to compute the class probabilities. The third way to perform semantic segmentation involves dilated convolutional networks. The resolution of the output is reduced by a fixed factor by changing the step size in the max-pooling layer and the convolutional layer, which is why bilinear interpolation is used to expand the class values of the output layer calculated in this way to the size of the input image.

The second method is object identification or object detection, in which the network recognizes previously trained objects in an image or video and indicates their position and location to the user by framing the identified object using a rectangle or circle. This approach is particularly useful for completeness checking or determining the location of multiple objects.

The third method is the single-image classification. Here, the CNN analyzes the complete image content by extracting relevant features whose assignment to the trained classes is already known. Subsequently, each single image is assigned to the class with the highest output layer score. The structure of these classification networks is similar and differs only by the number of consecutive convolutional layers, which is why it is exemplified by the VGG-16 network [19, 20, 21]. The input layer has a fixed size of 224×224 pixel RGB images. The image is passed through the stack of convolutional layers using very small 3×3 filters. The convolution step is fixed at one pixel, and the padding of the convolution layer input is chosen to maintain spatial resolution after convolution. Pooling is performed by the five max-pooling layers that follow some of the convolution layers. Max-pooling is performed on 2×2 pixels each with a step size of two. Three fully connected layers follow the convolutional layers, which have different depths in the different architectures (VGG-16, VGG-19, or ResNet-50). The first two have 4,096 classes each, while the third performs a 1,000-fold classification and therefore contains 1,000 classes. The last layer is the Softmax layer. The configuration of the fully connected layers is the same in all networks, only the last layer is adjusted to the actual number of classes. All hidden layers are equipped with a ReLU layer for normalization.

2.3 Methodical approach

2.3.1 Procedure for training the AI methods

There are several methods to use CNNs for the given computer vision task of counting weld spatters. Therefore, suitable methods shall first be determined and a maximum of three networks shall be selected for each method. Their suitability shall then be evaluated in a preliminary experiment with respect to the best training results. The network that achieves the best results for selected performance metrics for each method is then trained on the entire available data set and implemented in a computer analysis program for counting weld spatters in each case. Numerous pre-trained CNNs are available for semantic segmentation and single-image classification, so a preliminary experiment evaluates the extent to which the networks under investigation prove useful. The method of object identification is incompatible with numerous internal requirements and is therefore not pursued further in the present work. Since single-image classification alone cannot compute the required spatter volume, a static algorithm, which sequentially processes explicit procedural instructions to solve the given problem, is added to the classification method for evaluation, so that a combination of single-image classification and static algorithm is investigated.

2.3.2 Selection of networks for semantic segmentation

The first network investigated for semantic segmentation is DeepLab V3+ [22], which belongs to the class of dilated convolutional networks and is based on ResNet-18. The second network is U-Net, which features five encoders and decoders [21]. The third network is FCN 8s, a fully convolutional network based on VGG-16 [18]. The suffix 8s in this case means that the upsampling of the final feature map by a factor of 8 happens after the merging of the feature maps of the third and fourth max-pooling layers [18]. This additional information from earlier layers allows for more detailed segmentation despite the increased computational power required. The networks use the layers along with their pre-initialized weights from the specified base networks trained on the ImageNet dataset to enable a fast and effective training process.

2.3.3 Selection of networks for single-image classification

For single-image classification, the VGG-16 network is investigated [19]. On the other hand, the VGG-19 network is examined to investigate the effects of the depth of the networks on the results of the spatter detection. In previous work, the VGG networks were found to be suitable for spatter detection, so an investigation as part of a combination of classification and static algorithm is of particular interest. Furthermore, the 50-layer ResNet-50 is trained on the data for comparison purposes [23].

2.3.4 Obtaining the training data

To generate consistent and sufficient data for all in-process test methods, a welding experiment is carried out in which corresponding hairpin pairs of five stators are welded according to a fixed test plan and under identical conditions. The measured data are subsequently analyzed by both the static algorithm and the selected AI methods. The results thus obtained from the basis for the final comparison between the different analysis methods.

The aim of the welding experiment is to have sufficient sample data for as many factors as possible that can occur during the welding process. Accordingly, a detailed test plan with firmly predefined factors is developed in advance. This results in a total of 739 welds. Table 1 provides information on the selected factors and their factor levels [15].

Table 1. Experimental plan of the welding experiment [15]

Factor	Min. value	Max. value	Factor levels	Welds per factor level	Sum
Reference [ ]	-	-	1	64	64
Initial laser power [W]	3,150	5,400	7	24	168
Focal position shift [mm]	1	6	6	27	162
Height offset Δz [mm]	0.5	2.5	5	27	135
Gap Δx [mm]	0.2	1.2	6	27	162
Stripping quality [ ]	Class 1	Class 4	3	16	48

The reference values represent 64 welds that were produced under optimum conditions. For P_1 and P_2, this means a laser power of 4,500 W with a feed rate of 750 mm/s. P_3, in deviation from the first two, is welded with a laser power of 4,000 W and a feed rate of 500 mm/s. The first type of defect factor is generated by varying the initial laser power, which corresponds to a changed welding power. In this case, the minimum value of the first two process phases is 3,150 W and is increased in factor levels of 3,250, 3,400, 3,500, 4,500, 4,950 up to finally 5,400 W. The welding power of P_3 is adapted to maintain the reference ratio. For each factor, 24 hairpin welds are performed. The second type of error factor is the defocusing of the optics on hairpin A of figure 1 in the positive z-direction, which corresponds to an incorrect positioning of the stator in the system. The factor values are 1, 2, 3, 4, 5, and 6 mm, with 27 hairpin welds per factor level. The third type of defect factor is the height offset of the outer hairpin B in the camera image from figure 1 in the positive z-direction to the front hairpin A, which corresponds to an incorrect positioning of the hairpins in the clamping tool. The factor values are 0.5, 1.0, 1.5, 2.0 and 2.5 mm with 27 hairpin welds each. The penultimate factor is a gap between the hairpins, which can occur due to insufficient clamping force or jamming of the hairpins in the clamping tool. The factors are realized by means of specially made spacer plates and are 0.2, 0.4, 0.6, 0.8, 1.0 and 1.2 mm for each of 27 hairpin welds. The last type of defect realized in the welding experiment is contamination of the hairpin surfaces, due to poor stripping performance in the stripping process. The quality of stripping is classified into four classes using point fluorescence measurement, with a class 1 stripping quality corresponding to that of a reference pin and a class 4 stripping quality to that of a pin with most remaining insulation residues after the stripping process [24].

2.3.5 Configuration for training the AI methods

In the preliminary experiment, five videos per factor level from table 1 are labeled by factor and then trained by the networks. Then, the network with the best training results for each method is trained on the entire data set from the welding experiment. Table 2 lists the training parameters of the networks studied. The ADAM optimizer is used as the optimizer of the gradient descent algorithm. The optimal learning rate used by the optimizer has been empirically determined to be 1e-4 in several experimental runs. The number of training epochs has also been empirically determined. The size of the mini batches varies and is chosen to maximize the use of graphics memory. Each epoch, the complete training dataset is presented to the network. Furthermore, the benefit of synthetic data augmentation is investigated in the preliminary experiment, as it can reduce the time and cost spent on the welding experiment. The generation of a synthetic image by a GAN or NST requires too much time and computing power, so classical image processing techniques are used. With that, the existing training data set is extended by a factor of four by applying horizontal and vertical mirroring and rotation by ±90 degrees. Effects due to brightness changes are not investigated since they are already covered in the multilayer networks. The NVIDIA GeForce RTX 3060 graphics card with 12 GB of RAM is available during training. Processes that cannot be handled by the graphics card are calculated by an AMD Ryzen 5 PRO 4650G main processor with 6 cores and a 32 GB RAM. All training data is stored on a 12 TB external hard drive, which is connected to the computer via USB 3.0.

Table 2. Empirically determined training parameters for the AI networks

Parameter	Semantic Segmentation	Single-image classification
Optimizer	ADAM	
Learning rate	1e-4	
Epochs	100	50
Mini-batch size	32	16
Shuffle	Every Epoch	

3 Results

3.1 AI training results

3.1.1 Training results from the preliminary experiment

To ensure comparability of the results of semantic segmentation, relevant performance metrics are initially defined and used to evaluate the results. For semantic segmentation, these are the global accuracy, the mean accuracy, the mean BF1 score, the mean IoU, and the training duration in minutes. For classification, suitable performance metrics for interpreting the training results and comparing the networks are the global accuracy, the final loss, and the training duration in minutes.
The results discussed below were obtained with a test script based on the test data set and are shown in Table 3. The data shows that for all performance metrics except training duration, the FCN is the most suitable method for semantic segmentation of the present data. Especially in Mean IoU, the FCN has advantage of about 20% over the other two networks for Semantic Segmentation. Although the training time is the highest with 286 minutes, it is acceptable since it is only required once. For single-image

classification, the VGG-19 network has the highest accuracy with a global accuracy of 95.07% and the smallest final loss of 0.0115. As with semantic segmentation, the network with the best performance metrics requires the longest training time. It can be clearly seen that the 50-layer deep ResNet-50 offers no advantage over the VGG-19 network, while the former is more than twice as deep as the latter. Considering the results, the VGG-19 network is selected to be used for single-image classification in the remainder of this work. For all networks, the synthetic training data usage leads to overfitting while the training times increased.

Table 3. Training results of each network from the preliminary experiment

	Semantic Segmentation			Single-image classification		
Performance Metrics	**DeepLab V3+**	**FCN**	**U-Net**	**VGG-16**	**VGG-19**	**ResNet-50**
Global Accuracy [%]	99.75	99.96	99.20	94.82	95.07	93.67
Mean Accuracy [%]	83.23	88.14	81.95	-	-	-
Mean BF1 Score [%]	53.45	87.12	27.32	-	-	-
Mean IoU [%]	54.63	71.13	51.15	-	-	-
Final Loss [ ]	-	-	-	0.0118	0.0115	0.0164
Training duration [min]	15	286	123	138	140	104

3.1.2 Training results from the large data set

The semantic segmentation and single-image classification networks determined in the preliminary experiment are subsequently trained on a larger set of data from the welding experiment. The training parameters are taken from the preliminary experiment. The pre-trained networks from the preliminary experiment are also used. After the labeling process, a total of 95,245 training, 15,926 validation and 19,888 test images are available for semantic segmentation. For the single-image classification there are 91,698 training images, 17,596 validation images and 17,628 test images. There is a disparity between the two classes in the single-image classification. Images without spatters are represented more often than images with spatters by a factor of 3. Since spatters occur less frequently in reality in relation to the totality of all images than images in which no spatter occurs, balancing of the classes is dispensed with. The final training results are shown in table 4. In comparison with the preliminary experiment, none the performance metrics considered decreased significantly. Only the training times increased by a multiple of the preliminary times.

Table 4. Final training results and their delta values compared to the preliminary experiment (ΔPE)

Performance Metrics	FCN	ΔPE	VGG-19	ΔPE
Global Accuracy [%]	99.82	-0.14	95.00	-0.07
Mean Accuracy [%]	88.03	-0.11	-	-
Mean BF1 Score [%]	87.11	-0.01	-	-
Mean IoU [%]	70.09	-1.04	-	-
Final Loss [ ]	-	-	0.163	0.151
Training duration [min]	1,909	1,623	7,256	7,116

3.2 Comparison of method results

All presented in-process methods for the evaluation of the high-speed camera images are subsequently implemented in an analysis program, which is used to perform a spatter analysis for each video of the welding experiment. In this step, the single-image classification method is enhanced by a static algorithm that evaluates an image, identified as a spatter image by AI, with respect to the position and volume of the spatter. It should be noted that the static algorithm alone (without AI support) provides the fastest spatter analysis with a duration of 40 s. However, the combination of single-image classification and static algorithm, takes an additional 20 s. The longest analysis is done when combining the static algorithm with semantic segmentation, which takes just under six minutes.

3.2.1 Investigation of the global parameters

The objective is to compare the AI methods with the static algorithm and to evaluate them using suitable performance metrics. To evaluate the data of the welding experiment, all videos of the welding experiment are analyzed using the previously mentioned analysis programs. Thus, there is one database for each method containing spatter data for each high-speed camera video. The data in the databases include the total spatter per video, the number of spatters per video divided into the three relevant process phases P_1, P_2 and P_3, the number of spatters per video divided into four empirically determined volume ranges, and the number of spatters per video divided into the volume ranges per process phase. The process phases P_1 to P_3 are the same as those in section 2.1. The process phase P_4 is not considered in detail, since in this phase the cooling process takes place, in which no spattering occurs. The volume ranges are determined empirically and correlate with the visual impression of a small, medium, and large spatters as well as a melt ejection. The ranges are divided into $V_1 \in]0...15[\, e3\ \mu m^3$, $V_2 \in [15...10{,}000[\, e3\ \mu m^3$, $V_3 \in [10{,}000...100{,}000[\, e3\ \mu m^3$, and $V_4 \in [100{,}000...\infty[\, e3\ \mu m^3$. The factor e3 in front of the unit μm³ is used for better readability of the data.

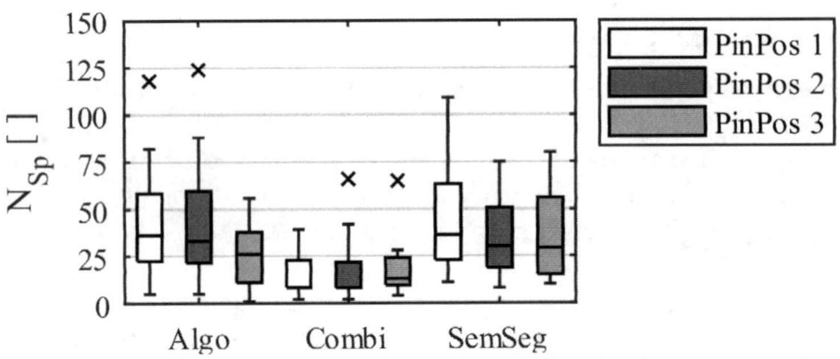

Figure 3. Investigation of a dependence of the pin position in the stator on the spattering result

The first factor examined is the pin position of the hairpin pairs in the stator. A total of three pin positions exist, with position 1 designating the inner hairpin row and position 3 the outer hairpin row, viewed from the center of the stator. For evaluation purposes, the spatter count of the reference welds, grouped by pin position is plotted above the analysis method. This is shown in figure 3. For the static algorithm (Algo) and the combination (Combi) of classification and algorithm, it can be observed that for pin position three, the scatter, i.e., the difference between the upper and lower whiskers is the smallest. However, the medians are similar for all methods across all three pin positions. To analyze whether there is a significant correlation between pin position and amount of spatters, the p-value of an analysis of variance can be used. The statistical analysis is performed with a significance level of $\alpha=0.05$, which yields a p-value of 0.42 for the static algorithm, a p-value of 0.86 for the combination, and a p-value of 0.94 for the semantic segmentation (SemSeg), so that it can be considered that there is no significant relationship between number of spatters and pin position. Also, the statistical analysis is performed for all welds welded with a predominant factor. All p-values are <1e-6, so that a significant dependency between factor and number of spatters is proven.
Subsequently, the spatters welded according to reference factors are evaluated by process phase and volume range. Narrowing down the range to be detected, especially for the process phase, would save computing time in the AI methods. There is a high number of spatters in P_3, which is accompanied by a wide scatter of results for all three methods. The constancy of the measured values of all methods is represented by the standard deviation of all three method medians per process phase. P_1 has the lowest value with a standard deviation of 10.19. The value for P_2 is not far greater at 10.68. The largest standard deviation is found in P_3 with a value of 43.10. The smallest scatters are also found in P_1. For this, the

sum of the individual scatters of each method per process phase is calculated. For P_1, the sum of the individual scatters is 202, for P_2 238 and for P_3 526. In view of these results, P_1 is considered more closely in the following evaluation regarding the consistency of the results and the lowest scatters.

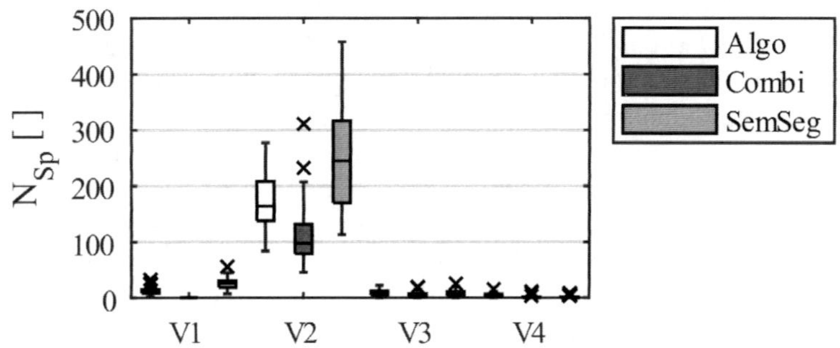

Figure 4. Comparison of the number of spatters as a function of the defined volume ranges

Figure 4 shows the evaluation of the volume ranges. V_2 contains the most spatters for every method investigated. The combination method has the smallest scatter in V_2 with a value of 161, followed by the static algorithm with a value of 193. The semantic segmentation has the largest scatter with a value of 345. Since V_2 contains the most data, it will be considered further below.

3.2.2 Investigation of the welding parameters

The following statements about the behavior of the median and scatter are made before evaluating the welds with factor present:

- Due to an increase in defocusing, the objects on the camera sensor become blurrier. Furthermore, it can be observed in the original images that from a defocus of 5 mm the hairpins are outside the range visible to the camera. Thus, the medians and the scatters must decrease with increasing defocus.
- Due to an increasing height offset, the laser must cope with a step between hairpin A and hairpin B, which favors spatter formation. Therefore, an increase in medians and scatters is to be expected here. [15, 25]
- Increasing the power increases the energy introduced into the copper material and thus the tendency to spatter formation. Therefore, an increase in the medians and the scatters is to be expected here. [15]
- Due to a larger distance between the hairpins, it is possible for the laser to weld the space between the two hairpins or that the two melting pools of hairpins A and B cannot be joined in P_2. Both factors favor spatter formation, which is why increasing medians and scatters are to be expected here. [15, 25]
- The higher the degree of contamination of the hairpins, the greater the tendency to spatter. An increase in the medians and, above all, in the scattering is to be expected since the contamination is not evenly distributed on the hairpin surface. [15, 24]

Finally, the results of the analysis methods are examined using the two criteria of median correlation and scatter for all factors. For this purpose, the correlation between the median of the examined method and the respective factor level per factor is computed, to be able to understand how well the median follows the established courses from the hypotheses for each method. On the other hand, the scatter is defined as the difference between the upper and lower whiskers of the individual methods per factor above all factor levels. The dependency is indicated by the coefficient of determination with respect to a linear model. The exact values of the coefficient of determination and the scatter are given in Table 5. R_M^2 denotes the coefficient of determination of the median with the respective factor levels of the selected factor and S the scatter. It can be clearly seen that large differences in the coefficients of determination and the scatters are prevalent, depending on the method and the factor. While the coefficients of determination for the gap factor are very low, the coefficients of determination for the initial power factor always reach values beyond 90%. The same can be observed for the scatters. The lowest scatters occur for the focus factor, while the highest scatters occur for the height offset and gap factors. If the best combination of the largest coefficient of determination and the lowest scatter is sought for each factor, the combination method must be used to detect the focus factor, the algorithm method for the height offset, the combination method for the power, the algorithm method for the gap and the combination method for the contamination. The results from this observation are presented qualitatively using scores in a combined diagram, as seen in figure 5. The highest score is given to the method with the highest coefficient of determination and the lowest scatter. The lowest score is given vice versa. Figure 5 clearly shows that the combination method most frequently achieves the best scores, although it does achieve low scatters in the height offset and gap factors, but also has very low coefficients of determination. If the individual scores per method are summed up and then the arithmetic mean is calculated, a value of 1.5 is obtained for the static algorithm, a value of 2.5 for the combination and a value of 2.1 for the semantic segmentation.

Table 5. Coefficient of determination and scatter values

	Focal position shift		**Height offset Δz**		**Initial laser power**		**Gap Δx**		**Stripping quality**	
	R_M^2	S	R_M^2	S	R_M^2	S	R_M^2	S	R_M^2	S
Algo	56.5%	110	64.2%	387	92.5%	421	42.9%	410	52.1%	249
Combi	85.2%	37	20.3%	110	95.6%	103	0.9%	148	58.1%	111
SemSeg	73.5%	110	65.9%	528	96.3%	266	0.0%	212	85.9%	205

It can be concluded that the combination of single-image classification of the images by an AI and subsequent evaluation by the static algorithm is a suitable method to detect spatters and the spatter deviation type. When looking at the medians in terms of their qualitative course as well as their correlation with the factor levels, disadvantages can be seen in the height offset and gap factors, which the static algorithm compensates, although this is accompanied by large scatters. The great advantage of the combination is the low scatter that this method consistently exhibits for the respective factors. The functionality of the semantic segmentation as a pure AI method is proven, but the achieved scatters are higher than those of the combination for the respective factors. For the factors height and gap, where the combination method has disadvantages, these cannot be compensated by the semantic segmentation.

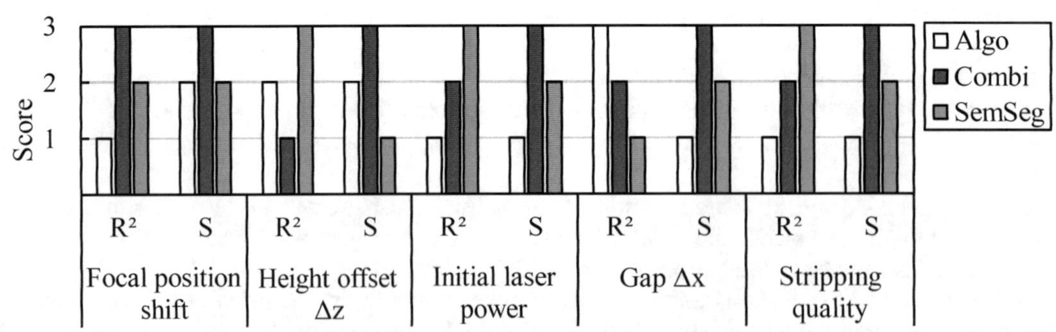

Figure 5. Final consideration of all methods, their median dependencies and scatters from best (3) to worst (1)

4 Conclusion and outlook

Spatter means loss of melt and damage to the component or the equipment, therefore the occurrence of spatter must be recorded quantitatively. The method for recording spatters according to [15] exists but is not applicable in the case of extrinsic changes. Therefore, a pure AI approach and a combined approach of static algorithm and AI were investigated. For this purpose, the AI methods of single-image classification and semantic segmentation were selected. For the single-image classification and semantic segmentation methods, a preliminary experiment was performed to evaluate effects of layer depth, training data texture, and training parameters, among others, in an indicator-based approach. VGG-19 for combination and FCN for semantic segmentation achieved the best training results and were then trained on a more extensive dataset generated in a welding experiment. No appreciably better training results were obtained. A subsequent static algorithm was added to the single-image classification, resulting in a combination of the two methods. Finally, an analysis of the videos of the welding experiment was performed. For this purpose, each video of the welding experiment was analyzed and evaluated using previously developed computer programs. Based on this data, a comparison between a static algorithm and AI methods could be made. The following results emerged from the investigation:

- The single image classification and semantic segmentation methods can reliably detect spatter as a deviation of the laser beam welding process.
- The combination of single image classification and algorithm reaches the best performance data (R_M^2) compared to the static algorithm and semantic segmentation.
- The pure use of artificial intelligence, here in the case of semantic segmentation, is not as productive as that of a static algorithm, which is also less time-consuming, labor-intensive, and computationally expensive to prepare and analyze.
- The main disadvantages of semantic segmentation are the high time and computational costs. For example, a fourfold increase in analysis time and a sixfold increase in computing power must be expected compared to the static algorithm.
- The combination of serially linked classification by means of AI and subsequent evaluation by the static algorithm uses the advantages of both methods and has an acceptable analysis time. In addition, the time required for labelling the training data as well as the actual training duration is less, since labelling and training is carried out per image and not per pixel.

Numerous suggestions for improvement can be derived for future research in this field. It is possible to use the positioning camera installed in the optical path of the welding system to train object identification networks on the positioning problems especially for height offset and gap, so that these can be eliminated in advance. If necessary, recurring patterns in conspicuous positions of several stators can be identified and thus incorrect positioning can be predicted by the networks. Furthermore, these nets can be used to influence the positioning of the system in case a hairpin pair is outside defined positioning limits. It is also interesting to examine individual image areas. Thus, it can be investigated whether the analysis of a certain ROI is sufficient to infer the factors, which would result in a speed improvement.

References

[1] Glässel, T.; Seefried, J.; Franke, J.: Challenges in the manufacturing of hairpin windings and application opportunities of infrared lasers for the contact-ing process. IEEE EDPC, 2017. (DOI 10.1109/EDPC.2017.8328150).

[2] Kamker, A.; Heimes, H.; Kawollek, S.; Treichel, P.; Kraus, A.: Produktionsprozess eines Hairpinstators. VDMA, 2019. (DOI 10.13140/RG.2.2.26518.09285).

[3] Bliedtner, J.; Müller, H.; Barz, A.: Lasermaterialbearbeitung. Grundlagen - Verfahren - Anwendungen – Beispiele. Hanser, Munich, 2013. (ISBN 978-3-446-42168-4).

[4] Dilthey, U.; v. Alvensleben, F.: Laserstrahlschweißen: Prozesse, Werkstoffe, Fertigung und Prüfung. DVS Verlag, 2000. (ISBN 978-3-87155-906-8)

[5] Hügel, H.; Graf, T.: Laser in der Fertigung. Strahlquellen, Systeme, Fertigungsverfahren. 2. Aufl., Wiesbaden, Vieweg + Teubner, 2009. (ISBN 978-3-8351-0005-3).

[6] TRUMPF SE + CO KG (pub.): BrightLine Weld - Perfect welding results in record time. White Paper, 2020.

[7] Gläßel, T.; Seefried, J.; Masuch, M.; Redel, A.; Mayr, A.; Kühl, A.; Franke, J.: Process Reliable Laser Welding of Hairpin Windings for Automotive Traction Drives. IEEE ICESI, 2019. (DOI 10.1109/ICESI.2019.8863004).

[8] Vater, J.: Präskriptive Automatisierung - Ein optisches Prüfsystem zur Klassizierung und automatisierten Nacharbeit mittels Cloud-/Edge-Computing. dissertation, Munich, mediaTUM, 2021. (https://mediatum.ub.tum.de/?id=1576629, 30.06.2023).

[9] Mayr, A.; Lutz, B.; Weigelt, M.; Glabel, T.; Kibkalt, D.; Masuch, M.; Riedel, A.; Franke, J.: Evaluation of machine learning for quality monitoring of laser welding using the example of the contacting of hairpin windings. IEEE EDPC, 2018. (DOI 10.1109/EDPC.2018.8658346)

[10] You, D.; Gao, X.; Katayama, S.: Monitoring of high-power laser welding using high-speed photographing and image processing. Mech. Sys. and Sig. Proc., Vol. 49, 2014, p. 39–52. (DOI 10.1016/j.ymssp.2013.10.024).

[11] Hartung, J.; Jahn, A.; Stambke, M.; Wehner, O.; Thieringer, R.; Heizmann, M.: Camera-based spatter detection in laser welding with a deep learning ap-proach. Forum Bildverarbeitung, 2020, p. 317-328. (DOI 10.5445/IR/1000129216).

[12] Hartung, J.; Jahn, A.; Bocksrocker, O.; Heizmann, M.: Camera-Based In-Process Quality Measurement of Hairpin Welding. Appl. Sci., Vol. 11(21), 2021. (DOI 10.3390/app112110375).

[13] Nicolosi, L.; Abt, F.; Blug, A.; Heider, A.; Tetzlaff, R.; Höfler, H. 2011. A novel spatter detection algorithm based on typical cellular neural network operations for laser beam welding processes. Measurement Science and Technology. 23. (DOI 10.1088/0957-0233/23/1/015401)

[14] Tan, Z., Fang, Q., Li, H., Liu, S., Zhu, W., & Yang, D. (2020). Neural network-based image segmentation for spatter extraction during laser-based powder bed fusion processing. Optics & laser technology, 130. (DOI 10.1016/j.optlastec.2020.106347)

[15] Omlor, M., Reith, J., Breitbarth, A., Hake, C., Dilger, K. (2022). Inline process monitoring of hairpin welding using optical and acoustic quality metrics. 12th International Electric Drives Production Conference (EDPC). Regensburg, Germany.

[16] Ronneberger, O., Fischer, P., & Brox, T. (2015). U-Net: Convolutional Networks for Biomedical Image Segmentation. arXiv. (DOI 10.48550/ARXIV.1505.04597)

[17] Brandenbusch, K. (2018). Semantische Segmentierung mit Deep Convolutional Neural Networks. Technische Universität Dortmund.

[18] Long, J., Shelhamer, E., & Darrell, T. (2014). Fully convolutional networks for semantic segmentation. arXiv. (DOI 10.48550/ARXIV.1411.4038)

[19] Tammina, S. (2019). Transfer learning using VGG-16 with Deep Convolutional Neural Network for Classifying Images. International Journal of Scientific and Research Publications (IJSRP), 9(10). (DOI 10.29322/ijsrp.9.10.2019.p9420)

[20] He, K., Zhang, X., Ren, S., & Sun, J. (2015). Deep residual learning for image recognition. arXiv. (DOI 10.48550/ARXIV.1512.03385)

[21] Simonyan, K., & Zisserman, A. (2014). Very deep convolutional networks for large-scale image recognition. arXiv. (DOI 10.48550/ARXIV.1409.1556)

[22] Chen, L.-C., Zhu, Y., Papandreou, G., Schroff, F., & Adam, H. (2018). Encoder-decoder with atrous separable convolution for semantic image segmentation. arXiv. (DOI 10.48550/ARXIV.1802.02611)

[23] Kaushik, A. (2020). Understanding ResNet50 architecture. OpenGenus IQ: Computing Expertise & Legacy. https://iq.opengenus.org/resnet50-architecture/

[24] Omlor, M., Racu, A.-M., Hensel, J., Dilger, K. (2022) "Quality analysis of stripped wire contacts in the production of hairpin windings for electric drives" 31st International Conference on electrical Contacts.

[25] Omlor, M., Seitz, N., Butzmann, T. et al. (2023) Quality characteristics and analysis of input parameters on laser beam welding of hairpin windings in electric drives. Weld World (2023). (DOI 10.1007/s40194-023-01500-y)

IOP Conf. Series: Materials Science and Engineering 1296 (2023) 012008 doi:10.1088/1757-899X/1296/1/012008

In-situ monitoring for PBF-LB/M processes: Does multispectral optical tomography add value in recognizing process deviations?

Tina Becker[1], Philipp P Breese[1], Christian Metz[2] and Simon J Altenburg[1]

[1] Bundesanstalt für Materialforschung und -prüfung (BAM),
Unter den Eichen 87, 12205 Berlin, Germany
[2] ThetaScan GmbH, Thyssenstraße 183a, 46535 Dinslaken, Germany

tina.becker@bam.de

Abstract. Laser powder bed fusion of metallic components (PBF-LB/M) is gaining acceptance in industry. However, the high costs and lengthy qualification processes required for printed components create the need for more effective in-situ monitoring and testing methods. This article proposes multispectral Optical Tomography (OT) as a new approach for monitoring the PBF-LB/M process. Compared to other methods, OT is a low-cost process monitoring method that uses long-time exposure imaging to observe the build process. However, it lacks time resolution compared to expensive thermographic sensor systems. Monochromatic OT (1C-OT) is already commercially available and observes the building process layer-wise using a single wavelength window in the NIR range. Multispectral OT (nC-OT) utilizes a similar setup but can measure multiple wavelength ranges per location simultaneously. By comparing the classical 1C-OT and nC-OT approaches, this article examines the advantages of nC-OT (two channel OT and RGB-OT) in reducing the false positive rate for process deviations and approximating maximum temperatures for a better comparison between different build processes and materials. This could ultimately reduce costs and time for part qualification. The main goal of this contribution is to assess the advantages of nC-OT compared to 1C-OT for in-situ process monitoring of PBF-LB/M.

1. Introduction

The process of Laser Powder Bed Fusion of metallic components (PBF-LB/M) is an Additive Manufacturing (AM) technique, which utilizes a high-powered laser to melt metal powder across a powder bed. It enables the production of components with a high degree of design freedom directly from a CAD file. Despite its growing applications and usage of AM technology in industry, there is still a significant lack of understanding of the fundamental processes involved. Especially for applications in highly regulated industries, e.g., aerospace, cost and time extensive post-process non-destructive and destructive testing is needed to ensure component quality. Reliable monitoring systems could enhance the component reliability and save processing time and costs.

The substantial attention towards in-process monitoring for AM is shown in the high number of publications [1]. McCann et al. [2], AbouelNour and Gupta [3] and Grasso et al. [1] provide a good overview of the current state of research on the topic.

The occurring defects in the PBF-LB/M process can be divided into the categories: porosity, cracking, delamination, balling, geometric defects, surface defects, microstructural inhomogeneities and impurities [4]. Most of the defects are therefore caused by the extreme thermal gradients in the process. Due to this thermal character of the PBF-LB/M process, the thermal radiation emitted by the printed surface, mainly by the melt pool, is a good indicator for the process stability. It can be monitored by camera systems or photo diodes positioned outside the process chamber (off-axis) or integrated in the laser optical path (on-axis).

A promising on-axis monitoring approach to detect the surface temperature is the quotient pyrometry. Already in 2010 Pavlov et al. [5] could show that induced changes of the process parameters, e.g.

laser power and layer thickness, can be monitored reliably by a setup of two photo diodes equipped with different spectral filters in the scanning laser path. Very detailed shape and temperature information of the melt pool could be captured by Paul A. Hooper [6]. He presented an elaborated quotient pyrometric setup consisting of two on-axis positioned high speed cameras and equipped with spectral filters. The maximum occurring temperatures in the PBF-LB/M process could be shown to be significantly higher than the ambient boiling temperature of the used material titanium. Furumoto et al. [7] monitored a one-layer PBF-LB/M process by means of one high-speed camera equipped with a two-color radiometric thermal imaging system. They also could determine temperatures well above the ambient boiling temperature of the used powder material 18NI maraging steel.

On-axis quotient pyrometry delivers a good insight into the thermal PBF-LB/M process. However, expensive equipment and on-axis access to the PBF-LB/M machine is needed. In addition, a high amount of data of up to 40 Gbits/s [6] needs to be handled.

Optical Tomography (OT) is an off-axis thermographic approach using relatively simple equipment. Here, highly spatially resolved layer-images of the manufacturing process are taken by an off-axis positioned camera operating in the near infrared spectrum (NIR). This approach, described by Bamberg et al. [8], Zenzinger et al. [9] and Ladewig [10] enables a process monitoring with comparably low data volume. Compared to elaborated quotient pyrometry, information in the time domain is lacking. Guerra et al. could show that OT data can be used for detection of distortion of thin walls and lattice structures [11, 12]. Also lack-of-fusion defects could be correlated to OT data. [8] However, the information value of the greyscale images of the OT is limited since process inherent factors influence the measurement result as well as real process deviations [13, 14].

In this paper, the authors present a novel approach for process monitoring on basis of maximum temperature measurement: the multispectral Optical Tomography (nC-OT). The nC-OT is a method that combines the benefits of spatially resolved optical tomography with quotient pyrometry for surface temperature measurement in the PBF-LB/M process. Two different approaches were realized and tested on the in-house developed PBF-LB/M machine "SAMMIE": the two-channel optical tomography (2C-OT) and the RGB optical tomography (RGB-OT). This is the first time that an actual physical quantity, the maximum temperature, was deduced from data of a low-cost one camera OT system.

2. Approach

For each layer of the construction process (duration t_{layer}), one OT image is generated for each camera. To this end, the camera is in long exposure mode and several camera images of an exposure time $t_{exp} < t_{layer}$ are added together to form a single layer image, since t_{exp} is usually limited. Thus, the time history of the melting and cooling processes is lost, but the resulting amount of data is greatly reduced. Each resulting pixel value of a single layer image represents the sum of all emitted radiation at that surface point captured at the sensor area A of the camera at the viewing angle α.

The specific spectral radiance $M_\lambda(\lambda, T)$ of the melt pool can be calculated according to Planck's law as the product of the specific spectral radiance of an ideal black body M_λ^0 with the spectral angular emissivity $\varepsilon(\lambda, T)$ of the melt pool, see equation 1.

$$M_\lambda(\lambda, T) \cdot dA\, d\lambda = \varepsilon(\lambda, T) \cdot M_\lambda^0(\lambda, T) \cdot dA\, d\lambda \tag{1}$$

The signal intensity $I(T)$ of a monitoring camera can be calculated as a function of the temperature by integrating $M_\lambda(\lambda, T)$. Also, the spectral transmittance $\tau(\lambda)$ of the optical setup, e.g., lenses and filters, the spectral sensitivity $S(\lambda)$ of the camera detector and a device specific correction factor c need to be considered. The real temperature curve of a surface point shows large temperature

IOP Conf. Series: Materials Science and Engineering 1296 (2023) 012008 doi:10.1088/1757-899X/1296/1/012008

gradients but can be assumed to be $T(t) = T_{max} \cdot \Delta t - t_{max}$ in a first approximation. Equation 2 shows this correlation for one camera sensor.

$$I(T) = c \int \int \varepsilon(\lambda, T)\, M_\lambda^0(\lambda, T(t)) \cdot \tau(\lambda) \cdot S(\lambda) \cdot dt\, d\lambda \quad (2)$$

Thus, the temperature of the molten pool can theoretically be determined by using the numerical inversion of equation 2. However, since the course of time of the appearing temperatures and exact emissivity of the molten pool in the PBF-LB/M process is unknown and prone to unpredictable influences such as the oxidation-level of the surface and keyhole depression, this approach is unfeasible.

Here, as a first approximation, the monitored melt pool in the PBF-LB/M process is regarded as gray body. That means, the influence of the emissivity ε is not completely neglected but is considered as wavelength independent. Having simplified this, we decided to go for the quotient pyrometric approach, by using at least two different wavelength ranges for the process monitoring. Considering Stefan-Boltzmann's law, which states the dependence of the emitted radiant power on the temperature to the power of four, we can assume that the maximum occurring temperature T_{max} has by far the greatest influence on the recorded grey value. This is particularly true in the visible (VIS) and near infrared (NIR) spectral regions, used in the tested setups. Combining the quotient pyrometric approach with the gray body assumption and assuming the main influence by the maximum temperature T_{max} occurring, the ratio of two camera signal intensities can be calculated by equation 3.

$$I(T) = c\,\frac{\varepsilon(T_{max}) \cdot \int M_\lambda^0(\lambda, T_{max}) \cdot \tau_1(\lambda) \cdot S_1(\lambda)\, d\lambda}{\varepsilon(T_{max}) \cdot \int M_\lambda^0(\lambda, T_{max}) \cdot \tau_2(\lambda) \cdot S_2(\lambda)\, d\lambda} \quad (3)$$

This formula can be used to calculate the temperature of the melt pool using curve fitting via look-up-table, without knowledge of the emissivity. Having n different cameras/channels, $\binom{n}{2}$ intensity ratios can be determined in the same way for one surface point. By including several measurement values in the determination of the maximum temperature, measurement inaccuracies in individual channels can be better compensated. The reliability of the measurement can be improved.

3. Material and method

3.1. *RGB Optical Tomography*

The introduced RGB Optical Tomography (RGB-OT) setup is shown in figure 1a. A camera equipped with an RGGB Bayer filter, with a resolution of 24.4 megapixels and a pixel size on chip of 2.74 µm x 2.74 µm was used. Due to Planck's Law, a significantly higher proportion of spectral radiance can be expected at longer wavelengths. Therefore, the camera was equipped with a color balancing filter, reducing the intensity towards longer wavelengths. To further adjust the incident light, a neutral density filter was used. The setup was focused on the top-layer of the AM build process with a lens. To protect the sensor from the laser light (1070 nm), an additional low-pass filter with a cut-off wavelength of 750 nm was used.

The exposure time t_{exp} for the long-time exposure images was optimized and set to $t_{exp} = 3$ s to avoid saturation of pixels with unusually high dark current. The red, blue and one green channel of the camera were used for data analysis. The matching of the three datasets were achieved by simple superimposition. To reduce the optical distortion of the final dataset, a perspective transformation was performed.

3.2. Two channel Optical Tomography

For the two channel Optical Tomography (2C-OT) setup presented in this paper, two identical monochromatic CMOS cameras (C1 and C2 in figure 1c) with a resolution of 24.4 megapixels and a pixel size of 2.74 µm x 2.74 µm were used. These were each equipped with narrow band-pass filters in the NIR range (around 700 and 905 nm), and the fields of view were matched using a dichromatic long-pass beam splitter. To avoid ghost images induced by the beam splitter and to enable the option to insert and remove different optical filters without inducing focus shifts, an optical setup with parallel beam path was designed using a set of plano-convex lenses.

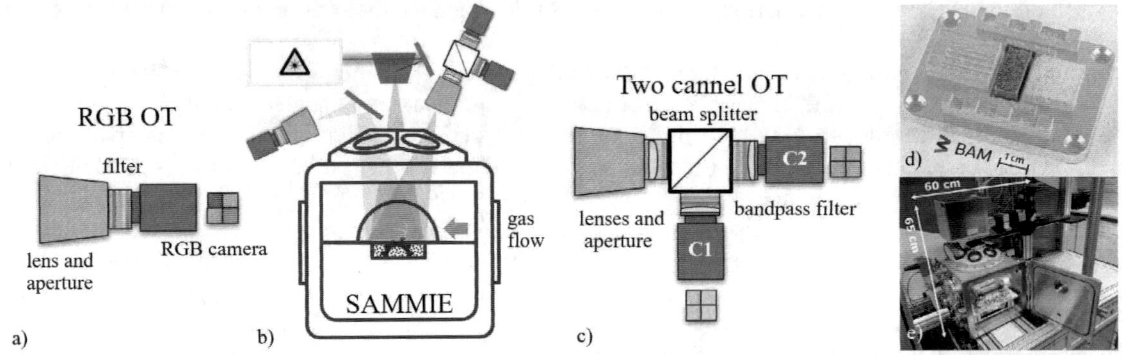

Figure 1. Setups used for the a) RGB-OT and c) 2C-OT presented, b) arrangement for the initial testing and e) image of the in-house developed PBF-LB/M machine "SAMMIE", d) monitored cuboid on the build plate

Similar to the RGB-OT setup, the exposure time t_{exp} for the long-time exposure images was optimized and set to $t_{exp} = 3$ s to avoid saturation of pixels with unusually high dark current. The two cameras were focused independently on the top-layer of the AM build process, accounting for the different optical focal lengths of the two different wavelength ranges. To further adjust the intensity of the incident light for both cameras, an aperture and several neutral density filters were used. The principle of the 2C-OT setup is illustrated in figure 1c. The generated datasets of both cameras were matched by affine perspective transformation.

The setup presented in this paper is a further development of a setup already described in [15] which has already been tested on a commercial PBF-LB/M system.

3.3. Calibration setup and calculation of the look-up table

Bias, dark, and flat frames corrections were performed for all images to minimize the influence of dark noise, readout noise and image errors such as vignetting [16]. The flat frames were created using an isotropic sphere (ED-ISS-100-MD, Gigahertz-Optik GmbH) and a calibrated halogen lamp (ISS-30-VA-V01, Gigahertz-Optik GmbH) with variable intensity setting, maintaining the colour temperature. The linearity of the used cameras was checked by means of the same setup.

The spectral transmittance $\tau(\lambda)$ of the used filters (bandpass, lowpass, temperature balancing filter, process windows) and lenses were measured by use of a spectrometer (OCEAN-FX-VIS-NIR-ES, Ocean Insight Inc.) in combination with the calibrated halogen lamp.

To correlate the measurement results of each setup to temperature values, the isotropic sphere with the calibrated halogen lamp was used as reference light source. The light source has a known spectral specific radiance $M_\lambda(\lambda)$ and a correlated color temperature $T_{cal} = 3190$ K in the visible range between 360 nm to 750 nm, when operated using the calibrated power settings. To have multiple supporting points, the light source was also operated with lower powers, resulting in lower color temperatures. By curve fitting of the spectral specific radiances $M_\lambda(\lambda)$, the corresponding color temperatures T_{color} were determined. These measurements provided the basis for the correction of the theoretically calculated correlations between measurement signals and temperature values.

3.4. Process setup

To test and compare both sensor setups at real process conditions, they were attached to the in-house developed sensor testing PBF-LB/M machine "SAMMIE", illustrated in figure 1b and shown in 1e. SAMMIE is equipped with an Ytterbium fiber laser with a wavelength λ_L of 1070 nm and maximum laser power P of 500 W, and has a build volume of 50 mm x 70 mm x 35 mm. The shielding gas flow direction of SAMMIE is marked with a blue arrow in figure 1b. Currently, four off-axis process windows and an on-axis access enable a simultaneous monitoring of the PBF-LB/M process with different optical sensor systems.

Both setups were connected to optical windows with a viewing angle of 22° to the build surface normal. The field of view of the RGB-OT was additionally guided by a silver mirror. For the 2C-OT, a sapphire process window was used, the RGB-OT was connected via a BK7 process window.

For the evaluation of the OT setups, the scanning process of a cuboid 30 mm x 8.5 mm x 25 mm sample, as shown in the middle of the build plate in figure 1d, was monitored. The component was built of 316L stainless steel and contains two sets of 12 cylindrical areas with diameters of d_{void} = 1.5 mm to 0.4 mm that were not illuminated to produce powder filled cavities as artificial defects. One set was built as a complete cylinder, on the other set the covering top was closed in a dome shape. The layer number 10 of the cuboid was scanned twice, to simulate double exposure scanning. Other components were printed in the same build job but are not focused on this paper.

A laser power P_L of 167 W, scanning speed v of 0.7 m/s, hatching distance l of 100 µm and layer thickness h of 40 µm were used as process parameters for the component. The laser spot diameter d_L was set to 80 µm. The monitored process was performed under argon atmosphere with a flow velocity of about v_{flow} = 3.5 m/s.

4. Results and Discussion

The generated look-up tables for the RGB-OT and 2C-OT are shown in figure 2a and 2b. Both figures show the correlation between measured surface temperature and detected intensity signal ratios. The continuous blue lines show the theoretically determined measured values. These were determined using Planck's law based on the measured transmission spectra of the optical elements and the sensitivity curves of the cameras.

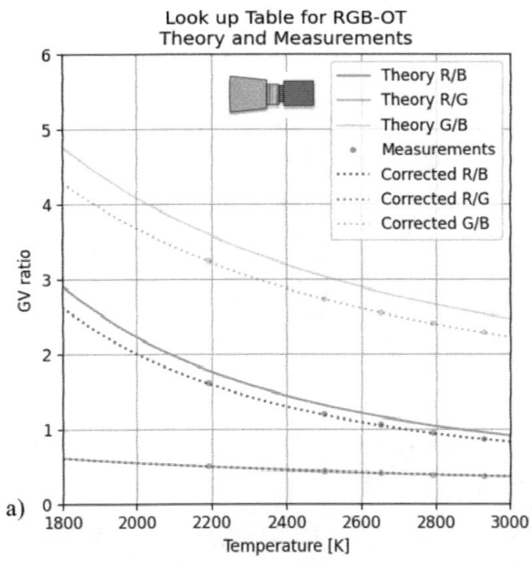

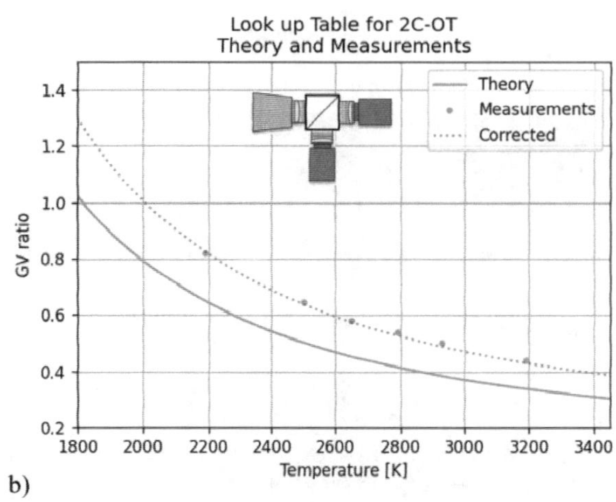

Figure 2. Look-up tables for the a) RGB-OT and the b) 2C-OT, both theoretical and corrected curves

The data points show the measured values obtained by means of the isotropic sphere with calibrated halogen lamp for different color temperatures. The purple dotted lines represent the final used corrected curves of the look-up table after the determination of the specific correction factors. The corrected curve for the 2C-OT shows a good consistency with the measured values with maximum residuals e of $e_{2C} = \pm 0.013$. The RGB-OT corrected curves also show good agreement with residuals of $e_{R/B} = \pm 0.01$, $e_{R/G} = + 0.003$ and $e_{G/B} = \pm 0.005$.

The deviation between the theoretical curve and the measured values of both RGB-OT and 2C-OT can be explained by the partly significant deviation D of the spectral composition of the radiation emitted by the calibrated lamp from the Planck spectrum, e.g. $D_{900nm} = 9$ % for the $T_{color} = 2290$ K. This deviation adds to the temperature measurement uncertainty and needs to be further investigated. However, due to the thermal calibration of the measuring systems, the measurement results can be considered correct.

A blackbody radiator as calibration source might give more reliable results. However, the emitted spectrum of many commercial blackbody radiators is only calibrated starting at about 800 nm and deviates significantly from the Planck spectrum in visible light range.

Figure 3 shows the measurement results of the two setups introduced in this paper and the underlying image data of the individual cameras/channels for an exemplary layer. The shown data was recorded during the production of the layer directly above the cylindrical cavities, i.e., the first completely exposed layer. The positions of the over-welds are marked on the sketch of the cuboid in figure 3c. The scanning lines of the laser are clearly visible in all images due to the high image resolution on the printing layer of 20 μm/pixel for the 2C-OT and 17 μm/pixel for the RGB-OT.

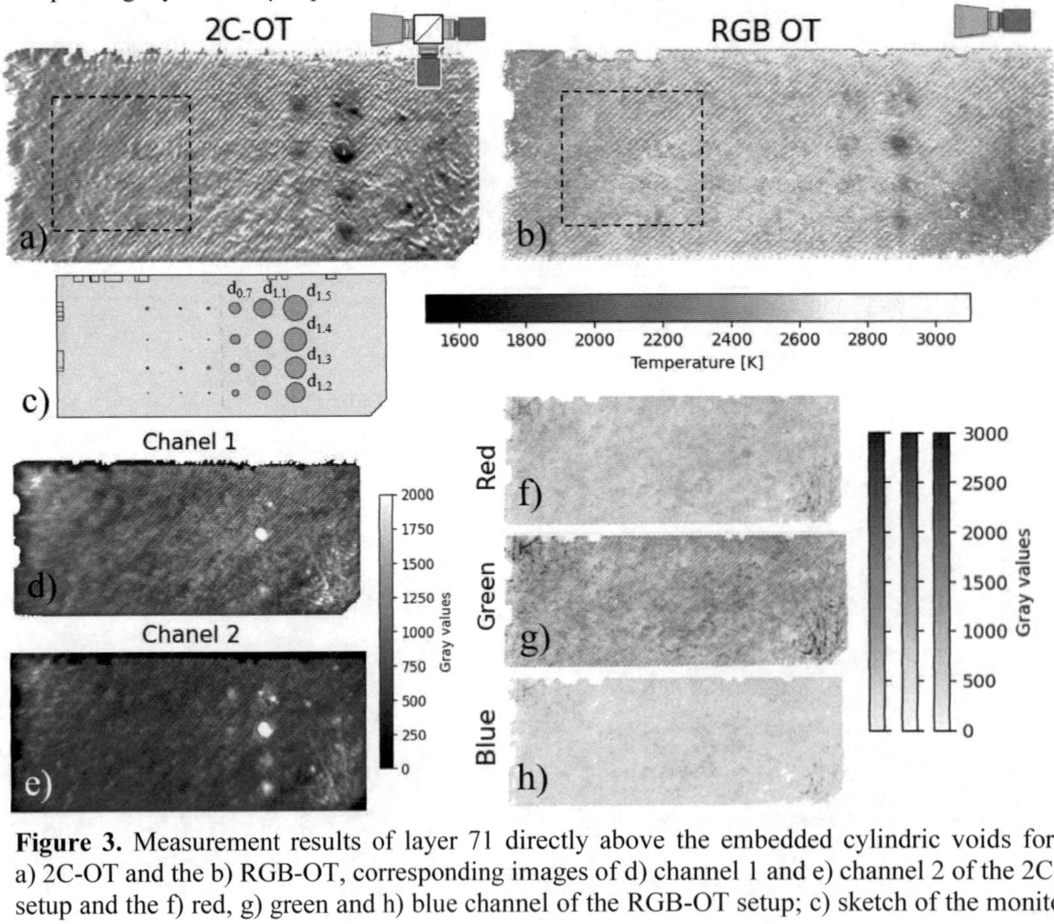

Figure 3. Measurement results of layer 71 directly above the embedded cylindric voids for the a) 2C-OT and the b) RGB-OT, corresponding images of d) channel 1 and e) channel 2 of the 2C-OT setup and the f) red, g) green and h) blue channel of the RGB-OT setup; c) sketch of the monitored cuboid

A comparison of the maximum temperature images of the two methods (figure 3a and 3b) shows a qualitatively good agreement. The averaged maximum temperature determined by the two methods are shown in table 1. The areas that were evaluated for this purpose are marked with a dotted contour in both images.

Table 1. Averaged maximum temperature determined by the RGB-OT and 2C-OT for the presented layer 91.

	Mean maximum temperature [K]	*2σ [K]*
RGB-OT	2550	370
2C-OT	2417	360

The maximum temperature measured for both systems is significantly above the melting temperature of the printed material 316L of about T_S= 1650 K but below the boiling temperature of about T_B = 2770 - 3086 K [17,18, 19]. The high values of the deviations of the measured maximum temperature of more than ± 370 K are expected due to the gaussian profile of the scanning laser, resulting in significantly higher energy input in the center of the scanned vector compared to the edges.

By use of a three-dimensional laser-keyhole welding model, Ki et al. [20] predict significantly higher maximum temperatures of 3600 - 4000 K, 500 - 900 K above the boiling temperature of the analyzed material iron in key-hole welding mode. Khairallah et al. [18] also suggest higher maximum temperatures of about 3000 K for a BPF-LB/M process of 316L using a fine-scale 3D simulation model. Ansari et al. [19] predicted temperatures of about 2800 K for a 316L PBF-LB/M process with similar process parameters using a Multiphysics simulation.

Using two-color pyrometric measurements, Pavlov el. Al. [5] determined surface temperatures in the range of 1173 – 2873 K for an PBF-LB/M IN625 process and Furumotoa et. Al. [7] reported surface temperatures between 1793 – 2083 K for an 18Ni stainless steel process. Pavlov et at. Used significant lower volumetric energy density of E_v = 44 J/mm³ compared to the E_v = 59 J/mm³ used in this paper. No detailed information is given for the process parameter experiment of Furumotoa et al.

To get an even deeper insight into the thermal history of the melt pool, more elaborate equipment, such as hyperspectral imaging, can be used. A detailed spatial temperature distribution and peak temperatures of over 2000 K for punctual laser welding of 316L stainless steel could be shown by Devesse et al. [21]. Similar temperature ranges could be shown for laser cladding of the same material by Lison et al. [22].

Thus, very different values for the maximum temperatures occurring in the PBF-LB/M process can be found in the literature. The authors of these papers agree that the maximum temperatures are highly dependent on the process parameters and the material used. The mean maximum temperature values determined with the setups presented here are very plausible and in the expected range. Still, further validation experiments are needed.

In figure 3a, the temperature image of the 2C-OT, most of the over-welded areas can be seen with the naked eye. In figure 3b, the temperature image of the RGB-OT, some of the over welds are visible as well. The over welds with diameters smaller than d = 0.9 mm, as well as the cavities with dome-shaped roofs cannot be distinguished. In the raw image data of the channel 2 (2C-OT), the image captured at the highest wavelength range, the over scanned areas are clearly visible down to a diameter of 0.9 mm. In all other raw images, just the defect with d = 1.4 mm diameter is clearly discernible. Consequently, for a monochromatic OT, the choice of the wavelength range in the upper NIR range is advantageous to observe these effects.

The mean temperature values of the two presented systems 2C-OT and RGB-OT over the height of the printed component are shown in figure 4b as well as the deviation between the two systems. The evaluated region of interest (ROI) on the part, a center positioned rectangle of 50 x 50 pixels is

IOP Conf. Series: Materials Science and Engineering 1296 (2023) 012008 doi:10.1088/1757-899X/1296/1/012008

marked in figure 4a. Correspondingly, the grey values of the underlying image data of the individual cameras/channels are shown in figure 4c and 4d. The temperatures detected by the RGB-OT are almost continuously higher than those of the 2C-OT. This small offset needs to be further investigated.

The fluctuations between the values of the different layers can probably be explained by the changing scanning angle with regard to the shielding gas flow. The grey values of the individual channels shown in Figure 4 have a variation of $2\sigma = \pm\ 38 - 48\ \%$. In contrast, the fluctuations of the presented nC-OT systems are only $2\sigma_{2C} = \pm\ 7.8\ \%$ and $2\sigma_{2C} = \pm\ 7.2\ \%$. Thus, the influence of the shielding gas flow direction is greatly reduced using nC-OT. This phenomenon and the influence of the shielding gas flow on the PBF-LB/M process has already been described for the 1C-OT [23] but needs to be investigated further for all OT approaches. The origin of the variations is expected in the different absorption of thermal radiation by the welding plume, which depends on the relative orientation between plume direction (gas flow direction) and the scanning direction. Thus, a reduction of this effect is expected using ratio-based analysis, as done for the nC-OT

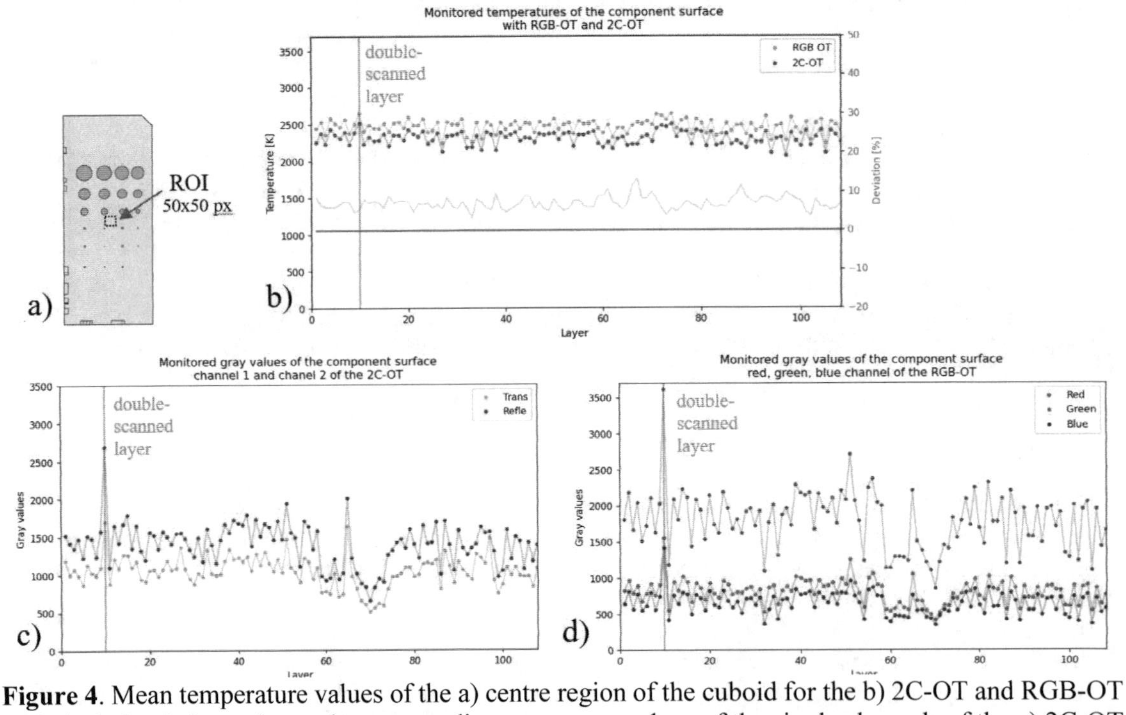

Figure 4. Mean temperature values of the a) centre region of the cuboid for the b) 2C-OT and RGB-OT with deviation information and corresponding mean grey values of the single channels of the c) 2C-OT and the d) RGB-OT

The red line indicates the layer with double exposure in figure 4b-d. The two presented systems did not record any excessive temperatures for the double-scanned layer ten. All single channels, on the other hand, show significantly higher values. This shows the ability of the two multispectral systems to distinguish between double-scanned areas and increased temperatures. This is especially advantageous, since local temperature deviations are expected to cause defects, while double exposures are frequently used, e.g., for contour scans, to improve the surface quality in standard PBF-LB/M processes.

The grey values of the individual channels also show deviating values in layers 60 to 70, the layers with the cavities, even though no changed scanning regime was applied at the ROI. In contrast, the temperature values of the two systems presented do not show any different behavior. It can be assumed that the systems presented in this paper represent the actual temperature distributions better than single-channel systems.

5. Conclusion and Outlook

Two approaches for multispectral optical tomography - two-channel OT and RGB-OT - were introduced and first results of a feasibility study were presented. The spatially resolved temperature OT images obtained with the presented setups still need to be validated by comparative experiments, e.g., with thermocouples or other IR imaging methods. However, good agreement with literature values indicates that the results are very plausible.

By means of the two-channel OT, the two used narrow wavelength ranges can be optimally selected for the PBF-LB/M process. Since both camera sensors are used independently of each other, maximum spatial resolution can be achieved. However, the optical design is challenging and an image registration between the two channels has to be performed for data analysis. On the other hand, the RGB-OT setup is much simpler and more robust, off-the-shelf optics can be used and the results are still very promising.

It has also been shown that the two nC-OT systems are robust against apparent overheating in areas of double exposure, such as additional contour scans or as used for better attachment of components to the building platform. This is a common problem with 1C-OT systems. A higher robustness against false detections due to process radiation from other areas of the part was also demonstrated. Validation of these results, i.e., correlation with actual defects, is planned by means of reference CT measurements. The offset between theoretically determined curves and measurements that occurred during calibration, as well as the offsets between the two systems presented, will be investigated in more detail.

Further work will also investigate the influence of scan angle and shielding gas flow, by-products above the melt pool such as smoke and plasma plume, on the detected signal and component quality. With a better understanding and consideration of the occurring emissivity and the temporal shape of the temperature development during heating and cooling of the melt pool, we aim to improve the measurement results of our systems even more.

Since the PBF-LB/M process is already widely used in industry, economic efficiency and an easy implementation into existing industrial machines is very important. A reduction of the initial and operating costs, e.g., for the data storage, move even more into focus. Thus, the promising results of the presented systems show a great potential for future industrial applications.

Acknowledgment

Subsidized by the initiative QI-Digital.

Many thanks for the support in setting up our PBF-LB/M machine SAMMIE to Simon Oster, Jens Barow, Mathias Röllig, Jolanda Schumann and Tsuyoshi Shimada.

References

[1] Grasso M, Remani A, Dickins A, Colosimo B M and Leach R K 2021 *Meas. Sci. Technol.* **32** https://doi.org/10.1088/1361-6501/ac0b6b
[2] McCann R et al. 2021 *Additve Manufacturing* **45** doi.org/10.1016/j.addma.2021.102058
[3] AbouelNour Y and Gupta N 2022 *Materials & Design* **222** doi.org/10.1016/j.matdes.2022.111063
[4] Grasso M and Colosimo B M 2017 *Meas. Sci. Technol.* **18** doi.org/10.1088/1361-6501/aa5c4f
[5] Pavlov M, Doubenskaia M and Smurov I 2010 *Physics Procedia* **5** doi.org/10.1016/j.phpro.2010.08.080
[6] Hooper P A 2018 *Additive Manufacturing* **22** doi.org/10.1016/j.addma.2018.05.032
[7] Furumoto T, Oishi K, Abe S, Tsubouchi K, Yamaguchi M and Clare A T 2022 *Journal of Materials Processing Technology* **299** doi.org/10.1016/j.jmatprotec.2021.117384

[8] Bamberg J, Dusel K-H and Satzger W 2015 *AIP Conference Proceedings* **1650** doi.org/10.1063/1.4914605
[9] Zenzinger G, Bamberg J, Ladewig A, Hess T, Henkel B and Satzger W 2015 *AIP Conference Proceedings* **1650** doi.org/10.1063/1.4914606
[10] Ladewig A 2019 Optische Tomographie *PhD Thesis* Karlsruher Institut für Technologie
[11] Guerra M G, Lafirenza M, Errico V and Angelastro A 2023 *Optics & Laser Technology* **162** doi.org/10.1016/j.optlastec.2023.109252
[12] Guerra M G, Errico V, Fusco A, Lavecchia F, Campanelli S L and Galantucci L M 2022 *Additive Manufacturing* **55** doi.org/10.1016/j.addma.2022.102850
[13] Mohr G, Altenburg J S, Ulbricht A, Heinrich P, Baum D, Maierhofer C and Hilgenberg K 2020 *Metals* **10** doi.org/10.3390/met10010103
[14] Ulbricht A, Mohr G, Altenburg J S, Oster S, Maierhofer C and Bruno G 2021 *Metals* **11** doi.org/10.3390/met11071012
[15] Becker T, Altenburg S J, Scheuschner N, Breese P P, Metz C, Hilgenberg K and Maierhofer C 2022 *Procedia CIRP* **111** doi.org/10.1016/j.procir.2022.08.035
[16] Tomsik H Kalibrierung mittels Bias-, Dark- und Flatframe 2011 https://astrofotografie.fg-vds.de/grundsatzartikel.php
[17] Fischer U, Gomeringer R, Heinzler M, Kilgus R, Näher F, Oesterle S, Paetzold H, Stephan A 2011 Tabellenbuch Metall (Haan-Gruiten: Verlag Europa Lehrmittel)
[18] Khairallah S A, Anderson A T, Rubenchik A, King W E 2016 *Acta Materialia* **108** dx.doi.org/10.1016/j.actamat.2016.02.014
[19] Ansari P, Rehman A U, Pitir F, Veziroglu S, Mishra J K, Aktas O C, Salamci M U 2021 *Metals* **11** doi.org/10.3390/met11071076
[20] Ki H, Mazumder J, Mohanty P S 2002 *Metall. Mater. Trans. A* **33** doi.org/10.1007/s11661-002-0191-5.
[21] Devesse W, Baere D, Guillaume P 2017 *Sensors* **17** doi:10.3390/s17010091
[22] Lison M, Devesse W, Baere D, Hinderdael M, Guillaume P 2019 *Journal of Laser Applications* **31** doi: 10.2351/1.5096129
[23] Becker T 2020 Statistical analysis of optical monitoring data for layer wise defect detection in the LPBF process, *MSc Thesis* Technische Universität Berlin

IOP Conf. Series: Materials Science and Engineering 1296 (2023) 012009 doi:10.1088/1757-899X/1296/1/012009

Towards data driven quality monitoring: alignment and correlation of photodiode-based co-axial melt pool monitoring signals to part quality in laser powder bed fusion

Joni Reijonen, Alejandro Revuelta, Hari P.N. Nagarajan

VTT Technical Research Centre of Finland Ltd., Advanced Manufacturing Technologies, Espoo FI-02044, Finland

Correspondance to: joni.reijonen@vtt.fi

Abstract. Online quality control of advanced manufacturing processes often utilize process monitoring as a qualification method to reduce or eliminate time consuming and costly destructive and non-destructive tests. The need for process monitoring based qualification becomes more apparent with the advent of laser based additive manufacturing for complex geometries and lot-size-one production. However, the usability of the process monitoring solution heavily relies on the accuracy of the established alignment and assignment of process signals to specific part and defects herein. This research proposes a data fusion strategy for data pre-processing to structure and align multi-source sensor data for improving correlation accuracy in process monitoring. A laser powder bed fusion system equipped with a photodiode-based co-axial melt pool monitoring is used for the study. First, the monitoring system collects photon emission data from the melt pool using photodiodes with high frequency (>100 kHz) which are aligned spatially and temporally using position data (xy-coordinates) of the galvanometer scanner and the laser ON/OFF signals. Next, a clustering approach is used to assign each photodiode signal with the individual parts on the build platform. Finally, correlation between melt pool monitoring signals and part level porosity is studied using micro-computed X-ray tomography (µCT) from a build job containing parts produced under varying process conditions.

1. Introduction

Additive manufacturing (AM) technologies has created a paradigm shift in manufacturing promoting a gradual shift towards mass customization and lot size one production by leveraging its advantages over traditional manufacturing. AM supports design freedom enabling customizable and complex product realization, selective deposition strategies for better material utilization, elimination of tools and fixtures simplifying the product supply chain by reducing/eliminating their dependence on location, material distribution, inventory holding, and machine setup and changeovers, and related challenges, through just-in-time manufacturing and build-to-order strategies [1]. However, in individualized production scenarios, the qualification of the products/processes to ensure reliable and repeatable production quality becomes challenging. Thus, online process monitoring is pursued as a qualification method to meet the stringent functional and quality requirements of complex products. The need for process monitoring based qualification becomes more apparent with the advent of laser based additive manufacturing (AM) for complex geometries and lot-size-one production. Laser powder bed fusion (PBF-LB) is the most established metal AM process to produce critical components in the aerospace, nuclear, and other high value industries that require extensive quality control measures. Recently, melt pool monitoring (MPM) systems which record the thermal signatures of the process emissions have been identified as a potential method for online qualification by the manufacturing community, supported also by standardization organizations [2].

Many of the MPM systems today use photodiodes as the main sensor for recording thermal data, due to their higher temporal frequency and lower cost than for example CMOS cameras. Early examples of using photodiodes to monitor laser processing (e.g., laser welding), date back more than 30 years [3]. Since then, multiple studies and development on photodiode-based MPM of laser processing have been conducted and our underlying understanding of what the signal response at different spectral ranges could tell us about the process has improved [4]. By applying the same method for PBF-LB process, many key developments were made at KU Leuven around 10-15 years ago [5][6], along with the method of mapping the sensor signal to a 2D pixelated grid to generate an image for every build layer [7] to go from time-based domain to position-based domain. This approach was introduced as a commercial product by Concept Laser GmbH and similar approaches were taken by other PBF-LB OEMs on their respective MPM systems in the following years. Around the same time, SigmaLabs Inc also launched their own MPM system PrintRite3D that was OEM-independent and could be retrofitted on different brands. They also use the same approach of mapping the sensor data to 2D-images per layer on their software product, however the system also provides the possibility and access for the user to work with the raw data.

Several studies have been conducted in literature for correlating the MPM signals to the process parameters and part quality using sensors operating at various different wavelengths [8][9][10][11]. However, to rely on MPM data for quality control in a production environment, variations in the signal needs to be correlated to stochastic defects that are formed in the PBF-LB process even when operating in the optimal, stable process parameter regime. Such research has been more scarce, but the recent study by Snow et al. for correlating spatter-induced stochastic defects to optical tomography based process monitoring signals in PBF-LB shows promise that this is achievable at least to some extent [12]. It must be noted that many of the actual defects revealed by the CT scan were not captured by their method, although for especially large defects >300 μm the correlation between spatter-induced defect and a corresponding indication in the monitoring signal was strong.

The main advantage of the image based MPM approach is to provide qualitative and real-time visual output to the machine operator after each scanned layer. However, the image data lacks labelling and assignment of the sensor signal to individual parts on the data structure level, which limits the possibilities of using the data for machine learning and other advanced data analytics to do quantitative studies on the correlation between MPM signal and part quality. Another drawback is evident when using very high frequency sensors (>100 kHz), which is necessary to capture the transient variations in melt pool and/or vapor plume dynamics during the PBF-LB process due to the high scanning speed (usually around 500-2000 mm/s depending on material). Mapping this data onto a 2D-pixelated grid will result in compromise between fidelity and image size, as the resolution would need to increase exponentially so as to not omit or average out the data. Maintaining high spatial resolution is also needed to increase the accuracy of correlating the MPM signal to actual defects on the part. Thus, to overcome these limitations, this research presents an approach for MPM where we remain in the time-based raw signal data domain but utilize a clustering approach to assign the signal to each individual part on the building platform. Then we transform and align the MPM signal with the coordinate system of the actual part geometry by fusing the data collected from multiple sources (MPM, μCT, and 3D-CAD model). This method has two key advantages over the image-based approaches used in commercial state-of-the art systems. Firstly, the capabilities of using the MPM data for quantitative data analytics and machine learning is enhanced. Secondly, the whole extent of the achievable spatial resolution with the very high frequency photodiode (10 μm at 100 kHz sampling rate and 1000 mm/s scanning speed) can be utilized due to more efficient data storage format (.h5 hierarchical data format is used) than 8 or 16 bit images. This improves the spatial resolution at which MPM signal can be mapped in relation to part geometry and defect locations as explained in detail in the next section.

IOP Conf. Series: Materials Science and Engineering 1296 (2023) 012009 doi:10.1088/1757-899X/1296/1/012009

2. Methods

2.1. Melt pool monitoring setup

The MPM setup used in this research, consists of the SigmaLabs Inc. PrintRite3D melt pool monitoring sensor pack retrofitted on a SLM Solutions SLM 125 HL powder bed fusion system, as detailed in Figure 1a). The MPM system consists of three photodiodes, named as TED, TEP LOW and TEP HIGH. The sensor pack is integrated on the Scanlab Camera adapter (bs1) between the scanner and the laser focusing unit, guiding the process emissions towards a 90° mirror and from there on to the Thorlabs BPD254S-G Polka Dot beam splitter (bs2) that divides the emissions to the sensors. A wide range of wavelengths (450-900 nm limited by the bs1) are captured by the TED sensor (Thorlabs PDA36A2) to monitor the total thermal emission density from the process. The two TEP sensors (Thorlabs APD440A) have narrow band filters (636±5 for low and 650±5 for high) before them and together the ratio of the sensors (TEP HIGH/TEP LOW) can be used as a two-color pyrometer. The filters are changeable and should be selected so that they are appropriate in relation to the emission spectrum of the material being melted. Selecting the filter wavelengths close to one another allows for assuming the emissivity to be equal and hence the TEP ratio metric can be used as a proxy for changes in temperature as per Planck's spectral radiance law. No attempt is made here to convert sensor signals to absolute temperature, as we are interested here in changes (i.e., disturbances) in the process. The typical spectral curve for melting 316L material in the system (Figure 1b) was measured with Ocean Insight HR4PRO Spectrometer.

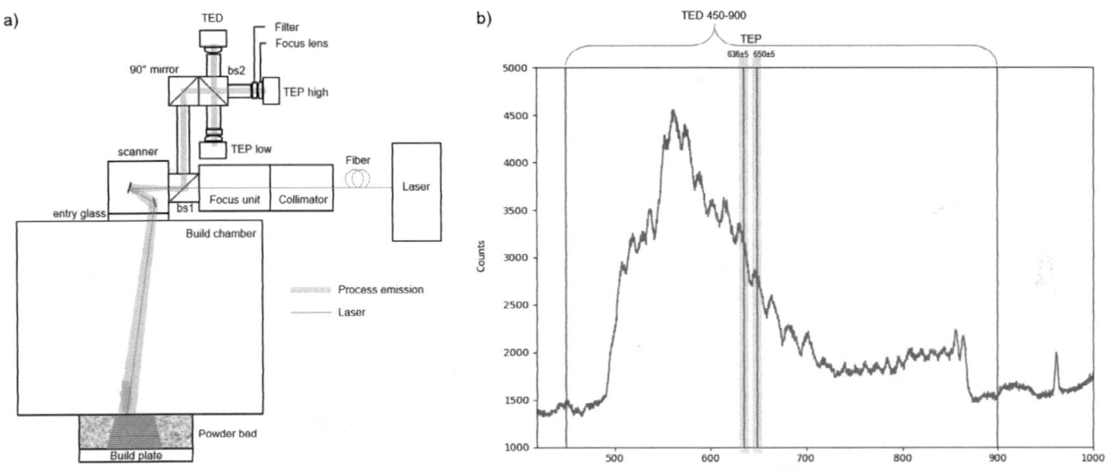

Figure 1. a) Schematic of melt pool monitoring setup at VTT and b) typical spectral curve when melting 316L stainless steel with the spectral ranges collected by TEP and TED sensors respectively.

2.2. PBF-LB experimental build description

SLM 125 HL PBF-LB system from SLM Solutions GmbH was used for the study. The optical configuration (seen in Figure 1a) consists of IPG YLR-400-SM fiber laser emitting at central wavelength of 1070 nm, followed by IPG fiber, IPG f=70, D=25 collimator, Scanlab Varioscan 40 dynamic focusing unit, Scanlab camera adapter and Scanlab Intelliscan 25 galvanometer scanner. This optical setup results in 80±6 μm focused spot size at the work plane with Gaussian intensity distribution. Argon 5.0 was used as the shielding gas, with the flow speed set to 7.0 m/s as measured in the 40 mm diameter inlet pipe, corresponding to a volumetric flow rate of ~528 l/min. Layer thickness was set at 30 μm, linear speed of re-coater was 95 mm/s and a soft (polymer composite) re-coater blade with a rounded tip profile was used. The parts were built directly onto a 316L base plate without supports and baseplate pre-heating. No post-process heat treatments were carried out and the parts were used in the as-built condition for defect analysis.

IOP Conf. Series: Materials Science and Engineering 1296 (2023) 012009 doi:10.1088/1757-899X/1296/1/012009

The experimental build layout is shown in Figure 2. The build included a total of 10 parts. The cylindrical parts 1 through 8 and part no. 10 were the experimental parts to be analyzed in the study while the rectangular part (number 9) was a dummy used to induce more spatters to the part number 1 located in its vicinity. Table 1 lists the process variables studied for each of the parts produced with PBF-LB. Cuts were made with a sharp knife to the rubber re-coater blade to simulate a damaged re-coater blade to affect specimens 5, 6 and 10. Gas flow was disturbed by placing a steel plate to block parts of the gas flow inlet nozzle to affect specimens 8,9 and 10 at the bottom of the build plate. The dashed lines in Figure 2 indicate the areas affected by these disturbances. Specimen 1 had excessive spattering induced to it by the nearby block (number 9). Specimen 2 was the nominal reference without any intentionally induced process disturbances. Specimen 3 had laser power decreased and specimen 4 increased by 100 W. The purpose of the process disturbances was to try and promote defect formation on those specimens affected.

The scanning strategy for all the specimens was stripes, with a stripe length of 10 mm with a 33° rotation between layers. The scanning sequency of parts proceeds from -x to +x (from left to right in Figure 2), against the shielding gas flow direction, with the purpose of scanning the down-stream first and then moving towards up-stream. The data collected from these experimental specimens were then used in the part defect analysis, the results of which is explained in Section 4.

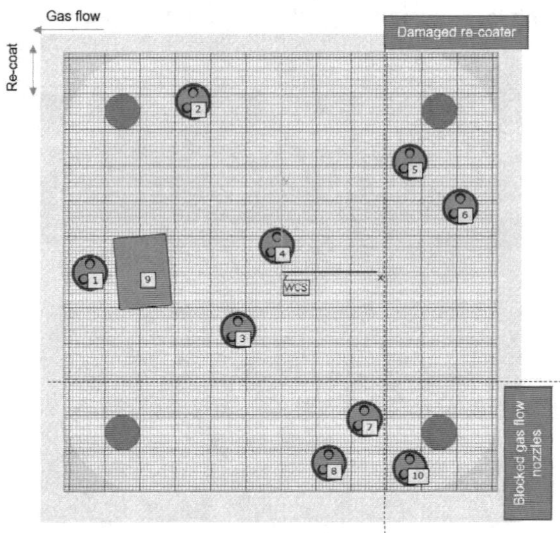

Figure 2. Layout of specimens on the building platform for the experiment. Red circles mark the non-build zones reserved for the platform attachment bolt holes.

Part no.	Power [W]	Induced process disturbance
1	200	increased spattering
2	200	nominal reference
3	100	decreased power
4	300	increased power
5	200	damaged re-coater
6	200	damaged re-coater
7	200	blocked gas flow
8	200	blocked gas flow
9	200	none (just dummy)
10	200	blocked gas flow and damaged re-coater
Scanning speed was 800 mm/s and hatch distance 0.12 mm for all parts		

Table 1. Laser parameters and induced process disturbances of each part.

2.3. MPM signal clustering and assignment to parts

The PrintRite3D MPM system used for this work has no a priori information about the number of parts included on a build job. As the build job proceeds, it collects the signal coming from the photodiodes, and the positional commands to the galvanometer scanner and the laser ON/OFF signal sent by the SLM machine simultaneously. Based on this information, it maps the time-based signal coming from the photodiodes into spatial coordinates and then provides aggregate information about the photodiode signal in the form of a picture with a spatial resolution of 100 um/px, with the value of the pixel being the average value of the photodiode signal at that location (~20 samples). Once this image is produced, the number of parts is calculated on a per layer basis. All further analysis on the PrintRite3D software product on the quality of the build job is based on this generated image. Although this approach is understandable in the context of real-time analysis of the build job, it presents a few drawbacks: The time-dependent information is lost in the conversion and the signal is converted to a down sampled version. In addition, when the geometry of the part is complex, e.g., with several branching features, the approach can lead to incorrectly identifying one part as several different ones in some layers.

In order to preserve as much information as possible, the authors have taken a different approach wherein, the identification of parts is done directly on the raw time-based signal, clustering every sample into a part. This is done in two stages. First, all adjacent non-zero samples belonging to a layer are grouped together and clustered into parts based on location (Figure 3). This signal is mapped into a 2D image for each layer, and then a flood fill algorithm is applied to the whole stack of images to ensure continuity in the part numbering across the different layers.

After the MPM signal is assigned to part numbers and aligned to the nominal 3D-CAD space coordinates, the next step is identifying and extracting defects and their locations from the μCT data and transforming it to the same coordinate system, as explained in the following sections.

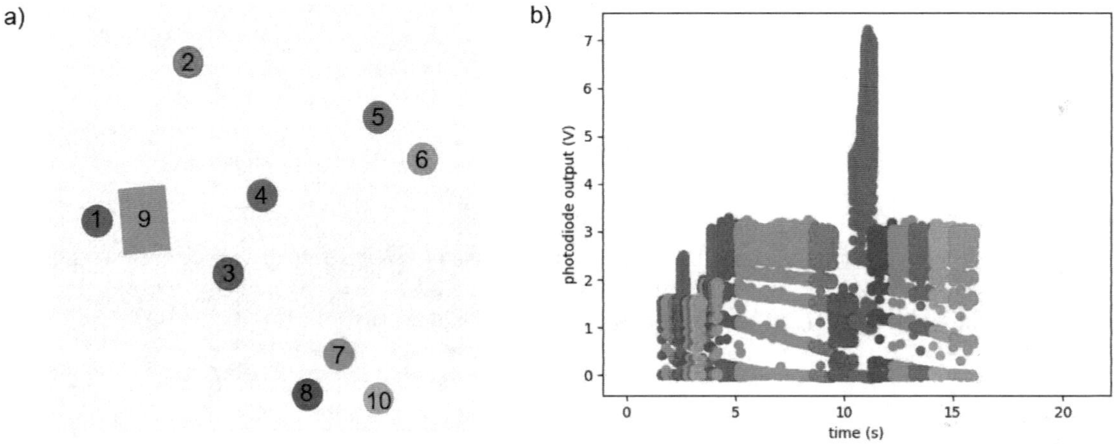

Figure 3. Photodiode signal clustering and assignment to part number shown for one layer TEP signal a) in xy-coordinates and b) on the time-based raw signal, visualized with different color for each part.

2.4. μCT scanning

GE Phoenix v|tome|x s μCT system with a 240 kV direct microfocus tube was used to scan the parts for defects. The output is an 8-bit image stack, with the achieved voxel resolution of 12 μm for the studied parts. The part design with main dimensions are shown in Figure 4. The specimen design included features such as the internal holes and the cones at the top of the specimen to allow for exact determination of the part orientation from the CT-scan data and relate it to the coordinate system of the xy-scanner to allow for mapping the defect locations to the MPM signal location. The two cones and the internal holes are aligned such that, their central axis converge at the rotation axis of the cylinder at a known point in space above the specimen surface. The volume between the bottom and top spheres was used for analyzing defect distributions with μCT (⌀10 mm cylindrical volume having height of 12 mm).

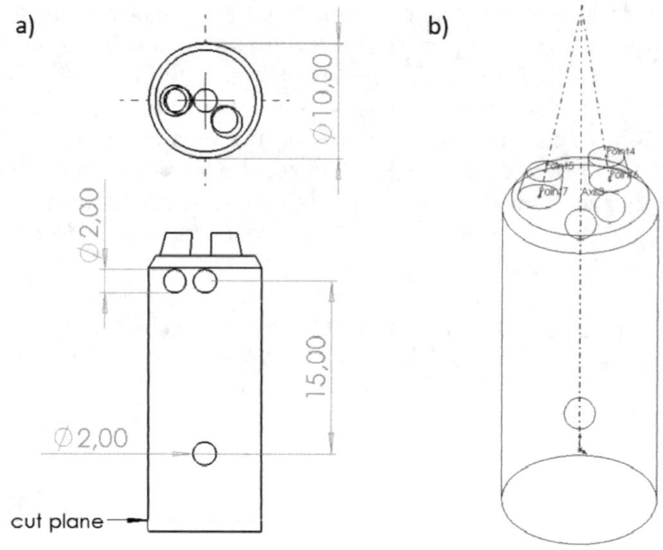

Figure 4. a) Part dimensions and b) design features of the μCT specimens.

2.5. μCT data processing and alignment

In order to calculate the defects on the μCT scan data, the Otsu [13] algorithm is applied for thresholding between solid and void, and all defects inside the cylinder body are calculated using a 3D flood fill algorithm. Due to the nature of the PBF-LB process, the actual manufactured part will have some deviations with respect to the nominal part geometry. Also, in the μCT scanning process, even if great care is taken to align the axis of rotation of the μCT scan with the generatrix axis of the physical cylinder, it is challenging to align these two with high accuracy due to the lack of precise reference points in the component. In order to produce this alignment, the OpenCV [14] image processing library was used to calculate the center of the boundary ellipse of the cylinder section for each μCT slice in the central area of the cylinder. Similarly, the centroids of the ellipses corresponding to the sections of the inclined cones on the top were also calculated. Using the information about the nominal angles between the axis of the cylinder and the cones as a constraint, it is possible to calculate mathematically the location of the convergence point between the three axes and the local coordinate system of the nominal geometry in the μCT scan image space (Figure 4b). Once the local coordinate system of the physical part is known, it is possible to calculate the rotation matrix to transform all μCT data into nominal CAD space, and thus, align it with the data obtained from the MPM monitoring system.

3. Results and Discussion

First, we have analyzed the macroscopic, part level defect size distributions in all of the studied parts and compared it with the MPM signal. Figure 5 shows boxplots of the TEP HIGH signal and identified defects >3 voxels on each specimen. From Figure 5b) it is clear that in specimen 3 (decreased laser power) there is significantly higher number of large defects than in the other specimens. Further, specimen 4 (increased laser power) also shows somewhat increased defect size (2x higher mean size) compared to all the rest (parts 1-2 and 5-10) where the defect distributions are almost equal. Comparing this to the TEP HIGH sensor signal (Figure 5a), it can be seen that the MPM signal level for part 3 is significantly reduced, whereas for part 4 it is slightly increased, the other parts having more-or-less similar signal levels. This correlates well with the macroscopic defect level found in the parts shown in Figure 5b).

The average MPM signal responses as a function of set laser power are further plotted in Figure 6. It can be seen, that on average, all the photodiode signals (TEP HIGH/LOW and TED) have strong linear correlation with changes in laser power. This makes sense, as higher laser energy input would result in stronger thermal emission from the melt area. However, as discussed later, focusing on local variations in the signal with respect to defect formation around the same area would be much more valuable than only looking at signal averages and part level porosities.

From the defect analysis it was clear that the attempt to induce increased number of defects to parts 5-8 and 10 by damaging the re-coater or blocking some of the gas flow nozzles was not successful, as there was no increase in defects found compared to the nominal reference (part 2). The cuts made to the re-coater blade were too thin to result in any observable disturbance to the powder spreading, and the blockage of few gas flow nozzles was perhaps not enough to severely alter the flow above the build area.

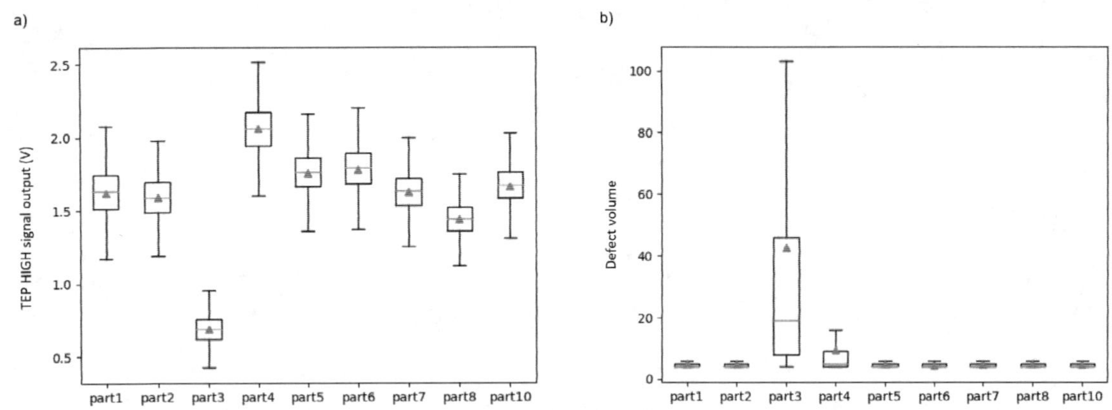

Figure 5. Part level correlation between a) TEP HIGH MPM signal intensity and b) defect distributions. Orange line marks the median and triangle the mean of the data.

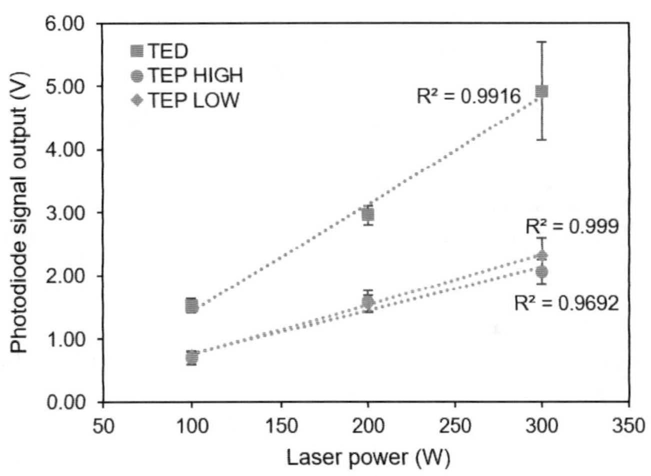

Figure 6. MPM photodiode signal intensities as a function of laser power.

To illustrate the defect level in the parts, Figure 7. shows a slice of the μCT scan data for parts 1, 3 and 4. In part 3 where the laser power was decreased by 100 W, the energy input is so low that severe lack-of-fusion defects are generated throughout the specimen. In part 4 with 100 W increased laser power, increased number of circular keyhole defects are found, often concentrated on a clear row, which implies they are formed at the hatch pattern overlap sections. Figure 6 a) shows part 1, which was produced with nominal laser parameters, but had another specimen close to it to expose it to increased spattering. Here we can see that in an otherwise flawless part with a part level density >99.98 %, there can be one or few stochastic defects caused by spattering or other stochastic process phenomena even as the process parameters are optimized and kept constant. This particular defect is irregular in shape with max dimension ~300μm, which alone could be detrimental to fatigue properties of the part. As was shown on Figure 5, plotting the MPM signal and μCT scan data on macroscopic (part level) scale can only reveal when something is severely and constantly off in the process, but is of little use to indicate the presence of such individual stochastic defects. Within all the seven parts where laser parameters were kept constant, only few such stochastic lack-of-fusion defects were found, all in specimen 1 where a nearby dummy part was placed in its vicinity to induce increased spattering.

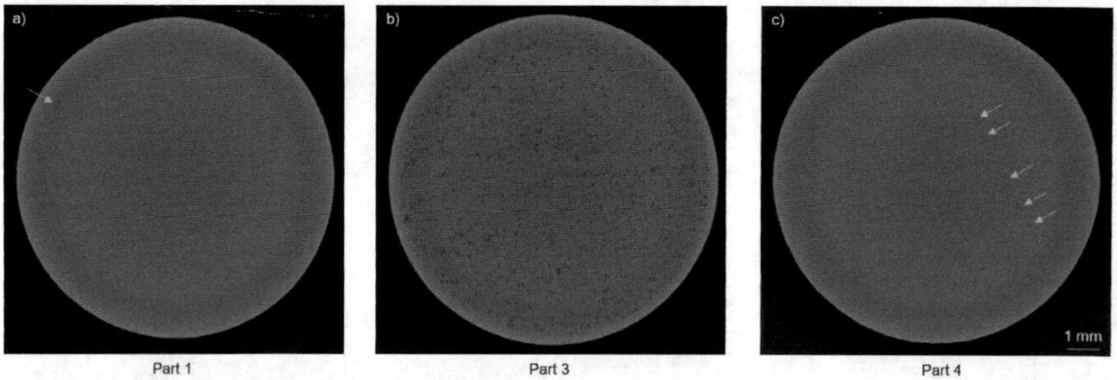

Figure 7. Cross-sectional slices of μCT scan showing examples of defects found for a) part 1 made with nominal laser parameters, b) part 3 with -100 W laser power and c) part 4 with +100 W. Yellow arrows mark defect locations in part 1 and 4.

Identification of individual, stochastic defects (such as those seen on part 1) which are generated in the process even when operating within the optimal process parameter window requires aligning the MPM signal and defect location in the same x-y-z coordinate system with high spatial accuracy, as described in Section 2. Figure 8 shows the result of this alignment. First in Figure 8a) a μCT slice with three defects are shown. In Figure 8b) a layer of MPM data at the same height (layer 717) is shown, aligned in the common coordinate system with the defects and their locations (in yellow) extracted from the μCT data, and with the nominal 3D CAD geometry of the part (shown in transparent grey). The color scale in the MPM data shows the photodiode signal intensity at each data point. In Figure 8c) is a close-up of the two defects found near the two internal hole features close to the top of the cylinder part (see Figure 4a).

It can be seen from this that these two defects are formed at a location where the core vectors and the fill contour vector overlap, around the borders of the internal hole feature. The standard scanning strategy for 316L developed by SLM Solutions was used, which consists of first scanning two border vectors, then a fill contour vector and finally the core hatching vectors. The fill contour and border vectors have a much lower MPM signal on average, as they are scanned with different laser power and speed than the core vectors. Apart from this, there is no clear anomaly in the MPM signal to be seen at the specific locations of the defects. This implies that a more advanced data analysis is needed to find out whether such defects leave an indication in the MPM signal. For example, a defect may form multiple layers below the actual layer being scanned, and hence the indication in the MPM signal would not be precisely at the same z-height as the defect. Furthermore, the defects seen here are quite small and it may be that they form as a result of such process phenomena that does not show as an anomaly in the MPM signal. In the study by Snow et al. [12], only 15% of μCT defects ≤ 100 μm were matched with a corresponding signal in the optical tomography based process monitoring. The defects seen in Figure 8c) have a maximum dimension around 100 μm.

It must be emphasized that regardless of attempts to generate defects to the parts included in the build job, only a handful of such defects were generated and identified in the μCT data. More MPM data from parts with stochastic defects would be needed for training advanced machine learning algorithms and to establish the level of correlation between the used MPM system and defect sizes and locations.

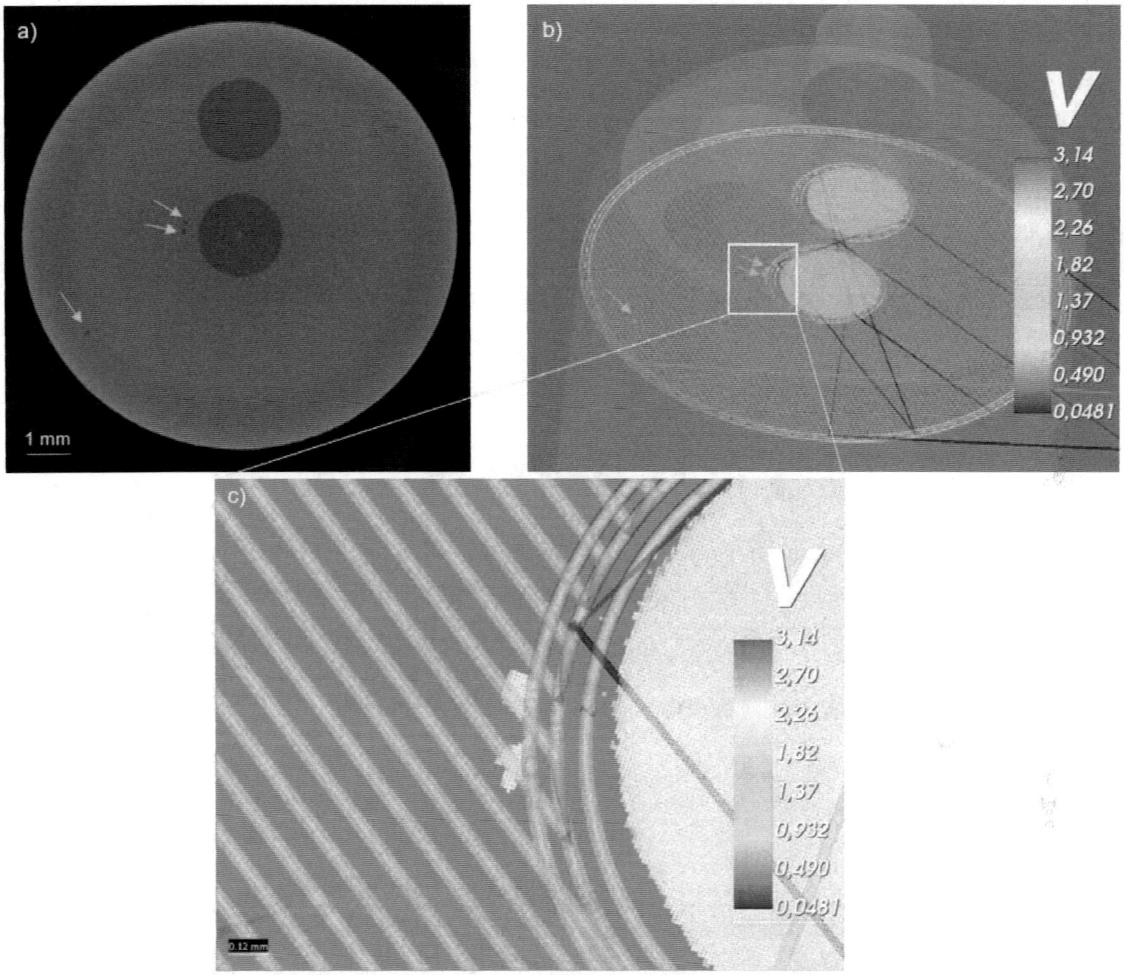

Figure 8. a) μCT slice of part 1 showing three defects, b) TEP HIGH MPM signal, μCT defects and CAD geometry aligned in common 3D coordinate system and c) close-up of the two defects formed at the contour and core vector overlap section.

4. Conclusions

Melt pool monitoring using co-axial photodiodes is proposed as a method to identify defects and move towards data-driven online quality monitoring of PBF-LB AM process. In this study we have demonstrated a method to work with the MPM sensor raw signals and assign them to each part on the building platform, and to reference spatial coordinates with the actual part geometry, instead of generating layered images of the data. Furthermore, we have analyzed defects from manufactured parts with μCT and transformed their size and locations to the same common coordinate system with rest of the data. This type of data alignment methodology presented here is a pre-requisite that enables quantitative data analysis and high accuracy mapping of the MPM signal to the produced parts and defect locations within. It was shown that on the macroscopic part level, the MPM signal intensity correlated well with defect generation in parts where the melting process was made non-optimal by adjusting laser power. However, to identify individual stochastic defects induced, e.g., by spattering that may be generated even as the process is operated with constant process parameters, the MPM signal must be aligned accurately in relation to the actual part geometry and possible defects within. Applying the alignment of all the data, defect formation at the overlap location of fill contour and core hatching vectors was demonstrated.

In the experimental build conducted within this study, only few stochastic defects were generated to the parts, which did not provide enough data needed for training advanced machine learning algorithms. Further work will focus on generating more stochastic defects (without altering the laser process parameters) with the MPM setup, to be used as data for advanced machine learning algorithms to establish the level of spatial correlation between MPM signal and defect location, with the aim of automatically identifying such defects based on the MPM signal.

References

[1] Gibson I, Rosen D and Stucker B 2015 *Additive Manufacturing Technologies: 3D Printing, Rapid Prototyping, and Direct Digital Manufacturing*

[2] ASTM International 2022 *E3353 − 22 Standard Guide for In-Process Monitoring Using Optical and Thermal Methods for Laser Powder Bed Fusion*

[3] Gatzweiler W, Maischner D and Beyer E 1989 On-Line Plasma Diagnostics For Process-Control In Welding With CO 2 Lasers *High Power CO2 Laser Syst. Appl.* **1020** 142

[4] Eriksson E A I, Norman P and Kaplan A F H 2009 Basic Study of Photodiode Signals From Laser Welding Emissions *12th NOLAMP Conf. Proc.* 1–14

[5] Mercelis P 2007 *Control of selective laser sintering and selective laser melting processes* (Katholieke Universiteit Leuven)

[6] Craeghs T 2012 *A Monitoring System for On-line Control of Selective Laser Melting* (Katholieke Universiteit Leuven)

[7] Clijsters S, Craeghs T, Buls S, Kempen K and Kruth J P 2014 In situ quality control of the selective laser melting process using a high-speed, real-time melt pool monitoring system *Int. J. Adv. Manuf. Technol.* **75** 1089–101

[8] Alberts D, Schwarze D and Witt G 2017 In situ melt pool monitoring and the correlation to part density of Inconel® 718 for quality assurance in selective laser melting *Solid Free. Fabr. 2017 Proc. 28th Annu. Int. Solid Free. Fabr. Symp. - An Addit. Manuf. Conf. SFF 2017* 1481–94

[9] Schwerz C and Nyborg L 2021 Linking in situ melt pool monitoring to melt pool size distributions and internal flaws in laser powder bed fusion *Metals (Basel).* **11**

[10] Jayasinghe S, Paoletti P, Sutcliffe C, Dardis J, Jones N and Green P L 2022 Automatic quality assessments of laser powder bed fusion builds from photodiode sensor measurements *Prog. Addit. Manuf.* **7** 143–60

[11] Taherkhani K, Sheydaeian E, Eischer C, Otto M and Toyserkani E 2021 Development of a defect-detection platform using photodiode signals collected from the melt pool of laser powder-bed fusion *Addit. Manuf.* **46** 102152

[12] Snow Z, Scime L, Ziabari A, Fisher B and Paquit V 2023 Observation of spatter-induced stochastic lack-of-fusion in laser powder bed fusion using in situ process monitoring *Addit. Manuf.* **61** 103298

[13] Otsu N 1979 A Threshold Selection Method from Gray-Level Histograms *IEEE Trans. Syst. Man. Cybern.* **9** 62–6

[14] Anon OpenCV *https://github.com/opencv/opencv*

IOP Conf. Series: Materials Science and Engineering 1296 (2023) 012010 doi:10.1088/1757-899X/1296/1/012010

In-process monitoring of track geometry as a control approach for laser metal deposition

A Schinderling [1], A Bohlen [1], T Seefeld [1,2], H Falldorf [3]

[1] BIAS – Bremer Institut für angewandte Strahltechnik GmbH, Bremen, Germany
[2] MAPEX Center for Materials and Processes – University of Bremen, 28359 Bremen, Germany
[3] Falldorf Sensor GmbH, Bremen, Germany

schinderling@bias.de

Abstract. Laser metal deposition (LMD) is an additive manufacturing process, which is used to generate large parts or for reconditioning machine components. During the process the welding tracks are affected by complex dynamic temperature-time curves that results in geometric deviations, which limits the build-up process. Therefore, complex structures are more difficult to manufacture with constant parameters. The present work describes an in-process monitoring system and first approaches of a control concept for LMD processes. It is set up using a newly developed laser triangulation sensor from Falldorf Sensor GmbH, which is trailing mounted. Due to the resistance to diffused radiation, e.g. from process light, it is possible to place the sensor close to the process zone which offers the possibility of a height control system. For the experiments, single tracks were welded from 316L powder by using a diode laser. The results show that the track width and height increase with rising laser power. Depending on the process parameter the height changes up to 29% and the width up to 58% if only the laser power is modified by 1800 W. The monitoring system forms a good basis for a later control concept, which can be used to make the process more stable in the case of complex structures.

1. Introduction

Laser metal deposition (LMD) is an additive manufacturing process, where metallic powder is melted layer by layer with a laser beam. Typically, an inert gas is used to deposit the powder through the nozzle and melt it onto the substrate with a coaxial laser beam. Additional coaxially supplied shielding gas prevents oxidation of the melt pool during solidification. By repeating the welding procedure in both horizontal and vertical planes, it can be used to generate large parts or for reconditioning machine components, especially because of its high built-up rate compared to powder bed-based processes [1]. LMD is a complex generative process, with many differently dependent parameters such as laser power, traverse speed and powder mass flow rate, which influence the thermal conditions of the process and lead to a changing melt pool size as well as resulting shape of the solid track [2]. Besides the geometry the dilution as well as defects, e.g. pores and cracks, are used to evaluate the quality of the welded tracks. The dilution is the ratio of molten substrate (welding depth) and the deposited powder material (track height), which influences the mechanical and corrosion properties of the welded layer. Therefore, the dilution should be small in order to obtain good bonding without influencing the material properties [3].

While the width deviation causes changing track overlap in surface coating, the height deviation changes the working distance between the nozzle and the substrate surface. Both effects can limit the build-up of parts and increase the amount of surface post-processing like milling [4]. For this reason, complex structures are more difficult to manufacture with constant parameters. A simple method to compensate the height offset is a manual adjustment of the optics after several layers, however this can lead to bonding defects due to the cooling down of the specimen. A better way is a continuous sensor-based real-time control system [5]. To monitor the process, some studies use pyrometers, thermal or digital cameras to get inline information about the thermal conditions of the melt pool and conclude about the theoretical track geometry. Other studies use height measurement sensors with optical coherence tomography (OCT), laser triangulation or structured light scanners to measure the height of the melt pool or the solid layer. With an OCT the melt pool height can be measured and used for an in-process numerical simulation of the solid track formation. Depending on the theoretical weld track height, the traverse speed is then adjusted. They are installed coaxially which reduces the interfering contour during the process but requires an adjustment of the optical components and increases the calibration effort. In addition, OCT only detect single tracks and cannot measure the valleys between side-by-side tracks during layer welding [6; 7].

Structured light scanners are used to generate a 3-dimensional point cloud from which the layer height can be determined and thus the depth of the valleys. Due to the stationary outstanding positioning, the measuring principle is not influenced by machine movements but increases the risk of collision. Furthermore, process emissions reduce their accuracy, so conclusions for a control system are limited [8].

The present publication focuses on height measurement methods based on the triangulation principle. Laser triangulation measurement systems have an emitting diode laser beam and an opposite detector. Both parts are at a defined angle (triangulation angle) to each other. At a certain working distance, the laser is focused on the surface of the substrate, where the beam is reflected. The reflection is then focused on a position of the sensitive detector. If the distance between the measured surface and the sensor varies, the position of the reflected beam on the sensor also changes and generates an output signal. The displacement of the reflected beam is thus proportional to the distance change and can be used to detect the track height [9; 10]. Due to interfering process emissions, triangulation sensors are used offline in most cases. This means that the part is scanned at the end of the process and between the layers, which reduces productivity and increases costs [11]. The sensor is often mounted on the side of the optics, which on the one hand increases the risk of collision, but on the other hand does not require a high calibration effort.

In addition to the lateral arrangement the triangulation sensor can also be integrated coaxially into the optics. Here, the triangulation laser is guided parallel to the processing laser. The focal lens of the processing head then angles the measuring beam. Due to the small nozzle opening, the triangulation angle and thus also the distance to the laser spot is limited [2]. Based on trigonometry, a small triangulation angle causes less shading and hidden points, but a larger angle allows a higher distance accuracy. [10] It is also desirable to detect the height profile along a line rather than the distance of a single point. For this reason, a laser line, longer than the nozzle aperture has to be projected by a projector outside the processing laser beam path. This measurement method is utilized by the triangulation sensor used in this work to acquire the height data of the track in-process. A newly developed sensor from Falldorf Sensor GmbH is being tested for the first time in the LMD process and verified with a confocal microscope.

2. Experimental & methodology

For the experiments a 3 kW diode laser LDM 3000-30 with a Zoom OTZ-5 VC optics, both manufactured by Laserline, were used. The wavelength of the laser is between 900 nm to 1080 nm and is guided to the processing optics with a 600 μm optical fiber. The processing head has the possibility to enlarge the spot size variable from 1.0 mm up to 3.6 mm without a focus shift. 316L powder was supplied by a TWIN-150 powder feeder made by Oerlikon Metco. Argon as the carrier gas was used to supply the powder to the six powder injectors of the GTV PN6625 nozzle, which was mounted on the processing head. Underneath, on the machining table of the 3-axis CNC machining station a substrate plate was placed in powder focus (25 mm below the nozzle).

2.1. Used material

The substrate material was an 60×60×8 mm³ structural steel S235, where 7 tracks next to each other of 316L powder made by DEW Deutsche Edelstahlwerke with a grain size between 45 µm and 106 µm was welded on. Table 1 shows the chemical composition of the powder material.

Table 1. Chemical composition of 316L (1.4404) by DEW.

C (%)	Si (%)	Mn (%)	P (%)	S (%)	Cr (%)	Mo (%)	Ni (%)
<0.03	0.7	0.7	0.01	0.008	16.7	2.4	12.6

2.2. Experimental Procedure

In order to determine the influence of the laser power and traverse speed on the track geometry, single welding tracks with a length of 40 mm were produced. For the experiments, the laser power was varied between 800 W and 2600 W and the traverse speed between 600 mm/min and 1200 mm/min.

The following parameters are kept constant for the experiments:

Table 2. Constant process parameters for single tracks.

Constant Parameters	
Powder mass flow	16.5 g/min
Carrier gas flow	10 l/min
Laser spot diameter	3.0 mm
Shielding gas flow	15 l/min
Track length	40 mm

To capture the height data of the weld track, a prototype laser triangulation sensor from Falldorf Sensor GmbH was attached to the processing head in trailing position. It was angled at 13° to reduce the horizontal distance to the laser spot to 10 mm. Figure 1 shows a schematic view of the LMD process on the left side and the real experimental setup with the trailing Falldorf Sensor on the right side.

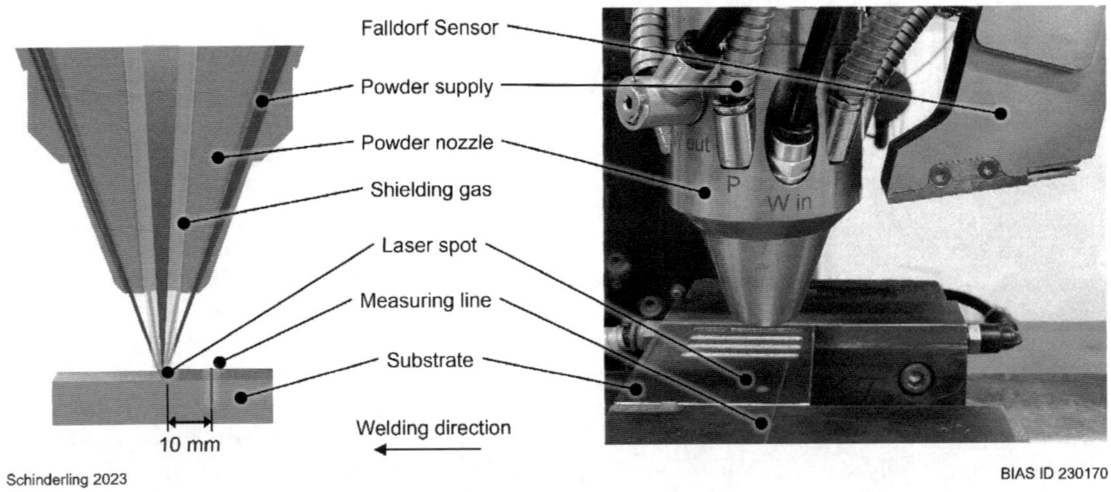

Figure 1. Schematic view of the LMD process (left) and real experimental setup with Falldorf Sensor (right).

The sensor was connected with an associated computer and JTrack Software, which converts the sensor data into height data of a single profile line. To verify the new Falldorf Sensor the height data is compared to an off-line measurement with a confocal microscope VK-X 3000 from Keyence. A lens with 5× resolution and the focus variation method was used resulting in a vertical resolution of 0.5 μm. In addition to the triangulation measurements, micrographs were also taken to examine imperfections such as pores or cracks, as well as to determine the welding depth and thus the dilution.

2.3. *Laser triangulation sensor*

The geometry sensor was designed to work as close as possible to the processing spot and at the same time should maintain a narrow outline to minimize obstructions. The lateral field of view, which corresponds to the length of the laser line is approximately 20 mm, the vertical range for height measurement is approximately 15 mm, the lateral and vertical resolution is 20 μm. The height profiles are acquired with a rate of 100 Hz. While profile features like thickness or voids may vary at a slower rate, a certain oversampling is advantageous to filter out spurious disturbance by stray particles.

The laser wavelength of 405 nm was chosen for a high depth of focus and the ability to work even on hot (red glowing) surfaces. Optical filters are used to suppress the disturbances from the process light emissions and to protect the internal components against back reflection from the procession laser. An exchangeable protective window blocks spatters, the influence of fumes is reduced by the integrated purge air/gas flow. An integrated cooling loop also keeps the sensor temperature low. The sensor system JTrack has its own Graphical User Interface (GUI) as well as the capability to send the profile data over a real-time ethernet interface for external evaluation as described in following. Furthermore, its large vertical working distance of 61 mm and slim design reduce the risk of collision.

2.4. *Monitoring Setup*

The JTrack Software transmitted the averaged values via ethernet protocol to an NI controller. The controller is a cRIO 9030 from National Instruments (NI) with a plug-in modul. The NI 9205 module is connected to a DC/DC Converter, which reduces the output voltage of the CNC control system to 5 V. Each time the laser is turned on and the process starts, the CNC sends a 24 V signal for better allocation of the recorded data to the track. For a future control system an additional slot for a NI 9263 module is available, which will be used to send a voltage signal between 1 V and 10 V to the laser to define the set point of the laser power. The PS-15 power supply for the controller is also made by NI and outputs a voltage of 24 V. In addition, a laptop is connected to the cRIO via another ethernet connection to display the raw data and filtering them as well as measuring the track height and width of the single track in a GUI, which was programmed with LabVIEW. Here, the raw data are leveled, smoothed and the projection distortion caused by the 13° angle of the sensor as well as the slop of the substrate is subtracted. By filtering the height values (Z-data), occurring measurement artifacts can be removed. Furthermore, a temporal filtering reduces strong fluctuations of the measured height and width values, for this, the last 5 measurements are observed.

3. Results

The above described filtering of the raw data with LabVIEW is shown in Figure 2. Here you can see the raw data (blue line), as well as the cut leveled data (green line) and the filtered profile (red line). While the left diagram shows a recording at 1400 W and 600 mm/min traverse speed, in the right picture the laser power and traverse speed has been increased to 2600 W and 1200 mm/min. Here a strong growth of measurement artifacts can be seen. The dashed lines also show the identified track height and sidewalls, which are used to calculate the track width.

At the beginning the in-line height monitoring of the Falldorf sensor were verified with the off-line measurement with the confocal microscope VK-X 3000. Therefore, a single track with 1400 W and 800 mm/min was welded on a substrate by using 316L powder. The distance between the laser spot and measuring line of the Falldorf sensor is 10 mm. The diagram in Figure 3 displays both measurement results of the whole track length. On the bottom the 3-dimensional track, which is measured with the confocal microscope, is shown in false colors.

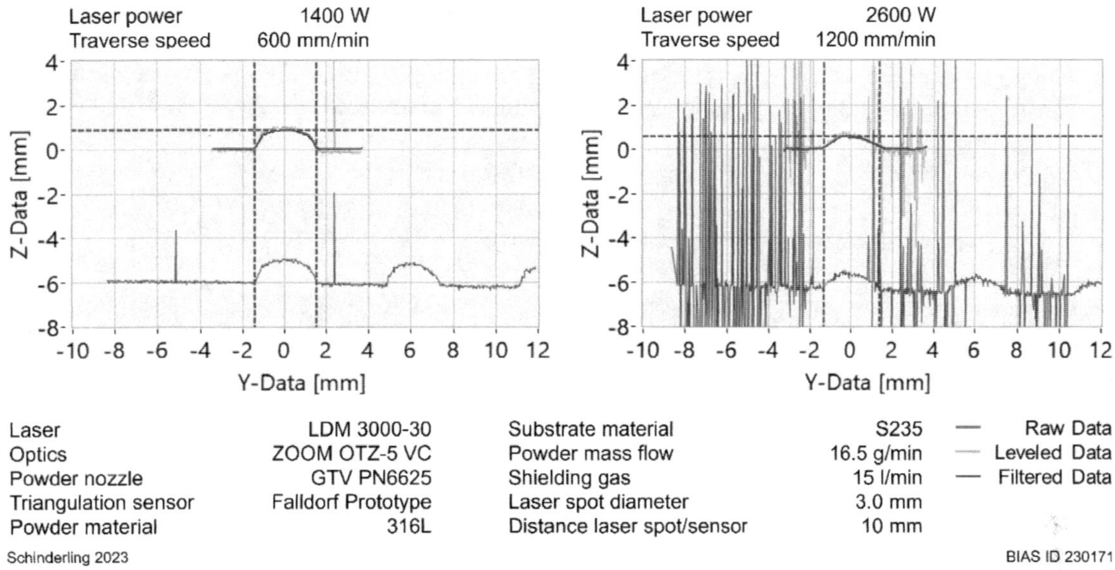

Figure 2. Raw data recordings and filtering with LabVIEW

The data in the diagram of Figure 3 as well as the picture below displays a slight decrease at the end of the track. Both height data, from the triangulation sensor and the confocal microscope show nearly the same course, which is particularly noticeable near 17.5 mm in x-position, where the height negligible drops. Only in the area of 5 mm a stronger deviation of both graphs can be seen. For the average value of the confocal microscope, 242 cross sections were evaluated at a distance of 165 µm, which corresponds to the same measurement interval as the triangulation data. While the average height of the confocal microscope is (661±111) µm, the average height of the triangulation sensor amounts to (657±100) µm. This results in a deviation of the measuring methods of 4 µm related to the average height. Considering all 28 welding tracks of the study, the deviation is about (24±11) µm.

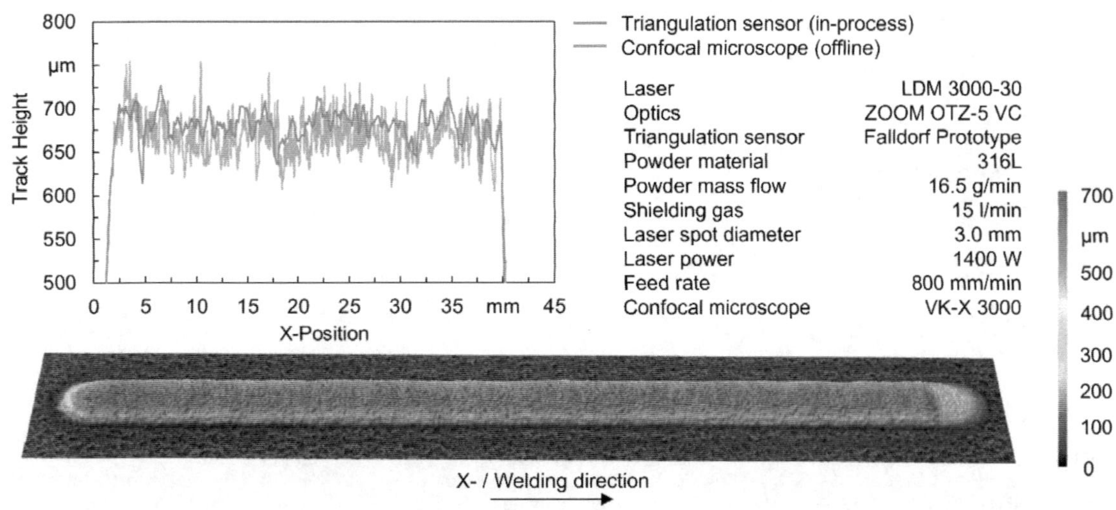

Figure 3. Height comparison between triangulation sensor and confocal microscope.

Figure 4 shows the detected height and width data of the single tracks with 1400 W and verifying traverse speed between 600 mm/min and 1200 mm/min. In both cases the number of data point of each welding track, as well as the height and width decrease with higher welding speed.

While the average height data changes between (438±70) µm (1200 mm/min traverse speed) and (826±138) µm (600 mm/min traverse speed), the average width varies between (2390±106) µm (1200 mm/min traverse speed) and (2837±282) µm (600 mm/min traverse speed). This results in a height difference of 388 µm and a width difference up to 447 µm related to the average value at twice the traverse speed.

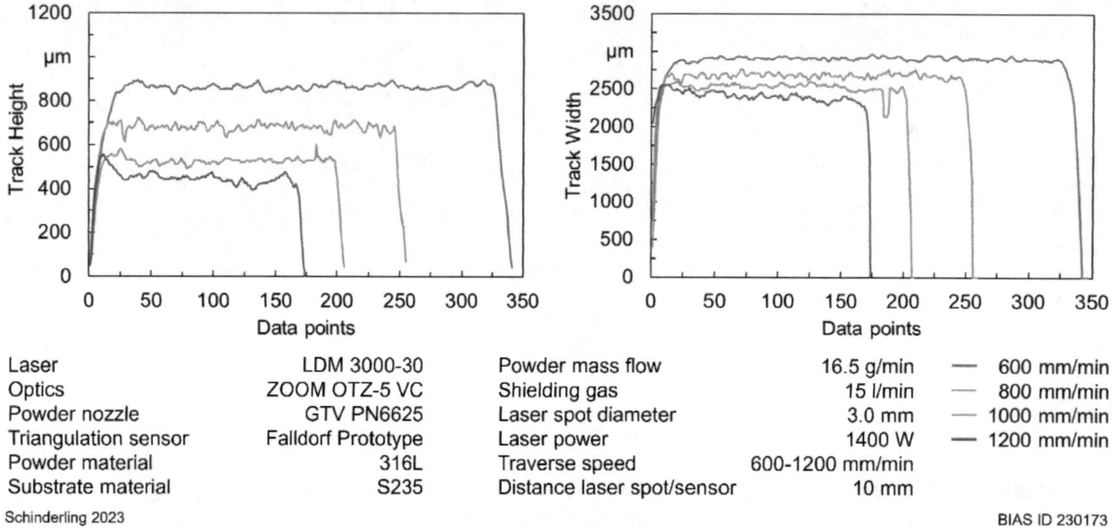

Figure 4. Height and width measurement of a single track with triangulation sensor.

Looking at the entire parameter study, shown in Figure 5, an increasing trend approaching saturation of the track height with rising laser power and decreasing traverse speed can be easily seen. This trend is also reflected in the track width. The largest change in track height up to 212 µm (29%) at constant traverse speed and rising laser power is observed at 600 mm/min, where the track height increases from (730±110) µm (600 W laser power) to (942±158) µm (2600 W laser power). At 1200 mm/min, the greatest change in track width up to 951 µm (58%) is seen. Here the width changes from (21645±206) µm (600 W laser power) to (2596±304) µm (2600 W laser power). If the speed of

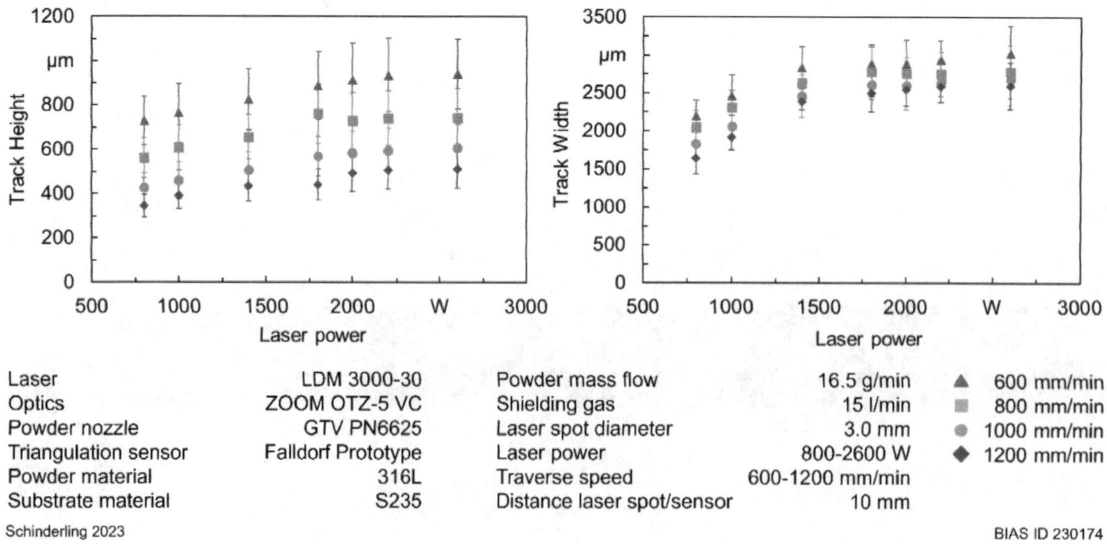

Figure 5. Height and width monitoring with triangulation sensor.

600 mm/min is also considered here, the width can be increased from (2194±205) µm to (3028±356) µm, which means a growth of 834 µm. In both diagrams the standard deviation rises with lower traverse speeds.

In addition to the triangulation measurements Figure 6 presented micrographs of the single tracks. Here, the significantly larger tracks, in terms of both height and width, with lower traverse speeds are particularly noticeable. At all traverse speeds, the micrographs show a sudden increase of the welding depth from 1400 W to 1800 W. Furthermore, the off-center position of the welding depth at 1800 W as well as the decreasing connection angle between welding track and substrate with higher traverse speed can be seen.

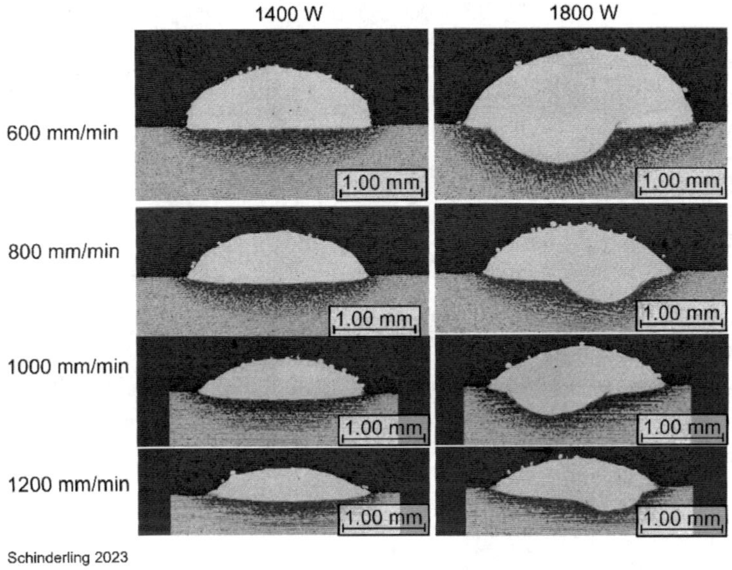

Laser	LDM 3000-30
Optics	ZOOM OTZ-5 VC
Powder nozzle	GTV PN6625
Substrate material	S235
Powder material	316L
Powder mass flow	16.5 g/min
Shielding gas	15 l/min
Laser spot diameter	3.0 mm
Laser power	1400 W, 1800 W
Traverse speed	600-1200 mm/min

BIAS ID 230175

Figure 6. Changing welding depth between 1400 W and 1800 W.

4. Discussion

The verification of the triangulation sensor in LMD processes with the confocal microscope in Figure 3 shows a good in-line detection of the track height. This can be seen from the small deviation between the two measurement methods, which is about (24±11) µm to the average height of all single track, as well as from the almost similar height profile in the diagram of Figure 3. Only in the area of 32.5 mm a small deviation of both graphs can be seen, which is probably caused by measurement artifacts. The differences between the two measurement methods in Figure 3 are the vertical resolution, which is 0.5 µm for the confocal microscope and 20 µm for the Falldorf sensor. Additionally, unmelted powder particles that remain on the consolidated weld track can be detected during the in-process monitoring, which fell down and are no longer present during the off-line measurement. The sensor was tested and validated by using 316L powder. Previous experience in welding and detecting joints with conventional Falldorf sensors shows that other metallic non-polished materials are also measured well. Only carbon black surfaces absorb too much of the laser light, thus gaps in the profile line can occur. Considering the reflectivity of aluminium powder at a wavelength of 405 nm the measuring line may be more deflected and therefore the distance to the powder stream has to be increased. [12]

The previous described measurement artifacts are especially noticeable at high laser powers in combination with high traverse speeds, which is shown in Figure 2. The reason for this could be powder particles rebounding from the substrate surface, moving into the direction of the sensor. Another possible reason could be the enlarged melt pool at higher powers leading to more radiation effects during recording. However, Figure 2 and the standard deviation in Figure 5 show, that the artifacts can be filtered out and the track geometry can be detected well.

Looking at the height and width measurement with constant laser power (Figure 4) or at the parameter study in Figure 5 the welding track decrease in both dimensions with increasing traverse speed. The reason could be the reducing energy density when using lower laser power and/or higher traverse speeds, which decrease the size of the melt pool in which the powder can be melted. Furthermore, as the traverse speed increases, the amount of powder fed per area also decreases, resulting in a smaller track geometry. The higher standard deviation of the track height at lower traverse speeds is due to the fact that more measuring points record both the rise and the fall of the track height. For both track height and width, a saturation of the trend with increasing power can be detected. This comes from keeping the laser spot size and powder mass flow constant, which limits the melt pool size and the track growth. It has already been shown by other studies, which identified the greatest influence on the track height in the traverse speed and the powder mass flow, while the laser power is more involved in formation of the welding depth [13]. However, if only the laser power is varied, it is possible to increase the track height by 212 μm (29%) and the track width by 834 μm (38%) at a traverse speed of 600 mm/min.

However, if the welding depth is also considered, which increases with rising laser power (Figure 6), the track height can be varied by up to 13% and the width up to 29%. Figure 6 shows an increasing weld depth or dilution above 1800 W. Without sufficient powder flow, higher laser power leads to an excess of laser energy being conducted into the substrate, causing a deeper welding track. [14] The reason for the off-center position of the welding depth in Figure 6 would be a non-symmetric powder intensity distribution coming from the nozzle. [15]

Generally, the results show that the new developed triangulation sensor is well suited for LMD processes. Even with the small distance between the laser spot and the measuring line of the sensor (10 mm), the track geometry can be detected well. Measurement artifacts that occur especially at high traverse speeds and laser powers can be filtered out. The small size, as well as the large working distance, also offers the possibility of generating complex geometries without the risk of collision.

5. Conclusion

The present work describes an in-line monitoring system for the track geometry with a new developed triangulation sensor from Falldorf Sensors GmbH. The results show that the sensor is well suited to monitor close to the process zone. Verification with an offline measurement using a confocal microscope results in a height deviation of (24±11) μm.

In addition, the influence of laser power and traverse speed on track geometry (height and width) was investigated. It was found that the track height and width increase constantly until saturation with increasing power, when keeping the powder mass flow and traverse speed constant. However, this also leads to an increasing welding depth above a certain laser power.

At higher traverse speeds and laser power, more measurement artifacts can be seen, but can be filtered out well. At a traverse speed of 600 mm/min, the highest change in track height with low weld depth up to 13% was achieved by increasing the laser power by 600 W. Here, the track width can vary up to 29%. The good measurement results in combination with the small distance between laser spot and triangulation sensor provide a solid basis for further control of the track geometry by varying the laser power.

6. Acknowledgements

The ZIM-project number KK 5274704KT1 was funded by the Federal Ministry for Economic Affairs and Climate Action (BMWK) via the German Federation of Industrial Research Associations (AiF) in accordance with the policy to support the Central Innovations of Medium-Sized Enterprises (ZIM) based on a decision by the German Bundestag and the swedish governmental agency for innovation systems (Vinnova). Furthermore, the authors gratefully acknowledge the collaboration with the members of the project affiliated committee regarding the support of knowledge, material and equipment over the course of the research. The "BIAS ID" numbers are part of the figures and allow the retraceability of the results with respect to the mandatory documentation required by the funding organization.

References

[1] Langebeck A, Bohlen A, Rentsch R and Vollertsen F 2021 Low-Defect AM of High Strength Aluminium Alloy by LMD *SSRN Journal*

[2] Donadello S, Motta M, Demir A G and Previtali B 2019 Monitoring of laser metal deposition height by means of coaxial laser triangulation *Optics and Lasers in Engineering* **112** pp 136–44

[3] Alvarez P, Montealegre M, Pulido-Jiménez J and Arrizubieta J 2018 Analysis of the Process Parameter Influence in Laser Cladding of 316L Stainless Steel *JMMP* **2** p 55

[4] Tyralla D, Köhler H, Seefeld T, Thomy C and Narita R 2020 A multi-parameter control of track geometry and melt pool size for laser metal deposition *Procedia CIRP* **94** pp 430–5

[5] Buhr M, Weber J, Wenzl J-P, Möller M and Emmelmann C 2018 Influences of process conditions on stability of sensor controlled robot-based laser metal deposition *Procedia CIRP* **74** pp 149–53

[6] Kogel-Hollacher M, Strebel M, Staudenmaier C, Schneider H-I and Regulin D 2020 OCT sensor for layer height control in DED using SINUMERIK® controller *Proc. SPIE* **11271** pp 59–63

[7] Baraldo S, Vandone A, Valente A and Carpanzano E 2020 Closed-Loop Control by Laser Power Modulation in Direct Energy Deposition Additive Manufacturing *Proc. of 5th Int. Conf. on the Industry 4.0 Model for Advanced Manufacturing* pp 129–43

[8] Garmendia I, Leunda J, Pujana J and Lamikiz A 2018 In-process height control during laser metal deposition based on structured light 3D scanning *Procedia CIRP* **68** pp 375–80

[9] Chen P H, Liu C H, Chen Y F, Chiu H S and Liu Y C 2015 A Height Measurement Based on Laser Triangulation Probe and a Two Dimensional Image Measurement System for Smart Manufacturing *AMM* **764-765** pp 1324–8

[10] Swojak N, Wieczorowski M and Jakubowicz M 2021 Assessment of selected metrological properties of laser triangulation sensors *Measurement* **176** p 109190

[11] Borovkov H, La Yedra A G de, Zurutuza X, Angulo X, Alvarez P, Pereira J C and Cortes F 2021 In-Line Height Measurement Technique for Directed Energy Deposition Processes *JMMP* **5** p 85

[12] Dahotre N B and Harimkar S Laser Fabrication and Machining of Materials. Springer eBook Collection Engineering. Springer US Boston, MA 2008

[13] Ebrahimnia M, Xie Y and Chi C 2021 Effect of cladding parameters on microstructure and defects in direct laser metal deposition of 24CrNiMo steel *Journal of Laser Applications* **33** p 12007

[14] Dalaee M, Cerrutti E, Dey I, Leinenbach C and Wegener K 2022 Parameters Development for Optimum Deposition Rate in Laser DMD of Stainless Steel EN X3CrNiMo13-4 *Lasers Manuf. Mater. Process.* **9** pp 1–17

[15] Zhong C, Pirch N, Gasser A, Poprawe R and Schleifenbaum J 2017 The Influence of the Powder Stream on High-Deposition-Rate Laser Metal Deposition with Inconel 718 *Metals* **7** p 443

IOP Conf. Series: Materials Science and Engineering 1296 (2023) 012011 doi:10.1088/1757-899X/1296/1/012011

Multi sensor monitoring of the wire-melt pool interaction in hot-wire directed energy deposition using laser beam

A Kisielewicz[1], Y Mi[1] and F Sikström[1] and A Ancona[1,2]

[1]Department of Engineering Science, University West, Gustava Melins Gata 2, 461-32 Trollhättan, Sweden

[2]Physics Department, University of Bari, via Amendala 173, 70126 Bari, Italy

E-mail: agnieszka.kisielewicz@hv.se

Abstract. This study investigates the combination of three sensors to improve in-process monitoring of the liquid bridge between the feedstock wire and melt pool in hot-wire Directed Energy Deposition using Laser Beam. The stability of the deposition process relies on the transfer of metal between the molten feedstock wire and melt pool. Therefore, monitoring the condition of the liquid bridge and the interaction between the feedstock wire and melt pool is crucial. By utilizing a laser-optics-integrated visible range optical spectrometer and electrical sensors measuring voltage and current, relevant process changes and indications of instabilities were detected. Combined information from the current sensor and the spectrometer provided a better understanding of the process and helped to identify deviations leading to unstable deposition modes.

1 Introduction

Hot-wire Directed Energy Deposition using Laser Beam (DED-LB) is a modified wire-based DED-LB additive manufacturing process. In the case of wire-based DED-LB a high-power laser beam (LB) is used to melt locally supplied feedstock wire and create a melt pool [1, 2]. While in hot-wire DED-LB, a combination of a high-power laser and a resitive pre-heating of a feedstock wire is utilized instead, for the same purposes [3]. Both processes are used to fabricate metallic, near-neat-shape parts by depositing the feedstock metal wire layer-by-layer. In comparison to powder-based DED-LB, utilizing a wire as a feedstock is a more sustainable and cost-efficient alternative as the material utilization can reach up to 100% [4]. The wire-based technologies are suitable for manufacturing large-scale elements of airplane engines [1, 5, 6] as they offer high material utilization and high deposition rates. In the case of hot-wire DED-LB, the resistive pre-heating of the feedstock wire gives the additional advantages of increased deposition rates [7], better heat input fine-tuning and improved material integrity [3].

For both processes, deposition stability is governed by the interaction between the wire molten front-end and the melt pool [8]. In stable process conditions a continuous metal transfer can be observed in a form of a liquid metal bridge. With changes in the heat input or the distance between the molten wire front-end, also referred to as the wire tip, and the surface of the melt pool, the metal transfer can undergo changes, leading to the occurrence of unstable deposition modes such as droplets and stubbing [8, 3]. As both of the unstable modes can result in material characterized by discontinuities [8], e.g. pores and lack of fusion, it is relevant to monitor the process, detect the instabilities and, if possible prevent their occurrence.

Various sensing technologies were investigated with the purpose of in-process monitoring and control of DED-LB processes as reported in the review article [9]. However, limited sources report monitoring and control setups measuring electrical current and potential difference in a circuit including the liquid bridge. Kisielewicz et al. [3] observed a correlation between the increased variation of voltage in the hot-wire circuit and the droplet mode. Hagqvist et al. [10] used resistance and wire feed rate to establish an empirical regression model relating the quantities to the feedstock wire molten front-end position in reference to the melt pool top surface. Based on the model a PI-controller was developed to adjust robots tool center point (directly related to wire molten front-end position). Kottman et al. [11] developed a system preventing arcing between the feedstock wire and the melt pool based on electrical power signal. Optical emission spectroscopy has been extensively researched for powder-based DED-LB. Ya et al. [12] used relative intensity of observed Fe I absorption lines present in the visible range spectra and correlated it with deposition quality. Song and Mazumder [13] used the optical emission spectroscopy to predict Chromium composition in the resulting built. Plasma optical emission was analyzed to construct calibration curves relating Cr and Fe emission lines intensity ratios and electron temperature/density to Chromium content in the deposits. Kisielewicz et al. [14] found a correlation between the presence of high-intensity Cr I emission lines and the vaporization of Chromium from deposited Alloy 718. The decreased Chromium content in the deposited material resulted in its impaired corrosion resistance. Ren et al. [15] developed a machine learning recognition model to evaluate the quality in powder-based DED-LB. Their approach involved a visible range spectra and an unsupervised model comprising a long short-term memory-based autoencoder and K-means clustering. The model accurately classified the deposition quality of Al7075 alloy under various process conditions. In all the stated examples the spectrometer was mounted off-axis. Despite being an effective way of monitoring plasma plume, off-axis positioning decreases flexibility of a deposition tool movement as well as is sensitive to collisions. The latter is of high relevance as even a slight displacement of the collimator has a large influence on the acquired spectral intensity. Only a few multi-sensor monitoring systems for wire-based DED-LB were reported. Stutzman et al. [16] assessed the cross-sectional area of the plasma plume above the melt pool using side-view images from a camera mounted off-axis, simultaneously extracting information on the state of the plasma plume from the optical emission spectra (OES). In the results, the authors proposed an in-process flaw detection system based on OES. Jamnikar et al. [17] developed in-process microstructure prediction model for wire-based DED-LB. The model utilizes melt pool characteristics extracted from top-view images and temperature data obtained from pyrometers.

In this paper a multi sensor setup for in-process monitoring of hot-wire DED-LB is proposed. Electrical signals from the hot-wire circuit and optical emission spectra acquired by a co-axially integrated spectrometer are used to detect process changes as well as to monitor interaction between the molten feedstock wire front-end and the melt pool. The acquired signals are assessed by comparison with reference signals acquired during nominal process conditions and their natural variation expressed by 3σ limits.

2 Experimental setup and procedures

2.1 Hot-wire DED-LB setup

The in-house hot-wire DED-LB setup consisted of a 6-axis industrial robot, a fiber laser, a DED tool, a wire feeding system, a hot-wire system, a Programmable Logic Controller (PLC) and an inflatable chamber with an exhaust system. Figure 1 shows the experimental setup. Depositions of single bead tracks were realized by movements of the DED tool attached to the 6 axis ABB

IOP Conf. Series: Materials Science and Engineering 1296 (2023) 012011 doi:10.1088/1757-899X/1296/1/012011

robot (mod. IRB-4400). To melt the substrate and the feedstock wire the IPG Ytterbium-doped fiber laser (mod. YRL-6000-S), working in continuous mode and producing a maximum output power of $6kW$, was used. The wavelength of the LB was $1070nm$. The spot size, measured at focal length of $300mm$, was approximately $1.5mm$ in diameter. To achieve stable processing conditions, the distance of the tool from the substrate was adjusted to obtain a larger spot diameter of approximately $3.2mm$. Therefore, the LB operated out-of-focus with a Gaussian power distribution. The Beckhoff PLC (mod. CX9020) governed movements of actuators and communication between all subsystems. On-line changes of wire tip position were possible since the wire nozzle was attached to the linear servomotor form Beckhoff (mod. AL2003) equipped with the absolute linear scale from Heidenhain (mod. LC 415). The servomotor was the integrated element of the DED tool. The copper nozzle guiding the feedstock wire was mounted off-axis. Wire feeding was performed by the wire feeder manufactured by EWM (mod. T drive 4 Rob 3) and Beckhoff servo-drive (mod. AM8131). The electrical power in the hot-wire system was supplied by the three power sources manufactured by Elektro Automatik (mod. EA-PS 8080-60T) connected in parallel, details of the system can be found here [3].

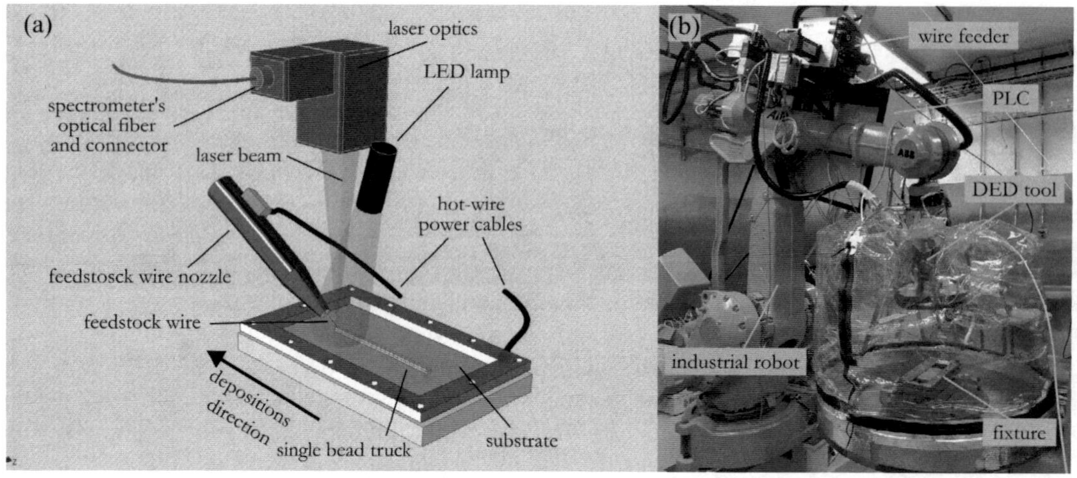

Figure 1. The hot-wire DED-LB setup: a) visualization of the DED tool including monitoring devices, b) overview of the DED-LB cell.

In the course of experiments, three selected key process parameters (electrical power ($P_{electrical}$), laser power (P_{laser}) and wire tip position(WTP)) were varied. Table 1 presents nominal, high and low levels of the varied parameters. The electrical power values are estimations of average power measured in continuous, liquid bridge conditions. Initially, $50mm$ long, single beads with nominal process parameters were deposited to obtain reference signals. The nominal process parameters were selected with the guidance of previous experiments [3]. Subsequently, the values of $P_{electrical}$ or P_{laser} were changed to obtain different stationary conditions. WTP in z-axis, aligned with the LB propagation axis, was changed dynamically by a Δz of $\approx 0.053mm$ per $0.01s$, which allowed to capture gradual changes in the process conditions. For this trial, the deposited bead was $170mm$ long to increase the resolution of the WTP change. Figure 2 present measured trajectory of the wire tip in z-axis. During all the experiments, the front feeding direction was used with the feeding angle of $\approx 18.6deg$. The longitudinal feedstock wire tip position was set so that the wire was in contact with the substrate in the center of the LB. Length of the feedstock wire sticking out of the nozzle was $\approx 6.5mm$ and the wire feed rate (WFR) was set to $33.3mm/s$. Traverse speed was $10mm/s$. Nickel-based Alloy 718 wire of the diameter equal to $1.2mm$ was deposited on $300 \times 100 \times 10mm$ Alloy 718 substrate. All experiments

Table 1. Nominal, low and high level of the key process parameters.

	$P_{electrical}$ [W]	P_{laser} [W]	WTP [mm]
Nominal	158	3500	0
High	378	4500	2
Low	47	2500	- 2

were conducted in a shielding, inert gas atmosphere to prevent the liquid metal from oxidizing. The shielding gas used was pure Argon.

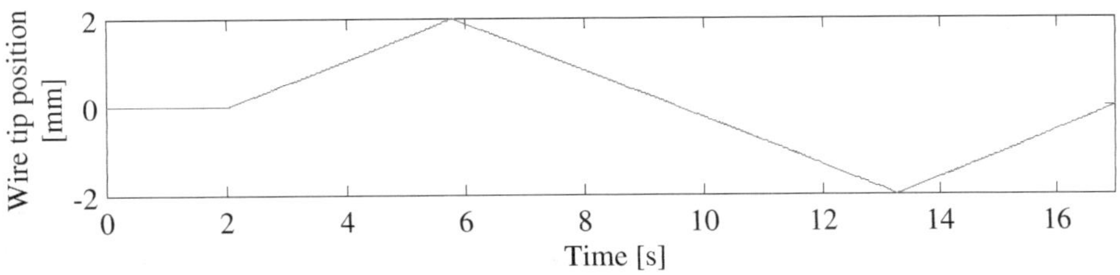

Figure 2. Measured trajectory of the wire tip position in z-axis.

2.2 Sensor setup and data processing
The monitoring system consisted of three different subsystems: a voltage and current measurement circuit and a visible rage spectrometer, see Figure 1(a).

The voltage and current were measured using two PLC modules (mod. EL3751) manufactured by Beckhoff. The modules have the accuracy of $\pm 0.01\%$. The voltage drop between the wire nozzle and the fixture, including the process load, was measured directly. Whereas, current was indirectly derived from the voltage drop over the current shunt manufactured by Thermovolt AB (mod. 1 6053-20) connected in series with the process load. A circuit diagram of the electrical monitoring system can be found here [3]. The sampling rate of the signals was 100 Hz. The electrical data were logged in a database.

The Ocean Optics spectrometer (mod. HR2000+) equipped with a 2048 pixel charge-coupled device (CCD) array detector and a $10\mu m$ entrance slit was co-axially integrated into the laser optics, see Figure 1(a). The spectrometer operated in the visible range of the electromagnetic spectrum between $400nm$ and $530nm$ with a spectral resolution of $0.07nm$. The process spectral emission was transmitted through $200\mu m$-core optical fiber to the detector. The spectrometer was operating with an exposure time of $20ms$. To accommodate for different levels and variations of spectral intensity depending on the wavelength, each acquired spectrum was integrated over the wavelength range. The signal is referred in this manuscript as integrated spectrum's intensity (ISI).

3 Results and discussion
3.1 Nominal process
Figure 3 presents signals acquired during depositions conducted with nominal process parameters. Average and standard deviation values of the signals are given in Table 2. All three signals were characterized by a relatively low variations, which corresponds well with stable liquid bridge observed during the deposition. Based on the average ($\bar{x}$) and standard

IOP Conf. Series: Materials Science and Engineering 1296 (2023) 012011 doi:10.1088/1757-899X/1296/1/012011

deviation (σ) values of the acquired signals reference limits were established to create a measure of comparison between the data acquired during different depositions. Considering that 99.7% of natural process variation is contained within the three sigma limits [18], thresholds of $\bar{x}+3\sigma$, called upper control limit (UCL) and $\bar{x}-3\sigma$, called lower control limit (LCL) were calculated for each signal.

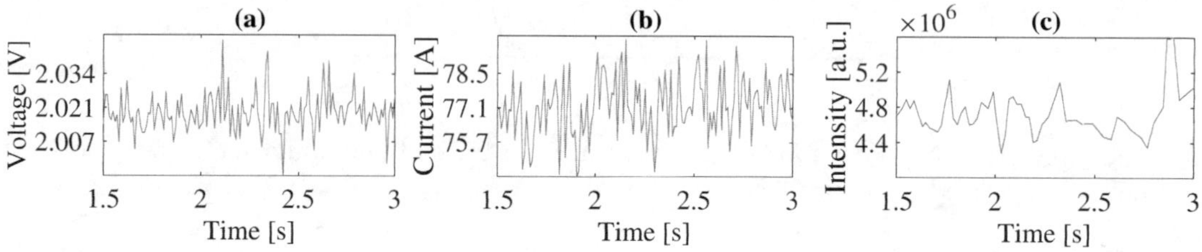

Figure 3. Signals acquired during deposition with nominal process parameters: (a) voltage, (b) current, (c) ISI signal.

Table 2. Average and standard deviation of measured in-process monitoring signals

Signal/ Settings	$\bar{U}$ [V]	σ_U [mV]	$\bar{I}$ [A]	σ_I [A]	$\bar{ISI}$ [Ma.u.]	σ_{ISI} [Ma.u.]
Nominal	2.02	7.7	77.0	1.3	4.75	0.21
High $P_{electrical}$	3.87	152.7	93.6	56.8	4.32	0.38
Low $P_{electrical}$	1.14	9.4	41.6	1.1	3.67	0.15
High P_{laser}	2.02	9.2	81.2	1.6	9.39	0.73
Low P_{laser}	2.02	10.1	79.2	2.1	2.52	0.06

3.2 Varying electrical power

Figure 4(a)-(c) presents signals acquired during depositions conducted with high electrical power. In the case of voltage, see Figure 4(a), the signal was characterized by a periodic, instantaneous increase of its level. Average value and standard deviation of the voltage signal are given in Table 2. Between the peaks, the signal stabilized at level of $\approx$2.94V with the standard deviation of 15.8mV. Despite, the temporal stability of the signal, both of the values were higher than in the case of nominal process and maintained levels considerably higher than UCL. Similar behaviour was observed for the current signal, see Figure 4(b). However, in stead of positive peaks, negative peaks were observed. Average value and standard deviation of the entire signal are given in Table 2. In the period between the peaks, signal stabilized at level of $\approx$126.9A with standard deviation of 2.5A. The given average and standard deviation values were higher than in the case of nominal process. ISI signal was characterized by a high variation, see Table 2, crossing multiple times LCL limit, see Figure 4(c). The high variation in the three acquired signals reflects well the unstable interaction between melt pool and the feedstock wire, as droplets formed and fall into the melt pool. During a single droplet formation, the liquid bridge between melt pool and molten wire front-end was disrupted, causing sudden increase in voltage and decrease in current. Thus, each observed peak of electrical signals relates to one droplet formation. Figure 4(d)-(f) presents signals acquired during deposition conducted with low electrical power. Both, voltage and current signals remained at the levels significantly lower than LCL limits, see Figure 4(d) and (e). The standard deviation of voltage was higher than

IOP Conf. Series: Materials Science and Engineering 1296 (2023) 012011 doi:10.1088/1757-899X/1296/1/012011

that of the nominal process, whereas the standard deviation of current was comparable to that of the nominal process conditions, see Table 2. The ISI signal maintained at a level significantly below LCL limit, see Figure 4(f). The signal was also characterized by a standard deviation lower than in the nominal process conditions, see Table 2. This deposition was characterized by stubbing, where irregular solid-to-solid feedstock wire to substrate interactions occurred. The observed higher standard deviation of the voltage signal can be related to this phenomenon, as the electrical connection between feedstock wire and melt pool changes its conditions.

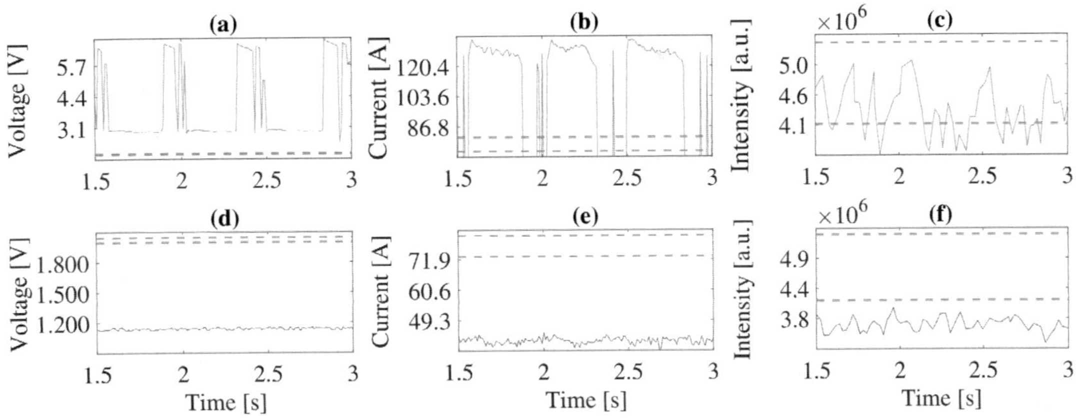

Figure 4. Signals acquired during deposition with varying electrical power: (a) voltage, (b) current, (c) ISI signal during high $P_{electrical}$ deposition; (d) voltage, (e) current, (f) ISI signal during low $P_{electrical}$ deposition. Dashed lines represent $\bar{x}\pm3\sigma$ limits.

3.3 Varying laser power

Figure 5(a)-(c) presents signals acquired during depositions conducted with high laser power. Major part of voltage signal remained within the control limits, see Figure 5(a). The standard deviation of the signal was higher than in the nominal case, see Table 2. Current signal was characterized by a noticeably higher average value, see Table 2, oscillating at the level of HCL. The ISI signal remained consistently above the HCL, see Figure 5(c), and its standard deviation was significantly higher than that corresponding to the nominal process conditions, see Table 2. For these process conditions, the liquid metal bridge was continuous. Yet, feedstock wire melted faster, creating a different liquid bridge profile and allowing higher current flow. Figure 5(d)-(f) presents signals acquired during deposition conducted with low laser power. Similarly to the high laser power conditions, major part of the voltage signal remained within the control limits, occasionally crossing both of them, see Figure 5(d). Variation of the signal was higher than in the nominal case, see Table 2. Current signal crossed HCL multiple times, see Figure 5(e), being characterized by higher average and standard deviation values, see Table 2. ISI signal maintained distinctively below the LCL limit, see Figure 5(f). The standard deviation of the signal was of a magnitude lower to that of the nominal process, see Table 2. The observed higher variation of electrical signals can be explained by the observed, unstable metal transfer through stubbing.

3.4 Varying wire tip position in z-axis

Figures 6(a)-(c) present the acquired signals when values of WTP were change dynamically increasing above the nominal value. Numbered continuous vertical lines indicate different process modes observed. Between $t\approx2.0s$ and $t\approx3.9s$ of the deposition (line 1), voltage signal remained

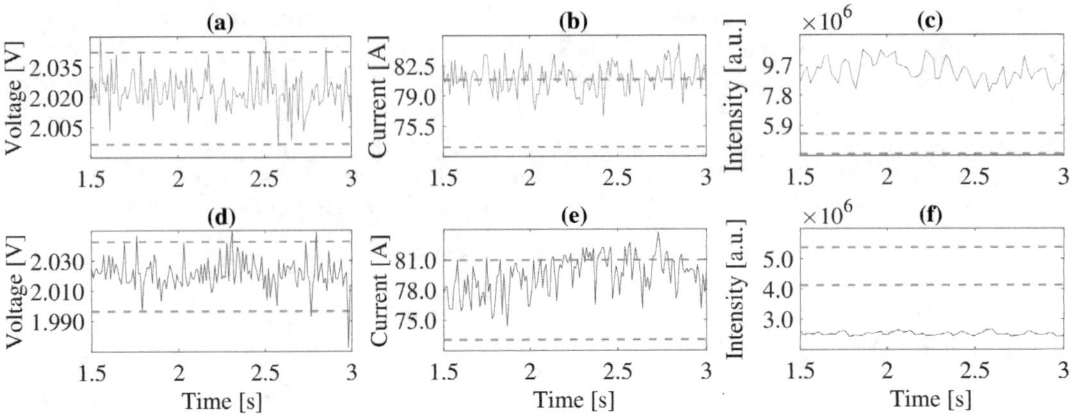

Figure 5. Signals acquired during deposition with varying laser power: (a) voltage, (b) current, (c) ISI signal during high P_{laser} depositions; (d) voltage, (e) current, (f) ISI signal during low P_{laser} depositions. Dashed lines represent $\bar{x}\pm3\sigma$ limits.

within the control limits for the vast majority of the measurement points with average value of $2.02V$ and standard deviation of $1.0mV$, see Figure 6(a). The current experienced much more variation. From $t\approx3.0s$ its value started to decrease, crossing LCL at $t\approx3.4s$, see Figure 6(b). The values of ISI were within the control limits until $t\approx2.9s$, when the signal crossed LCL for the first time. From the imaging system it was observed that until $t\approx3.9s$ the liquid metal bridge was continuous. As WTP increased, a necking of the liquid bridge was observed. Thus, effective cross-section area of the liquid bridge perpendicular to the current flow decreased causing the decrease in the current flow. Between $t\approx3.9s$ (line 1) and $t\approx7.7s$ (line 2), all of the acquired signals exhibited a significantly higher level of variation, see Figure 6(a)-(d). For this period metal transfer changed from continuous to droplet mode. As the liquid bridge was disrupted and electrical circuit became open, value of voltage increased rapidly, while the current value decreased rapidly. As a drop was forming, the tip of the feedstock wire was exposed to the LB to a larger degree causing ISI signal to increase, see Figure 6(d). A release of a drop was accompanied with a sudden signal intensity decrease. Number of peaks observed in all signals corresponds to the number of droplets formed. Between $t\approx7.7s$ (line 2) and $t\approx9.5s$ (line 3), the continuous liquid bridge was restored. Voltage signal oscillated at the level of HCL with average value of $2.04V$ and standard deviation of $12.9mV$, see Figure 6(a). Current signal exhibited positive trend, increasing over HCL at $t\approx8.9s$, see Figure 6(b). The behaviour of the signal was related to the increase of effective cross-section of the liquid bridge. ISI signal increased to finally stabilize within the control limits at $t\approx8.4s$ with average value equal to $4.8Ma.u.$ and standard deviation of $0.19Ma.u.$. Figures 6(d)-(f) present the acquired signals when values of WTP were change dynamically decreasing below the nominal value. Until $t\approx11.5s$ (line 4) voltage values maintained a steady level above the HCL and current signal exhibited constant steady increase, see Figure 6(d) and (e). ISI signal maintained values within the control limits, see Figure 6(f), with average value of $4.7Ma.u.$ and standard deviation of $0.24Ma.u.$. From $t\approx9.5s$ (line 3) to $t\approx11.5s$ (line 4) stubbing mode was observed as the feedstock wire got in direct contact with the solid substrate. Between $t\approx11.5s$ (line 4) to $t\approx13.0s$ (line 5), considerable decrease of voltage and increase of current was observed, see Figure 6(d) and (e). Voltage values decreased below $1V$ and current saturated at the level of $180V$, corresponding to the maximal current supplied by the hot-wire system. A decrease of the integrated spectral intensity signals was also observed, see Figure 6(f). The signal crossed for the first time LCL at $t\approx12.3s$ and stabilized below the LCL between $t\approx13.0s$ and $t\approx14.0s$. The behaviour of all three signals reflects well the highly unstable process conditions as the feedstock wire was in contact with solid substrate before reaching the melt pool. Thus, large area of the wire was pressing on the solid substrate changing significantly characteristics of the electrical connection. At this point, the hot-wire system was saturated as the system could not supply enough current to maintain voltage at the reference level. Between $t \approx 14.0s$ (line 5) and $t \approx 15.2s$ (line 6), as the WTP increased, stubbing occurred again. From $t \approx 14.7s$ voltage maintained stable level above HCL with average value equal to $2.08V$ and standard deviation of $9.5mV$, see Figure 6(d). Current signal exhibited positive trend maintaining values above HCL, see Figure 6(e). Value of spectral integrated intensity remained within the control limits for the vast majority of signal samples, see Figure 6(f), with average value equal to $4.4Ma.u.$ and standard deviation of $0.21Ma.u.$. The deposition in this period was characterized by a heavy stubbing, which is reflected in level and variation of the electrical signals. After $t \approx 15.2s$ (line 6), deposition was disrupted as the feedstock wire was pushed out of the melt pool.

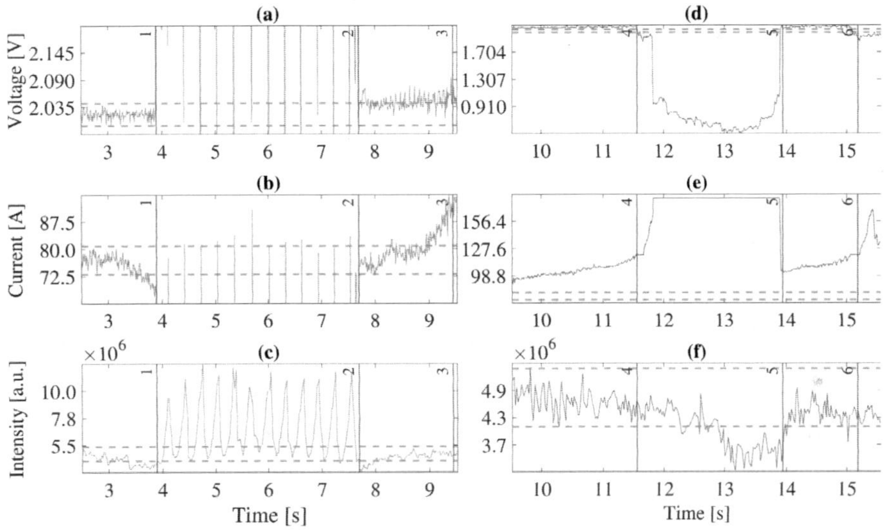

Figure 6. In-process monitoring signals acquired during deposition with varying WTP: (a) voltage, (b) current, (c) integrated spectrum's intensity over wavelengths signals during deposition with WTP above nominal; (d) voltage, (e) current, (f) integrated spectrum's intensity over wavelengths signals during deposition with WTP below nominal. Dashed lines represent $\bar{x} \pm 3\sigma$ limits.

3.5 Detection of process changes and wire-melt pool interactions

Based on the presented results, monitoring of voltage level in hot-wire circuit gave limited information on process changes. For example, in the cases of depositions with high and low electrical power as well as initial increase of WTP, the signal maintained within the control limits. Yet, one can observe its higher variation for the unstable deposition modes.

In the case of current, more pronounced level changes were observed when process conditions differed from nominal. However, when varying the laser power, the two distinctly different conditions were associated with a similarly elevated signal level. Increased current level was also caused by the decreased WTP and higher electrical power. Thus, used individually current signal can indicate process deviation form nominal, yet does not allow to interpret the cause of the change.

In the case of both electrical signals, droplet mode was easily detected as the drop formation was related to distinguishable peaks of signals. However stubbing mode proved to be harder to distinguish from cases of continuous liquid bridge in non-nominal conditions, i.e. high laser power depositions, as they were characterized by similar signal behaviour, i.e. increased variation of voltage and increased level of current.

Changes of ISI signal proved to give a good indication of varying laser power conditions as the signal maintained on distinctively different levels. Slightly less pronounced differences were observed in the case of varying electrical power experiments, where both of the signals had lower intensity than the nominal. In all cases of increased heat input either into the feedstock wire or into the wire-melt pool interaction zone, ISI signal variation increased, which relates to the stronger heat radiation over the interaction zone. The opposite behaviour was observed for the depositions performed with low heat input conditions. A clear indication of droplet formation was observed as the signal intensity rapidly increased. The phenomenon was more pronounced during varying WTP experiments when a molten droplet was exposed to the LB for longer time. In the case of stubbing, it was not possible to distinguish the stubbing mode from nominal when it was caused by changes of WTP. As the feedstock wire experienced vibrations, the ISI signal remained within the control limits having comparable standard deviation to the nominal process conditions.

4 Conclusions

This study investigated a multi sensor system for in-process monitoring of hot-wire Directed Energy Deposition using Laser Beam of Alloy 718. The system was based on sensors integrated with the DED tool to increase the robustness of the solution. By varying three selected key process parameters (laser power, electrical power and wire tip position) different deposition modes were provoked. In relation to the observed process changes, signals from three different sensors were evaluated. Based on the findings the following conclusions have been drawn.

Increased variation in voltage and changes in current strongly correlate to changes in the liquid bridge and the unstable deposition modes. However, it proves to be difficult to distinguish stubbing from other non-nominal process conditions considering only the indications from electrical signals.

The ISI signal obtained by the optical spectrometer provides good results in detecting process changes associated with varying heat input. Additionally, the proposed setup offers a more robust alternative to the off-axis mounted collimator. By locating the hardware further away from the interaction zone, the setup becomes less susceptible to unintentional position displacements and unwanted process-generated heat.

Simultaneous in-process monitoring of current and variation of ISI signal indicates whether the process is deviating from the nominal conditions. It also enables to distinguish whether the change is related to a variation of the heat input into feedstock wire and melt pool interaction zone or to other factors like e.g. changes of wire tip position. These findings provide a valuable input for a future multiple-input, multiple-output control solutions as the combination of signals could allow to determine if the control action should influence parameters related to heat input or feedstock wire position.

Acknowledgments

The authors would like to thank Dr. Petter Hagqvist from Procada AB as well as Mr. Jimmy Johansson and Mr. Janardhanan Swathanandan from GKN Aerospace Sweden AB for valuable discussions and support in data analysis. This work was funded by the Swedish Knowledge Foundation within the project SAMw [20170060] as well as Sweden's innovation agency Vinnova within the project InAIRwire [2019-02752].

IOP Conf. Series: Materials Science and Engineering 1296 (2023) 012011 doi:10.1088/1757-899X/1296/1/012011

References

[1] Özel T, Shokri H and Loizeau R 2023 *J. Manuf. Mater. Process.* **7**(1) 45

[2] Ahn D G 2021 *Int. J. of Precis. Eng. and Manuf.-Green Tech.* **8** 703–742

[3] Kisielewicz A, Thalavai Pandian K, Sthen D, Hagqvist P, Valiente Bermejo M A, Sikström F and Ancona A 2021 *Metals* **11**(4) 634

[4] Ding D, Pan Z, Cuiuri D and Li H 2015 *Int J Adv Manuf Technol* **81** 465–481

[5] GKN Aerospace expands RRSP contract with Pratt & Whitney and injects new Additive Manufacturing and Composite technologies in key components URL `https://www.gknaerospace.com/en/newsroom/news-releases/2019/gkn-aerospace-expands-rrsp-contract-with-pratt--whitney-and-injects-new-additive-manufacturing-and-composite-technologies-in-key-components/`

[6] Blakey-Milner B, Gradl P, Snedden G, Brooks M, Pitot J, Lopez E, Leary M, Berto F and de Plessis A 2021 *Mater. Design* **209** 110008

[7] Bambach M, Sizova I, Silze F and Schnick M 2018 *Proc. CIRP* **74** 206–209

[8] Heralic A 2012 *Monitoring and Control of Robotized Laser Metal-Wire Deposition* (Göteborg: Chalmers Reproservice) ISBN 9789173856553

[9] Tang Z, Liu W, Wang Y, Saleheen K M, Liu Z, Peng S, Zhang Z and Zhang H 2020 *Int J Adv Manuf Technol*

[10] Hagqvist P, Heralić A, Christiansson A K and Lennartson B 2014 *Opt. Lasers Eng.* **54** 62–67

[11] Kottman M, Zhang S, McGuffin-Cawley J, Denney P and Narayanan B K 2015 *JOM* **67** 622–628

[12] Ya W, Konuk A R, Aarts R, Pathiraj B and Huis in 't Veld B 2015 *J. Mater. Proces. Tech.* **220** 276–284

[13] Song L and Mazumder J 2012 *IEEE Sens. J.* **12** 958–964

[14] Kisielewicz A, Sadeghi E, Sikström F, Christiansson A K, Palumbo G and Ancona A 2020 *Mater. Design* **186** 108317

[15] Ren W, Wen G, Zhang Z and Mazumder J 2022 *Mater. Manuf. Proces.* **37** 1339–1346

[16] Stutzman C B, Nassar A R and Reutzel E W 2018 *Addit. Manuf.* **21** 333–339

[17] Jamnikar N D, Liu S, Brice C and Zhang X 2022 *J. Manuf. Process.* **79** 803–814

[18] Montgomery D C 2020 *Introduction to Statistical Quality Control* (New York: John Wiley & Sons) ISBN 978-1-119-72309-7

IOP Conf. Series: Materials Science and Engineering 1296 (2023) 012012 doi:10.1088/1757-899X/1296/1/012012

Numerical modelling of high-power laser spot melting of thin stainless steel

I Bunaziv[1,*], M H Danielsen[1], M Eriksson[1], X Ma[2], X Ren[2], O E Godinez Brizuela[3], P Skjetne[3]

[1] Materials and Nanotechnology, SINTEF Industry, P.O. Box 4760 Torgarden, NO-7465 Trondheim, Norway

[2] Metal Production and Processing, SINTEF Industry, NO-0314 Oslo, Norway

[3] Process Technology, SINTEF Industry, Trondheim, Norway

*E-mail: ivan.bunaziv@sintef.no

Abstract. Numerical modelling is an important scientific tool in laser materials processing to study different melting, evaporation, and solidification phenomena. It can assist in understanding why certain defects are formed, and thus may provide solution paths for how to decrease certain defects and to increase the quality of a product. Laser spot melting in heat conduction mode represents a simple case and is an excellent first step to build and test solvers prior to moving on to more complicated cases such as laser keyhole welding of thick plates. Modelling of laser spot melting requires only a limited computational domain and allows more complex physics to be added gradually. In this work, a thin 3.0 mm thick stainless steel plate was melted with high-power fiber laser and numerically simulated using a native and custom-build solver based on the VOF method within OpenFOAM. The process was captured with a high-speed imaging camera and simulation results are compared with the experimental results. It was found that temperature-dependent surface tension plays a vital role in controlling melt flow directions within melt pool.

1. Introduction

Laser materials processing plays a significant role in modern industry, encompassing a diverse range of applications and technologies. Its processing range is staggering and is used from nanostructuring for medical applications to thick plate welding or cutting. After many decades of research in laser technology, substantial progress has been made in understanding the interaction between laser-beam and matter. However, there are still many uncertainties, and no universal comprehensive model has been developed to capture all possible phenomena. In processing with high power lasers of metals, there are melting/solidification and evaporation/condensation phenomena with non-linear effects. The latter is among the most challenging phenomenon to model since it includes complex physics and both large spatial and temporal gradients resulting in computationally intensive models. Here we adopt a numerical model based on the finite volume method to study laser spot melting.

Otto et al. **[1]** in 2011 developed an advanced model using OpenFOAM which comes close to representing a universal laser beam model, and is still one of the most notable models. Following publication of their model, numerous other numerical models were published using different software and solution strategies, especially related to laser-powder bed fusion (L-PBF). One of most advanced models was published recently by Gan et al. **[2]** using Ansys Fluent in numerical modelling of keyhole formation in L-PBF where vapour dynamics, ray tracing and complex tangential surface tension forces were included. Recently advanced models for deep laser welding have also been published **[3-6]**.

However, few high-quality models are available specifically for laser spot welding. It is simpler since no keyhole is formed but it contains most of the physics involved in both keyhole formation and deep laser welding apart from multiple reflections. Thus, laser spot welding is an excellent starting point for testing numerical models and solution strategies before progressing to the more advanced cases. From an operational point it is of value to be able to predict the transition from a heat conduction dominated operational regime to a keyhole forming regime without having to perform expensive experiments. In heat conduction mode, surface tension plays a key role in predicting melt flow dynamics. The melt pool dynamics is to a large extent driven by a balance of viscous forces, surface tension forces and buoyancy forces. All these forces are sensitive to changes in temperature, resulting in complex temporal and spatial gradients in these forces which drive flow. Typically melt will flow from areas with low surface tension to areas with high surface tension. Such gradients can form e.g., due to compositional differences or temperature gradients. In the latter case the effects are known as thermocapillary effects. Likewise, thermal expansion will induce a change in density resulting in varying buoyancy relative to the average density in the melt. Only a few codes allow for surface tension gradients called as Marangoni forces to be accurately represented in the modelling framework. Some of those that do allow for inclusion of Marangoni forces do so at a large computational cost. Despite a large and growing body of published work, several challenges remain before laser materials processing will have access to efficient and validated general purpose modelling tools. Progress towards this goal is hampered by lack of transparency with regards to details of both modelling and implementation as well as physical data.

OpenFOAM (OF) is an open-source general purpose toolbox for computational fluid dynamics based on the finite volume method. It has a growing community of developers and allows developers to implement and contribute new models and solvers. Due to its open-source nature, running OF is more complicated than running a proprietary GUI driven commercial solver, since it requires general programming skills (in this case the C++ programming language) and Linux OS/WSL system understanding since it cannot be natively run on Windows OS. Therefore, OF has a higher barrier to entry for development compared to most commercial software. OF comes in two distributions, one managed by the ESI-group and one by the OpenFOAM foundation. One of the most promising solvers for laser materials processing is found in the ESI-group's distribution of OF. One of its solvers known as icoReactingMultiphaseInterFoam includes models for phase change and radiative heat transfer, which are key to enabling simulating laser beam energy. It was developed and available in 2018 but is rarely mentioned in the published works. According to the authors' knowledge only a few works have been published which clearly indicated that it was used, e.g., works by Wirth et al. **[7]** and Leung et al. **[8]**. It seems the solver so far has not integrated tangential surface tension forces. The solver was heavily modified in work by Leung et al. where level set method was incorporated, and a laser heat source was added externally. Recently, Flint et al. **[9]** published an alternative model with open-source code where the surface temperature gradient effects in the surface tension has been implemented allowing for accurately accounting for the Marangoni forces. This is achieved by use of an explicit ray tracing model. However, according to the authors' experience, the computational overhead associated with the ray tracing model necessitates the use of high-performance computing to be practical for 3D applications. Moreover, it has limited number of phases involved. The advantages of the native OF solver is that it utilizes the Lagrangian particles for the radiative heat transfer and ray tracing is implicitly implemented which provides faster calculation time. This is discussed by Otto et al. **[10]** where a similar approach was used. Moreover, the multiple reflections can be turned off when they are not required, such as laser spot welding and this will boost the computational speeds immensely. In addition, many phases/species can be included into a model.

In this work, we investigate high power laser beam spot melting on a thin 3.0 mm thick stainless steel plate. Laser beam spot melting is a simple process since it operates in heat conduction mode, however it contains most of the physical phenomena needed to model more complex processes. In heat conduction mode, surface tension plays an important role and capturing the resulting dynamics is one of the main objectives of this work. The process was characterized using a high speed imaging (HSI) camera which allowed direct comparison of the overall dynamics with the numerical modelling. The native OF solver was used and has provided promising results in prediction of melt flows inside the weld pool and partially correlates well with the observed physical phenomena gained from HSI.

IOP Conf. Series: Materials Science and Engineering 1296 (2023) 012012 doi:10.1088/1757-899X/1296/1/012012

However, the solver needs to be modified to include tangential surface tension forces to achieve more realistic melt flow behaviour. The results are compared with works from the literature and critical issues that need to be resolved to improve the solver are elucidated.

2. Methodology

2.1. Materials, equipment and setup parameters

For laser spot welding, a 3.0 mm thick 316L stainless steel was used. In the experiments, a Yb:fiber laser of the IPG Photonics YLS-16000-S2 model was utilized. This continuous wave multi-mode laser operates at a maximum power of 16 kW, with a wavelength of 1070 nm. It features a fiber core diameter of 300 μm and a beam parameter product of 12 mm·mrad. The laser beam exhibited a heat distribution resembling a Gaussian profile. It possessed the following optical parameters: a focal length of 300 mm, with the smallest focused spot size measuring 512 μm in diameter (as measured). Additionally, the Rayleigh length was determined to be 5.6 mm (as measured). To safeguard the spot melt pool from atmospheric conditions, pure argon gas was employed at a flow rate of 10 L/min.

In the initial experiment, a laser power output of 1.0 kW and an irradiation time of 1.5 seconds were utilized. The laser beam was focused at the +60 mm focal point position (*FPP*), resulting in a laser beam spot diameter of 4.1 mm. This configuration enabled the creation of a weld with partial penetration depth (case I). Subsequently, the laser power output was increased to 2.0 kW, leading to a full penetration case, and expanding the model to more complex scenarios (case II). Additionally, a different weld was performed at the +30 mm *FPP* to achieve greater penetration depth and further expand the capabilities of the model (case III).

Spot welds were recorded using a Photron Fastcam Nova S6 type 800K-M-64GB high speed imaging camera at 20 000 frames per second. An illumination diode laser (Cavilux HF 640nm) was used to improve the lighting of the spot welds resulting in improved video quality. The complete melting process was recorded to gain insights into the underlying physics and establish a meaningful correlation with numerical modeling.

2.2. Numerical model description

In this work, both the original and modified *icoReactingMultiphaseInterFoam* was used and tested using OF v2006/v2012/v2106 versions. A few assumptions and simplifications have been made for initial laser spot remelting cases: flow is incompressible and laminar, fluid is Newtonian, and most of thermo-physical parameters are temperature-independent, no multiple reflections are considered/activated, and no evaporation physics was considered. The native solver is designed for N-incompressible phases for transient non-isothermal applications and many phases can be added. The interface tracking was performed using the Volume-of-Fluid (VOF) method which is integrated in OF by default and no modification were used to VOF. Material properties were weighted as a mixture (or volume fraction) of the three primary phases, solid/liquid (stainless steel) metal and gaseous phase (argon). The native solver solves the following equations considering the stated assumptions for mass (equation 1), momentum (equation 2), and energy conservation (equation 3):

$$\frac{\partial \rho}{\partial t} + \nabla \cdot \left(\rho \vec{U}\right) = 0 \, , \tag{1}$$

$$\frac{\partial \rho \vec{U}}{\partial t} + \nabla \cdot \left(\rho \vec{U}\vec{U}\right) = -\nabla p + \nabla \cdot \left(\mu \nabla \vec{U}\right) - C_u \left[\frac{(1-f_l)^2}{f_l^3 + 0.001}\right] \vec{U} + \rho g + F_{st}, \tag{2}$$

$$\frac{\partial (\rho C_p T)}{\partial t} + \nabla \cdot \left(\rho C_p T \vec{U}\right) = \nabla \cdot (k \nabla T) + S_{laser} - h_c\left(T - T_{ref}\right) - \sigma_{sb} \varepsilon_e \left(T^4 - T_{ref}^4\right) , \tag{3}$$

Here, the following terms are (every term description from the left to the right consequently): for equation 1: (I) density change with time, and (II) gradient of density change with velocity; for equation 2: (I) density and velocity change with time, (II) divergence of momentum flux, (III) hydrostatic pressure (p) gradient, (IV) viscous stress, (V) the Darcy dampening force for solidification as porous interface which is modelled through the Voller-Prakash model, (VI) buoyancy force (g is gravitational constant), and (VII) surface tension forces; and for equation 3: (I) change energy per volume with time and (II) the divergence of the thermal energy flux, (III) gradient of heat transfer, (IV) laser heat source, (V) convective losses, and (VI) radiative losses. The nomenclature of symbols is indicated in **Table 1**.

The processing complexity was categorized into three levels, to introduce different physical phenomena gradually for the sake of clarity: (i) Solid body remelting: This level involved only melting and solidification, without any gas phase. Its aim was to establish the Lee constants and provide a quick assessment of the development of the thermal cycle during remelting. (ii) Addition of blanketing gas layer on top: At this level, a gas layer was introduced on top of the solid plate, incorporating surface tension physics. This configuration was typical for partial penetration welding and additive manufacturing scenarios. (iii) Gas layer addition at the bottom: This level featured the inclusion of a gas layer at the bottom, allowing for melt dropout in full penetration cases.

The size of the computational domain was 15×15×(5–7) mm with a base resolution of the computational mesh of 80×80×35 elements. Additional mesh refinement was employed at the centre of the plate. The smallest element size was 0.065 mm and grew to 0.112 mm at the outer edge of the melt pool. To keep the Courant number < 1.0 for fluid velocities in the range of 500–600 mm/s the time step was set to $10e^{-5}$ s. In this work, we consider only modelling which is affordable to run on a standard laptop computer with a few cores (4–6 cores with a frequency of 2.7–4.2 GHz).

The thermos-physical properties used for numerical modelling are shown in **Table 1**.

Table 1. Thermo-physical properties (316L stainless steel) and coefficients used for initial numerical simulations (dimensions provided are aligned with OF conventions). Subscript s is for solid (base metal), l is for liquid (stainless steel melt), and g is for gas (Ar shielding gas).

Density of solid/liquid/gas	$\rho_s/\rho_l/\rho_g$	7900/7500/1.70 kg m^{-3}
Thermal conductivity of solid/liquid/gas	$k_s/k_l/k_g$	18/25/0.018 W m^{-1} K^{-1}
Specific heat of solid/liquid/gas	$C_{ps}/C_{pl}/C_{pg}$	600/775/520 kg^{-1} K^{-1}
Solidus/solidification temperature	T_s	1650 K
Liquidus/melting temperature	T_l	1727 K
Initial surface tension coefficient (liquid/gas)	σ	1.7 N m^{-1}
Latent heat of fusion	H_f	247e3 J kg^{-1}
Absorption coefficient of solid/liquid	a_s/a_l	7e5/6e6 m^{-1}
Dynamic viscosity of liquid/gas	μ_l/μ_g	6.0e-3/2.09e-5 J kg m^{-1} s^{-1}

3. Results and discussion

3.1. Experimental observations

The columns in **Figure 1** show the evolution of the melt pool for the three cases discussed in this section. Laser spot remelting using a 1.0 kW highly defocused laser (+60 mm *FPP*, case I) representing a pure heat conduction welding mode is shown in **Figure 1**a. The depth of the melt pool is shallow and typically exhibits an ellipsoidal shape. This type of melting represents a pure heat conduction welding mode. According to the HSI video, 50 ms is required to form an incipient melt pool. It takes about 500 ms to reach the maximum size of the melt pool. Beyond this time there is a slight increase of the melt pool size between 500 to 1500 ms (the maximum irradiation time). In the beginning of the melt pool formation, there is a clear indication of low outward melt flow velocities. After ~580 ms, a melt stagnation occurs with puls-like melt behaviour with a periodic change in the surface flow between inward and outward flow, with outward melt flow behaviour dominating.

An increase of the laser power to 2.0 kW provided a full penetration of the plate at +60 mm *FPP* (see **Figure 1**b, case II). Due to higher energy input per time, the time scales were shorter. During inception of melting, a slow outward melt flow directions occurred with flow speeds measured in the range of 35–45 mm/s, based on a few floating oxides grains on the surface. With melt stagnation occurring around ~100 ms again followed by puls-like behaviour. The flow then settles into a mode with strong inward surface flow. Such complex behaviour with flow reversals was also reported by Zhao et al. **[11]**. Moreover, a small vortex formed in the middle of the melt pool that exhibited a counterclockwise rotation. A portion of the melt near the edges continued to have outward melt flows with some periodic pulsing and rotation-like flows around the edges.

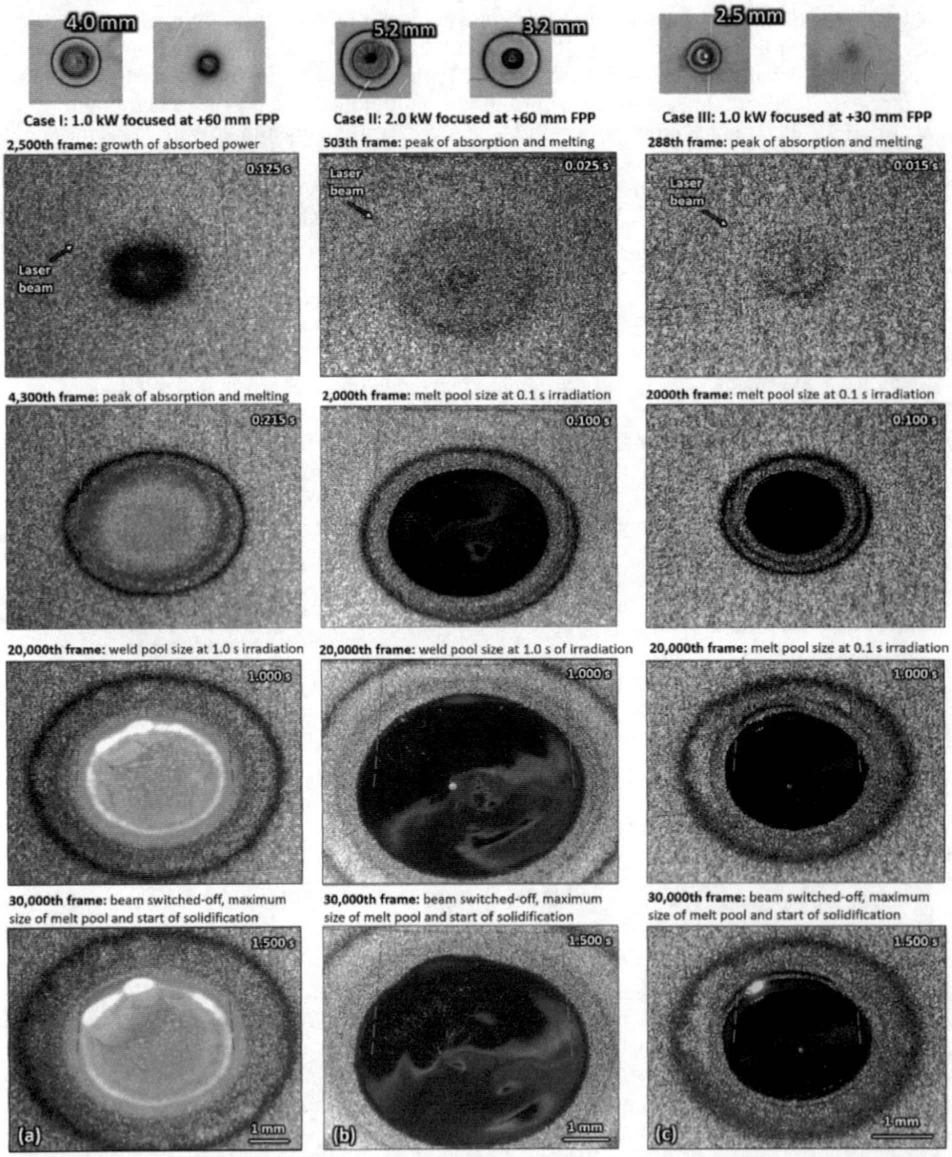

Figure 1. Experimental observation of melt pool formation, growth, and dynamics of melt flows of selected test cases.

A more focused laser beam (+30 *FPP*) with 1.0 kW (see **Figure 1**c, case III), the melt pool dynamics was almost identical to case II. In this case, the vortex is more pronounced and with similar velocity magnitudes of the surface melt flow. Flow reversals occurred slightly earlier at ~60 ms. Cases II and III

IOP Conf. Series: Materials Science and Engineering 1296 (2023) 012012 doi:10.1088/1757-899X/1296/1/012012

were selected for numerical modelling due to their similarity. A complete overview of melt flow behaviour for each case is illustrated in **Figure 2**.

There are a few reasons for the complex melt pool behaviour which is related to surface tension phenomena. In heat conduction laser welding gradients in surface tension dominate the physics resulting in surface flows from areas of low surface tension to areas of high surface tension. It is well known that in pure iron melting by a heat source, only outward melt flows occur (cf. refs in **Figure 3**), and surface tension decreases linearly with temperature, see **Figure 3**a. The temperature is at a maximum at the centre of the melt pool and decreases towards the edges, thus resulting in an outward surface flow. With addition of oxygen or sulphur into the bulk metal, the surface tension no longer exhibits a linear dependence with temperature. Starting from the melting point (for steel) it will first *increase* with increasing temperature until it reaches a maximum after which it will start to *decrease* with a further increase in temperature, cf. **Figure 3**. For sulphur the surface tension goes through a minimum instead of a maximum producing an opposite chain of events **[12]**. In these cases, the initial surface flow will be directed inwards in the pool until the maximum temperature in the pool reaches the temperature corresponding to the maximum value in surface tension. At this point there will be a period where the surface tension gradient is no longer monotonous, and stagnation and pulsing may occur. Once the temperature of the whole melt pool has crossed this maximum temperature the surface melt flow will reverse and flow outwards. If the melt pool becomes larger than the irradiation area of the laser the melt temperature at the edges can drop below the temperature corresponding to the maximum in surface tension and complex edge flows can result. Therefore, based on the performed experimental observations, the surface tension at melting point is high, with no oxygen, and rapidly decreasing with temperature according to the **Figure 3**. In such case, the surface tension temperature gradient **Figure 2**b is negative in the beginning and becomes more positive.

The surface tension temperature gradient causes Marangoni flows and includes the tangential components for the surface tension force that may strongly affect the melt flow behaviour and its direction. Such behaviour can be explained in our case by increase of the oxygen from the surface oxides (the used plates were not cleaned mechanically) present on the plates into the melt pool during the growth of the melt pool. There is also additional explanation of chaotic-like behaviour of the melt pool which can be related to: uneven temperature distribution of melt surface, evaporation, and chemical anisotropy with thermo-physical properties. Due to these phenomena, a realistic spot welding simulation represents a significant challenge in predicting the complex melt flow characteristics and capturing such quasi-stable behaviour which also may require inclusion of turbulent modelling.

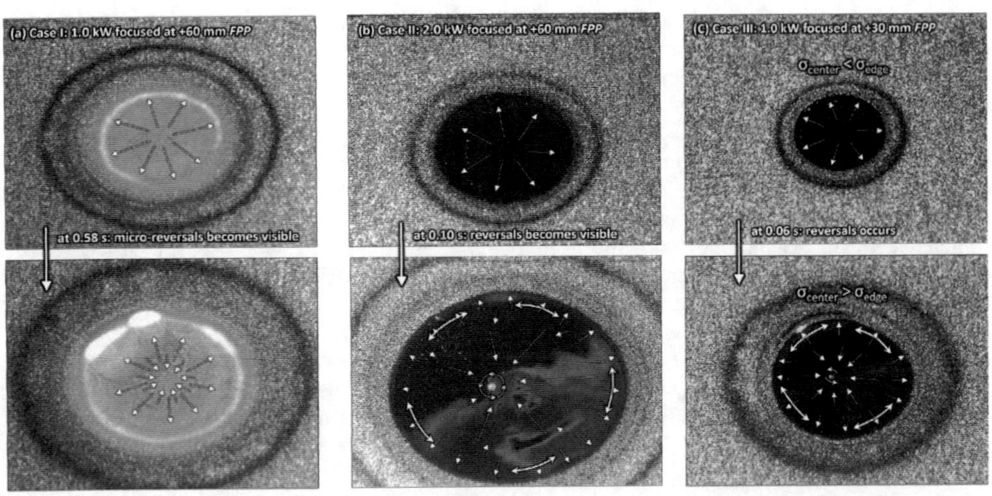

Figure 2. Experimental observation of melt pool formation, growth, and dynamics of melt flows of selected test cases.

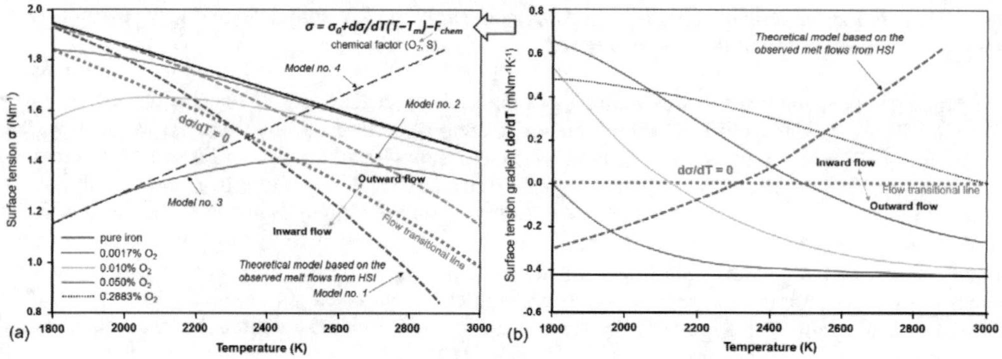

Figure 3. Effect of (a) temperature dependent surface tension and (b) surface tension temperature gradient on melt flow behaviour. Graphs constructed based on **[11]**. Note the curve employed for this work is shown only schematically and approximated with tabulated data in the numerical modelling part.

3.2. Numerical modelling

High speed imaging is an excellent tool to observe dynamic process behaviour. However, it is primarily limited to surface phenomena and internal melt flows cannot be directly observed. Therefore, numerical modelling is a great tool to infer a more complete understanding of the system. Building credible models is a gradual process, starting from first principles accounting for the dominant phenomena in succession. To build credible models requires significant skills both in formulation of the underlying physics and numerical solution of the governing equations. As a first step, we have investigated a simple plate with two phases (solid and liquid) to test the implementation of melting, and to establish approximate melting parameters. In this case, the Lee model of melting was used assuming a porous interface between the two phases. Next, a gas layer was added on top of the plate resulting in a three-phase system. This significantly increased the computational time.

Initially, only the normal component of the surface tension force was included assuming a constant value for the surface tension. From these simulations the melting parameters of the Lee model were tuned to give good agreement with melting times from the experimental part. Good agreement with the fusion zone lines were achieved, cf. **Figure 4**. For case II full penetration occurred with a slight dropout at the root but surface tension prevents formation of a hole, cf. **Figure 4**b.

The melt flows behaviour in all three cases is outward flows. However, the remelt with 1.0 kW at +60 mm *FPP* had a melt flow reversal to inward flow after 800 ms (0.8 s) which occurs only slightly later than experimental observations. At this point such behaviour is unknown but it was not expected to provide flow reversals at constant surface tension. Notably, the numerical model was not able to capture a fully correct melting evolution of the molten pool (considering the liquid fraction of > 0.8) which formed at a slightly later time with delay of 50 ms. Generally, the weld pool evolution times are in good agreement with the experimental observations.

The next step is to include the temperature dependent surface tension which is employed as tabulated data (or polynomial) corresponding to represent three selected cases of surface tension behaviour on temperature. As the prototype, the case with 2.0 kW defocused at +60 mm *FPP* and up to 0.40 s simulation time was considered since the flow reversals occurs early in the beginning and at such time the melt pool is large enough. Furthermore, the bottom gas layer was omitted since no full penetrated is needed for proof of concept and the mesh can be further refined to increase the accuracy.

Initially, two monotonically decreasing models for surface tension as a function of temperature were investigated (Model no. 1 and Model no. 2, cf. **Figure 3**a). Both models started out with a relatively high surface tension of ~1.95 N/m at 1800 K. Model no. 1 exhibited a more rapid decrease in surface tension with temperature compared with Model no. 2. Additionally, both exhibited a linear increase in reduction of surface tension with temperature. Model no. 1 would not converge, and we are still investigating the reasons for the numerical problems seen in these simulations. Model no. 2 had a smaller reduction in surface tension with temperature and would converge. The outward melt from in the beginning of melting was achieved and only partly reversed into the inward melt flows, see **Figure 5**a. Therefore, the proposed model only partially in agreement with the experimental observations.

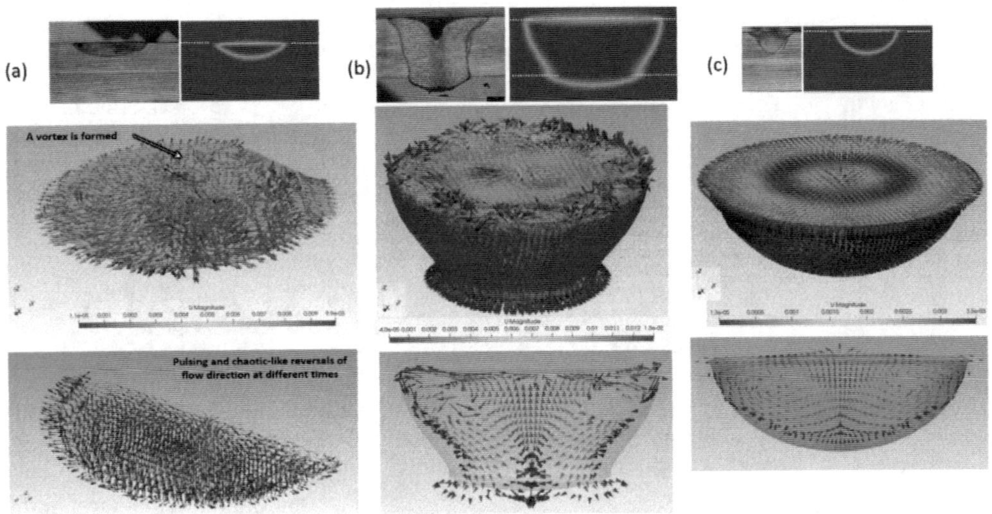

Figure 4. Numerical modelling results of the selected cases with constant or temperature independent surface tension. Comparison of fusion zone geometry and melt flows in case of (a) defocused 1.0 kW laser at +60 mm *FPP* at different timings, (b) defocused 2.0 kW laser at +60 mm *FPP* at 1.5 s (end of welding time), and (c) defocused 1.0 kW at +30 mm *FPP* at 1.5 s (end of welding time).

The next model, Model no. 3 cf. **Figure 3**, represented a steady increase of the surface tension with temperature and a slight decrease after 2400 K. The inward melt flow was achieved at start of melting and transited to outward flow in a short time (see **Figure 5**b) which is in a good agreement with the theoretical assumptions (see **Figure 3**) and the model responded correctly to the temperature dependency of the surface tension coefficient. However, no vortex was formed in the centre and flow behaviour was not realistic exhibiting random waviness upon enlargement of the melt pool. The final surface tension model investigated, Model no. 4 cf. **Figure 3**, exhibited a gradual (linear) increase in surface tension with temperature. Unlike Model no. 3, it only exhibited inward flow in the main melt pool. Close to the edges a thin annular region exhibited outward flow accompanied by small tangential flows creating a number of eddies along the edge. A large vortex was gradually formed at the centre of the melt pool which represent mostly inward melt flows as shown in **Figure 5**c. This is in a good agreement with the experimental observations except that the melt flow did not exhibit flow reversal at the start of melting. Intermittently, smaller vortices were formed around the large central vortex having short lifetimes dissipating rapidly. The central vortex became quasi-stable state after > 0.30 s. The melt flow velocities increased significantly compared to the cases with constant surface tension (see **Figure 4**) and were at a maximum at the centre of the melt pool.

According to theoretical assumptions, more oxides enter the weld pool as it grows, hence the surface tension should increase with temperature. To exhibit flow reversal the surface tension needs to pass through a maximum as temperature increases. This could result from increasing in oxides concentration but warrants further investigation and access to data for stainless steel as **Figure 3** is only valid for iron systems. Hence, some discrepancy between theoretical considerations based on **Figure 3** and our experimental observations of stainless steel should be expected.

Addition of the tangential component of the surface tension force was included using the code provided by Wirth et al. **[7]**. However, no difference was found as shown in **Figure 5**d. It may be related to: (*i*) rough mesh size which is unable to properly capture tangential surface tensions; (*ii*) for the presented case using higher laser power than for typical L-PBF, these forces may have a low effect;

and/or (*iii*) the native solver somehow derives the tangential surface component by itself using the temperature-dependent surface tension which cannot be confirmed at this point. A further mesh refinement yielded similar results, **Figure 5**e-f, for both unmodified and modified solvers. In general, the VOF method struggles to capture effects such as the Marangoni forces due to the dissipative nature of the surface reconstruction. Based on the provided modelling results, the significance of the tangential surface tension component is low. A non-monotonic temperature dependent surface tension which included only the normal component was all that was needed to capture flow reversal in the melt pool as it heated up. Unfortunately, the proposed models were not able to fully coupe with all the details observed experimentally using HSI. The future work will employ more refined mesh models which will require more powerful computational workstations for higher efficiency.

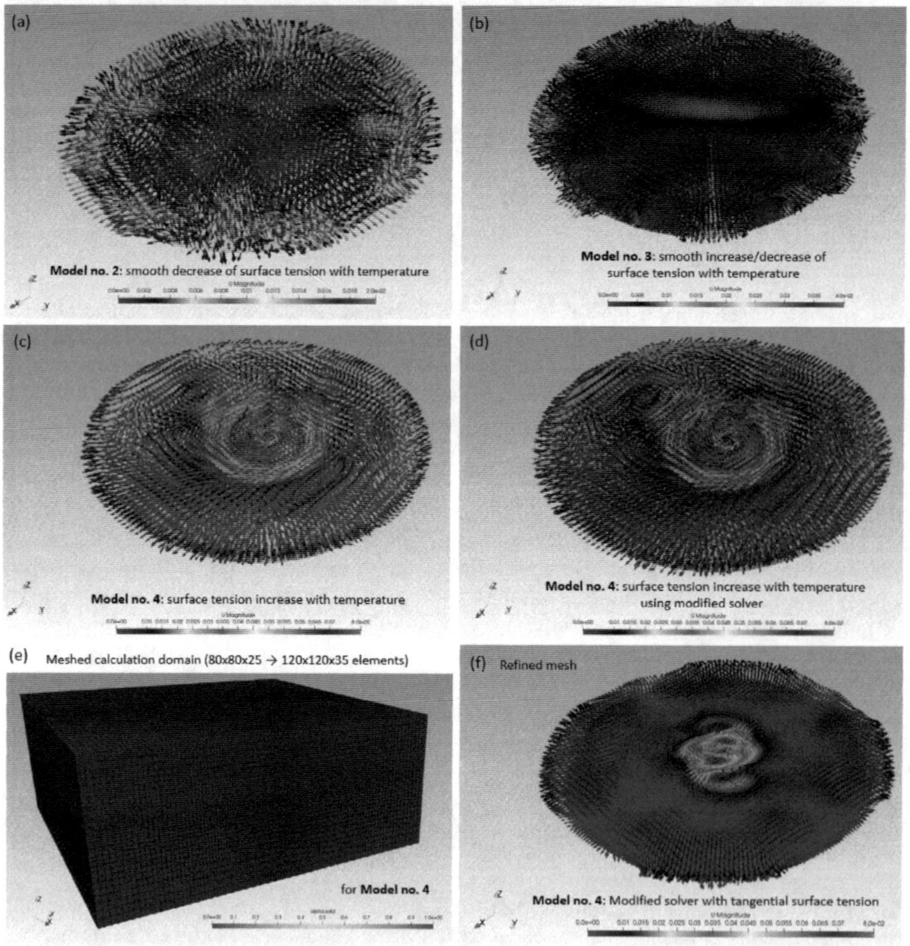

Figure 5. Numerical modelling results of the selected case (2.0 kW focused at +60 mm *FPP*) using different temperature dependent surface tension models at 0.40 seconds.

In this work it has been demonstrated that the identification of the melt flow directions and their magnitudes is challenging to estimate based on the experimental observations. The HSI equipment is unable to show the melt characteristics inside the melt pool, thus it is limited to the observations on the surface. In fact, very accurate measurement and numerical estimated melt flow velocities are not required, hence only a general understanding of the flow patterns is needed and is enough to understand the process behaviour with possible formation of weld defects (or penetration depth) which is the main purpose of the numerical modelling and experimental observations. In general, numerical modelling is a complex procedure and to achieve similar results is challenging since precise thermo-physical parameters of the base metal must be known. Moreover, they are temperature dependent. Absorption coefficients are difficult to estimate and cannot be readily measured which also depends on the temperature and phase of the absorbing media. The relationship of temperature-dependent surface tension and flow direction is difficult to measure practically and rarely provided in the literature for a certain alloy system. Other thermo-physical parameters may also affect the fluid flow such as temperature-dependent viscosity and vaporisation which were not tested here. A high-fidelity model may reduce the knowledge gap and explain the melt flow behaviour more correctly.

4. Summary and conclusions

The experimental data and the numerical modelling showed a good agreement with regard to weld pool growth and the general nature of the melt flows. The native OpenFOAM solver was tested for the simplest case at low power heat conduction laser spot welding. A successful implementation of the temperature-dependent surface tension was performed with testing of different models to control directions of the melt flows. The normal surface tension component was sufficient to model and control the flow reversals. Therefore, the selected solver may provide a reliable and realistic behaviour of the melt pool. At this point it is not clear if the tangential surface tension component is absent in the native solver or should be introduced and tested properly for the future works.

Acknowledgments

The experimental work was conducted in MANULAB (Norwegian Manufacturing Research Laboratory) infrastructure, funded by the Norwegian Research Council (Grant No: 269898).

References

[1] Otto A, Schmidt M 2010 Towards a universal numerical simulation model for laser material processing. *Phys. Procedia* **5** 35-46
[2] Gan Z, Kafka O L, Parab N, Zhao C, Fang L, Heinonen O, Sun T, Liu W K 2021 Universal scaling laws of keyhole stability and porosity in 3D printing of metals. *Nat. Commun* **12**(1) 2379
[3] Sun Y, Li L, Hao Y, Lin S, Tang X, Lu F 2022 Numerical modeling on formation of periodic chain-like pores in high power laser welding of thick steel plate. *J. Mater. Process. Technol* **306** 117638
[4] Wang Y, Jiang P, Geng S, Xu B, Zhao J 2022 Influence of the wrinkle surface structures on the vapor flow and keyhole stability in 20 kW high power laser welding. *Int. J. Heat Mass Transf.* **193** 122958
[5] Artinov A, Meng X, Bachmann M, Rethmeier M 2022 Study on the transition behavior of the bulging effect during deep penetration laser beam welding. *Int. J. Heat Mass Transf.* **184** 122171
[6] Geng S, Yang W, Jiang P, Han C, Ren L 2022 Numerical study of keyhole dynamics and porosity formation during high - power oscillating laser welding of medium - thick aluminum alloy plates. *Int. J. Heat Mass Transf.* **194** 123084
[7] Wirth F, Tonn T, Schöberl M, Hermann S, Birkhofer H, Ploshikhin V 2022 Implementation of the Marangoni effect in an open-source software environment and the influence of surface tension modeling in the mushy region in laser powder bed fusion (LPBF). *Model. Simul. Mater. Sci. Eng.* **30**(3) 034001
[8] Leung C L A, Luczyniec D, Guo E, Marussi S, Atwood R C, Meisnar M, Saunders B, Lee P D 2022 Quantification of Interdependent Dynamics during Laser Additive Manufacturing Using X-Ray Imaging Informed Multi-Physics and Multiphase Simulation. *Adv. Sci.* **9**(36) 2203546
[9] Flint T F, Robson J D, Parivendhan G, Cardiff P 2023 laserbeamFoam: Laser ray-tracing and thermally induced state transition simulation toolkit. *SoftwareX* **21** 101299
[10] Otto A, Vázquez R G 2018 Fluid dynamical simulation of high speed micro welding. *J. Laser Appl.* **30**(3) 032411
[11] Zhao C X, Kwakernaak C, Pan Y, Richardson I M, Saldi Z, Kenjeres S, Kleijn C R 2010 The effect of oxygen on transitional Marangoni flow in laser spot welding. *Acta Mater.* **58**(19) 6345-57
[12] Pitscheneder W, DebRoy T, Mundra K, Ebner R 1996 Role of Sulfur and Processing Variables on the Temporal Evolution of Weld Pool Geometry during Multikilowatt Laser Beam Welding of Steels. *Weld J.* **75** 71-80

Defect detection in laser-based powder bed fusion process using machine learning classification methods

Jan Akmal[1, 2], Mihai Macarie[3], Roy Björkstrand[1], Kevin Minet[2], Mika Salmi[1]

1. Department of Mechanical Engineering, Aalto University, Puumiehenkuja 3, 02150 Espoo, Finland
2. EOS Metal Materials, Electro Optical Systems Finland Oy, Lemminkäisenkatu 36, 20520 Turku, Finland
3. Faculty of Electrical Engineering, Mathematics and Computer Science, University of Twente, Drienerlolaan 5, 7522 NB Enschede, The Netherlands

Email: jan.akmal@aalto.fi, jan.akmal-ext@eos.info

Abstract. The aim of this study is to deploy machine learning (ML) classification methods to detect defective regions in additive manufacturing, colloquially known as 3D printing, particularly for the laser-based powder bed fusion process. A custom-designed test specimen composed of 316L was manufactured using EOS M 290 machine. Multinomial logistic regression (MLR), artificial neural network (ANN), and convolutional neural network (CNN) classification techniques were applied to train the ML models using optical tomography infrared images of each additively manufactured layer of test specimen. Based on the trained MLR, ANN, and CNN classifiers, the ML models predict whether the manufactured layer is standard or defective, yielding five classes. Defective layers were classified into two classes for lack of fusion and two classes for keyhole porosity. The supervised approach yielded impeccable accuracy (>99%) for all three classification methods, however CNN inherited the highest degree of performance with 100% accuracy for independent test dataset unfamiliar to the model for unbiased evaluation. The high performance and low cost of computing observed in this work can have the potential to detect and eliminate defective regions by tuning the processing parameters in real time resulting in significantly decreased costs, lead-time, and waste. The proposed quality control can enable mass adoption of additive manufacturing technologies in a vast number of industries for critical components that are design- and shape- agnostic.

1. Introduction

Additive manufacturing (AM), colloquially known as 3D printing, is a novel and disruptive mode of manufacturing that allows creation of parts on a layer-by-layer basis using digital 3D model data [1]. The layer-by-layer mechanism of the additive process enables tool-less production of parts with unprecedented geometric freedom in one build cycle [2]. Comparatively, the subtractive process involves selective removal of material via a tool and formative process involves pressurizing a body of raw material into a mould to obtain a part [3]. In both cases, part-specific tools and moulds favour economies of scale and limit design freedom. On the other hand, the tool-less layer-by-layer mechanism is emerging into a general purpose technology akin to computers and dynamos serving a plethora of applications [4]. AM can be categorized into seven group of technologies [1] depending on the layer-wise addition and fusion mechanisms of feedstock material. Today, it is used for tooling [2,5], prototyping [6,7], and end-use part [8–11] applications in a wide range of industries, e.g., medical, industrial machines, defence, automotive, and aerospace industries.

More specifically, the laser-based powder bed fusion (PBF-LB) process has been gaining traction within the industrial landscape to create lightweight metallic parts on demand [4,12] using generative design, topology optimization, and lattice formation [2,7]. During the PBF-LB process, the impinging laser energy selectively fuses regions of powder bed [2] to create the intended part on a layer-by-layer principle. The underlying process dynamics are complex and not fully understood [13]. The process is inherent to large thermal gradients, heat absorption, local melting and re-solidification of particles, phase change, plumes, spatter, and Marangoni convection [14]. To mitigate these inherent characteristics, the parts are manufactured using support structures for heat dissipation and anchoring the parts to the build platform [10].

Nevertheless, the process is highly sensitive to energy density, i.e., laser power, scan speed, hatch distance, and layer thickness, which can cause under- or over- heating. A slight deviation from the thresholded volumetric energy density caused by process and environment uncertainties can induce lack of fusion and keyhole porosity resulting in structural defects, i.e., porosity, support rupture, warpage, residual stresses, cracking, balling and geometric defects, in the part [15]. Further, the under- or over-heating can cause disturbances in powder bed, recoater crashes, and eventually premature failure of intended parts.

The aim of this study is to create PBF-LB process variation-dependent defective regions on a layer-by-layer basis and to use three machine learning (ML) methods for detection. This study overcomes the lack of understanding in the transitional phase of process parameters and temporal thermal fingerprint emitted during the process. This allows for engineering functional end-use parts by reducing the uncertainty in passing stringent quality standards and easing the widespread adoption in various applications in a wide range of precision industries.

2. Methods

2.1. Additive manufacturing

The EOS M 290 (EOS Gmbh, Krailling, Germany) machine was used to manufacture a tailor-made test specimen using the PBF-LB process. The volumetric energy density (VED) function is the critical independent variable for the design of experiment, which is formulated as follows:

$$VED = \frac{P_L}{v \times h \times L_t} \tag{1}$$

where P_L is the laser power [W], v is the exposure scan speed [mm/s], h is the hatch distance [mm], and L_t is the layer thickness [mm]. The design of experiment involving five volumetric energy densities is shown in Table 1. EOSTATE Exposure optical tomography (OT) monitoring system (EOS Gmbh, Krailling, Germany), comprising of a scientific complementary metal-oxide-semiconductor (sCMOS) sensor, was used to capture near-infrared images of each layer. The test specimen was manufactured using 316L as feedstock powder ($D_{v(25)}$=24μm, $D_{v(50)}$=37μm, $D_{v(75)}$=56μm) at the centre of the build platform using the EOS 316_SurfaceM291_1.10 parameter set. Argon was used as inert shielding gas. Table 1 shows the number of OT images as a function of exposure nomenclature for hatch infill adopted in this study. The test specimen and the segmented exposure strategies are shown in Figure 1. Owing to such a small selectively melted cross-section region per layer of the build, a minimum exposure time of 18 s was added to ensure the likelihood of a successful print.

Table 1. Additive manufacturing parameters

Exposure Nomenclature	Power [W]	Scan speed [mm/s]	Hatch Distance [mm]	Layer Thickness [mm]	Energy Density [J/mm3]	No. of OT Images [-]
Δ-20% VED	156	1083	0.09	0.02	80.02	249
Δ-10% VED	175.5	1083	0.09	0.02	90.03	247
Standard VED	195	1083	0.09	0.02	100.03	1248
Δ+10% VED	214.5	1083	0.09	0.02	110.03	249
Δ+20% VED	234	1083	0.09	0.02	120.04	247

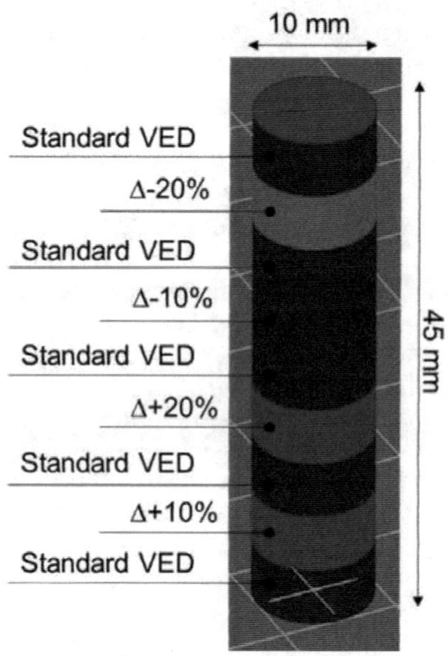

Figure 1. Dimensions and exposure strategies of the test specimen.

2.2. Machine learning

2.2.1. Dataset. The data comprises of 2238 OT images in JPG format after removing the first ten and the last two OT images owing to the variation in the light emissions. For the first ten layers, the variation can be caused by for example reduced layer thickness of first layer, reflectivity of the build platform, and double exposure of the first two layers. For the last two layers, the variation is caused by up-skin parameters. Each image represents process radiation in the near infrared spectrum of the laser-based selectively melted layer over the entire build platform. Each image corresponds to the thermal history of one layer. The original dimension of each image is 2000 x 2000 pixels. After visualizing the dataset, the insignificant regions of the OT images were removed. Subsequently, each OT image is cropped to focus on the region of interest resulting in 100 x100 pixels (width x height) and 10000 pixel squared. Each pixel represents a shade of grey as a 24-bit integer (range 0-255). The number of OT images as a function of exposure nomenclature are shown in Table 1.

2.2.2. Features. This study uses deep learning in which features of each OT image can be depicted as a vector $\boldsymbol{x}^{\boldsymbol{i}} \in \mathbb{R}^{10000}$ composed of integers, where every **i**-th entry represents a grayscale value in the range of 0 to 255. A grayscale value of 0 depicts black and 255 denotes white. Further, OT images were mapped with RYB colour model for visualization. To assist the activation function in deep learning, this range is normalized between 0 and 1 [16]. Each image corresponds to the 100 x 100 tensor format containing multidimensional feature input. The first 100 entries of the feature vector depict the top row of each OT image, the second 100 entries denote the second row from top, and similarly the remaining entries are filled until the bottom-most row with an index of 100. This is the prescribed tensor format for Tensorflow [16] which is adopted in this work.

2.2.3. Labels. The label of each datapoint, i.e., each OT image, is a multi-class classification, $y \in \{0,1,2,3,4\}$, denoting a Δ-10% VED layer $\hat{y}^i = 0$, a Δ-20% VED layer $\hat{y}^i = 1$, a Δ+10% VED layer $\hat{y}^i = 2$, a Δ+20% VED layer $\hat{y}^i = 3$, and a Standard VED layer $\hat{y}^i = 4$. Supervised approach is used to manually label the datapoints yielding five classes.

2.2.4. Classification methods. This study adopts three ML classification methods, i.e., multinomial logistic regression (MLR), artificial neural network (ANN), and convolutional neural network (CNN), to classify five classes of OT images as shown in Table 1. Open-source Jupyter Notebook (6.4.8, Project Jupyter) operating on Python (3.9.12, Python Software Foundation) was used to build the architecture of the ML models using an off-the-shelf mobile workstation.

The MLR applies the parameters of the linear map $h(x)=w^Tx$ that classifies datapoints, i.e., OT images, into label value $\hat{y}^i$. Sparse categorical cross-entropy loss is used to quantify the performance of the linear map because the five label classes comprise of integers. A dense layer (fully connected layer) is created with 5 neurons. Softmax is used to transform logits to valid probabilities [16].

The ANN architecture is built on a feedforward neural network that classifies datapoints, i.e., OT images, into label value $\hat{y}^i$. In this case, two dense layers, comprising of 2048 and 5 neurons respectively, are created with rectified linear unit (ReLU) as an activation function. Further, a dropout layer (regularization layer) is added with a rate of 0.2 to prevent overfitting. The model also uses sparse categorical cross-entropy as the loss function.

Finally, the CNN architecture adopts a feedforward neural network that classifies datapoints, i.e., OT images, into label value $\hat{y}^i$. It consists of a convolutional base, which includes convolutional layers (i.e., spatial convolutional over each OT image), and pooling layers (i.e., to produce lower resolution of the feature map) [17]. Finally, two dense layers, comprising of 64 and 5 neurons respectively, are added to perform classification using ReLU. Sparse categorical cross-entropy loss is used to quantify the performance of the network.

2.2.5. Data splitting. For this supervised approach, OT images were manually sorted into five classes, $y \in \{0,1,2,3,4\}$, as shown in Table 1. The dataset is randomly split in a manner that represents 70% for training the ML models and 20% for validating them. This means 1571 OT images are used to train the ML models and 442 OT images are used for validating. The remaining 10% of the dataset were used for testing the ML models. The five classes were split identically, representing similar percentages. This training, validation, and testing ratio is within the range suggested by empirical studies obtaining the optimal results [18].

3. Results and discussion

The additively manufactured test specimen with segmented exposure strategies according to Figure 1 is shown in Figure 2. When observed closely, the segmentation regions are visible on the surface of the manufactured test specimen. A highlight of processed OT image for each VED is shown in Figure 3.

IOP Conf. Series: Materials Science and Engineering 1296 (2023) 012013 doi:10.1088/1757-899X/1296/1/012013

Figure 2. Additively manufactured test specimen (316L) with segmented exposure strategies.

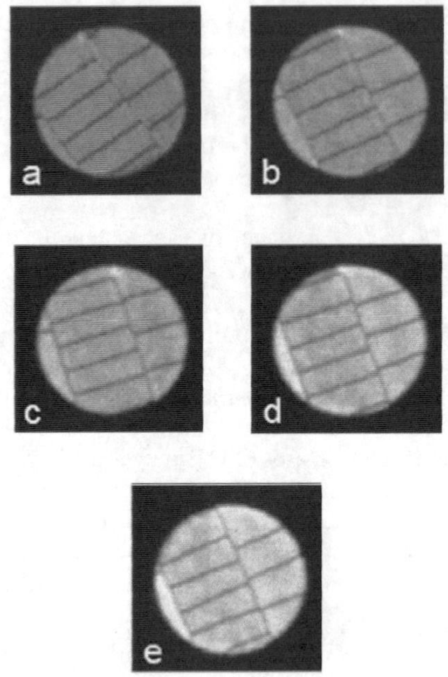

Figure 3. A processed OT image for a) Δ-20%, b) Δ-10%, c) Standard, d) Δ+10%, and e) Δ+20% VED layers.

The accuracy and loss functions of MLR, ANN, and CNN are shown in Figure 4. The MLR model was able to obtain a 100% training and validation accuracy at the 57th epoch with loss values of 0.026 and 0.035 respectively. Comparatively, the CNN model was able to achieve 100% training and validation accuracy at the 5th epoch with loss of 0.007 and 0.004. On the other hand, the ANN model was not able to converge training and validation accuracy to a value of 100%. In this case, the highest values observed were 96% for training and 98% for validation with 0.146 and 0.081 as loss values. As depicted in Figure 5, the CNN model was able to train and validate the model 42% faster than MLR and 66% faster than ANN.

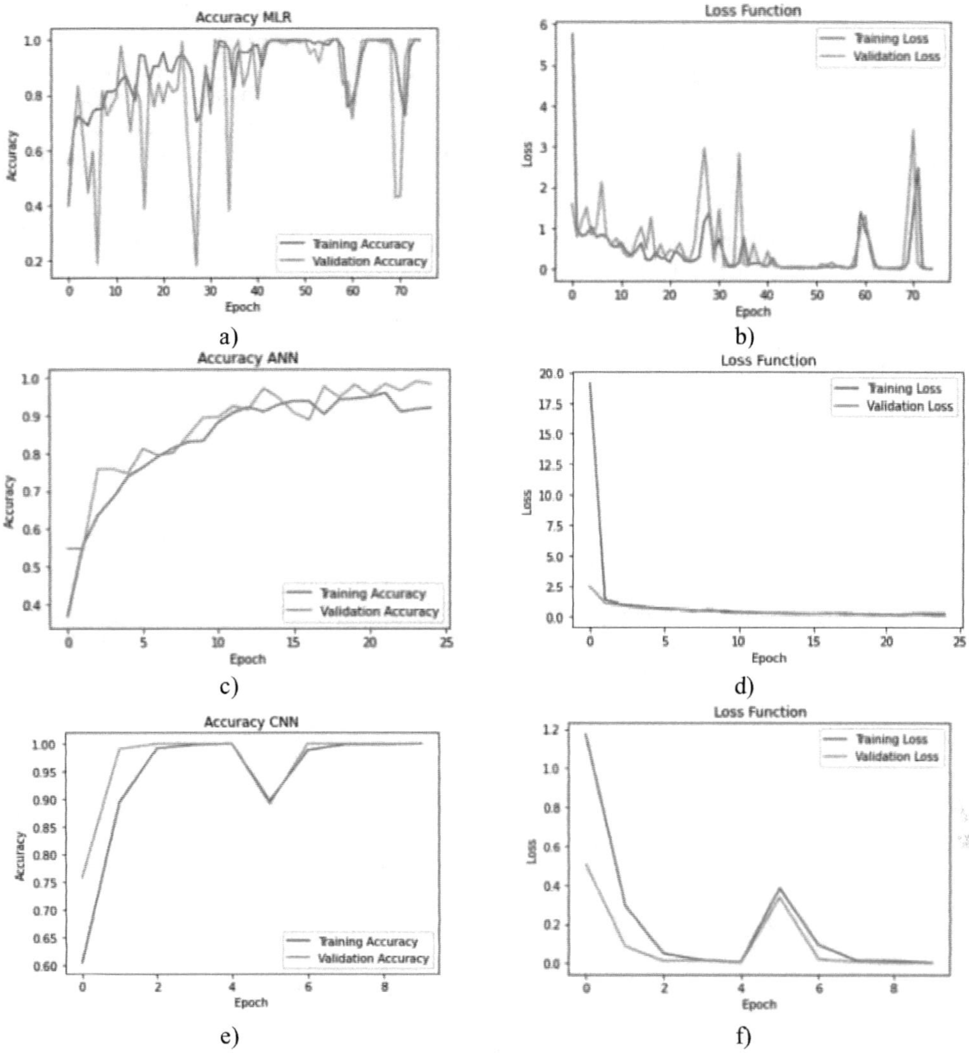

Figure 4. Accuracy as a function of epoch for a) MLR, c) ANN, and e) CNN, and results of loss function with respect to epoch for b) MLR, d) ANN, and f) CNN.

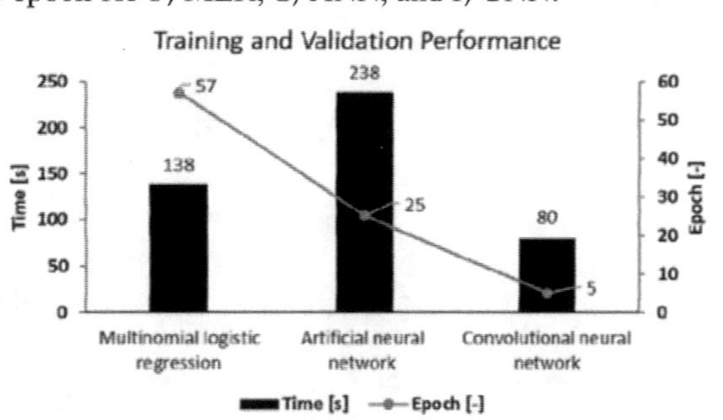

Figure 5. Training and validation performance of the ML models.

Confusion matrices of each ML classification method for test dataset are shown in Figure 6. As observed from Figure 6a and 6c, the MLR and CNN models yielded impeccable accuracy of 100% for evaluating the test dataset. Considering that the test dataset was not used for training nor validating the ML models, this is a remarkable outcome. The loss values for MLR and CNN were 0.023 and 0.002 respectively. In contrast, the ANN model yielded 99.6% accuracy for the test dataset with a loss value of 0.082. As observed from Figure 6b, the ANN model predicted one Δ+10% (Pos-10) VED layer as Standard VED layer. The testing performance of each ML model is denoted in Figure 7.

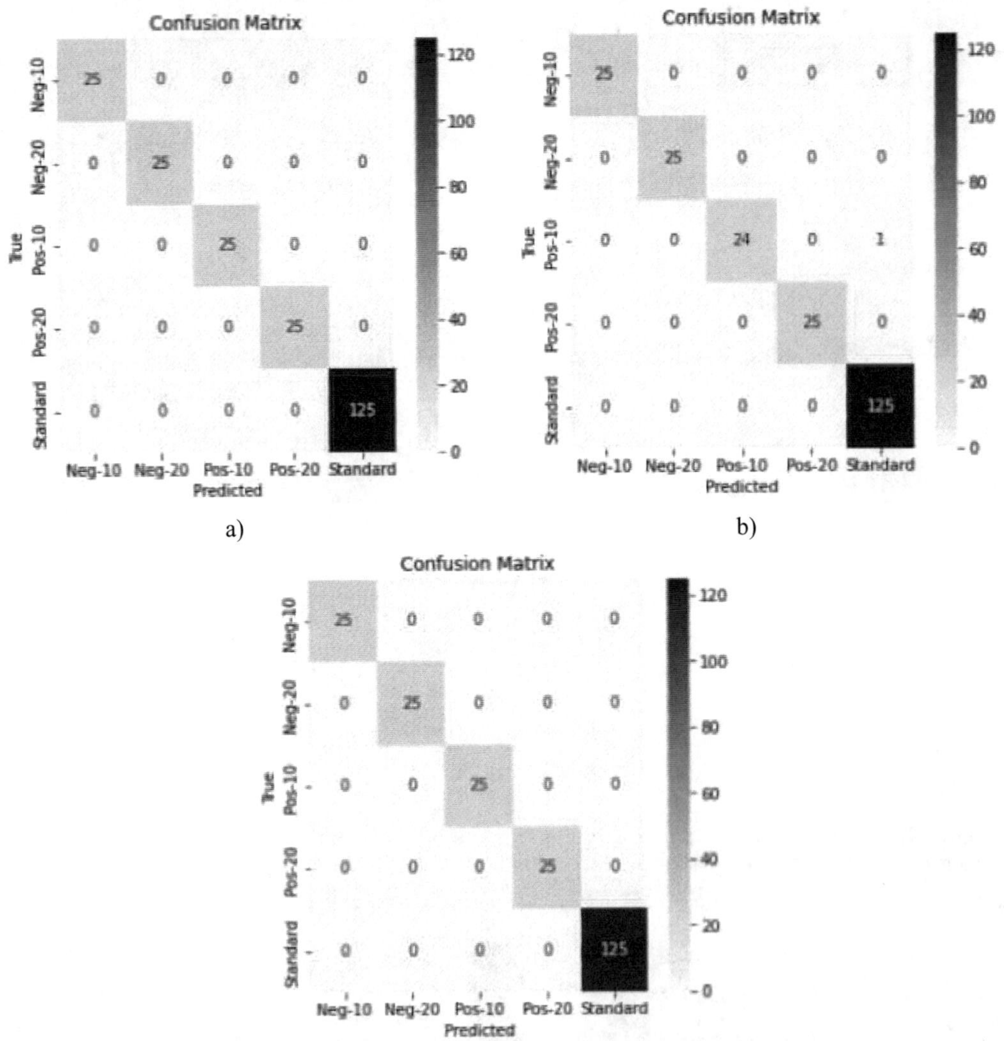

Figure 6. Confusion matrix of a) MLR, b) ANN, and c) CNN models for test dataset.

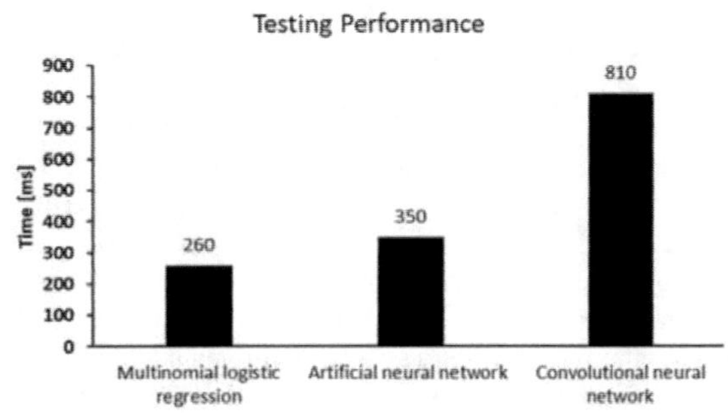

Figure 7. Testing performance of the ML models.

Extant literature has identified that a slight deviation from the thresholded VED can induce structural defects in the part [13,19–22]. This is also evidenced in this study as shown in Figure 8, which includes micrographs of polished cross-section with a magnification of 3.15x, highlight of OT images and their mean gray values as a function of each exposure strategy (Figure 1) in the build direction (Z-height).

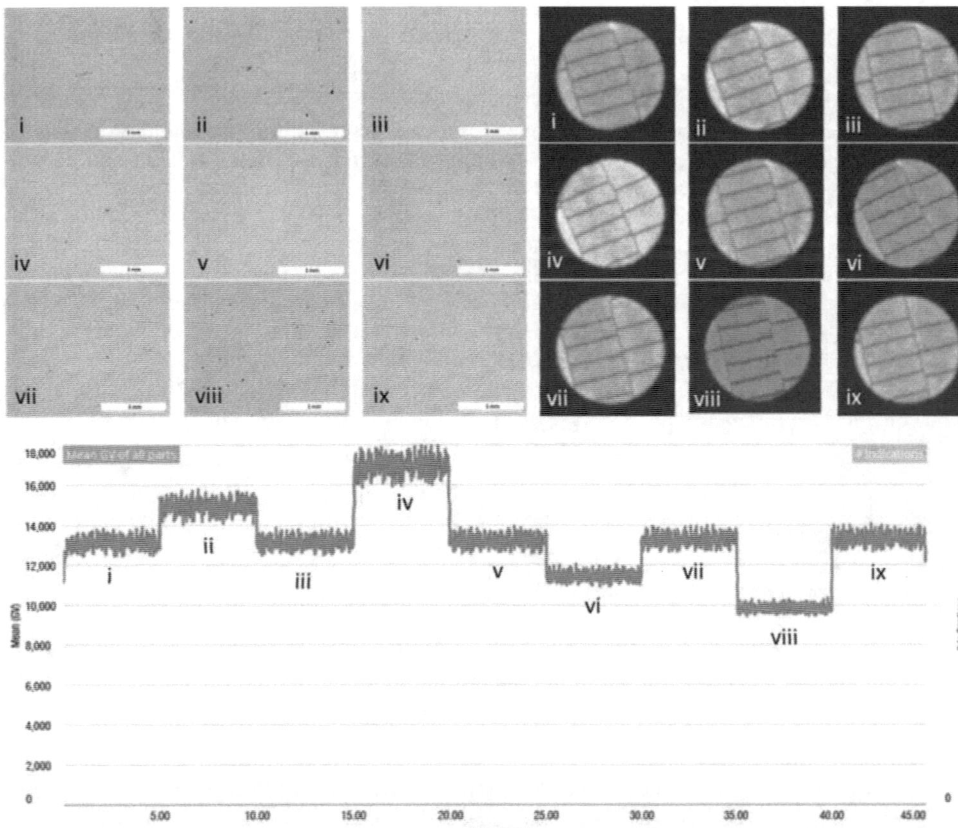

Figure 8. Micrograph (3.15x), OT image, and mean gray values of each exposure strategy in the build direction of the test specimen.

This study created defective regions using build parameter manipulation that are concomitant to lack of fusion and keyhole porosity. Two classes of keyhole porosity were created by an increase in Standard VED by 10% (Figure 8*ii*) and 20% (Figure 8*iv*). Similarly, two classes of lack of fusion were created by a reduction in Standard VED by 10% (Figure 8*vi*) and 20% (Figure 8*viii*). Generally, the micrographs presented in Figure 8 yielded higher defect percentage for deviated VED compared to the preceding Standard VED. The Δ-20% VED yielded the highest defect percentage of 0.24% compared to the average Standard VED defect percentage of 0.096% observed in this study. Additional samples representing a statistically sound dataset should be collected for conclusive outcome. Compared to the literature [13,22–24], this study develops and evaluates the accuracy and performance of three ML classification methods for detecting four defective classes and one standard class. The outcome confirms high performance (up to 100% accuracy) for training, validation, and test datasets. No indications of over- or under- fitting were observed because the validation error was similar to the training error. The high speed of detection (within 100s of millisecond for test dataset of 225 OT images) observed in this study enables opportunities to detect defective regions in real time and even compensate them in the current or subsequent layers by tuning the VED using a closed feedback loop. Considering such a high performance outcome for detecting layer-based defective regions with conventional computing power, this study opens opportunities to detect multiple vector-based defective regions in real-time for future work. Future studies should focus on further destructive and non-destructive testing for verifying the degree of defect generation and compensation also considering the degree of self-healing phenomena.

4. Conclusion

Additive manufacturing is shifting towards general-purpose manufacturing technology analogous to dynamos and computers that offer a vast number of applications in a wide range of industries. However, uncertainty in passing stringent quality standards hinders its adoption. To assist this transition, this study leverages the layer-by-layer mechanism of PBF-LB to develop three ML classification methods to detect defective regions inside the parts. A tailor-made test specimen is additively manufactured with five segmented exposure strategies emulating defective regions composed of two classes of lack of fusion and two classes of keyhole porosity. The artificial intelligence-based classifiers yielded flawless accuracy (up to 100% accuracy for MLR and CNN) for classifying OT images of each exposure strategy composed of independent test dataset for unbiased evaluation metrics. The high performance and low cost of computing of this layer-based approach opens a direction for detecting and compensating scan-based anomalies in real time for shifting the paradigm of uncertainty in quality of additively manufactured parts.

Funding: This research has been executed with Tandem Industry Academia funding from the Finnish Research Impact Foundation.

Conflict of interest: The authors declare no conflict of interest.

Acknowledgements: Special acknowledgements are extended to Harald Krauss, Jouni Partanen, Olli Nyrhilä, Tatu Syvänen, Pilvi Ylander, and Niko Raitanen for valuable discussions.

References

1. ISO/ASTM 52900. ISO/ASTM 52900:2021(en) Additive Manufacturing - General principles - Terminology. 2nd ed. ISO/ASTM International 2015; 2021.
2. Akmal JS. Digital Unique Component Manufacturing Through Direct and Indirect Additive Manufacturing [Internet]. Aalto University; 2017. Available from: https://aaltodoc.aalto.fi/handle/123456789/28509
3. Akmal JS. Switchover to additive manufacturing: Dynamic decision-making for accurate,

personalized and smart end-use parts [Internet]. Aalto University; 2022. Available from: http://urn.fi/URN:ISBN:978-952-64-1013-5
4. Akmal JS, Salmi M, Björkstrand R, Partanen J, Holmström J. Switchover to industrial additive manufacturing: dynamic decision-making for problematic spare parts. Int J Oper Prod Manag. 2022;42(13):358–84.
5. Salmi M, Partanen J, Tuomi J, Chekurov S, Björkstrand R, Huotilainen E, et al. Digital spare parts [Internet]. 2018. Available from: http://urn.fi/URN:ISBN:978-952-60-3746-2
6. Akmal JS, Salmi M, Mäkitie A, Björkstrand R, Partanen J. Implementation of industrial additive manufacturing: Intelligent implants and drug delivery systems. J Funct Biomater. 2018 Jun 29;9(3):41.
7. Wohlers. Wohlers Report. Colorado: Wohlers Associates Inc.; 2021.
8. Kukko K, Akmal JS, Kangas A, Salmi M, Björkstrand R, Viitanen AK, et al. Additively manufactured parametric universal clip-system: An open source approach for aiding personal exposure measurement in the breathing zone. Appl Sci. 2020;10(19):6671.
9. Salmi M, Akmal JS, Pei E, Wolff J, Jaribion A, Khajavi SH. 3D printing in COVID-19: Productivity estimation of the most promising open source solutions in emergency situations. Appl Sci [Internet]. 2020 Jun 9;10(11):4004. Available from: https://www.mdpi.com/2076-3417/10/11/4004
10. Akmal JS, Salmi M, Hemming B, Teir L, Suomalainen A, Kortesniemi M, et al. Cumulative inaccuracies in implementation of additive manufacturing through medical imaging, 3D thresholding, and 3D modeling: A case study for an end-use implant. Appl Sci [Internet]. 2020 Apr 24;10(8):2968. Available from: https://www.mdpi.com/2076-3417/10/8/2968
11. Ullah R, Lian J, Akmal J, Wu J, Niemi E. Prediction and validation of melt pool dimensions and geometric distortions of additively manufactured AlSi10Mg. Int J Adv Manuf Technol [Internet]. 2023;(0123456789). Available from: https://doi.org/10.1007/s00170-023-11264-w
12. Oettmeier K, Hofmann E. Impact of additive manufacturing technology adoption on supply chain management processes and components. J Manuf Technol Manag [Internet]. 2016 [cited 2021 May 19];27(7):944–68. Available from: https://www.emerald.com/insight/content/doi/10.1108/JMTM-12-2015-0113/full/html
13. Wang P, Yang Y, Moghaddam NS. Process modeling in laser powder bed fusion towards defect detection and quality control via machine learning: The state-of-the-art and research challenges. J Manuf Process [Internet]. 2022;73(November 2021):961–84. Available from: https://doi.org/10.1016/j.jmapro.2021.11.037
14. Soundararajan B, Sofia D, Barletta D, Poletto M. Review on modeling techniques for powder bed fusion processes based on physical principles. Addit Manuf [Internet]. 2021;47(April):102336. Available from: https://doi.org/10.1016/j.addma.2021.102336
15. Bauereiß A, Scharowsky T, Körner C. Defect generation and propagation mechanism during additive manufacturing by selective beam melting. J Mater Process Technol [Internet]. 2014;214(11):2522–8. Available from: http://dx.doi.org/10.1016/j.jmatprotec.2014.05.002
16. Abadi M, Agarwal A, Barham P, Brevdo E, Chen Z, Citro C, et al. TensorFlow: Large-Scale Machine Learning on Heterogeneous Distributed Systems. 2016 Mar 14 [cited 2022 Oct 8]; Available from: https://arxiv.org/abs/1603.04467
17. Albawi S, Mohammed TA, Al-Zawi S. Understanding of a convolutional neural network. Proc 2017 Int Conf Eng Technol ICET 2017. 2018;2018-Janua:1–6.
18. Gholamy A, Kreinovich V, Kosheleva O. Why 70/30 or 80/20 Relation Between Training and Testing Sets : A Pedagogical Explanation. Dep Tech Reports. 2018;1–6.
19. Luo Q, Yin L, Simpson TW, Beese AM. Effect of processing parameters on pore structures, grain features, and mechanical properties in Ti-6Al-4V by laser powder bed fusion. Addit Manuf [Internet]. 2022;56(February):102915. Available from: https://doi.org/10.1016/j.addma.2022.102915
20. Gong H, Rafi K, Gu H, Starr T, Stucker B. Analysis of defect generation in Ti-6Al-4V parts

made using powder bed fusion additive manufacturing processes. Addit Manuf [Internet]. 2014;1:87–98. Available from: http://dx.doi.org/10.1016/j.addma.2014.08.002

21. ISO/ASTM/TR 52906. Additive manufacturing. Non-destructive testing. Intentionally seeding flaws in metallic parts (ISO/ASTM/TR 52906:2022). 2022.
22. Schwerz C, Nyborg L. A neural network for identification and classification of systematic internal flaws in laser powder bed fusion. CIRP J Manuf Sci Technol [Internet]. 2022;37:312–8. Available from: https://doi.org/10.1016/j.cirpj.2022.02.010
23. Baumgartl H, Tomas J, Buettner R, Merkel M. A deep learning-based model for defect detection in laser-powder bed fusion using in-situ thermographic monitoring. Prog Addit Manuf [Internet]. 2020;5(3):277–85. Available from: https://doi.org/10.1007/s40964-019-00108-3
24. Kunkel MH, Gebhardt A, Mpofu K, Kallweit S. Quality assurance in metal powder bed fusion via deep-learning-based image classification. Rapid Prototyp J. 2020;26(2):259–66.

Laser welding monitoring with multisensory data fusion: A brief review

Li-Wei Hsu, Antti Salminen

liwhsu@utu.fi

Abstract As digital manufacturing is being implemented across industries, the automation of the laser welding process is a crucial step to enhance production efficiency. To monitor the in-situ welding process, there are several approaches to detect the electromagnetic and mechanical waves on various frequencies for comprehending laser beam-material interaction. Five sensing techniques, namely the optical microphone, welding camera, inline coherent imaging, infrared camera, and heat flux sensor, can be employed to identify distinct features in the laser welding process. These features include pore formation, melt pool geometry, weld bead topography, keyhole depth, and thermal distribution. The discussion of a proposed welding system designed with compatibility for multisensory data fusion is included, both on its capabilities and potential challenges, to offer guidance of welding monitoring.

Keywords: Laser welding monitoring, inline coherent imaging, surface topography, keyhole depth, welding camera, acoustic emission

1 Introduction

Laser welding is a versatile joining method with high precision and flexibility applied in various industries, and the operational scope spans from a few millimeters in thickness to several meters in length. The high-power density contributes to a high production rate, compared to arc welding and plasma welding [1]. As the laser welding process involves highly complex physical mechanisms like laser absorption, conductive and convective heat transfer and hydrodynamic melt pool flow, it takes substantial research effort to understand these phenomena and to enhance the stability and repeatability [2]. Consistent quality is crucial for ensuring structural safety, so reduction of defects is the main goal. This review paper explains how a laser welding is conducted and how to improve the quality with the monitoring data from various types of sensors. A welding system is proposed to outline a structured hierarchy of data and information and to demonstrate how these elements can improve the current system.

2 Mechanism

The common types of lasers used for welding are fiber laser, disk laser, and carbon dioxide laser with the wavelength of near-infrared or far-infrared spectrum, depending on its laser media. Coherent stimulated emission has a fixed frequency as monochromatic radiation, excluding the presence of ionizing radiation.

2.1 Laser beam-metal interaction

To understand the laser welding process, light-matter interaction, or to be specific, laser beam-metal interaction is discussed here in the microscopic manner. When the laser beam reaches the material surface, the radiation provides the energy to heat up the metals into liquid state and vapor state, then the thermal energy which conducts to the heat-affected zone (HAZ) to change the mechanical properties on the base metal. Depending on energy input and interaction time, lasers can do ablation, welding and cutting [3] [4]. If the penetration depth exceeds the thickness of the base metal, the melt pool falls from the spot, which is called drop-through or piercing [5]. For in-situ observation of the laser welding process, the X-ray phase contrast method can be used for identifying keyhole and melt pool [6] [7]. The dynamic radiography, or the X-ray imaging with high frame rate, can reach 50 thousand frames per second, meaning the increment change in 0.02 millisecond can be analyzed [8]. The classical problem of laser-induced cavitation bubble dynamics can be investigated attributed to the highly resolved images [9] [10] [11]. Note that these experiments with X-ray phase contrast imaging are conducted e.g., in synchrotrons, the large-scale facilities for high brightness and fast acquisition time. From the in-situ observation of laser welding, the keyhole, consisting of gaseous metal, can be seen at the laser beam-metal interaction point, and its geometry should be analyzed to ensure the full penetration without bubbles trapped during the keyhole wall closing stage [12] [13]. In some research papers, the keyhole is called vapor capillary, and its geometry is discussed in the manner of shape, size, and depth [14]. Since the keyhole behaves as an efficient point to absorb the laser energy, the relation of absorption rate between laser energy and welding speed is investigated with water-calorimetric method or integrated sphere radiometry [15] [16] [17]. Ideally the metals joining method aims to form the metallic bonds on the interface; however, in less favorable instances, it results in some defects like pores, undercuts, cracks, deformation and oxidation to a certain context [18] [19]. Due to the unstable keyhole movement, high thermal gradient, and insufficient melt flow, the laser welding speed has an upper process limit which prevents it from faster production [20] [21] [22] [23]. Therefore, it is important to monitor the keyhole, melt pool and weld bead to ensure the weld quality. Meanwhile, the mechanical properties of the weld joints must be suitable to avoid e.g., the fatigue crack initiation [24] [25]. The monitoring results can be incorporated to correlations of crack initiation location and fatigue strength.

2.2 Sensor development

Over the last few decades, sensor technology has made significant advancement in terms of accuracy, speed and versatility thanks to the miniaturization of integrated circuits and improvement in optic systems. The enhanced signal processing techniques with the novel algorithms and the development of materials in sensors with better sensitivity also contribute to the sensor development. In addition, the high computing power enables the real-time monitoring and control system with visualization and user interface to evaluate the quality and make informed decisions. The system with these sensors can extract meaningful information and identify patterns or anomalies, which are beneficial for developing a smart system to enhance the weld quality. These sensors include low-coherence interferometry, infrared cameras, high-speed welding cameras with active illumination, pyrometers, photodiodes, acoustic emission sensors, and heat flux sensors, which will generate data to be interpreted and integrated to understand the weld quality. Before the processed data can be combined to evaluate the weld, the data alignment is needed to ensure the time frame is compatible with each other and synchronized to have a common starting point and finishing line.

IOP Conf. Series: Materials Science and Engineering 1296 (2023) 012014 doi:10.1088/1757-899X/1296/1/012014

3 The trend in laser welding technology

This review paper briefly categorizes the welding techniques into three stages, depending on the involvement of machines on implementation and perception, meanwhile the ability for the algorithm to improve the data interpretation quality based on the experience is also discussed here as it is the trend of developing the smart welding system. Below is the table to show the three stages of the welding system.

Table 1. The three stage of welding system based on the perception and implementation.

Welding system	Human perception	Machine perception
Human implementation	1 Conventional welding training	Not in discussion
Machine implementation	2 Predefined welding automation	3 Future adaptive system

3.1 Conventional welding training

When welding is performed by human workers, the training includes the types of welding symbols and materials properties. While doing the hand-held laser welding, there is pictorial drawing well defined by the mechanical design for the welders to follow. Thus, hand-eye coordination is important, but there is limited endurance and recognition capability. Restricted by nature, the visible light spectrum spans from 380 nm to 760 nm and the welding helmet or welding goggles filter out the radiation beyond the visible light, which may result in a loss of detailed information about weld quality [26].

3.2 Predefined welding automation

As the automation gains prevalence on the production line, the welding equipment installed on the robotic arm can enhance the production rate. The trajectory of the weld line is determined by predefined welding parameter values and is executed using the robotic arm, so the quality is more consistent and uniform than the preceding stage. Real-time monitoring equipment can be installed to measure the quality, but the adjustment is not made in time. Typically, laser welding is carried out in this category. Post-weld X-ray inspection is run to examine the pores and cracks after solidification of the melt pool. Though this process can verify the quality, the delay of the defects formation and discovery is prolonged, and there is no integrated feature for perturbation cancellation in the system. To spot the anomalies on time and to mitigate the defects, the future adaptive system is needed [27].

3.3 Future adaptive system

To develop an adaptive system, both advanced sensing and control technologies are used for adjusting the laser welding parameters during the welding process. There are several types of real-time feedback that can be used to analyze the welding process, and the first one is the seam trajectory. The seam tracking capability relies on the ahead sensing line and control architecture to adjust the position and orientation for a curved line or a straight line [28]. If the robotic arm is installed on a mobile platform, the working area of the seam tracking welding can be extended [29]. Next, the second type of the feedback is the melt pool geometry, and it can exhibit the thermal condition on the surface. When excessive energy from the laser causes a larger melt pool, it is prone to drop-through. Therefore, either the laser power should be reduced or the welding speed should be increased before the piercing takes place. The real-time feedback from these non-destructive testing can reduce the delay between defect formation and defect discovery, and it can be applied in other types of welding like hybrid laser welding too. After discussing the feedback, there is one trend that many researchers start to investigate, which is the deep learning algorithm to correlate the signature of defects and predict the crack initiation [30] [31]. Thus, the application of these features will be considered to develop the adaptive system for better weld quality control.

IOP Conf. Series: Materials Science and Engineering 1296 (2023) 012014 doi:10.1088/1757-899X/1296/1/012014

4 Proposed welding system

Based on the trace of signal from the laser welding process, there are three types of inspection to be discussed in this system, which are ultrasound wave signal, electromagnetic wave signal, and heat flux on the surface. The corresponding sensors tailored to each signal type are involved in collecting data, subsequently interpreting it into observable phenomena. For example, a welding camera can capture visible light to provide insights of melt pool geometry and surface topology. Similarly, a thermal camera can detect the infrared light to show the thermal distribution across the surface. These phenomena are beneficial for researchers to comprehend the properties after a considerable number of samples are tested.

With the identified properties over time, the domain knowledge can be cultivated to inform future decision on welding parameters. As deep learning algorithm has increasing prevalence in the manufacturing industry, it becomes imperative to assess its rule in the overall process too. If the human learning represented by the blue box in Figure 1 is replaced by machine learning, the system cannot function because the algorithms for computational statistics lack the capability to understand the properties or to establish the domain knowledge that includes the relevant theories, principles and relationships within the domain. Therefore, in this review paper, only the connection between sensing level and features level, marked with purple arrows, would incorporate the algorithms to interpret the certain phenomena from the welding process. Noteworthy that some phenomena utilize multiple sensors in parallel, which is planned as the engineering redundancy for enhancing system robustness. One sensing technique that has yet to receive substantial attention in the community is the heat flux monitoring during welding, and this aspect is integrated in the proposed system discussed in this review paper.

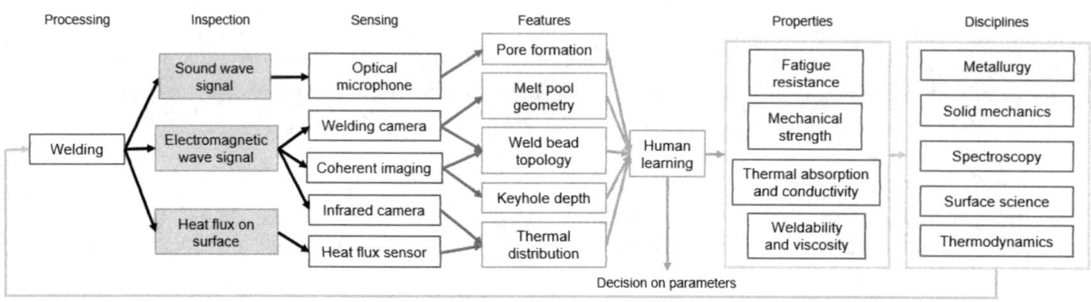

Figure 1. The proposed welding system showing the level of data and information.

4.1 Ultrasonic signal processing

During the welding process, the formation of spatters and pores emits distinctive acoustic signals detectable by a microphone [32]. As acoustic emission is essentially a mechanical wave propagating through the air, both mechanical microphone and optical microphone can be used for monitoring. However, these two types of microphones operate on the different mechanisms to generate the data. In this proposed system, the optical microphone is preferred as it has high sensitivity and significantly broader frequency response.

Since the laser sensor detects the air density change during the welding process, there is no mechanical moving part in the microphone. The reason that it can cover a wider frequency range from 2 kHz to 2 MHz is that the signal is processed through Fourier transform to convert the time-domain signal to frequency-domain signal. In this way, the frequency range is not constrained by membranes as in conventional mechanical microphones. Subsequently, the acoustic emission signals can be analyzed by filtering and correlation, or convolutional neural network, which is a machine learning algorithm designed for pattern recognition [33].

4.2 Multispectral image processing

Regarding electromagnetic waves, or optical signals, there are four aspects to be discussed, firstly, low-coherence interferometry; secondly, active pulsed laser illumination; thirdly, the structured light; and fourthly, radiative emission from the surface based on its temperature.

First part, optical coherence tomography, utilizing low-coherence interferometry, enables the real-time measurement of keyhole depth, and the imaging technique can generate point cloud representing the targeted surface area [34]. The path length is derived from the time difference of the sample point and the reference point. With a high sampling rate reaching several hundreds of kilohertz, its interested area includes not only keyhole depth, but also the seam profile and weld bead profile.

Second part, rapid pulsed laser illumination enables high-speed photography with the welding camera [35]. As the narrow band-pass filter can concentrate the detection of the monochromatic laser light source, the process radiation would hardly cause overexposure in the images. The acquisition rate of the high-quality welding images can easily reach thousands of frames per second, given the short exposure time that can capture illumination from the brief laser pulse, preferably on the microsecond scale.

Third part, structured light can present the surface profile when it traverses along the welding direction. The calculation of the surface profile involves knowledge of the angle and position between the bar pattern and the sensor. Noteworthy that the technique of photogrammetry can be used to construct the weld profile and to measure the misalignment without the structured light, but they are conducted as the post-weld inspection, rather than in-situ monitoring [36] [37]. Hence, if photogrammetry can be integrated to the welding camera without relying on structured light, it would be a notable achievement as it overcomes challenges associated with varying radiation from the melt pool.

Fourth part, the electromagnetic radiation emitted from the surface of the base metal and heat-affected zone can reveal the surface temperature. As laser energy is absorbed in the keyhole and extends to the surrounding area, the thermal gradient can be measured with the infrared thermal camera [38]. Other than the infrared thermal camera, photodiodes can be used to measure the surface temperature, but the drawback is that it only shows the overall emission rather than the thermal gradient [39]. It is important to note that all four parts rely on electromagnetic signals, and the potential overlapping of wavelengths may lead to interference among these different sources. In case the situation worsens, the spatial and temporal relations need to be taken into consideration to avoid constructive or destructive interference.

4.3 Heat flux monitoring

During the laser welding process, the plume and fume are the undesirable byproducts, containing metal particulates and metal oxides, that can impede optical sensors from receiving signals. Therefore, heat flux sensors would be valuable to measure the heat flow without potential interference from the fume or plume. These heat flux sensors should be mounted on the surface of the base metal for the thermal conduction from the keyhole and heat-affected zone, so there will be multiple sensors on the both sides of the workpiece to form an array to calculate the changing temperature [40] [41]. Once the temperature at the contact points is determined, it can be used to analyze the thermal phenomena and phase transformation in the solid state with the numerical model according to the metallurgical composition [42].

4.4 Multisensory data fusion

After each sensor receives the signal, it will be routed to the processing module for feature extraction, and the classification algorithms are used to evaluate the weld quality based on the significance of these features [43]. The algorithms for feature extraction may use conventional statistical filtering model or convolutional neural networks model, and the defects such as pores, undercuts and cracks are identified when specific features are derived from the deployed sensors. To establish an efficient monitoring system, the detection rates from various sensor combinations will be compared for a comprehensive evaluation of performance. This iterative process can optimize the monitoring system for enhanced weld quality assessment.

5 Potential problems

Challenges ahead, there are some common problems existing in previous experiments and will persist in the future experiments, such as the plume formation, shielding gas, and the beam shift [44] [45]. The solution of the plume formation may be incorporating the inert gas like argon to prevent oxidation, but it also decreases the keyhole stability, which may lead to increased pores formation. Moreover, as the plume is present within the atmospheric environment, the particulates can actually cause the laser beam shift [2] [46]. This type of associated dilemma needs more comprehensive approaches to solve in the future research. Additionally, from the computational perspective, the substantial volume of data from high-speed cameras and other high frame rate sensors demands considerable computing power, so achieving real-time result is a challenge to the algorithms on both sensors and computers in the adaptive welding system.

6 Conclusion

Digitalization in the manufacturing industries will increase the prevalence, and it can contribute to a safer workspace and more efficient production. This review paper presents the current research on the welding monitoring techniques and proposes an adaptive welding system with the brief introduction of each type of sensing method. Positioned as a foundation for future welding system development, more experiments are scheduled to validate the proposed system structure and to affirm the increased stability and repeatability. The research team is committed to achieving a more autonomous manufacturing system as a key objective.

7 Acknowledgement

Authors of this review paper acknowledge the support from CaNeLis project (Carbon-neutral lightweight ship structures using advanced design, production, and life-cycle services). The project is funded by Business Finland with the duration of project is from October 2022 to September 2025, and it includes Meyer Turku Oy, SSAB AB, Cavitar Oy, Mobimar Oy, Refamo Oy, Suisto Engineering Oy together with Aalto University and LUT University. Authors would like to thank partners of this project for their contribution to this article.

8 CRediT authorship contribution statement

Li-Wei Hsu: Methodology, Formal analysis, Investigation, Validation, Visualization, Writing – original draft, Writing – review & editing. Antti Salminen: Conceptualization, Writing – review & editing, Supervision, Resources, Funding acquisition

9 Reference

[1] S. Katayama, "Introduction: fundamentals of laser welding," in *Handbook of Laser Welding Technologies*, 2013, pp. 3-16. DOI: 10.1533/9780857098771.1.3

[2] S. Katayama, A. Yohei, M. Mizutani and Y. Kawahito, "Development of Deep Penetration Welding Technology with High Brightness Laser under Vacuum," vol. 12, 2011, pp. 75-80. DOI: 10.1016/j.phpro.2011.03.010

[3] H. Wang, M. Nakanishi and Y. Kawahito, "Effects of welding speed on absorption rate in partial and full penetration welding of stainless steel with high brightness and high power laser," vol. 249, 2017, pp. 193-201. DOI: 10.1016/j.jmatprotec.2017.06.014

[4] H. Roozbahani, M. Alizadeh, H. Handroos and A. Salminen, "Design and Development of a Multisensory Real-Time Monitoring Platform for Ultrafast Laser Engraving Process," vol. 10, 2022, pp. 66529-66544. DOI: 10.1109/ACCESS.2022.3185131

[5] J. Lind, N. Weckenmann, C. Hagenlocher, R. Weber and T. Graf, "Analysis and optimization of the piercing process in laser beam cutting by means of high-speed X-ray imaging," vol. 69, 2021, pp. 303-310. DOI: 10.1016/j.jmapro.2021.07.048

[6] Y. Kawahito and H. Wang, "In-situ observation of gap filling in laser butt welding," vol. 154, 2018, pp. 73-77. DOI: 10.1016/j.scriptamat.2018.05.033

[7] J. Wagner, C. Hagenlocher, M. Hummel, A. Olowinsky, R. Weber and T. Graf, "Synchrotron X-ray Analysis of the Influence of the Magnesium Content on the Absorptance during Full-Penetration Laser Welding of Aluminum," vol. 11, 2021, p. 797. DOI: 10.3390/met11050797

[8] Z. Wu, G. Tang, S. J. Clark, A. Meshkov, S. Roychowdhury, B. Gould, V. Ostroverkhov, T. Adcock, S. J. Duclos, K. Fezzaa, C. Immer and A. D. Rollett, "High frequency beam oscillation keyhole dynamics in laser melting revealed by in-situ x-ray imaging," vol. 4, 2023, p. 5. DOI: 10.1038/s43246-023-00332-z

[9] G. T. Bokman, L. Biasiori-Poulanges, B. Lukić, C. Bourquard, D. W. Meyer, A. Rack and O. Supponen, "High-speed x-ray phase-contrast imaging of single cavitation bubbles near a solid boundary," vol. 35, 2023, p. 013322. DOI: 10.1063/5.0132104

[10] W.-S. Chung, A. Häusler, M. Hummel, A. Olowinsky, A. Gillner, F. Beckmann and J. Moosmann, "In-situ x-ray phase contrast observation of the full penetration spot welding on limited aluminum material thickness," vol. 34, 2022, p. 042019. DOI: 10.2351/7.0000772

[11] M. Boley, F. Abt, R. Weber and T. Graf, "X-Ray and Optical Videography for 3D Measurement of Capillary and Melt Pool Geometry in Laser Welding," vol. 41, 2013, pp. 488-495. DOI: 10.1016/j.phpro.2013.03.105

[12] M. Vänskä, F. Abt, R. Weber, A. Salminen and T. Graf, "Effects of Welding Parameters Onto Keyhole Geometry for Partial Penetration Laser Welding," vol. 41, 2013, pp. 199-208. DOI: 10.1016/j.phpro.2013.03.070

[13] S. M. Robertson and A. F. Kaplan, "Multi-keyhole separation during multi-spot laser welding of duplex steel," vol. 143, 2021, p. 107382. DOI: 10.1016/j.optlastec.2021.107382

[14] E. N. Reinheimer, R. Weber and T. Graf, "Influence of the capillary geometry on the weld seam quality during high-speed laser welding," vol. 111, 2022, pp. 431-434. DOI: 10.1016/j.procir.2022.08.181

[15] Y. Kawahito, N. Matsumoto, Y. Abe and S. Katayama, "Relationship of laser absorption to keyhole behavior in high power fiber laser welding of stainless steel and aluminum alloy," vol. 211, 2011, pp. 1563-1568. DOI: 10.1016/j.jmatprotec.2011.04.002

[16] H. Wang, Y. Kawahito, R. Yoshida, Y. Nakashima and K. Shiokawa, "A model to calculate the laser absorption property of actual surface," vol. 118, 2018, pp. 562-569. DOI: 10.1016/j.ijheatmasstransfer.2017.11.023

[17] T. R. Allen, T. G. Fleming, T. J. Krause and J. M. Fraser, "Simultaneous high-speed keyhole depth and absorptance measurements in laser spot welding of dissimilar metals," vol. 111, 2022, pp. 5-9. DOI: 10.1016/j.procir.2022.08.041

[18] I. Eriksson, J. Powell and A. F. H. Kaplan, "Surface tension generated defects in full penetration laser keyhole welding," vol. 26, 2014, p. 012006. DOI: 10.1016/j.procir.2022.08.041

[19] J. Frostevarg and A. F. Kaplan, "Undercuts in Laser Arc Hybrid Welding," vol. 56, 2014, pp. 663-672. DOI: 10.1016/j.phpro.2014.08.071

[20] M. Mazar Atabaki, J. Ma, W. Liu and R. Kovacevic, "Pore formation and its mitigation during hybrid laser/arc welding of advanced high strength steel," vol. 67, 2015, pp. 509-521. DOI: 10.1016/j.matdes.2014.10.072

[21] M. Bachmann, X. Meng, A. Artinov and M. Rethmeier, "Evaluation of narrowed weld pool shapes and their effect on resulting potential defects during deep penetration laser beam welding," vol. 34, 2022, p. 042005. DOI: 10.2351/7.0000733

[22] Y. Hu, S. Wu, Y. Guo, Z. Shen, A. M. Korsunsky, Y. Yu, X. Zhang, Y. Fu, Z. Che, T. Xiao, S. Lozano-Perez, Q. Yuan, X. Zhong, X. Zeng, G. Kang and P. J. Withers, "Inhibiting weld cracking in high-strength aluminium alloys," vol. 13, 2022, p. 5816. DOI: 10.1038/s41467-022-33188-x

[23] E. N. Reinheimer, R. Weber and T. Graf, "Process limit imposed by the occurrence of undercuts during high-speed laser welding," vol. 34, 2022, p. 032003. DOI: 10.2351/7.0000621

[24] R. Zhang, C. Buchanan, V.-P. Matilainen, D. Daskalaki-Mountanou, T. B. Britton, H. Piili, A. Salminen and L. Gardner, "Mechanical properties and microstructure of additively manufactured stainless steel with laser welded joints," vol. 208, 2021, p. 109921. DOI: 10.1016/j.matdes.2021.109921

[25] A. Niraula, H. Remes and P. Lehto, "Local weld geometry-based characterization of fatigue strength in laser-MAG hybrid welded joints," vol. 67, 2023, pp. 1527-1544. DOI: 10.1007/s40194-023-01488-5

[26] Harris M, Welding Health and Safety: A Field Guide for OEHS Professionals, M. K. Harris, Ed., 2700 Prosperity Ave., Suite 250 Fairfax, VA 22031: American Industrial Hygiene Association, 2002. ISBN-10 : 1950286037

[27] Y. ZHANG, Q. WANG and Y. LIU, "Adaptive Intelligent Welding Manufacturing," vol. 100, 2021, pp. 63-83. DOI:10.29391/2021.100.006

[28] M. de Graaf, R. Aarts, B. Jonker and J. Meijer, "Real-time seam tracking for robotic laser welding using trajectory-based control," vol. 18, 2010, pp. 944-953. DOI: 10.1016/j.conengprac.2010.04.001

[29] X. Chen, A. G. Dharmawan, S. Foong and G. S. Soh, "Seam tracking of large pipe structures for an agile robotic welding system mounted on scaffold structures," vol. 50, 2018, pp. 242-255. DOI: 10.1016/j.rcim.2017.09.018

[30] S. Shevchik, T. Le-Quang, B. Meylan, F. V. Farahani, M. P. Olbinado, A. Rack, G. Masinelli, C. Leinenbach and K. Wasmer, "Supervised deep learning for real-time quality monitoring of laser welding with X-ray radiographic guidance," vol. 10, 2020, p. 3389. DOI: 10.1038/s41598-020-60294-x

[31] W. Huo, N. Bakir, A. Gumenyuk, M. Rethmeier and K. Wolter, "Strain Prediction Using Deep Learning during Solidification Crack Initiation and Growth in Laser Beam Welding of Thin Metal Sheets," vol. 13, 2023, p. 2930. DOI: 10.3390/app13052930

[32] L. Schmidt, F. Römer, D. Böttger, F. Leinenbach, B. Straß, B. Wolter, K. Schricker, M. Seibold, J. Pierre Bergmann and G. Del Galdo, "Acoustic process monitoring in laser beam welding," vol. 94, 2020, pp. 763-768. DOI: 10.1016/j.procir.2020.09.139

[33] K. Wasmer, R. Drissi-daoudi, G. Masinelli, T. Quang-le, R. Loge and S. A. Shevchik, "When am (additive manufacturing) meets ae (acoustic emission) and AI (artificial intelligence)," vol. 28, 2023. DOI: 10.58286/27606

[34] M. Werner, J. Wagner, F. Ribbeck, S. Hensel, K. Goth, T. Graf and G. Meschut, "Influence of the incident angle on the OCT measurement during remote laser beam welding," vol. 111, 2022, pp. 513-517. DOI: 10.1016/j.procir.2022.08.081

[35] I. Eriksson, P. Gren, J. Powell and A. F. Kaplan, "New high-speed photography technique for observation of fluid flow in laser welding," vol. 49, 2010, p. 1. DOI: 10.1117/1.3502567

[36] M. Rodríguez-Martín, P. Rodríguez-Gonzálvez, S. Lagüela and D. González-Aguilera, "Macro-photogrammetry as a tool for the accurate measurement of three-dimensional misalignment in welding," vol. 71, 2016, pp. 189-197. DOI: 10.1016/j.autcon.2016.08.016

[37] E. R. d. Oña, M. Rodríguez-Martin, P. Rodríguez-Gonzálvez, R. Mora and D. González-Aguilera, "WELDMAP: A Photogrammetric Suite Applied to the Inspection of Welds," vol. 12, 2022, p. 2553. DOI: 10.3390/app12052553

[38] X. Gao, D. You and S. Katayama, "Infrared image recognition for seam tracking monitoring during fiber laser welding," vol. 22, 2012, pp. 370-380. DOI: 10.1016/j.mechatronics.2011.09.005

[39] G. Chianese, P. Franciosa, J. Nolte, D. Ceglarek and S. Patalano, "Characterization of Photodiodes for Detection of Variations in Part-to-Part Gap and Weld Penetration Depth During Remote Laser Welding of Copper-to-Steel Battery Tab Connectors," vol. 144, 2022. DOI: 10.1115/1.4052725

[40] E. Baygildina, L. Smirnova, R. Juntunen, K. Murashko, A. V. Mityakov, M. Kuisma, O. Pyrhönen, P. Peltoniemi, K. Hynynen, V. Y. Mityakov and S. Z. Sapozhnikov, "Condition Monitoring of Wind Power Converters Using Heat Flux Sensor," vol. 11, 2016, p. 239. DOI: 10.15866/iree.v11i3.8404

[41] V. Mityakov, A. Mityakov, A. Vintsarevich and D. Gerasimov, "Gradient heat flux measurement as monitoring tool for the diesel engine," vol. 245, 2018, p. 14001. DOI: 10.1051/matecconf/201824514001

[42] M. Kubiak and W. Piekarska, "Comprehensive model of thermal phenomena and phase transformations in laser welding process," vol. 172, 2016, pp. 29-39. DOI: 10.1016/j.compstruc.2016.05.014

[43] C. Knaak, J. von Eßen, M. Kröger, F. Schulze, P. Abels and A. Gillner, "A Spatio-Temporal Ensemble Deep Learning Architecture for Real-Time Defect Detection during Laser Welding on Low Power Embedded Computing Boards," vol. 21, 2021, p. 4205. DOI: 10.3390/s21124205

[44] C. Brock, R. Hohenstein and M. Schmidt, "Mechanisms of vapour plume formation in laser deep penetration welding," vol. 58, 2014, pp. 93-101. DOI: 10.1016/j.optlaseng.2014.02.001

[45] F. Tenner, C. Brock, F. Klämpfl and M. Schmidt, "Analysis of the correlation between plasma plume and keyhole behavior in laser metal welding for the modeling of the keyhole geometry," vol. 64, 2015, pp. 32-41. DOI: 10.1016/j.optlaseng.2014.07.009

[46] M. Sokolov, A. Salminen, S. Katayama and Y. Kawahito, "Reduced pressure laser welding of thick section structural steel," vol. 219, 2015, pp. 278-285. DOI: 10.1016/j.jmatprotec.2014.12.026

IOP Conf. Series: Materials Science and Engineering 1296 (2023) 012015 doi:10.1088/1757-899X/1296/1/012015

Influence of buoyancy on phase separation during laser melting of Silicon and Iron ore

Dan Yang*, Himani Naesstroem*, Joerg Volpp

Department of Engineering Sciences and Mathematics, Luleå University of Technology, 97187 Luleå, Sweden

*Corresponding author; Email: dan.yang@ltu.se, himani.naesstroem@ltu.se

Abstract. Phase separation is a well-known effect in liquid-liquid interactions, which can also occur during laser-related processes such as laser melting and laser alloying. However, the mechanism of phase separation between immiscible liquids on a millimeter scale during rapid laser processes has not been fully investigated, in which the extent of buoyancy's contribution to it has remained unclear. Therefore, this investigation focused on the effect of buoyancy on liquid phase separation during the laser melting of silicon (Si) and iron ore. A simplified 2D numerical model was established to simulate the motion of a single Si liquid droplet in iron ore melt with and without the impact of gravity, respectively. The rise velocity of the droplet was calculated and analyzed under the effect of gravity. In addition, the phenomena of simultaneous laser remelting of Si and iron ore in layers were recorded with a high-speed camera, and the element and phase distributions of the target nugget were analyzed by Scanning Electron Microscopy/Energy Dispersive Spectroscopy (SEM/EDS). By combining the simulation, high-speed imaging, and SEM/EDS analysis, the effect of buoyancy on phase separation has been qualitatively analyzed. This investigation revealed that buoyancy is not the main driving force of liquid-liquid phase separation during rapid laser processing.

1. Introduction

Liquid-liquid phase separation is a widely observed phenomenon between two immiscible solutions with different chemical compositions, which can also occur in rapid laser material processing. Some major factors contribute to this phenomenon, such as buoyancy, Marangoni motion, and miscibility gap [1]. In some cases, buoyancy is the main driving force, for example, rapid separation of metal components of different densities may occur by precipitation or flotation on the micrometer and nanoscale[2–4]. The behavior of two immiscible liquids depends on their physical properties, such as density, viscosity, and the interfacial tension between the liquid-liquid interface and surface tension in the gas-liquid interface, some of which are temperature-dependent. However, the mechanism of phase separation at the millimeter scale has not been fully investigated in the fast laser processing of metals and its main driving force remains unclear.

The work presented in this paper is part of the project work. The aim of the project is to investigate the potential reduction reactions between reducing agents, such as Si [5], and iron ore. The main goal of this paper is to verify to what extent phase separation is affected by buoyancy on a millimeter scale. An experimental setup was designed with two independent Si-rich and Fe-rich phases coexisting in a melt pool. Based on the level set method[6], a numerical model was established to simulate the motion of liquid Si in iron ore melt, and the simulation result was validated by high-speed imaging and SEM/EDS.

2. Methods and materials

2.1 Experiments

In order to design an experiment with two independent phases coexisting in a controlled manner, two layers of powder were manually applied on a copper substrate and the iron ore powder was spread on top of the silicon powder. The thickness of each layer was 0.5 mm and the processing speed was 600 mm/min. The iron ore powder and Si powder layers were simultaneously melted by a traveling laser beam, see Figure 1. A continuous wave Yb: fiber laser (wavelength 1070 nm) with a maximum power of 5 kW. The defocused spot radius on the powder bed surface is 0.66 mm. Argon was supplied as shielding gas (flow rate 13 L/min) from a laterally placed tubular copper nozzle (20 mm diameter).

The laser melting process was observed by a high-speed camera at an angle of 60°. The images were preferably recorded at a frame rate of 10000 fps, using a shutter time of 9 μs. Two diode laser beams illuminated the process with a wavelength of 810 nm. Therefore, a narrow bandpass filter with a wavelength of 810±10 nm was used in front of the camera objective, blocking most of the process light. To determine the composition of two independent phases that coexist in a generated nugget, the nugget was prepared by mounting and polishing for SEM/EDS analysis. The distribution of elements and phases in the cross-section of the generated nugget was then examined using SEM/EDS, with emphasis on detecting the primary elements of each powder layer, namely Si and Fe.

The iron ore powder (grain size: < 63 μm) was provided by LKAB, which primarily consisted of hematite (Fe_2O_3). Additionally, the Si powder utilized had a grain size of < 45 μm.

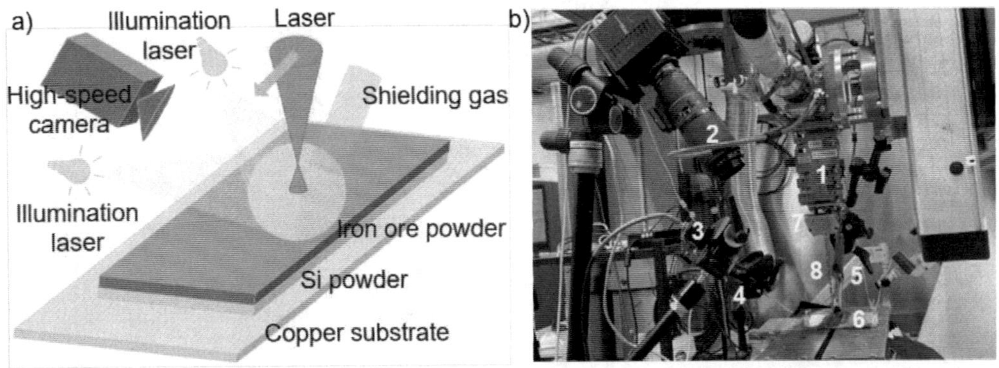

Figure 1. Designed experimental set-up a) schematic diagram; b) experimental setup (1: Beam focusing system, 2: High-speed camera, 3: Illumination laser 1, 4: Illumination laser 2, 5: Gas tube, 6: Sample, 7: Cross jet, 8: Laser beam)

2.2 Numerical solution

In this work, the commercial software COMSOL Multiphysics 6.0 was used to analyze the influence of buoyancy forces on phase separation during rapid laser processing. The phase and velocity field distributions were calculated by the level set method. Assumptions were made to simplify the simulation: a) the potential chemical reactions between iron ore melt and Si droplets were not simulated; b) the simulation was performed under isothermal conditions, and differences in heat transfer and physical properties due to temperature-dependency or content dependency were neglected; c) the fluid motion in the calculation domain is an incompressible fluid.

The physical effect of liquid Si interacting with liquid iron ore is a two-phase flow problem. Accordingly, a simplified two-dimensional axisymmetric model was developed to calculate the single droplet motion. The computational domain, as depicted in Figure 2, was designed with constraints on the length and width to keep computational efficiency. The initial droplet radius was chosen based on the size of the Si droplet measured via SEM, and the solidification time was measured by High-speed imaging. The typical size of the Si droplet in the track measured via SEM was adopted as the initial condition for the simulation.

Standard non-slip boundary conditions were set on the boundaries of the upper, side, and lower walls, while symmetric boundary conditions were set on the symmetry axis. Additionally, the pressure constraint point was placed in the lower right corner. The temperature for both liquids was set to 1550°C, which is the melting temperature of iron ore. The physical properties of iron ore melt and Si liquid at 1550°C are given in Table 1. The initial shape of the liquid droplet was assumed to be a circle. A finer mesh was created using the default physics-controlled meshing algorithm in COMSOL Multiphysics 6.0.

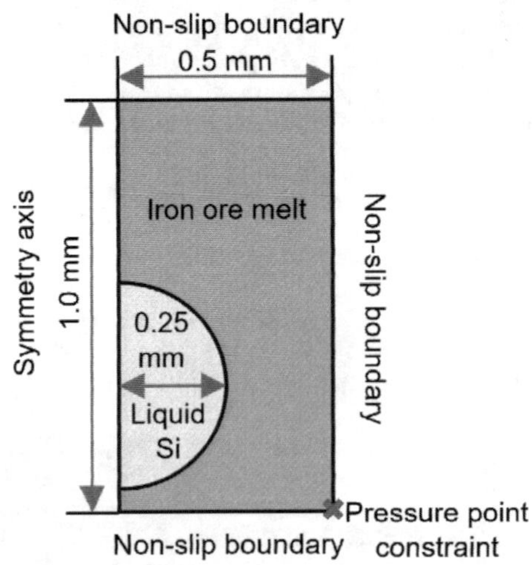

Figure 2. Schematic diagram of the computational domain and initial geometry and boundary conditions

Table 1. Physical properties of materials

Properties	Iron ore melt	Si liquid
Density (kg/m^3)	4000 [7]	2520 [8]
Viscosity (Pa·s)	0.99 [7]	0.000541 [8]
Interface tension with iron ore melt (N/m)	--	0.0078 [9]

3. Results and discussion

The track generated from the process showed two solidified phases, with the darker phase (Solid phase I) distributed rather uniformly along the track and the brighter phase (Solid phase II) distributed randomly along the track. Also, the darker phase has uneven sizing. A high-speed image of the laser remelting process in Figure 3 shows a partially solidified track with a melt pool, and the morphology of the solidified track was obtained by optical microscopy. During laser processing, two different liquid phases were found in the melt pool, while the solidified track consisted of two solidified phases. As shown in Figure 3, the length of the melt pool was 3 mm. Accordingly, the solidification time was 0.3 s, which was set as a logical time constraint for the simulation. After that, the liquid was considered to be solidified.

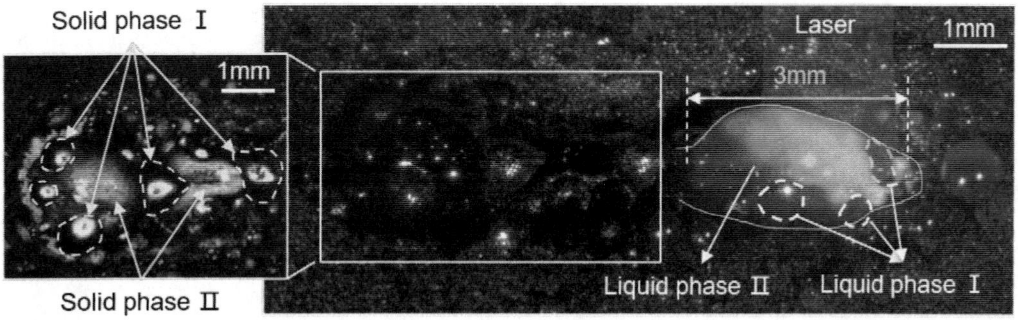

Figure 3. High-speed image and morphology of the solidified track

The generated track from the laser melting process was analyzed by SEM/EDS. Figure 4 shows the SEM/EDS image of the cross-section of a part of the track. There were some Si droplets in the track with a typical size of 0.25mm. Those larger droplets, shown in Figure 4, are not investigated in this paper because their size exceeds the average height of the melt pool. The dark spots were common pores in laser material processing, as verified in Figure 4 b) and c), where fewer elements were distributed. The gray phase inside the nugget indicated by the white dashed line was a Si-rich phase, while the brighter phase was a Fe-rich phase, as verified in Figure 4. Some Si-rich droplets appear ellipsoidal in shape, indicating that the droplet was not uniformly stressed in the liquid and show a tendency of moving upward. Besides, there are some small domains attached to the bottom of the track, which might be caused by uneven heat penetration, less heat at the bottom of the track, and faster cooling at the bottom due to the high thermal conductivity of the copper substrate. For this paper, the larger Si droplet was analyzed in detail.

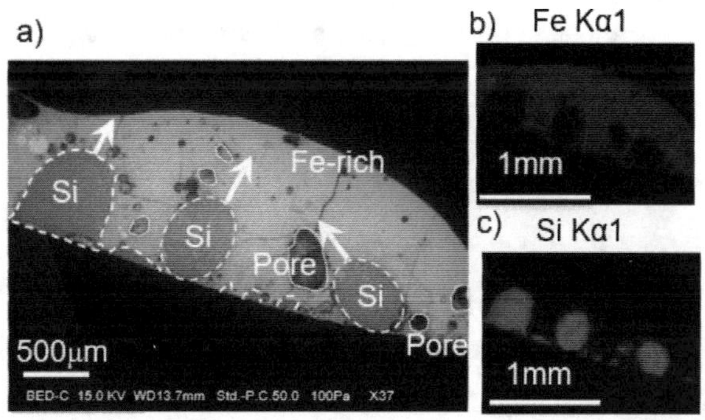

Figure 4. a) Cross-section polish of an exemplary longitudinal view of the track and SEM/EDS analysis of the area b) Fe- and c) Si-contents

High-speed imaging and SEM/EDS analysis showed that the powder layers were melted under laser irradiation, forming two immiscible Si-rich and Fe-rich liquids that formed phase separation. The Si-rich liquid showed an accumulation and upward-moving tendency in the Fe-rich melt. Subsequently, these immiscible liquids solidified into the Si-rich and Fe-rich phases within 0.3 s.

To further verify the impact of buoyancy on the motion of a single droplet was simulated. The phase-field distribution within the solidification time with and without gravity is shown in Figure 5 a) and b). Under gravity impact, the Si droplet moved upward by 0.025 mm in 0.3 s, while it remained still in the absence of gravity. It is indicated that buoyancy force does not make a significant contribution to the upward motion of Si droplets in this case. This is due to the relatively high viscosity of the iron ore melt, which provides a relatively high resistance for the silicon droplet within the melt to move slowly upward. It can also be seen by the SEM analysis in Figure 4, that the droplet in the middle shows an upward-moving tendency.

The final shape of the Si-rich domain is influenced by the balance of buoyancy force and flow drag resistance[10]. In the actual case, the shape of the Si droplet did not distort as it was expected. Two main reasons account for it: 1) the interfacial tension between the Si droplet and iron ore melt at their melting point is relatively high, which inhibits the deformation of the droplet. 2) the value of interfacial tension decreases with increasing temperature and the maximum temperature is around 1550 °C in this case. The interfacial tension employed in the simulation is 7.8 mN/m at 1550°C. For comparison, the interfacial tension between oil and water is 23-26 mN/m at 25 °C[11]. On the other hand, according to high-speed imaging frames, the solidification time is 0.3 s, which is too short for the droplet to distort. Another possibility is that the initial shape and size of the droplet also affect its terminal state, and they are closely related to the mechanism of Si droplets formation. This will be investigated in future work.

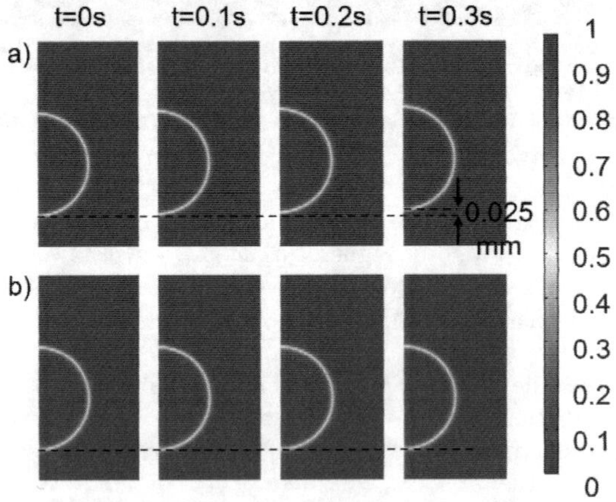

Figure 5. Phase-field distribution of the computational domain a) with gravity, b) without gravity

4. Conclusions

Based on the simulation and experimental results, investigating the phase separation of Si and iron ore under laser treatment, it has been proven that buoyancy does not make a significant contribution to the upward motion of the Si droplet during short-time laser treatment and is not the main driving force for millimeter-scale phase separation in rapid laser processing.

Acknowledgment

The authors acknowledge the project AMORF (No. P2022-00202) funded by Energimyndigheten, the Swedish Energy.

References

[1] Lappa M, Piccolo C and Carotenuto L 2004 Mixed buoyant-Marangoni convection due to dissolution of a droplet in a liquid–liquid system with miscibility gap *European Journal of Mechanics - B/Fluids* **23** 781–94

[2] Dai X, Zhou S, Wang M, Lei J, Wang C and Wang T 2017 Microstructure evolution of phase separated Fe-Cu-Cr-C composite coatings by laser induction hybrid cladding *Surf Coat Technol* **324** 518–25

[3] Zhang L, Yu G, He X, Li S, Ning W and Wang X 2019 Phase separated characteristics affected by cooling rate of immiscible Cu-Cr alloy by laser surface melting *J Alloys Compd* **772** 209–17

[4] Yue S, Qu J, Li G, Liu S, Guo Z, Jie J, Guo S and Li T 2022 A comprehensive investigation of carbon micro-alloying on microstructure evolution and properties of metastable immiscible Cu-Fe alloy *J Alloys Compd* **921** 166163

[5] Kaplan A F, Fedina T, and Brückner F 2022 Study of Si-domains enabling local reduction of laser-melted iron ore for iron-making during 3D-printing *Procedia CIRP* **111** 377-380.

[6] Osher S and Fedkiw R P 2001 Level Set Methods: An Overview and Some Recent Results *J Comput Phys* **169** 463–502

[7] Liu D, Loo C E and Evans G 2016 Flow characteristics of the molten mix generated during iron ore sintering *Int J Miner Process* **149** 56–68

[8] Assael M J, Armyra I J, Brillo J, Stankus S V., Wu J and Wakeham W A 2012 Reference Data for the Density and Viscosity of Liquid Cadmium, Cobalt, Gallium, Indium, Mercury, Silicon, Thallium, and Zinc *J Phys Chem Ref Data* **41** 033101

[9] Veksler I V., Kähn J, Franz G and Dingwell D B 2010 Interfacial tension between immiscible liquids in the system K2O-FeO-Fe2O3-Al2O3-SiO2 and implications for the kinetics of silicate melt unmixing *American Mineralogist* **95** 1679–85

[10] Hua J and Lou J 2007 Numerical simulation of bubble rising in viscous liquid *J Comput Phys* **222** 769–95

[11] Fisher L R, Mitchell E E and Parker N S 1985 Interfacial Tensions of Commercial Vegetable Oils with Water *J Food Sci* **50** 1201–2

IOP Conf. Series: Materials Science and Engineering 1296 (2023) 012016 doi:10.1088/1757-899X/1296/1/012016

Thermal post-treatment and material characterization of laser powder bed fusion additively manufactured Ti-6Al-4V

Abhinav Anand[1,2], Nagarajan Devarajan[2], Rohit Kumar Gupta[3], Nikhil Kamboj[1,4], Ashish Ganvir[1]

[1] Research group of Digital Manufacturing and Surface Engineering, Department of Mechanical and Materials Engineering, University of Turku, FI-20500 Turku, Finland

[2] Department of Metallurgical and Materials Engineering, National Institute of Technology, Tiruchirappalli, Tamil Nadu, 620015, India

[3] Vikram Sarabhai Space Centre, Thiruvananthapuram 695022, India

[4] Department of Biomaterials Science, Faculty of Medicine, Institute of Dentistry, Turku Clinical Biomaterials Centre, University of Turku, 20014, Turku, Finland

E-mail: abhinitt521@gmail.com

Abstract. Laser powder bed fusion of Ti-6Al-4V (PBF-LB/Ti-6Al-4V) alloy results in the formation of non-equilibrium microstructures due to very high cooling rates, causing them to be unfit for direct applications; therefore, post-heat-treatment operations are required. This paper investigates the effects of different heat treatment operations on the microstructure, hardness and phase composition of as-built PBF-LB/Ti-6Al-4V samples. Six sets of heat-treatment operations have been designed in which samples have been first subjected to either above or below the beta (β) transus zone of 980 °C, then subjected to air cooling or water quenching followed by ageing at 510 °C. The novelty of this article is performing double quenching on as-built Ti-6Al-4V parts. Optical microscopy, Vickers microhardness testing, and XRD analysis have been performed on heat-treated samples for material characterization. Microstructural studies have revealed that depending upon the cooling medium and subjected temperature during heat treatment, the extent of dissolution of martensitic (α′) needle-like phases present in as-built samples into α-β phase varies, resulting in variation of hardness values. XRD analysis confirmed the presence of β-phase along with the α-phase in the matrix in air-cooled heat-treated samples. Maximum hardness was obtained in the case when the sample was solutionized at 1050 °C, followed by water quenching and ageing due to the formation of intermetallic precipitates.

Keywords: Heat treatment, Ti-6Al-4V, Selective laser melting, Double quenching, Phase transformation

1. Introduction

Ti-6Al-4V is one of the most sought-after and widely used titanium alloys in aerospace and biomedical applications due to its high strength-to-weight ratio, good corrosion resistance, and biocompatibility. It is also referred to as alpha(α)+beta(β) titanium alloy because it consists of a combination of both phases in its microstructure. The α-phase is a hexagonal close-packed crystal structure stabilized by aluminium (6 weight %), whereas the β-phase is a body-centred cubic crystal structure stabilized by vanadium (4 weight %) [1]. The production of Ti-6Al-4V alloys by conventional manufacturing is challenging due to its high melting temperature, low thermal conductivity, and high chemical reactivity. Moreover, the high cost of manufacturing, longer lead time and suboptimal use of raw materials are major barriers to traditional manufacturing routes. Additive manufacturing (AM) solves these issues and produces parts with complex geometries that are difficult to achieve through conventional routes, along with design freedom for specific end-use applications [2,3]. Laser powder bed fusion (PBF-LB) is the most used AM process for this purpose.

The microstructure of Ti-6Al-4V obtained via casting or forging consists of a mixture of alpha and beta phases, i.e., lamellar, bimodal or equiaxed, which can be later altered by adjusting the thermal history or processing routes, such as thermo-mechanical processing [1]. However, in the case of PBF-LB, due to the very high cooling rate of the melt pool (range of 10^3 to 10^8 kelvin/second), the obtained microstructure consists of mostly acicular (α′) martensitic phase, which results in high-strength but minimal ductility [4]. Also, due to very short interaction time and localized heat inputs, the parts fabricated by PBF-LB have high thermal gradients resulting in thermal stress build-up. The as-printed microstructure consists of non-equilibrium phases accompanied by segregation and is anisotropic due to rapid solidification and complex heating-cooling cycles involved during AM. Therefore, a post-heat treatment operation is required to eliminate residual stress and enhance the mechanical properties of the as-built part by obtaining a microstructure that can yield an optimum combination of ductility and enhanced structural stability [1].

In the case of Ti-6Al-4V, the α-β phase matrix transform entirely into β phase above 980 °C, which is referred to as β-transus temperature [5]. In this study, several post-heat treatment operations, including air cooling and water quenching followed by ageing, have been performed on PBF-LB/Ti-6Al-4V samples to understand the effect of cooling or quenching from above or below the β transus temperature. Also, the effects of double quenching, which has not been reported before, have been investigated in this study.

2. Materials and Methods

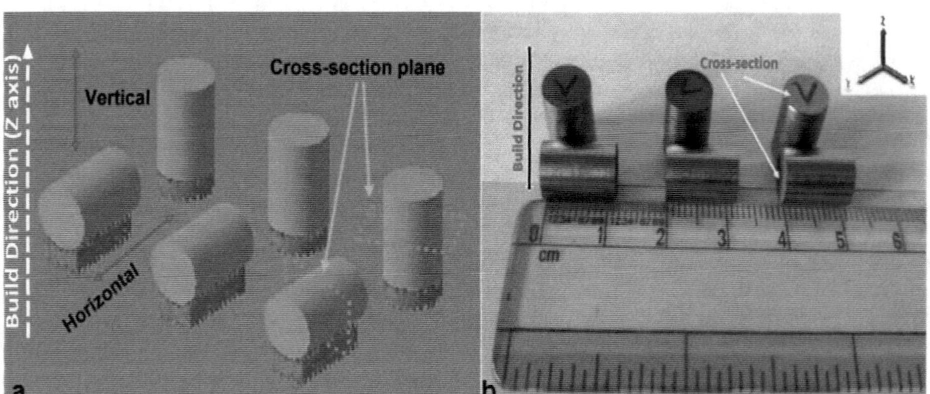

Figure 1. a) CAD image showing specimens orientation before printing and highlighting cross-section planes used for study b) PBF-LB/Ti-6Al-4V specimens with build direction along the Z-axis.

Two sets of samples with vertical (V) and horizontal (H) build having a diameter of 8 mm and height of 12 mm were printed using EOS M290 PBF machine as shown in **Figure** 1. The powder size used during the fabrication lied between 20 to 60 μm. "Chess-board pattern" scanning strategy was used during printing, i.e., 90 degrees rotation of laser beam for every other layer of the print. This strategy was adopted to improve print quality by evenly distributing the heat across the entire print and minimizing the risk of warping during printing. A laser power of 340 W with a scan speed of 1200 mm/s and a hatch distance of 0.1 to 0.14 mm was used. The powder layer thickness during printing was 60 μm. Each as-built sample was cut into equal halves across the cross-section planes via wire cut EDM to obtain a total of twelve specimens (six horizontal, six vertical) out of three, each H and V, with dimensions of diameter 8 mm and height 6 mm.

Six different heat treatment operations, as shown in **Table 1**, were performed on different sets of samples, each set containing one horizontal and one vertical sample. The heating temperatures were chosen in accordance to study the effect of air-cooling and quenching followed by ageing from the above/below transus temperature of 980 °C. All heat treatments were performed in a box furnace without vacuum manufactured by Nano Tec Pvt. Ltd. with a heating rate of 7 °C/minute.

Table 1. Details on the six distinct heat treatments employed for each set of PBF-LB specimens.

Treatment number	Heat treatment details
1 // 2 // 3	Holding for 30 min at 730 °C // 950 °C // 1050 °C, followed by AC to room temperature
4 // 5	Holding for 30 min at 950 °C // 1050 °C, followed by WQ + Ageing at 510 °C for 8 h and then AC to room temperature
6	Holding for 30 min at 1050 °C followed by WQ + Solution treatment at 950 °C for 30 min followed by WQ + Ageing at 510 °C for 8 h and then AC to room temperature

Note: here AC and WQ denote air cooling and water quenching, respectively.

Microscopic analysis was performed after every heat treatment step and also on the as-printed samples. Samples were ground to a fine 2000-grit size using SiC paper and were polished using alumina suspension. Kroll's reagent with a composition of 2 ml H.F., 6 ml HNO_3 and 50 mL distilled water was used to reveal the microstructure of polished samples. Microscopy was performed using Olympus optical microscope with an image analyzer. Microhardness testing was performed on hot-mounted samples using Vickers microhardness testing with an applied load of 1000 gram-force with a ten-second dwell time. Eight indentations were marked during testing, and an average hardness value and mean standard deviation were calculated. Further, XRD was performed on heat-treated samples to identify the type of phase formed using Cu-Kα radiation for recording the data in a 2θ range of 20°to 90°.

3. Results and discussion

3.1. As-built PBF-LB/Ti-6Al-4V

The observed microstructure for the as-built parts is shown in **Figure 2.** The unetched surface was observed under the microscope to reveal the porosities formed during the printing process, possibly due to lack of fusion, partial melting or due to entrapment of gases. As shown earlier in **Figure 1**, the horizontally printed samples are cross-sectioned in such a way that the plane of microstructural examination lies along the build direction, whereas in the case of vertical samples, the cross-section plane is along the laser-scanning direction. The slightly higher average hardness of the horizontal sample (356.4 ± 6.9 HV) compared to the vertical sample (350.6 ± 7.9 HV) is due to the faster cooling of the material resulting from the higher thermal flows present in the former build configuration [6].

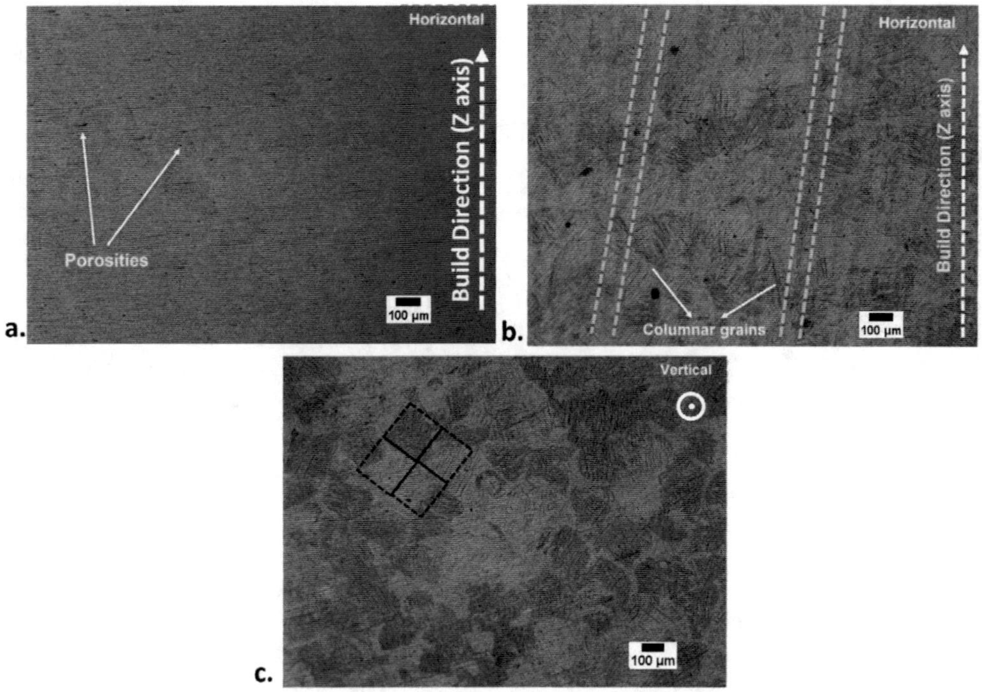

Figure 2. a) Polished unetched surface of the as-built part (horizontal) b) Microstructure of horizontal sample revealing columnar grains along the build direction c) Chess-board pattern observed in the cross-sectioned vertical sample (perpendicular to build direction)

Rapid cooling during laser powder bed fusion process resulted in the formation of acicular martensite (α′), transforming from the original columnar β grains. The formation of a columnar shape is attributed to the epitaxial growth of the initial β phase caused by successive layer deposition and directional cooling in the build direction, as shown in **Figure 2**(b). Chess-board patterns, as highlighted in **Figure 2**(c), have been formed due to the adopted laser scanning strategy used during the printing.

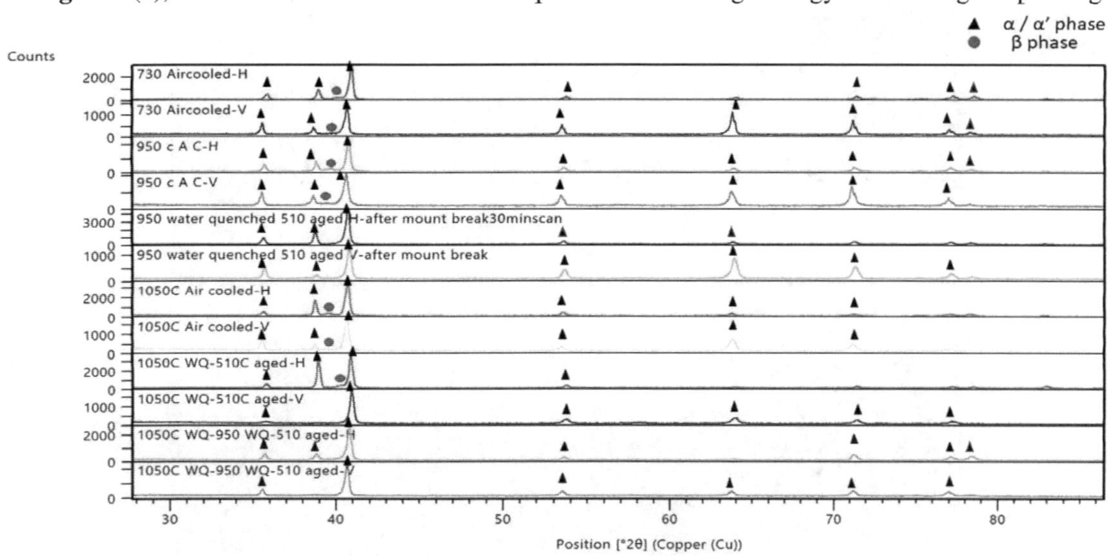

Figure 3. The XRD patterns of the heat-treated samples (H-horizontal, V-vertical, AC-air cooled, WQ-water quenched).

3.2. Effects of air cooling on PBF-LB/Ti-6Al-4V

X-ray diffraction (XRD) patterns of heat-treated samples have been shown in **Figure 3**. Since both α' and α phases have the same hexagonal close-packed crystalline structure, they produce XRD peaks at the same angular position, so they have been indicated as α/α' in the XRD pattern [7]. The presence of the β phase, as indicated by the XRD study and the shifting of XRD peaks towards the left with an increase in soaking temperature has proved that the air cooling-based heat treatment results in the decomposition of acicular martensitic (α') into $\alpha+\beta$, and the transformation of α into β phase increases with increase in solution temperature. According to previous studies, the equilibrium α-fractions get reduced to 23 % at 950 ° C from around 90 % at 730 ° C [8].

As highlighted in **Figure 4**, the microstructural studies indicate the same outcome. Acicular martensitic (α') are no longer visible in samples cooled from 950 °C, and a lamellar $\alpha+\beta$ phase is formed. Also, the prior β grains begin to disappear during heat treatment and get diffused at 950 ° C. When the sample is heated above the β-transus temperature of 980 °C, i.e. at 1050 °C, semi-equiaxed β grains are formed by replacing the columnar β grains, resulting in an α-Widmanstätten microstructure or basket weave structure. Also, the chess-board pattern in the microstructure of vertical samples begins to disappear by air cooling from 950 ° C and vanishes at 1050 ° C.

The measured Vickers microhardness of heat-treated samples has been represented in **Figure 5**. The decrease in the hardness of as-built specimens upon air cooling is due to the acicular martensitic (α') phase decomposition. With an increase in solution temperature from 730 ° C to 950 ° C, a further decrease in values was observed due to the increased β phase in the matrix. But at 1050 ° C, an increase in hardness was observed due to formation of Widmanstätten microstructure [9].

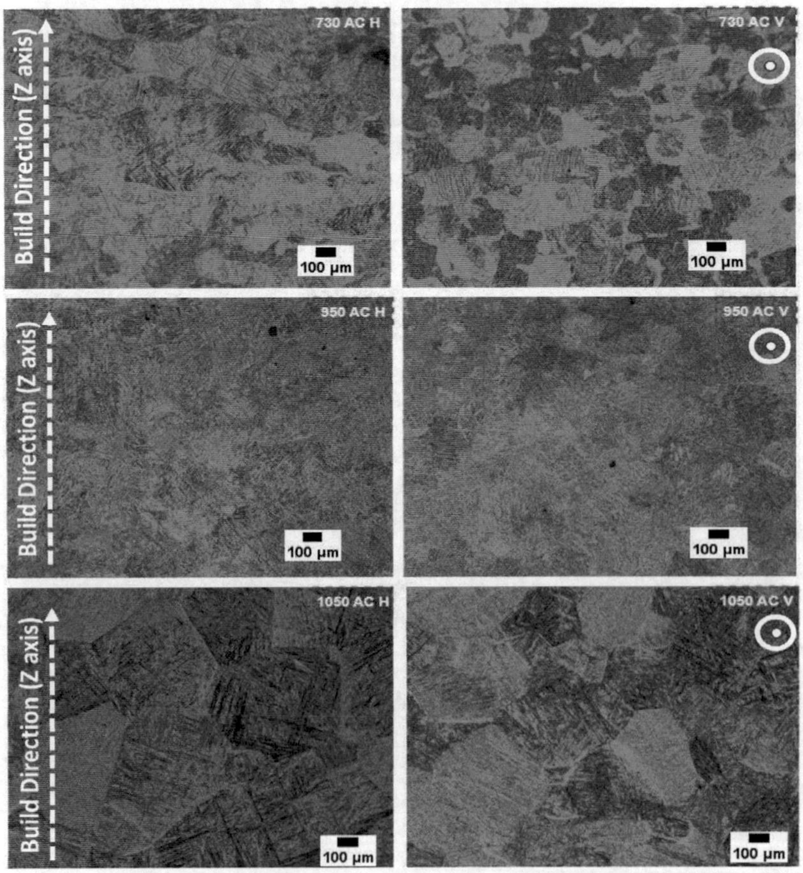

Figure 4. The observed microstructures of the samples after being heat-treated for 30 minutes at a) 730 ° C, b) 950 ° C, c) 1050 ° C, followed by air-cooling (AC-air cooling, H-horizontal, V-vertical)

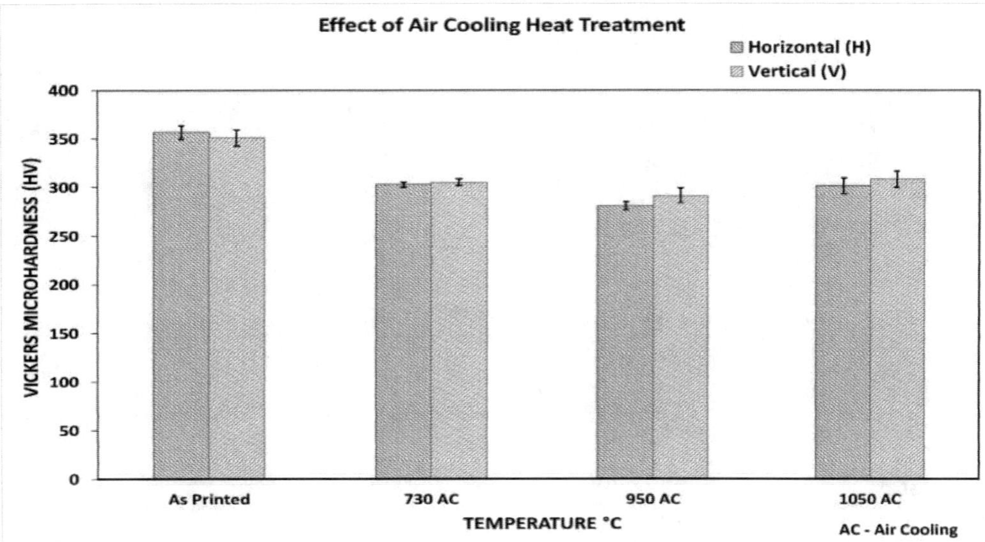

Figure 5. Vickers microhardness of heat-treated specimens (air cooled) compared to as-built specimens.

3.3. Effects of quenching and ageing on PBF-LB/Ti-6Al-4V

Since the martensite start temperature (M_s) for Ti-6Al-4V is ~900 °C [10], upon quenching above this temperature, martensitic phases were observed in the matrix. Upon quenching from 950 °C, a microstructure containing lamellar α phase along with acicular martensite (α′) needles were observed, as shown in **Figure 6**(a, b). Upon subjecting the specimens to a solution temperature below the β-transus temperature, almost all the previous existing acicular martensite (α′) (formed during PBF-LB) get decomposed into α and β phases, and when the quenching is performed from above M_s, the β phase transform into martensitic phase, leaving behind the primary α into the matrix. The prior β microstructure containing the laser scan pattern, as shown in **Figure 6**(b), was retained in this case. The decomposition of martensitic needles was noticed upon ageing the quenched samples for 8 hours. However, the amount of β phase formed after ageing was too low to get detected in the XRD pattern (**Figure 3**).

IOP Conf. Series: Materials Science and Engineering 1296 (2023) 012016 doi:10.1088/1757-899X/1296/1/012016

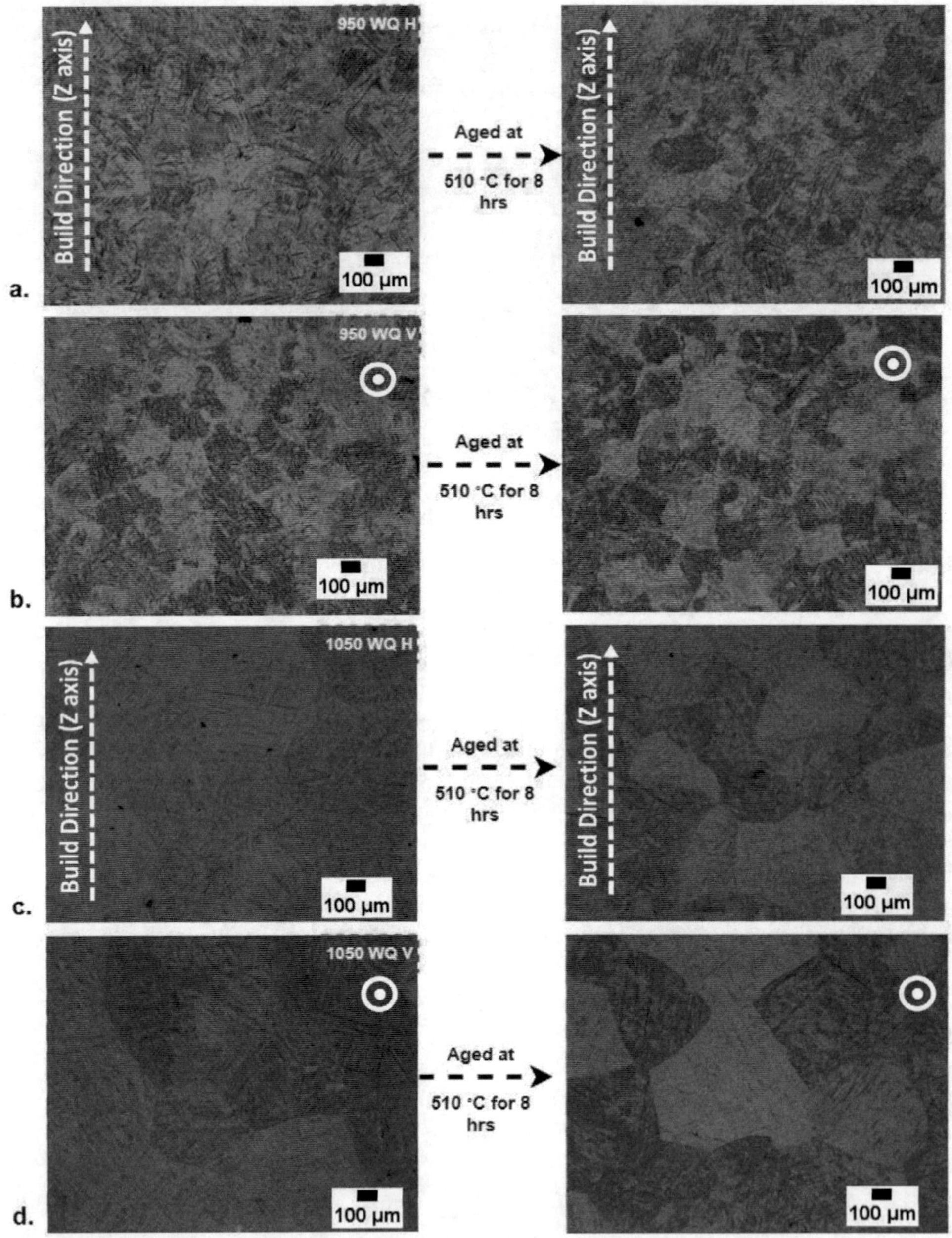

Figure 6. Observed microstructures after as-built specimens were subjected to quenching from different temperatures followed by ageing a,b) at 950 °C c,d) at 1050 °C (H-horizontal, V-vertical, WQ-water quenched)

In the case of quenching from above the β-transus temperature, i.e., from 1050 °C, a 100 % martensitic phase originating from the new equiaxed β grains was observed (**Figure 6** (c, d)). Shearing of previously existing columnar grains generated during PBF-LB took place, and the previously existing acicular martensitic phase completely transformed into β phases upon solutionizing at 1050 ° C. Upon quenching, this resulted in the formation of weave-type acicular (α′) martensite inside equiaxed β grains. Also, the microstructure containing the chess-board laser scan pattern vanished in this case. Furthermore, when the samples were aged, the β phase was detected in the XRD (**Figure 3**), verifying the decomposition of acicular α′ martensite upon ageing above 500 ° C.

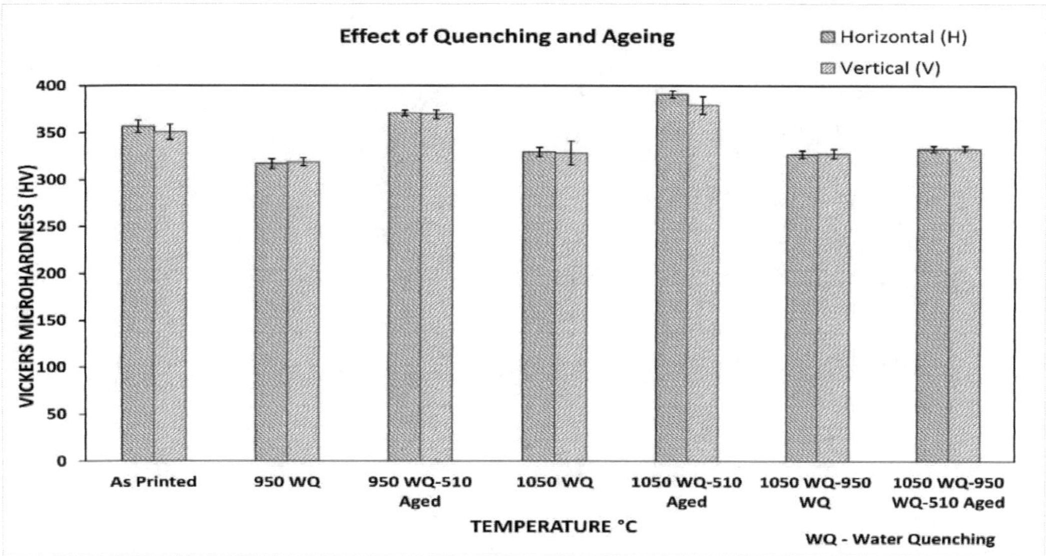

Figure 7. Vickers microhardness values calculated after quenching and ageing.

Figure 7 shows a decrease in hardness value when the samples were quenched from 950 ° C and 1050 ° C compared to as-built specimens. Since during the water-quenching, the quenching rate is much lower than the quenching due to rapid solidification in PBF-LB, thicker martensitic needles were observed (**Figure 6**) compared to as-built specimens (**Figure 2**). Since the finest microstructure gives more hardness [11], a dip in hardness value is justified. Also, the samples quenched from 1050 °C showcased higher hardness than those quenched from 950 °C because the former has a 100 % martensitic structure, whereas the latter has acicular martensite accompanied by primary α, which leads to a softer matrix. Moreover, upon ageing, a significant increase in hardness was noticed for both quenching operations. This is possibly contributed by the precipitation of intermetallic Ti-Al phases in the matrix. A previous study has shown that when a quenched Ti-6Al-4V is subjected to ageing between 500 to 900 °C, the intermetallic phases get precipitated in the matrix due to the low solubility of Al in Ti in this temperature range [12].

3.4. Effect of double quenching followed by ageing on PBF-LB/Ti-6Al-4V

During double quenching, the samples were first solutionized above β -transus temperature. i.e., at 1050 ° C. This resulted in the formation of equiaxed β grains containing weave-type acicular α′ martensite. When the samples were again solutionized at below β-transus temperature. i.e., at 950 °C, most of acicular α′ martensite transformed into α+β phases and upon quenching, the β phases in the matrix transformed again into acicular α′ martensite. However, this time, compared to a normal 950 °C quenching (**Figure 6**(a,b))**,** the observed microstructure (**Figure 8**(a,b)) has primary α and martensite originating inside equiaxed β grains, which were formed when the sample was initially heated to 1050 ° C. The difference in microstructures resulted in a noticeable change in hardness values of 950 °C WQ and double-quenched samples. Also, the softer primary α phase contributed to a lower hardness of double-quenched samples compared to as-built and 1050° C WQ.

However, unlike the ageing of 950 °C WQ and 1050 °C WQ samples, no significant changes in hardness values were observed upon ageing the double-quenched samples (**Figure 7**). Also, no β phase was detected this time in the XRD pattern after ageing. A possible explanation is that double quenching can reduce the amount of precipitates (based on TiAl) that can form in the matrix by disrupting the diffusion of alloying elements and the nucleation of precipitates. Nevertheless, more research needs to be done in this regard to derive a conclusion.

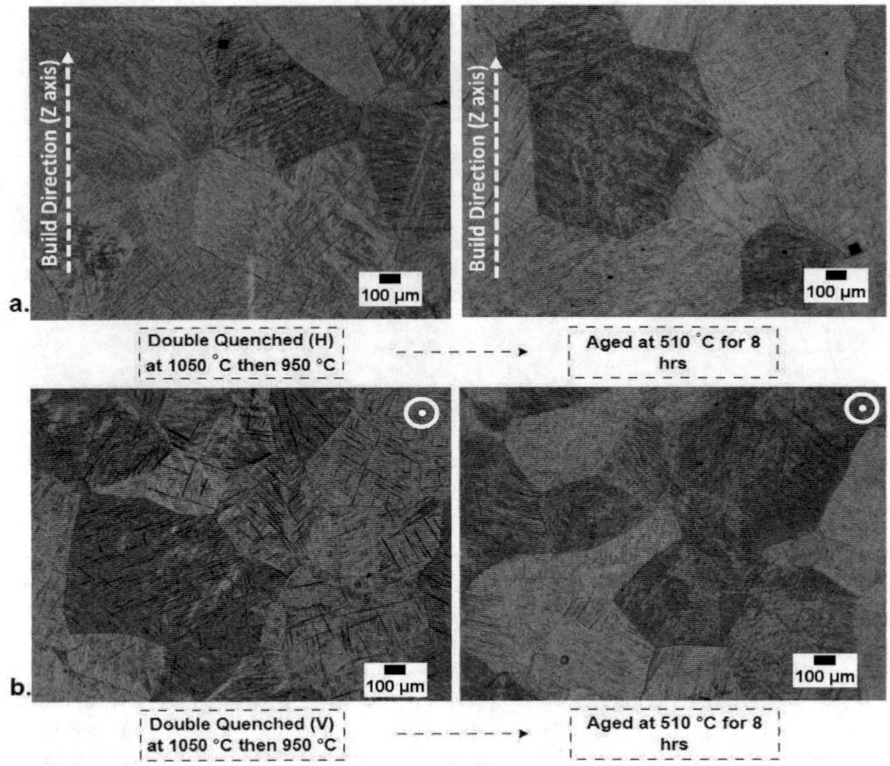

Figure 8. Observed microstructures when the as-printed samples get quenched at 1050 ° C, followed by quenching again at 950 ° C and then aged at 510 ° C

4. Conclusions

In this paper, a detailed study was conducted on the heat treatment of as-built PBF-LB/Ti-6Al-4V samples. Optical microscopy, Vickers microhardness test and XRD were performed for material characterization. The following conclusions can be derived from the study:

- As-built horizontal (H) samples constitute prior β columnar grains containing acicular martensitic phase in the matrix, whereas microstructure highlighting chess-board laser scanning pattern was observed in as-built vertical (V) samples.
- Air cooling from 730 °C, 950 °C, and 1050 °C decreases hardness values due to the decomposition of the acicular martensitic phase into α and β phases.
- Air cooling from above β-transus temperature of 980 °C results in α-Widmanstätten microstructure, whereas from below β-transus results in lamellar α+β phase.
- Water quenching from above β-transus temperature results in weave-type acicular α′ martensite inside equiaxed β grains, whereas from below beta-transus results in lamellar α phase along with acicular martensite.
- Water quenching results in a decrease in hardness values as compared to as-built specimens due to an increase in the thickness of the martensite needle due to a slower cooling rate than PBF-LB.
- Upon double quenching, the microstructure contains primary α and martensitic phases originating inside equiaxed β grains and ageing the double-quenched samples results in no significant change in hardness.

Acknowledgement-

The authors thank the Group Director of Materials and Metallurgy Group and Deputy Director of Materials and Mechanical Entity for their encouragement and support. They further thank the Director of Vikram Sarabhai Space Centre for permission to publish this work.

References:

[1] Polishetty A, Nomani J and Littlefair G 2022 Effect of transus based heat treatment on material characterisation of wrought and additive titanium alloy Ti-6Al-4 V *Mater. Today Proc.* **59** 1749–53

[2] Lu S L, Zhang Z J, Liu R, Qu Z, Wang B, Zhou X H, Eckert J and Zhang Z F 2022 Prior β grain evolution and phase transformation of selective laser melted Ti6Al4V alloy during heat treatment *J. Alloys Compd.* **914**

[3] Anand A, Nagarajan D, El Mansori M and Sivarupan T 2022 Integration of Additive Fabrication with High-Pressure Die Casting for Quality Structural Castings of Aluminium Alloys; Optimising Energy Consumption *Trans. Indian Inst. Met.*

[4] Sun S, Zhang D, Palanisamy S, Liu Q and Dargusch M S 2022 Mechanical properties and deformation mechanisms of martensitic Ti6Al4V alloy processed by laser powder bed fusion and water quenching *Mater. Sci. Eng. A* **839**

[5] Liu S and Shin Y C 2019 Additive manufacturing of Ti6Al4V alloy: A review *Mater. Des.* **164** 107552

[6] Palmeri D, Buffa G, Pollara G and Fratini L 2021 The Effect of Building Direction on Microstructure and Microhardness during Selective Laser Melting of Ti6Al4V Titanium Alloy *J. Mater. Eng. Perform.* **30** 8725–34

[7] Ju J, Zhao C, Kang M, Li J, He L, Wang C, Li J, Fu H and Wang J 2021 Effect of heat treatment on microstructure and tribological behavior of Ti–6Al–4V alloys fabricated by selective laser melting *Tribol. Int.* **159**

[8] Al-Bermani S S, Blackmore M L, Zhang W and Todd I 2010 The Origin of Microstructural Diversity, Texture, and Mechanical Properties in Electron Beam Melted Ti-6Al-4V *Metall. Mater. Trans. A* **41** 3422–34

[9] Vrancken B, Thijs L, Kruth J P and Van Humbeeck J 2012 Heat treatment of Ti6Al4V produced by Selective Laser Melting: Microstructure and mechanical properties *J. Alloys Compd.* **541** 177–85

[10] Wu S Q, Lu Y J, Gan Y L, Huang T T, Zhao C Q, Lin J J, Guo S and Lin J X 2016 Microstructural evolution and microhardness of a selective-laser-melted Ti-6Al-4V alloy after post heat treatments *J. Alloys Compd.* **672** 643–52

[11] Vilaro T, Colin C and Bartout J D 2011 As-fabricated and heat-treated microstructures of the Ti-6Al-4V alloy processed by selective laser melting *Metall. Mater. Trans. A Phys. Metall. Mater. Sci.* **42** 3190–9

[12] Liang Z, Sun Z, Zhang W, Wu S and Chang H 2019 The effect of heat treatment on microstructure evolution and tensile properties of selective laser melted Ti6Al4V alloy *J. Alloys Compd.* **782** 1041–8

IOP Conf. Series: Materials Science and Engineering 1296 (2023) 012017 doi:10.1088/1757-899X/1296/1/012017

Digital manufacturing in laser powder bed fusion of aluminium alloy safety feature for EV battery cell

Heikki Saariluoma[1,2], **Kalevi Vesterinen**[1], **Antti Salminen**[1,2]

1 Department of Mechanical Engineering, Turku University of Applied Sciences, Turku, Joukahaisenkatu 3, FIN-20520, Finland; heikki.saariluoma@turkuamk.fi, kalevi.vesterinen@turkuamk.fi

2 Department of Mechanical Engineering, University of Turku, Turku, FI-20014, Finland; heikki.v.saariluoma@utu.fi, antti.salminen@utu.fi

Corresponding author's email: heikki.saariluoma@turkuamk.fi

Abstract. Recently, a new way to produce prismatic Li-ion battery cell's lid part for Battery Electric Vehicles (BEV) was presented, with integrated cell thermal management, based on the Laser Additive Manufacturing (LAM) method. In the solution, the safety vent protects the battery cell from unpredictable explosions caused by the increasing internal pressure. Additive manufacturing allows integrating the feature to the cell lid part and significantly decreases the process steps in manufacturing and increases the production efficiency. Aluminium powder materials for the used Laser Powder Bed Fusion (LPBF) method have relatively high tensile strength, which causes the need for minimum layer thickness for the safety vent. Reliability of producing such a thin layer as a gas tight structure is challenging and several procedures were studied. Material with lower tensile strength and higher elongation ability was tested during the second phase. The challenge is to produce such a thin wall thickness, which breaks at the pressure level of one MPa and opens the valve. The design is based on FEM analysis with constant equivalent strain theory to optimize safety vent dimensions. The pressure test was performed to measure the rupturing pressure of the thin aluminium LPBF layer as materials in normal mode and in annealed mode. The constant equivalent strain theory for estimate rupture pressure was in fine agreement with the experimentally measured ones, even with a maximum error rate. The results show that the manufacturing method is possible for such automotive mass production parts and when comparing different stacking strategies, it showed a great influence in the productivity of such parts' production.

Keywords: Additive Manufacturing; LPBF; Battery cell; Ductile fracture; Finite Element Method; Integrated safety vent; Lithium-ion battery; Rupture pressure; BEV

1. Introduction

The additive manufacturing (AM) process gives enormous possibilities to design complex structures for product functionality. This study concentrates on the laser powder bed fusion (LPBF) process with aluminium AlSi10Mg material's possibilities to be used in the laser welded battery cells and automate the disassembly process. Waste management has partially been ignored in current electric vehicles' (EV) battery designs. That should be a major concern when designing lithium-ion (Li-ion) battery packs in the future. In 2017, the global production volume of EV's passed one million vehicles in a year for the first time. That number means half a million cubic meters and quarter million tons of partially toxic waste without properly planned recycling [1]. Year 2018 doubled the production of EVs and is expected to rise as high as to 12–22 million vehicles by the year 2025. Then the yearly battery mass to be produced increases to four million tons per year [2]. Today, a typical lifetime for the EV battery pack is ten years.

After that life cycle, the battery cells and modules may be used in other purposes than EV energy sources like energy storages [3]. The so-called second life of batteries requires dismantling the battery packs in most cases. Disassembling, storing or re-manufacture generates major challenges, on the other hand, the Lithium-Ion Batteries (LIBs') recycling market is projected to be USD 18.1 billion by 2030 [1,4,5,6]. As a comparison, in Powys, Wales a stock of rubber tyres caught fire and smoked for fifteen years up to the year 2004; today, stored batteries include a similar or higher risk of uncontrollable fire. Without an effective waste-management strategy for LIBs', the greater dangers associated with stockpiling of end-of-life LIBs' is present [1]. Batteries are also a valuable resource as battery raw materials and therefore should be considered already in EV's battery packs' designing phase as a material and energy efficient 3R system basis (reduce, re-use, recycle) [1]. Greenhouse gas (GHG) emissions reduce substantially by using EVs instead of combustion engine vehicles. At the end of year 2018, 11 countries pledged to increase their annual EV sales to 30% of all vehicle sales by 2030. If this target is achieved, the global EV amount increases 70 times over the next 12 years. That means 35 million tons of EV battery material [3]. Such a huge amount of battery material should be re-used or recycled effectively and environmentally friendly by properly designing EV battery packs in the first place. In the five-level waste management hierarchy, it means the first level, "Prevention", by design means. In the second level, "Re-use", packs are typically disassembled for modules and cells used as components in other assemblies. The third level "Recycling" means material separation in shredding or pyrometallurgy processes to new raw material. In the fourth level, "Recovery", the material's energy burns in pyrometallurgy processes. In the final fifth level, "Disposal", no valuable material ends up in a landfill [1]. Design of integrated functions like a safety vent in the battery lid simplifies and enables lowering the cost of production. In the disassembly process, the cell lid part, housing and battery cell jellyroll or stack should be easily separated to their own pins. By combining the LPBF part's clever design and laser welding's low heat input, the effective re-use of a battery cell becomes possible. The 3R thinking should be used already in the design phase and taken to the battery module and pack level, especially when larger cells are used in the cell to pack battery designs.

Joining the LPBF manufactured AlSi10Mg components with a safety feature to aluminium sheet metal parts is one of the key objectives of the study. Designing a safety feature i.e., safety vent dimensions, with the finite element method (FEM) is also a key objective of this study. This kind of AM part can be porous material, which is a disadvantage for high-quality laser weld [7,8]. Electron beam welding can produce high quality hybrid welds while the suitability for mass production is limited [9]. A fillet lap joining configuration has been studied for similar LPBF part joint with successful joints [10, 11]. Prismatic battery cell components and modules for the EV's can be laser welded in the assembly phase, which is the most common method for overlap join in sheet metal prismatic battery cell construction. The freedom of geometry in AM parts gives a big advantage, enabling the integration of many features in one part and should increase assembly competitiveness compared to conventional sheet metal prismatic battery cell housing and module design. The newest multi-laser metal AM machines with high productivity can create a great advantage also for a high production volume of automotive parts. Prismatic battery cells are typically located in the modules side by side [12]. Electrodes are connected by a busbar which is typically a sheet metal part joined with laser welding. The LPBF process creates an advantage to design free forms in the battery cell lid. Disassembly of the busbars in the battery module is difficult to do automatically in current battery module designs. The battery cell and module design should allow automatic disassembly. In using an LPBF -produced lid for prismatic battery cells, there are two joints to consider: first, lid to battery cell cup and second, lid to busbar. Both joints are laser welds between an AlSi10Mg printed part and a 1050 Al sheet metal part. To benefit of free forms in an additively manufactured lid, reliable welds of described laser welded joints are enabled [12]. Conventional welding methods may produce porosity to the weld when welding the LPBF parts.

2. Specimen material, manufacturing and preparation

The additively manufactured tensile test specimens were prepared using a SLM 280 HL (SLM Solution Group AG, Germany) metal printing system and gas atomized AlSi10Mg powder (SLM) with spherical particle size of 20 to 63 μm. The 400/700 W machine is equipped with two Yttrium fiber lasers with beam diameters 80–115 μm. All plates were fabricated under the same processing conditions in argon atmosphere with the concentration of O_2 controlled below 200 ppm. Used material properties are shown in table 1 and 2.

Table 1. Chemical properties of the material.

Chemical composition [Mass fraction in %]											
Al	Si	Fe	Cu	Mn	Mg	Zn	Ti	Ni	Pb	Sn	Other total
Balance	9.00-11.00	0.55	0.05	0.45	0.20-0.45	0.10	0.15	0.05	0.05	0.05	0.15

Table 2. Mechanical properties of the material.

Build-up rate		[cm3/h]	35.6 cm3/h			
Component density		[%]	>99.0%			
Tensile test			M [a]	SD [b]	M [c]	SD [d]
Tensile strength	Rm [Mpa]	H [e]	443	6	264	5
		V [f]	432	28	273	5
Offset yield strength	$R_{p0.2}$ [Mpa]	H	277	7	151	8
		V	259	5	154	6
Elongation at break	A [%]	H	8	1	19	3
		V	5	1	14	3
Reduction of area	Z [%]	H	7	2	30	5
		V	4	1	19	6
Young's modulus	E [GPa]	H	72	5	57	14
		V	71	3	58	14
Hardness test			M	SD	M	SD
Vickers hardness	HV5		127	4	80	1

[a] Median, as-built
[b] Standard Deviation, as-built
[c] Median, material annealed, 2 hours in 300° C, air quenching
[d] Standard Deviation, material annealed, 2 hours in 300° C, air quenching
[c] Horizontal direction
[d] Vertical direction

The following chapters describe the key objectives of the study and testing those characteristics.

2.1 Specimen fabrication methods for the safety area

The safety vent should break if the pressure corresponds to 1 MPa in the battery cell. Dimensioning of the model of the safety vent was based on material AlSi10Mg and are shown are in figure 1.

IOP Conf. Series: Materials Science and Engineering 1296 (2023) 012017 doi:10.1088/1757-899X/1296/1/012017

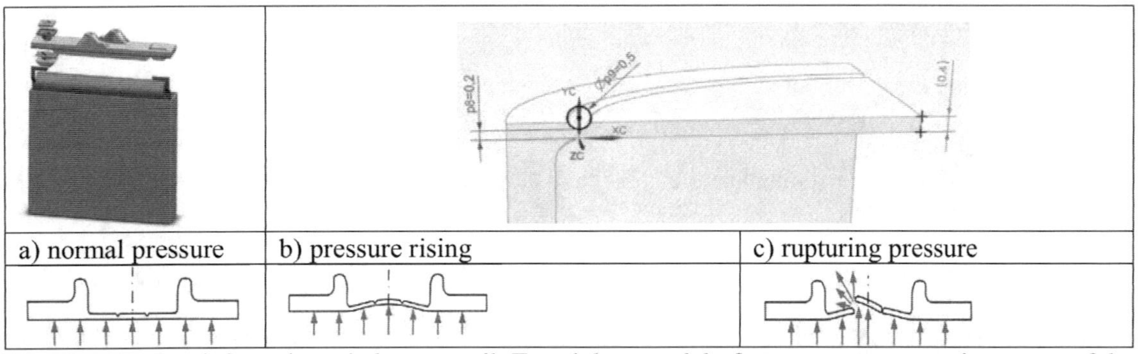

Figure 1. On top left a prismatic battery cell. Top right a model of a two-way symmetric quarter of the safety vent and used dimensions for FEM modelling and calculation. The ellipse's bigger diameter was 15mm and half of it is 7.5mm. Below a, b and c show the rupturing mechanising in different pressures.

2.2 Tensile test specimens for joint strength test

The material properties needed on the nonlinear FEM simulation came from tensile tests. Test specimens were produced with an SLM-280 3D metal printer machine and tensile tests were made with the Shimadzu-AGS-X tensile test machine. The thickness of the produced test specimens was chosen to be 0.4 mm to get the real properties of the thin layer of material produced with the LPBF process (figure 2). In printing the specimen, the middle section had supports with thickness of 1mm. Before testing, the supports were removed by milling from the middle section of specimens.

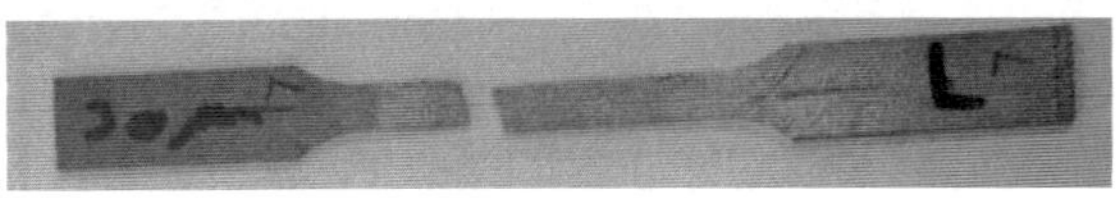

Figure 2. In the test, the middle part of the specimen was closely corresponding the safety vent thickness of 0.4mm.

The joining method and principle is shown in figure 4. The joint type is an overlap joint from the sheet metal side to the AM part. The material thickness of sheet metal varies 0.5 to 1.5 mm in the test series and for AM parts, 1.0 to 2.0 mm. The tensile test method was used in this study as the test method.

In the first trial, the produced specimens were detached from supports. The support structure has an important role in achieving the correct tolerances to a part. When the structure does not hold the layers strong enough, the ends of the part twisted to the printing direction due to residual stress. Maximum deviation of planned geometry was measured at 2.1 mm, which is not acceptable. Figure 3 shows the difference to planned dimensions of the AM produced specimen. The produced samples were measured for dimensional accuracy and deviations using Zeiss T20 3D-scanner.

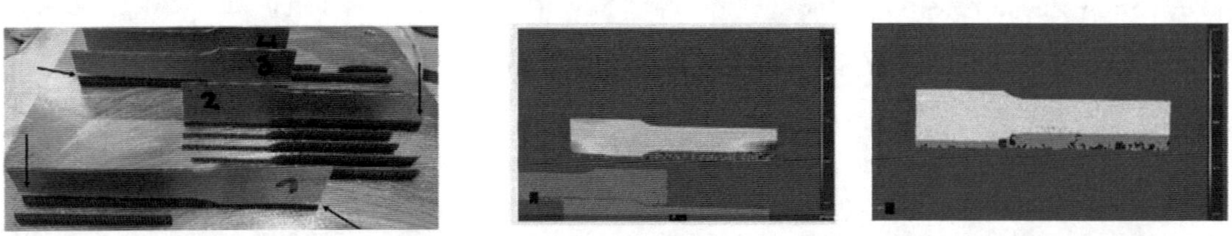

Figure 3. Left – faulty first print; middle – Deformation bending of specimen number 2, shown as red colour corresponding more than 0.2 mm, this case 1.2 mm displacement measured with a calliper. The picture on the right shows that with a right support, structure distortion can be avoided and specimen deviation is very small, less than 0.1 mm, and well acceptable.

IOP Conf. Series: Materials Science and Engineering 1296 (2023) 012017 doi:10.1088/1757-899X/1296/1/012017

2.3 Sheet metal samples

The upper part of tensile test specimens was cut from 1 mm thick Al 1050 sheets with Bodor i7 fiber laser cutting machine. The edges of the specimens were left in as is condition (not grinded after cutting). The oxide layer covering the joint forming top surface of the specimens was not removed prior to welding, thus, oxide layer with thickness of 1 μm and electrical conductivity 1/(Ω·m) was evaluated as acceptable.

2.4 Laser welding process

Welding of the specimens was performed with the IPG YLS-2000-SM fiber laser with wavelength of 1064 nm and 2000 W power with ARGES 3D Elephant galvanometer driven laser beam scanner for beam delivery. The setup delivered a beam with focused spot diameter of 120 μm and 585 mm focal length. The welding length was 20 mm.

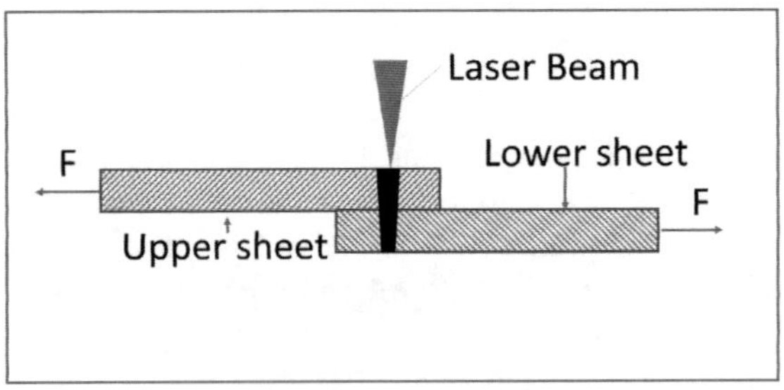

Figure 4. Drawing of overlap joint. An additively manufactured lower sheet and upper part is sheet metal material.

Figure 5 shows the used tensile test specimens' dimensions and specimens are following the ISO 6892-1 standard type B.1. The welding line goes over the black dot in the middle of the welding clamp.

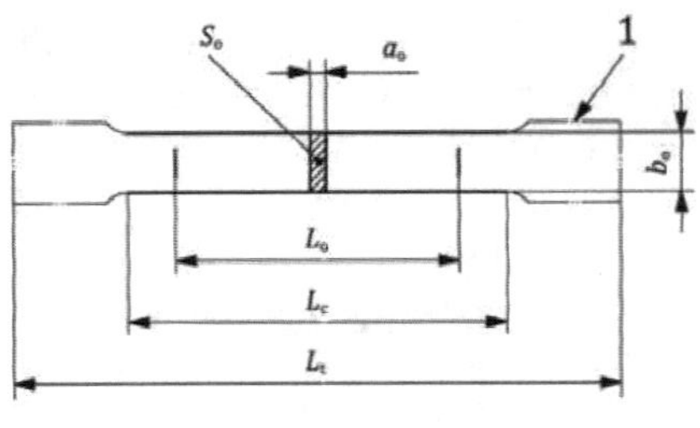

L_0=80mm
L_c=120mm
L_t=250mm
b_0=20mm
a_0=1mm

Figure 5. Shows the dimensions of the specimen and S_o marks the weld position in the welded specimen.

Welding parameters' varieties according to the following table 3.

Table 3. Used welding parameters. During welding, the Al 1050 sheet metal part was on top and AlSi10Mg part below it.

No	P [kW]	s [mm/s]	Al 1050 [mm]	AlSi10Mg [mm]
1–5	1.0	50	1.5	2.0
7–9	1.0	50	1.0	2.0
11–13	1.0	50	1.0	1.0
15–17	1.0	50	1.0	1.0
23–25	1.0	200	1.0	1.0
27–29	1.0	200	0.5	2.0
31–33	1.0	200	0.5	1.0
35	1.0	100	0.5	1.0
36	1.0	150	0.5	1.0
38	1.0	200	0.5	1.0

Specimens for the weldability test were produced according to ISO 6892-1 standard type B.1 so that one side of the specimen was manufactured in the LPBF process and other side as a laser cut sheet metal part. In the LPBF process, the used material was AlSi10Mg and the sheet metal part was laser cut of Al 1050 material. Then the weld was made in the middle of the standard-length specimen. First, before welding, the strength of the AM produced material was tested to verify that it is homogenous AM material. The same characterisation was made to complete the sheet metal material specimen produced via laser cutting. Table 4 shows the base material test results.

Table 4. Tensile strength of the welded specimen's material.

Specimen number	Material	Thickens	Width	Cross section	Rm force	Tensile strength
1 (LPBF)	AlSi10Mg	1 mm	20 mm	20 mm2	5 520N	276 MPa
2 (sheet metal)	Al 1050	1 mm	20 mm	20 mm2	1 300 N	65 MPa

2.5 Specimen fabrication methods for the safety vent area

The safety vent should break if the pressure corresponds to 1 MPa in the battery cell as shown earlier in figure 1. Dimensioning of the model of the safety vent was based on material AlSi10Mg. The dimensions of the FEM model were shown in figure 1 as quarter of the valve while the feature is symmetric in two directions. To estimate the rupture pressure of the safety vent, a nonlinear FEM model was made and solved with a Simcenter Nastran Multi-Step solver SOL401.

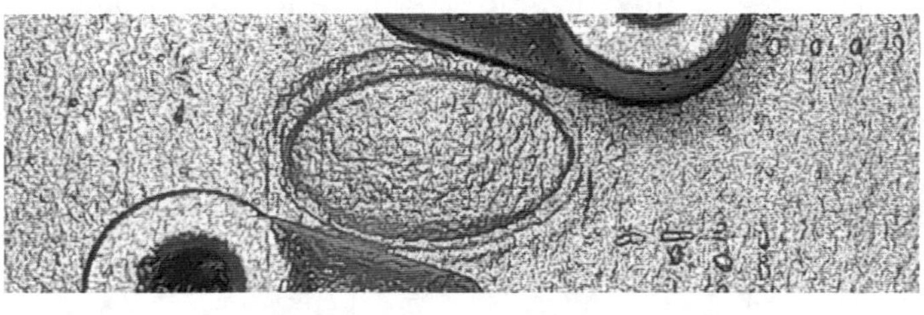

Figure 6. FEM model is a quarter of the safety vent, the oval feature in the picture (dimensions differ).

3. Results

Calculations show that it is difficult to produce a safety vent feature of AlSi10Mg material to open in 1 MPa rupturing pressure due to it's the strength. One possibility to solve this problem is to anneal the material on the safety vent area. The safety vent tensile strength properties measured annealed and unannealed material and the results are shown in figures 7 and 8 as a stress-strain curve.

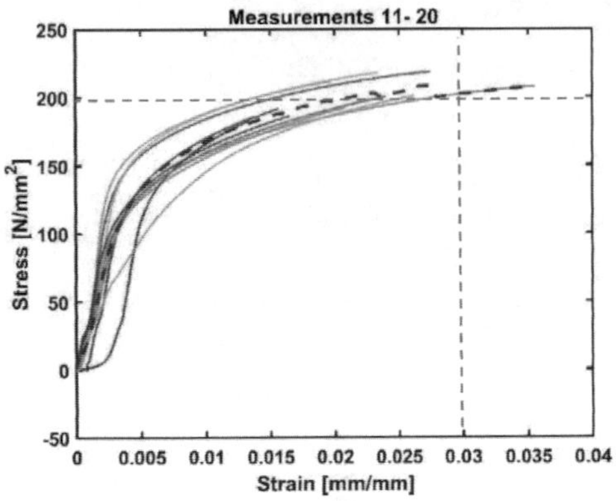

Figure 7. Without annealing, the tensile strength in level of 200 N/mm2 as shown in the stress-strain curve.

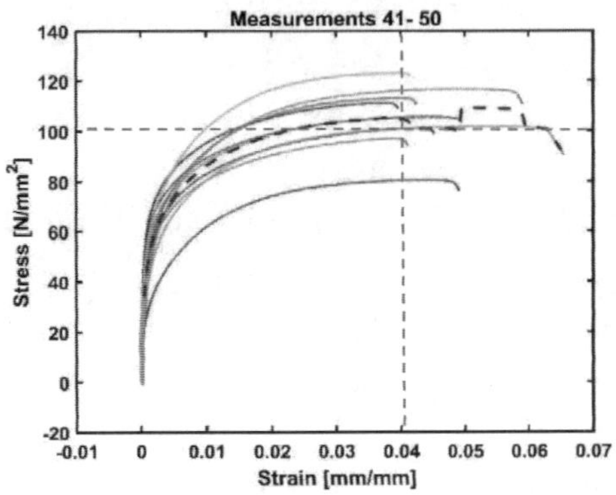

Figure 8. After annealing, the tensile strength drops in level of 100 N/mm2 as shown in the stress-strain curve.

The calculation shows that rupture pressure with the material without annealing was about 2.4 MPa and with the material annealed about 1.6 MPa. The aim of 1.0 MPa rupture pressure could not be achieved. The strength of the materials is high, which leads to that such a thin vent membrane with current LPBF processes is difficult to produce reliably. The results of the FEM simulation are shown in Figures 9 and 10.

Figure 9. Stresses of the safety vent model with no annealing with 2.4 MPa pressure.

Figure 10. Stresses of the safety vent model with annealed material with 1.6 MPa pressure.

The scale is showing stress under the mentioned 2.4 MPa and 1.6 MPa pressure in figures 11 and 12. Annealed material stress is close to the required rupturing pressure 1 MPa, but still too high. In table 5, several pressure levels are calculated for both unannealed and annealed material. The results show that these material properties and used membrane thicknesses are not completely meeting the set requirement 1 MPa rupturing pressure.

3.6 MPa – Unannealed	4.2 MPa – Unannealed	4.8 MPa – Unannealed	5.4 MPa – Unannealed
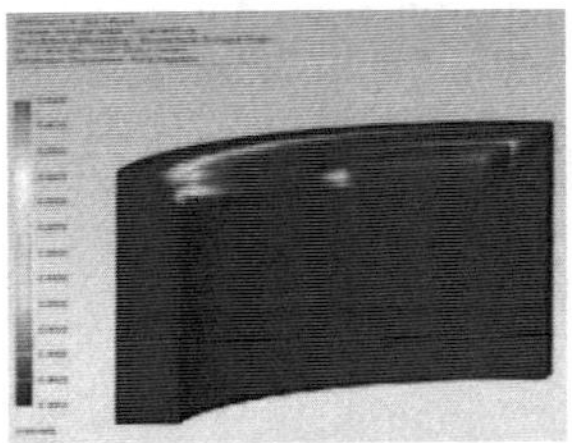	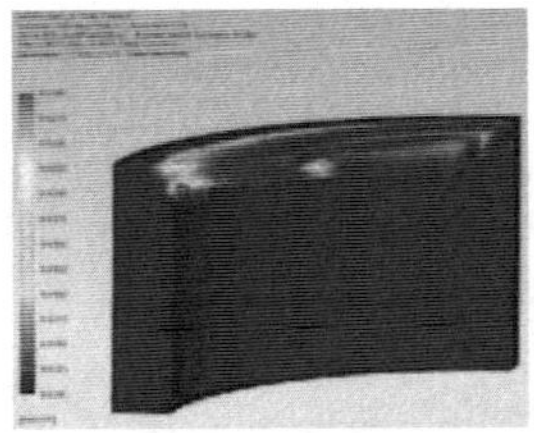	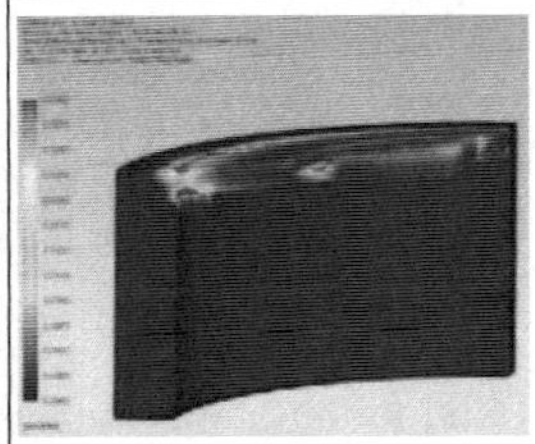	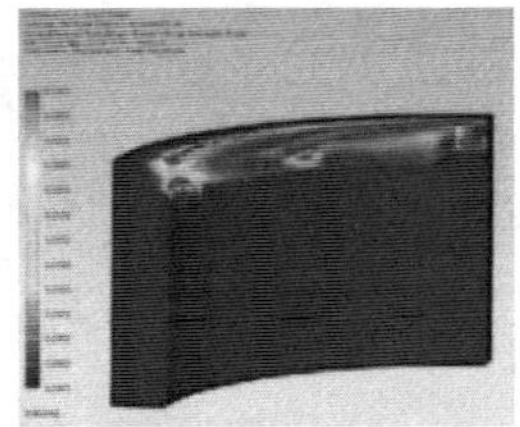
3.6 MPa – Annealed	4.2 MPa – Annealed	4.8 MPa – Annealed	5.4 MPa – Annealed
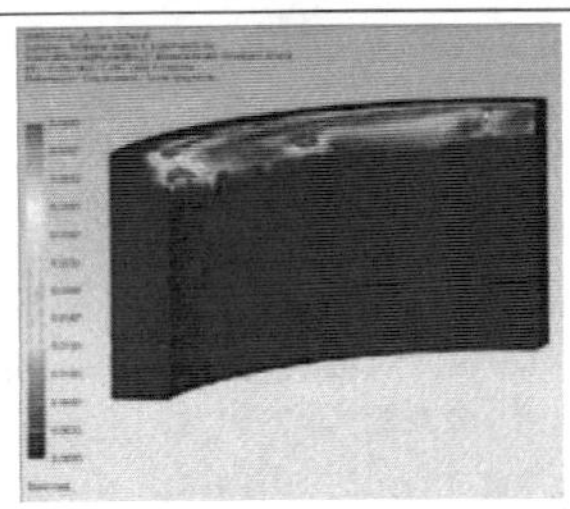	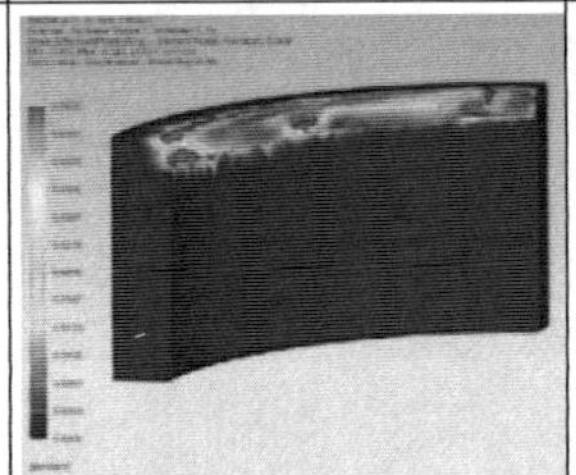	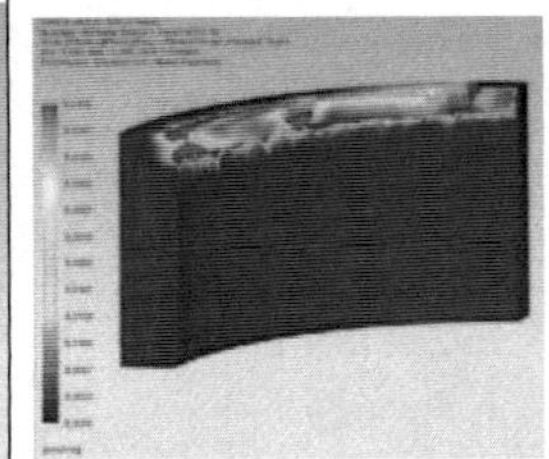	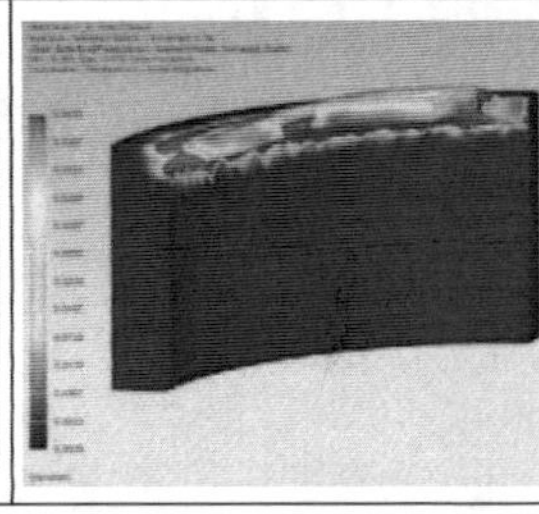

Figure 11. In the FEM simulations different pressures are used to simulate membrane rupturing for unannealed and annealed materials. The red color indicates that the strain is exceeding maximum elongation at break of the material. The maximum used elongation values are based on the conducted test results. When the red areas meet from below and above at membrane outer edge, the rupturing pressure has been met, as in the picture at below right (5.4 MPa – Annealed).

The membrane designed for 1.0 MPa rupturing pressure and 0.3 mm thickness is shown in figure 12. The FEM analysis showed that AlSi10Mg material requires a minimum 30 mm diameter to exceed 400N/mm^2 tensile strength. Preliminary rupturing test made for the 0.3mm thick and 30 mm diameter membrane which is shown in Figure 13.

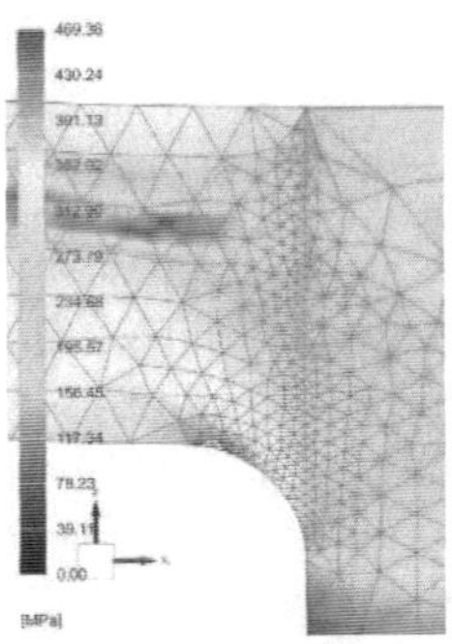

Figure 12. FEM modelling of 0.3 mm membrane and 1 MPa pressure. Tensile strength 400N/mm^2 can exceed and rupturing of the safety vent can be met.

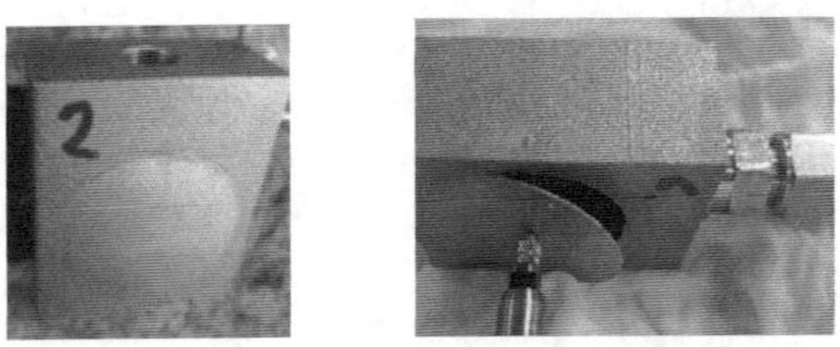

Figure 13. Test specimens of rupturing test at 1.0 MPa (on left before and on right after). The pressure was produced with a hand pump, diesel oil is used as the medium.

IOP Conf. Series: Materials Science and Engineering 1296 (2023) 012017 doi:10.1088/1757-899X/1296/1/012017

The laser welding test results are shown as a graph in the figure 14. According to the tensile test results in the table 4, the AM part has higher tensile strength than sheet metal Al 1050. Therefore, it is assumed that welded specimens will break at the force level of sheet metal part tensile strength. That will fulfil the strength demand of the construction. Figure 14 shows the welded specimens' tensile strength and calculated heat input during welding. Three or more identical specimens were tested per welding parameter set. Variation of the test results is relatively small per set.

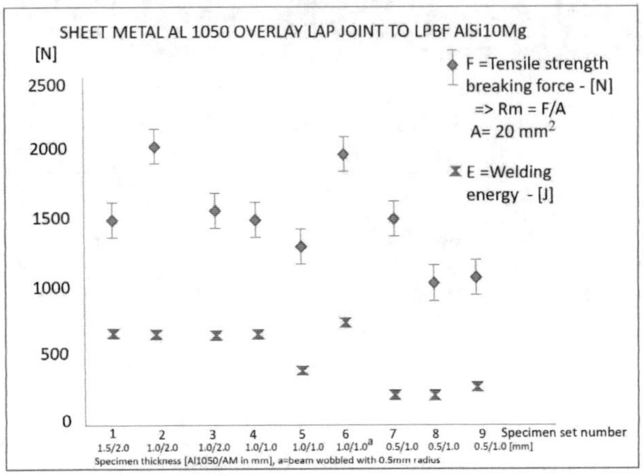

Figure 14. Tensile strength test results of the welded specimens' y-axis breaking force in [N] and heat input [J]. The specimen set number is presented on the x-axis.

Breaking force 1500 N corresponds 75 N/mm^2 tensile strength, which is higher than sheet metal specimens' nominal tensile strength 65 N/mm^2. Therefore, the parameter set for specimens 27, 28 and 29 is giving best-compromised result for strength and minimised heat input. Low heat input can be explained also due to the thinner Al 1050 sheet metal part and welding speed 200 mm/s.

4. Conclusions

The main findings of the study are the following. The safety vent area material properties are critical for the reliability of the battery cell safety. The results of this study indicate that a wider study should be conducted. Also the results indicates that the safety vent area of the lid part should be annealed, or dimensioning should be changed to meet the requirement. The measured tensile strength properties of annealed and unannealed material revealed that the maximum elongation at break values of the specimens, are lower than the material supplier's given values. One conclusion is that the main part of the lid may remain in normal condition after printing and the safety vent area may annealed locally e.g., by a laser beam. It would be optimal to make the annealing even during the printing process. Advanced LPBF machines, e.g., which Aconity3D GmbH has introduced [Martin Buscher comment's, 15.6.2022 Industrial 3D printing seminar at Turku], should be capable of that, but that was not realised in this study. The preliminary rupturing test result in 1.0 MPa pressure met the requirements and indicates that a full-scale study with this larger dimension safety vent should be carried out. While the safety vent is a part of the lid component and lid joining strength to cell housing part is critical a compromised solution should be found. Battery cells in EV must made with mass production methods. LPBF machines and systems are developing and machines for continuous process, like SLM 500 and EOS M 400-4 machines, are available in the market. In those machines, the chamber can be changed rapidly, and the process is almost continuous. That indicates LPBF to be a possible method for manufacturing automotive parts like in this study. Laser welding results of AM lid part to sheet metal aluminium indicates that the AM part can combine battery cell assembly as an industrial process.

IOP Conf. Series: Materials Science and Engineering 1296 (2023) 012017 doi:10.1088/1757-899X/1296/1/012017

5. References

[1] Harper G, Somerville R., Driscoll L, Slater P, Stolkin R, Walton A, Christensen P, Heidrich O, Lambert S, Abbott A, Ryder K, Gaines L, Anderson P, Recycling lithium-ion batteries from electric vehicles, *Nature* | Vol 575 | 7 November 2019

[2] D'Souza R, Patsavellas J, Salonitis K, Automated assembly of Li-ion vehicle batteries: A feasibility study, *Procedia CIRP*, Volume 93, 2020, Pages 131-136, ISSN 2212-8271

[3] Kwade, A, Haselrieder, W, Leithoff, R. et al. Current status and challenges for automotive battery production technologies. *Nat Energy 3*, 290–300, 2018

[4] Dai Q, Kelly J, Gaines C, L.; Wang, M. Life Cycle Analysis of Lithium-Ion Batteries for Automotive Applications. *Batteries* 2019, 5, 48. https://doi.org/10.3390/batteries5020048

[5] Meng K, Xu G, Peng X, Youcef-Toumi K, Li J, Intelligent disassembly of electric-vehicle batteries: a forward-looking overview, *Resources, Conservation and Recycling*, Volume 182, 2022, 106207, ISSN 0921-3449, https://doi.org/10.1016/j.resconrec.2022.106207.

[6] Velázquez-Martínez, O., Valio, J., Santasalo-Aarnio, A., Reuter, M., Serna-Guerrero, R. A Critical Review of Lithium-Ion Battery Recycling Processes from a Circular Economy Perspective. *Batteries,* |Volume 5| Issue4, 2019. 10.3390/batteries5040068

[7] Schwarz A., Schleser M., Gerhards B., Popoola P., Gebhardt A. Welding of Additive Manufactured ALSI10MG: Using Laser Welding in a Vacuum for High Quality Weld Seams — A New Approach to Welding LPBF Manufactured ALSI10MG. *South African Journal of Industrial Engineering* Vol. 32, No. 4, 2 Jan 2022.

[8] Dimatteo V., Liverani E., Ascari A., Fortunato A. Weldability and mechanical properties of dissimilar laser welded aluminum alloys thin sheets produced by conventional rolling and Additive Manufacturing. *Journal of Materials Processing Technology*, Volume 302, April 2022, 117512.

[9] Mäkikangas J., Rautio T., Mustakangas A., Mäntyjärvi K. Laser welding of AlSi10Mg aluminium-based alloy produced by Selective Laser Melting (SLM). *Procedia Manufacturing*, Volume 36, 2019, Pages 88-94

[10] Sun T., Ferguson N., Liu C., et al. Application of adjustable ring mode laser in remote laser welding of additive manufactured AlSi10Mg alloy, *J. Laser Appl.* 34, 042007, 2022, https://doi.org/10.2351/7.0000794

[11] Michler M, Hollmann P, Zenker R, Buchwalder A. Investigation of Pore Reduction in Hybrid Joints of Conventionally and Additively Manufactured AlSi10Mg Using Electron Beam Welding. *Advanced Engineering Materials,* First published: 11 March 2021. https://doi.org/10.1002/adem.202001325

[12] Saariluoma H, Piiroinen A, Unt A, Hakanen J, Rautava T. and Salminen A, Overview of Optical Digital Measuring Challenges and Technologies in Laser Welded Components in EV Battery Module Design and Manufacturing, *Batteries*, 2020, *6*, 47. https://doi.org/10.3390/batteries6030047

Authors contribution: Methodology, H.S., K.V., and A.S.; formal analysis, H.S., K.V. and A.S.; acquisition, H.S. and K.V.; interpretation of data, H.S., and K.S.; writing—original draft preparation, H.S., K.V. and A.S.; writing—review and editing, H.S., K.V. and A.S.; visualization, H.S. and K.V.; supervision, A.S.; project administration, H.S. and A.S. All authors have read and agreed to the published version of the manuscript.

Acknowledgments: Authors of this Module (funded by Business Finland from January 2022 to article acknowledge projects of New Battery December 2023 as a Research to Business -project) and DREAMS (Database for Radically Enhancing Additive Manufacturing and Standardization, the project was funded by Business Finland from April 2022 to February 2024). Authors would like to thank partners of this project for their contribution to this article. Authors wish to acknowledge the assistance of laboratory engineers P Törnqvist and K Metsämäki from Turku University of Applied Sciences in the rupture testing of safety valves.

IOP Conf. Series: Materials Science and Engineering 1296 (2023) 012018 doi:10.1088/1757-899X/1296/1/012018

Development of a build volume reduction kit for studying epitaxial re-solidification in laser powder bed fusion

M. Norouzi-Inallu[1*], V. Laitinen[1], I. Poutilainen[2], and K. Ullakko[1]**

[1]Material Physics Laboratory, Lappeenranta-Lahti University of Technology LUT, Lappeenranta 53850, Finland
[2]Research Group of Laser Material Processing and Additive Manufacturing, Lappeenranta-Lahti University of Technology LUT, Lappeenranta 53850, Finland

*Corresponding author e-mail: meysam.norouzi.inallu@lut.fi
**Corresponding co-author e-mail: ville.laitinen@lut.fi

Abstract. Laser powder bed fusion (PBF-LB/M) is a promising additive manufacturing process that enables the production of complex and high-performance parts. However, the high cost of materials and the need for large quantities of powder in conventional industrial-grade systems pose challenges for experimental materials development and testing activities. This study focuses on the development of a modular build volume reduction kit for an existing EOS EOSINT M-series PBF-LB/M machine. The proposed build volume reduction kit can be customized and adapted for specific research needs, expanding the capabilities of existing infrastructure without significant capital investment. This study describes the design and characterization of the build volume reduction kit and a detachable Pt-heater module, which allows for preheating of the substrate material above 500 °C. The kit's operation was validated by manufacturing simple cuboid samples using EOS 316L stainless steel powder on a 316L stainless steel substrate. The results demonstrate the feasibility of using the reduction kit for cost-effective experimental investigations, as well as highlighting its potential for studying the epitaxial solidification of PBF-LB/M-built functional materials.

1. Introduction

Laser powder bed fusion (PBF-LB/M) is a type of additive manufacturing (AM) process that uses a focused laser beam to selectively melt and fuse layers of powders (polymer or metal) to create complex three-dimensional (3D) structures with high geometrical precision [1]. PBF-LB/M enables the production of complex parts with intricate geometries, to an extent that is not possible with conventional manufacturing processes. Additionally, in some cases, PBF-LB/M can produce parts with superior mechanical properties in comparison to those that have been conventionally manufactured, which results from having the capacity to control parts microstructure during the manufacturing process [2]. Such capabilities have made PBF-LB/M an attractive option for industries requiring complex and high-performance parts, such as for sports cars, aerospace components, and biomedical implants [2,3].

A large quantity of the research published on PBF-LB/M is focused on materials development, e.g. characterization of the properties of PBF-LB/M-built materials or the development of new materials suitable for PBF-LB/M [4,5]. Indeed, PBF-LB/M has proven particularly useful in manufacturing metallic alloys that are otherwise difficult to achieve using traditional manufacturing methods, e.g. tungsten alloys [5] or Ni-Mn-Ga-based magnetic shape memory alloys [6,7] that, due to their high brittleness, are generally difficult to machine. Furthermore, PBF-LB/M also enables the processing of new materials with unique chemical compositions and properties [8]. Although modern materials development significantly benefits from the computational materials engineering, developing new alloys and processes still heavily relies on experimental materials development and testing activities. The majority of the commercially available PBF-LB/M machines have build volumes above 150×150×150 mm^3 [9]. Different research institutes and universities have typically purchased these types of larger industrial-scale machines due to the capabilities they offer in the manufacture of larger parts and patch sizes [10]. However, although this is advantageous for general mechanical engineering and application-oriented studies, a significant challenge remains when it comes to experimental materials development and testing activities. Most of these industrial-scale machines require the minimum usage of dozens of kilograms of metal powder to fill the build chamber and successfully operate the device. This scenario is problematic because the relatively high cost [11] of new experimental materials and precious alloys can make it difficult for researchers to conduct experimental research with these systems.

The costs associated with material usage within the manufacturing process can be reduced by exploring ways to reuse and recycle powder (minimizing waste), reducing the cost of used powders, and powder consumption too. These actions can also help achieving time savings and sustainability-related goals, such as minimizing material waste and the amount of used energy, during experimental materials development activities. In cases where the experimental development of new materials or expensive precious alloys is required, reducing the amount of required powder is the most effective approach. One option is using an PBF-LB/M system with a smaller build volume [12], although this may require significant capital investment if a new system needs to be purchased and installed. Another option is to develop a customized testbed PBF-LB/M system to enable precise control over the build volume, as well as to optimize the system for specific research tasks [8,12]. There is a major disadvantage with this strategy, however, in that these types of testbed systems tend to be tailored for specific experiments, making them generally less reliable than commercially sold systems. A third option is to purchase or develop a build volume reduction kit for an existing PBF-LB/M machine [13,14]. This kit would enable the machine to use less powder, hence, reducing the cost of materials involved with experimental materials development and testing activities. The advantage is that this option allows for expanding the capabilities of existing (reliable) infrastructure while reducing the need for large capital investment.

The aim of this study was to develop a modular build volume reduction kit with which experimental investigations using exotic or precious alloys can be conducted in a cost-effective manner. The kit was designed to fit into an existing (modified) EOS EOSINT M-series PBF-LB/M machine. The modular design of the developed reduction kit was specifically tailored to enable system customization and adaptation to align with research needs. By manufacturing simple cuboid samples using EOS 316L stainless steel powder, the operation of the manufactured reduction kit was finally validated. In this study, the reduction kit was designed to feature an additional detachable Pt-heater module, allowing for preheating the used substrate material above 500 °C. Taking this step enables the future studies on epitaxial solidification of PBF-LB/M-built functional materials.

2. Experimental setup and materials

2.1. PBF-LB/M system

The PBF-LB/M machine used in this study is a modified research machine that represents the EOSINT M-series – see [14] for specific machine details. The machine consists of a laser source, a galvanometric scanner, movement and laser exposure control software, and a build chamber with a powder re-coater system. The system was retrofitted with a new single-mode IPG YLS 200W SM CW ytterbium fiber laser with a wavelength of 1,070 nm and a maximum continuous wave average power output of 200W. The laser beam is transported from the laser to the galvanometric scanner optics (Scanlab hurry-SCAN 20 with F-theta lens) via optical fiber. Figure 1 shows the power intensity distributions of the laser beam at its focal point. The laser beam produced by the system was capable of near-Gaussian power distribution with the intensity in the center of the beam. The resulting focal point diameter of the laser beam is 83 µm at laser power of 80 W and 92 µm at laser power of 200 W. The atmosphere control system in the build chamber allows for the use of inert process gas, such as argon or nitrogen. The system has an inbuilt nitrogen generator with an approximate productivity of 20 m^3/h. 3D file manipulation (including slicing and generation of scanning vectors) and laser beam manipulation were conducted using SCAPS SAMLight scanner software with 3D functionality, whereas the physical movements of the systems axis (re-coater, powder feed, and the build platform position) were controlled externally using a G-code based software.

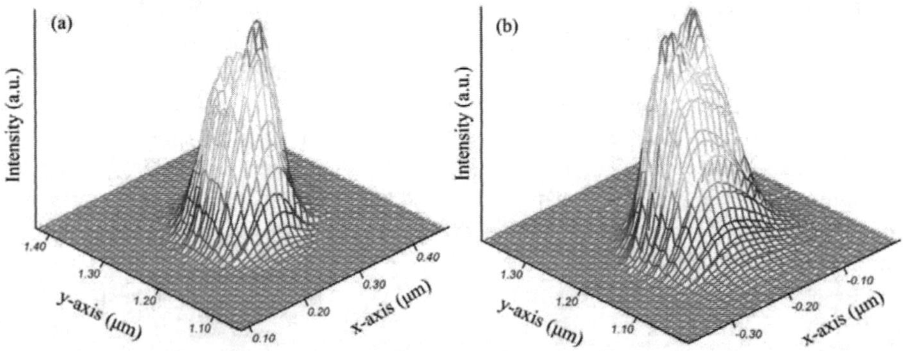

Figure 1. Power intensity distribution of the near-Gaussian-shaped laser beam in focal point at (a) 80 W and (b) 200 W.

2.2. Design and characterization of the build volume reduction kit and detachable Pt-heater module

The build volume reduction kit and the additional detachable Pt-heater module were designed using Dassault Systèmes SolidWorks. Based on discussions and the subsequent manufacture of prototypes for initial testing, the design was gradually iterated to the final setup presented in this study. Whereas the main frame, some of the other smaller bulk components, and the final assembly were all implemented at the LUT machine shop and the LUT laser processing laboratory. Characterization of the detachable heater module performance was implemented using a calibrated IR camera (Fluke Ti10, USA) with the uncertainty of ± 2 °C or 2 % (-20 – 250 °C) and a basic electronic multimeter equipped with K-type thermocouples with the uncertainty of ± 2.2 % (~ 0 – 293 °C) and ± 0.75 % (~ 293 – 1250 °C). By applying voltage to the Pt-heater module, using a DC-supply Tenma 72-10495 (0-30 V ± 0.01, 5 A ± 0.01), circulating current was measured to study the electric current vs. electric voltage characteristics.

2.3. Manufacturing of EOS 316L stainless steel test samples

The developed build volume reduction kit was tested by manufacturing test samples using EOS 316L stainless steel powder. The samples were melted on a 316L stainless steel substrate that was cut from a circular rod and machined down to the final dimensions of Ø45×15 mm^3. Table 1 shows the chemical compositions of the used powder and the substrate. Figure 2(a) shows that the volume-weighted particle size distribution ($d_{0.1}$ = 29.6 µm, $d_{0.5}$ = 46.2 µm, and $d_{0.9}$ = 63.3 µm) of the powder was measured using Panalytical Morphologi G3S as well as Figure 2(b) shows the SEM secondary electron image of the powder particles using Hitachi SU3500 electron backscatter microscope.

Table 1. Compositions of the used materials in weight percentage (wt.%).

Material	Fe	Ni	Mo	Cr	Mn	Si	N	O	Al	C
EOS 316L powder	Balance	12.4	2.4	19.0	2.4	0.3	0.1	0.1	0.2	0.1
316L substrate	68.1	7.1	0.2	20.9	1.7	1.1	0.1	-	0.7	0.1

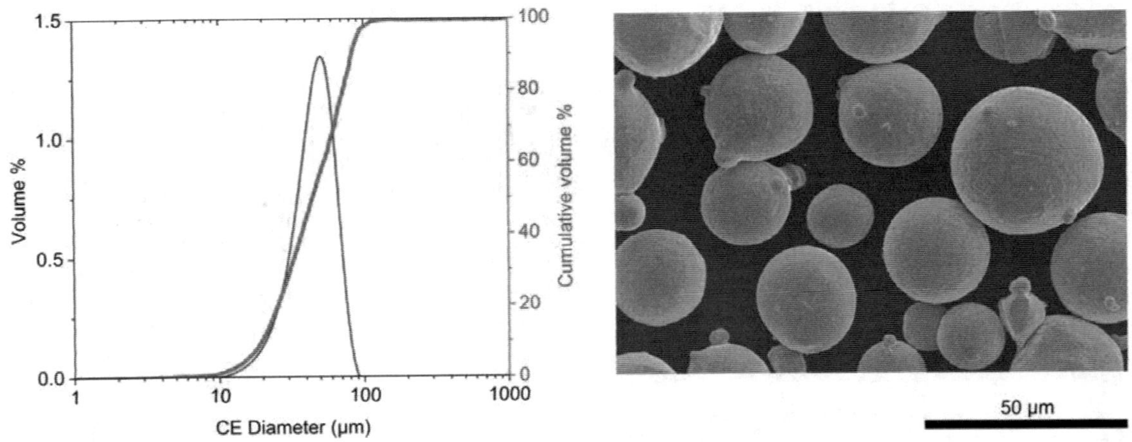

Figure 2. (a) Powder particle distribution of the used EOS 316L stainless steel powder; (b) SEM secondary electron image of the powder particles.

The process parameters used for manufacturing the EOS 316L test samples are summarized in Table 2. The samples themselves were cuboids with approximate target dimensions of 7×7×1 mm^3. All samples were manufactured using a unidirectional scanning strategy (aligned along one of the side facets of each) without rotation of the scanning pattern from layer to layer. A fixed laser power of 200 W and layer thickness of 20 µm was also utilized for manufacturing while varying the applied scanning speed and hatch distance in three levels. Contour scans were melted using the same parameters as the hatched areas. The center-point values were determined based on the known optimal values for this material, as previously used in the same system without the reduction kit. During the experiments, the laser beam was focused on the powder bed surface. The test samples were deposited in a single patch (nine samples) on the 316L substrate, leaving a 1 mm wide gap between each to minimize thermal interactions during the process. The melted cuboids were ground incrementally along the build direction using SiC abrasives and then polished using 1 µm alumina particle suspension and a napless cloth. Polished cross-sections of the samples were inspected using a customized Zeiss Axio Scope.A1 optical polarized light microscope, and the relative densities of the manufactured samples were determined optically by measuring the area fraction of the pores within each sample. To facilitate comparison, a 316L sample manufactured using the known optimal values (scanning speed of 1000 mm/s and hatch distance of 100 µm) in the same PBF-LB/M system without the reduction kit was also ground, polished, and measured.

Table 2. The applied process parameters for manufacturing the EOS 316L cuboids.

Parameter	Value
Laser power (W)	200
Scanning speed (mm/s)	950, 1000, 1050
Hatch distance (µm)	90, 100, 110
Layer thickness (µm)	20
Shielding gas type (-)	Nitrogen (99.998%)

3. Results and discussion

3.1. Design and characterization of the modular build volume reduction kit

The developed build volume reduction kit's structure and main components are shown in Figure 3. The kit is enclosed in the shielding gas chamber of the initial system, which includes inlet and exhaust ports for filling the shielding gas and providing a gas flow across the powder bed during builds. The focused laser beam passes through a window located on the top surface of the shielding gas chamber.

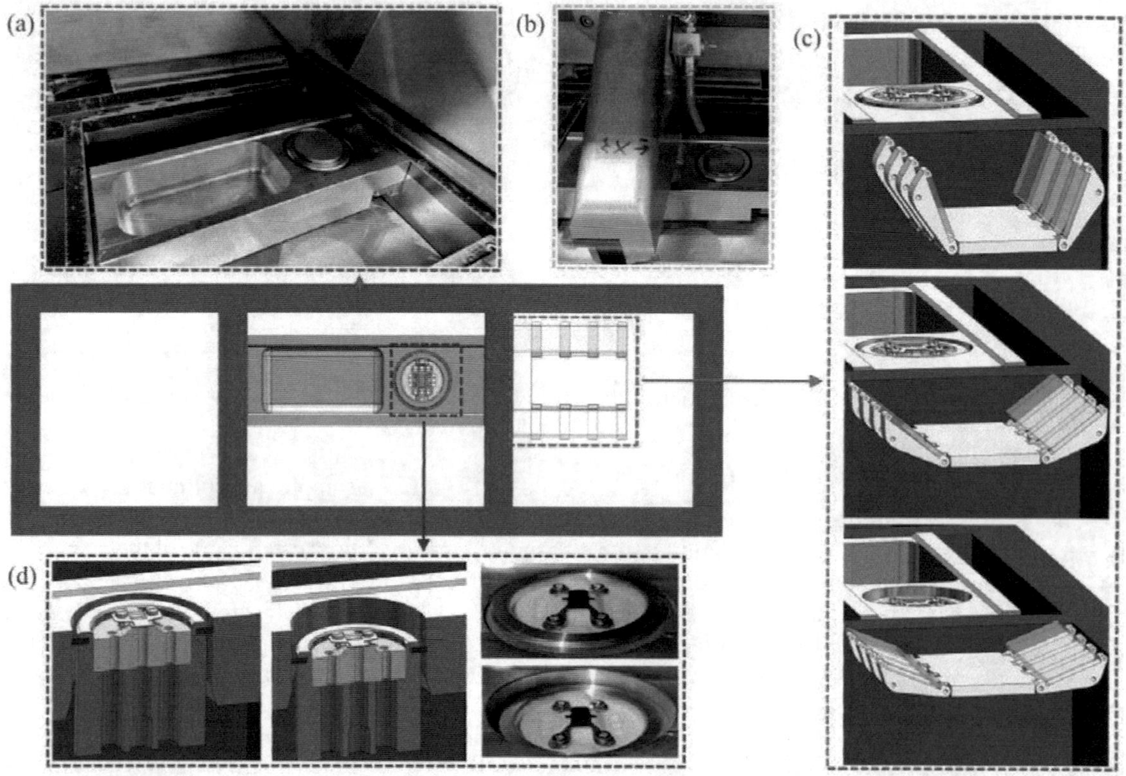

Figure 3. Schematic of the designed build volume reduction kit. (a) the excess powder reservoir and the build platform with an installed stainless-steel substrate, (b) secondary (Argon or Nitrogen) gas inlet and the powder re-coater, (c) operation of the main powder reservoir, and (d) build platform assembly with the z-axis piston and the detachable Pt-heater module. The flat grey sections in the central image show the structure and positions of the original PBF-LB/M system's main chambers.

The build volume reduction kit comprises two main components: the main powder reservoir and the 'manufacturing zone' with an integrated build platform and fixed depth reservoir for excess powder. The main powder reservoir (powder feed) and the build platform (powder layer height) were designed to be operated using the original in-built movement axis (z-direction) of the PBF-LB/M machine. Adopting this design approach effectively eliminated the need for building a complex setup of external electrical drives to produce the required motions. The powder feed mechanism selected for the build volume reduction kit was a type of simple cantilever hinge mechanism. This mechanism is directly mounted on the top surface of the original powder feed plate, serving as a relatively inexpensive solution to be used in the initial device setup. This structure is separate and fully modular, allowing for technical design iterations and possible setup replacements with a more conventional piston-style approach in the near future. By moving the machine's original powder re-coater linearly in one direction, the powder is spread from the main reservoir to the build platform. The powder layers of the desired thickness are produced by lowering the build platform and, thereby, restricting the spread powder layer thickness to a fixed value. The build plate is installed on a circular piston head that is mechanically connected to the initial machine's larger build platform surface. A Teflon ring around the piston's top edge prevents powder from falling through the gap between the frame of the assembly and the piston itself. The build plate is attached to the head of the piston using a screw.

3.2. Design and characterization of the detachable Pt-heater module

The main motivation for designing and building a detachable Pt-heater module can be explained by a special research need involving experimental investigations on the epitaxial solidification of PBF-LB/M-built functional materials. To significantly lower the thermal gradients during melting in PBF-LB/M [15], this approach requires the use of substantial substrate heating (above 500 °C). In the case of Ni-Mn-Ga-based magnetic shape memory alloys, the used substrate may be either polycrystalline or, in specific solidification studies, also single crystalline, which sets limits to the maximum size of the used substrate – typical Ni-Mn-Ga single crystals are less than 20 mm in diameter [16] The detachable Pt-heater module was designed, therefore, to facilitate the use of small (below the aforementioned size) substrates.

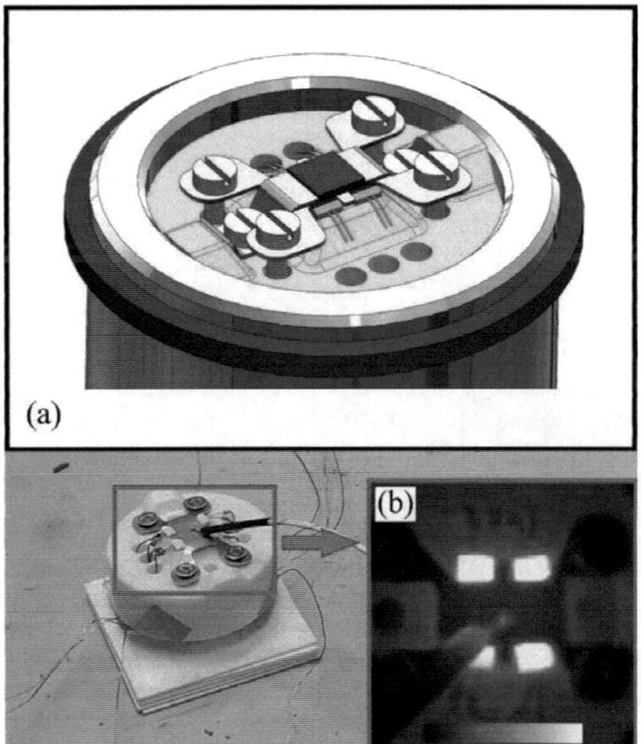

Figure 4. (a) 3D model of the designed detachable Pt-heater module, and (b) a manufactured heater module with an externally mounted K-type thermocouple used for characterizing the module's thermal behaviour. The inset shows a capture from the IR camera data used for monitoring thermal distributions during heating.

Figure 4 shows the designed detachable Pt-heater module, which includes a clamp with a top surface that is heated linearly using a PID controller. PID (Proportional-Integral-Derivative) is a control algorithm commonly used in engineering systems to achieve precise and efficient control of a heating process. The main body of the heater module consists of a Ø45×70 mm^3 alumina block that acts as a thermal isolator preventing the flow of heat from the Pt-heaters into the build platform piston and the reduction kit frame. The main heater assembly is placed on top of the alumina block and consists of a rectangular clamp measuring 20×10 mm^2 and four Pt heater elements sized 5.2×3.9×1 mm^3. The build plate, which is a 7×7×1.5 mm^3 substrate (single crystal or polycrystal), is fixed to a rectangular support assembly that is connected rigidly to the vertical stage (z-axis) using two screws. This module can also be mounted on top of the build platform piston – again, also by using screws. The piston (and the alumina block) requires holes for transferring the heater element and thermocouple wires through the piston, thus, preventing the wires from interacting with the powder re-coater or the powder itself during builds.

The temperature of the Pt microheaters is controlled by adjusting the voltage, whereas a K-type thermocouple is used to measure the temperature of the clamp and monitor the temperature increase. This facilitates compensating for any error caused by the physical separation between the clamp and the Pt-heaters. By monitoring the temperature of Pt heater elements using a calibrated IR camera and establishing a linear proportional relationship with the clamping surface, the temperature measurement of the samples was enhanced. Figure 5(a) shows the electric current vs. electric voltage characteristics of the used Pt-heater elements the heating rate of 55 °C min^{-1} (test 1) and 110 °C min^{-1} (test 2) to reach maximum temperatures of ~272 °C, and ~540 °C for in a direct current mode.

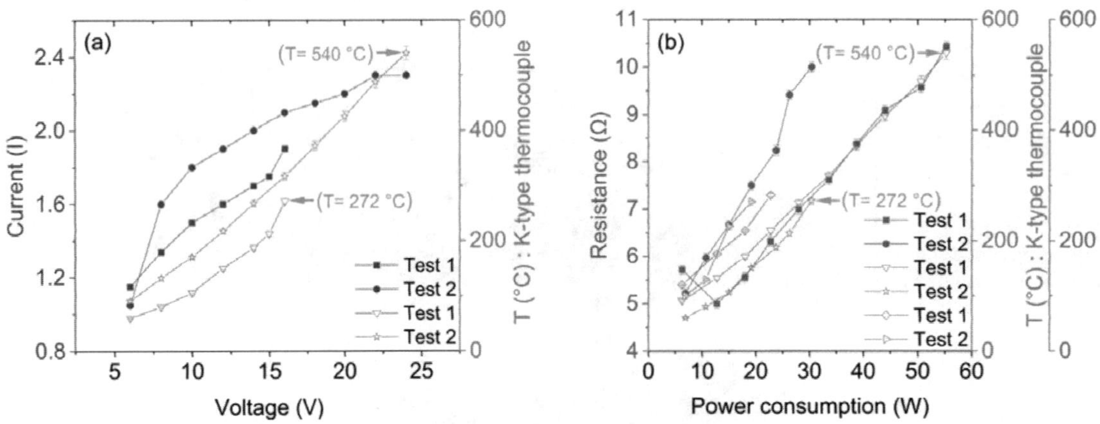

Figure 5. Characterization of the performance of the used Pt-heater elements during test 1 (maximum temperature ~272 °C), test 2 (maximum temperature ~540 °C). (a) electric current vs. electric voltage and the electric current vs. heater temperature (*T*), and (b) electric resistance vs. power consumption and the electric resistance vs. heater temperature.

The investigation of the current (*I*) versus voltage (*V*) characteristic of the sample reveals that the Pt heater elements exhibit nonlinear behaviour, as well as showing that the curve in Figure 5(a) can be divided into two distinct regions: 1) a nonlinear zone wherein the electrical conductivity of the Pt heater elements undergoes an abrupt change from a high value to a low value, resulting in a reduction of electrical resistance, and 2) an upturn zone that exhibits a moderate increase in electrical resistance. As the Pt heater elements can be utilized without experiencing a 24 V breakdown voltage, employing them for heating the sample as a substrate with low overvoltage is suggested. Figure 5(a) demonstrates that any rise in clamp temperature can be controlled by adjusting the *I* - *V* characteristics. As illustrated in Figure 5(b), the electrical resistance of the Pt heater elements was measured by applying input power. The resistance (*R*) of the Pt heater elements resulting from Joule heating with input power (power consumption) is plotted. The temperature of the clamp and Pt heaters, as well as their relationship with the input power, exhibit nearly identical linear behaviour – consistent with the data obtained from an IR camera. These measurements indicate that some heat losses occur due to the physical gap between the clamp and Pt heater elements.

3.3. Fabrication of the EOS 316L stainless steel cuboids

By manufacturing test samples using EOS 316L stainless steel powder on a 316L stainless steel substrate, the developed build volume reduction kit was tested. During this, two process deviations were observed that affected the manufacturing results. Firstly, it was found that the build platform piston – responsible for the substrate movement – was not securely attached to the assembly. As a result, the piston started rotating during the build process, causing slight rotational drifting of the scanning pattern. By modifying and reinforcing the piston mounting to the main platform of the system, this issue can easily be resolved to ensure stable and precise movement without any rotational deviations. Secondly, the powder spreading process resulted in the powder being spread over two wide areas, causing some of the powder to flow sideways over the edges of the designed reduction kit frame. This resulted in the loss of powder that, from an economic standpoint, is undesirable. To address this issue, reducing the width of the active section of the powder re-coater that comes into physical contact with the powder is recommended. By reducing the width, the flow of powder to the sides can be minimized, thus, reducing wastage, and optimizing the powder utilization.

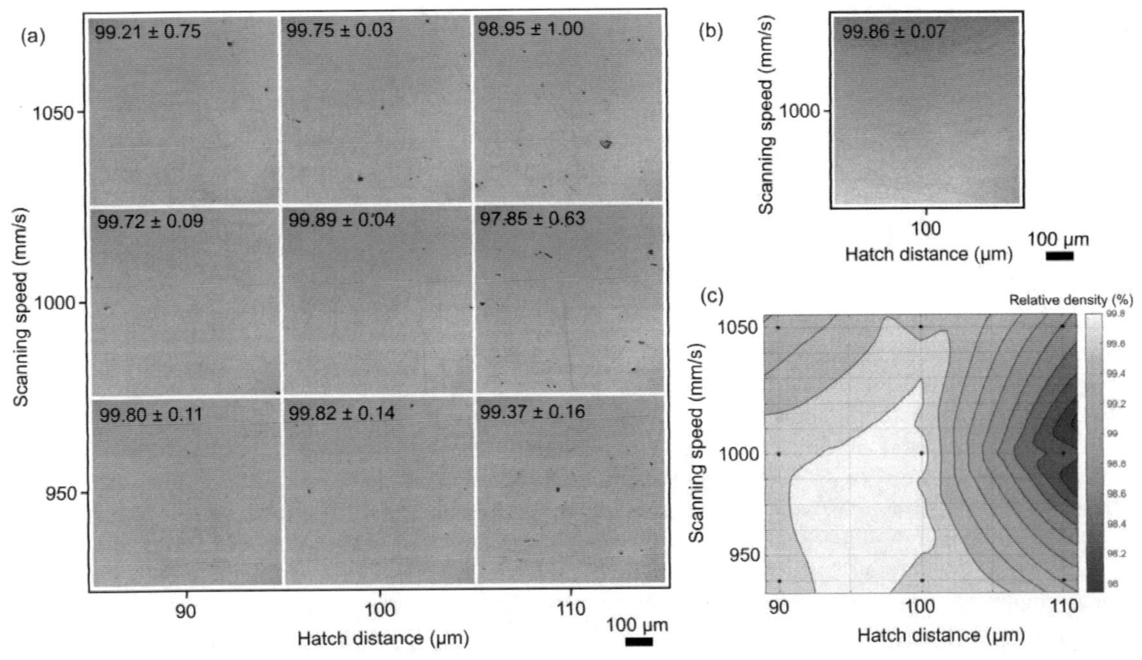

Figure 6. (a) Optical images of the EOS 316L test samples built using the reduction kit. The inset markings indicate relative densities for each sample, wherein the shown error corresponds to the standard measurement deviation. (b) Optical image of the EOS 316L reference sample built without using the build volume reduction kit. (c) Lowess surface fit showing the main effects of the applied parameters on relative density of the built samples.

Despite these minor process deviations, the results indicate that the build volume reduction kit was effective in controlling thermal interactions during the build. Figure 6(a, b) shows example cross-sections and relative densities of each test sample and the reference sample. The shown errors correspond to the standard deviations of the measurements. As can be observed from the figure, the relative densities of the test samples manufactured using scanning speeds of 950 and 1000 mm/s and hatch distances of 90 and 100 μm were relatively close to the reference sample manufactured using the same PBF-LB/M machine without the build volume reduction kit installed. However, the samples manufactured using scanning speeds of 1100 mm/s and hatch distances of 110 μm exhibited significantly lower relative densities – see Figure 6(c). These samples also exhibited larger standard deviations of relative density. The observed drop of relative density values is drastic considering the relatively small change in the applied processing parameters, which implies that the optimal processing window for the used material shifts slightly towards lower scanning speed and hatching distance values when using the reduction kit. The observed shifting of the optimal processing window likely occurs due to the different thermal diffusion and heat accumulation characteristics between the substrates in these two setups (with and without the reduction kit). Besides influencing relative density, this effect may also induce minor microstructural differences between the samples manufactured using the initial system and the samples manufactured using the reduction kit. Future studies using this system will focus on benchmarking these differences and conduct more thorough microstructural investigations on the produced materials.

However, the successful test samples highlight the suitability of the designed build volume reduction kit for manufacturing precise and dimensionally accurate components, thus, facilitating the intended experimental materials development and testing activities. However, addressing the observed deviations and implementing the recommended modifications will further enhance the system's performance and efficiency. Overall, the results demonstrate the successful implementation of the build volume reduction kit.

Conclusions and further studies

Laser powder bed fusion (PBF-LB/M) serves as a promising additive manufacturing process that enables the production of complex and high-performance parts. Nevertheless, the high material costs and the need for large quantities of powder in conventional industrial-grade systems pose challenges for experimental materials development and testing activities. To address these challenges, this study developed a modular build volume reduction kit that was then integrated into an existing (modified) EOS EOSINT M-series PBF-LB/M machine. The reduction kit allows for the use of smaller quantities of powder, thus, reducing material costs and enabling cost-effective research on new exotic alloys or precious metals. The kit was successfully designed and characterized before its operation was validated by manufacturing test samples using EOS 316L stainless steel powder. The results demonstrate the feasibility of using the reduction kit to conduct experimental investigations in a cost-effective manner. Moreover, a detachable Pt-heater module was developed to facilitate studies on the epitaxial solidification of PBF-LB/M-built functional materials. This module enables precise control of the substrate temperature above 500 °C, which is crucial for the formation of specific microstructures. The module's thermal behaviour was characterized using an IR camera and K-type thermocouples, with results showing that the module effectively provided the desired temperature control and could be used for experiments involving the solidification of materials like Ni-Mn-Ga-based shape memory alloy. In summation, the developed build volume reduction kit and detachable Pt-heater module offer valuable tools for researchers working with PBF-LB/M, specifically for enabling cost-effective materials development and precise temperature control for specialized studies of functional materials.

Acknowledgments

This work was financially supported by the Strategic Research Council of Finland (grant number 313349), the Academy of Finland (grant number 325910), and the AMBI (Analytics-based Management for Business and Manufacturing Industry) research platform of LUT University.

Reference

[1] Gibson I, Rosen D, Stucker B, Khorasani M, Gibson I, Rosen D, Stucker B and Khorasani M 2021 Development of additive manufacturing technology *Additive manufacturing technologies* 23–51

[2] Wang D, Liu L, Deng G, Deng C, Bai Y, Yang Y, Wu W, Chen J, Liu Y, Wang Y, Lin X and Han C 2022 Recent progress on additive manufacturing of multi-material structures with laser powder bed fusion *Virtual Phys Prototyp* **17** 329–65
[3] Land W S, Zhang B, Ziegert J and Davies A 2015 In-Situ Metrology System for Laser Powder Bed Fusion Additive Process *Procedia Manufacturing* vol 1 pp 393–403
[4] Ahmed N, Barsoum I, Haidemenopoulos G and Al-Rub R K A 2022 Process parameter selection and optimization of laser powder bed fusion for 316L stainless steel: A review *J Manuf Process* **75** 415–34
[5] Krakhmalev P and Kazantseva N 2021 Microstructure of L-PBF alloys *Fundamentals of Laser Powder Bed Fusion of Metals* pp 215–43
[6] Laitinen V, Sozinov A, Saren A, Salminen A and Ullakko K 2019 Laser powder bed fusion of Ni-Mn-Ga magnetic shape memory alloy *Addit Manuf* **30** 100891
[7] Laitinen V, Saren A, Sozinov A and Ullakko K 2022 Giant 5.8% magnetic-field-induced strain in additive manufactured Ni-Mn-Ga magnetic shape memory alloy *Scr Mater* **208** 114324
[8] Sing S L, Huang S, Goh G D, Goh G L, Tey C F, Tan J H K and Yeong W Y 2021 Emerging metallic systems for additive manufacturing: In-situ alloying and multi-metal processing in laser powder bed fusion *Prog Mater Sci* **119** 100795
[9] Vlasea M L, Lane B, Lopez F, Mekhontsev S and Donmez A 2020 Development of powder bed fusion additive manufacturing test bed for enhanced real-time process control *Proceedings - 26th Annual International Solid Freeform Fabrication Symposium - An Additive Manufacturing Conference, SFF 2015*
[10] Sow M C, De Terris T, Castelnau O, Hamouche Z, Coste F, Fabbro R and Peyre P 2020 Influence of beam diameter on Laser Powder Bed Fusion (L-PBF) process *Addit Manuf* **36** 101532
[11] Colosimo B M and Grasso M 2020 In-situ monitoring in L-PBF: Opportunities and challenges *Procedia CIRP* vol 94 pp 388–91
[12] Bidare P, Maier R R J, Beck R J, Shephard J D and Moore A J 2017 An open-architecture metal powder bed fusion system for in-situ process measurements *Addit Manuf* **16** 177–85
[13] Matilainen V, Piili H, Salminen A, Syvänen T and Nyrhilä O 2014 Characterization of process efficiency improvement in laser additive manufacturing *Physics Procedia* vol 56 pp 317–26
[14] Järvinen J P, Matilainen V, Li X, Piili H, Salminen A, Mäkelä I and Nyrhilä O 2014 Characterization of effect of support structures in laser additive manufacturing of stainless steel *Physics Procedia* vol 56 pp 72–81
[15] Toman J, Pagan D C, Müllner P and Chmielus M 2021 Epitaxial re-solidification of laser-melted Ni-Mn-Ga single crystal *Acta Mater* **219** 117236
[16] Ullakko K, Sozinov A and Järvinen J 2016 Is it possible to make single crystalline Ni-Mn-Ga devices by 3d printing? *ICFSMA'16 Fifth Int. Conf. Ferromagn. Shape Mem, Alloy.* (Sendai, Japan) p 51

IOP Conf. Series: Materials Science and Engineering 1296 (2023) 012019 doi:10.1088/1757-899X/1296/1/012019

Optimized Inconel 718 pressure vessel manufactured with laser powder bed fusion

V Tepponen, S Westman, P Nyamekye, I Poutiainen

vesa.tepponen@lut.fi, sami.westman@lut.fi, patricia.nyamekye@lut.fi, ilkka.poutiainen@lut.fi

Corresponding author's vesa.tepponen@lut.fi

Abstract. Laser powder bed fusion (L-PBF) is one of the most novel additive manufacturing methods used for a wide range of industrial grade metallic materials. The process can produce end-use metal parts with desirable qualities and mechanical properties. L-PBF however, remains a complicated and expensive manufacturing method. Design for additive manufacturing (DfAM) is a key aspect leveraging the uptake of advantages and possibilities offered by AM in augmenting its competitiveness against conventional manufacturing (CM) methods. Inconel 718 (IN718) is a nickel-based superalloy boasting high temperature strength, good oxidation, and corrosion resistance at elevated temperatures. IN718 is commonly used for high performance applications, such as power and process industry parts, and gas turbine components. High inherent toughness, hardness, work hardening, and low thermal conductivity properties make the material difficult to manufacture through conventional machining methods. The layer-by-layer building of powder metals via L-PBF makes it possible to build different geometrical intricacy. The offered manufacturing flexibility for complex high-end metal structures for variety of applications makes L-PBF an alternative manufacturing method for high performance metals. This study investigates use of DfAM for a small-scale pressure vessel with predefined geometry, dimensions, design space and load condition. The aim is to introduce and exploit contemporary design optimization methods and their feasibility with AM. Structures, such as lattices and stress field driven geometries based on finite element analysis are investigated in this study. The designs are virtually tested under predefined pressure load of 50 bar. All four design options are manufactured on EOS M290 and IN718 powder. The result of the study shows the different optimizations decrease weight and improve material savings without compromising the linear load capacity. Optimized designs could also be made in such a way that it does not increase the manufacturing duration or add additional steps to it.

1. Introduction

Additive manufacturing (AM) has in recent years become one of the established industrial manufacturing methods offering several benefits including resource efficiency, production flexibilities [1–3] . In AM, CAD geometries of intended products are sliced into layers, which are built in sequential layer by layer with successive joining of material on AM machine [4]. AM allows unprecedented design possibilities in the designs of new products and redesigns of existing products. Material catalogue for AM has fast become broad, including several types of plastics, metals, ceramics, and glass. The mechanical properties of AM end products may be identical with their conventionally manufactured counterparts, however, the utilized AM method together with chosen material have an impact in the outcome [5]. AM unlike the conventionally manufacturing methods combines the capabilities of modern manufacturing software (e.g., computer-based and, simulations-assisted modelling) to allow design flexibility along digital thread for innovative production method in design, manufacturing, and distribution to end-users. [6]. AM through the digital thread allows the simultaneous consideration of product design, manufacturing planning, postprocessing, quality assurance and data management prior and post build. This approach allows to optimize product designs, process parameters, and consequently reduce manufacturing time, resource inefficiency and unforeseeable errors that may occur. Product design optimization methods including lattice, topology and variable thickness can be digitally generated and fine-tuned based on design for additive manufacturing (DfAM) rules to suit specific AM subcategory prior physical commitments. Different studies have investigated how optimized structures including honeycombs, lattice, variable thickness, build plan, process parameters (mainly from material deposition, beam shapes or spots) influence functional properties, energy intensities and related costs of high-performance metals [7–13].

1.1. Prospects of additively manufactured Inconel 718 pressure vessel

The inability of one conventional manufacturing (CM) method to manufacture complex or net-shape geometries [14,15] and the limitations associated to conventionally manufacturing IN718 can increase manufacturing steps and time. CM methods often create parts with sacrificial volume that do not contribute to performance and thereby increasing material and energy consumption in applications that require lightweight [14,15]. In this regard, net-shape pressure vessels are possible via combination of CM methods, wrought, casting, forging, cutting, machining, welding, to produce the final part geometry [16–19]. This usually lengthen manufacturing time and steps [16] and produce a large amount of scrap [14]. AM facilitates first-time-right manufacturing of optimized and efficient products. L-PBF can create net-shape, intricate nickel-based superalloy structures with minimal finishing machining depending on application [20]. Studies have highlighted the benefits of L-PBF and paired emerging product design methods and tools contribute to the creation of high valued products [21,22]. The study specifies contributions from four main aspects of product design or products engineering, namely digital design management, topological and parametric design optimization, generative design, material selection, and uncertainty quantification [22].

Making pressure vessel with Inconel 718 (IN718) is promising to enhance performance, one which require structural modifications to surpass the functioning of comparable composite or ceramic vessel to ensure integrity. AM is offering manufacturing sectors innovative approach to manufacture optimized, lightweight and high performing structures (e.g., bionic) [15,23,24] without direct increasing to time and costs [25]. Exemplary to studies [18,26–28] AM is shown to improve performance and/or offer economic benefits for pressure vessels. Studies [19,29,30] on the other hand have demonstrated how different AM specially PBF and DED enhance performance (microstructure, mechanical properties) of additively manufactured IN718 parts. The studies according to [19,20,30] demonstrate the capability of L-PBF to manufacture fully dense nickel-based alloys components. L-PBF is capable to bond IN718 powder layers with regular microstructural without major defects.

L-PBF for IN718 require the right product design optimization to create highly optimized structures void of defects and uncertainty quantification. Making first-first-right pressure vessels via L-PBF is promising to life cycle resource and cost efficiencies which are potential to boasting adoption confidence. The geometric/microstructural alteration benefits enabled by L-PBF may be compromised for industrial applications without guaranteed repeatability and quantified properties. The digital thread of AM enables virtual optimization of the geometric design and building process prior the actual build. Inconel 718 is a precipitation-hardening nickel-chromium alloy. The material offers high temperature strength, good oxidation, and corrosion resistance at elevated temperatures. Inconel 718 has numerous applications spanning various industrial sectors including aerospace, marine, oil and gas [31] due to their high chemical and temperature resistance. Some of the end use applications of Inconel 718 are pressure vessels, jet engine exhausts, marine propeller blades, aerospace and marine fasteners, exhaust couplings, heavy machinery parts [31,32]. Main mechanical characteristics include good tensile, fatigue, creep, and rupture strength properties at temperatures up to 700°C [32]. Heat treatment can be applied to IN718 parts produced to further enhance the mechanical properties and relieve residual stresses [33,34]. IN718 alloy offer inherent toughness, hardness, work hardening, and low thermal conductivity properties [32,35]. These properties make the processing of IN718 difficult to machine to as due to tendency of tool wear, poor surface integrity and low material removal rate [19,32]. With the rising demand of various complex structures and high-end mechanical properties in beforementioned industries, AM enables competent production methods of IN718 parts [19,35].

2. Experimental study

Initial pressure vessel had dimensions of 100x60x15 mm (see figure 1). Wall thickness was set to 0.5 mm, which is close to the guideline value for minimum wall thickness set by EOS, the material manufacturer for IN718. This was used as base for each upcoming iterative design variation.

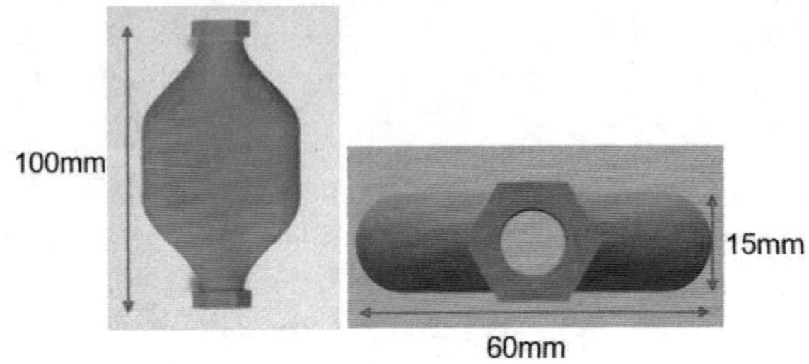

Figure 1. Pressure vessel baseline geometry and dimensions with 0.5 mm wall thickness.

As seen in figure 1, the case vessel geometry is highly inefficient for an internal pressure bearing structure, as ball shape is ideal for such applications. Aim of the study was to optimize a non-ideal shape with high surface area to volume ratio. Optimization of such non-ideal vessel shape with AM specific design possibilities is aimed to provide novelty in this study.

2.1. Design procedure

Hexagon shape was added to each end to help in attaching of the vessel to necessary components e.g., piping and pressure gauge. Both hexagons had fillets to remove possible overhangs and notches at the corners to decrease displacement caused by residual stresses. All the vessels were designed to be printed in upright position with supports only at the bottom of the part. Initial model planning was carried out with Dassault Systèmes SolidWorks and optimized designs were finalized with nTopology. FE-analysis carried out alongside design optimization was based on a linear load condition with static 50 bar pressure in each vessel iteration. Boundary condition for optimization was chosen as close exceeding of material UTS of 1500 MPa stress value in notable sections of vessel geometry.

2.1.1. *Benchmark design.* The most basic design had uniform wall thickness where the thickness was increased outwards until the load condition induced stresses approached ultimate tensile strength (UTS) of the heat-treated material. The overall geometry remained comparable to the basic model seen in figure 1, as only changeable variable was the wall thickness according to FE-analysis. The benchmark design was aimed to provide a comparison target for more optimized design variations. Increasing the uniform wall thickness throughout the vessel is a simplistic approach in strengthening the vessel model, but it is expected to induce excess material in various sections of the structure, therefore increasing the weight and material consumption compared to upcoming design iterations.

2.1.2. *Internal lattice design.* The baseline wall thickness of 0.5 mm was used for the vessel design variation with internal lattices. Two rows of periodic lattice structures were introduced to both sides, crossing the vessel chamber longitudinally as seen in figure 2.

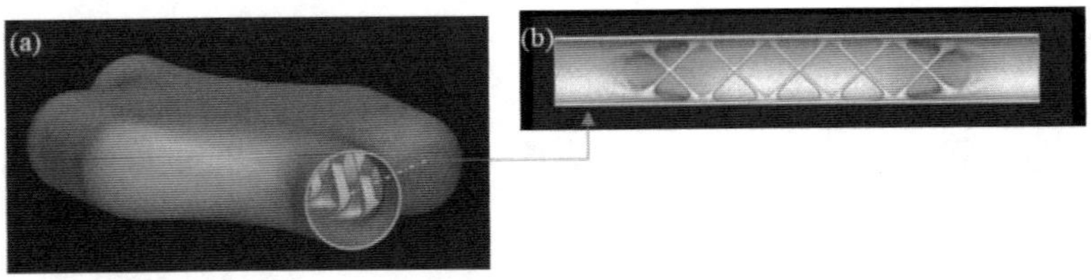

Figure 2. Internal lattice design: (a) external geometry and (b) cross-sectional view of the internal lattice.

Vessel wall thickness was kept uniform while lattice thickness was adjusted according to von Mises stresses caused by the load case. Lattice structures are typically challenging to model and more notably to optimize with common CAD software, however specialized software such as nTopology can be utilized for the task in an efficient manner. Compared to other design iterations in this study, the internal features will influence internal volume of the vessel and flowability dynamics, which would have to be considered in end-use of relevant designs. For this case study, such investigation is left out of research scope.

2.1.3. *Variable shell thickness design.* Field driven design was used to create vessel with variable wall thickness (see figure 3). FE-analysis was run on the original vessel to create von mises stress field. Offset was used on the wall with stress field as modifier. This means wall is thicker when stress is higher. Material added was between 0-1.5mm.

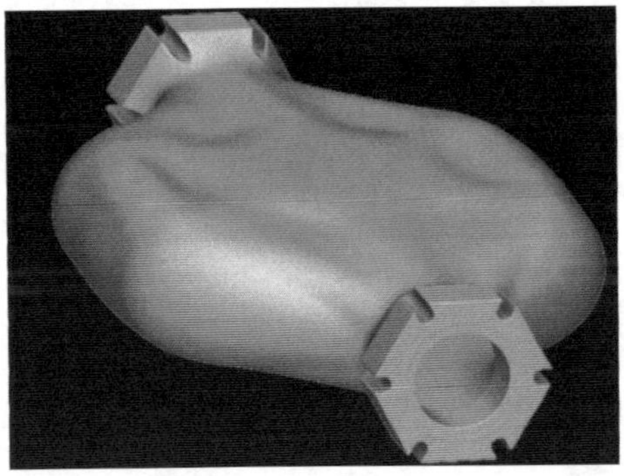

Figure 3. Representation of external geometry of the variable shell thickness design.

2.1.4. *Ribbed design.* The original 0.5 mm wall thickness model was used as base, ribs were added on top of the surface at 1.0 mm offset. The ribs were created with cell map of the surface of the vessel. The map bad 28 cells around and 12 cells across. Lattice used for ribs was rotated honeycomb. Wall thickness of the ribs were adjusted based on stress analysis results varying between 0.5-2.0 mm.

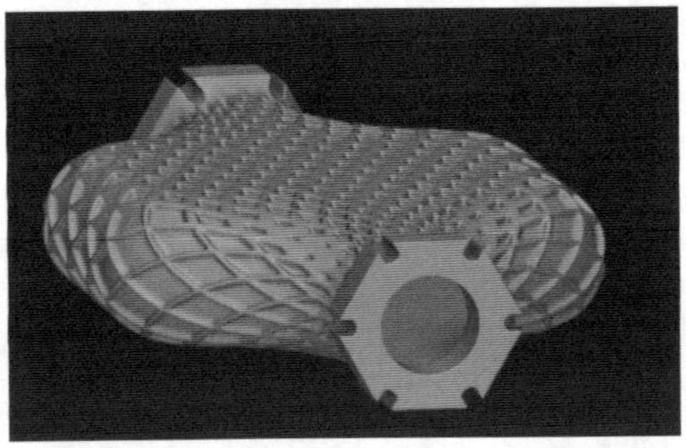

Figure 4. Ribbed design: external geometry.

2.2. Materials and manufacturing

EOS M290 L-PBF machine was used to manufacture the pressure vessels in this study. The feedstock EOS Nickel Alloy IN718 particle size range of 20 to 55 microns (20-55 μm). The used powder is characterised to have spherical shape with negligible pores due to the used atomization method. The study uses the predefined process parameters (IN718_040_PerformanceM291 2.11) according to manufacturer specification.

The manufacturing phase started with the build planning and machine setup using Materialise Magics and EOSPRINT. An upright orientation conforming to design position were defined as the build direction. The designed geometries and build direction were pre-planned during CAD modelling to omit the need of support structures. One end of the vessels was planned to serve as an attachment to the build plate. Build planning considered build orientation, part placement, possible support, and manufacturing time. Pre-machine set up considered the processes parameters, required chamber conditions. The actual building of the parts started with setting build chamber to the required inert environment (argon gas) and temperature (80°C). The different layers were built under consistent build environment with defined process parameters layer upon layer according to designed G-codes of the sliced layers. Each build layer consisted of (1) layering of powder according to defined thickness (40 μm), (2) scanning and melting of layer geometry, (3) lowering down the powder bed and levelling the next layer of powder according to defined layer thickness (40 μm) and (4) commence of successive layer accordingly. These steps were repeated layer by layer until the completion of the parts. A total of 8 samples were additively manufactured via L-PBF, which consisted of 2 samples per design iteration. Figure 5 shows the build chamber with the as-built parts post vacuuming, and the final post-processed parts after heat-treatment.

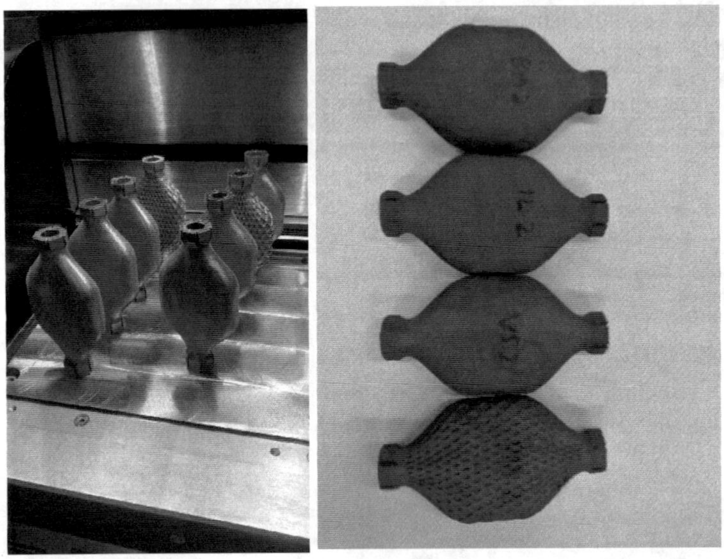

Figure 5. Representation of the as-printed parts attached to the build plate in the build chamber (left) and final heat-treated samples (right).

The as built parts were post heat-treated according to EOS recommendations consisting of two steps. Step 1 involved solution annealing where parts are heated to 954 degrees Celsius (954°C) with one hour (per 25 mm of thickness) and subsequently air cooled. Second step involved ageing treatment where the parts are held at 718 degrees Celsius (718°C) for eight hours (8 hr). Then furnace is cooled to 621 degrees Celsius (621°C) for 18 hours (18 hr). Similar air cooling to earlier step.

3. Results and discussion

All vessel samples were successfully manufactured with no defects noted in visual inspection. The as-built samples were weighed and compared to estimations provided by nTopology (Digital models). Weights of the different designs are shown in Table 1.

Table 1. Weight of the different vessels from simulations and weighted parts.

Design option	Weight (g)	
	Digital model	As-built part
Benchmark	160	158
Inner lattice	85,4	82
Ribbed	133	132
Variable thickness	128,9	126

As seen in table 1, the weight estimations of digital models and as-built parts match closely with very little deviation. Using benchmark model as comparison target, a weight reduction of 48% with inner lattice, 16% with ribbed and 20% with variable shell thickness variations were achieved respectively.

3.1. Basic benchmark design

The most basic design had uniform wall thickness where the thickness was increased outwards until the vessel would break at about 50 bar pressure. The model design was effortlessly generated as the wall thickness acts as the only variable. Manufacturability for the design is also excellent, as risky geometries regarding L-PBF and AM in general remain minimal. The load case was analysed with nTopology and the tensile strength value of heat-treated material was closely exceeded with a uniform wall thickness of 1,5 mm in the stress area shown in red in figure 6.

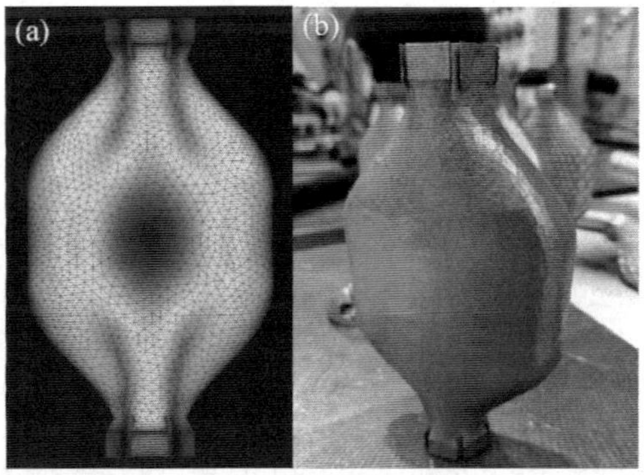

Figure 6. Representation of (a) simulation result of stress field distribution with 50 bar pressure and (b) as-built pressure vessel with uniform wall thickness (benchmark) design.

Stresses distributed unevenly along the vessel and high stress gradient could be observed. The heaviest stresses were observed in the center of the flat side sections of the model as seen in figure 7a. The uniform wall thickness was the easiest to design as there is only one variable to change in wall thickness. The simplicity of the design does not take advantage of the strengths of AM process. It was also the heaviest design as seen in table 1.

3.2. Internal lattice design

The only design option that had structures inside the vessel. Inner lattices decrease the inner volume of the vessel and could affect the flow of the liquid/gas through the vessel. The inner lattice was however the lightest design. Design-wise optimization of the inner lattices are difficult with CAD software as the lattices need to be created and modified manually. nTopology software alternatively easily allow such optimizations with few parameter changes.

IOP Conf. Series: Materials Science and Engineering 1296 (2023) 012019 doi:10.1088/1757-899X/1296/1/012019

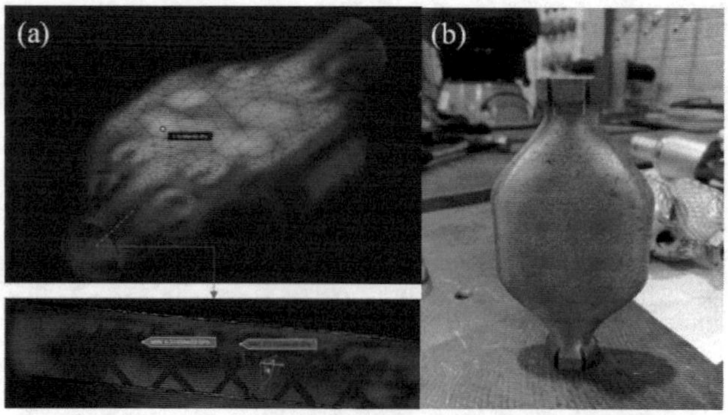

Figure 7. Representation of (a) simulation result of surface stress field with 50bar pressure with cross-section and (b) as-built pressure vessel with internal lattice design.

UTS (1500 MPa) exceeding stresses were noted in the internal lattice structures in the FE analysis phase, while stresses applied to the vessel shell were notably low. Initial failure is therefore expected to happen in the lattice structure.

3.3. Variable shell thickness design

The variable wall thickness comparatively exhibited the highest strength to weight ratio with outer changes. The designing was comparably difficult with other designs as there were multiple different variables affecting the end results. The design method used also tends to create some weak spots as the method did not consider stress distribution changes.

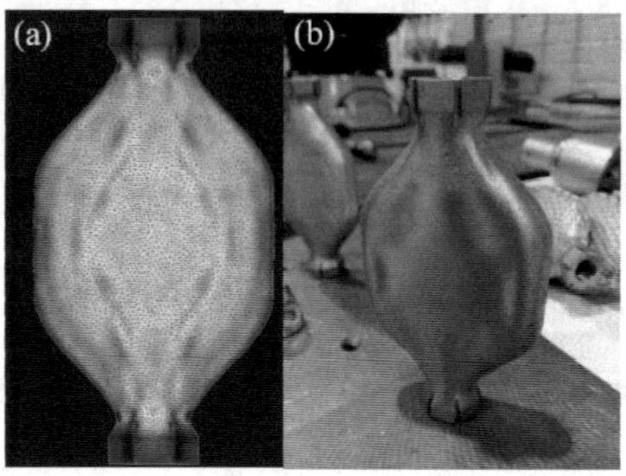

Figure 8. Representation of (a) simulation result of surface stress field with 50bar and (b) as-built pressure vessel with variable shell thickness design.

Variable wall thickness affects the stress distribution which is not considered as only the results from unedited model were used. High stress concentrations were observed at the neck of the opening. This created small bumps when thickened which increased the stress concentration, so the thickness was manually thickened there. Optimization process was relatively difficult due to the number of variables affecting the result. Topology optimization could be easier to use since the optimization process is automatic.

3.4. Ribbed design

The ribbed optimization was effortlessly achieved with nTopology software. This primarily involved the creation of cell map from the desired surface and using it as the base for the lattices. Non-linear cell map allows creation of rounded lattices. There were challenges with FE analysis since there are problems in meshing of the part due to small geometries. This creates stress spikes in some places where the mesh was rough. This created difficulties with the optimization process.

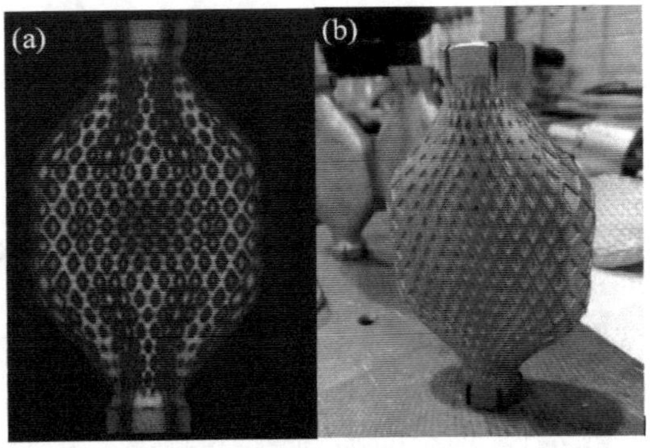

Figure9. Representation of (a) simulation result surface stress field with 50 bar and (b) as-built pressure vessel with ribbed design.

Stress distribution was better than in the benchmark model but not as good as variable shell or inner lattice. High stress concentration can still be seen in the middle of the vessel.

4. Conclusions

In conclusion, the characteristics, and applications of high-performance metals (e.g., titanium alloys and nickel alloys) with L-PBF are gaining significant interest in various industries. These metals possess exceptional properties that make them suitable for high-stress applications, and L-PBF allows the production of complex components with high precision and efficiency. The aviation, biomedical, and automotive sectors are some examples of industries that are utilizing L-PBF of high-performance metals in their manufacturing chains. However, there are also some challenges and limitations associated with L-PBF for metal parts. One of the biggest challenges is the high cost of AM machines and materials, which can impede the adoption of smaller and medium size companies. Additionally, the need of support structures, post-processing can be costly as result of time and labour intensiveness. Finally, the quality control and certification of AM-produced parts are still evolving, which may impact their acceptance in highly regulated industries. However, AM allows optimization of the designs which can decrease the weight of the product and used material without decreasing the capacity or affecting the manufacturing process. This does could make the designing process somewhat more difficult depending on the selected optimization type and used software. Results from this case study showed that different AM applicable structures e.g., lattice and stress field driven geometries can be utilized effectively to topologically optimize small-scale pressure vessel structures in a linear load case. It is estimated that with a non-linear load analysis, more accurate final vessel geometries would have been achieved.

5. References

[1] A. Vafadar, F. Guzzomi, A. Rassau, K. Hayward, Advances in metal additive manufacturing: A review of common processes, industrial applications, and current challenges, Appl. Sci. 11 (2021) 1–33. https://doi.org/10.3390/app11031213.

[2] H. Lee, C.H.J. Lim, M.J. Low, N. Tham, V.M. Murukeshan, Y.J. Kim, Lasers in additive manufacturing: A review, Int. J. Precis. Eng. Manuf. - Green Technol. 4 (2017) 307–322. https://doi.org/10.1007/s40684-017-0037-7.

[3] D. Gu, X. Shi, R. Poprawe, D.L. Bourell, R. Setchi, J. Zhu, Material-structure-performance integrated laser-metal additive manufacturing, Science (80-.). 372 (2021). https://doi.org/10.1126/science.abg1487.

[4] ISO/ASTM, ISO/ASTM 52900:2021 Additive manufacturing — General principles — Fundamentals and vocabulary, 2021.

[5] F. de Oliveira Campos, A.C. Araujo, A.L. Jardini Munhoz, S.G. Kapoor, The influence of additive manufacturing on the micromilling machinability of Ti6Al4V: A comparison of SLM and commercial workpieces, J. Manuf. Process. 60 (2020) 299–307. https://doi.org/10.1016/j.jmapro.2020.10.006.

[6] M. Mehrpouya, A. Dehghanghadikolaei, B. Fotovvati, A. Vosooghnia, S.S. Emamian, A. Gisario, The potential of additive manufacturing in the smart factory industrial 4.0: A review,

Appl. Sci. 9 (2019). https://doi.org/10.3390/app9183865.

[7] G. Kasperovich, R. Becker, K. Artzt, P. Barriobero-Vila, G. Requena, J. Haubrich, The effect of build direction and geometric optimization in laser powder bed fusion of Inconel 718 structures with internal channels, Mater. Des. 207 (2021) 109858. https://doi.org/10.1016/j.matdes.2021.109858.

[8] T.A. Le Néel, P. Mognol, J.Y. Hascoët, Design methodology for variable shell mould thickness and thermal conductivity additively manufactured, Weld. World. 62 (2018) 1059–1072. https://doi.org/10.1007/s40194-018-0598-2.

[9] T. Shi, B. Lu, T. Shen, R. Zhang, S. Shi, G. Fu, Closed-loop control of variable width deposition in laser metal deposition, Int. J. Adv. Manuf. Technol. 97 (2018) 4167–4178. https://doi.org/10.1007/s00170-018-1895-z.

[10] A.J. Wildgoose, K.A. Thole, P. Sanders, L. Wang, Impact of Additive Manufacturing on Internal Cooling Channels With Varying Diameters and Build Directions, J. Turbomach. 143 (2021). https://doi.org/10.1115/1.4050336.

[11] L. Yang, O. Harrysson, H. West, D. Cormier, Mechanical properties of 3D re-entrant honeycomb auxetic structures realized via additive manufacturing, Int. J. Solids Struct. 69–70 (2015) 475–490. https://doi.org/10.1016/j.ijsolstr.2015.05.005.

[12] H. Zhang, L. Zhu, P. Xue, Laser direct metal deposition of variable width thin-walled structures in Inconel 718 alloy by coaxial powder feeding, Int. J. Adv. Manuf. Technol. 108 (2020) 821–840. https://doi.org/10.1007/s00170-020-05434-3.

[13] X. Zhang, R. Tiwari, A.H. Shooshtari, M.M. Ohadi, An additively manufactured metallic manifold-microchannel heat exchanger for high temperature applications, Appl. Therm. Eng. 143 (2018) 899–908. https://doi.org/10.1016/j.applthermaleng.2018.08.032.

[14] M.D. Bambach, M. Bambach, A. Sviridov, S. Weiss, New process chains involving additive manufacturing and metal forming - A chance for saving energy?, in: Procedia Eng., No longer published by Elsevier, 2017: pp. 1176–1181. https://doi.org/10.1016/j.proeng.2017.10.1049.

[15] Z. Wang, N. Wu, Q. Wang, Y. Li, Q. Yang, F. Wu, Novel bionic design method for skeleton structures based on load path analysis, Appl. Sci. 10 (2020) 1–17. https://doi.org/10.3390/app10228251.

[16] K. Basinger, C. Webster, C. Keough, R. Wysk, O. Harrysson, Advanced Manufacturing Using Linked Processes: Hybrid Manufacturing, in: Mass Prod. Process., 2020. https://doi.org/10.5772/intechopen.88560.

[17] M.H. Hajmohammad, A. Tabatabaeian, A.R. Ghasemi, F. Taheri-Behrooz, A novel detailed analytical approach for determining the optimal design of FRP pressure vessels subjected to hydrostatic loading: Analytical model with experimental validation, Compos. Part B Eng. 183 (2020). https://doi.org/10.1016/j.compositesb.2019.107732.

[18] X. kou He, C. sheng Xie, L. jun Xiao, Y. Luo, D. Lu, Z. dong Liu, X. tao Wang, Microstructure and impact toughness of 16MND5 reactor pressure vessel steel manufactured by electrical additive manufacturing, J. Iron Steel Res. Int. 27 (2020) 992–1004. https://doi.org/10.1007/s42243-020-00467-0.

[19] Q. Jia, D. Gu, Selective laser melting additive manufacturing of Inconel 718 superalloy parts: Densification, microstructure and properties, J. Alloys Compd. 585 (2014) 713–721. https://doi.org/10.1016/j.jallcom.2013.09.171.

[20] Y.M. Arısoy, L.E. Criales, T. Özel, B. Lane, S. Moylan, A. Donmez, Influence of scan strategy and process parameters on microstructure and its optimization in additively manufactured nickel alloy 625 via laser powder bed fusion, Int. J. Adv. Manuf. Technol. 90 (2017) 1393–1417. https://doi.org/10.1007/s00170-016-9429-z.

[21] M. Leary, Digital design for AM, in: Des. Addit. Manuf., Elsevier, 2020: pp. 33–90. https://doi.org/10.1016/b978-0-12-816721-2.00003-8.

[22] M. Leary, D. Downing, B. Lozanovski, J. Harris, Design principles, in: Fundam. Laser Powder Bed Fusion Met., Elsevier, 2021: pp. 119–154. https://doi.org/10.1016/B978-0-12-824090-8.00013-5.

[23] C. Emmelmann, P. Sander, J. Kranz, E. Wycisk, Laser additive manufacturing and bionics:

Redefining lightweight design, Phys. Procedia. 12 (2011) 364–368. https://doi.org/10.1016/j.phpro.2011.03.046.
[24] D. Deng, J. Moverare, R.L. Peng, H. Söderberg, Microstructure and anisotropic mechanical properties of EBM manufactured Inconel 718 and effects of post heat treatments, Mater. Sci. Eng. A. 693 (2017) 151–163. https://doi.org/10.1016/j.msea.2017.03.085.
[25] N. Raghavan, R. Dehoff, S. Pannala, S. Simunovic, M. Kirka, J. Turner, N. Carlson, S.S. Babu, Numerical modeling of heat-transfer and the influence of process parameters on tailoring the grain morphology of IN718 in electron beam additive manufacturing, Acta Mater. 112 (2016) 303–314. https://doi.org/10.1016/J.ACTAMAT.2016.03.063.
[26] B.P. McNelly, R.L. Hooks, W.R. Setzler, C.S. Hughes, Additive manufacturing of pressure vessels (with plating), in: Am. Soc. Mech. Eng. Press. Vessel. Pip. Div. PVP, 2017. https://doi.org/10.1115/PVP2017-65888.
[27] H.M.A. Abdalla, D. Casagrande, F. De Bona, T. De Monte, M. Sortino, G. Totis, An optimized pressure vessel obtained by metal additive manufacturing: Preliminary results, Int. J. Press. Vessel. Pip. 192 (2021). https://doi.org/10.1016/j.ijpvp.2021.104434.
[28] H. Li, T. Yue, F. Zhu, Y. Peng, Y. Li, C. Huang, Y. Zhang, Research on application of additive manufacturing technology in nuclear fuel assembly field, in: Int. Conf. Nucl. Eng. Proceedings, ICONE, 2021. https://doi.org/10.1115/ICONE28-65776.
[29] S.H. Sun, Y. Koizumi, T. Saito, K. Yamanaka, Y.P. Li, Y. Cui, A. Chiba, Electron beam additive manufacturing of Inconel 718 alloy rods: Impact of build direction on microstructure and high-temperature tensile properties, Addit. Manuf. 23 (2018) 457–470. https://doi.org/10.1016/j.addma.2018.08.017.
[30] Z. Wang, K. Guan, M. Gao, X. Li, X. Chen, X. Zeng, The microstructure and mechanical properties of deposited-IN718 by selective laser melting, J. Alloys Compd. 513 (2012) 518–523. https://doi.org/10.1016/j.jallcom.2011.10.107.
[31] G. Mahalle, O. Salunke, N. Kotkunde, A.K. Gupta, S.K. Singh, Neural network modeling for anisotropic mechanical properties and work hardening behavior of Inconel 718 alloy at elevated temperatures, J. Mater. Res. Technol. 8 (2019) 2130–2140. https://doi.org/10.1016/j.jmrt.2019.01.019.
[32] E. Hosseini, V.A. Popovich, A review of mechanical properties of additively manufactured Inconel 718, Addit. Manuf. 30 (2019) 100877. https://doi.org/10.1016/j.addma.2019.100877.
[33] V.A. Popovich, E.V. Borisov, A.A. Popovich, V.S. Sufiiarov, D.V. Masaylo, L. Alzina, Impact of heat treatment on mechanical behaviour of Inconel 718 processed with tailored microstructure by selective laser melting, Mater. Des. 131 (2017) 12–22. https://doi.org/10.1016/j.matdes.2017.05.065.
[34] EOS GmbH, Eos Nickel alloy IN718 Material Data Sheet, 2021.
[35] M. Machno, Investigation of the Machinability of the Inconel 718 Superalloy during the Electrical Discharge Drilling Process, Materials (Basel). 13 (2020) 3392. https://doi.org/10.3390/ma13153392.

Acknowledgments
The authors would like to express their gratitude to all partners and companies of Verkostoitumisella voimaa 3D-tulostukseen (VERKOTA) -project for supporting this study. Authors would also like to acknowledge Laboratory of Laser Materials Processing and Additive Manufacturing and Analytics-based Management for Business and Manufacturing Industry (AMBI) platform at LUT University for supporting this study.

IOP Conf. Series: Materials Science and Engineering 1296 (2023) 012020 doi:10.1088/1757-899X/1296/1/012020

Validation of powder layering simulation via packing density measurement for laser-based powder bed fusion

E Haapa[1], A Gopaluni[1], H Piili[1], A Ganvir[1], A Salminen[1] and J Ottelin[2]

[1] Department of Mechanical Engineering, Faculty of Technology, Research group of Digital Manufacturing and Surface Engineering (DMS), University of Turku, Turku, 20520, Finland
[2] EOS Finland Oy, Turku, 20520, Finland

erik.r.haapa@utu.fi

Abstract. Powder bed fusion using a laser beam (PBF-LB/M) is considered one of the most versatile additive manufacturing methods as the parts printed have high resolution thanks to the low layer thickness used. The powder packing density (*PD*) of the powder layer has a significant impact on the density, surface roughness and other mechanical properties of the built parts. Due to the difficulty of characterizing the powder bed in situ, simulation has often been used to study the powder behavior on the powder bed. However, in order for the simulation to have practical value, there must be some way of confirming the results via experimental methods, also called validation. The aim of this study was to develop a powder packing density-based validation method for a powder bed simulation. The developed method featured a simplistic "open cup" style sample which traps powder inside for *PD* measurement. The samples were built with an EOS M 290 PBF-LB/M system using Alloy 718 (also known as "IN718" or "Inconel") powder. Average *PD* over the five built samples was 52.4 %, with a standard deviation of 0.2 %. The method was used to successfully validate a powder bed simulation with four recoated powder layers, modelled using FLOW-3D DEM simulation software from Flow Science Inc. Similar methods for *PD* characterization were found in literature, but in many cases the method does not fully correspond to the conditions of a simulated powder bed, the scale is very small, or the reliability of the *PD* measurement is not confirmed. The method presented in this study corresponds to typical powder bed simulation conditions, while retaining high reliability and repeatability of results.

1. Introduction

Laser-based powder bed fusion (PBF-LB/M) is the most applicable of all the standardized methods of additive manufacturing (AM) in terms of industrial applications, due to its capability of fabricating complex parts with high dimensional accuracy [1–3]. The PBF-LB/M process uses a laser beam to melt thin (<0.1 mm) layers of metal powder on top of previously melted and fused layers until the required design is complete. However, building solid, accurate, and strong parts with PBF-LB/M is only possible if the powder material properties (e.g. particle size distribution or PSD [4] and particle morphology [5]), processing parameters (e.g., laser power [6], scanning speed [6], and layer thickness [7]), and post-process heat treatments [8] are carefully optimized. This is often done by practical experiments, by comparing the mechanical properties of finished parts resulting from each combination of powder material or process parameters, which is labor-intensive, time consuming, and expensive, [9–11]. For this reason, it is an attractive proposition to replace experiments with computer simulation of the powder bed and the laser-material interaction.

Powder bed simulations (PBS) commonly use the discrete element modelling (DEM) technique which models the contact forces and trajectories of individual powder particles [12]. DEM has been used for observing powder packing behavior [13,14] as well as for creating realistic powder beds for the simulation of laser-material interaction [15]. Packing behavior is typically described with the packing density (*PD*) of the particles in the powder bed, which is the portion of volume in the powder bed occupied by particles. For example, a bed with a *PD* of 60 % consists of 60% of actual powder material and 40% empty interstitial space. As is shown in Figure 1, high *PD* correlates with reduced porosity in finished parts [4,16,17] and improved absorptivity of energy from the laser beam [18,19]. Higher *PD* also reduces the effective powder layer thickness, as the difference of powder layer volume and melt pool volume is smaller [4], which has been found to reduce surface roughness of parts [20].

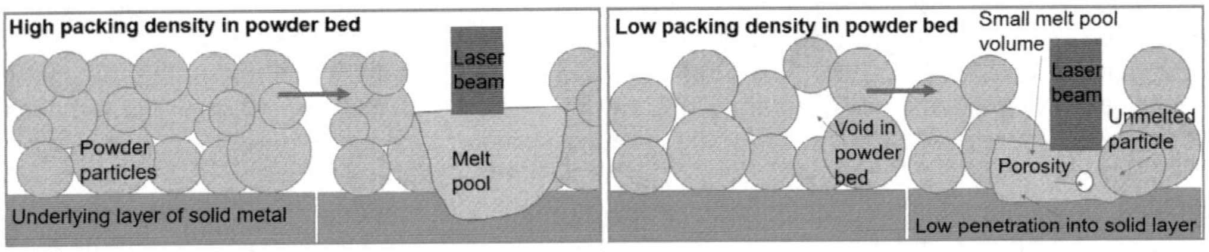

Figure 1. Effect of powder packing density in PBF-LB/M, adapted from: [20], and data in [18].

It is therefore important that PBS models accurately represent the powder packing behavior. This can be confirmed by a validation experiment, where the powder packing density is measured from the real powder bed and compared to the packing density of the simulated powder bed, these validation methods are referred to as "*PD*-based validation" in this study. Such validation experiments have been developed across many studies conducted by various authors [11, 13, 22, 23, 25]. In [21] the "staircase" method, which was developed in [22], was used to validate a DEM recoating simulation. In this method, a 10 x 25 mm rectangle is built to a height of about eight layers, and the build continues with five zones of stepped height increase incremented by one layer height, creating an object resembling a staircase. Powder was removed from around the staircase, and the single layer of powder left on the second "step" of the staircase was weighed to find its packing density. The thickness of this single powder layer was measured using a high depth of field digital microscope, enabling the calculation of *PD*. In [23] the PBF-LB/M process was used to build one layer tall 15 x 15 mm rectangular enclosures in three lines along the length of the build platform. The surrounding powder was removed, and the sampled powder inside the enclosures was weighed to determine *PD*. This style of sample is referred to as an "open cup" sample in this study. Both of these *PD*-based validation methods investigate a small area and single layer of powder. Such a small sample volume increases the relative impact of measurement error, and cannot be considered a representative sample of the whole powder bed. This is problematic if a DEM-simulated powder bed is intended to be used as a "digital twin" representation of the actual powder bed of an PBF-LB/M system.

Simulations of powder packing behavior are also sometimes validated with other parameters, e.g. the angle between the build platform and the powder pile during recoating, the dynamic repose angle (*DRA*) [24], or the slope angle of a powder pile formed under gravity, the angle of repose (*AOR*) [24,25], or by visual comparison between the simulated and real powder bed [13]. More research is needed to confirm if parameters not based on *PD* can be used to validate a powder bed simulation of powder packing behavior.

There are also multiple methods developed for measuring *PD,* which are not intended for validation, but could be applied for validation purposes. A non-exhaustive list of these methods include the closed container samples [26], open cup samples [22,27–29], the sample picker device [30], impact penetration into the powder bed [31], or light-curable liquid polymer to extract powder bed samples [32]. The sample picker, impact penetration test and liquid polymer methods require purpose-built equipment, and have a seemingly high risk of measurement error (e.g. the sample picker is hand-held). The closed container method was found in preliminary tests to be difficult to perform repeatably, in addition to the relatively large amount of material waste and long build times, making the method expensive. The high powder column also does not correspond to the simulated thin powder beds. Therefore, only the open cup sample was chosen for further development. In this study, a *PD*-based validation method which uses a large area open cup sample was designed. Five samples were built and measured to confirm the repeatability of the method. Sources of uncertainty were recognized and quantified. The method was used to validate a PBS model based on *PD*, and the significance of uncertainty in the validation method was compared to inherent uncertainty in the simulation.

2. Aim and purpose of study

Simulation of powder packing and especially the validation of these simulations form a modest portion of total research on PBF-LB/M simulation, as is shown in Figure 2. This figure shows the number of yearly publications between the years 2003-2023 with keywords related to the topic of this study, which were extracted from the Elsevier SCOPUS database in May 2023. Unless otherwise mentioned, all terms are separated by AND, and were searched in all article fields.

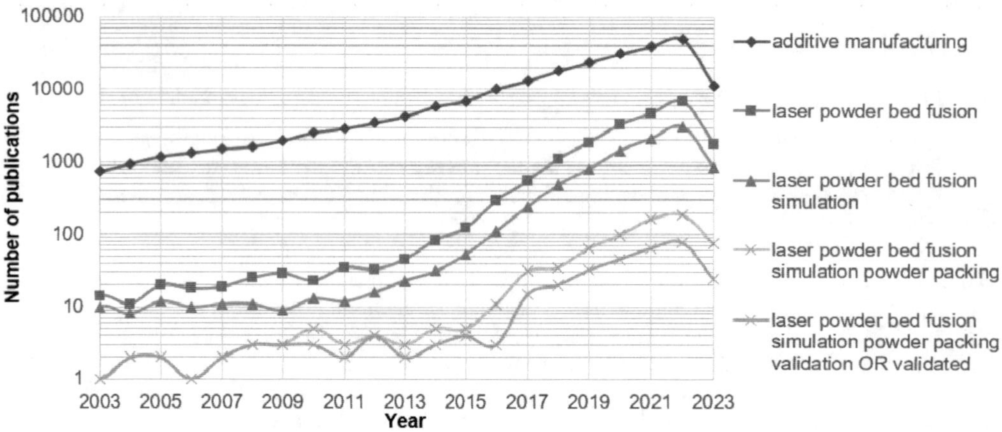

Figure 2. Yearly publications for search terms related to PBF-LB/M, simulation, and validation.

The total number of articles containing "additive manufacturing" was over 230 000, and over 20 000 for "laser powder bed fusion". Adding "simulation" yields nearly 9 300 results, and adding "simulation powder packing" and "simulation powder packing (validation OR validated)" yields 701 and 312 results respectively. Though simulation is a relatively popular topic in research related to PBF-LB/M, less than 10 % of it is related to the powder packing behavior despite it being crucial for building a realistic base for simulating the laser-material interaction. Of the studies on simulation of powder packing behavior, less than half mention validation by definition. One possible reason is the expense of testing a validation method for PBS. Furthermore, the lack of well-established standardized powder characterization routines and tests during and pre- or post-PBF-LB/M processing also makes validation of PBS models less popular in literature.

This study aims to develop a repeatable, simple, and inexpensive validation method which corresponds to the conditions of a typical PBS model; low powder layer thicknesses and high repeatability. The scientific relevance of this study is the new understanding of the significance of uncertainty in the validation method compared to uncertainty in the simulation model. Also, the inexpensive and repeatable *PD*-based validation method can enable the accurate validation of PBS models which increase the understanding of powder packing behavior in PBF-LB/M. The industrial

IOP Conf. Series: Materials Science and Engineering 1296 (2023) 012020 doi:10.1088/1757-899X/1296/1/012020

relevance of this study is that the presented large area open cup sample is well suited for validating PBS models which are extrapolated to the entire build area. PBS has potential for replacing expensive experiments, but only if the models can be reliably validated. The presented validation method can also be used as a method for characterizing *PD* during PBF-LB/M processing.

3. Experimental setup and procedure

3.1. Experimental setup

An EOS M 290 PBF-LB/M system by EOS GmbH (Germany) was used to build the open cup samples. The powder material was the gas-atomized EOS NickelAlloy IN718 (Alloy 718). The powder PSD is between 22-50 μm, and particles are mostly spherical with occasional satellites, see supplier specification in [33]. Alloy 718 was chosen for this study as it is commonly processed with PBF-LB/M in industry [34,35] and its behavior in PBF-LB/M is well documented [36]. A validated parameter set could be provided by EOS, which reduced the risk of error from build defects. The samples were weighed using a Mettler Toledo ICS425 (Switzerland) scale (measuring accuracy 0.1 g). The length and width of the open cup sample were measured with a Mitutoyo 500-173-30 Absolute Digimatic (Japan) caliper (measuring accuracy 0.01 mm). The height of the open cup walls was measured using a Mitutoyo ID-C112X Absolute Digimatic (Japan) depth gage (measuring accuracy 0.001 mm). The simulation model of the powder bed was created with FLOW-3D DEM software by Flow Science Inc. (USA).

3.2. Experimental procedure

Only the topmost powder layers are affected by the laser-material interaction, so the sampled powder bed should be relatively thin, but still multiple layers thick to reduce the impact of measuring error. Of the build platform area, only the central region near the dispenser would be sampled, because the edges and the end of the powder bed may have higher variation in *PD* due to the possible segregation of different sized particles [32] and the unlikely possibility that the recoater does not pick up enough powder from the dispenser. Thus, the sample should be shallow and follow the shape of the platform, but it should be centered and placed near to the dispenser, as seen in Figure 3.

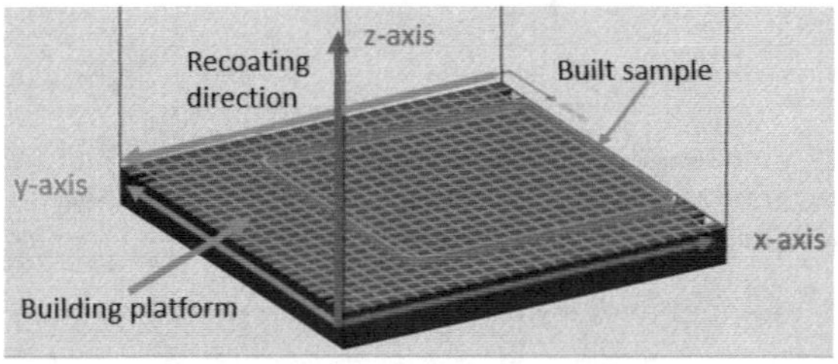

Figure 3. A 3D-model of the "open cup" *PD* validation sample (green) on the build platform of an EOS M 290. The recoater picks up powder from the dispenser at end of x-axis, and spreads the powder over the building platform towards the origin, i.e., right to left.

The outside width of the sample is 178 mm. The corners were rounded with a 10 mm radius to ease the removal of powder from the open cup sample. The height was 5 mm and wall thickness was 2.5 mm. Both values could still be lower, though this will make it more difficult to remove surrounding powder without disturbing the sampled powder. The samples were built using the 40 μm layer height Alloy 718 process for the M 290 [33]. After printing, powder outside the sample walls was removed with a paint brush, and the build platform with the sample was weighed as seen in Figure 4 A. This weighing yields m_{full} which is the mass of the platform, sampled powder, and any powder residue left on the platform after initial brushing. After this, the sampled powder was moved from the walled area into a plastic bag with a scoop and paint brush, and weighed. The weight of the bag was tared. This gives the minimum mass of sampled powder, $m_{powder\text{-}min}$. Then, the sample area inside the wall was cleaned with a cloth, but the area outside the wall was left untouched, as is seen in Figure 4 B. After this the build platform was weighed, yielding $m_{emptied}$ which was deducted from m_{full} to find $m_{powder\text{-}measured}$ which is the likely most accurate value for mass of sampled powder.

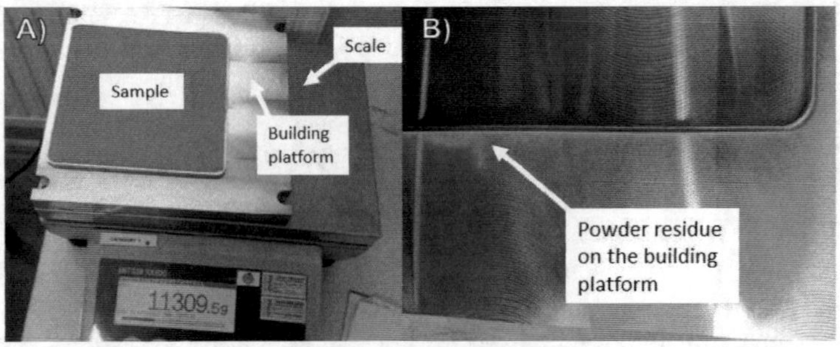

Figure 4. The open cup sample during initial weighing (A) and after emptying the sampled powder (B), where the arrow indicates powder residue outside the sampled area.

Measuring $m_{powder\text{-}measured}$ this way results in a value where only the mass of sampled powder is included and the mass of residual powder outside the sampled area is excluded. Finally, the entire platform was vacuum cleaned and wiped with a cloth and isopropyl alcohol, including the bottom side and the screw holes. After this the platform was again weighed, yielding $m_{cleaned}$, which was deducted from m_{full} to find $m_{powder\text{-}max}$, which is the likely maximum value for mass of sampled powder.
After this weighing procedure, the inner width and length of the sample rectangle were measured at three locations (at midpoint and both edges) and averaged. Wall height was measured at the midpoint of each rounded corner and averaged. Sample volume V_{sample} was calculated based on the average width w, average length l, and the average height h, using Equation 1. The rounded corners were assumed perfectly round with a 10 mm radius r.

$$V_{sample}\,[cm^3] = \frac{(w-2r)*l*h+\pi*r^2*h+(l-2r)*2r*h+r^2*\pi*h}{1000} \tag{1}$$

Next, the volume occupied by powder particles was calculated for each three powder mass values with skeletal density ρ_s of particles (measured by supplier according to ASTM B923-22), using Equation 2.

$$V_{powder} = \frac{m_{powder}}{\rho_s} \tag{2}$$

With V_{powder} and V_{sample}, it is possible to calculate PD using Equation 3.

$$PD = \frac{V_{powder}}{V_{sample}} \tag{3}$$

In addition, a nominal PD ($PD_{nominal}$) is calculated using the nominal volume of the CAD model and $V_{powder\text{-}measured}$ (the volume of powder calculated from $m_{powder\text{-}measured}$). As is shown in Figure 5, a PD value was calculated for each measured powder mass (min, measured, max) and their calculated powder volumes, using the measured sample volume V_{sample}.

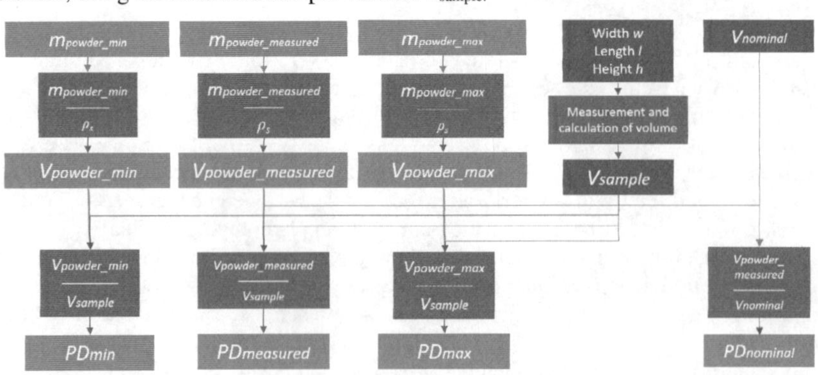

Figure 5. Different *PD* values and their calculation logic.

3.3. Simulation model

The model has approximately 260 000 particles recoated in four 40 μm layers. The shape of recoater and recoating speed were the same as in the real M 290. The total simulation area including the dispenser was 17.5 x 1.5 mm, of which the powder bed was 6.5 x 1.5 mm. The simulated PSD was based on data from dynamic image analysis (ISO 13322-2) and particle density on pycnometry results (ASTM B923-22), both done by the powder supplier. Friction coefficients were approximated as 0.3 and 0.6 for particle-particle dynamic and static friction respectively and 0.4 for particle-wall friction, based on rheometry experiments done according to standard ASTM D7891. A suitable particle stiffness was found by calibration as 10 N/m, and the coefficient of restitution was approximated from [37] as 0.6.

The *PD* of the simulated powder bed was measured in three ways, since the choice of method has significant effect on the result. Measuring the fraction of volume occupied by particles from a sampling volume completely embedded in the powder bed will yield higher *PD* than if the sampling volume is defined by the extents of the powder bed. The former measurement in this study is referred to as $PD_{max\text{-}sim}$ and the latter $PD_{min\text{-}sim}$. A measurement taken between these two is $PD_{measured\text{-}sim}$. *PD* was sampled from both the full length of the simulated powder bed and a central stable area, where the thickness of the powder bed did not vary. The "stable area" measurement is considered more useful, as varying powder bed thickness distorts the *PD* measurement. The simulation was then repeated with identical parameters to quantify the effect of randomized particle generation pattern on *PD*.

4. Results and discussion

4.1. Results of validation experiment

The PD_{min}, $PD_{measured}$, PD_{max} results of each sample as well as their mean and standard deviation are shown in Table 1. The samples are labeled OL (Open cup, Large area) and numbered in build order.

Table 1. Measured and calculated *PD* of all samples.

PD	OL1	OL2	OL3	OL4	OL5	Mean	Standard deviation
PD_{min}	51.9 %	51.9 %	52.2 %	52.6 %	52.5 %	52.2 %	0.29 %
$PD_{measured}$	52.2 %	52.2 %	52.3 %	52.6 %	52.6 %	52.4 %	0.20 %
PD_{max}	52.2 %	52.2 %	52.3 %	52.7 %	52.7 %	52.4 %	0.22 %
$PD_{nominal}$	54.0 %	52.4 %	53.0 %	52.8 %	53.5 %	53.2 %	0.56 %

As it can be seen in Table 1, there is slight variance between the *PD* values, but all values of PD_{min}, $PD_{measured}$, and PD_{max} are between 51-53 %. $PD_{measured}$ is considered the most accurate measurement since the values of PD_{min} and PD_{max} are likely lower and higher than the true *PD* respectively, and are not intended as a characterized value, but only for quantifying uncertainty from the cleaning method. The mean value of $PD_{measured}$ was 52.4 %, and it is considered the definitive *PD* value characterized with this method. The $PD_{measured}$ is very uniform across samples, with a standard deviation of 0.20, likely due to the large mass of sampled powder which reduces the relative significance of small measurement errors. The only variation in $PD_{measured}$ is a slight increase between the second and fourth samples, after which there is no change between OL4 and OL5. This could be due to increasing proficiency in sample handling, leading to less sampled powder being lost before weighing. This increase is not present in $PD_{nominal}$, which alternates between 52.4-54 % without a clear trend. No correlation between $PD_{nominal}$ and other *PD* values was found, which indicates that variations in sample volume which are not accounted for in the sample measurement (e.g. variation in the surface of the build platform or the built wall) do not impact *PD* and are not a significant source of uncertainty. The built parts were slightly taller (0.02-0.18 mm taller) than the nominal CAD model, which was unexpected as PBF-LB/M-built parts with a short z-dimension are typically slightly shorter than nominal due to instability of layer thickness in the first ~10 layers [38]. This is possibly due to the

manually coated first layer being slightly thicker than the nominal layer height. For this reason, $PD_{nominal}$ is always slightly higher than $PD_{measured}$ or PD_{max}.
The PD_{min} and PD_{max} values were measured to quantify the variation resulting from different sample cleaning methods or error in sample cleaning. Greatest difference between PD_{min} and $PD_{measured}$ values was in samples OL1 and OL2, with a difference of 0.3 percentage points. The PD_{max} values have at most 0.1 percentage point difference to $PD_{measured}$ values, indicating that the samples have low error due to residual powder and other debris on the build platform, and that the error from accidentally removing residue from outside the sampled area before weighing is lower than the error caused by trying to remove only the sampled powder by brushing and scooping. Therefore it is better practice to e.g. vacuum clean the sample and entire build platform before weighing. Overall, the results show that the used cleaning method is an insignificant source of uncertainty regardless of the used technique.

4.2. Application of validation method to simulation
As seen in Figure 6, the method of measuring *PD* from the simulation has significant effect on the *PD*.

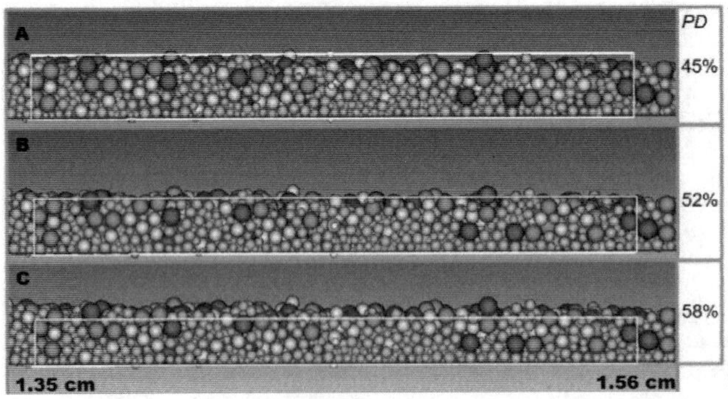

Figure 6. *PD* measurement from the stable region of the simulated powder bed; sampling volumes marked with white rectangle, and *PD* on the right.
A: sampling volume following extents, $PD_{min\text{-}sim}$
B: $PD_{measured\text{-}sim}$, where the sampling volume height is defined at midpoint between $PD_{min\text{-}sim}$ and $PD_{max\text{-}sim}$.
C: sampling volume embedded into the powder bed so that maximum PD is found, $PD_{max\text{-}sim}$.

The sampling volume heights for the extent-following $PD_{min\text{-}sim}$ and fully embedded $PD_{max\text{-}sim}$ were 150 μm and 210 μm respectively, with the $PD_{measured\text{-}sim}$ being defined as the midpoint between these extremes (180 μm). The nominal powder bed thickness was 200 μm. $PD_{measured\text{-}sim}$ measured at a stable region of the powder bed, between 1.35-1.56 cm of simulated volume (full length of powder bed being 1.1-1.75 cm of the total volume) is considered the best analogue to the measurement method in the validation experiment, which also sampled a central stable region. $PD_{measured\text{-}sim}$ was 52 %, which agrees with the validation experiment. $PD_{min\text{-}sim}$ and $PD_{max\text{-}sim}$ were 45 % and 58 % respectively, which means the choice of measurement method is a significant source of uncertainty, much larger than any uncertainty associated with the validation method.
The $PD_{min\text{-}sim}$, $PD_{measured\text{-}sim}$, and $PD_{max\text{-}sim}$ measured from the full length of the simulated powder bed (1.1-1.75 cm) were 37 %, 49 %, and 58% respectively. The lower results were expected due to lower powder bed thickness at the end of the bed resulting in more empty volume being included in the sampling volume. However, $PD_{max\text{-}sim}$ was identical in both measurements, which is significant since $PD_{max\text{-}sim}$ is fully embedded in the simulated powder bed and thus should not be affected by any variation in powder bed thickness, indicating that as long as the sampling volume is equally embedded in the powder bed, its size and thus the change in wall-powder interface area does not affect the *PD* result. If the length of the sampling volume affected *PD*, a change in $PD_{max\text{-}sim}$ should have been seen.
The results of a repeated simulation run were nearly identical, except $PD_{measured\text{-}sim}$ in the stable region of the repeat run was 53 % compared to 52 % in the first run. This is not a sufficient sample size to make conclusions, but it demonstrates that the randomized particle generation is a potential source of uncertainty, though further research on the effect of randomized particle generation on the variance of *PD* is needed to quantify how significant a source of uncertainty it is, and to confirm that the variation in *PD* observed in this study is not an outlier.

The open cup *PD*-based validation method can be considered sufficiently repeatable and precise for the purposes of validating DEM simulations of powder bed recoating. The validation result matching the carefully parametrized simulation indicates that the sampled powder bed in the open cup method is not significantly compressed by gravity compared to the simulation despite being about 50 times taller, and that the *PD* of a ~9 mm^2 sampling area of the simulation is comparable to the average *PD* of the entire center area of the build platform.

5. Conclusions

The aim of this study was to develop a powder packing based validation method which matches the typical conditions of a powder bed simulation and is suitable for their validation. Uncertainties related to the measurement process were identified and quantified. The results of the validation experiment and its sources of uncertainty were compared to a powder bed simulation model and the sources of uncertainty related to it. The main conclusions of the study are:
Multiple validation methods based on different powder bed parameters exist in literature, as well as many methods for measuring *PD* for other purposes. The presented open cup *PD*-based validation method with Alloy 718 powder material had a mean value of $PD_{measured}$ over five samples of 52.4 % with a standard deviation of 0.20 %. Despite the sample being orders of magnitude larger, the *PD* result is a good match to a simulated powder bed, which was carefully parametrized by practical experiments, literature review, and calibration. The sources of uncertainty quantified from the simulation are more impactful than those of the validation method, making the validation method sufficiently accurate and reliable for validating powder bed simulation models.

6. Future work

The validation model built in the scope of this research could be utilized as a basis to develop a thermomechanical CFD/FEM model of laser-material interaction as an initial step in the PBF-LB/M simulation methodology. In addition, the powder bed model could be a part of a powder characterization sequence. Traditional powder characterization for PBF-LB/M includes the characterization of PSD, particle morphology, packing density, and flowability. With the rapid development of simulation software, it is a natural progression that simulation of the powder bed in the build chamber becomes a part of the powder characterization sequence.

7. References

[1] Godec D, Gonzalez-Gutierrez J, Nordin A, Pei E and Ureña Alcázar J 2022 *A Guide to Additive Manufacturing* (Cham: Springer Nature)
[2] Gibson I, Rosen D, Stucker B and Khorasani M 2021 *Additive Manufacturing Technologies* (Cham: Springer International Publishing)
[3] Willkomm J, Kuhlbach K, Merget D, Wagner M, Reich S, Ziegler S and Schleifenbaum J H 2021 Design and manufacturing of a cylinder head by laser powder bed fusion *IOP Conf. Ser.: Mater. Sci. Eng.* **1097** 012021
[4] Spierings A B, Herres N and Levy G 2011 Influence of the particle size distribution on surface quality and mechanical properties in AM steel parts ed D Bourell and B Stucker *Rapid Prototyping Journal* **17** 195–202
[5] Guzmán J, de Moura Nobre R, Rodrigues Júnior D L, de Morais W A, Nunes E R, Bayerlein D L, Falcão R B, Sallica-Leva E, Oliveira H R, Chastinet V L and Landgraf F J G 2021 Comparing Spherical and Irregularly Shaped Powders in Laser Powder Bed Fusion of Nb47Ti Alloy *J. of Materi Eng and Perform* **30** 6557–67
[6] Kumar P, Farah J, Akram J, Teng C, Ginn J and Misra M 2019 Influence of laser processing parameters on porosity in Inconel 718 during additive manufacturing *Int J Adv Manuf Technol* **103** 1497–507
[7] Nath S D, Gupta G, Kearns M, Gulsoy O and Atre S V 2020 Effects of layer thickness in laser-powder bed fusion of 420 stainless steel *Rapid Prototyping Journal* **26** 1197–208

[8] Al-Maharma A Y, Patil S P and Markert B 2020 Effects of porosity on the mechanical properties of additively manufactured components: a critical review *Mater. Res. Express* **7** 122001

[9] Aversa A, Marchese G, Saboori A, Bassini E, Manfredi D, Biamino S, Ugues D, Fino P and Lombardi M 2019 New Aluminum Alloys Specifically Designed for Laser Powder Bed Fusion: A Review *Materials* **12** 1007

[10] Asnafi N 2021 Application of Laser-Based Powder Bed Fusion for Direct Metal Tooling *Metals* **11** 458

[11] Budinoff H D and Shafae M 2022 Connecting part geometry and cost for metal powder bed fusion *Int J Adv Manuf Technol* **121** 6125–36

[12] Zohdi T I 2018 PART II—Discrete Element Method (DEM) Approaches: Dynamic Powder Deposition *Modeling and Simulation of Functionalized Materials for Additive Manufacturing and 3D Printing: Continuous and Discrete Media: Continuum and Discrete Element Methods* Lecture Notes in Applied and Computational Mechanics ed T I Zohdi (Cham: Springer International Publishing) pp 83–119

[13] Shaheen M Y, Thornton A R, Luding S and Weinhart T 2021 The influence of material and process parameters on powder spreading in additive manufacturing *Powder Technology* **383** 564–83

[14] Lee Y S, Nandwana P and Zhang W 2018 Dynamic simulation of powder packing structure for powder bed additive manufacturing *Int J Adv Manuf Technol* **96** 1507–20

[15] Tian Y, Yang L, Zhao D, Huang Y and Pan J 2020 Numerical analysis of powder bed generation and single track forming for selective laser melting of SS316L stainless steel *Journal of Manufacturing Processes* **58** 964–74

[16] Spierings A and Levy G 2009 Comparison of density of stainless steel 316L parts produced with Selective Laser Melting using different powder grades *Proceedings for the 2009 International Solid Freeform Fabrication Symposium* 20th Annual International Solid Freeform Fabrication Symposium (Austin) pp 342–53

[17] Pleass C and Jothi S 2018 Influence of powder characteristics and additive manufacturing process parameters on the microstructure and mechanical behaviour of Inconel 625 fabricated by Selective Laser Melting *Additive Manufacturing* **24** 419–31

[18] Boley C D, Khairallah S A and Rubenchik A M 2015 Calculation of laser absorption by metal powders in additive manufacturing *Applied Optics* **54** 2477–82

[19] Pal S, Gubeljak N, Bončina T, Hudák R, Toth T, Zivcak J, Lojen G, Leben N and Drstvenšek I 2021 The effects of locations on the build tray on the quality of specimens in powder bed additive manufacturing *Int J Adv Manuf Technol* **112** 1159–70

[20] Shi X, Ma S, Liu C, Chen C, Wu Q, Chen X and Lu J 2016 Performance of High Layer Thickness in Selective Laser Melting of Ti6Al4V *Materials* **9** 975

[21] Wu S, Lei Z, Jiang M, Liang J, Li B and Chen Y 2022 Experimental investigation and discrete element modeling for particle-scale powder spreading dynamics in powder-bed-fusion-based additive manufacturing *Powder Technology* **403** 117390

[22] Wischeropp T M, Emmelmann C, Brandt M and Pateras A 2019 Measurement of actual powder layer height and packing density in a single layer in selective laser melting *Additive Manufacturing* **28** 176–83

[23] Chen H, Wei Q, Zhang Y, Chen F, Shi Y and Yan W 2019 Powder-spreading mechanisms in powder-bed-based additive manufacturing: Experiments and computational modeling *Acta Materialia* **179** 158–71

[24] Xiang Z, Zhang M, Yan R, Yin Q and Zhang K 2021 Powder-spreading dynamics and packing quality improvement for laser powder bed fusion additive manufacturing *Powder Technology* **389** 278–91

[25] Cao L 2019 Study on the numerical simulation of laying powder for the selective laser melting process *Int J Adv Manuf Technol* **105** 2253–69

[26] Jacob G, Donmez A, Slotwinski J and Moylan S 2016 Measurement of powder bed density in powder bed fusion additive manufacturing processes *Meas. Sci. Technol.* **27** 115601
[27] Liu B, Wildman R, Tuck C, Ashcroft I and Hague R 2011 Investigation the effect of particle size distribution on processing parameters optimisation in selective laser melting process *Proceedings for the 2011 International Solid Freeform Fabrication Symposium* 22nd Annual International Solid Freeform Fabrication Symposium - An Additive Manufacturing Conference (Austin) pp 227–38
[28] Haferkamp L, Haudenschild L, Spierings A, Wegener K, Riener K, Ziegelmeier S and Leichtfried G J 2021 The Influence of Particle Shape, Powder Flowability, and Powder Layer Density on Part Density in Laser Powder Bed Fusion *Metals* **11** 418
[29] Rogalsky A, Rishmawi I, Brock L and Vlasea M 2018 Low cost irregular feed stock for laser powder bed fusion *Journal of Manufacturing Processes* **35** 446–56
[30] Lampitella V, Trofa M, Astarita A and D'Avino G 2021 Spreading of Powders in Powder Bed Additive Manufacturing: an Experimental Approach *ESAFORM Proceedings 2021* 24th International Conference on Material Forming (Liège)
[31] Li J, Gao G, Zhang S and Du Y 2019 Calibration and Measurement Study on Density of Selective Laser Sintered 3D Printing Powder Bed *Journal of Physics: Conference Series* **1267** 012072
[32] Ali U, Mahmoodkhani Y, Imani S, Esmaeilizadeh R, Liravi F, Sheydaeian E, Huang K, Marzbanrad E, Vlasea M and Toyserkani E 2018 On the measurement of relative powder-bed compaction density in powder-bed additive manufacturing processes *Materials & Design* **155** 495–501
[33] EOS GmbH 2023 EOS NickelAlloy IN718 Material Data Sheet. Available from: https://www.eos.info/en/3d-printing-materials/metals/nickel-alloys [Accessed 10 May 2023]
[34] Kelkar R, Andreaco A, Ott E and Groh J 2018 Alloy 718: Laser Powder Bed Additive Manufacturing for Turbine Applications *Proceedings of the 9th International Symposium on Superalloy 718 & Derivatives: Energy, Aerospace, and Industrial Applications* The Minerals, Metals & Materials Series ed E Ott, X Liu, J Andersson, Z Bi, K Bockenstedt, I Dempster, J Groh, K Heck, P Jablonski, M Kaplan, D Nagahama and C Sudbrack (Cham: Springer International Publishing) pp 53–68
[35] Caiazzo F, Alfieri V, Corrado G and Argenio P 2017 Laser powder-bed fusion of Inconel 718 to manufacture turbine blades *Int J Adv Manuf Technol* **93** 4023–31
[36] Sanchez S, Smith P, Xu Z, Gaspard G, Hyde C J, Wits W W, Ashcroft I A, Chen H and Clare A T 2021 Powder Bed Fusion of nickel-based superalloys: A review *International Journal of Machine Tools and Manufacture* **165** 103729
[37] Li X, Dunn P F and Brach R M 1999 Experimental and numerical studies on the normal impact of microspheres with surfaces *Journal of Aerosol Science* **30** 439–49
[38] Chen H, Cheng T, Li Z, Wei Q and Yan W 2022 Is high-speed powder spreading really unfavourable for the part quality of laser powder bed fusion additive manufacturing? *Acta Materialia* **231** 117901

Acknowledgements
Authors acknowledge the project of DREAMS (Database for Radically Enhancing Additive Manufacturing and Standardization) funded by Business Finland and managed by Finnish Additive Manufacturing Ecosystem (FAME), University of Oulu, Lappeenranta University of Technology, and industrial partners. Authors would also like to express gratitude for EOS Finland Oy and their staff for essential help and assistance for the study. Lastly, authors thank also staff of Mechanical Engineering in Faculty of Technology at University of Turku for their knowledge and help.

CRedit: Haapa: main contributor to the planning and execution of experiments and simulation, data gathering and analysis and in writing. Gopaluni: planning of experiments and simulation, feedback and advice on data analysis and contribution in writing. Piili: supervision of research and findings, feedback and advice in planning of experiments and on data analysis, feedback and advice in writing. Ganvir: feedback, and advice on writing the paper. Salminen: feedback, and advice on writing the paper. Ottelin: practical expertise for the experimental set-up and methods, supervised research and findings

IOP Conf. Series: Materials Science and Engineering 1296 (2023) 012021 doi:10.1088/1757-899X/1296/1/012021

Effects of powder recycling on laser-based powder bed fusion produced SS316L parts

Aditya Gopaluni[1], Chinmayee Nayak[1], Aki Piironen[1], Tuomas Kantonen[1], Antti Salminen[1]

1 Digital Manufacturing and Surface Engineering, Department of Mechanical and Materials Engineering, University of Turku, Joukahaisenkatu 3, Turku, Finland 20520,

Abstract. Laser-based powder bed fusion (PBF-LB/M) is one of the extensively used additive manufacturing (AM) methods as the parts printed by PBF-LB/M have high resolution due to low layer thickness. Recycling and Reusing of the powder in this process has a significant impact on the surface roughness, mechanical properties such as hardness, elastic modulus and fracture strength etc. of the manufactured parts. The aim of the present study is to understand the effects of powder recycling on the properties of SS316L parts such as hardness and surface roughness. The developed method featured the use of SS316L powder five times over with weight percentage adjusted as required for the build job. The printed parts were checked for surface roughness across different surfaces with respect to the build direction along with Vickers hardness test. The surface roughness of the parts before polishing showed a steady increasing trend of about 40% with the recycling count while there was no significant effect on hardness as the values stayed in the range of 230+/-5 HV. Powder morphology studies with SEM displayed visible changes in terms of satellite formations, broken particles etc. with the increase in counts of recycling and particle size distribution showed a linear increase with the increase in recycling counts. This study was performed within the limited scope of a bigger study for which a detailed methodology of powder recycling will be developed.

Keywords: Additive Manufacturing, Laser powder bed fusion, powder recycling, characterization, hardness, surface roughness, digital manufacturing

1. Introduction

Laser-based Powder Bed Fusion (PBF-LB/M) is one of the extensively used additive manufacturing methods (AM) as the parts manufactured have high resolution due to low layer thickness [1]. In the industry, it is a common practice to reuse the metal powder used for manufacturing parts in the PBF-LB/M process for the purpose of cutting costs and ensuring sustainability [2]. Reusing and recycling metal powder has an effect on the part properties such as surface roughness, density, hardness, tensile strength, fatigue, fracture toughness etc. and powder properties such as particle size distribution (PSD), packing density and morphology [3][4]. The SS316L stainless steel material is often used in the medical and aerospace industries where corrosion resistant parts are required [5]. The surface characteristics of such parts are important in the said fields as the PBF-LB/M manufactured parts are often used as a part in assemblies [6].

In PBF-LB/M process, various parameters such as raw material parameters- particle size distribution (PSD), powder flowability, packing density, process parameters- powder spreading speed, powder layer thickness and build parameters- laser energy density, laser power, laser scan speed have an effect on the powder properties and the part properties [7]. When it comes to powder reuse, the research is focused on how the reusing and recycling phenomenon affects the part properties[8]. Emminghaus et al. have studied the effect of residual oxygen content and powder recycling on the part properties of PBF-LB/M produced Ti6Al4V [3]. The study was conducted by studying the effect of residual oxygen on virgin and recycled powders through understanding the part properties such as porosity, surface roughness and density. It was concluded from the study that the surface roughness of the parts increased steadily with the counts of recycling for the top surface and the side surface with respect to the build direction. A standard deviation of 4.01 µm was observed for the top surface and a standard deviation of 2.79 µm. The reason was that there was an increase in the PSD of the powder particles with the counts of recycling. There was no significant change in relative density of the parts due to powder recycling.

Ahmed et al. studied the effect of powder recycling of 17-4PH Stainless steel powder on the part properties without adding any virgin powder in the recycling iterations. The powder was recycled ten times over after starting with the virgin powder. In each recycling stage, the powder characteristics were investigated by performing powder rheology, morphology, microstructure and chemical compositions. The morphology study showed that there was an increase in the powder particle size by 3% with each iteration of recycling. This is due to the spatter and partially molten particles getting attached to the spherical particles of the virgin powder and the subsequent powders from there. The difference in tensile strength was also studied between the recycled and the virgin powders where the parts manufactured with recycled powder showed no significant difference. The reason for the observation was suggested to be the lack of austenite phase present in the samples [9]. Gorji et al. conducted a synchrotron x-ray study on SS316L powder to understand the effect of recycling on the oxidation of powders. This study revealed that there was an increase in the concentration of metallic oxides on the surface of the powder particles as the recycling iterations increased. This was supported by the phenomenon seen in the SEM images of powder particles where there were indications of cracks and porosity within the powder particles as it is an effect of oxygen weakening the bonds existing in the powder particles [2].

Quinn et al. studied the effect of recycling SS316L powder on the parts manufactured using an EOS M280 machine. The design of experiments for the research included the measurement of hardness, surface roughness, and porosity for SS316L samples for each iteration of recycling. It was concluded from the study that the surface roughness of the parts manufactured increased by 50% by the time the number of recycling runs were finished while no significant difference in the hardness was observed. The increase in the surface roughness was attributed to the increase in the PSD of powder particles with the number of recycling runs. An increased PSD means that there is less than ideal fusion of the powder particles, which causes the formation of rough surfaces [10].

Different studies have been conducted to understand the effect of powder recycling on powder degradation and how the powder properties change throughout the recycling process [11]–[13]. Research was also carried out on how the additively manufactured part properties are affected by powder recycling as well. The scope of the present work is on how the number of recycling iterations affects the PSD and powder morphology of the powder that is used for printing. In addition, an investigation of the change in part properties such as surface roughness and hardness is also conducted. The scope of the present work is part of a bigger study where the aim is to develop a methodology to qualify metal powder for part production where recycling of powder is the key point of consideration. The study will include simulation from the point of view of powder bed and the effects of recycling will be simulated. The idea to include simulation is motivated by the need to reduce the dependency on the experimental data and procedures.

2. Materials and Methods

2.1. Laser Powder Bed Fusion

Stainless steel powder of grade SS316L has been used in Aconity MIDI+ PBF-LB/M printer from Aconity3D GmbH for this study. The SS316L powder used for printing was provided by SLM Solutions GmbH and has a PSD of 10-45μm and consists of mostly spherical shaped particles. Figure 1 shows the set of samples printed using SS316L which belong to the template defined for the project which funds the research. These samples included the powder canisters, bending test samples, tensile test samples and 10x10x10mm cubes.

Figure 1. Print layout for the project, sample of interest highlighted

The scope of the study presented in this paper however only covers the cubic samples that can be seen in different locations across the build plate. The 10x10x10mm cubes were printed with a laser beam scan speed of 800mm/s, laser power of 150W and laser beam size of 80μm. Hatch distance was 80μm. Laser was 400W continuous wave single mode fiber laser with Gaussian beam intensity profile and wavelength of 1070nm. The used shielding gas was argon.

2.2. Powder Recycling

The powder recycling methodology for this study involved recycling the powder five times over after starting with the virgin SS316L powder. For each build, the virgin powder was added to the remaining unmelted powder after printing to match the required weight of the powder to carry out the printing process. For the given set of samples that were printed using the pre-defined layout, the ratio between the used powder and re-added virgin powder was 3:2. This method was repeated five times to understand the effect of recycling on the produced parts and also the powder properties. In this methodology, the unmelted powder was not sieved to completely understand the behaviour of untouched and unmelted powder.

2.3. Powder properties

The powder properties that were studied in the scope of this research are particle size distribution and powder morphology. Powder samples were collected from each build starting with the virgin powder build until the 5th recycling iteration. PSD was measured using a Malvern3000 laser diffraction particle size analyzer from Malvern Panalytical Ltd. The PSD measurements were performed in the dry powder analysis method for all the powder samples from the five recycling iterations. The experiments were repeated three times for each sample for repeatability and were conducted a total of 15 times. Scanning Electron Microscopy (SEM) was performed on all the powder samples from the first four builds with high resolution FE-SEM Thermo Scientific Apreo 5 from Thermo Fisher Scientific Ltd. For SEM, only one samples of powder from each build was taken and studied.

2.4. Surface Roughness

Surface Roughness measurements were performed on the as-printed samples on the top surface of the cubic samples with respect to the build direction and also on the diamond polished samples of the same surface. The equipment used to measure surface roughness is the Bruker Alicona Infinity G6 from Bruker Alicona, which measures the form and surface roughness of samples. The machine performs an optical 3D scanning of the sample surface and provides values of R_a. The measurement for R_a was performed parallel to the hatch direction and perpendicular to the build direction. For the surface roughness measurement of the as-printed parts, a magnification of 800WD17 was used which indicates the working distance of 17mm and a resolution of 50nm. For the measurement of surface roughness of the same surfaces which have been diamond polished, a magnification of 400WD19 has been used which indicates a working distance of 19mm and a resolution of 30nm. The roughness measurements were conducted for a distance of 15μm perpendicular to the build direction on the sample surface. This procedure was repeated for both polished and unpolished samples.

2.5. Hardness

Vickers Hardness test was performed on all of the samples of interest which were diamond polished. The equipment used to perform the tests is the Innovatest Falcon 600G2 from Innovatest Europe. For all the samples, HV10 scale was used to measure the Vickers hardness, which means that the dwell time for the indentor was 10 seconds for a force of 1kgf. For each sample 4 indents were made across the diagonal of the top surface with respect to the build direction. The spacing between each indent was 1mm.

3. Results and Discussion

3.1. Powder properties

Figure 2 indicates the powder morphology of the SS316L starting from the virgin powder until the 4th recycling iteration as measured with SEM . The back scattered electron images show that there is a visible change in the morphology of the 2nd build powder sample when compared with the sample from the 4th recycling iteration. Powder from the 5th recycling iteration was not characterized as the gas flow stoped in the chamber during build time, leading to the powder attaining a burnt texture.

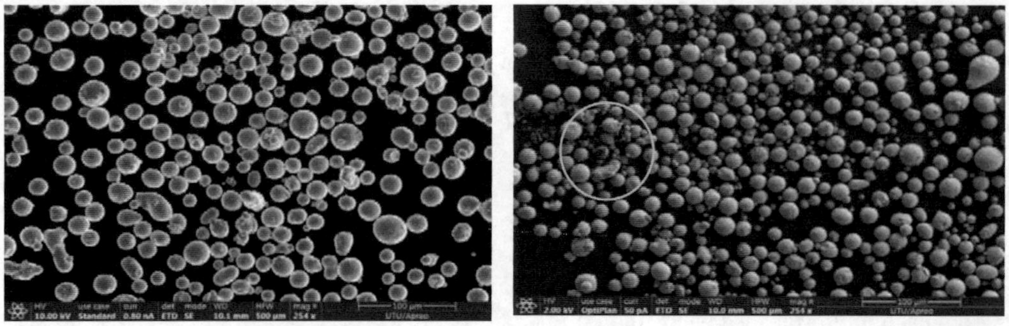

Figure 2. SEM images of powder samples. Build 2 (left) and Build 4 (right, particles with different morphologies)

The powder from the second build presents a morphology that is mostly spherical in nature for the powder particles present as the powder has not undergone significant changes within the build chamber of the PBF-LB/M machine. The particle size from the SEM images for build 2 is observed to be about 46μm while for the 4th build, it is observed to be at 49 μm. When the recycled powder is considered, it can be noticed from the image on the right that there are different types of morphologies present in the powder samples that belong to the recycled powders. The presence of satellite particles can be seen on the larger spherical particles of powder along with smaller nanosized particles. Other morphologies such as cracked and broken particles can also be observed. Partially fused particles, which occur due to the improper fusion of the melt pool can also be noticed in this highlighted area. The appearance of cracks and broken particles can be seen in figure 3.

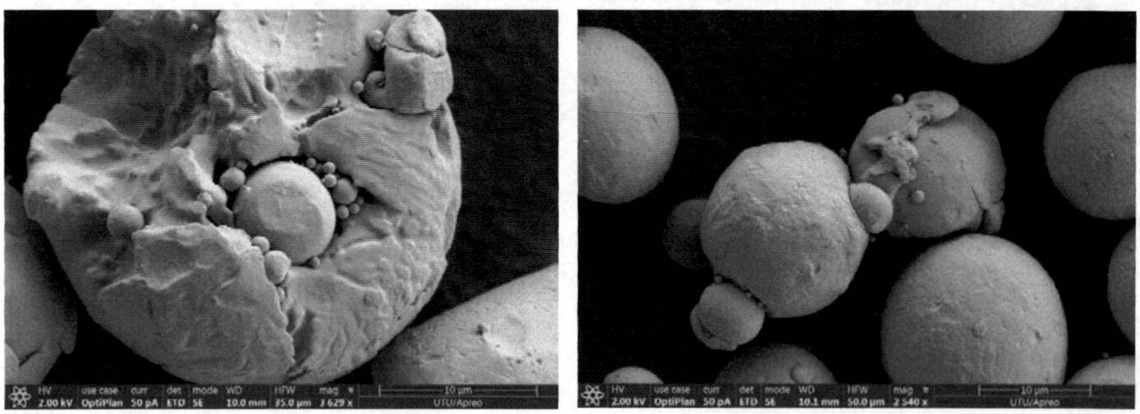

Figure 3. Broken powder particle (left) from build 4 and partially fused powder particles (right) from build 4

The cause for the formation of satellites on the larger powder particles is e.g. due to the fact that there is partially melted powder particles resulting from the laser beam and material interaction. These particles are formed inside the build chamber due to different phenomena such as spatter, particle breakage upon laser impact, improper melting of the particles etc. In this case, it is visible that the nano-size particles which are attached to the larger powder particles are spatter. The change in morphology is also supported by the PSD data accumulated for all the four powder samples.

The bar diagram in the figure 4 indicates a steady increase in PSD of the powder samples as the number of counts of recycling is increased.

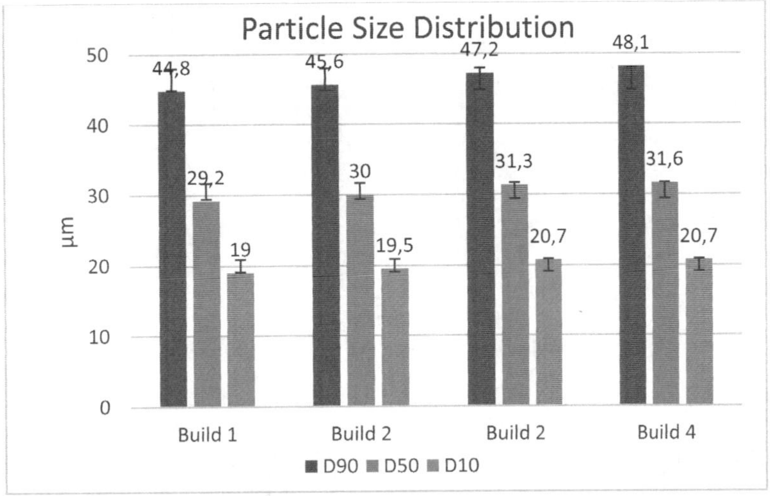

Figure 4. PSD of powder samples from each build

The D90 for Build 1 in the figure 4 shows that 90% of the powder particles in the given samples are of the minimum size of 45µm or above. D50 is an indication that 50% of the particles in the sample are of the minimum size of 29,2 µm and D10 shows that 10% of the samples in the build are of the minimum size of 19 µm. The increase in the PSD with the increase in the count of recycling is due to the formation of satellites on most of the powder particles within the sample and also due to lack of fusion between the two regular-sized powder particles which ultimately form a bigger particle.

3.2. Surface Roughness

The graph in the figure 5 depicts the average surface roughness of the samples for each build that has been measured on the top surface of the samples, perpendicular to the build direction. The surface roughness here corresponds to samples which were in the as built condition, meaning without polishing or any other surface treatment performed on the sample surface.

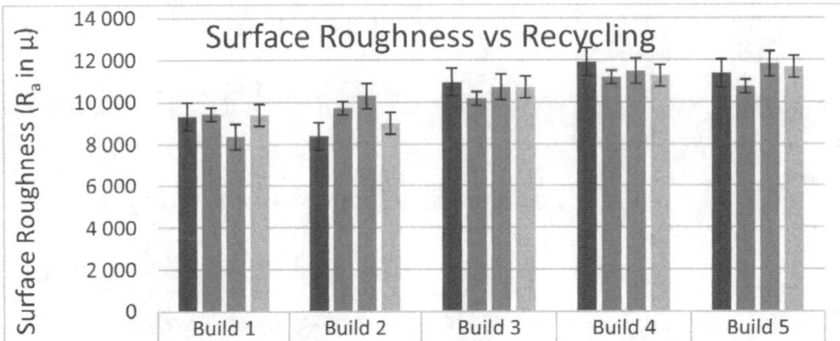

	Build 1	Build 2	Build 3	Build 4	Build 5
Series1	9 312	8 368	10 938	11 906	11 369
Series2	9 425	9 715	10 174	11 171	10 744
Series3	8 355	10 290	10 710	11 464	11 813
Series4	9 389	8 997	10 702	11 248	11 663

Figure 5. Surface Roughness of the pre polished samples

A trend of steady increase in the surface roughness of these samples is observed to be a direct effect of powder recycling. In PBF-LB/M, the surface characteristics of the parts produced are of high interest as they have an effect on the functionality of the part and as seen in the figure, there is a significant difference between the surface roughness of the first build and the surface roughness of the 4th build. The significant difference comes down to the surface characteristics that are affected by the lack of fusion and partial fusion of the powder particles on the surface of the sample while printing. For the preset laser parameters, the increase in the particle size of the powder samples means that there is not enough fusion in the powder bed when the laser beam interacts with the material. The fusion of the powder particles progressively becomes worse as the counts of recycling are increased and it is visible in the increasing values of the surface roughness of the samples which is an increase of 20%.

Figure 6 is a representation of the surface roughness of the diamond polished samples across all the builds. When diamond polishing is done, all the irregularities in the surface are removed. This can be seen in the values of the surface roughness across all the builds. The average surface roughness (R_a) of all the samples in the four builds is about 0.25µm.

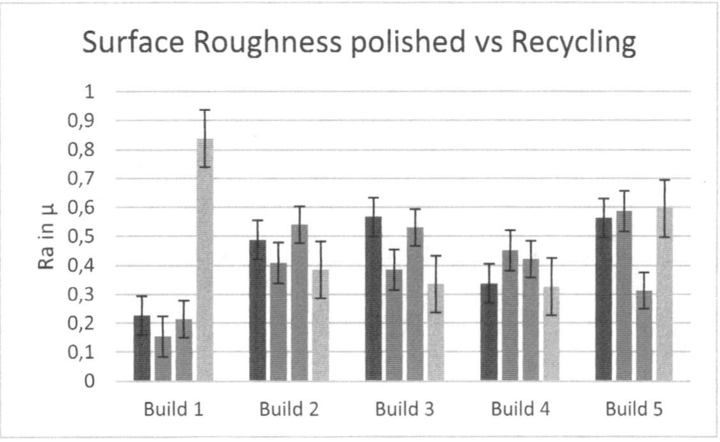

Figure 6. Surface Roughness of the polished samples

The similar values of surface roughness across all the builds suggest that the effect of recycling is not as significant after the polishing in comparison to before polishing. Once the removal of the irregularities and the contaminants from the surface occur, the recycling does not seem to have enough effect to alter the surface characteristics of the cubic samples printed. A polishing anomaly for the 4th sample in the 1st build incurred a really high value for surface roughness which is 300% higher than the average value of surface roughness. In both the cases of polishing and non-polished samples surface roughness is also affected by various factors such as packing density, powder flowability and melting behaviour. A uniform PSD means that the powder bed which is melted is smoother, meaning there is a uniform packing of the particles in a given layer. The nature of PSD for the powder obtained from the recycled builds is non-uniform, which means that the powder layer is irregular and causes improper melting of the powder particles and thereby promoting higher surface roughness.

3.3. Hardness

Figure 7 shows the Vickers Hardness values of the cubic samples. The values correspond to the average hardness of the build for which the measurements were done.

IOP Conf. Series: Materials Science and Engineering 1296 (2023) 012021 doi:10.1088/1757-899X/1296/1/012021

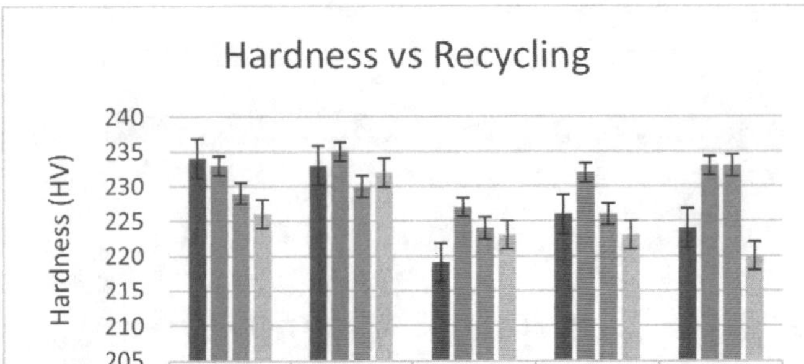

	Build 1	Build 2	Build 3	Build 4	Build 5
Series1	234	233	219	226	224
Series2	233	235	227	232	233
Series3	229	230	224	226	233
Series4	226	232	223	223	220

Figure 7. Hardness of the SS316L samples

Hardness values were measured to find out the existence of a possible correlation between them and powder recycling. As it can be noticed, the mean hardness of the samples falls in the category of 230+/-5 HV for all the 5 builds. This suggests that there is no significant effect of powder recycling on the hardness values. The recycled powder in the first and second iterations showed a higher hardness as compared to the cubic samples printed with the virgin powder. This is due to the oxidation of the powder which localizes the powder particles closer to the surface and influences the hardness. A similar phenomenon was studied and reported in the research of Gorji et al. where nano-indentation was performed to measure hardness [2]. The expected behaviour usually in the case of hardness of the samples is correlated to porosity present in the samples. The non-uniform PSD of the recycled powder causes the improper melting of the powder which leads to porosity and higher hardness. However, this phenomenon could not be seen in this case. The plausible reason for this could be that the laser parameters used to print the samples were at an optimum which would play down the effects of powder recycling.

4. Conclusion

The aim of this study was to understand the effect of powder recycling without sieving on part properties such as surface roughness and hardness has been studied with respect to the counts of recycling of powder. 5 builds have been printed according to the main project template out of which the cubic samples have been used for optical depth scanning to measure the surface roughness (for both as-printed and diamond polished samples) and Vickers hardness measurements. The powder samples collected only from the first four builds were analyzed because of handling issues. Powder morphology was analyzed by SEM and PSD was measured using laser diffraction method.

- The surface roughness of the as built samples showed an increasing trend of until 50% with the counts of recycling which suggests there was improper fusion of powder particles closer to the surface of the samples
- The trend in the hardness values for the samples is very uniform with the values being in the range of 230+/-5HV which suggests that there is no significant effect of recycling of the powder on hardness of the samples
- The trend in surface roughness of the samples shows that the packing density and PSD of the powder have a significant effect on melt behaviour of the powder, however, this effect is not visible in case of

hardness of the samples. This could be attributed to optimized laser parameters which offset the porosity effects.

5. Future work

Powder recycling has a significant effect on the powder properties and the parts produced by PBF-LB/M. The work in this direction will continue with the characterization of the samples outside the scope of this study. Tensile testing and Bending testing will be carried out on the samples that were printed along with the samples that were printed for this study to understand the effect of powder recycling on part properties such as tensile strength and fatigue.

In addition to the finish the remaining part property characterization, new materials will be printed and the parts will be characterized with a similar methodology. Simulation will also be introduced at a certain point to try and understand the changes in part properties.

6. Acknowledgement

Authors of this article acknowledge project of DREAMS (Database for Radically Enhancing Additive Manufacturing and Standardization). This project was funded by Business Finland and the duration of project is from April 2022 to February 2024. The Participant of project were the Finnish Additive Manufacturing Ecosystem (FAME). Aim of the project was to establish standards for AM and develop a powder material database for AM. Authors would like to thank partners of this project for their contribution to this article.

7. Conflict of interest

All the authors involved in the article report no conflict of interest.

8. References

[1] F. Veron, F. Lanoue, V. Baco-Carles, K. Kiryukhina, O. Vendier, and P. Tailhades, “Selective laser powder bed fusion for manufacturing of 3D metal-ceramic multi-materials assemblies,” *Addit Manuf*, vol. 50, Feb. 2022, doi: 10.1016/j.addma.2021.102550.

[2] N. E. Gorji, R. O’Connor, A. Mussatto, M. Snelgrove, P. G. M. González, and D. Brabazon, “Recyclability of stainless steel (316 L) powder within the additive manufacturing process,” *Materialia (Oxf)*, vol. 8, p. 100489, Dec. 2019, doi: 10.1016/J.MTLA.2019.100489.

[3] N. Emminghaus, C. Hoff, J. Hermsdorf, and S. Kaierle, “Residual oxygen content and powder recycling: Effects on surface roughness and porosity of additively manufactured Ti-6Al-4V,” *Addit Manuf*, vol. 46, p. 102093, Oct. 2021, doi: 10.1016/J.ADDMA.2021.102093.

[4] V. V. Popov, A. Katz-Demyanetz, A. Garkun, and M. Bamberger, “The effect of powder recycling on the mechanical properties and microstructure of electron beam melted Ti-6Al-4 V specimens,” *Addit Manuf*, vol. 22, pp. 834–843, Aug. 2018, doi: 10.1016/J.ADDMA.2018.06.003.

[5] B. Blakey-Milner *et al.*, “Metal additive manufacturing in aerospace: A review,” *Mater Des*, vol. 209, p. 110008, Nov. 2021, doi: 10.1016/J.MATDES.2021.110008.

[6] H. Yeung, B. M. Lane, M. A. Donmez, J. C. Fox, and J. Neira, "Implementation of Advanced Laser Control Strategies for Powder Bed Fusion Systems," *Procedia Manuf*, vol. 26, pp. 871–879, Jan. 2018, doi: 10.1016/J.PROMFG.2018.07.112.

[7] S. Mounsey, B. Hon, and C. Sutcliffe, "Performance modelling and simulation of metal powder bed fusion production system," *CIRP Annals*, vol. 65, no. 1, pp. 421–424, Jan. 2016, doi: 10.1016/J.CIRP.2016.04.065.

[8] Y. Sun, M. Aindow, and R. J. Hebert, "The effect of recycling on the oxygen distribution in Ti-6Al-4V powder for additive manufacturing," *https://doi.org/10.1080/09603409.2017.1389133*, vol. 35, no. 1–3, pp. 217–224, May 2017, doi: 10.1080/09603409.2017.1389133.

[9] N. Ahmed, I. Barsoum, G. Haidemenopoulos, and R. K. A. Al-Rub, "Process parameter selection and optimization of laser powder bed fusion for 316L stainless steel: A review," *J Manuf Process*, vol. 75, pp. 415–434, Mar. 2022, doi: 10.1016/J.JMAPRO.2021.12.064.

[10] P. Quinn, S. O'Halloran, J. Lawlor, and R. Raghavendra, "The effect of metal EOS 316L stainless steel additive manufacturing powder recycling on part characteristics and powder reusability," *https://doi.org/10.1080/2374068X.2019.1594602*, vol. 5, no. 2, pp. 348–359, Apr. 2019, doi: 10.1080/2374068X.2019.1594602.

[11] S. Giganto, S. Martínez-Pellitero, J. Barreiro, and P. Zapico, "Influence of 17-4 PH stainless steel powder recycling on properties of SLM additive manufactured parts," *Journal of Materials Research and Technology*, vol. 16, pp. 1647–1658, Jan. 2022, doi: 10.1016/J.JMRT.2021.12.089.

[12] P. Muthuswamy, "Influence of powder characteristics on properties of parts manufactured by metal additive manufacturing," *Lasers in Manufacturing and Materials Processing*, vol. 9, no. 3, pp. 312–337, Sep. 2022, doi: 10.1007/S40516-022-00177-3/TABLES/2.

[13] Q. B. Nguyen, M. L. S. Nai, Z. Zhu, C. N. Sun, J. Wei, and W. Zhou, "Characteristics of Inconel Powders for Powder-Bed Additive Manufacturing," *Engineering*, vol. 3, no. 5, pp. 695–700, Oct. 2017, doi: 10.1016/J.ENG.2017.05.012.

Bioinert ceramics scaffolds for bone tissue engineering by laser-based powder bed fusion: a preliminary review

N Kamboj [1,2,3*], H Piili [1], A Ganvir [1], A Gopaluni [1], C Nayak [1], N Moritz [2] and A Salminen [1]

[1] *Research group of Digital Manufacturing and Surface Engineering, Department of Mechanical and Materials Engineering, University of Turku, FI-20500, Turku, Finland*

[2] *TCBC– Turku Clinical Biomaterials Centre, Department of Biomaterials Science, Faculty of Medicine, Institute of Dentistry, University of Turku, FI-20014, Turku, Finland*

[3] *Tallinn University of Technology, Department of Mechanical and Industrial Engineering, Ehitajate tee 5, 19086 Tallinn, Estonia*

Email – nikhil.kamboj@utu.fi

Abstract. The implementation of laser powder bed fusion (PBF-LB) on ceramics is far more demanding than their metallic and polymeric counterparts for bone tissue engineering (BTE). The review will shed light on bioinert ceramics-based biomaterials manufacturing through PBF-LB incorporating alumina and yttria-stabilized zirconia as oxide-based ceramics and nitride-based ceramics as non-oxide-based ceramics with particular prominence on their properties and requirements for biomedical devices and BTE. The review paper will also classify bioinert scaffolds processed through PBF-LB as a medium to manufacture drug delivery systems (DDS) and to ameliorate critical-sized bone defects based on the fracture site length of the bone with the various modes of functionalization through the incorporation of drugs, stem cells, and growth factors for personalized medicine.

Keywords: bioinert ceramics; bone tissue engineering; critical-sized defects; digital manufacturing; laser powder bed fusion; drug delivery systems;

1. Introduction

Osseous tissue, widely known as bone, provides structural integrity to the living being and has a profound role in shielding the organs, biomechanical movement, and anchoring several blood cells. Bone tissue presents a hierarchy at the macro, micro, and nano-level with the schematic representation depicted in Figure 1. Bone tissue has a peculiar property to dynamically heal and remodel itself from micro-level damages by the activity of osteoblasts and osteoclasts. However, critical orthopedic injuries cannot heal by themselves, and henceforth usually needs engineered structures i.e. biomaterials [1].

Biomaterials can be elucidated as natural or synthetic supporting structures which can assist biomechanical movements. Biomaterials are made of assorted types of materials encompassing metals, ceramics, and polymers. Metallic biomaterials orchestrate and severe for long periods of time without any significant rejection with the characteristics like high corrosion and wear resistance, suited mechanical properties for large bone defects, and biocompatibility. However, the several downsides to using metallic biomaterials which include stress shielding [2] of the mangled bone, bioactivity, and

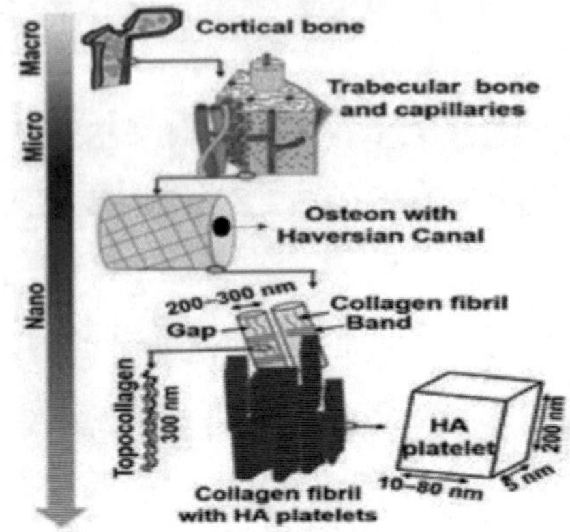

Figure 1. The bone hierarchy at the macro, micro, and nano-level, adapted from [3]

revision of the surgery at the orthopedic defect. On the contrary, polymeric biomaterials have the largest batch of materials used in hard tissue restoration but the disadvantages of the polymeric materials include lower elastic modulus and tensile strength, time-dependent degradation based on the complexity of the polymeric chains, and, maintenance after the sterilization before surgery [4].

Ceramic biomaterials have peculiar attributes and present several advantages which can fulfill the shortcomings and gaps of metallic and polymeric biomaterials since ceramics being inorganic compounds are made up of metals and nonmetals linked by ionic and/or covalent bonds with forming crystal structures. The bestowed property on ceramics imparts them crystallographic and texturing properties coupled with chemical functionalization of the ceramics to fine-tune the degradation and bioactivity with the physiological medium making them excellent candidates for BTE. Ceramic biomaterials or bio-ceramics can be broadly classified into three different categories which include bioactive, bioresorbable, and bioinert ceramics. Bioactive and bioresorbable ceramics have the potential to bond with the bone tissue in the vicinity due to the similar inorganic composition of bone through the nucleation [5,6] and crystal growth of hydroxyapatite (HA) [7]. Therefore, there is a chemical reaction between the bone tissue and bioactive and bioresorable ceramics. On the other hand, bioinert ceramics remains inert and therefore there is no chemical reaction between the surrounding defective bone tissue and bioinert ceramics since the chemical composition of the bioinert ceramics is different from the bone. Most of the articles in the literature are devoted to bioactive and bioresorbable ceramic scaffolds. However, on the contrary, there is a single review [8] devoted to bioinert ceramics which was our prime motivation for writing the present review article to fulfill the gap in the literature.

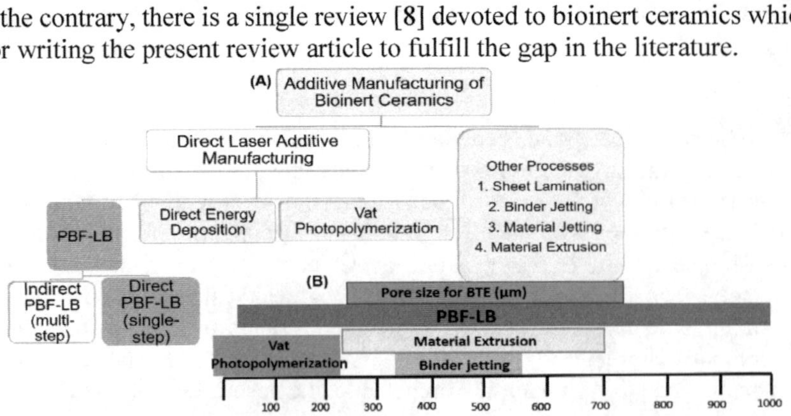

Figure 2. (A) Current additive manufacturing methods for bioinert ceramics scaffolds and process categories highlighted in green are the ones discussed in this review paper, (B) Advantage of PBF-LB over other AM techniques used for BTE, adapted from [3]

The additive manufacturing (AM) of bioinert ceramics scaffolds encloses techniques like PBF-LB, material extrusion, directed energy deposition, binder jetting, vat-photopolymerization, sheet lamination, and material jetting as depicted in Figure 2(a). However, the most commonly used techniques from Figure 2(a) for bioinert scaffold manufacturing are PBF-LB, material extrusion, binder jetting, and vat-photopolymerization. Amongst all the commonly used techniques used for BTE, PBF-LB has a major advantage in the fabrication of the scaffolds since it presents a wider window for the fabrication of pore size of scaffolds ranging from 100 – 1000 μm as shown in Figure 2(b). The functionality of scaffold manufactured by PBF-LB can further be enhanced by the incorporation of hydrogels [9]. They have the characteristics of absorbing and retaining water into their structure (20–100% water relative to the total mass), without dissolution of the material into the pores of the scaffold. Secondly, the most favored way of enhancing the functionality of the scaffolds is the incorporation of DDS on the powder bed for personalized medicine. DDS encompasses the incorporation of a drug, an active molecule, and cells in the scaffolds fabricated by PBF-LB with the more detailed information presented in section 4. Additionally, PBF-LB has the capability to manufacture customized shapes to fulfill the defect at the anatomical site of the defective bone based on the critical size defects. Therefore, taking the advantage of PBF-LB process as the most applicable AM technique for bioinert ceramics from the perspective of BTE is the second motivation to write the review article.

The authors have described the manufacturing of bioactive ceramic scaffolds for BTE through the PBF-LB technique [3]. The review article sheds light on silicate-based ceramics, hydroxyapatite, and bioactive glass-based ceramics processed through direct PBF-LB. However, as far as bioinert ceramics scaffolds the number of articles published in the last 13 years are depicted in Figure 3.

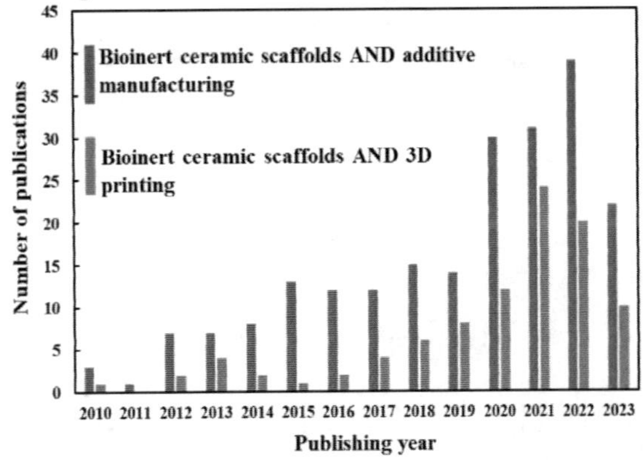

Figure 3. The number of articles published in the last 13 with different keywords years

Two different sets of keywords are used as shown in Figure 3. Furthermore, there was not even a single review article in Scopus with the keywords "Bioinert ceramics scaffolds" AND "laser powder bed fusion" Henceforth, this was our prime motivation to write the review article. This review article will only review direct PBF-LB articles without any use of binders which were used to manufacture bioinert ceramic scaffolds since the scaffolds were not subjected to any geometrical constraints, no additional post-processing treatment like debinding and firing process, single-step technology, unlike most in indirect PBF-LB process of bioinert ceramic scaffolds [10].

The aim of this study is to review the lasers used to consolidate bioinert ceramics materials which are mostly used for orthopedic load-bearing devices (spinal defects, bone defects, and femoral head fabrication) in the form of oxides ceramics as Al_2O_3, and YSZ (yttria-stabilized zirconia), and non-oxides as Si_3N_4 manufactured by direct PBF-LB. Lastly, the manufacturing of bioinert ceramic scaffolds for critical-sized defects of the bone defects ranging from 1-4.5 cm and DDS through PBF-LB are presented.

2. PBF-LB of bioinert ceramics

PBF-LB process is one of the AM techniques as it is used to manufacture the scaffold as shown in Figure 4(a), from a Computer Aided Design/Manufacturing (CAD/CAM). The powder particles on the powder bed are laser melted or laser sintered. The binding mechanism unraveling in the PBF-LB process comprises solid state sintering (Al_2O_3, YSZ), liquid phase sintering/partial melting (Al_2O_3, YSZ), full melting (Al_2O_3), and chemically induced binding (Si_3N_4). However, the process of manufacturing bioinert ceramic coupons through PBF-LB is far more challenging since ceramics usually have higher melting point, brittleness (eventually no plasticity), and poor thermal shock resistance. Furthermore, a rapid heating and cooling process in powder bed results in cracking, and unrequired porosity on the scaffolds [11].

The process parameters of PBF-LB have a profound effect on the functionality of the manufactured scaffolds in-terms of porosity and surface roughness. The main process parameters of PBF-LB for the manufacturing of bioinert ceramic scaffolds encircles around the source of lasing media, the wavelength of laser beam, laser power, laser scanning speed, laser scanning strategy, particle shape and size in the powder bed, and layer thickness as shown in Figure 4(b). Nowadays, ceramic machines are usually implemented through two different lasers CO_2 (wavelength is 10.6 μm) and Nd:YAG (λ:1.064 μm) for the manufacturing of ceramic scaffolds. Furthermore, some ceramic machines in the industry for manufacturing ceramic scaffolds are equipped with both CO_2 (for pre-heating the ceramic powders) and Nd:YAG lasers (for consolidation). Laser adsorption for the oxide bioinert ceramics can be different for both the CO_2 and Nd: YAG lasers. Besides adsorption, laser beam diameter plays a significant role in the amount of energy transferred to the ceramic powder bed and can be expressed as

$$E=P_{eff}/(V_s \cdot h_d \cdot d) \quad (1)$$

where E is the laser energy density, P_{eff} is the laser power, V_s is the laser beam scanning speed, h_d is the hatch distance and d is the layer thickness. Higher scanning speed will lead to poor sintering/melting of the layer and vice versa. Also, authors have shown in recent studies that modulating the scanning speed can eventually fine-tune or customize the desired phases in the ceramic scaffold [12].

Particle morphology also has a perspicacious effect on the final microstructure of the manufactured scaffold as they are directly related to the packing density of the powder bed and flowability. As the basic criterion and preferred AM norm is the powder of narrow particle size distribution with a spherical size distribution, the better will be the flowability. The advantages of using multimodal powders as the feedstock since the packing bimodal size distribution can fulfill the interstitial sites/voids of the powder bed. The preferable coarse-to-fine particle size ratio is (1:10) with the weight fraction of the coarser particles being 70% to accomplish maximum packing [10]. Therefore, all the parameters depicted in Figure 4(b), of the PBF-LB have direct repercussions on the printability of the scaffold which can have an influence on performance customization of the scaffold.

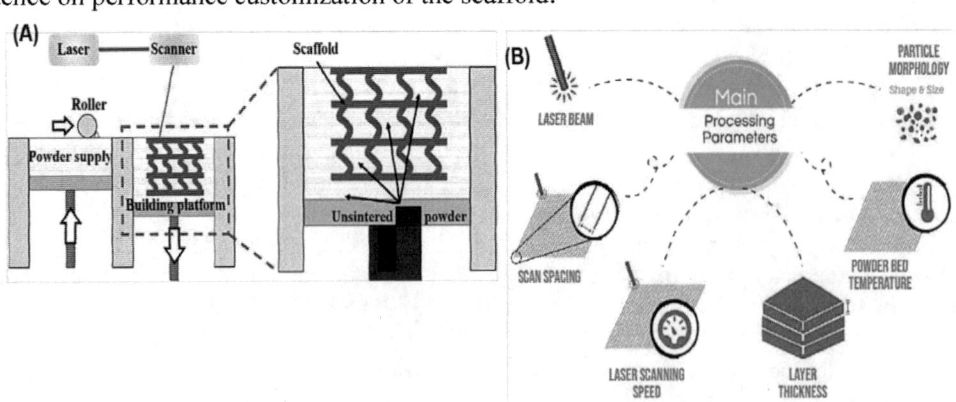

Figure 4. (a) Schematic of PBF-LB process, (b) parameters to be taken into consideration for PBF-LB process, adapted from [13,14]

3. Bioinert ceramic scaffolds processed by PBF-LB

Mostly, the applicability of bioinert ceramics concerning BTE can be on joint prostheses, spine fixtures, and femoral head manufacturing. The most frequently used oxide bioinert ceramics are alumina, YSZ, and non-oxide bioinert ceramics including Si_3N_4. They offer several advantages like less oxidation and corrosion resistance in the body fluids along with low levels of friction and wear. However, on the downside, they are too brittle with low elasticity. The materials i.e. alumina, YSZ, and Si_3N_4 will be presented in detail and are processed through direct PBF-LB.

3.1 Alumina

Alumina is a bioceramic used as a potential alternative to metal implants for a total hip prosthesis (load-bearing applications) [15]. The rationale for using alumina for load-bearing applications can be attributed to the low friction coefficient, high hardness and abrasiveness, chemical inertness, and good thermal conductivity which makes alumina a potential candidate for BTE [16]. It has been also proven from the clinical point of view that the alumina-on-alumina bioceramic on the femoral with the acetabular side leads to minimal osteolysis, which can be associated with superior wear resistance [17,18].

However, alumina has a drawback concerning BTE. It is limited to the capability of the formation of non-adherent fibrous mass or membrane at the bio-interphase when implemented as the implant material. As a result, it can lead to implant failure at the bio-interphase [19]. Nevertheless, the alumina implant failure at the bio-interphase can be averted by removing surface defects (surface modification) or even modifying the surface features at the nano-scale to reduce the pro-inflammatory macrophage response [20]. All these aforesaid surface modifications could facilitate the strong interfacial bonding between alumina and defected bone and can be prevented from loosening [21].

The first and foremost implementation of alumina ceramics by PBF-LB dates back to 1995 when Subramanian et al [22] first implemented alumina in the powder feedstock. Various researchers have used various additives for alumina bioceramic processing. The additives include diboron trioxide [23], zirconia [24], and polyamide [25] with the summary of alumina bioinert ceramics shown in Table 1.

Table 1. Summary of PBF-LB of alumina bioinert ceramics.

Powder feedstock	Laser type	Scanning speed	Reference
Al_2O_3 (80 wt%)+ stearic acid	Nd:YAG laser	2-60 mm/s	[26]
Al_2O_3 + 6YSZ	CO_2 laser for pre-heating, Nd:YAG laser for sintering/melting	200 mm/s	[24]
Al_2O_3 + 6YSZ	CO_2 laser for pre-heating, Nd:YAG laser for sintering/melting	200 mm/s	[27]
Al_2O_3 + polyamide	CO_2 laser	60 mm/s	[25]
Pure Al_2O_3	Fibre laser	30-1000 mm/s	[28]
Al_2O_3 + additives (Na_2O + Fe_2O_3 + SiO_2 + MgO + TiO_2 + CaO)	Fibre laser	90 mm/s	[29]
Compressed Al_2O_3 powder at different compressive pressure (0, 50, 100 MPa)	CO_2 laser	50-1500 mm/s	[30]

As table 1 illustrates, most of the studies focused on CO_2 and Nd:YAG laser for laser sintering/melting of Al_2O_3. Porous alumina can expedite the bone regeneration and healing process by presenting a non-toxic surface at the bio-interphase [16]. Also, the superior mechanical properties of alumina make it a suitable implant for both open defects and long bone defects (load-bearing application) [31].

3.2 YSZ

YSZ (ZrO_2 - Y_2O_3) has been chiefly used for orthopedic implants/scaffolds [32]. The attractive properties, which make it suitable for orthopedic (hip, knee) encircles around good wear resistance, biocompatibility superior mechanical properties especially fracture toughness, and durability [33]. 3 mol% yttria, the most common and stabilized biomedical grade zirconia possess superior fracture toughness to be used for BTE. YSZ was used for the first time in 1990 to manufacture the femoral head for load-bearing applications. This material was in practice for the femoral head until steady fractures were reported and suggested by the authors in their study [34]. The catastrophic failure was not attributed to the mechanical properties but instead to the phase transformation from tetragonal to monoclinic zirconia [35]. Also, the low thermal degradation by occupying the anionic variances (oxygen) in water molecules, or called ageing could also evoke this catastrophic transformation in zirconia. Alumina-reinforced or toughened zirconia and cerium oxide incorporation have comparatively shown better and enhanced ageing resistance when compared to the YSZ counterpart.

PBF-LB of zirconia and zirconia toughened ceramics are also used for dental implants apart from the load-bearing bone implants and are explored by various researchers as it depends on the limited processing steps (no debinding stage) with various examples by the researchers in Table 2 of direct PBF-LB process on YSZ.

Table 2. Summary of PBF-LB of YSZ bioinert ceramics.

Powder feedstock	**Laser type**	**Scanning speed**	**Reference**
ATZ (Alumina (80 wt%) toughened zirconia (20 wt%))	Nd:YAG laser	200 mm/s	[36]
YSZ (ZrO_2-8Y_2O_3)	Nd:YAG laser	60-75 mm/s	[37]
Eutectic mixing ratio of 41.5 wt% YSZ and 58.5 wt% of Al_2O_3	CO_2 laser for pre-heating, Nd:YAG laser for sintering/melting	200 mm/s	[24]
ZrO_2 (90 wt%) + Y_2O_3 (10 wt%)	Nd:YAG laser	2-60 mm/s	[26]
ZYP30 (Zircar)	Fibre laser	1250-2000 mm/s	[38]
ZrO_2 – Al_2O_3 various blends	CO_2 laser for pre-heating, Nd:YAG laser for sintering/melting	200 mm/s	[27]

As Table 2 illustrates the majority of the studies focused on CO_2 and Nd:YAG laser for laser sintering/melting of YSZ. The limitation of zirconia concerning laser interaction is very limited due to the unstable thermal behavior of zirconia, which can induce crack formation and ultimately lead to the failure of the implant. One way of avoiding the cracks is introducing a dual laser system (CO_2 and Nd:YAG) operating at different wavelengths. The formation of the cracks can be avoided by heating the powder bed with a CO_2 laser operating at a longer wavelength and shorter frequency (CO_2 laser) and followed by melting the ceramic powder with Nd:YAG operating at a shorter wavelength and higher frequency to fabricate the bioceramic dense part without any cracks and superior strength.

3.3 Silicon nitride

Silicon nitride ceramics are admired for their corrosion resistance, hardness, fracture toughness, and wear performance. The lifetime of the oxide ceramics (Al_2O_3, YSZ) is limited due to its oxygen activity in the simulating body fluids (hydrothermal environment). However, on the contrary, Si_3N_4 has a high wear rate initially followed by the formation of hydroxylated silicon oxide will be formed as the primary product of the thermomechanical wear and resulting in decreasing friction level which is a mandate for high-load medical applications [39].

Si_3N_4 also complies well with biomedical applications such as dental and orthopedic implants/scaffolds. Si_3N_4 meets the satisfactory requirements of orthopedic surgery displaying several attractive features like low solubility in the aqueous solutions, high wear resistance, and proper mechanical properties for BTE. The other complemented properties with regards to BTE include anti-bacterial properties, biocompatibility, and excellent radiographic imaging, thus making them strong candidates to support high loads for deformed or mangled bone. The specific targets of Si_3N_4 as the biomaterial focuses on the locations like intervertebral spacers, spinal surgery, joint arthroplasty, otorhinolaryngology, dental, osteo-fixation system, and orthopedic scaffolds [40]. Table 3 shows the summary of the direct PBF-LB on Si_3N_4. As Table 3 shows scarce studies have been devoted to PBF-LB of Si_3N_4 ceramics since the binding mechanism type is called chemically induced binding which tends to dissociate Si_3N_4 ceramics.

Table 3. Summary of PBF-LB of Si_3N_4 bioinert ceramics.

Powder feedstock	Laser type	Scanning speed	Reference
Si_3N_4 hollow microspheres	CO_2 laser	150 mm/s	[41]
Si_3N_4 powder consolidation on pure titanium	Nd:YAG laser	pulse time 1-20 ms	[42]
Silicon powder as the powder bed and further nitriding the green part	Nd:YAG laser	80 mm/s	[43]

4. Manufacturing of bioinert ceramic scaffolds for critical-sized defects and drug delivery system through PBF-LB

Critical-sized bone defects, commonly with a length irrespective of the dimensions and usually in the range of 1-4.5 cm can be repaired by synthetic bioinert grafts/scaffolds. The scaffolds which can be implemented on bone defects encompass the femur, shoulder, hip, wrist, tibia, ankle, and spinal cord. Synthetic bioinert bone scaffolds play an imperative role in BTE manufactured by PBF-LB by repairing spinal defects, bone defects, femoral head fabrication, and heart valves by facilitating and providing a 3D network for regenerating the bone for critical defects. PBF-LB process can be also used as a DDS [44] as shown in Figure 5. In formulations, I and II, the drug and active biomolecules can be directly blended with a powder bed of ceramic powders. However, in these types of formulations, the active ingredient is directly subjected to the thermal energy imposed by laser which can degrade them or destroy the functionality of the drug or active biomolecules. In order to diverge from the aforementioned formulations, III and IV types of formulations can be implemented on the powder bed by necessitating the incorporation of the drug molecules by surface functionalization onto the bioinert scaffolds. Henceforth, PBF-LB process presents a very flexible approach to manufacturing orthobiologic materials as a mediator to deliver drug or active biomolecules. In recent studies, authors proved that manipulating the scan spacing can fine-tune drug release in the in-vitro conditions by PBF-LB process [45].

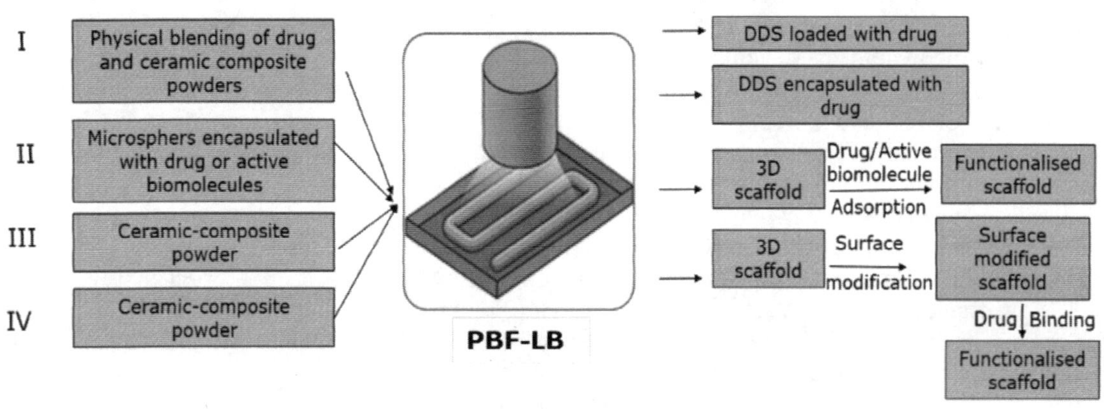

Figure 5. Schematic of developing DDS through PBF-LB process

The first class of synthetic bioinert scaffolds which can affect the functionality PBF-LB allows custom-designed bioinert scaffolds suitable to the defect by having superior control over pore size, porosity, and surface roughness. The second class of synthetic bone scaffolds manufactured by PBF-LB can also be integrated with active biomolecules to have DDS directly on the powder bed. The second class of bioinert scaffolds is mainly used for cervical and lumbar spine, wrist, ankle, and grim fractures of the femur and tibia, where the fracture size is between 2 to 4 cm. The third and the last class of bioinert ceramic scaffolds are usually implemented in the femoral neck fracture of the hip joints and joint arthroplasty by imparting functionality through stem cells by embodying onto the scaffold as shown in Figure 6.

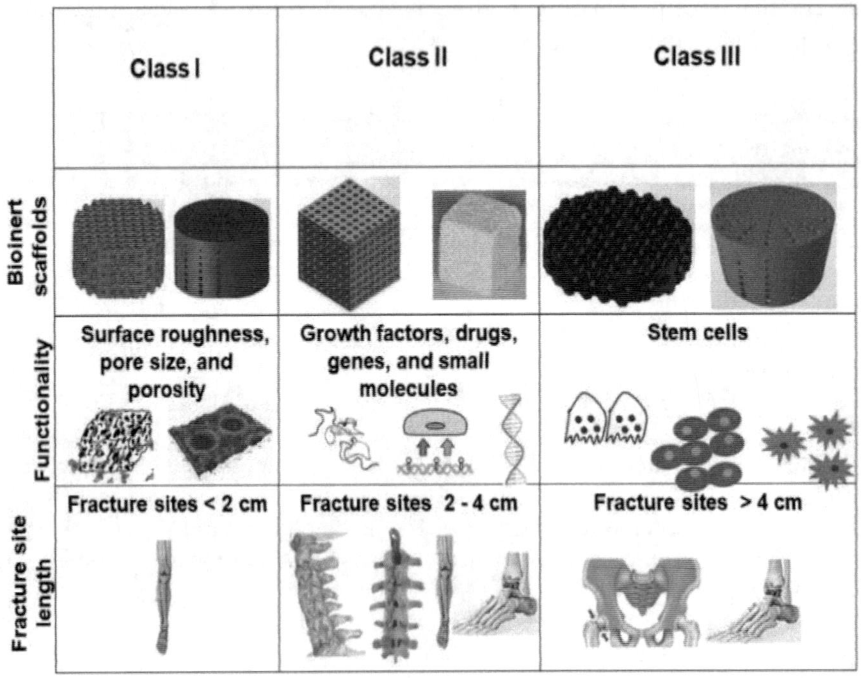

Figure 6. Three major classes to enhance the functionality of the bioinert ceramic scaffolds for bone repair by the PBF-LB process, adapted from [3]

5. Discussion and potential developments

Ideally, it is a very strenuous task to fabricate an ideal scaffold, which can fit both the dense and the porous parts of the bone. One solution can be to fabricate a triple periodic minimal surface (TPMS) by improving the architectural design through ceramic materials to withstand mechanical loads and significantly increase the permeability of body fluids. The metal porous TPMS can be easily fabricated and are well known for BTE [46]. However, on the contrary, there is very limited progress and development of TPMS bioceramics surface, although there is a recent study existing in the literature. Furthermore, recently, ceramics powders are incorporated into metal lattices by consolidating through spark plasma sintering [47] for biomedical applications [48].

Usually, bioceramics are not well suited for shorter wavelengths and higher frequencies. A probable solution can be the mixing of the additive with a bioceramic powder bed. The additive itself should be good orthobiologic material that can coalesce actual ceramic particles in the powder bed to sinter or melt. For instance, silicon can act as an efficient absorber of the fiber laser to manufacture the scaffold for drug delivery [49]. The future recommendation of PBF-LB for ceramics can be focused on the customized machine laser designs and process parameters.

6. Conclusions

In this review, we presented the different ceramic materials used to produce bioinert scaffolds through PBF-LB. Alumina and YSZ are well versed with PBF-LB but on the contrary Si_3N_4 ceramics have to be studied more by the PBF-LB process. Si_3N_4 ceramics have a binding mechanism type called chemically induced binding which tends to dissociate Si_3N_4 ceramics. One solution to tackle this problem would be to have an over-pressure of nitrogen in the powder bed atmosphere in order to prevent the dissociation of Si_3N_4. Henceforth, this area can also be explored in further studies. Furthermore, the most used lasers for bioinert ceramics are CO_2 and Nd:YAG lasers with a very limited number of studies on fibre lasers. This can be attributed to the fact that fibre lasers are more expensive than the CO_2 and Nd:YAG lasers which are used in bioinert ceramics. Finally, it was presented that PBF-LB has the potential to fabricate scaffolds by incorporating drugs, and biomolecules straightway in the powder bed. The customized scaffolds fabricated through PBF-LB can be a solution for personalized medicine for bone repair along with gateways to DDS.

Credit authorship contribution statement

N Kamboj - Writing – review & editing, Visualization, Investigation, Answering reviewer comments. **H Piili, A Ganvir, A Gopaluni, C Nayak, N Motiz, A Salminen** – Editing, Visualization, Investigation.

Acknowledgment

This research was done as part of a project funded by the Academy of Finland with Grant Number 325003 (Recovery of gold from secondary resources by novel electrochemical reactors realized with additive manufacturing, ReGold-AM). The project started on 01/09/2019 and will end on 31/08/2023. Project partners are the University of Turku and LUT University. The authors would like to thank the project partners and their assistance in enabling conducting of this study. The authors would like to thank also to the staff of the Department of Mechanical Engineering in the Faculty of Technology at the University of Turku for their knowledge and help to be able to execute this study. This work was also supported by the Estonian Research Council grant PUTJD1169.

References

[1] Kulkarni S V., Nemade AC, Sonawwanay PD. An Overview on Metallic and Ceramic Biomaterials, 2022, p. 149–65. https://doi.org/10.1007/978-981-16-7787-8_11.

[2] Rezapourian M, Kamboj N, Jasiuk I, Hussainova I. Biomimetic design of implants for long bone critical-sized defects. J Mech Behav Biomed Mater 2022;134:105370. https://doi.org/10.1016/j.jmbbm.2022.105370.

[3] Kamboj N, Ressler A, Hussainova I. Bioactive Ceramic Scaffolds for Bone Tissue Engineering by Powder Bed Selective Laser Processing: A Review 2021. https://doi.org/10.3390/ma14185338.

[4] S. A. Bento C, Gaspar MC, Coimbra P, de Sousa HC, E. M. Braga M. A review of conventional and emerging technologies for hydrogels sterilization. Int J Pharm 2023;634:122671. https://doi.org/10.1016/j.ijpharm.2023.122671.

[5] Kamboj N, Aghayan M, Rodrigo-Vazquez CS, Rodríguez MA, Hussainova I. Novel silicon-wollastonite based scaffolds for bone tissue engineering produced by selective laser melting. Ceram Int 2019;45:24691–701. https://doi.org/10.1016/j.ceramint.2019.08.208.

[6] Kamboj N, Kazantseva J, Rahmani R, Rodríguez MA, Hussainova I. Selective laser sintered bio-inspired silicon-wollastonite scaffolds for bone tissue engineering. Materials Science and Engineering: C 2020;116:111223. https://doi.org/10.1016/j.msec.2020.111223.

[7] Ressler A, Žužić A, Ivanišević I, Kamboj N, Ivanković H. Ionic substituted hydroxyapatite for bone regeneration applications: A review. Open Ceramics 2021;6:100122. https://doi.org/10.1016/j.oceram.2021.100122.

[8] Buj-Corral I, Tejo-Otero A. 3D Printing of Bioinert Oxide Ceramics for Medical Applications. J Funct Biomater 2022;13:155. https://doi.org/10.3390/jfb13030155.

[9] Ressler A, Kamboj N, Ledinski M, Rogina A, Urlić I, Hussainova I, et al. Macroporous silicon-wollastonite scaffold with Sr/Se/Zn/Mg-substituted hydroxyapatite/chitosan hydrogel. Open Ceramics 2022;12:100306. https://doi.org/10.1016/j.oceram.2022.100306.

[10] Grossin D, Montón A, Navarrete-Segado P, Özmen E, Urruth G, Maury F, et al. A review of additive manufacturing of ceramics by powder bed selective laser processing (sintering / melting): Calcium phosphate, silicon carbide, zirconia, alumina, and their composites. Open Ceramics 2021;5:100073. https://doi.org/10.1016/j.oceram.2021.100073.

[11] Rodrigo-Vázquez CS, Kamboj N, Aghayan M, Sáez A, Aza AH De, Rodríguez MA, et al. Manufacturing of silicon – Bioactive glass scaffolds by selective laser melting for bone tissue engineering. Ceram Int 2020. https://doi.org/10.1016/j.ceramint.2020.07.171.

[12] Ressler A, Kamboj N, Ivanković H, Hussainova I. Optimisation of trabecular bone mimicking silicon-hydroxyapatite based composite scaffolds processed through selective laser melting. Open Ceramics 2022;10:100252. https://doi.org/10.1016/j.oceram.2022.100252.

[13] Ly M, Spinelli S, Hays S, Zhu D. 3D Printing of Ceramic Biomaterials. Engineered Regeneration 2022;3:41–52. https://doi.org/10.1016/j.engreg.2022.01.006.

[14] Awad A, Fina F, Goyanes A, Gaisford S, Basit AW. 3D printing: Principles and pharmaceutical applications of selective laser sintering. Int J Pharm 2020;586:119594. https://doi.org/10.1016/j.ijpharm.2020.119594.

[15] Zeng P, Rainforth WM, Stewart TD. Characterisation of the wear mechanisms in retrieved alumina-on-alumina total hip replacements. Wear 2017;376–377:212–22. https://doi.org/10.1016/j.wear.2016.11.045.

[16] Bose S, Darsell J, Hosick HL, Yang L, Sarkar DK, Bandyopadhyay A. Processing and characterization of porous alumina scaffolds. J Mater Sci Mater Med 2002;13:23–8. https://doi.org/https://doi.org/10.1023/A:1013622216071.

[17] Cogan A, Nizard R, Sedel L. Occurrence of noise in alumina-on-alumina total hip arthroplasty. A survey on 284 consecutive hips. Orthopaedics & Traumatology: Surgery & Research 2011;97:206–10. https://doi.org/10.1016/j.otsr.2010.11.008.

[18] Hamadouche M, Boutin P, Daussange J, Bolander ME, Sedel L. Alumina-on-Alumina Total Hip Arthroplasty: A Minimum 18.5-Year Follow-up Study. J Bone Joint Surg 2002;84:69–77. https://doi.org/10.2106/00004623-200201000-00011.

[19] Caravaca CF, Flamant Q, Anglada M, Gremillard L, Chevalier J. Impact of sandblasting on the mechanical properties and aging resistance of alumina and zirconia based ceramics. J Eur Ceram Soc 2018;38:915–25. https://doi.org/10.1016/j.jeurceramsoc.2017.10.050.

[20] Khang D, Liu-Snyder P, Pareta R, Lu J, Webster TJ. Reduced responses of macrophages on nanometer surface features of altered alumina crystalline phases. Acta Biomater 2009;5:1425–32. https://doi.org/10.1016/j.actbio.2009.01.031.

[21] Rahmati M, Mozafari M. Biocompatibility of alumina-based biomaterials–A review. J Cell Physiol 2019;234:3321–35. https://doi.org/10.1002/jcp.27292.

[22] Subramanian PK, Marcus HL. Selective Laser Sintering of Alumina Using Aluminum Binder. Materials and Manufacturing Processes 1995;10:689–706. https://doi.org/10.1080/10426919508935060.

[23] Lee IS. Full densed alumina-glass composites by SLS and ceracon forging process. Metals and Materials 1997;3:203–9. https://doi.org/10.1007/BF03025964.

[24] Yves-Christian H, Jan W, Wilhelm M, Konrad W, Reinhart P. Net shaped high performance oxide ceramic parts by selective laser melting. Phys Procedia 2010;5:587–94. https://doi.org/10.1016/j.phpro.2010.08.086.

[25] Deckers J, Kruth J-P, Shahzad K, Vleugels J. Density improvement of alumina parts produced through selective laser sintering of alumina-polyamide composite powder. CIRP Annals 2012;61:211–4. https://doi.org/10.1016/j.cirp.2012.03.032.

[26] Shishkovsky I, Yadroitsev I, Bertrand P, Smurov I. Alumina–zirconium ceramics synthesis by selective laser sintering/melting. Appl Surf Sci 2007;254:966–70. https://doi.org/10.1016/j.apsusc.2007.09.001.

[27] Wilkes J, Hagedorn Y, Meiners W, Wissenbach K. Additive manufacturing of ZrO 2 -Al 2 O 3 ceramic components by selective laser melting. Rapid Prototyp J 2013;19:51–7. https://doi.org/10.1108/13552541311292736.

[28] Fan Z, Lu M, Huang H. Selective laser melting of alumina: A single track study. Ceram Int 2018;44:9484–93. https://doi.org/10.1016/j.ceramint.2018.02.166.

[29] Zheng Y, Zhang K, Liu TT, Liao WH, Zhang CD, Shao H. Cracks of alumina ceramics by selective laser melting. Ceram Int 2019;45:175–84. https://doi.org/10.1016/j.ceramint.2018.09.149.

[30] Qiu Y-D, Wu J-M, Chen A-N, Chen P, Yang Y, Liu R-Z, et al. Balling phenomenon and cracks in alumina ceramics prepared by direct selective laser melting assisted with pressure treatment. Ceram Int 2020;46:13854–61. https://doi.org/10.1016/j.ceramint.2020.02.178.

[31] Yin L, Peng HX, Yang L, Su B. Fabrication of three-dimensional inter-connective porous ceramics via ceramic green machining and bonding. J Eur Ceram Soc 2008;28:531–7. https://doi.org/10.1016/j.jeurceramsoc.2007.07.006.

[32] Leib EW, Vainio U, Pasquarelli RM, Kus J, Czaschke C, Walter N, et al. Synthesis and thermal stability of zirconia and yttria-stabilized zirconia microspheres. J Colloid Interface Sci 2015;448:582–92. https://doi.org/10.1016/j.jcis.2015.02.049.

[33] DENRY I, KELLY J. State of the art of zirconia for dental applications. Dental Materials 2008;24:299–307. https://doi.org/10.1016/j.dental.2007.05.007.

[34] Roy ME, Whiteside LA, Katerberg BJ, Steiger JA, Nayfeh T. Not All Zirconia Femoral Heads Degrade In Vivo. Clin Orthop Relat Res 2007;465:220–6. https://doi.org/10.1097/BLO.0b013e318158b4d3.

[35] Chevalier J, Gremillard L, Virkar A V, Clarke DR. The Tetragonal-Monoclinic Transformation in Zirconia: Lessons Learned and Future Trends. Journal of the American Ceramic Society 2009;92:1901–20. https://doi.org/10.1111/j.1551-2916.2009.03278.x.

[36] Koopmann J, Voigt J, Niendorf T. Additive manufacturing of a steel–ceramic multi-material by selective laser melting. Metallurgical and Materials Transactions B 2019;50:1042–51.

[37] Ferrage L, Bertrand G, Lenormand P. Dense yttria-stabilized zirconia obtained by direct selective laser sintering. Addit Manuf 2018;21:472–8. https://doi.org/10.1016/j.addma.2018.02.005.

[38] Bertrand P, Bayle F, Combe C, Goeuriot P, Smurov I. Ceramic components manufacturing by selective laser sintering. Appl Surf Sci 2007;254:989–92. https://doi.org/10.1016/j.apsusc.2007.08.085.

[39] Du X, Lee SS, Blugan G, Ferguson SJ. Silicon Nitride as a Biomedical Material: An Overview. Int J Mol Sci 2022;23:6551. https://doi.org/10.3390/ijms23126551.

[40] Sainz MA, Serena S, Belmonte M, Miranzo P, Osendi MI. Protein adsorption and in vitro behavior of additively manufactured 3D-silicon nitride scaffolds intended for bone tissue engineering. Materials Science and Engineering: C 2020;115:110734. https://doi.org/10.1016/j.msec.2020.110734.

[41] Wei Z-H, Cheng L-J, Ma Y-X, Chen A-N, Guo X-F, Wu J-M, et al. Direct fabrication mechanism of pre-sintered Si3N4 ceramic with ultra-high porosity by laser additive manufacturing. Scr Mater 2019;173:91–5. https://doi.org/10.1016/j.scriptamat.2019.07.046.

[42] Zanocco M, Boschetto F, Zhu W, Marin E, McEntire BJ, Bal BS, et al. 3D-additive deposition of an antibacterial and osteogenic silicon nitride coating on orthopaedic titanium substrate. J Mech Behav Biomed Mater 2020;103:103557. https://doi.org/10.1016/j.jmbbm.2019.103557.

[43] Minasyan T, Liu L, Aghayan M, Kollo L, Kamboj N, Aydinyan S, et al. A novel approach to fabricate Si3N4 by selective laser melting. Ceram Int 2018. https://doi.org/10.1016/j.ceramint.2018.04.208.

[44] Mohammed A, Elshaer A, Sareh P, Elsayed M, Hassanin H. Additive Manufacturing Technologies for Drug Delivery Applications. Int J Pharm 2020;580:119245. https://doi.org/10.1016/j.ijpharm.2020.119245.

[45] Awad A, Fina F, Goyanes A, Gaisford S, Basit AW. Advances in powder bed fusion 3D printing in drug delivery and healthcare. Adv Drug Deliv Rev 2021;174:406–24. https://doi.org/10.1016/j.addr.2021.04.025.

[46] Rezapourian M, Kamboj N, Hussainova I. Numerical study on the effect of geometry on mechanical behavior of triply periodic minimal surfaces. IOP Conf Ser Mater Sci Eng, vol. 1140, IOP Publishing; 2021, p. 12038.

[47] Rahmani R, Karimi J, Kamboj N, Kumar R, Brojan M, Tchórz A, et al. Fabrication of localized diamond-filled copper structures via selective laser melting and spark plasma sintering. Diam Relat Mater 2023;136:109916. https://doi.org/10.1016/j.diamond.2023.109916.

[48] Rahmani R, Kamboj N, Brojan M, Antonov M, Prashanth KG. Hybrid metal-ceramic biomaterials fabricated through powder bed fusion and powder metallurgy for improved impact resistance of craniofacial implants. Materialia (Oxf) 2022;24:101465. https://doi.org/10.1016/j.mtla.2022.101465.

[49] Kamboj N, Rodríguez MA, Rahmani R, Prashanth KG, Hussainova I. Bioceramic scaffolds by additive manufacturing for controlled delivery of the antibiotic vancomycin. Proceedings of the Estonian Academy of Sciences 2019;68:185. https://doi.org/10.3176/proc.2019.2.10.

IOP Conf. Series: Materials Science and Engineering 1296 (2023) 012023 doi:10.1088/1757-899X/1296/1/012023

Laser polishing of PBF-LB manufactured stainless steel surfaces

J Karlsson[1], I Väisänen[2], E Tourunen[3], A Piiroinen[2], A Gopaluni[2], A Salminen[2]

1 Department of Technology and Business, School of Machine Technology, Turku University of Applied Sciences, Turku, 20520, Finland
2 Department of Mechanical and Materials Engineering, Faculty of Technology, University of Turku, Turku, 20520, Finland
3 Department of Mechanical Engineering, Faculty of Technology, University of Turku, Turku, 20520, Finland

jussi.karlsson@turkuamk.fi

Abstract. Laser powder bed fusion (PBF-LB) is additive manufacturing where material is added layer by layer. This technology has revolutionized the manufacturing during last few recent years by enabling more efficient material usage and more functions to each component. However, for many industrial cases, the PBF-LB surfaces need an additional finishing operation like the machining or polishing operation. Laser polishing process was studied to solve the problems of traditional technologies to improve the surface quality and to enable automation of it. In this study parts were manufactured by PBF-LB of austenitic stainless steel (AISI 316L) into shapes of mechanical property testing samples in different orientations and locations in building platform. The surface roughness was measured prior and after polishing. The experiments showed that considerable surface smoothening even up to 85% (Sa from 48 μm to 11 μm) was reached in two treatments.

1. Introduction

Laser powder bed fusion is additive manufacturing where material is added layer by layer. The metal powder is fused according to 3d-model sliced to thin layers on powder layers. This technology has revolutionized the manufacturing during last few recent years by enabling more efficient material usage and more functions to each component. Even though this technology is growing in use gradually there is some major drawbacks slowing down the development. (ISO/ASTM 52900:2021)
However, as commonly known by engineering practices and studies there are a lot of applications where the surface quality, e.g. due to fatigue, aesthetic or tribological functions; accuracy or low roughness requirements, requirements are such that the L-PBF manufactured surfaces need an additional finishing operation like machining or polishing operation. Traditional polishing processes usually are based on abrasive or chemical techniques. These technologies have several drawbacks like dimensional constraints in case of complex shape or variations in material thickness, high environmental impact, long processing time and high cost. Typically, these also involve issues with awkward work phase with health risks for operators, which cannot be solved due to difficulties in automation of the process.
The research on laser polishing has started in early 2000 and has been growing on volume during last few years by Scopus. Typically, it is used for smoothening the surface of metals, polymers, ceramics and glasses, so the technology is pretty versatile. Naturally different materials require different

technologies and the process itself can be carried out with some different techniques e.g. via melting or via ablation. The process can be carried out with various different lasers reaching form CO_2 and Nd:YAG lasers to currently more common fiber and disk lasers.
Even though the process is not brand new the research on topic has increased considerably during laser years. A Scopus search in May 15th 2023 with search text: "laser polishing" AND "metals" and "laser polishing" AND "stainless steel", see Figure 1, found totally 117 and 47 publications, since 1994 when first publications were published.

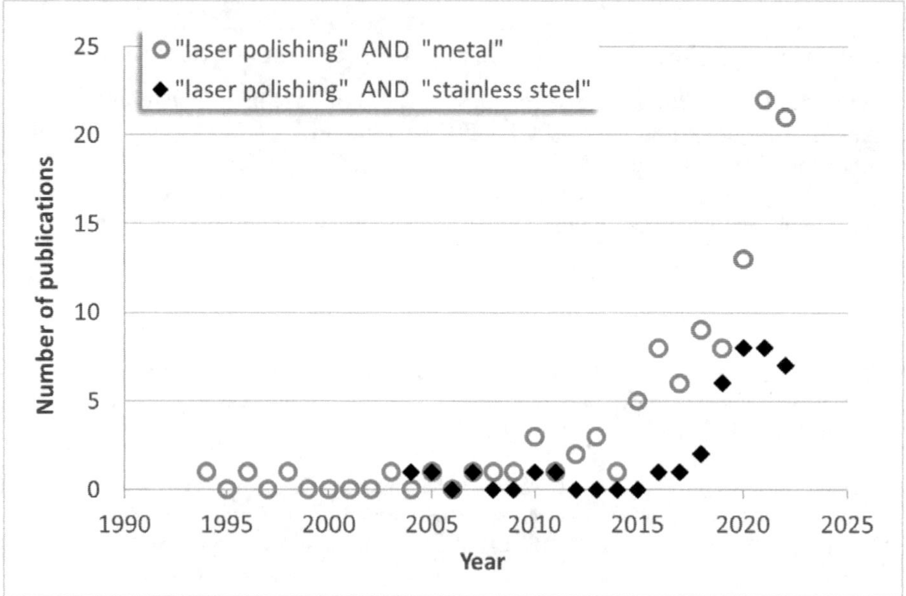

Figure 1. Number of publications per annum by Scopus for searches "laser polishing" AND "metals" and "laser polishing" AND "stainless steel".

The process can be done both with continuous wave or pulsed laser beam. Laser polishing for metals is typically based on melting a thin surface layer of the workpiece followed with the surface smoothening by interfacial tension. In this case beam is using the existing material of "hills" to fill the "valleys" in the surface texture. Figure 2 shows the principle of laser polishing in practice, whereas figure 3 shows typical result of laser polishing.

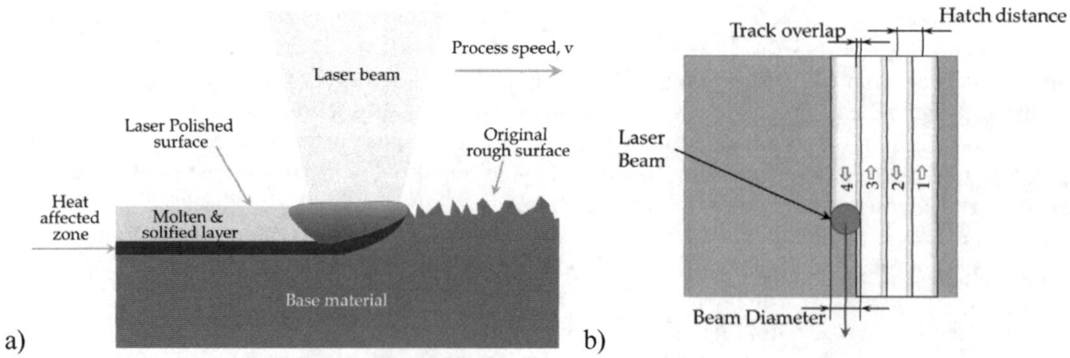

Figure 2. Laser polishing principle a) side view and b) top view.

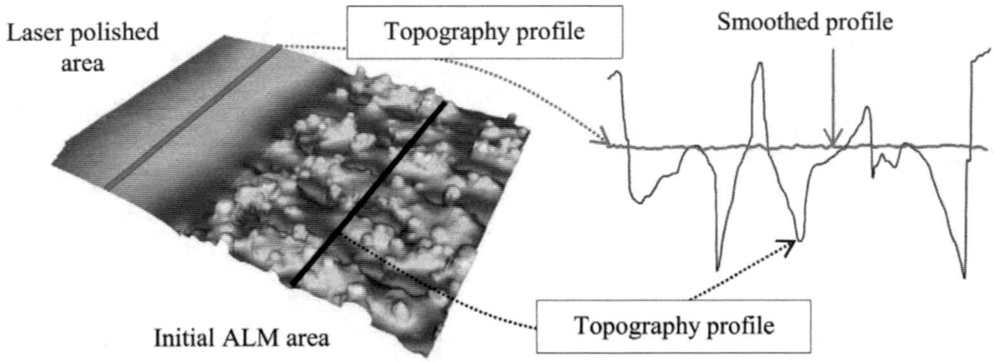

Figure 3. Laser polishing is melting the material which is then smoothened by surface tension and leaves a totally different surface roughness. [1]

The technology has actually been used also for deburring of parts after machining operations and rounding and deburring of sheet metal parts [2]. In these cases, naturally there is usually not need to do more than one track. There are some results reported of polishing of austenitic stainless steel of AISI 316L standard. Table 1 has collection some of them. As from table 1 shows one problem in comparison is formed by the type of roughness informed. Ra is conventional line roughness measurement procedure, but it will bring out different results as Sa which is roughens of an area. In any case typically the polishing is done with solid state laser fiber or Nd:YAG, which will result typically to roughness decrease of 84.6 – 96.2%. The only results with CO_2 laser show the lowest improvement which still is of quite high value.

Table 1. Results from literatures of polishing experiments for AISI 316L type stainless steel manufactured with PBF-LB process. Unfortunately, the references did not follow rules of accuracy.

Ref.	Material	Laser type		Ra µm	Ra µm	Sa µm	Sa µm	St µm	St µm	Roughness decrease
1	316 L	Fiber	CW			21.0	0.79			96.2 %
1	316 L	Fiber	CW			18.0	2.34			87.0 %
3	316 L	Fiber	pulsed					9.77	1.5	84.6 %
6	316 L	Fiber	CW	10.0	1.4					86.0 %
7	316 L	Nd:YAG	pulsed	12.0	1.5					87.5 %
8	316 L	Fiber	pulsed	4.75	0.4					91.6 %
9	316 L	CO_2	CW	10.4	2.7					74.0 %

A pulsed laser can be used also for surface polishing. This makes process a bit more complicated since together with track overlap the pulse overlap must also be considered. However, the process seems to work pretty well e.g. according to Bhaduri et al., 2017[3] who used MOPA-based Yb-doped nanosecond (ns) fiber laser source, maximum average power 50 W and wavelength of 1064 nm. They polished austenitic stainless steel SS316L parts made with binder jetting process called Digital Metal®(DM). In this study it was found that the energy density and pulse overlap along are together with the beam scanning direction the most important factors in laser polishing. A maximum reduction in areal surface roughness of over 94% was achieved at the optimized LP settings. [3]
Ability of laser polishing to treat selected locations of part is presented e.g. by Bhaduri et al., who presented results of polishing specific locations of a machine component [3]. They improved the surface roughness from 2.41 µm to 0.24 µm. The energy input plays an important role in polishing process such that lower energy input result in insufficient melting while that beyond the optimized value the higher energy input led to ablation and surface over-melting of material in case of pulsed laser. [3]

IOP Conf. Series: Materials Science and Engineering 1296 (2023) 012023 doi:10.1088/1757-899X/1296/1/012023

The aim of this study is to find out whether can laser polishing is suitable for post-processing to improve surface quality to level required by engineering applications. The second aim was to study can normal high-power multimode laser be used for polishing. Laser polishing process has potential to solve the problems of mechanical tools in improvement of surface quality and finally to enable automation of it. Automation would be the final goal and laser beam as non-wearing tool, no force process and flexibility in shape and form can be much more flexible tool and set less restrictions to product design than e.g. machining of grinding.

2. Experimental procedure

In this study the test samples were manufactured by PBF-LB/M/ of austenitic stainless steel (AISI 316L) as a sheet in 45-degree angle to the building platform. The sheet was the cut with laser beam into samples of suitable size for testing. The surface roughness of the samples was measured prior an after the laser polishing experiment with optical measuring system. Polishing was carried out with fiber laser with different process parameters. There were three sub goals for the experimental procedure:

A. study the effect of laser power (B1 vs B2 vs B4)
B. study the effect of energy input at various speed and various laser power (B1 vs B2 vs B3 vs B4)
C. study the effect of number of treatments on the polishing result (A5 vs B1)

The material used was EOS StainlessSteel 316L (according to UNS S31673 standard) material, with general powder particle size distribution of 20 – 65 µm [9]. The nominal analysis of material is presented in table 2. The printing was carried out with EOS290M powder bed fusion printer with EOS standard parameter values for this material (316L_040_FlexM291_1.).

Table 2. Powder chemical nominal composition (wt.-%)

Cr	Ni	Mo	C	N	Fe
17.00 - 19.00	13.00 - 15.00	2.25 - 3.00	- 0.03	- 0.10	bal.

2.1. The polishing procedure

The polishing experiments were performed with high power laser system typically used for e.g. welding and hybrid welding experiments. This was intentionally selected laser, since the availability of such a laser is good and the beam intensity profile, top hat, would provide quite even heat input to whole process zone. The laser used can be tuned to use only 1-2 modules thus the low laser power used was attainable.

- Laser: IPG YLS 10000 CW fiber laser
- Wavelength: 1070 nm
- Processing fiber core diameter: 400µm
- Optical processing head: Precitec YW52
- Collimation length: 200 mm
- Focal length of focusing: 300 mm
- Beam intensity profile: top hat
- Beam diameter at focal point: 0.8 mm.
- Focal point position: on the sample surface.

The experiments were performed with Yaskava / Motoman UP50N Industrial robot. With this movement systems the constant processing speed for polishing experiments reached by starting process on top of the masking plate such that in beginning while system was accelerating the beam did not reach the test sample.

The laser power used was varied between 50 – 200W (to avoid too high heat input on lower speeds, only low power levels were used). The processing speeds used were between 20 – 130 mm/s (defined by previous experience to be suitable for robot use). Most of experiments were carried out with two round treatment rounds. Table 3 summarizes the used parameter value combinations in different experiments.

Table 3. The used parameter value combinations in experiments.

Experiment	laser power, W	Process speed, mm/s	Treatment rounds
A	50, 100, 160	20 - 70; interval 10	2
B	50, 100, 160	20 - 70; interval 10	2
	200	80 – 130; interval 10	2
C	100	20 - 70; interval 10	1/2

2.2. The surface roughness measurement procedure

The measurements have been made with a Bruker Alicona Infinite Focus G6 optical 3D surface quality measuring device, the software of which is MetMaX 2.1 and its LaboratoryMeasurementModule 9.2 software. When measuring the surface roughness of the printed, unpolished plates, a 5x magnification was used, because a reproducible result was not obtained from the image at a higher magnification. In the measurements of the polished surfaces, 10x magnification was used, which was the highest possible accuracy for the surface quality in question. The measurement processes were carried out according to the device manufacturer's instructions.

Surface roughness measurements were made on metal printed (PBF-LB/M/) pieces before polishing and after laser polishing. Since no established method has been defined for the implementation of the measurements, the profile (2D) and area (3D) surface roughness of all samples were measured. In 2D measurements of surface roughness, the measuring device complies with standards ISO 4287 and ISO 4288. In 3D surface roughness measurements of the area, the instructions of the ISO 25178 standard series were followed.

The optical measurement used is well suited for measuring the surface quality of metal products printed with the PBF-LB/M/ method. The surface of the printout is so rough that a traditional tactile measuring probe may stick. In addition, the printed surface has sharp valleys that the touching measuring probe cannot reach.

The measurement parameters are defined according to the device manufacturer's instructions. First, a sample of one image has been taken from the surface to be measured, from which a rough estimate of the Ra and Rz values of the surface has been obtained, and the structure of the surface structure has been determined from the profile of the surface. With these values, the vertical resolution (Vr) and lateral resolution (Lr) could be calculated for the measurement, the lens suitable for the measurement was confirmed, and the length of the measurement area (Lc). The factors resulting from the properties of the surface of the object to be measured affect the resolutions used in this study as follows:

$$Vr = \frac{Rz}{(15...30)}$$

where:
Vr = Vertical resolution
Rz = Average height difference between the highest peaks and the lowest valleys of the surface

The definition provides a range of Vr values suitable for the shape of the surface, from which a certain value is assigned to the measuring device.

$$Lr = \frac{Structural\ lenght}{10...40}$$

where:
Lr = lateral resolution
Structural length = defined from the profile measured from the sample

IOP Conf. Series: Materials Science and Engineering 1296 (2023) 012023 doi:10.1088/1757-899X/1296/1/012023

The definition provides a range of Lr values suitable for the shape of the surface, from which a certain value is assigned to the measuring device.

The definition of surface roughness according to the 2D standard is based on five measurement areas, and the length of individual measurement areas is defined in five categories according to surface roughness. In this study, the Ra values of the surface roughness were in the categories 2 – 10 μm and 10 – 80 μm, in which case the length of the measuring profile should be 12.5 mm (Lc=2500 μm) or 40.0 mm (Lc=8000 μm).

In these measurements, it was not possible to describe the measurement profile length (Lc) according to the standard, so the measurement profile was made as a continuous back and forth line on the surface described. Using this method, all measurements were made with standard Lc values.

3. Results and discussion

3.1. The effect of laser power on the polishing result

The effect of laser power on polishing was tested with three power levels (parameter values B) 50, 100 and 160 W. The processing speed with each power was varied from 20 to 70 mm/s at intervals of 10 mm/s. All samples were treated twice with same parameter set. After each polishing, the change in surface roughness was determined, in which the obtained surface roughness parameters after treatment were compared to the untreated surface. The graphs in figure 4 contain samples B1, B2 and B4 parameters ΔRa, ΔRz, ΔSa and ΔSz as functions of process speed. As the figure 4 shows typically, the decrease of roughness is decreased with the increase in speed with all power levels. Also, the decrease of roughness was also decreased with decrease in power. This leads to the conclusion that the selected power levels were quite good for first experiments since the optimum power range is probably in between the currently used power levels with this beam set-up.

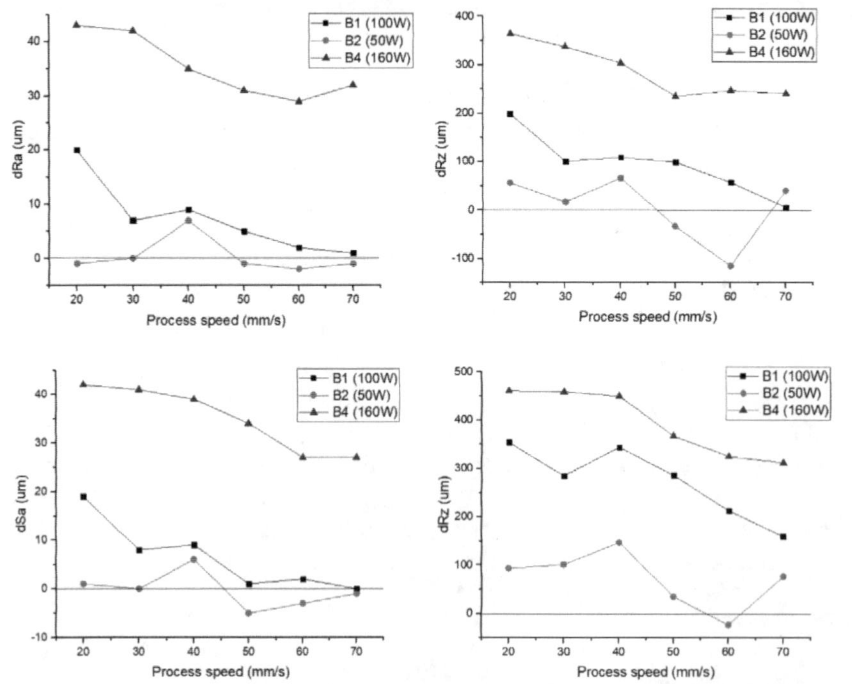

Figure 4. The effect of process speed and laser power for surface roughness values. Different surface roughness change values in function of the process speed with different laser polishing powers.

Based on the results, at a power of 50 W, the average surface roughness (Ra, Sa) does not significantly change at the speeds used. In contrast, at a power of 50 W, the average peak-to-valley roughness profile (Rz, Sz) has changed somewhat. It can be assumed that at this power and at these speeds, the highest peaks of the surface to be polished will be lowered without significant polishing taking place.

At higher powers, the changes in surface roughness are already significant. As the power increases and the process speed slows down, the polishing process improves as expected. It can also be seen that the Rz and Sz values change more strongly than the average surface roughness values Ra and Sa. Even in these cases the highest peaks melt and descend more easily.

3.2. The effect of energy input on polishing result

The heat input is calculated by dividing the laser power with process speed. Typically, this is suitable way to analyse the effect of parameter values on the process. In addition to the results of the previous experiment (50W, 100W and 160W), the evaluation of the effect of the energy input was performed with the parameter 200 W with each line varied speed from 80 to 130 mm/s at intervals of 10 mm/s. Each sample was measured prior and after the treatment. All samples had two treatment rounds on top of each other. The average of each prior roughness parameter was calculated to measure the difference between the prior and after values to represent the amount that the treatment affected the roughness. The graphs contain all the samples parameters ΔRa, ΔRz, ΔSa and ΔSz as functions of energy input. Naturally the values of Rz and Sz are always higher than the more common value in mechanical engineering the Ra and Sa. Typically for conventionally manufactured surfaces the Ra and Sa are suitable since the surface roughness is often quite even e.g. in after machining. See figure 5.

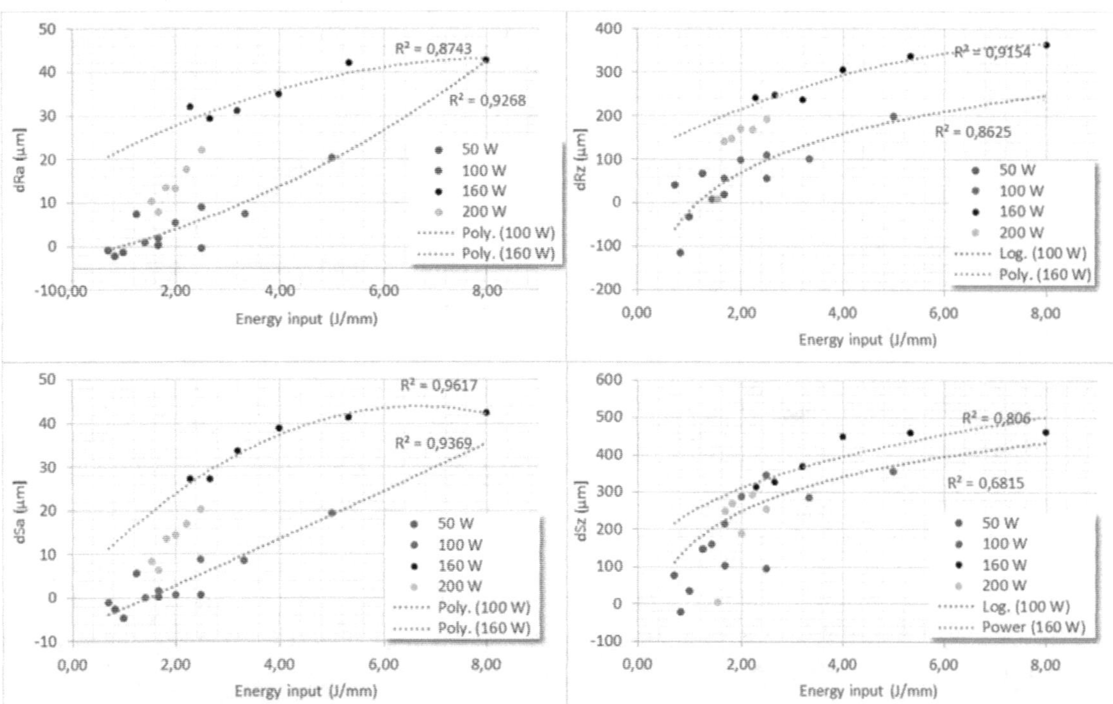

Figure 5. Different surface roughness change values in function of the energy input with different laser polishing powers and speed.

Based on the obtained results, it can be observed that at the lowest energy inputs used in the experiments, polishing did not occur regardless of the used laser power. With the energy inputs of the experiment 5 - 8 J/mm at 160 W laser power, the surface roughness changes were almost the same, i.e. increasing the energy input did not improve the obtained results. In contrast, with 100 W laser power, changes occurred in that range. Even with 200W laser power, changes in surface roughness were obtained with a lower energy input, but the results do not yet show an optimal energy input, whereby increasing the energy input no longer improves the change in surface roughness. The risk of increasing the energy input is the possible unwanted changes to the shape and properties of the piece. With a higher power of 200 W, the surface melts, even if the speed is increased. Based on the results, by increasing the laser power, polishing takes place with a lower energy input.

3.3. Effect of repeated treatment

To study the effect of carrying out several polishing rounds by second polishing treatment on top of first treatment two sample lines were selected for secondary treatment study. A5 and B1 sample lines were made with using the single laser power (100W). The processing speed with each line varied from 20 to 70 mm/s at intervals of 10 mm/s. The surface roughness of each sample was measured prior and after the treatment. Sample A5 was treated once and sample B1 was treated twice. The average of each prior roughness parameter was calculated to measure the difference between the prior and after values to represent how did the treatment affect to the roughness of the surface. The graphs in fig 6. contain both samples A5 and B1 parameters ΔRa, ΔRz, ΔSa and ΔSz as functions of process speed.

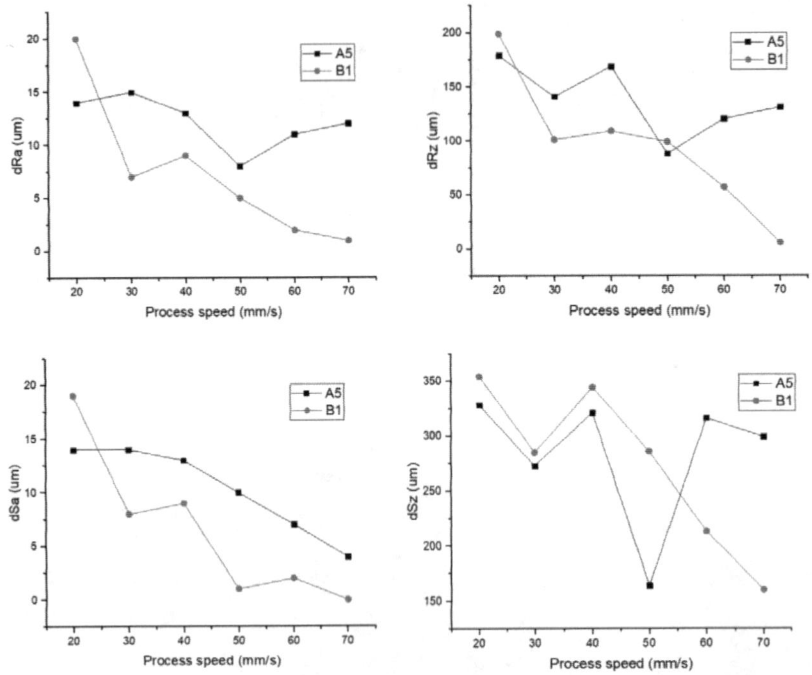

Figure 6. Different surface roughness change values in function of the process speed with different laser polishing powers after two polishing.

Based on the results, the changes in surface roughness measured at all values for both polishing cycles are significant at lower process speeds. As the process speed increases, the effect of the second polishing time ends. From the results, it can be concluded that during the first polishing, the highest peaks are smoothed out and at higher process speeds, the energy is not enough to melt larger surfaces.

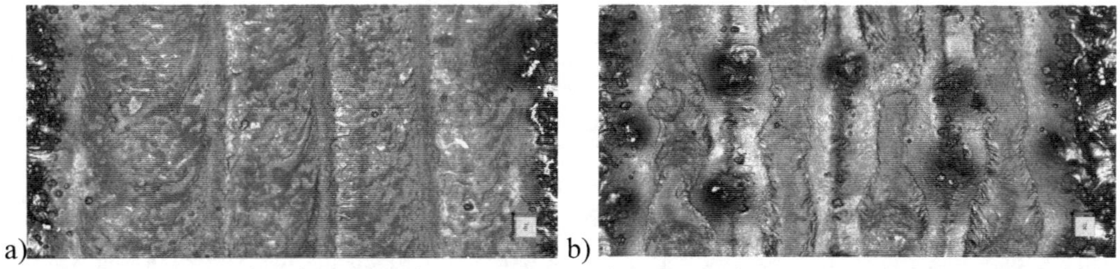

Figure 7. An example of LPF-LB/M manufactured surface after a) one polishing treatment (A5) and b) two polishing treatments (B2).

The results of two polishing rounds were interesting. The second polishing treatment made the surface rougher in some cases from the surface of first polishing treatment. From the images in Figure 7 a, the first polishing time created a smooth surface, but after the second polishing time, the surface has deep valleys (Figure 7 b). This is visible more clearly in Sz and Rz values, which are measuring the maximum difference in surface roughness and only with high, over 60 mm/s speed, the second round was smoothening the surface more. Also, in case of Sz which measures the area this phenomenon is more obvious. It can be assumed that the first polishing session has left pores under the layer, which will open with the second polishing session.

4. Conclusions

The experiments showed that considerable surface smoothening was reached in all part surface orientations and location. It was also found that this technology can provide a considerable improvement to surface finish of PBF-LB/M/ manufactured parts and that the technology is potential for automation.

The results clearly show that laser polishing improves surface quality when the power and speed of the process are suitable for polishing. There is a lot of scatters in the results, but the trend clearly shows the effect. The number of tests is too small to determine more precise polishing power, speed and other values.

Based on the obtained measurement results, measurements of the surface quality of PBF-LB metal prints should be made using optical measurement methods. The quality of the surface of the prints is so rough that you cannot get a reliable result with a touch measuring device, or you don't get a result at all.
In terms of evaluating the efficiency of the polishing process, it does not matter if the parameters of surface roughness Ra, Rz, Sa or Sz are used. All values or their changes can be used to describe the effect of the process on the surface quality. When a measuring device is available to measure the regional surface roughness parameters (Sa, Sz), then their use we recommended.

5. References

[1] Rosa B, Mognol P, Hascoët J-Y. Laser polishing of additive laser manufacturing surfaces. J. Laser Appl., Vol. 27, No. S2, February 2015
[2] Kumstel J, Lüken S. Deburring an dounding of sheet metals and functionally relevant edges of parts using laser radiation. MATEC Web of Conferences 190 02003 (2018). https://doi.org/10.1051/matecconf/201819002003
[3] Bhaduri D, Penchev P, Batal A, Dimov S, Leung Soo S, Sten S, Harrysson U, Zhang X, DongH. Laser polishing of 3D printed mesoscale components. Applied Surface Science 405 (2017) 29–46.
[4] Ermergen T, Taylan F. Review on Surface Quality Improvement of Additively Manufactured Metals by Laser Polishing. Arabian Journal for Science and Engineering (2021) 46:7125–7141. https://doi.org/10.1007/s13369-021-05658-9
[5] Alrbaey K, Wimpenny D, Tosi R, Manning W, Moroz A. On Optimization of Surface Roughness

of Selective Laser Melted Stainless Steel Parts: A Statistical Study. Journal of Materials Engineering and Performance, Volume 23(6) June 2014—2139. DOI: 10.1007/s11665-014-0993-9

[6] Yasa E, Kruth J-P, Deckers J. Manufacturing by combining Selective Laser Melting and Selective Laser Erosion/laser re-melting. CIRP Annals - Manufacturing Technology 60 (2011) 263–266. https://doi.org/10.1016/j.cirp.2011.03.063.

[7] Chen L, Richter B, Zhang X, Ren X Pfeffefkorn FE. Modification of surface characterictics and electrochemical corrosion behavior of laser powder bed fused stainless-steel 316L after laser polishing. Addit. Manuf. (2020). https:// doi. org/ 10. 1016/j. addma.2019. 101013

[8] Obeidi MA, McCarthy E, O'Connell B, Ul Ahad I, Brabazon D. Laser Polishing of Additive Manufactured 316L Stainless Steel Synthesized by Selective Laser Melting. Materials 2019, 12, 991. https://doi.org/10.3390/ma12060991

[9] https://www.eos.info/03_system-related-assets/material-related-contents/metal-materials-and-examples/metal-material-datasheet/stainlesssteel/material_datasheet_eos_stainlesssteel_316l_en_web.pdfMore references

Acknowledgements

Authors of this article acknowledge project of DREAMS (Database for Radically Enhancing Additive Manufacturing and Standardization). This project was funded by Business Finland and the duration of project is from April 2022 to February 2024. The Participant of project were the Finnish Additive Manufacturing Ecosystem (FAME). Aim of the project was to establish standards for AM and develop a powder material database for AM. Authors would like to thank partners of this project for their contribution to this article.

IOP Conf. Series: Materials Science and Engineering 1296 (2023) 012024 doi:10.1088/1757-899X/1296/1/012024

Additive manufacturing systems integration

J Heilala[1*], S Parchegani[1], H Piili[1]

[1] Department of Mechanical Engineering, Faculty of Technology, University of Turku, Turku, 20520, Finland
[*] janne.p.heilala@utu.fi

Abstract. This research explores real-time database systems' evolution, focusing on unique features and the addressed challenges. It examines the role of multi-material additive manufacturing quality domain databases in innovation and maintaining standards. It also looks at the challenges of implementing quality manufacturing systems from a technology, organization, and people of European Manufacturing research perspective. The study offers a new perspective on reconfigurable intelligent surfaces with multi-material additive manufacturing with system integration, discussing its applications and digital products' transformative potential. Enhancing multi-material additive manufacturing capabilities redefines the industries, creating a strong communication culture by adopting quality integration of quality and robotics while exploring the future of mathematics complexity in optimizing manufacturing education. The paper also explains the statistical classification of historical technology in manufacturing engineering education by flagging the platform's role of opportunities in secure research collaboration. The manufacturing horizontal is essential for effectively managing the multi-material additive manufacturing system through advanced technology to quality management integration. Certified advanced training and competency development econophysics show the multi-material additive manufacturing systems development influence on the production theories and mechanics of complex electronics. To enable, for example, communication links and cellural multi-material additive manufacturing integrations research for new technologies.

1. Introduction

The evolution of intelligent product development has been impacted by advancements in additive manufacturing (AM), a topic this study focuses on with a particular emphasis on artificial neural network algorithms developed for the analog neural networks via arithmetic-logic units [1] to be utilized physically. With AM at the forefront of modern manufacturing, effective quality management for maintaining standardized multi-material AM production is A requirement [2]. Understanding the importance of organizational integration, this research navigates its ties to econophysics that is in intersection of expanding statistical mechanics sciences. Computing physics could be involved in comparative operating systems performance constraints [3]. The mathematical modeling foregrounding of this study leverages data from the Finnish European Manufacturing Survey 2022 (EMS22) in Finland, a collective performance evaluation of firms based on perspectives concerning non-digitally or digitally offered product related services, digital elements, and cybersecurity [4]. Equally, virtual training and professional development are binding research organizations [5], this is important to note for studying quantum physics within manufacturing horizontal integrations. Research on spherical harmonic expansion, originate in geophysics [6], provides essential theoretical insights with applications extending from optics to additive manufacturing and laser fusion droplets into mechanics of feedstocks elements. The study emphasizes the interconnectedness of quality

IOP Conf. Series: Materials Science and Engineering 1296 (2023) 012024 doi:10.1088/1757-899X/1296/1/012024

management, organizational integration, and virtual training, all having roles in intelligent product development with AM. Embracing database-driven operations management and artificial neural network algorithms can strengthen the AM processes [8][9]. While in its historical cutting-edge research background, advanced manufacturing shows promising potential for future innovations, especially considering the implications of frequencies in various fields [10]. The growth of AM is being propagated by ML algorithms and efficient database management, harbingering a new generation of technological advancement in the field [11]. The practical application of these findings is evident in the new guidelines for data systems [12]. This research also addresses strategies for managing manufacturing data, considering its under-utilization [13]. Improvements in artificial neural network algorithms have led to significant risk assessment enhancements and resulted in redesigning system components, such as operation systems for sustainability [14]. The exploration of intelligent system simulation design on wireless network systems for multi-material additive manufactured products have a role, using the technology, organization processes, and people to artifact, user, task, organization, and situation framework in system design [15]. This study provides a foundation for future research in the AM industries, aiming to expand the boundaries of contextual review for advanced manufacturing.

2. Aim and purpose of this study

2.1. Motivation

The motivation for this study is to elaborate various physic constraints for manufacturing success. The mixed methods approach explores cybersecurity's impact on system integration competitiveness [16] by the open design. This integration encapsulates human systems from all societal bodies and support providers, leading to enterprise level physics constraint analyses. The aim is to understand how a integrated system can support manufacturing [17] through quantum level by quantifying a survey elements.

2.2. Research question

This paper examines specific variables in a theoretical manufacturing management framework using unexplored variables from EMS22 [4]. These variables offer a multidisciplinary perspective on factory structures. Employee numbers indicate factory size and demographic insights, while manufacturing capacity utilization reflects operational efficiency by showing how much of a firm's manufacturing physical production capacity is used. The return principle highlights profitability by comparing revenue to product structure costs. The study also contrasts non-digitized and digitized product-related services, indicating factory levels as a rare analogy to tectonophysics, showing insights from the Earth's modernizing construction processes. Digital maturity and cybersecurity effectiveness are analyzed, providing insights into a firm's digital and security tensor constants. The research formulates hypotheses to categorize companies using these variables and establishes a correlation matrix (R) to understand their relationships. Correlations (with a p-value above 0.05) could be considered insignificant and omitted in the Hessian matrix ∇, equally assuming that p is not changeable. This approach leads to an augmentation; how do we visualize matrice's ∇ significant variable relationships?

3. Methodology

3.1. Experimental set-up: Industry Surveying and Literature Review

This research is two-parted where first, conducted through a database (Scopus) search, synthesized findings from an industry EMS of Finnish factories relying upon basic assumption forming production theory in experimental econophysics, employed a multimethod-literature processing embedded into correlation model evaluation from EMS and evaluating it on based on learning from the broad surveying [18], and then conducted a literature review to support the industry's needs [19]. Attention was directed to fundamental factory smart systems physics. This included number of employees, manufacturing capacity utilization, to the extent of training physics system integration in factories' digital elements strategies [4]. Evaluation extended to non-digitally or digitally offered product related services and cybersecurity practices, highlighting the relevance of both digital and non-digital services for enhancing manufacturing digital changing spectral constants [4]. Parallelly, a (Scopus) search examined literature trends in general AM. Tables 1 and 2 show that themes like "Quality Management" and "Artificial Neural Network Algorithm" were widely explored. At the same time, "Organization Integration," "Smart Products" for intelligent product development, and "Smart Frameworks" received less attention, indicating potential areas for future research spectrum. A correlation was found between theme complexity and article quantity with topics for intelligent product development to be integrated together experimentally [20].

Table 1 shows the count of studies from the Scopus database related to each keyword combination, facilitating comparative analysis of different research areas within AM.

Used keywords	Number of publications													
	2023	2022	2021	2020	2019	2018	2017	2016	2015	2014	2013	2012	2011	2010
"additive manufacturing" OR "3d printing"	22791	52252	41051	32091	23601	16928	11455	7782	4502	2930	1418	836	543	437
"simulation" AND "additive manufacturing" OR "3d printing"	7616	17019	12838	9514	6643	4531	2974	1896	963	613	262	137	84	80
"artificial neural network algorithm" AND "additive manufacturing" OR "3d printing"	23	41	16	4	0	5	2	0	3	1	0	0	0	0
"quality management" AND "additive manufacturing" OR "3d printing"	232	560	366	267	182	105	54	45	23	14	7	3	1	3
"organization integration " AND "additive manufacturing" OR "3d printing"	17	44	0	1	1	20	20	12	7	5	2	1	0	0
"virtual training" AND "additive manufacturing" OR "3d printing"	8	27	17	16	11	6	2	5	1	0	0	0	0	0
"smart frameworks" AND "additive manufacturing" OR "3d printing"	2	8	6	10	1	4	3	1	1	0	0	0	0	0
"smart products" AND "additive manufacturing" OR "3d printing"	95	205	136	111	63	46	14	9	4	0	0	2	0	1

As seen in Table 1, an overview of the total number of articles for each keyword helps identify the most discussed topics within the field. Conversely, Table 2 offers a more narrowed focus, indicating the number of articles that discuss these keywords within a specific context, offering a more detailed breakdown of the thematic concentration within the literature for Quality.

Table 2 shows the count of studies narrowed to the recent articles of the Scopus database related to each keyword combination for precision.

Used keywords	Number of publications					
	2022	Relation*	2020	Relation*	2018	Relation*
"additive manufacturing" OR "3d printing"	52252	100,00	32091	100,00	16928	100,00
"simulation" AND "additive manufacturing" OR "3d printing"	17019	32,57	9514	29,65	4531	26,77
"artificial neural network algorithm" AND "additive manufacturing" OR "3d printing"	41	0,08	4	0,01	5	0,03
"quality management" AND "additive manufacturing" OR "3d printing"	560	1,07	267	0,83	105	0,62
"organization integration " AND "additive manufacturing" OR "3d printing"	44	0,08	1	0,00	20	0,12
"smart products" AND "additive manufacturing" OR "3d printing"	27	0,05	16	0,05	6	0,04

* Number of publications with keywords "additive manufacturing" OR "3d printing" are given value 100.00, and then number of publications with other keywords are compared with this resulting a comparison factor which is here mentiones as "Relation".

Table 2, refined from Table 1, shows that AM and related topics are increasingly prominent in research, indicating it as a relatively unexplored industry. There is a rising trend in academic articles on AM, highlighting a growing interest in this area. Figure 1 A illustrates the prevalence of specific keywords in AM literature, while Figure 1 B shows the dominance of certain themes in academic studies, emphasizing the significance of research in uncovering gaps to reach to the topmost to close the gap. As the AM industry evolves, research is expected to deepen, fostering technological advancements that enhance manufacturing efficiency and customization, leading to manufacturing success. The research also highlights a notable focus on the convergence of artificial neural networks, simulations, quality management, organizational integration, and smart – frameworks in AM.

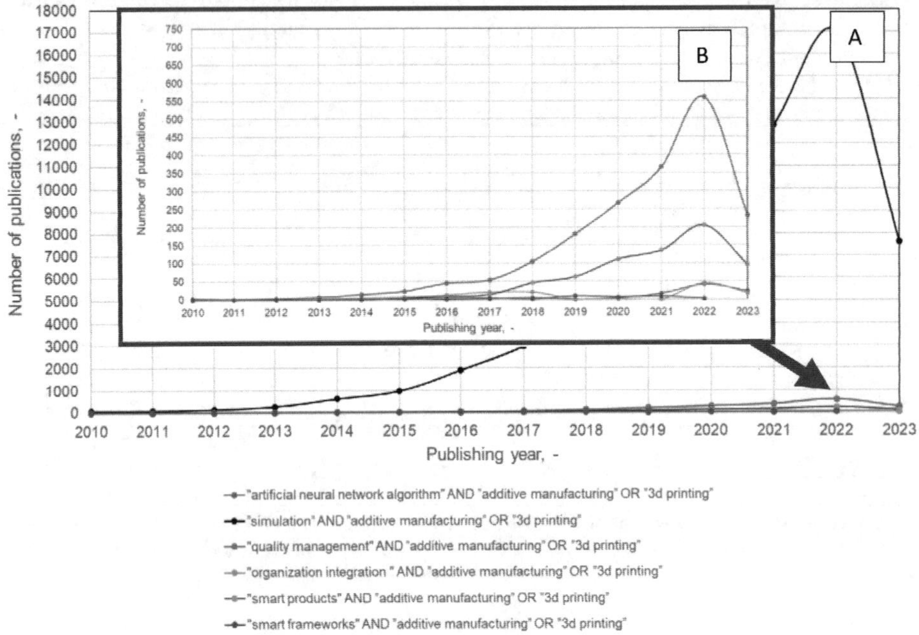

Figure 1: Distribution of (Scopus) studies based on keyword combinations; this figure consists of two graphs depicting research distribution using specific keyword combinations. Graph A displays the annual frequency of studies on topics like artificial neural network algorithms, simulations, and quality management in the AM context. Graph B, positioned above Graph A, offers a related distribution within Graph A's coverage area, with its details only clear through specific scoping and scaling.

3.2. Experimental analysis procedure within Finnish manufacturing

3.2.1. Total human system integration from elements of EMS and embedded correlation

This methodology, influenced by mathematics to gain the matrice's ∇, integrates human factors engineering from smart frameworks as outlined in [21]. It employs a unique approach, the Ergonomics, Axiomatic Design, and Complexity theory (EAC) from [22], for simplifying complex systems. Research gaps identified in [23] and [24] challenge the completeness of EAC theory. This approach has been successfully applied in the partially ∇ organizations, as seen incorporating independence and information axioms [22]. A matrix, developed using mathematical modeling [22], aligns function with design. The integration of these principles with the Naïve Bayes-based Theorem (NBT) selective ML to the prior probabilities from inspired [25] to the grounding theories [26] ensures that the design objectives are achieved [22]. For system integration, a variety of sustainability techniques are utilized.

Data presented in Figure 2 shows matrice's ∇ correlations between safety variables, product performance, and service security. Positive correlations are observed among the number of employees, manufacturing capacity utilization, non-digitally offered product-related services, and digital elements. Figure 2 also econophysics constants suggests that while institutions may lack robust cybersecurity physics, presenting collaboration opportunities, they compensate with fusion process of digital elements and product-related services. This tends to increase manufacturing capacity utilization and aids in intelligent product development. Contrarily, digitally offered product-related services show negative correlation, except with digital elements with factory scaler. The uniformity in cybersecurity practices across companies' physics statistical variabilities is hypothetically static.

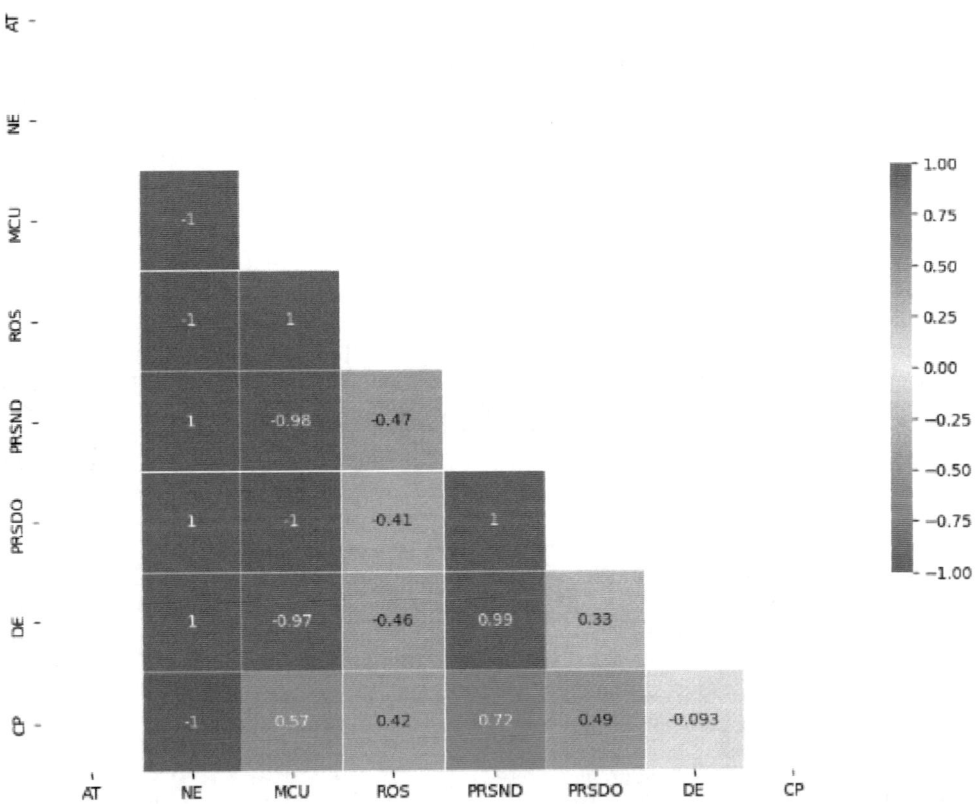

Figure 2. presents a weighted correlogram of EMS respondents, displaying cross-correlations of various factors: AT (annual turnover), NE (number of employees), MCU (manufacturing capacity utilization), ROS (return on sales), PRSND/PRSDO (product-related services not offered/offered digitally), DE (digital elements), and CP (cybersecurity practices), using data from 2019-2021. It shows alike squared tectonophysics plates of manufacturing context tensors. The visualization was created using immersive Python augmentation with its excellent modules [20].

3.2.2. Building a framework for intelligent production management

The growth of intelligent product development and frameworks using AM has notably accelerated, spanning applications such as antenna structures corresponding electromagnetics in dependence to the components exclusively designed to be additively manufactured [27]; [28]. Various AM methods have been utilized for electromagnetic structures steering, using materials like filament, resin, powder, and nanoparticle inks adaptable to smart composites [27], characterized by lightness and ease of use. AM has evolved from a prototyping method to a valuable tool for manufacturing end-use components, presenting unprecedented design freedom and capabilities. However, the advantages of AM currently extend beyond design possibilities, with operational benefits like improved delivery times and reduced manufacturing costs playing an integral role [28]. In optimizing flow-level understanding of systems development opportunities on AM, it is recommended (also by [28] to classify end-user applications based on design case for each category segregated physics, what composite must be manufactured. In conjunction with AM advancements, artificial neural network algorithms such as learning vector quantization have emerged as potent tools for processing vast data volumes and enhancing system predictive abilities. These algorithms can be applied across various domains for intelligent product development, to optimize information flow and lean management [29] [30].

3.2.3. Adopting AM principles from aviation industry
Fascinating research findings [31] and [32] underscored the importance of enterprise resource planning and customer requirement management systems physics within the infrastructure, rather than seeing this as load, in supporting restructuring the effects of the constant variable for the company performance and seeking the balance between statistical mechanics on customers. These systems, equipped with advanced processing capabilities, can facilitate econophysics to the production parameters. Interdisciplinary analyses of unstructured textual data from manufacturing processes to communication [31]; [32] is less conventional when the econophysics environmentally ought to change. The use of artificial neural network algorithms, such as learning vector quantization, optimizes information flow and promotes lean management across various manufacturing domains, including AM (adapted to [29]; [30]. To the creation of metal matrix composites, for example, reconfigurable intelligent surfaces (RIS) [33]; [34]. The manufacturing of smart reconfigurable intelligent surfaces requires robust quality assurance and control systems [35], melt pool monitoring [36], layer control systems [37], and optical coherence tomography [38]; [39] without forgetting the effective integration and factory digitization elements to manage planning of the frequency domain to extending toward multibody simulation [40]. This is a place for efficient information management [41], [42], cross-functional focus, creative learning [43], [44], and data management for the industries' [45] factory-level constants. Remote operations success requires adequate training against whirring intelligent surfaces to enhance capabilities and domain knowledge accessibility [46]. Real-world examples of virtual operations involve multi-agent model immersion across various industries [47]. Attaining appropriate competencies enables the design of optimal ML models. By synchronization with examining hierarchical structures in industry, benefits society [48]. The use of biosignals in optimizing and controlling industrial operations is critical [49], as is the programming for ML-based data acquisition in industrial operations and management systems constants [50]; [51]. Paired mathematical modeling provides a solid basis for exploring algorithmics for quality management but not without horizontal integration of the physics of security constants.

3.2.4. Real-time database systems integration
The integration of real-time database systems into manufacturing is optimal when employing a high-speed star design. When triangulated from the database, all relevant entries represent the similarity between queries within Scopus. This emphasizes the importance of scalable system-level integration in this rapidly developing field of control systems development [52]. An extensive investigation and comparison of different generations of real-time database systems, including framework and real-time to real-time database systems mobile backend as a service for frontend as a service line combination, is crucial. The database systems cannot explain the human systems complex design [53]. This necessitates physics systems constantly exhibit a variety of strategies to manage real-time data and provide diverse architectural models, data modeling tactics, and query processing technique constants. The role of effective caching in boosting system performance and the complexity of transaction management in distributed environments are also meticulously examined [54]. Potential enhancements and open research areas are identified for future directions, including distributed transactions with workload management aligning to the persistent factory constants. The study concludes with the significance and implications of real-time database systems for modern applications, along with suggestions for future research and development in this field [53] by following the optimal system

architecture integration through education [54] [55], deploying the safety of systems integration, and for example, launch the successful system online to test evolutionary algorithms with regression test validating the commission of the virtual twin before use in manufacturing.

4. Results and discussion

The study identified a gap in advancing econophysics for the manufacturers' education, particularly in integrating ML with AM and adopting digital elements for intelligent product development using artificial neural network algorithms. This approach brings a new dimension to the science, especially in exploring real-time system implementation, which enhances performance and customer relations. The study reveals a philosophical shift towards new technology and manufacturing operations in cyber-physical systems, suggesting new directions for system integration and performance measurement. Using conditional probability in NBT classification has shown promise in predicting manufacturers' education for adjusting physics. The study [14] indicates that improved cybersecurity benefits factories, where product quality is crucial. NBT is instrumental in enhancing manufacturing optimization, network utilization, and efficiency, as well as in quality management.

In addition to aerospace [56], innovations such as augmented interfaces are transforming industries adopting training-based ML in transdisciplinary spectrum. Real-time communication and integration of modern digital elements into factories enhance efficiency across sectors [57]. The study confirms the effectiveness of manufacturing engineering evaluation methods in manufacturing include advancement in mechanics for the quantifying and finding a solution with studying algorithms for these problem-based phenomena, highlighting the importance of mathematical modelling with NBT in optimizing systems. Advanced ML enhances AM cybersecurity constants, including NBT and artificial neural network algorithms' responsive solutions. This extends to system-level, material, electromagnetic, and organizational simulations for adjusting monitoring, sensing, and controlling processes. The industry's demand for next-generation systems is met with enhanced communication capabilities and versatility. This makes demand constant suitable for diverse environments and manufacturing applications statistically. These advancements, driven by industry-academia collaborations, represent the future direction of the sector's numerical boundaries [58].

5. Conclusions and future directions

This study investigates the integration of ML with artificial neural network algorithms in AM to enhance manufacturing processes' constant parameters within factories that would contribute to the mainstream of the reconfigurable intelligent metasurfaces (RIS) development software constant adjustment for next-generation 6G. The study finds that combining digital technologies and non-digital services, underpinned by mathematical modeling, improves competitiveness in manufacturing quality systems. A fundamental discovery is the vital role of RIS in identifying system errors early on. Mathematical models have been developed within cyber-physical systems, adhering to mathematic-physics principles. These models are essential for assessing frame quality innovations for intelligent product development on beamforming quality monitoring. The research underscores the effectiveness of RIS, showing promise in enhancing system reliability on airing new technologies' quality propagation. However, the study emphasizes the necessity for more profound research into manufacturing defects and finding a balance between technological innovation and its practical application. It advocates for further exploration of mathematical modeling extending to the RIS technologies with enhanced AM. Looking to the future, the paper suggests refining algorithms focusing on sustainable physics. Emphasizing eco-friendly practices, there is a shift towards model-based systems engineering that considers malfunctioning RIS impacts [59]. The paper also indicates a trend towards autonomous operations to revise training methods for the evolving industry [60][61]. The study highlights the importance of advanced materials in the biomedical field, advocating for systems integration with optimized RIS for training [64]. It anticipates developments in experimenting with RIS within multi-material AM systems that do not require interventions to address cellular challenges [65].

6. References

[1] Sackinger E, Boser B, Bromley J, Lecun Y, Jackel L. Application of the ANNA Neural Network Chip to High-Speed Character Recognition. Neural Networks, IEEE Transactions on. 1992.

[2] Mousapour, M., Salmi, M., Klemettinen, L., & Partanen, J. (2021). Feasibility study of producing multi-metal parts by Fused Filament Fabrication (FFF) technique. Journal of Manufacturing Processes, 67, 438-446. https://doi.org/10.1016/j.jmapro.2021.05.021

[3] Saab A, Nakhle V, Al Hajj Chehade G, Moussawi H. Testing and Comparing the Performances of Windows Server 2022, Ubuntu 20.04, and CentOS 8 under DDoS Attacks.

[4] EMS. European Manufacturing Survey Finland 2019-2021. 2022.

[5] Norman, Zoncita. (2023). Exploring Arts-Based Learning as an Effective Instructional Approach to Teaching ESL- STEAM (Science, Technology, Engineering, Arts, and Mathematics) Education. SSRN Electronic Journal. 10.2139/ssrn.4445329.

[6] Fukushima T. Rectangular rotation of spherical harmonic expansion of arbitrary high degree and order. Journal of Geodesy. 2017;91(8):911-933.

[7] Manogharan G, Wysk R, Harrysson O. Additive manufacturing-integrated hybrid manufacturing and subtractive processes: Economic model and analysis. International Journal of Computer Integrated Manufacturing. 2016;29(3):251-267.

[8] Liu Z, Wang G. Additive manufacturing: Current landscape, potential opportunities, and research topics. International Journal of Precision Engineering and Manufacturing-Green Technology. 2018;5(3):239-256.

[9] Mourtzis D, Vlachou E, Milas N, Xanthopoulos N. A cloud-based approach for maintenance of machine tools and equipment based on shop-floor monitoring. Procedia CIRP. 2016;41:292-297.

[10] Li Z, Chang B, Zhang D, et al. Effect of post-heat treatment on Ti2AlNb-based alloy fabricated by twin-wire alternating dual-electron beam additive manufacturing technology. Journal of Materials Science & Technology. 2023;157:130-143. doi:10.1016/j.jmst.2023.02.021

[11] Dutordoir V, Durrande N, Hensman J. Sparse Gaussian Processes with Spherical Harmonic Features. In: Proceedings of the 37th International Machine Learning Research. 2019;119:2793-2802.

[12] Davitkova A, Gjurovski D, Michel S. Compressing Learned Bloom Filters. ArXiv. 2022.

[13] Martinelli M, Pandolfi S, Martins CJ, Vielzeuf PE. Probing dark energy. ArXiv. 2012.

[14] Heilala, J., Kwegyir-Afful, E., Kantola, J. (2023). Training and competency development on virtual safety training. In: Tareq Ahram and Christianne Falcão (eds) Human Factors in Virtual Environments and Game Design. AHFE (2023) International Conference. AHFE Open Access, vol 96. AHFE International, USA. http://doi.org/10.54941/ahfe1003875

[15] Boy GA. Human-Systems Integration: from Virtual to Tangible. CRC Press, Taylor & Francis Group. 2020.

[16] Menichinelli M. A data-driven approach for understanding Open Design. Mapping social interactions in collaborative processeson GitHub. The Design Journal. 2017;20(Sup1).

[17] Chai K-H, Zhang J, Tan K. A TRIZ-based method for new service design. Journal of Service Research - J SERV RES. 2005;8:48-66. doi:10.1177/1094670505276683

[18] Creswell J, Clark V. Designing and conducting mixed methods research. Sage Publications. 2007.

[19] Creswell J. A Concise Introduction to Mixed Methods Research. Los Angeles: SAGE. 2015:33-62.

[20] Edmonds W, Kennedy T. Embedded Approach. In: An Applied Guide to Research Designs. Thousand Oaks: SAGE Publications. 2019:189-195.

[21] Karwowski W, Zhang W. The discipline of human factors and ergonomics. In: Handbook of Human Factors and Ergonomics. 2021.

[22] Nam P. Suh (2007) Ergonomics, axiomatic design and complexity theory, Theoretical Issues in Ergonomics Science, 8:2, 101-121, DOI: 10.1080/14639220601092509

[23] Suh N. Designing-in of Quality Through AD. Reliability, IEEE Transactions on. 1995;44:256-264.

[24] Salonitis K. Design for additive manufacturing based on the axiomatic design method. International Journal of Advanced Manufacturing Technology. 2016;87(1-4):989-996.
[25] Heilala, J., Nevalainen, P., Toivonen, K. (2023). AI-based sentiment analysis approaches for large-scale data domains of public and security interests. In: Salman Nazir (eds) Human Factors in Management and Leadership. AHFE (2023) International Conference. AHFE Open Access, vol 92. AHFE International, USA. http://doi.org/10.54941/ahfe1003738
[26] Setzler B. Maximum Likelihood Estimation (MLE) In Julia: The OLS Example.
[27] Tiismus H, Kallaste A, Vaimann T, Rassõlkin A. State of the Art of Additively Manufactured Electromagnetic Materials for Topology Optimized Electrical Machines. Additive Manufacturing. 2022;55:102778. doi:10.1016/j.addma.2022.102778
[28] Chekurov S. Industrial Opportunities of AM - Workflow planning and decision making of additively manufactured end-use components. Aalto University, Finland. 2019.
[29] Mumali F. Improving Value Stream Mapping Using Learning Vector Quantization: A Conceptual Framework. Poznan University of Technology. 2022.
[30] Mumali F. Artificial neural network-based decision support systems in manufacturing processes: A systematic literature review. Computers & Industrial Engineering. 2022;165:107964.
[31] Bataineh A, Abu-AlSondos I, Salhab H, Al-Abbas L. A structural equation model for analyzing the relationship between enterprise resource planning and digital supply chain management. Uncertain Supply Chain Management. 2022;10:1289-1296. doi:10.5267/j.uscm.2022.7.011
[32] Verhoef P, Kamakura W, Mela C, et al. Choice Models and Customer Relationship Management. Marketing Letters. 2005;16:279-291. doi:10.1007/s11002-005-5892-2
[33] Dicandia, Francesco & Genovesi, Simone. (2020). Characteristic Modes Analysis of Non-Uniform Metasurface Superstrate for Nanosatellite Antenna Design. IEEE Access. 8.
[34] Nturanabo F, Masu L, Kirabira J. Novel Applications of Aluminium Metal Matrix Composites. 2019.
[35] Ding C-F, Lee M-S, Li K-M. Development of on-line laser power monitoring system. SPIE. 2016.
[36] Vallabh C, Zhao X. Melt pool temperature measurement and monitoring during laser powder bed fusion based AM via single-camera two-wavelength imaging pyrometry (STWIP). Journal of Manufacturing Processes. 2022;79:486-500. doi:10.1016/j.jmapro.2022.04.058
[37] SLM. Selective Laser Melting Machine. 2023. Retrieved from https://novapax.gr/wp-content/uploads/2019/06/150EN181102_03_SLM800_WEB.pdf
[38] Steen WM, Mazumder J. Laser Material Processing. 4th ed. Springer London. 2010.
[39] Sacher S, Wahl P, Weißensteiner M, et al. Shedding light on coatings: Real-time monitoring of coating quality at industrial scale. Int J Pharm. 2019;566:57-66.
[40] Herôncio, Roberta & Santos, João & Bessa, Wallace & de Paula, Aline & Savi, Marcelo. (2022). A fuzzy feedback linearization scheme applied to vibration control of a smart structure.
[41] Objective 3D (O3D). (2022). Additive Manufacturing – a structured approach to get started. unisa.edu.au/contentassets/ec3778bc0f5d4da8a8093654d8b7d593/additive-seminar-q1-2022---handouts.pdf
[42] Tryashchenko V, Tytar T. MANAGEMENT OF INFORMATION SECURITY OF THE ENTERPRISE. Economy and society. 2022. doi:10.32782/2524-0072/2022-44-109
[43] Pérez-López R, Olguín Tiznado J, García-Alcaraz J, et al. Integrating and Controlling ICT Implementation in the Supply. Mathematics. 2021;9:1234. doi:10.3390/math9111234
[44] Swanston WJ, Biggar GT. Developing a framework for establishing cross-functional integration within a product development project. In: Project Management Institute Annual Seminars & Symposium, Houston, TX. Newtown Square, PA: Project Management Institute. 2000.
[45] Aspire. Building a Robust Cloud Security Architecture for IaaS. Retrieved from https://www.aspiresys.com/ebooks/building-a-robust-cloud-security-architecture-for-iaas.pdf
[46] Dillen A, Steckelmacher D, Efthymiadis K, et al. Deep learning for biosignal control: insights from basic to real-time methods with recommendations. J Neural Eng. 2022;19(1).
[47] Germine M. The Holographic Principle of Mind and the Evolution of Consciousness. Journal of Conscious Evolution. 2018;3. https://digitalcommons.ciis.edu/cejournal/vol3/iss3/1

IOP Conf. Series: Materials Science and Engineering 1296 (2023) 012024 doi:10.1088/1757-899X/1296/1/012024

[48] Dwivedi, Y. K., Hughes, L., Baabdullah, A. M., Ribeiro-Navarrete, S., Giannakis, M., Al-Debei, M. M., Dennehy, D., Metri, B., Buhalis, D., Cheung, C. M. K., Conboy, K., Doyle, R., Dubey, R., Dutot, V., Felix, R., Goyal, D. P., Gustafsson, A., Hinsch, C., Jebabli, I., Janssen, M., ... Wamba, S. F. Metaverse beyond the hype. International Journal of Information Management. 2022.

[49] Sarker IH. Machine Learning: Algorithms, Real-World Applications and Research Directions. SN COMPUT SCI. 2021;2(160). doi:10.1007/s42979-021-00592-x

[50] Köllőd C, Adolf A, Márton G, Ulbert I. Deep comparisons of Neural Networks from the EEGNet family.

[51] Zhang W, B Zhang C, Xiao Y, Wang M. An EEG/EMG/EOG-Based Multimodal Human-Machine Interface to Real-Time Control of a Soft Robot Hand. Frontiers in Neurorobotics. 2019;13:7.

[52] Das S, Tripathy D, Raheja JL. Real-Time Bci System Design to Control Arduino Based Speed Controllable Robot Using Eeg. Singapore: Springer. 2019. doi:10.1007/978-981-13-3098-8

[53] Roy-Hubara, Noa, Arnon Sturm, and Peretz Shoval. "Designing NoSQL Databases Based on Multiple Requirement Views." Data & knowledge engineering 145, 102149–, 2023.

[54] Wingerath W, Gessert F, Ritter N. NoSQL & Real-Time Data Management in Research & Practice – Part 1. BTW 2019, Rostock.

[55] Singh K. Optimization of polarizers and metasurfaces using evolutionary algorithms, Academic dissertation, School of Engineering Faculty of Science and Engineering Macquarie University Sydney, Australia. 2021.

[56] Esselle K, Singh K, Thalakotuna D, Koli MNY, Ahmed F. Beam-steering antenna technologies for space-related applications. School of Electrical & Data Engineering, UTS, Sydney, Australia.

[57] Keebler, J. R., & Fausett, C. M. (2023). Human factors in aviation and aerospace: An overview. Human Factors in Aviation and Aerospace (Third Edition), 1-9.

[58] Davison AC. Statistical Models. Cambridge: Cambridge University Press. 2003.

[59] Tabesh, Nikta. (2023). A Model-Based System Engineering Approach to Support System Architecting Activities in Early Aircraft Design. 10.13140/RG.2.2.25688.67842.

[60] Calzavara, Martina & Faccio, Maurizio & Granata, Irene. (2023). Multi-objective task allocation for collaborative robot systems with an Industry 5.0 human-centered perspective. The International Journal of Advanced Manufacturing Technology. 128. 1-18.

[61] Dekker, S., Zimmermann, V., & Woods, D. D. (2023). High-reliability organizations invest in resilience. Human Factors in Aviation and Aerospace (Third Edition), 41-57.

[62] Shaw, David M., & J. W. Harrell. "Integrating Physiological Monitoring Systems in Military Aviation: a Brief Narrative Review of Its Importance, Opportunities, and Risks." Ergonomics ahead-of-print.ahead-of-print (2023): 1–13. Web.

[63] Antonante, P., Veer, S., Leung, K., Weng, X., Carlone, L., & Pavone, M. (2023). Task-Aware Risk Estimation of Perception Failures for Autonomous Vehicles. ArXiv. /abs/2305.01870

[64] Brunila, K. & Nehring, D. (2023). Precision education governance and the high risks of fabrication of future-oriented learning humankinds. Research Papers in Education. 1-16.

[65] Uchida, D.T. & Bruschi, M.L. 3D Printing as a Technological Strategy for the Personalized Treat ment of Wound Healing. AAPS PharmSciTech, 2023; 24(1).

Author contributions

Piili has conducted a comprehensive meta-analysis, while Parchegani has provided expert validation of the materials' laser engineering aspects. As part of the 2022-2023 Doctoral Programme in Technology at the University of Turku, Heilala's manuscript was aligned toward a range of selective HORIZON-CL-Z-aligned partner selection to non-commercial research; draws inspiration from the 2023 Virtual Training Certification Business Finland project and Finland's 2022 European Manufacturing Research; aligns with the EU Directives 2022/2464 and 2014/95/EU; to contribute to the Degree of Technology, focusing on sustainable development aligned for supporting administrative research. Grammarly enhanced the handwriting of the manuscript to prevent grammatical mistakes.

IOP Conf. Series: Materials Science and Engineering 1296 (2023) 012025 doi:10.1088/1757-899X/1296/1/012025

Effects of laser welding speed on the microstructure and microhardness of ultra-high strength steel S1100

S Afkhami[1] M Ghafouri[1] K Lipiäinen[1] I Poutiainen[2] M Amraei[3] T Björk[1]

[1] Laboratory of Steel Structures, LUT University, Lappeenranta, 53850, Finland
[2] Laboratory of Laser Processing and Additive Manufacturing, LUT University, Lappeenranta, 53850, Finland
[3] Department of Mechanical and Materials Engineering, University of Turku, Turku 20520, Finland

E-mail: Shahriar.Afkhami@lut.fi

Abstract. Ultra-high strength steels strengthened via thermomechanical processing and low alloying constituents are currently quite popular for industrial applications. The popularity is primarily due to these steels being relatively economical compared with their high-alloy rivals, making them feasible to provide significant strength-to-weight ratios with reasonable costs in steel structures. However, welded joints in most steel structures cannot be avoided, and ultra-high strength steels, due to their thermomechanical strengthening mechanisms, are prone to loss of strength, toughness, or ductility after welding. Also, due to the ultra-high strength levels of the base metals, providing a suitable filler material with mechanical properties comparable to those of the base metals for filling the joint gap seems challenging. Accordingly, laser welding can be a relatively more suitable approach than arc welding to weld ultra-high strength steels, considering its easier control over the weld heat input, joint shape, and the possibility of welding without the requirement of filler metals. Therefore, this study investigates the influence of travel speed and its subsequent change in heat input of laser welding on the joint shape, microstructure, and hardness of the S1100 low-alloy carbon steel, one of the most commonly used ultra-high strength steels in modernized steel structures.

1. Introduction

Combinations of welding, as the fabrication process, and ultra-high strength steels (UHSS), as the raw material, can benefit manufacturers from the advantages of both sides. However, UHSSs achieve their unique combination of strength and ductility via the complex microstructural features achieved via thermomechanical processing; therefore, compared to conventional carbon steels with lower strength values, UHSSs are relatively more prone to losing their favorable mechanical properties when exposed to welding's extremely high heating energy. This deterioration is due to the welding heat input (HI) triggering microstructural transformations in the heat-affected zone (HAZ), changing intricate combinations of bainite and martensite, as hard metastable and supersaturated iron-carbon mixtures in UHSSs, into relatively softer ferritic alternatives. Such transformations can yield decreased strength, hardness, and ductility in the welded UHSSs, especially their HAZ. This phenomenon is so common in welded UHSSs that it is typically known as HAZ softening in the literature.

Utilizing laser welding to minimize the HI can be a suitable remedy to significantly reduce the adverse effects of welding on UHSSs and prevent HAZ softening. However, laser welding is not expected to make the disadvantages of welding as a joining process completely disappear since, like other welding procedures, it suffers from its own unique drawbacks. For example, the intensified focused heating source subjects the base metal to severe thermal cycles of highly rapid heating and cooling sequences. Hence, although the HAZ area is expected to be small compared to arc welding, the fusion zone of laser welded joints, especially in carbon and low-carbon steels like UHSSs, is prone to forming brittle microstructural features and, subsequently, deterioration in mechanical properties or even cracking [1, 2].

Due to the complexities mentioned above, the welding of UHSSs has been the subject of numerous studies. For example, Amraei et al. [3] investigated the effects of HI on the mechanical properties and HAZ softening of high-strength and ultra-high strength steels S700, S960, and S1100 welded via gas-metal arc welding and confirmed the presence of soft features like granular bainite as the main reason behind the HAZ softening. Also, the microstructural transformations along the HAZ area of S960 and S1100 were thoroughly investigated via thermomechanical simulations of HAZ subzones in [4] and [5] to precisely address local softenings in such areas and accurately associate the fluctuations in the mechanical properties to their respective microstructural transformations (with considering the thermal history that each subzone had endured through the simulations). Also, the combination of cold-forming and welding on the mechanical performance of UHSSs has been investigated in [6] to evaluate the weldability of hollow sections potentially made of UHSSs since the future of sustainable steel structures relies on them. In addition to HAZ softening, defects such as cold cracking, toughness reduction, brittleness, and lack of strength are other common difficulties associated with welded UHSSs [7-11]. Further, the general effects of laser welding on direct-quenched UHSSs have been investigated in studies like [2], [7], [8], [10], [12], [13], and [14]. However, having a comprehensive understanding of the influence of laser welding parameters on the UHSSs' welded joints and evaluating the laser welding of quenched and tempered UHSSs (as new variants to their direct-quenched peers) still require further research.

Consequently, this study investigated the effects of laser power and traveling speed (as influential parameters upon HI) and base metal thickness on the microstructural features of S1100 as a quenched and tempered UHSS, which has been commercially available for the past few years for structural and industrial applications. Also, correlations between microstructural features (as direct results of transformations caused by welding intense heat input) and changes in mechanical properties have been established, given fluctuations in Vickers hardness values.

2. Experimental procedure

Plates of the S1100 UHSS with rectangular cross-sections procured from SSAB were used as base metals in this study. The length and width of each plate were 400 mm and 150 mm, respectively. Also, two different thickness values of 4 mm and 8 mm were considered for the base metal in this research to investigate the influence of material thickness on the welded joint properties. A fiber laser welding system with a 10 kW total capacity mounted on a robotic arm to accurately control the welding parameters was used to perform fully automated welding. A total number of six welding trials were carried out to consider the simultaneous effects of base metal thickness and welding speed, as summarized in Table 1. Linear heat inputs in the table were approximated according to [15], considering an absorption rate of 90% for keyhole laser welding. Although all the welding parameters except power, welding speed, and the base metal thickness were set to be constant, the focal position for the thinner base metals had to be changed to zero mm to keep the melt pool profile consistent. Finally, the welds were done using a square edge joint design without applying filler material and an opening gap.

IOP Conf. Series: Materials Science and Engineering 1296 (2023) 012025 doi:10.1088/1757-899X/1296/1/012025

Table 1. Laser welding parameters.

Sample code	Thickness	Spot radius	Focal position	Laser power	Welding speed	Linear heat input
4H	4 mm	0.2 mm	0 mm	6 kW	35 mm/s	154 J/mm
4M	4 mm	0.2 mm	0 mm	6 kW	50 mm/s	108 J/mm
4L	4 mm	0.2 mm	0 mm	6 kW	120 mm/s	45 J/mm
8H	8 mm	0.2 mm	-1 mm	9 kW	21 mm/s	386 J/mm
8M	8 mm	0.2 mm	-1 mm	9 kW	30 mm/s	270 J/mm
8L	8 mm	0.2 mm	-1 mm	9 kW	75 mm/s	108 J/mm

After welding, welded joints were sectioned and cut into desired dimensions of 30 mm by 20 mm for hardness measurements and microstructural analysis. Vickers hardness values were then measured with a Struers DuraScan 70 machine by applying a 500 g load (HV 0.5). Two sets of measurements were carried out for every sample; the first set was 1 mm away from the top surface of the base metal, while the second set was carried out along the middle of the section. Next, the specimens were ground and polished sequentially for the microstructural analysis until a mirror-like polish was reached. Then, the polished specimens were etched with Nital for 15 seconds. Finally, scanning electron microscopy (SEM) was performed to capture microstructural features with a Hitachi SU3500 microscope.

Finally, the Finite element (FE) approach was used to simulate the thermal cycles experienced by the material during the welding procedures. The FE models were developed via SimuFact Welding software version 2022.0.1. The meshing was done using MSC Apex 2023.1 with solid hexagonal elements; the element size range was between 0.125 μm to 500 μm (a mesh convergence study was also done to affirm the mesh size reliability). The laser heat source was modeled as a conical geometry with a radius equal to the actual laser spot used in the welding procedures (0.2 mm) and non-Gaussian heat distribution, accompanied by a surface heat source with a radius of 1.5 mm and a Gaussian heat distribution ratio of 1.5. The total heat exposure from the two heat source geometries was achieved by a 60% division ratio between the conical and surface geometries. The required material data was imported from S960QL-SPM-sw data set for the base metal due to the similar thermal characteristics between the chosen data set (as a quenched-tempered multiphase steel with minimum bainite content of 79%) from the software's library and the actual base metal. Also, a pair of bearing and clamping were set on the top and bottom sides of the base plates to apply the geometrical boundary conditions used for the actual welds. Further, a symmetry plane was used along the longitudinal section of the base metal to decrease the simulation time. Finally, the models were calibrated according to the final geometries of the actual welds, shown in Figure 1.

3. Results and Discussion

Samples welded using the minimum HIs from Table 1 did not have fully penetrated joints, as shown in Figure 1. This Figure shows a linear relationship between the maximum weld depth and the applied HIs in these samples (~5 mm for 108 J/mm for the thick and ~2 mm for 45 J/mm for the thin plates). Regarding the 4 mm-thick samples, even the relatively medium HI resulted in a negative crown bead shape and sagging, indicating a too high HI to achieve a desirable bead geometry in such welds. However, reaching the optimum bead geometry is out of the scope of this research, and this issue will not be further addressed in the current study. Further, increasing the HI increased the width ratios of top crowns and mid-sections of both 8 mm- and 4 mm-thick samples, turning the welded joint cross-sections into a wine glass shape. In addition, enabling/encouraging the wine glass geometry by increasing the HI resulted in hardness inconsistencies in size of HAZ areas along the top and middle profile lines, especially in samples welded with the highest HIs.

IOP Conf. Series: Materials Science and Engineering 1296 (2023) 012025 doi:10.1088/1757-899X/1296/1/012025

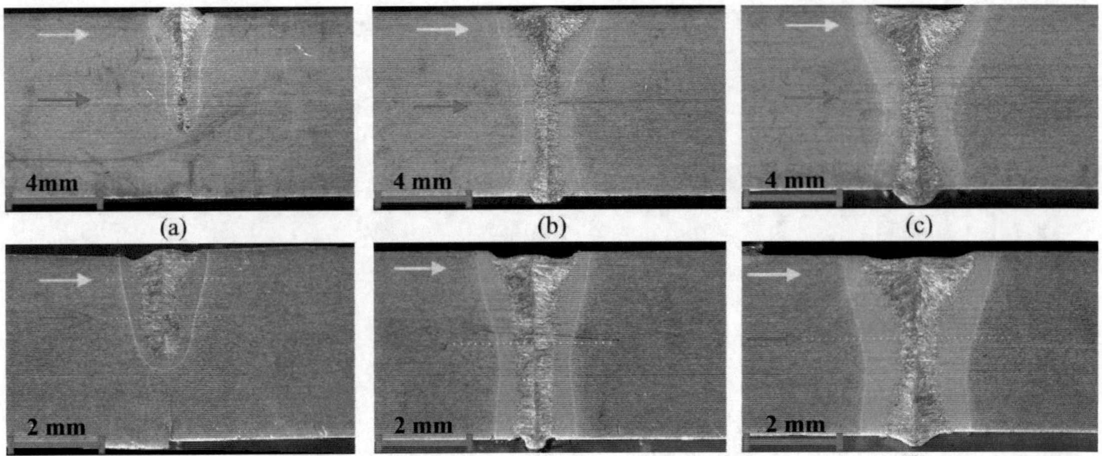

(a) (b) (c)
(d) (e) (f)

Figure 1. Cross-sectional macrographs of the welded specimens and locations of the linear hardness measurements carried out on the top (cyan arrow) and middle (red arrow) of the sections: (a) 8L, (b) 8M, (c) 8H, (d) 4L, (e) 4M, and (f) 4H.

The hardness profiles of the samples are shown in Figure 2. Considering the average value of 385 HV for the hardness of the base metal in the Figure, the HAZ softening did not occur in the samples welded with the lowest HIs, as presented in Table 2. Furthermore, no significant HAZ softening (> 10%) that can cause premature failure in the softened area [1] was detected in the welded samples. This observation agrees with the other studies on the same quench and tempered UHSS welded by gas-metal arc welding as another fusion welding procedure with a less intensified heat source than the one used in the current study [3-6]. Also, comparing the hardness values based on the applied HI in samples with the same thickness (Figure 3) showed that the lowest HIs caused slightly higher hardness peaks in the HAZ areas in the middle sections of both 8 mm and 4 mm specimens. However, the hardness peaks seemed more unified along the top sections, especially in the 4 mm-thick specimens. Finally, the minor softenings did not occur in the samples with the lowest HIs, as mentioned earlier; however, a negligible 1.6% hardness drop was recorded for the top side of 8L (which will be discussed later via FE analysis).

Further investigation on the hardness profiles requires establishing a correlation between the hardness changes and microstructural features in laser welded S1100. According to Figure 4, the microstructure of S1100 as the base metal consisted of a mixture of lath bainite, martensite, and bainitic ferrite. Also, the microstructural features of the weld metal and HAZ sub-zones (accompanied by their hardness values) in the 8H specimen are shown in Figure 5. According to this Figure, the weld metal mainly consisted of lath bainite and martensite combinations, aligned toward the weld pool thermal gradient, depending on the position in the weld metal [compare Figure 5(f) at the weld centerline and Figure 5(g), for example]. Further, minor hardness fluctuations in the scale of 20 HV were observed along the weld metal, although the area showed similar microstructural features throughout its section. Such minor fluctuations in the weld metal can be caused by minuscule discrepancies in the chemical composition originating from, for example, (welding) microsegregation or the base material in its as-received condition. However, it is not feasible to experimentally address such issues any further with the apparatus used in this study.

IOP Conf. Series: Materials Science and Engineering 1296 (2023) 012025 doi:10.1088/1757-899X/1296/1/012025

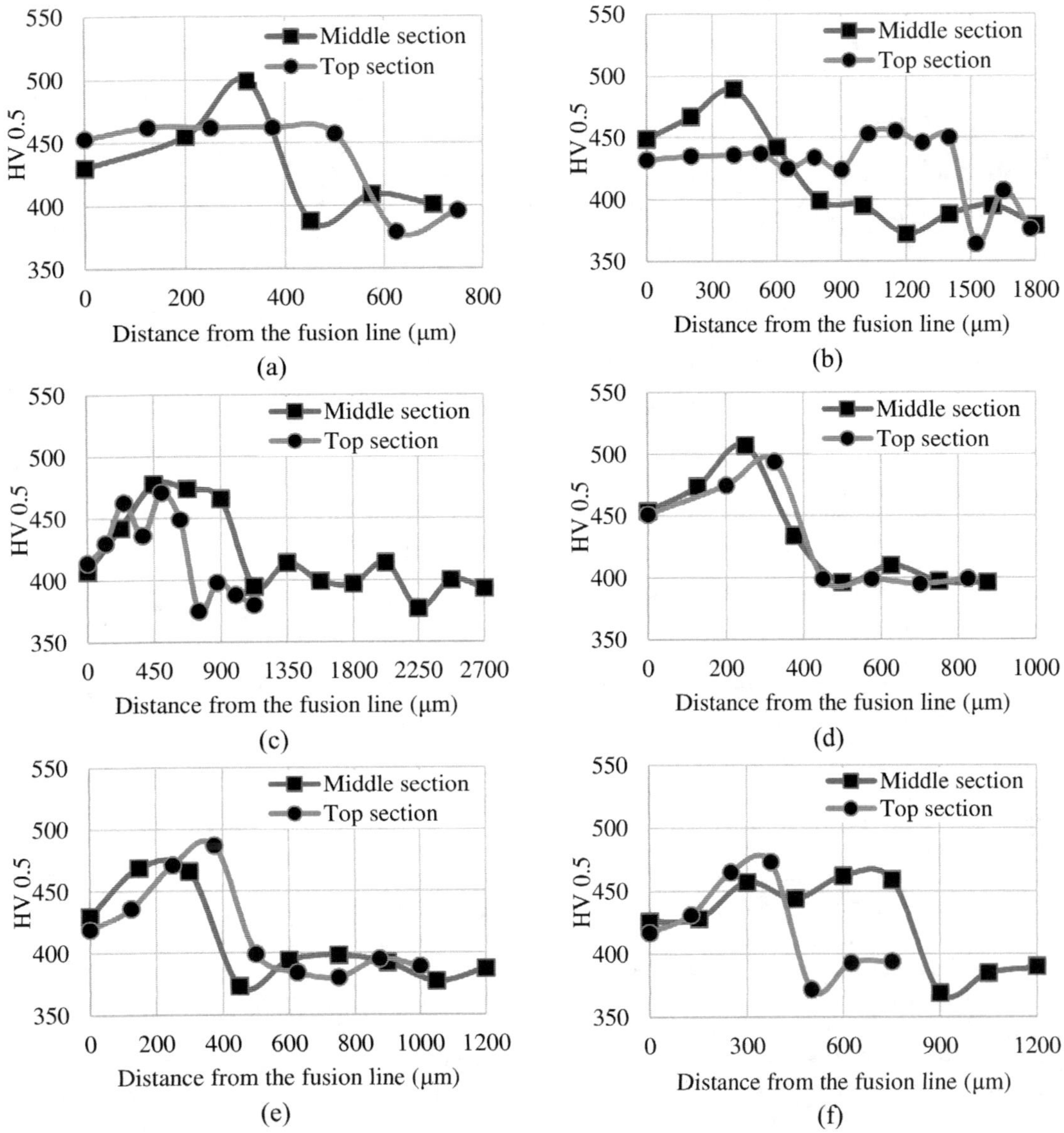

Figure 2. Hardness profiles of the samples: (a) 8L, (b) 8M, (c) 8H, (d) 4L, (e) 4M, and (f) 4H.

Table 2. Calculated HAZ softenings

Sample code	8L	8M	8H	4L	4M	4H
Softening % (top)	1.6	5.5	2.6	0.0	1.3	3.4
Softening % (mid)	0.0	3.4	2.1	0.0	3.1	4.2

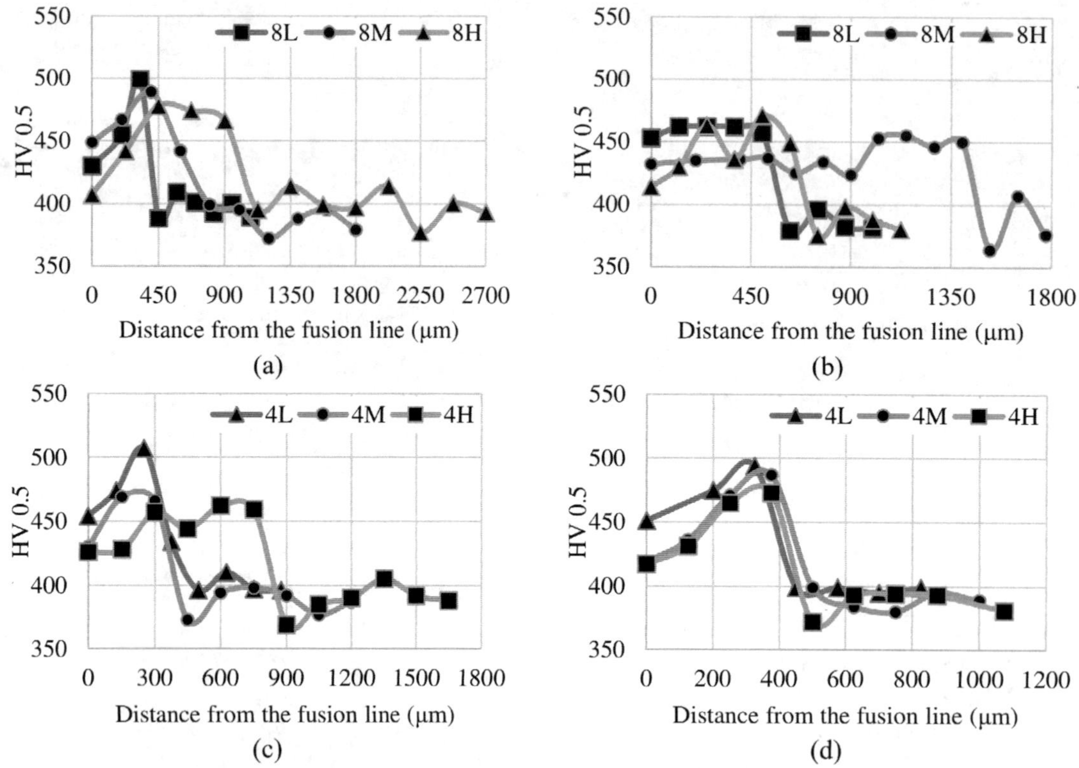

Figure 3. Hardness comparisons along the middle and top sections of 8mm-thick [(a): middle and (b): top] and 4mm-thick [(c): middle and (d): top] specimens.

Moving from the weld metal into the HAZ, the microstructure maintained its bainitic-martensitic mixture along the fusion line, but the weld pool thermal gradient did not impose any directionality in this area since it is outside the weld metal (it was not molten and solidified through the welding procedure). The occasional appearance of Widmenstäten ferrite was also observed in this region [Figure 5(h)]. Consequently, the hardness value along the fusion line was similar to the weld metal. However, further movement from the fusion line to the base metal showed a sudden hardness increase (~10%). According to Figure 5(c), the appearance of a mixture of martensite and acicular ferrite, accompanied by sporadic small bainitic features, caused such an increase in the hardness. Figure 5(a) shows that the hardness fluctuated between 435 HV to 470 HV in the hardened HAZ sub-zone (around 2500 μm away from the weld centerline). The relatively small drop to ~435 HV in this sub-zone was caused by a decrease in the martensite content and dominance of relatively coarse acicular ferrite [Figure 5(d), compared to Figure 5(c)]. However, the small hardness drop was compensated by a decrease in the size of these microstructural features, shown in Figure 5(e), resulting in a hardness value of ~470 HV as the peak hardness of the welded material. The mentioned changes in Figures 5(c), (d), and (e) can be the direct results of the competition between the effects of peak temperature and cooling rate in these areas. Next, the softening occurs by getting closer to the base metal with a minimum hardness value of 375 HV. Microstructural analysis of this HAZ sub-zone [Figure 5(b)] indicates the appearance of granular features as the main contributor to the resultant hardness. However, the size of this (microstructural) transition zone was relatively narrow (~20 μm) compared to the specimens' dimensions, and its hardness drop was so small (~4% max.) that its influence upon the mechanical performance of laser welded S1100 could be deemed insignificant.

Finally, samples with different thickness values but exposed to the same HI (8L and 4M) had similar microstructural features and hardness profiles (the microstructural features of the rest of the specimens are not presented here for the sake of brevity). These similarities indicate the dominant influence of the welding parameters compared to the material thickness, at least in this thickness range, over the material's characteristics.

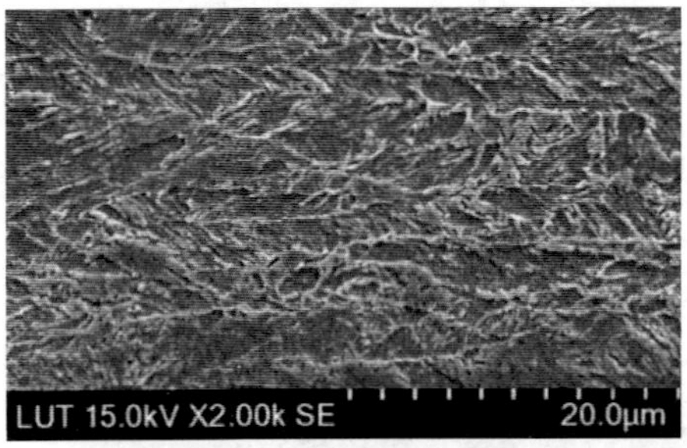

Figure 4. Microstructure of the base metal.

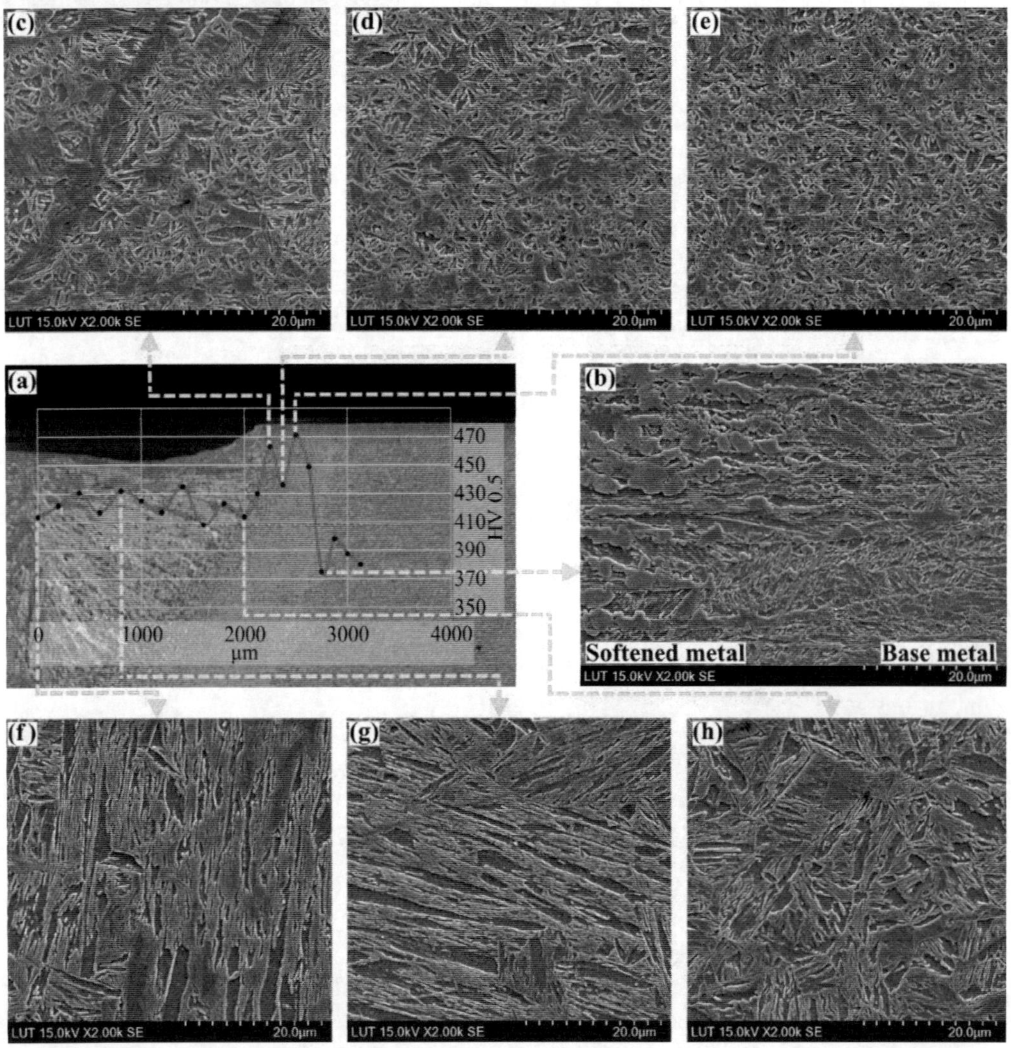

Figure 5. Cross-sectional view of the 8H specimen and the microstructural features of its sub-zones: (a) overall view of the joint accompanied by its hardness profile; (b) granular bainite and ferrite causing minimum hardness; (c, d, and e) mixtures of martensite, acicular ferrite, and bainite causing a range of maximum hardness; (f and g) combinations of lath bainite and martensite resulting in the relatively high hardness of the weld metal; and (h) microstructural transitions along the fusion line.

To consider the influence of thermal gradients on the microstructure, the CCT diagram of S1100, shown in Figure 6, can be compared to the time-temperature data from the thermal gradients modeled via FE. The thermal gradients of the 8H, 8M, and 8L specimens are shown in Figures 7(a, b), (c), and (d), respectively. The time-temperature diagrams in Figure 7 are plotted based on the peak temperatures and are not based on the distance from the fusion line. According to Figure 6, any cooling rate resulting in temperatures lower than 400 °C in less than 80 seconds causes the martensitic microstructure. Such a cooling regime matches all the simulated zones shown in Figure 7; and provides a reason behind the presence of martensite and acicular ferrite in the HAZ sub-zones with peak hardness values. Also, from the time-temperature data in Figure 7, it can be concluded that diffusional austenitic transformations during the cooling stage cannot contribute to the minor softening and granular feature in Figure 5(b). Hence, these features should be caused by the tempering effect of such thermal gradients imposed on adjacent regions resulting in the microstructure in Figure 5(b), with possibly a ferritic nature. Finally, the slightly higher hardness peak values along the sections' middle can be attributed to the slightly more rapid thermal gradients along these regions, as shown in Figure 7 [especially comparing the middle and top in Figure 7(d) shows the effect at its extremes, considering the 1.6% softening on the top and no softening in the middle (Table 2) of 4L].

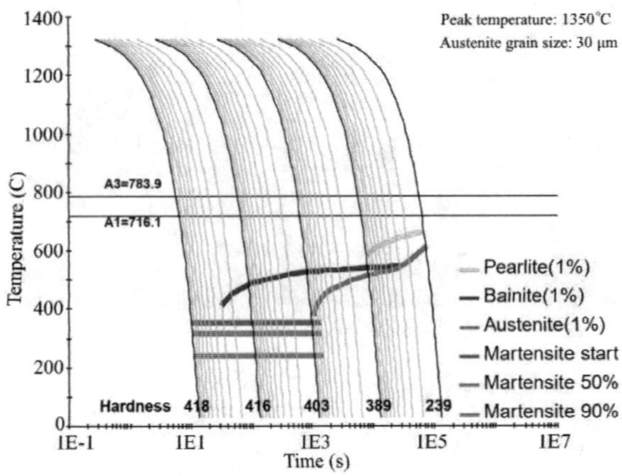

Figure 6. CCT diagram of an arbitrary S1100 specimen plotted by JMatPro [4].

4. Conclusions

The hardness and microstructural features of quenched and tempered S1100 steel welded via laser welding were investigated in this research. Following the results and discussion provided in this study, these conclusions were made:

- The microstructure of quenched and tempered S1100 was a mixture of bainite, martensite, and bainitic ferrite.
- No significant softening was detected in the laser-welded S1100.
- Samples welded using different beam powers or welding speeds showed similar trends in their hardness profiles and microstructural features.
- Hardness peaks in the HAZ sub-zones were mainly caused by the presence of martensite.
- Relatively small softening in the zero to 4% range was caused by the tempering effect of welding thermal gradients on their adjacent areas. However, the dimensions and softening degrees of such sub-zones were so minor that their influence on the mechanical properties could be considered insignificant.

CRediT authorship contribution statement

Shahriar Afkhami: Methodology, Fromal analysis, Software, Investigation, Validation, Visualization, Writing – original draft, Writing – review & editing. **Mehran Ghafouri:** Conceptualization, Methodology, Investigation, Writing – review & editing. **Kalle Lipiäinen:** Methodology and Data curation (SEM and Hardness measurements), Investigation, Writing – review & editing. **Ilkka**

Poutiainen: Methodology, Data curation, Investigation, Writing – review & editing, Resources, Project administration. **Mohsen Amraei:** Investigation, Validation, Writing – review & editing. **Timo Björk:** Conceptualization, Writing – review & editing, Supervision, Resources, Funding acquisition.

Acknowledgment

This study was carried out at LUT University; the authors of this work would like to acknowledge the financial support of the FOSSA project, funded by Business Finland. All industrial and academic partners of this project are highly appreciated.

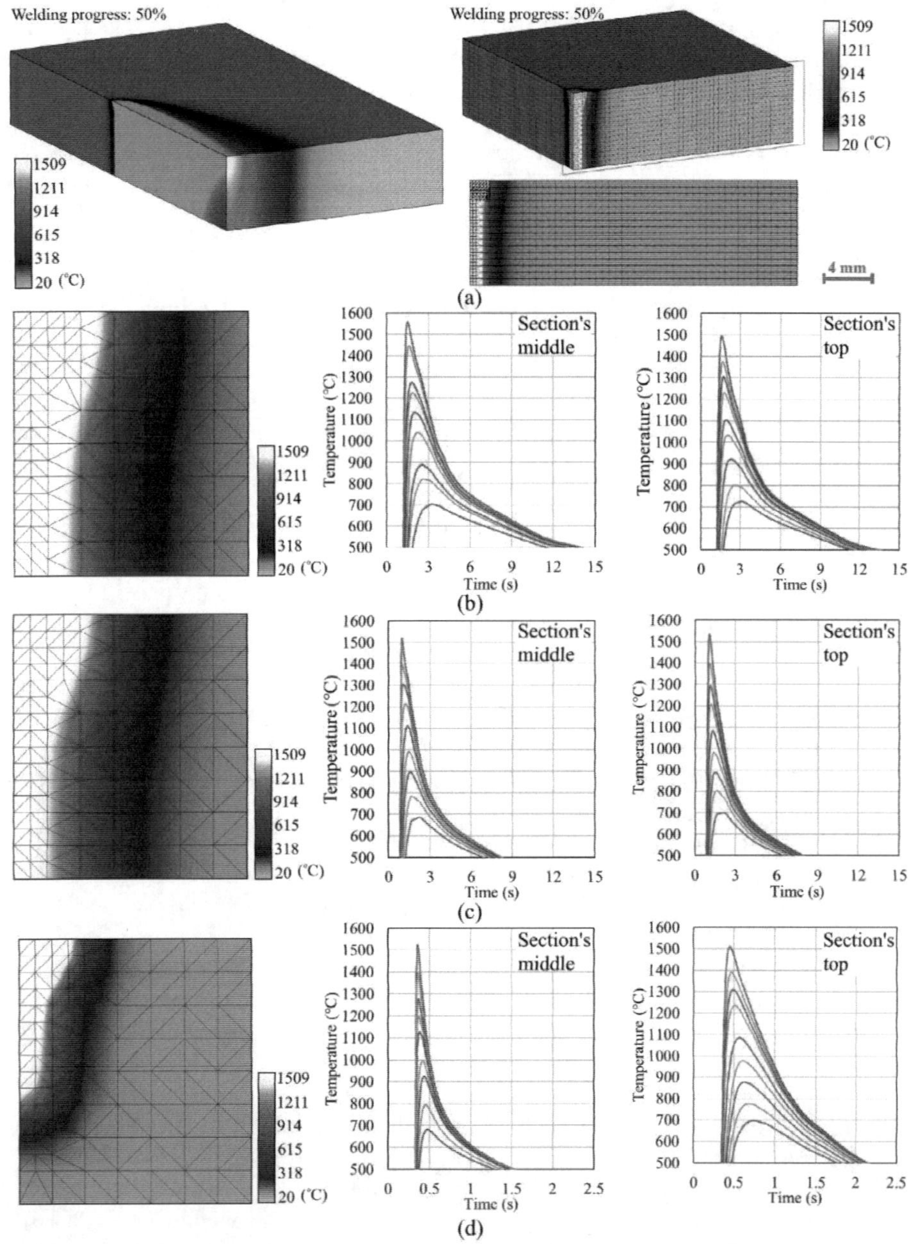

Figure 7. Thermal gradients simulated via FE: (a) 2D and 3D views of the welding procedure of 8H; (b) time-temperature data of 8H; (c) time-temperature data of 8M; and (c) time-temperature data of 8L. The welding progress is 50% in all the presented results.

IOP Conf. Series: Materials Science and Engineering 1296 (2023) 012025 doi:10.1088/1757-899X/1296/1/012025

References

[1] M. Amraei, M. Dabiri, T. Björk, T. Skriko, Effects of Workshop Fabrication Processes on the Deformation Capacity of S960 Ultra-high Strength Steel, Journal of Manufacturing Science and Engineering, Transactions of the ASME. 138 (2016) 121007. https://doi.org/10.1115/1.4033930.

[2] F. Farrokhi, J. Siltanen, A. Salminen, Fiber Laser Welding of Direct-Quenched Ultrahigh Strength Steels: Evaluation of Hardness, Tensile Strength, and Toughness Properties at Subzero Temperatures, Journal of Manufacturing Science and Engineering, Transactions of the ASME. 137 (2015) 061012. https://doi.org/10.1115/1.4030177.

[3] M. Amraei, A. Ahola, S. Afkhami, T. Björk, A. Heidarpour, X.L. Zhao, Effects of heat input on the mechanical properties of butt-welded high and ultra-high strength steels, Eng Struct. 198 (2019) 109460. https://doi.org/10.1016/j.engstruct.2019.109460.

[4] S. Afkhami, V. Javaheri, M. Amraei, T. Skriko, H. Piili, X.-L. Zhao, T. Björk, Thermomechanical simulation of the heat-affected zones in welded ultra-high strength steels: Microstructure and mechanical properties, Mater Des. 213 (2022) 110336. https://doi.org/10.1016/j.matdes.2021.110336.

[5] M. Amraei, S. Afkhami, V. Javaheri, J. Larkiola, T. Skriko, T. Björk, X.L. Zhao, Mechanical properties and microstructural evaluation of the heat-affected zone in ultra-high strength steels, Thin-Walled Structures. 157 (2020) 107072. https://doi.org/10.1016/j.tws.2020.107072.

[6] S. Afkhami, T. Björk, J. Larkiola, Weldability of cold-formed high strength and ultra-high strength steels, J Constr Steel Res. 158 (2019) 86–98. https://doi.org/10.1016/j.jcsr.2019.03.017.

[7] W. Guo, L. Li, S. Dong, D. Crowther, A. Thompson, Comparison of microstructure and mechanical properties of ultra-narrow gap laser and gas-metal-arc welded S960 high strength steel, Opt Lasers Eng. 91 (2017) 1–15. https://doi.org/10.1016/j.optlaseng.2016.11.011.

[8] W. Guo, L. Li, D. Crowther, S. Dong, J.A. Francis, A. Thompson, Laser welding of high strength steels (S960 and S700) with medium thickness, J Laser Appl. 28 (2016) 022425. https://doi.org/10.2351/1.4944100.

[9] I. Garašić, A. Ćorić, Z. Kožuh, I.S.-T. vjesnik, Occurrence of cold cracks in welding of high-strength S960 QL steel, Hrcak.Srce.Hr, (2010) 327-335. https://hrcak.srce.hr/59645 (accessed April 18, 2023).

[10] A. Kurc-Lisiecka, … J.P.-A. of M. and, undefined 2017, Laser welding of new grade of advanced high strength steel STRENX 1100 MC, Bibliotekanauki.Pl. https://bibliotekanauki.pl/articles/351943.pdf (accessed April 18, 2023).

[11] H. Tasalloti, P. Kah, J. Martikainen, Effect of heat input on dissimilar welds of ultra high strength steel and duplex stainless steel: Microstructural and compositional analysis, Mater Charact. 123 (2017) 29–41. https://doi.org/10.1016/j.matchar.2016.11.014.

[12] W. Guo, D. Crowther, J.A. Francis, A. Thompson, Z. Liu, L. Li, Microstructure and mechanical properties of laser welded S960 high strength steel, Mater Des. 85 (2015) 534–548. https://doi.org/10.1016/j.matdes.2015.07.037.

[13] W. Guo, D. Crowther, J.A. Francis, A. Thompson, L. Li, Process-parameter interactions in ultra-narrow gap laser welding of high strength steels, International Journal of Advanced Manufacturing Technology. 84 (2016) 2547–2566. https://doi.org/10.1007/S00170-015-7881-9/METRICS.

[14] J.H. Lee, S.H. Park, H.S. Kwon, G.S. Kim, C.S. Lee, Laser, tungsten inert gas, and metal active gas welding of DP780 steel: Comparison of hardness, tensile properties and fatigue resistance, Mater Des. 64 (2014) 559–565. https://doi.org/10.1016/J.MATDES.2014.07.065.

[15] J. Pekkarinen, V. Kujanpää, The effects of laser welding parameters on the microstructure of ferritic and duplex stainless steels welds, Phys Procedia. 5 (2010) 517–523. https://doi.org/10.1016/j.phpro.2010.08.175.

IOP Conf. Series: Materials Science and Engineering 1296 (2023) 012026 doi:10.1088/1757-899X/1296/1/012026

Impact of laser beam shaping on weld pool and seam geometry in butt-joint welding with filler wire

Y Mi[1], F Sikström[1], M Nilsen[1], S Joshi[1], P Guglielmi[2], G Palumbo[2] and A Ancona[1,3]

[1]Department of Engineering Science, University West, Gustava Melins Gata 2, 461-32 Trollhättan, Sweden

[2]Department of Mechanics, Mathematics and Management, Politecnico di Bari, via Orabona 4, 70125 Bari, Italy

[3]Physics Department, University of Bari "Aldo Moro", via Orabona 4, 70125 Bari, Italy

E-mail: yongcui.mi@hv.se

Abstract.
Beam shaping offers a promising solution to address various challenges in laser-based processes. This study explores the use of an adaptive beam shaping solution using a deformable mirror for high-power continuous wave lasers. Specifically, the impact of three different elliptical beam shapes on weld pool geometry, weld cross section in butt-joint welding with a filler wire is investigated. The results reveal that beam shaping significantly affects the weld pool and cross section, with elliptical beam shapes resulting in wider fusion zones in two tested cases and more uniform weld bead compared to a typical circular beam. These findings highlight the potential of beam shaping for enhancing consistency and quality in laser beam welding.

1 Introduction

Laser beam (LB) shaping (referring to changing the power density distribution), has been considered as a very promising solution to tackle many issues in laser beam welding (LBW) and laser directed energy deposition (DED-LB). It has been recognized that this technology holds significant potential for addressing challenges associated with welding of dissimilar metals [1], gap bridging in autogenous welding of butt-joints [2], as well as enhancing productivity [3] [4].

Several beam shaping methods have been developed and applied in LBW, including diffractive optical elements (DOE) [5], scanning mirrors [1], double cladded delivery fibers [6], coherent beam combining [7] and deformable mirrors [8]. In a prior [5], a DOE was utilized for beam shaping during butt-joint welding of multiple sheets. The LB was shaped into multiple spots arranged in a row perpendicular to the welding travel direction. The results revealed an increase in gap bridging ability and robustness meeting specified tolerances compared to multi-pass welding. In yet another study [2], the authors employed a beam splitting unit to create four separate spots arranged in either a row or a 2x2 matrix with varying power distribution among the spots. Improved gap bridgability was demonstrated compared to a single spot. This improvement was attributed to the influence of beam shaping on the melt flow. The high-intensity spots produced keyholes that pushed the melt towards the center of the gap. Simultaneously, the additional low-intensity spots facilitated the reduction of spatter by stabilizing the dynamics of the weld pool. There have been reports [1] of a LB being shaped by oscillating a mirror during butt-joint welding of Ti/Al dissimilar metals. The microstructure and tensile properties of the resulting seams were evaluated, and the findings revealed that with proper oscillating frequency and amplitude, the seams exhibited excellent intermetallic compound distribution and mechanical properties. Researchers have also employed a double cladded delivery fiber for beam shaping to investigate its impact of on dissimilar lap-joint welding, consisting of steel and copper with copper located on the top [6]. The weld seam appearance, topography, and cross-sectional characteristics were evaluated. The findings indicated that proper beam profile adjustments can effectively prevent spatter. In other studies [7], the authors have also applied beam shaping using coherent beam combining (CBC) with an optical phased array (OPA) during overlap welding of aluminum alloys with partial penetration. The effects of various beam shaping conditions on interface width, penetration, and interface defects were evaluated. The study suggested that by shaping the beam at a frequency exceeding 10 kHz, the desired weld seam width and penetration could be achieved while simultaneously expanding the process window, thereby minimizing the occurrence of defects, and ensuring a stable seam geometry. Existing beam shaping technologies have their own limitations when applied to production lines. For instance, applying a DOE is limited by the flexibility and alignment sensitivity [9], while galvoscanners cannot achieve the necessary scanning speeds for high-speed LBW applications. The utilization of double-cladded fiber for beam shaping is limited by the available range of different beam shapes, while coherent beam combining usually requires a relatively bulky system. Moreover, the recent development of a deformable mirror system, composed of piezoelectric actuators bonded to a flexible glass membrane, enables shaping of a multi-kilowatt continuous wave beam without active cooling [8]. The system can produce a huge range of aberration modes and enable change in beam shape at a rate of 100 Hz. This represents a real breakthrough for laser metal processing since it would allow adapting the heat input to the workpiece geometry or to varying process conditions only when and where it is needed.

Although a number of beam shaping research studies have been conducted for laser based manufacturing processes, very few have explored the impact of beam shaping on LBW and DED-LB using a filler wire. The incorporation of a filler wire in LBW and DED-LB offers several notable advantages, such as improvement in joint fit-up tolerance by accommodating air gaps and mismatches and improved weld surface profile by eliminating undercut [10]. Similarly, DED-LB with a filler wire provides advantages such as clean work environment and up to 100% material's usage [11]. However, these processes present challenges that require better control of heat input to achieve process stability and minimize defects. One current solution involves using additional electrical power to pre-heat the wire (hot-wire) [12] [13], which provides additional manipulation of the heat input. Therefore, beam shaping technologies would represent a valuable solution to adapt the heat input to laser based manufacturing processes that utilize a filler wire.

This study investigates the influence of beam shaping using a deformable mirror on the weld pool and seam geometry in butt-joint welding with a filler wire. The aim is to provide insights in the benefits of this technology in laser-based manufacturing processes where a filler wire is applied. Although the experiments were restricted to ellipsoidal beam shapes with different orientations of the major axis with respect to the traveling and the wire feeding direction, the deformable mirror can in principle produce a large number of different beam shapes.

2 Experimental setup and procedures

The welding setup comprises of a CNC gantry, fixtures, an imaging system, a wire feeding system, shielding gas supply equipment, and a water-cooled high-power ytterbium-doped fiber laser, along with a deformable mirror system in the welding head. The experimental procedure includes welding parameter definition, weld pool top surface analysis, and methods for analyzing the seam cross-sections and reinforcement height of the weld bead.

2.1 Experimental setup

The LBW setup used is shown in Figure 1. The welding head comprised a collimator lens with $f = 76$mm and a focusing lens with $f = 400$mm, and was manipulated by a CNC gantry from Isel® Germany (mod. M40). The laser was generated by a 6 kW IPG system (mod. YRL-6000-S) with a wavelength of 1070 nm and guided to the welding head through a 200 μm core diameter optical fibre. The welding head was mounted on to the gantry with an inclination angle of 5° (in the welding travel direction with respect to the workpiece surface normal) to avoid back reflections into the optics. The filler wire was guided by a copper nozzle with a 45° angle (with respect to the welding travel direction), as shown Figure 1(a) and Figure 1(b). Argon was supplied for shielding from both the top and root side of the workpiece as well as in front of the optics of the welding head to protect it from spatter.

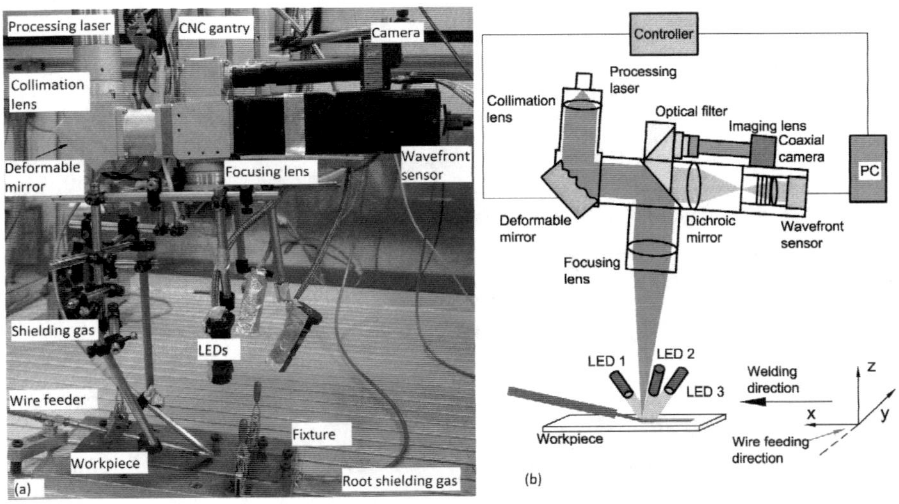

Figure 1. Experimental setup: (a) physical arrangement; (b) illustration of all components.

An imaging system was integrated coaxial with the LB in the welding head to capture the weld pool top surface. It consisted of a camera (SP5000M-CX4, from JAI), an imaging lens with $f = 150$mm, a hard-coated band pass filter (centre wavelength at 450 nm and full width at half max of 10 nm) and three illuminating LEDs. The LEDs (450 nm centre wavelength) were mounted with a steep angle (less than 30° with respect to the workpiece surface normal) to increase image contrast. With this configuration, the vision system had a resolution of u=14 μ m/pixel, obtained by calibration using a checkerboard with minor square size of 1 mm^2. The acquisition exposure time and frame rate were set to 500 μs and 200 frames per second, respectively.

The beam shaping system consisted of a deformable mirror, a Shack-Hartmann wavefront sensor and a control unit. As shown in Figure 1(b), the deformable mirror was positioned at an angle of 45° to the incident beam inside the welding head, after the collimating lens and before the dichroic mirror. It consisted of a thin glass flexible mirror and 19 piezoelectric actuators. The mirror was made of fused silica and coated with a multilayer dielectric layer, with high reflectivity ($R > 99.99$) and low absorptivity ($A < 10^{-5}$) at 1070 nm. Thanks to the high reflectivity and low absorptivity, no active cooling was needed. The actuators were designed to induce curvature changes, thereby introducing aberrations in the wavefront of the beam. The actuators were disposed in two concentric oval rings to match the incident LB diameter and the stroke of each actuator was ±5 μm. The wavefront sensor was placed opposite the deformable mirror and behind the dichroic mirror, as shown in Figure 1(b). A small fraction of the laser power that was reflected by the deformable mirror and leaked through the dichroic mirror was then captured by the wavefront sensor. This arrangement enabled real-time measurement of the aberrations introduced by the deformable mirror. Once a desired beam shape was defined, the feedback unit actuates in real-time to achieve and maintain the desired shape. A more detailed description of the system can be found in [8].

2.2 Welding parameters

Figure 2 visualizes the power density distributions of the four beam shapes investigated: TE, LE, and DE denote elliptical Gaussian shapes with the major axes oriented transverse, along, and at 45° to the travel direction (along the wire feeding), respectively. C denotes a circular Gaussian power distribution resulted from a flat deformable mirror. The beam shapes were measured by PRIMES focus monitor (mod. FM) in the focal planes, and the corresponding geometric parameters and power densities are given in Table 1.

S355MC steel plates with dimensions 300.0 mm × 40.0 mm × 2 mm were tack welded at four positions distributed along the joint. Keyhole mode welding with a filler wire and a total power of 2.1 kW was then carried out with beam shapes TE, LE, DE, and C. The wire (AWS: A5.28 ER70S-X, EN 14341(-A) G 42X) has a diameter of 1.2 mm. The wire feeding direction was with a 45° side angle with respect to the travel direction. The welding travel speed was 15 mm/s and the wire feeding rate was 6.8 mm/s. The length of each weld was 180 mm. Argon was supplied at flow rates of 10 L/min on the top and 15 L/min at the root side of the workpiece.

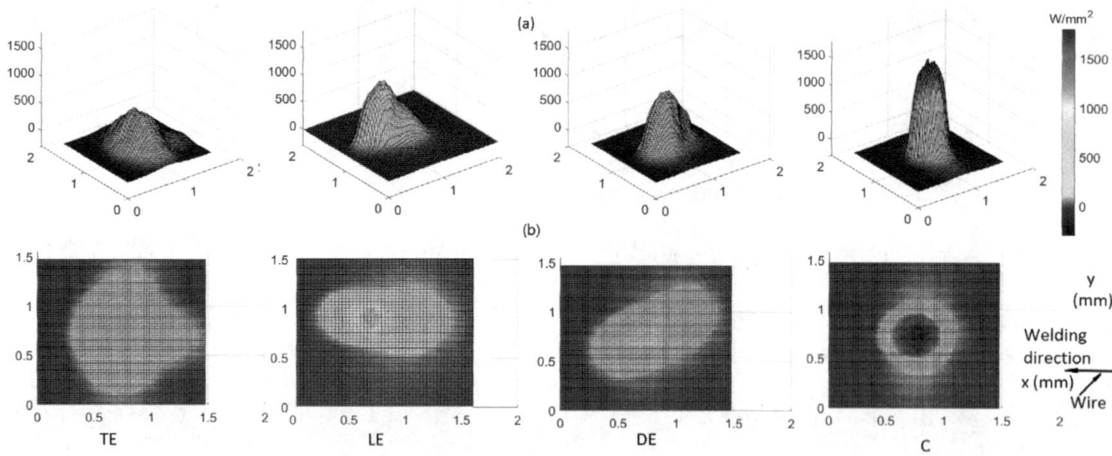

Figure 2. Measurements of the four beam shapes under investigation (with 500 W power), where (a) shows a 3-D visualization and (b) shows the top view.

Table 1. Geometrical characteristics of the beam shapes.

Beam shape	TE	LE	DE	C
R_x (mm)	0.425	0.619	0.566	0.333
R_y (mm)	0.554	0.420	0.382	0.399
Power density ($W/(mm^2)$)	1364	1494	1735	1909

2.3 Experimental data analysis

The weld pool contour was quantified by its area and shape. The contours were first extracted by manually selecting points from 10 image frames out of 1800 frames captured during a quasi-stationary process condition. The boundary of the weld pool was identified by the difference in surface textures between the liquid and solidified metal. The contour points were then transferred from unit in pixel to mm using the camera calibration results, as shown in Figure 4. Based on this, the width, length, and area were calculated. Finally, two straight lines were fitted to the edge of the weld pool tail part to estimate the angle of the weld pool rear curvature.

The obtained weld cross-sections were examined at a distance of 30 mm from the start of the weld to analyze the weld-bead geometries and the associated dimensions of interest. For this purpose, the weld beads were carefully cut at the desired location using a precision slow-speed saw, the transverse sections of the bead hot-mounted, and standard grinding and polishing steps followed for metallographic sample preparation to obtain a scratch-free, mirror finished surface. Light optical microscopy (OM) and scanning electron microscopy (SEM) were utilized to investigate the cross-sectional weld-bead microstructures at suitable magnifications. After acquisition, the images were analyzed in ImageJ software to quantify the weld cross-sectional geometry, including fusion zone width, fusion zone area, heat-affected zone area, and excess metal height. For each weld, three cross sections were used for the analysis.

To measure the weld bead reinforcement, a laser line scanner (scancontrol LLT2810-25) was used, and the maximum height of excess metal was found. A quantitative analysis of the weld appearance was performed using a histogram of the height measurements and the interquantile range as metrics. For each case, approximately 1000 measurements (taken over about 120-millimeter length of the weld bead with tack welds region excluded) were used for the evaluation.

3 Results and discussion

This section presents the results, focusing on three aspects of analysis: weld pool top surface contour, weld cross section geometries and along weld bead measurements.

3.1 Weld pool top surface visualization and measurement

Figure 3 shows top view images of the weld pool acquired during welding with beam shapes TE, LE, DE, and C, respectively.

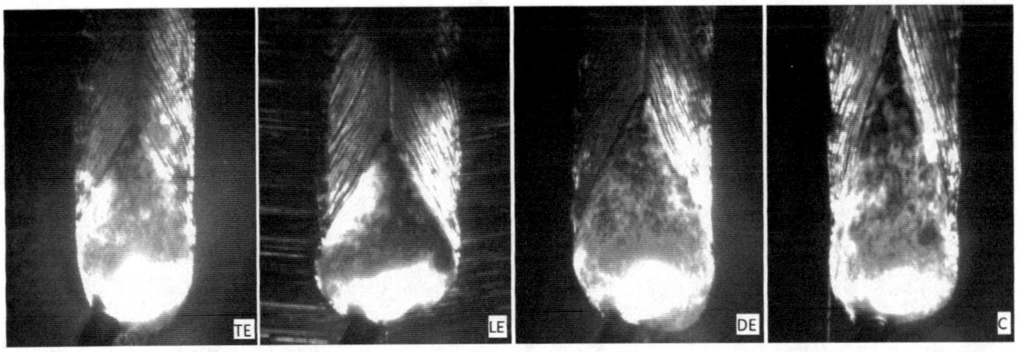

Figure 3. Weld pool top view images resulted from the four different beam profiles.

There is a noticeable difference in the geometry among the cases, with C beam shape (reference) resulting in the longest weld pool and LE exhibiting the shortest. This observation is anticipated, since both the temperature distribution and the fluid flow in the weld pool are governed by the distribution of the heat input [14].

Figure 4 shows weld pool contours, each one extracted from 10 randomly selected frames and plotted together for each of the different beam shapes. The effect of beam shaping on weld pool geometry was quantitatively analyzed using these contours.

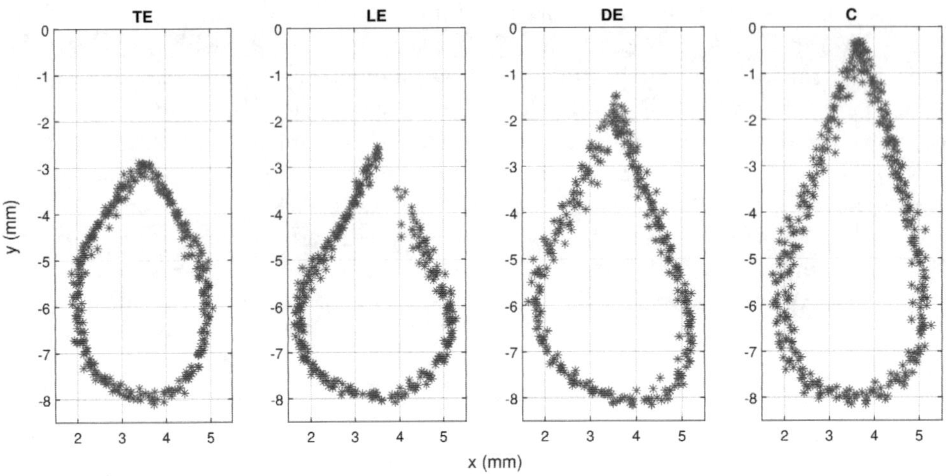

Figure 4. Weld pool contours extracted from 10 randomly selected image frames.

Figure 5 gives the dimensions of the weld pool, including width, length, and area measurements, and the angle formed at weld pool tails, resulting from the different beam shapes.

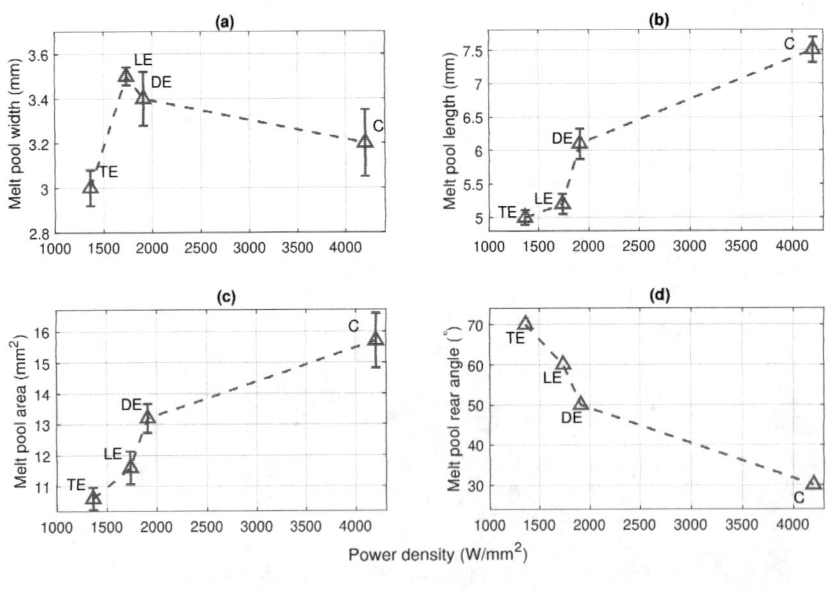

Figure 5. Quantitative data of the weld pool geometries in relation to the power density: (a) width; (b) length; (c) area; (d) angle of rear curvature.

The weld pool width, as given in Figure 5 (a), is compared to the reference case C. The LE beam shape resulted in a 9.0% wider weld pool while for beam shape TE the weld pool was 6.0% narrower. Taking the measurement uncertainty into consideration, beam shape DE resulted in a comparable width as LE. This means that even though the beam shape TE has the largest radius R_y (oriented perpendicular to the travel direction), as shown in Table 1, it still produced the narrowest weld pool among the four shapes. This finding is contrary to the conclusions drawn from earlier research on conduction mode welding [8] [14], which stated that larger beam radius R_y would result in a wider weld pool. This also implies that the crucial factors influencing the weld pool width are likely to be the laser's interaction time with the metal and the laser power density, since in this study, beam shapes LE and DE had larger beam radius R_x, which means longer interaction time, while beam shape C had the highest power density. Figure 5 (b) displays the measured length of the weld pool. Compared to beam shape C, the weld pools for beam shapes TE, LE, and DE were shorter by 33.3%, 30.7%, and 8%, respectively. One possible explanation for a longer weld pool for beam shape C is that the higher power density leads to an elevated temperature, causing a longer cooling time compared to the other three beam shapes. Another possibility is that, for beam shape C, the flow of molten metal is directed towards the end of the weld pool, exerting pressure on the liquid melt, and driving it away from the welding travel direction, resulting in a thinner and longer tail. Figure 5 (c) illustrates the area of the weld pools, showing the same trend as that for weld pool length. The largest weld pool area of beam shape C was due to the largest weld pool length. Figure 5 (d) shows the angle of the weld pool tail curvatures. Narrow weld pool tail (smaller angle of the curvature) is observed in the case of beam shape C. The narrow and long tail implies a slow solidifying rate at the center of the joint. The narrow tail could also affect the flow inside the weld pool, which eventually affects the process stability and the final weld seam properties. These geometrical measurement analysis demonstrate that the weld pool geometries were significantly influenced by beam shaping.

3.2 Weld seam cross section

Figure 6 shows the cross sections of the weld seams obtained from beam shapes TE, LE, DE, and C.

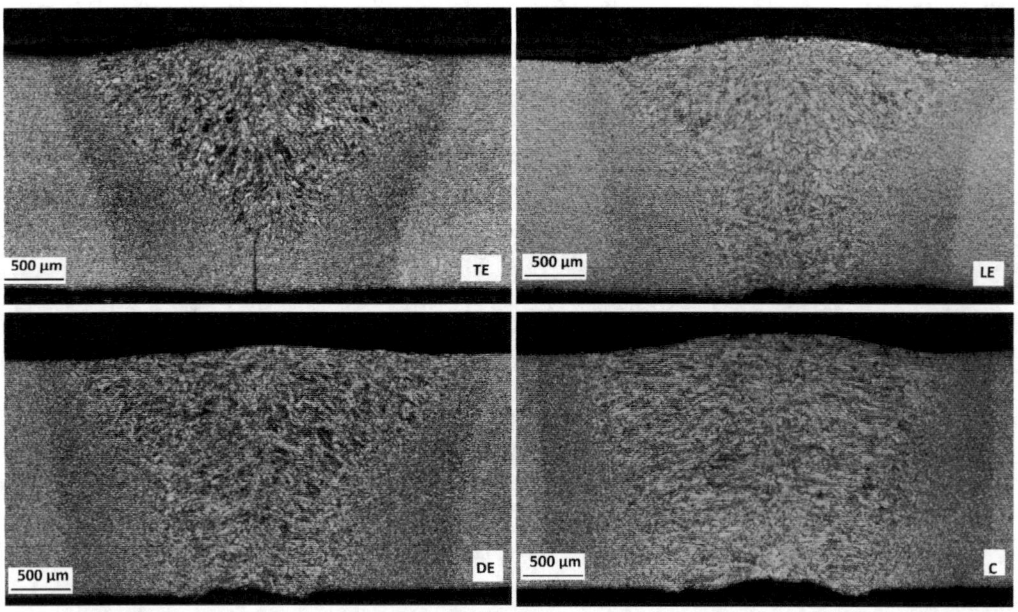

Figure 6. Cross section of the weld seam resulted from beam shapes TE, LE, DE, and C.

It can be seen that full penetration was obtained for all beam shapes except TE. The welding mode might also have been modified by the manipulation of the beam shape. Figure 6 (a) evidently shows that the weld produced by the TE beam does not represent a keyhole welding, while for DE and LE beams, the welding mode might be in transition mode or closer to keyhole mode. Besides, root concavity is observed in the seam obtained from beam shape C. The cross section images were further analyzed using software imageJ to quantify the fusion zone and the heat affected zone, as well as the height of the weld reinforcement. Figure 7 shows the quantitative measurements of the weld seam cross sections.

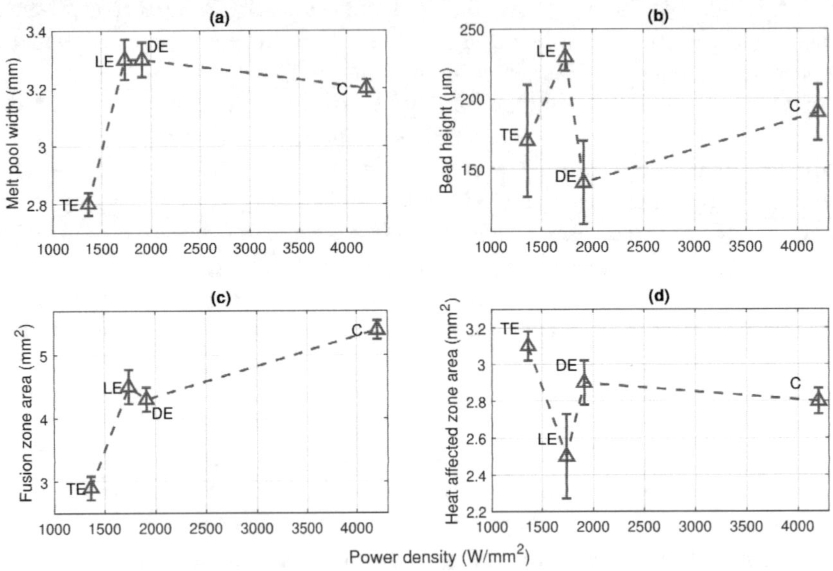

Figure 7. Cross section measurements in relation to the power density: (a) fusion zone width; (b): weld reinforcement; area of (c) the fusion zone and (d) the heat affected zone.

Figure 7(a) displays the fusion zone width measurement which is in good agreement with the weld pool width measurement presented in Figure 5(a). Figure 7(b) presents the weld bead reinforcement height. Beam shape LE yielded the highest weld bead reinforcement, whereas the lowest were obtained from the weld seam resulting from the DE beam shape. This is unexpected, since the wire feeding rate was constant, and both the top-view melt surface measurement and fusion zone width measurement confirmed that DE and LE resulted in comparable weld seam widths, so that a similar weld bead reinforcement is expected. However, one possible explanation could be that the shape of the weld bead reinforcement obtained by LE shape is more convex than that obtained by DE. This could also be observed from cross section images in Figure 6. This finding also reveals that by beam shaping, it is possible to modify the contact angle [15] (the convexity of the weld bead reinforcement), which is also an important factor in laser directed energy deposition with a filler wire. The weld bead reinforcement obtained by all the beam shapes meets the requirement even for the highest quality level, according to the standard (ISO13919-1). Figure 7(c) shows that the largest fusion zone area was obtained when the C beam shape was applied, while the fusion zone was smallest for the TE beam shape. This could be attributed to the larger fusion zone width at the bottom of the workpiece, which can be observed in Figure 6. Figure 7(d) shows that the largest heat affected zone was produced by beam shape TE, this might be due to the fact that less portion of the power contributed to melting when this beam shape was used.

3.3 Weld bead topology

Figure 8 presents the histogram of the maximal reinforcement height distribution measured from the weld bead scanning along the welded seam.

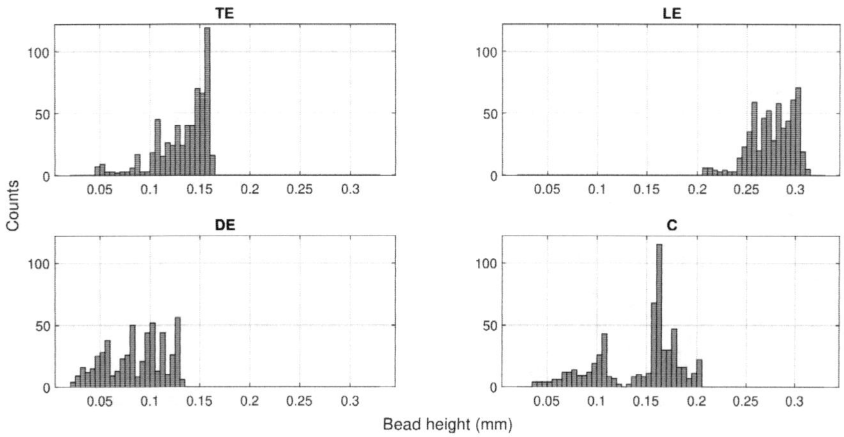

Figure 8. Histogram of measurements of the maximal weld bead reinforcement height produced by the beam shapes under investigation.

It can be observed that LE beam shape resulted in the highest bead and DE resulted in the lowest, which is in agreement with the measurement from the cross sections in Figure 7(b). It can also be noted that the height measurement distributes more widely for the beam shape C and this was quantified by the interquartile range, presented in Figure 9. It can be seen that beam shapes TE, LE and DE, all produced reinforcement height with lower interquartile range than that produced by beam shape C. This indicates that more stable process conditions can be achieved by shaping the beam, as concluded by previous research [16] [3]. It is also interesting to notice that the interquartile range of the height measurement is strongly correlated with the weld pool rear curvature angle. The smaller the angle, the larger the range.

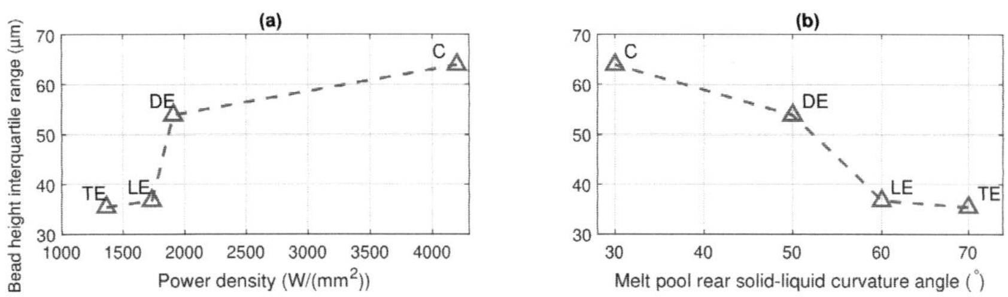

Figure 9. Interquartile range of weld bead reinforcement maximal height measurements as a function of (a) LB power density and (b) weld pool rear angle.

4 Conclusions and future work

This work presents a deformable mirror system that enables beam shaping of multi-kilowatt continuous wave lasers. Keyhole welding of butt-joints of S355MC steel plates with a filler wire with four different LB shapes has been investigated. The effect of beam shaping on weld pool geometries, welds cross sections and top surface appearance were quantified. Based on this investigation, the following conclusions can be drawn:

By beam shaping using a deformable mirror,

- the weld pool width (fusion zone with) can be expanded by 9% or contracted by 6%;
- the weld pool length can be diminished by a maximum of 33.3%;
- a more uniform weld bead appearance could also be obtained.

Acknowledgments

The authors gratefully acknowledge the Swedish Knowledge Foundation for granting the project SAMw (20170060) and Dedicate (20210094), which funded the activities of this work. Special thanks go to Daniel Sthen from GKN Aerospace in Trollhättan (Sweden) and to Kjell Hurtig from University West for their support during the experiments as well as to Dr. Stefano Bonora, from the CNR (Consiglio Nazionale delle Ricerche) Institute for Photonics and Nanotechnologies in Padova (Italy) for the valuable discussion and help with the deformable mirror system.

References

[1] Chen X, Lei Z, Chen Y, Han Y, Jiang M, Tian Z, Bi J, Lin S and Jiang N 2019 *Materials* **12** 4165
[2] Volpp J 2022 *J. Laser Appl.* **34** 042002
[3] Feuchtenbeiner S, Hesse T, Speker N, Haug P, Seebach J, Dubitzky W, Havrilla D and Hermani J P 2019 *High-Power Laser Materials Processing: Applications, Diagnostics, and Systems VIII* ed Kaierle S and Heinemann S W (San Francisco, United States: SPIE) p 31
[4] Galbusera F, Caprio L, Previtali B and Demir A G *J. Manuf. Process.* **85** 1024–1036
[5] Hansen K S, Olsen F O, Kristiansen M and Madsen O 2015 *J. Laser Appl.* **27** 032011
[6] Rinne J S, Nothdurft S, Hermsdorf J, Kaierle S and Overmeyer L 2020 *Procedia CIRP* **94** 661–665
[7] Prieto C, Vaamonde E, Diego-Vallejo D, Jimenez J, Urbach B, Vidne Y and Shekel E 2020 *Procedia CIRP* **94** 596–600
[8] Mi Y, Mahade S, Sikström F, Choquet I, Joshi S and Ancona A 2022 *Opt. Laser Technol.* **148** 107718
[9] Hansen K, Kristiansen M and Olsen F 2014 *Physics Procedia* **56** 467–476
[10] Nilsen M, Sikström F and Christiansson A K 2019 *J. Adv. Manuf. Technol.* **102** 3667–3676
[11] Heralic A 2012 *Monitoring and control of robotized laser metal-wire deposition* (Göteborg: Chalmers Reproservice) ISBN 9789173856553
[12] Phaoniam R, Shinozaki K, Yamamoto M, Kadoi K, Nishijima A and Yamamoto M 2014 *Welding in the World* **58** 469–476
[13] Kisielewicz A, Thalavai Pandian K, Sthen D, Hagqvist P, Valiente Bermejo M A, Sikström F and Ancona A 2021 *Metals* **11** 634
[14] Abadi S N R, Mi Y, Sikström F, Ancona A and Choquet I 2021 *Int. J. Therm. Sci.* **166** 106957
[15] Dass A and Moridi A 2019 *Coatings* **9** 418
[16] Volpp J and Vollertsen F 2019 *Opt. Laser Technol.* **112** 278–283

IOP Conf. Series: Materials Science and Engineering 1296 (2023) 012027 doi:10.1088/1757-899X/1296/1/012027

Effect of post-weld heat treatment on heat affected zone microstructure of additively manufactured Inconel 718

M Lindqvist[1], E Kivirasi[1], K Lipiäinen[2], H Piili[3] and A Salminen[3]

[1]Delva Oy, Hämeenlinna 13430, Finland

[2]Laboratory of Steel Structures, LUT University, Yliopistonkatu 34, Lappeenranta 53850, Finland

[3]Department of Mechanical Engineering Faculty of Technology, University of Turku, Turku 20520, Finland

markku.lindqvist@delva.fi

Abstract. Laser powder bed fusion for metals is a method of producing end use components for industrial use. Powder bed fusion machines are relatively small, and are usually used to create only the critical part of the larger assembly. Therefore, L-PBF manufactured parts must be attached to each other for example by welding. The industrial world needs to be able to join the printed superalloy components to the traditionally manufactured components to reach better corrosion, wear and/or heat resistance in selected parts in an assembly. The problem is that there is limited amount of information about the suitable welding parameter values for these applications. This study examines how the standard heat treatment cycles affect to the quality of the weld, and if the post-heat treatment is reducing undesired phases in the heat affected zone. Test has shown that post heat treatment highlights cuboidal shaped niobium rich carbides throughout the material to the heat affected zone grain boundaries.

1. Introduction

Laser based powder bed fusion (PBF-LB) which is one of the mostly wide used technology for additive manufacturing of metals is a mainstream technology used to manufacture metal components. It is used to replace traditional manufacturing usually in complex structures**[1,2]**. As products can be designed more efficiently many industries for example automotive, medical, and aerospace can benefit from it. It is also possible to cost-effectively use traditionally expensive or hard to manufacture materials. Inconel 718 is one of these materials. It is commonly used in demanding applications for example high temperature components**[3]**. Inconel 718 also has good fatigue and corrosion properties and it is weldable. **[3–6]** Even if traditionally manufactured Inconel welding is common in certain industries welding of AM Inconel is not widely researched. Process parameter values need to be carefully considered during laser welding to avoid undesired microstructures, mainly Laves phase **[5,7–9]**. Undesirable Laves phase is mainly generated due to the high heat input in the welding process and slow cooling rates**[10,11]**. Niobium rich Laves phase in Inconel 718 can reduce the material properties**[7]** in heat affected zone (HAZ)**[12,13]**. This is the reason for lower heat input when welding Inconel 718. Laser welding could be an effective method because its low heat input and faster cooling rates**[14,15]**. Higher cooling rates greatly reduce the Laves phase formation in the part**[16]**. Due to the AM microstructure, the phase formation during welding differs from traditionally manufactured Inconel 718. Post-weld heat treatments can be used to improve the situation after welding, and the research shows that the standard 980°C heat treatments significantly reduce the Laves phase and improve the properties of the weld**[5,12,17,18]**. Higher solution temperatures can be more beneficial.**[19]** Welding is often needed when combining benefits from additive manufacturing and traditional structures, welding of the smaller AM components to a larger structure can often be used to optimize benefits of different technologies. Especially in cases of large structures, AM is not feasible as a sole manufacturing method. **[20]** In these large structures, it is not always possible feasible to do the post-weld heat treatments, which can be problematic if a wrong microstructure is formed during the welding operations. Therefore, it is important to study what effect welding has in AM Inconel 718.

2. Aim and purpose

Authors have executed a study of laser welding and effect of post-weld treatment of additively manufactured materials. The whole study consists of IN718 – AISI316L and IN718 – IN718 material pairs. This study focuses on one material pair and heat treatment, laser welding of IN718-IN718 and post-weld heat treatment of it. This is done to have a deeper analysis for different cases. Structure of whole study is shown in Figure 1.

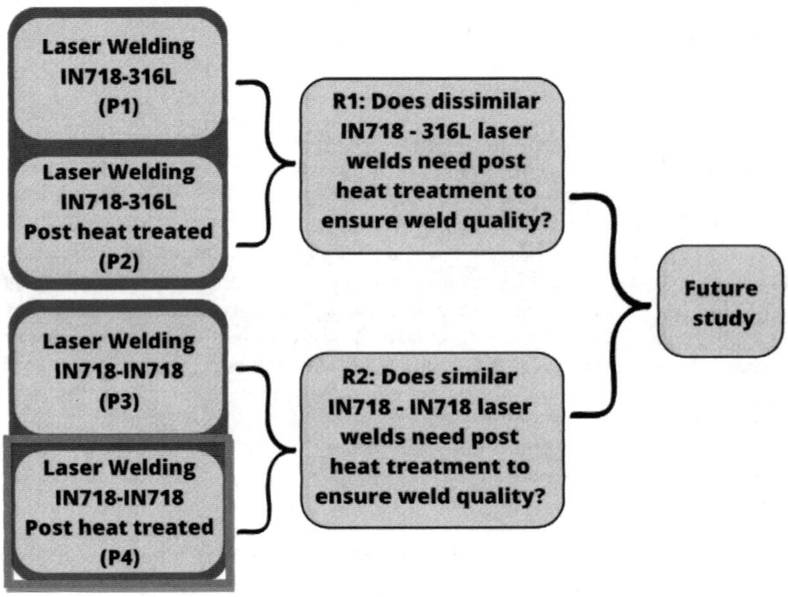

Figure 1. Structure of whole study on effects on laser welding of additively manufactured (PBF-LB) material pairs dissimilar IN718-AISI316L and IN718-IN718. Topic of the current study marked with red frame.

Aim of this study is to generate data on behaviour of studied materials and material pairs in laser welding and subsequent heat treatment. This data eases the use of different additively manufactured materials in industrial use.

3. Material and methods

3.1. Materials

Test samples for this test were manufactured with EOS M290 machine. Sample size was 10,5 x 80 mm round bar. All the samples were printed using standard EOS parameters. IN718_PerformanceM291_2.11 for PBF-LB Inconel 718. Samples for this test were printed in vertical position. Layer thickness of 40 micron was used. Powder was supplied by EOS GmbH. Chemical composition can be found in Table 1.

Table 1. Material composition of EOS NickelAlloy Inconel 718.

Element	Min [wt.-%]	Max [wt.-%]
Ni	50	55
Cr	17.0	21.0
Nb	4.75	5.5
Mo	2.8	3.3
Ti	0.65	1.15
Al	0.20	0.80
Co	N/A	1.0
Cu	N/A	0.3
C	N/A	0.08
Si, Mn	N/A	0.35
P, S	N/A	0.015
B	N/A	0.006
Fe	N/A	Balance

After printing, all the samples were heat treated according to standard AMS5662 heat treatment cycle. Heat treatment is performed by heating the samples to 954°C for 30 minutes followed by air cooling. After that double aging is done by holding the samples in 718°C for 8 hours followed by furnace cooling to 621°C at a rate of 50°C/h. The samples are then held at 621°C for 8 hours followed by cooling to room temperature. Welding parameters of tests are shown in Table 2.

3.2. Experimental setup

Laser welding tests were done by varying the welding speed with constant laser power. Ten different samples were welded with ten different speeds ranging from 1000 mm/min to 2350 mm/min. Welding parameters can be found in table 2. With this variation, the optimal energy input was studied. Welding tests for the samples were done at LUT University using IPG YLS-10000 fibre laser which has maximum power of 10 kW. Welding head in these tests was Precitec YW50 with 300 mm focal length focusing lens and 150 mm collimator lens. Used fibre core diameter was 600 μm. The optical arrangement give approximate diameter of focal point of 1200 μm. Shielding gas was not used during the welds. Focal point of the laser was set on the surface of the test samples. Test setup can be seen in Figure 2.

Figure 2. Test setup for welding.

Welds were performed by butt welding the two circular rods together. Samples are cut half from the middle of the sample to form suitable sample size for the microscopic analysis. Cut was done along the welded rods to form two semicircular samples. (Figure 3) All test samples were photographed after cutting. Samples were polished and etched with Adlers Etchant. Microscopic images were taken from all the samples, form fusion zone, HAZ, and base materials (Figure 3). SEM-imaging was used for some to get more information from the weld.

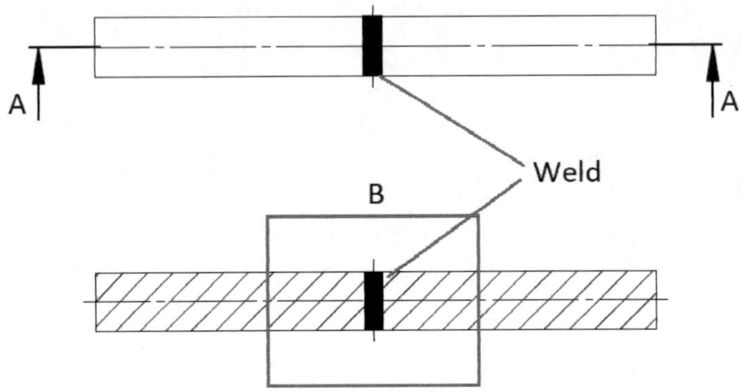

Figure 3. Welded rod was cut along axis A-A. Images were taken from samples from area B.

Welding parameters used in experiments were carried out by varying the welding speed thus the energy input changes. The laser power used was 2500W. The used welding speed and energy input for each test is shown in Table 2.

Table 2. Test welding variables in test for Inconel 718.

	Welding Speed [m/min]	*Energy input [J/mm]*
20-21	1.00	150
22-23	1.15	130
24-25	1.30	115
26-27	1.45	103
28-29	1.60	94
30-31	1.75	86
32-33	1.90	79
34-35	2.05	73
36-37	2.20	68
38-39	2.35	64

The energy input of laser welding was calculated by using equation:

$$E = P/v$$

Where E is energy input (J/mm), P is the laser power (W) and v is the welding speed (mm/s).

After the welding, samples were cut, and first half was considered as welded condition and second half was heat treated with standard AMS 5662 Inconel 718 heat treatment. Samples were polished and etched with Adlers Etchant. And microscopic images were taken. Some samples were also chosen for SEM-imaging to study further the test piece. Hardness measurements were done for the samples, from around 1 mm from the top of the sample across the weld zones (from base material to base material) to see the behaviour of the material across the weld.

4. Results and discussion

When comparing the welds between different heat inputs, effects can be seen. Rapid cooling leads to dendritic structures in the weld. Laves phase is formed in the HAZ which shows as a dark region in the optical microscopy images. This area becomes more visible and wider as the heat input is increased. This can be seen in Figure 4.

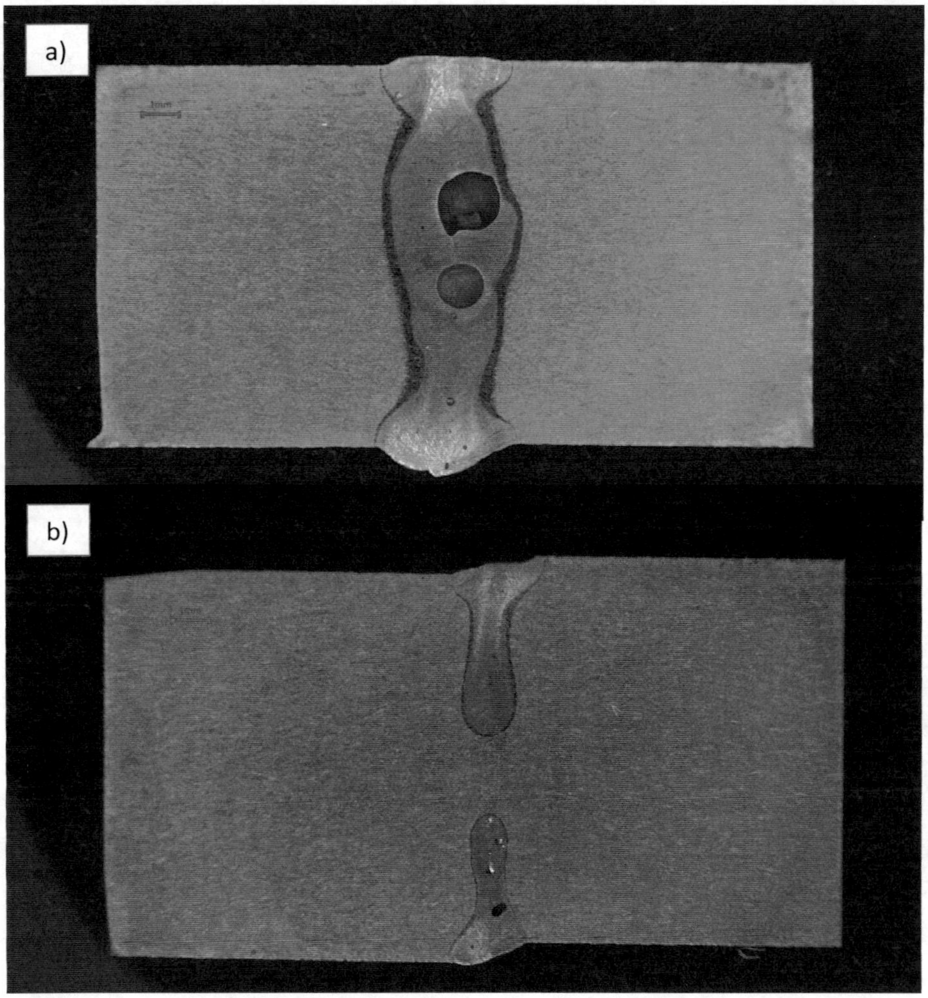

Figure 4. (a) High energy input weld test (samples 20-21). (b) low energy input weld test (samples 38-39). Comparison between highest and lowest energy input.

Weld penetration increases with energy input. It also increases the width of the weld. Width of the weld increases from around 1-1,5 mm (lowest energy input) to over 3 mm (highest energy input). At the same time penetration depth changes from around 4 mm where the roots are not touching to over 6 mm where the roots are overlapping each other. This causes some defects in the higher energy input samples. These welding defects are mainly porosity (ISO 6520-1 reference 200 as stated in standard ISO/DIS 13919-9) in the weld. Pre-heating was not used and it may help to reduce welding defects. On the other hand, pre-heating will introduce more Laves phase formation as the cooling rate after welding decreases.

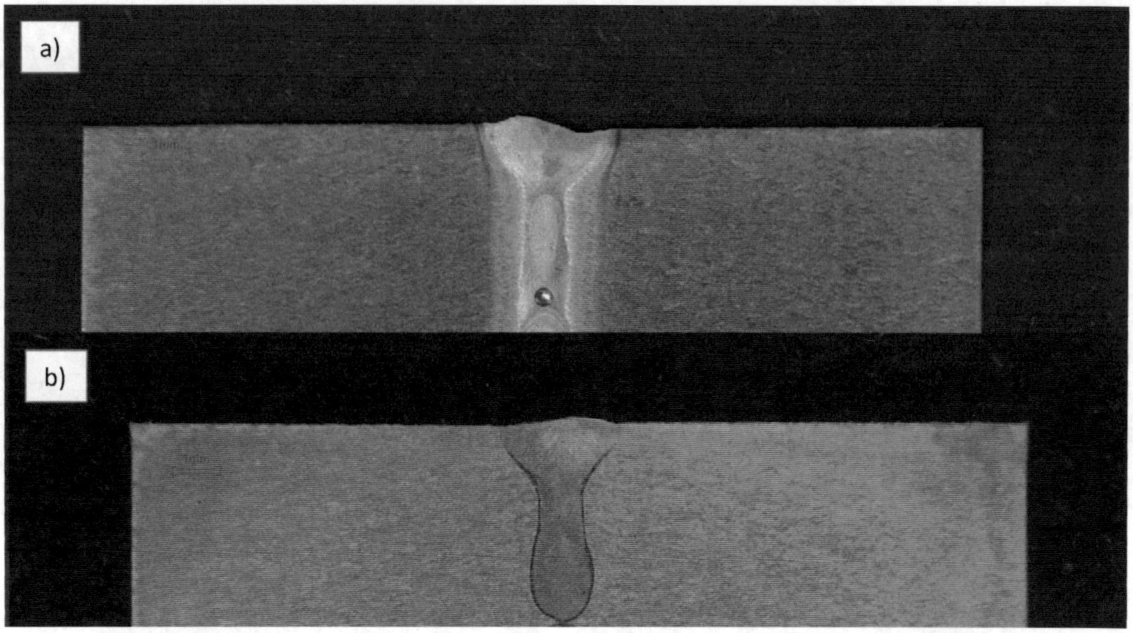

Figure 5. Differences in the HAZ can be seen in as welded (b) and post-weld heat treated parts (a). Samples 30-31.

Heat affected zone with undesired Laves phase can clearly be seen in as welded sample. Its visibility is greatly reduced with heat treatment (Figure 5). The standard AMS 5662 solution treatment + double aging is enough to modify the microstructure. The material microstructure becomes more uniform and separate region with Laves phase near the weld is removed.

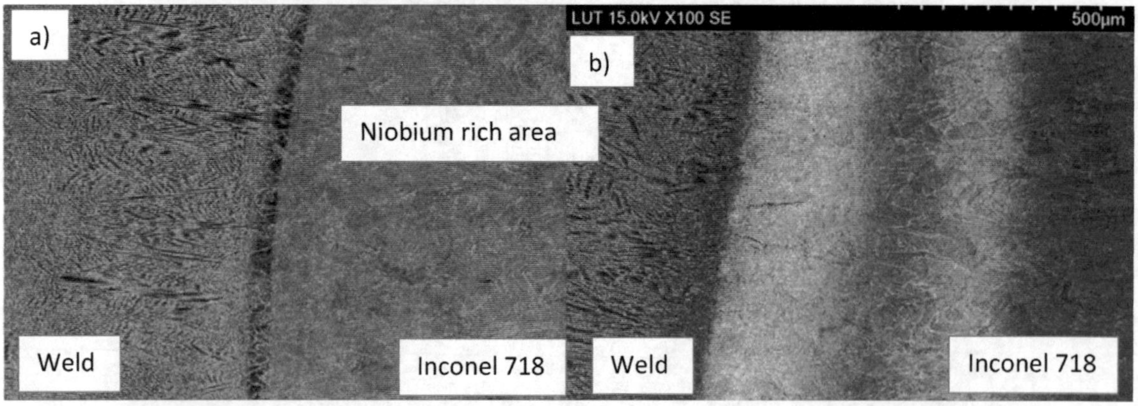

Figure 6. SEM image of the welds before heat treatment (a) and after heat treatment (b). Samples 30-31.

SEM images highlights the dark niobium rich zone between the weld and base material. SEM imaging shows Laves phase as white concentrations in the dark zone.

In addition to visual comparison, hardness tests were performed. Hardness profiles were measured across the welds in both post-weld heat treated and as welded states. Comparison of the hardness profiles can be seen in Figure 7. Sample 30-31 was chosen for this figure as the weld was visually the best looking. Other welds behaved similarly when comparing hardness's in as welded and post-weld heat treated state.

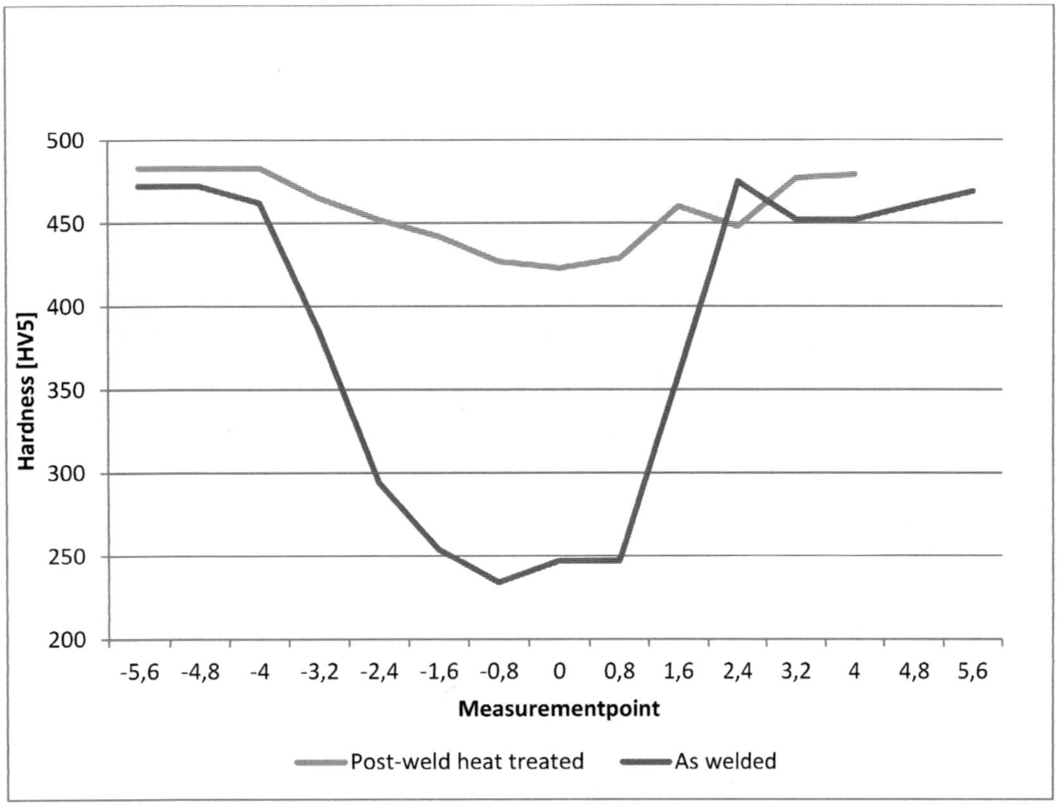

Figure 7. Hardness profile of the samples before post-weld heat treatment and after. Sample 30-31.

In these results post-weld heat treatment greatly increases the hardness of the weld. In as welded state, hardness observed were around 250 HV in the weld. After heat treating the parts, hardness of weld increased to 400-450 HV. This hardness range is lower than expected in printed and heat treated IN718 alloy.

5. Conclusions

Additively manufactured PBF-LB Inconel 718 welds were studied. According to the tests laser welding of Inconel 718 PBF-LB is very parameter value dependent. With suboptimal welding parameters, defects in the weld were observed. Also, hardness drop in the weld is significant compared to printed and heat-treated Inconel 718. This hardness can be increased by performing the AMS 5662 heat treatment for the welded component. The weld consists of columnar grains which transform into irregular grains in the melt pool edges. Heat treatment removes Laves phase and improves the material properties. Resulting hardness is not as high as non-welded Inconel 718. Both welding and heat-treating parameters should be optimized to find optimal solution for joining additively manufactured Inconel 718. Some cracking and porosity in the weld were observed. The microstructures and phases in the samples need to be examined further. Follow up Gleeble tests are planned to examine the right cooling time and adjust the welding parameters accordingly.

Credit authorship contribution

All authors contributed to the publication by either conceptualizing the idea, by analyzing the data or providing testing possibilities. Publication was commented and modified by all authors.
All authors contributed to chapters 2. Aim and purpose of the study and 3. Material and methods. Markku Lindqvist was main responsible for chapters 1. Introduction and 4. Results and discussion. All authors participated in conclusions and final analysis.

References

[1] Zhang R, Gardner L, Buchanan C, Matilainen V-P, Piili H, Salminen A. Testing and analysis of additively manufactured stainless steel CHS in compression. Thin-Walled Structures 2021;159:107270. https://doi.org/10.1016/j.tws.2020.107270.
[2] Karabulut Y, Tascioglu E, Kaynak Y. Heat treatment temperature-induced microstructure, microhardness and wear resistance of Inconel 718 produced by selective laser melting additive manufacturing. Optik 2021;227:163907. https://doi.org/10.1016/j.ijleo.2019.163907.
[3] Sonar T, Balasubramanian V, Malarvizhi S, Venkateswaran T, Sivakumar D. An overview on welding of Inconel 718 alloy - Effect of welding processes on microstructural evolution and mechanical properties of joints. Materials Characterization 2021;174:110997. https://doi.org/10.1016/j.matchar.2021.110997.
[4] Jiang R, Mostafaei A, Wu Z, Choi A, Guan P-W, Chmielus M, et al. Effect of heat treatment on microstructural evolution and hardness homogeneity in laser powder bed fusion of alloy 718. Additive Manufacturing 2020;35:101282. https://doi.org/10.1016/j.addma.2020.101282.
[5] Xu L, Chai Z, Peng B, Zhou W, Chen X. Effect of heat treatment on microstructures and mechanical properties of Inconel 718 additively manufactured using gradient laser power. Materials Science and Engineering: A 2023;868:144754. https://doi.org/10.1016/j.msea.2023.144754.
[6] Tucho WM, Cuvillier P, Sjolyst-Kverneland A, Hansen V. Microstructure and hardness studies of Inconel 718 manufactured by selective laser melting before and after solution heat treatment. Materials Science and Engineering: A 2017;689:220–32. https://doi.org/10.1016/j.msea.2017.02.062.
[7] Kouraytem N, Varga J, Amin-Ahmadi B, Mirmohammad H, Chanut RA, Spear AD, et al. A recrystallization heat-treatment to reduce deformation anisotropy of additively manufactured Inconel 718. Materials & Design 2021;198:109228. https://doi.org/10.1016/j.matdes.2020.109228.
[8] Kumara C, Balachandramurthi AR, Goel S, Hanning F, Moverare J. Toward a better understanding of phase transformations in additive manufacturing of Alloy 718. Materialia 2020;13:100862. https://doi.org/10.1016/j.mtla.2020.100862.
[9] Górka J, Jamrozik W, Kiel-Jamrozik M. The effect of TIG welding on the structure and hardness of butt joints made of Inconel 718. Heliyon 2023;9. https://doi.org/10.1016/j.heliyon.2023.e13175.
[10] Anbarasan N, Gupta BK, Prakash S, Muthukumar P, Oyyaravelu R, Kumar RJF, et al. Effect of Heat Treatment on the Microstructure and Mechanical Properties of Inconel 718. Materials Today: Proceedings 2018;5:7716–24. https://doi.org/10.1016/j.matpr.2017.11.448.
[11] Janaki Ram GD, Venugopal Reddy A, Prasad Rao K, Reddy GM, Sarin Sundar JK. Microstructure and tensile properties of Inconel 718 pulsed Nd-YAG laser welds. Journal of Materials Processing Technology 2005;167:73–82. https://doi.org/10.1016/j.jmatprotec.2004.09.081.
[12] Liu B, Ding Y, Xu J, Gao Y, Wang X, Zhang H, et al. Outstanding strength-ductility synergy in Inconel 718 superalloy via laser powder bed fusion and thermomechanical treatment. Additive Manufacturing 2023;67:103491. https://doi.org/10.1016/j.addma.2023.103491.
[13] Keivanloo A, Naffakh-Moosavy H, Miresmaeili R. The effect of pulsed laser welding on hot cracking susceptible region size and weld pool internal geometry of Inconel 718: Numerical and

experimental approaches. CIRP Journal of Manufacturing Science and Technology 2021;35:787–94. https://doi.org/10.1016/j.cirpj.2021.09.001.
[14] Sharma SK, Biswas K, Nath AK, Manna I, Dutta Majumdar J. Microstructural change during laser welding of Inconel 718. Optik 2020;218:165029. https://doi.org/10.1016/j.ijleo.2020.165029.
[15] Li L, Peng G, Wang J, Gong J, Li H. Experimental study on weld formation of Inconel 718 with fiber laser welding under reduced ambient pressure. Vacuum 2018;151:140–7. https://doi.org/10.1016/j.vacuum.2018.02.008.
[16] Hernando I, Renderos MA, Cortina M, Ruiz JE, Arrizubieta JI, Lamikiz A. Inconel 718 laser welding simulation tool based on a moving heat source and phase change. Procedia CIRP 2018;74:674–8. https://doi.org/10.1016/j.procir.2018.08.045.
[17] Gruber K, Stopyra W, Kobiela K, Madejski B, Malicki M, Kurzynowski T. Mechanical properties of Inconel 718 additively manufactured by laser powder bed fusion after industrial high-temperature heat treatment. Journal of Manufacturing Processes 2022;73:642–59. https://doi.org/10.1016/j.jmapro.2021.11.053.
[18] Wang X, Chou K. Effects of thermal cycles on the microstructure evolution of Inconel 718 during selective laser melting process. Additive Manufacturing 2017;18:1–14. https://doi.org/10.1016/j.addma.2017.08.016.
[19] Liu F, Lyu F, Liu F, Lin X, Huang C. Laves phase control of inconel 718 superalloy fabricated by laser direct energy deposition via δ aging and solution treatment. Journal of Materials Research and Technology 2020;9:9753–65. https://doi.org/10.1016/j.jmrt.2020.06.061.
[20] Zhang R, Buchanan C, Matilainen V-P, Daskalaki-Mountanou D, Britton TB, Piili H, et al. Mechanical properties and microstructure of additively manufactured stainless steel with laser welded joints. Materials & Design 2021;208:109921. https://doi.org/10.1016/j.matdes.2021.109921.

Acknowledgements

The authors would like to acknowledge all relevant personnel in Delva Oy for enabling the experiments carried out for this study and thank for their assistance and knowledge. Authors would also like to gratitude LUT University for their assistance and knowledge.

Manufacturing 4.0 (MFG4.0) project funded by the Strategic Research Council of Finland, 1.1.2018 – 31.12.2024. The Participant of project were MFG4.0 project is carried out by University of Turku, University of Jyväskylä, University of Helsinki and LUT University. The project aims to investigate on how the digital manufacturing will change fabrication in Finland.

DREAMS (Database for Radically Enhancing Additive Manufacturing and Standardization). This project was funded by Business Finland, 1.4.2022 – 28.2.2024. The Participants of project were the Finnish Additive Manufacturing Ecosystem (FAME). University of Turku, University of Oulu and LUT University. Aim of the project was to establish standards for AM and develop a powder material database for AM. Authors would like to thank partners of this project for their contribution to this article.

IOP Conf. Series: Materials Science and Engineering 1296 (2023) 012028 doi:10.1088/1757-899X/1296/1/012028

Process advantages of laser hybrid welding compared to conventional arc-based welding processes for joining thick steel structures of wind tower

Christian Brunner-Schwer[1], Ömer Üstündag[2,*], Nasim Bakir[2], Andrey Gumenyuk[1,2] and Michael Rethmeier[3,1,2]

[1] Fraunhofer Institute for Production Systems and Design Technology, Pascalstraße 8-9, 10587 Berlin, Germany
[2] Bundesanstalt für Materialforschung und -prüfung, Unter den Eichen 87, 12205 Berlin, Germany
[3] Institute for Machine Tools and Factory Management, Technische Universität Berlin, Pascalstraße 8-9, 10587 Berlin, Germany

*Corresponding author: oemer.uestuendag@bam.de

Abstract. The most common welding processes when joining thick-walled steels in the industry are arc-based welding processes such as GMAW or SAW. For this purpose, the sheets are joined in multi-layer technique, which can lead to productivity losses due to high welding times. The process-specific challenges in welding thick steels using multi-layer technique relate to the high heat input from the process. Therefore, alternative welding processes are being actively sought. A suitable alternative is provided by beam-based welding processes such as the laser hybrid welding processes, which are characterized by deep penetration welds and lower heat input. With implementation of the laser hybrid welding process in the heavy industry, such as the wind tower industry, economic benefits can be reached such as the increase in productivity by reducing the layer number, and the lower consumption of filler material and energy. When comparing SAW welded 25 mm thick steels in five to six layers and single-pass laser hybrid welding, the welding time can be reduced more than 80 % and the costs of filler material, flux and energy can be saved up to 90 %. However, the industrial use of the laser hybrid welding process is still limited to applications, where the material thickness does not exceed 15 mm due to some process-specific challenges such as the sagging, sensitivity to manufacturing tolerances such as gaps and misalignment, limited filler wire mixing, and deteriorated mechanical properties resulting from high cooling rates. To overcome these challenges, a contactless electromagnetic backing based on an externally applied AC magnetic field was used. Eddy currents are induced due to the oscillating magnetic field, and an upward-oriented Lorentz force is generated to counteract the droplets formed due to gravitational forces. It allows to weld up to 30 mm thick structural steels in a single-pass with a 20-kW fiber laser system. Additionally, the gap bridgeability and the misalignment of edges were increased to 2 mm when welding 20 mm thick steels. With the aid of the AC magnetic field, a vortex was formed in the weld root, which had a positive effect on the filler wire mixing. A further significant advantage of the EM backing was the possibility to expand

IOP Conf. Series: Materials Science and Engineering 1296 (2023) 012028 doi:10.1088/1757-899X/1296/1/012028

the process parameter window to maintain desired cooling times and mechanical properties, without suffering adverse effects concerning the root quality of the weld.

1. Introduction

The demand for clean energy has been steadily increasing in recent decades. According to data from eurostat, the share of renewable electricity generation in the European Union (EU) in total was approx. 38% in 2021. Wind energy accounts for 13.3% of the EU's total electricity generation. [1]. With regard to Germany, the federal government is aiming for an installed capacity of 115 GW (EEG 2023) of onshore wind energy by 2030 [2]. Some of the wind turbines (WT) have already reached the technical lifetime of about 20 years. This currently affects about 5,000 WTs that have reached the 20-year subsidy under the EEG, with another 8,000 WTs to follow by 2025 [3]. The expected decommissioning of old plants by 2030 is about 20 GW. Based on about 55 GW of currently installed capacity and an expected deconstruction of about 20 GW by 2030, an annual addition of 11 GW to 12 GW is necessary to achieve the targets for an installed capacity of 115 GW by 2030. The Federal Network Agency's tender volumes are 5.12 GW for 2022 and 12.84 GW for 2023 [4]. With a current nominal capacity of newly installed onshore WTs of about 4 MW per turbine, the requirement corresponds to an annual addition of about 3,200 WTs.

Figure 1 shows the basic structure of a tower schematically and the general production steps for one tower section. It can be seen that the prepared steel plates with a plate width of approx. 3 m and a plate thickness of up to 30 mm are first bent into a ring using a round bending machine. They are then joined with a tack weld and a longitudinal seam. After the rings have been aligned, around 10 rings are joined to form a section of approx. 30 m with the aid of a total of 9 circumferential weld seams. In the further course, flanges are attached, bolts welded on, door or hatch prepared, non-destructive tests carried out, painted, fitted inside with ladders and electrical lines and finally prepared for transport. The active German production sites for this kind of onshore steel wind towers are Leipzig, Schwerin, Gräfenhainichen and Fürstenwalde. According to the companies' own information, the production capacity of one plant is up to 10 sections per week. Subsequently, 4 sections each with an individual length of approx. 30 m are flanged on both sides and bolted to the tower at the installation site. The total production capacity of steel towers in Germany using conventional production methods, especially welding methods, is thus at most one sixth of the number of new towers planned.

2. Comparison of conventional welding processes with alternatives

Submerged arc welding (SAW) is the most common welding process in the thick plate sector, especially in the manufacture of steel towers for wind turbines. Currently, the plates, which are on average 25 mm thick, are welded in several layers both internally and externally using two-wire SAW. In a section of approx. 30 m length, this results in approx. 30 m longitudinal welds and, for a tower diameter of approx. 5 m, in approx. 141 m circumferential welds. Table 1 shows an example of the dimensions of a 120 m long steel tower and assumptions on the welding processes.

IOP Conf. Series: Materials Science and Engineering 1296 (2023) 012028 doi:10.1088/1757-899X/1296/1/012028

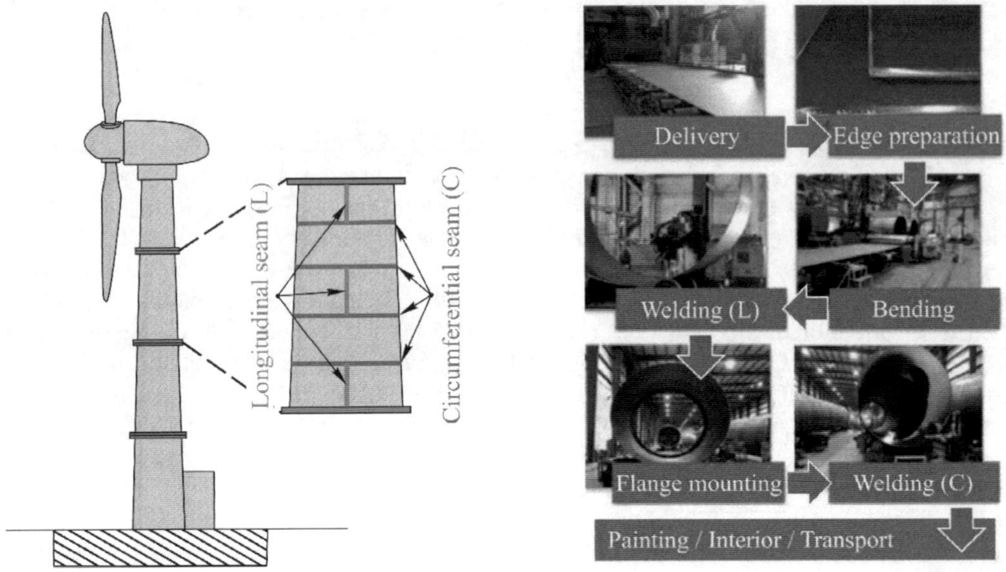

Figure 1. Schematic illustration of the structure of a tower and the general production steps.

Table 1. Assumptions for the manufacture of a teel tower for wind turbines.

Parameter	Value
Tower height L (m)	120
Average diameter of the tower D (m)	5
Number of sections A_s	4
Section length S_l each (m)	30
Panel width B (m)	3
Number of circumferential seams per section A_C	9
Number of longitudinal seams per section A_L	10
Sheet thickness d (mm)	25

	SAW	Laser-Hybrid
Welding speed (m/min)	$v_{SAW} = 0{,}6$	$v_l = 1$
Number of layers	$L_{SAW} = 5$	$L_l = 1$
Longitudinal seam	$L_n = L$	
Circular seam	$C_n = \pi * D * A_C$	
Total weld length L_T	$L_T = (L_n + C_n) * L_{SAW}$	$L_T = (L_n + C_n) * L_l$
Total welding time (h)	$t = L_T / v_{SAW}$	$t = L_T / v_l$

For a 120 m long tower consisting of 4 sections, this results in a weld seam length of just under 700 m, of which 120 m are longitudinal welds and approx. 564 m are circumferential welds. Currently, the towers with an average plate thickness of 25 mm are joined by SAW in 5 layers, bringing the total weld length to approx. 3.5 km. At a typical welding speed of 0.6 m/min for SAW, this results in a welding time of approx. 95 h. The alternative laser hybrid welding enables single-layer welding at a welding speed of approx. 1 m/min for the aforementioned plate thickness, so that the welding time for one tower can be reduced to approx. 12 h. Figure 2 shows the comparison of the welding times for multilayer SAW and single-layer laser hybrid welding for a tower with a length of 120 m, a diameter of 5 m and a plate thickness of 25 mm.

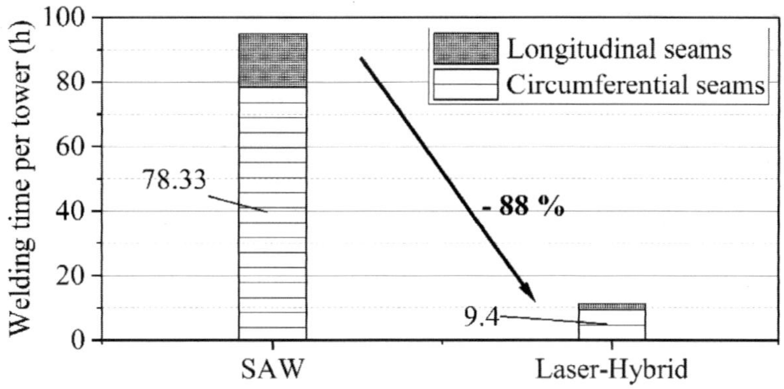

Figure 2. Comparison of welding times for multilayer submerged-arc welding and laser hybrid welding.

The Figure 2 shows that the laser hybrid welding process can significantly reduce the welding time for a tower by approx. 88% and significantly increase productivity by a factor of 8. In addition to the time factor, laser hybrid welding also offers other advantages, such as savings in filler material costs or powder costs. Typical seam preparations and for each case a schematic representation of the number of layers for submerged arc welding and laser hybrid welding of 25 mm thick sheets are shown in Figure **3**. From the seam preparations, it can be seen that in SAW a volume of approx. 231 mm² must be filled by means of molten filler material. For a weld length of approx. 700 m per tower, this corresponds to a volume of 0.17 m³, which means a consumption of approx. 1,250 kg of filler material per tower. The comparison of the costs for filler metal and powder or shielding gas for the examples already mentioned are listed in Table 2. The data refer to the production of one tower. The calculations made in this paragraph to compare the two welding processes are based on the formulas and the market prices for the materials used at that time (shielding gas, powder, etc.), which can be found in Table 2.

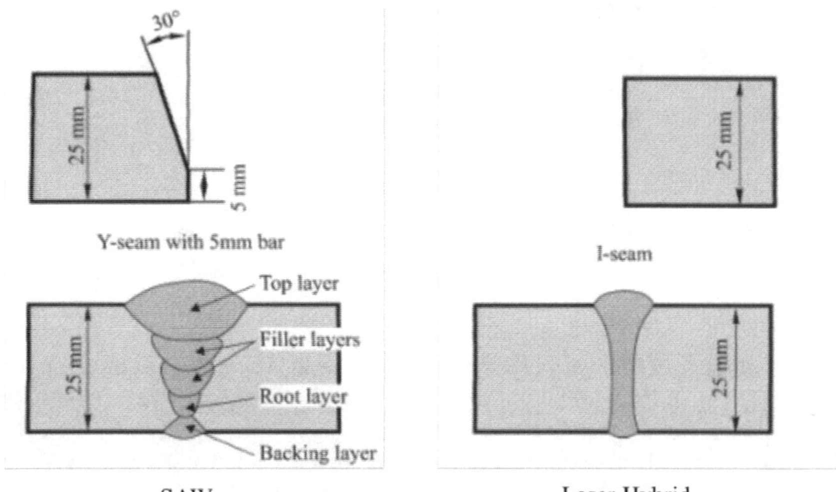

Figure 3. Comparison of edge preparations and number of layers for submerged arc welding (left) and laser hybrid welding (right) for welds with a plate thickness of 25 mm.

Table 2. Comparison of costs for filler metal and powder or shielding gas for submerged arc welding processes and laser hybrid welding.

	SAW		Laser-Hybrid
Wire consumption (kg)	$v_{joint} * \rho$		$v_{wire} * t * \rho$
Wire cost (€/kg)		1.50	
Powder/wire ratio		1.3:1	
Powder costs (€/kg)	$P_C = 1.30$		-
Inert gas consumption (l)	-		$20 \frac{l}{min} * t$
Inert gas costs (€)			35 € (50 l bottle at 300 bar)
Power P (kW)	19.5 – 24		25
Electrical efficiency η (%)	95		40
Power consumption (kWh)		$\frac{P}{\eta} * t$	
Electricity price S (Cent/kWh)		36.19	

Based on the assumptions and market prices, SAW results in approx. 1,860 € wire costs, 2,096 € powder costs and 1,592 € electricity costs per tower. With laser hybrid welding, the costs can be reduced by about 90% to less than 500 € per tower. The savings potential is over €5,000 per tower, Figure 4. With an annual production of approx. 100 steel towers, this corresponds to an annual cost reduction of 500,000 €. Thus, the investment costs of the laser system can be amortized within a few years.

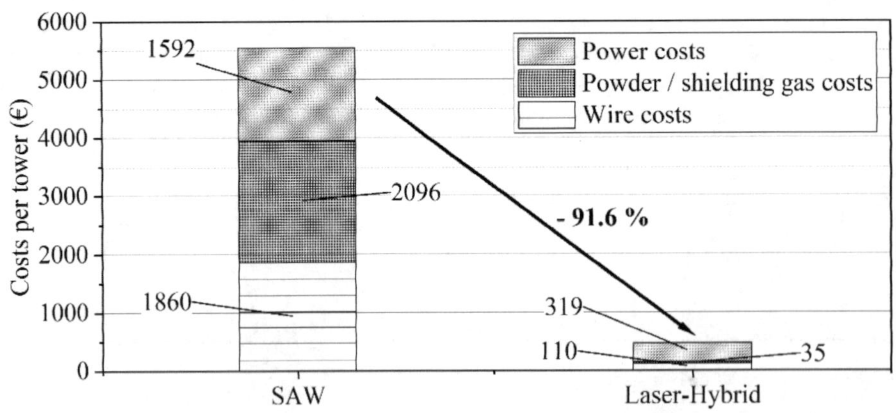

Figure 4. Comparison of costs for multilayer submerged-arc welding and laser hybrid welding.

In addition to all the economic benefits, laser hybrid welding also offers ecological advantages. For example, a study has shown that laser hybrid welding has a better ecological balance sheet than arc-based welding processes. For this purpose, measurements were carried out in the categories of climate change, acidification, eutrophication as well as photochemical ozone formation, showing that the laser hybrid welding process performed better in all categories compared to the conventional processes [5, 6]. Thus, it offers a high potential for sustainable manufacturing.

3. If the laser has so many advantages, why isn't it being used?

Recent developments in laser technology have brought a new generation of high-power laser systems in the power range between 10 kW and 100 kW onto the market. The use of high power lasers in welding thick-walled components is nevertheless severely limited. Successful examples are known from shipbuilding, where sheets with a thickness of up to 15 mm are joined using laser hybrid welding. However, an industrial breakthrough of high power lasers for joining higher sheet thicknesses, as in wind turbine towers, has not yet been achieved. The reasons for this are some process-specific limitations. First, it was found that the stability of the process decreased at higher laser powers. A scaling of the welding depth and the required laser power is not given above about 20 kW laser power. Other limitations are as follows:

- Droplet formation, especially during single-layer welding of thick sheets
- Sensitivity to manufacturing tolerances such as gaps or edge misalignment, which is why the panels would ideally have to be milled,
- deteriorated mechanical-technological properties, due to inhomogeneous distribution of the filler metal over the entire weld depth.

The drop formation on the root side occurs due to the high hydrostatic pressure as soon as this cannot be compensated by the Laplace pressure or the surface tension. The hydrostatic pressure increases linearly with the plate thickness, with the surface tension decreasing successively with an increase in the weld width. For molten steel, the stability criterion is dependent on the weld width and the plate height, Figure 5.

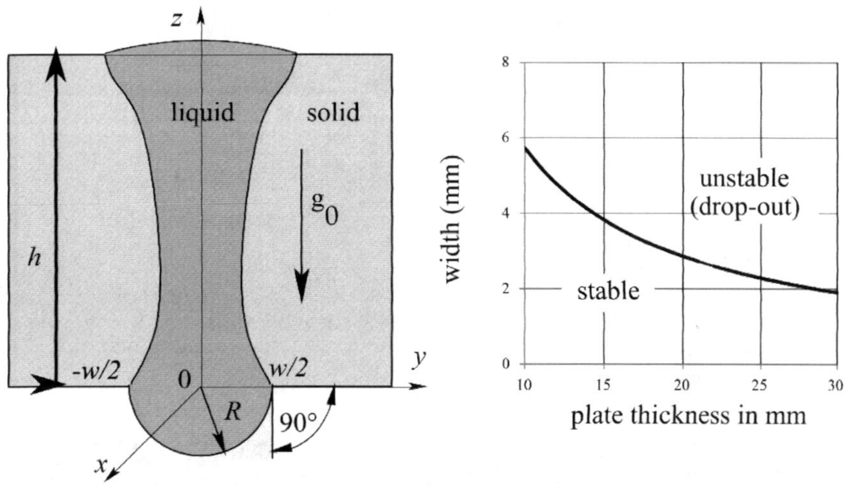

Figure 5. Stability criterion with regard to droplet formation during single-pass welding.

It can be seen that with a wall thickness of approx. 25 mm on the root side, from a weld width of over 2.5 mm, the molten metal can no longer be compensated by surface tension alone. A narrow weld seam can only be achieved at fast welding speeds. Increasing the welding speed leads to higher cooling rates and thus to martensite formation, hardening and deterioration of the mechanical-technological properties, in particular notched impact strength. Since the mixing and material transport of the filler metal is severely restricted at high wall thicknesses, the mechanical properties, especially in the root area, cannot be influenced by suitable filler metals.

4. The solution: magnet as a bath support

Thick-walled components can be welded at lower welding speeds in horizontal position (PC) or in PA position with weld pool supports or in multilayer technique. The more difficult handling of the components due to their dimensions makes welding in PC position not possible in most cases. Therefore, the use of a weld pool support is usually unavoidable. Conventional mechanical weld pool supports such as ceramic bath supports or copper rails are used. These require mechanical attachment, detachment and possibly reworking of the root area, which costs additional time and money. An alternative patented solution, which has been developed at the BAM, is the use of oscillating electromagnetic fields to prevent root drops. This solution is based on the generation of Lorentz forces in the weld pool. An oscillating magnetic field B perpendicular to the welding direction is generated by an AC magnet and produces eddy currents in the material. The electric current density j is parallel to the welding direction. The resulting Lorentz force F_L is directed upwards and counteracts the hydrostatic pressure. A schematic diagram of the non-contact electromagnetic weld pool support is shown in Figure 6.

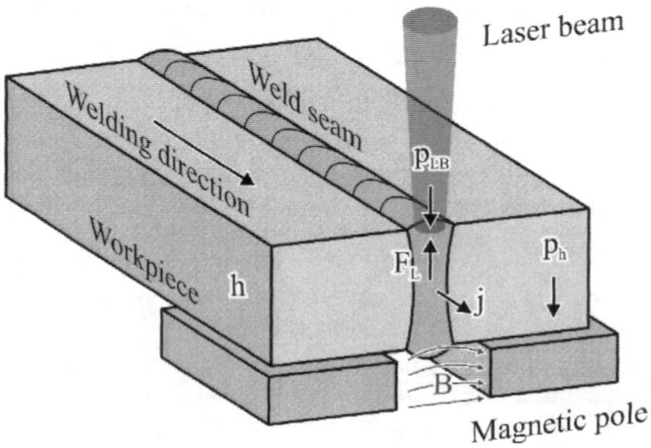

Figure 6 Schematic representation of the contactless electromagnetic weld pool support.

The non-contact electromagnetic weld pool support has already been successfully used in single-layer laser hybrid welding of S355J2 sheets up to 30 mm thick [7, 8]. The experiments were carried out using a 20-kW laser system. Sheets up to a thickness of 25 mm can be welded in square groove butt-joint. For higher plate thicknesses, a single V seam preparation with a total seam opening angle of about 30° - 45° and a root face of about 22 mm - 25 mm is preferred. These results also show that high sheet thicknesses can be joined in a single layer with lower laser powers.

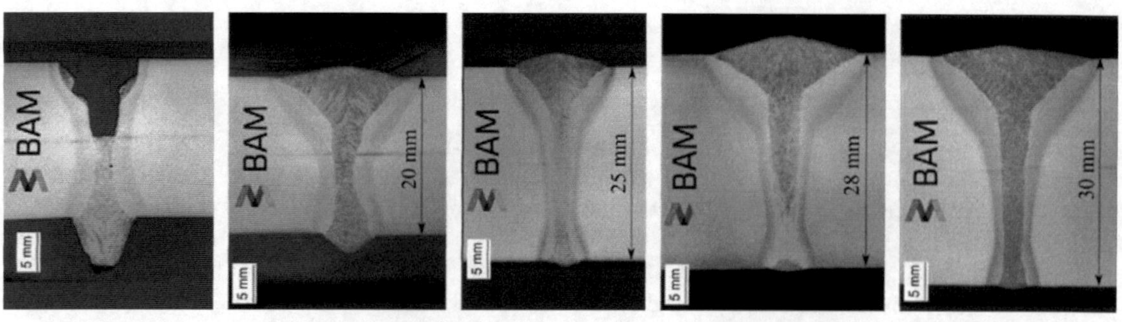

Figure 7. Single-layer laser hybrid welding tests on 20 mm to 30 mm thick sheets of S355J2; reference weld without non-contact electromagnetic weld pool support (left).

A weld seam of a single-layer laser hybrid welded 25 mm thick specimen with a total length of 1.2 m is shown as an example, in Figure 8.

Figure 8. Single-pass laser hybrid welded 25 mm thick specimen with use of the non-contact electromagnetic weld pool support.

5. Lower sensitivity to manufacturing tolerances due to magnet system

It is well known that laser beam welding or laser hybrid welding are sensitive to manufacturing tolerances such as gap and edge misalignment, especially when welding thick-walled steels due to the narrow beam diameter. Therefore, the joining parts must preferably be milled. Experimental investigations were carried out on single-pass laser hybrid welding of 20 mm thick steels made of S355J2 with edge preparation as laser cut, flame cut, water jet cut and plasma cut. The self-developed non-contact electromagnetic bath support was used as the weld pool support. The experiments show that the sensitivity of the laser hybrid process to a gap and edge misalignment could be reduced. A plasma cut quality is sufficient to weld the joining partners without internal and external defects by laser hybrid welding. For the laser cut and plasma cut specimens, high gap bridgeability could be achieved with a preset gap of up to 1 mm and 2 mm [9]. Compared to the state of the art, this corresponds to a threefold increase in gap bridgeability. Even with 25 mm thick sheets, a gap of up to 1 mm could be safely bridged.

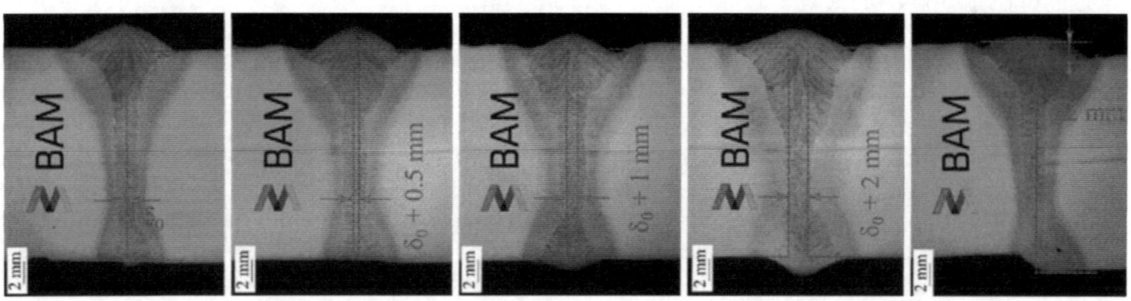

Figure 9. Transverse sections of a single-layer laser hybrid welded 20 mm thick plasma-cut specimen with different gap and edge offset.

6. Improved mixing

Another challenge is the inhomogeneous distribution of the filler metal over the entire weld depth. Some investigations to improve the mixing in laser hybrid welding have already been carried out. In these, a pulsed or modified arc process [10], a small air gap of about 0.3 mm between the joining partners [11], a trailing arc orientation [12], and an addition of about 2% oxygen in the shielding gas [13] were recommended for improved mixing. However, the experiments were carried out on sheets with a wall thickness of 10 mm to 15 mm. The noncontact electromagnetic bath support presented in this paper allows a more homogeneous distribution of the filler material on sheets up to 20 mm thickness. The electromagnetic fields cause a vortex-like flow in the molten bath, so that the distribution is positively influenced [14]. Figure 10 shows a comparison of a reference weld without electromagnetic bath support (15 mm wall thickness) and a single-layer welded 20 mm thick specimen with electromagnetic bath support. It can be seen that due to the external magnetic fields, alloying elements of the filler metal are homogeneously distributed even at a depth of up to 18 mm.

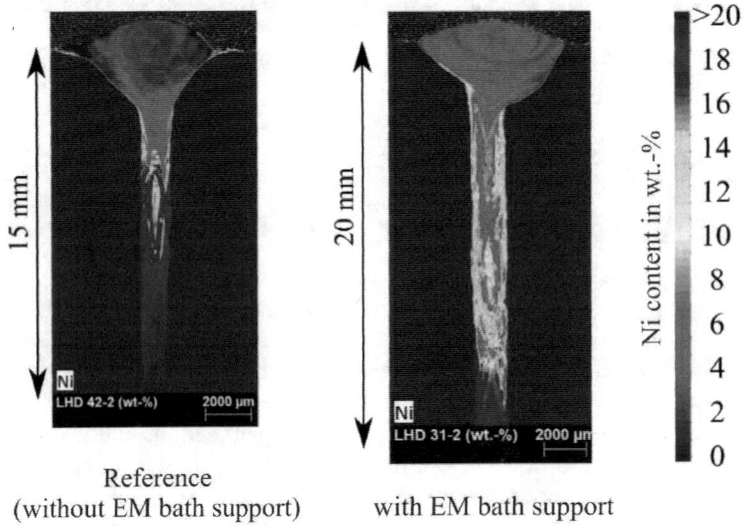

Figure 10. Element distribution of a single-layer laser hybrid welded specimen without electromagnetic weld pool support (left) and with electromagnetic weld pool support.

7. More control over welding parameters for highest welding quality

The choice of welding parameters is limited in laser or laser hybrid welding, especially for thick walled components. The reason for the limitations is, on the one hand, the formation of droplets at welding speeds that are too slow and, on the other hand, the deterioration of mechanical properties such as notched impact strength and hardening at welding speeds that are too high. This can result in poorer weld quality compared to conventional welding. With the help of noncontact electromagnetic bath support, this problem can be eliminated. This extends the parameters of laser or laser hybrid welding so that the user is no longer forced to weld at high speed. Since the risk of droplet formation is eliminated, a desired cooling time and thus adjustment of the welding speed can be made to maintain the required mechanical properties. For 25 mm thick plates of S355J2, it was found that the cooling time from 800 °C to 500 °C (t8/5 -time) below 1 s leads to poor notched impact strengths due to martensite formation and hardening, and the minimum requirement of 34 J/cm² is not achieved at a test temperature of -20 °C. At a t8/5 time of about 8 s (welding speed of 0.7 m/min to 0.9 m/min), notched impact strengths of about 150 J/cm² to 200 J/cm² could be achieved. At slower welding speeds and longer cooling times, the notched impact strength decreases due to grain coarsening. The impact strength of a laser hybrid welded 25 mm thick specimen made of S355J2 as a function of welding speed is shown in Figure 11.

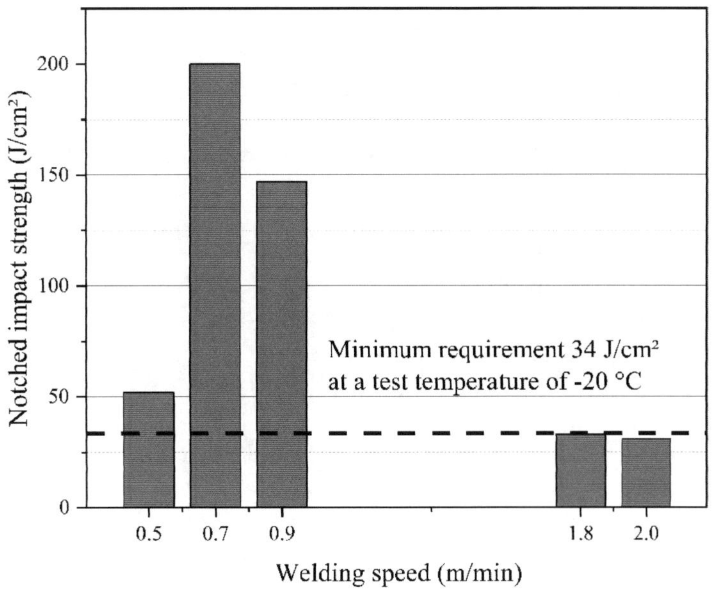

Figure 11. Notched impact strength of a single-layer laser hybrid welded specimen at different welding speeds.

8. Summary

Laser hybrid welding is an alternative to SAW of thick plates in multilayer technology and offers many advantages. One of the advantages is that the welding time of a wind turbine tower can be reduced by more than 85% or productivity can be increased by a factor of 8. In addition, with the help of laser hybrid welding, the cost of filler metal (wire and powder) and electricity costs can be reduced by about 90%. In addition to the economic advantages, the laser hybrid welding process is also the ecologically cleaner alternative compared to SAW. The use of the process in industry is limited by process-specific challenges. The patented non-contact electromagnetic bath support described in this paper serves to expand the potential of laser hybrid welding in the thick plate range > 15 mm. The advantages of the electromagnetic bath support includes:

- works contactless, so mechanical attachment and removal is not necessary
- sheets with a wall thickness of up to 30 mm can be joined in a single layer with a laser power of only 20 kW, drop formation is avoided
- time-consuming seam preparation can be dispensed with, a plasma or flame cut is sufficient
- gaps and edge offsets of up to 2 mm can be safely bridged
- more homogeneous distribution of the filler metal over the entire weld depth can be achieved
- process parameters can be extended to meet desired cooling times; positive influence on mechanical properties, especially notched impact strength.

References

[1] European Commission. Eurostat. Shedding light on energy in the EU. Publications Office. 2023.

[2] Wirtschaft und Klimaschutz, BMWK - Bundesministerium für. Erneuerbare Energien. 2023. https://www.bmwk.de/Redaktion/DE/Dossier/erneuerbare-energien.html. Abrufdatum 29.03.2023.

[3] Windräder: ab 2021 droht Rückbauwelle. 2023. https://www.din.de/de/din-und-seine-partner/presse/mitteilungen/windraeder-ab-2021-droht-rueckbauwelle-721594. Abrufdatum 29.03.2023.

[4] Bundesnetzagentur - Ausschreibungen. 2023. https://www.bundesnetzagentur.de/DE/Sachgebiete/ElektrizitaetundGas/Unternehmen_Institutionen/Ausschreibungen/start.html. Abrufdatum 29.03.2023.

[5] Sproesser, G., Chang, Y.-J., Pittner, A., Finkbeiner, M., and Rethmeier, M. Life Cycle Assessment of welding technologies for thick metal plate welds. In: Journal of Cleaner Production. 2015, S.46–53.

[6] Sproesser, G., Chang, Y.-J., Pittner, A., Finkbeiner, M., and Rethmeier, M. 2017. Sustainable technologies for thick metal plate welding. In *Sustainable Manufacturing: Challenges, Solutions and Implementation Perspectives*, R. Stark, G. Seliger and J. Bonvoisin, Eds. Springer Nature, Cham, S.71–84.

[7] Üstündağ, Ö., Fritzsche, A., Avilov, V., Gumenyuk, A., and Rethmeier, M. Hybrid laser-arc welding of thick-walled ferromagnetic steels with electromagnetic weld pool support. In: Welding in the World, 4. 2018, S.767–774.

[8] Üstündag, Ö., Avilov, V., Gumenyuk, A., and Rethmeier, M. Full penetration hybrid laser arc welding of up to 28 mm thick S355 plates using electromagnetic weld pool support. In: Journal of Physics: Conference Series. 2018, S.12015.

[9] Üstündağ, Ö., Bakir, N., Gook, S., Gumenyuk, A., and Rethmeier, M. Hybrid laser-arc welding of laser- and plasma-cut 20-mm-thick structural steels. In: Welding in the World, 3. 2022, S.507–514.

[10] Gook, S., Gumenyuk, A., and Rethmeier, M. Hybrid laser arc welding of X80 and X120 steel grade. In: Science and Technology of Welding and Joining, 1. 2014, S.15–24.

[11] Karhu, M., Kujanpää, V., Gumenyuk, A., and Lammers, M. 2013. Study of filler metal mixing and its implication on weld homogeneity of laser-hybrid and laser cold-wire welded thick austenitic stainless steel joints. In *Congress proceedings / ICALEO, 32nd International Congress on Applications of Lasers & Electro-Optics, October 6 - 10, 2013, Hyatt Regency, Miami, Florida, USA ; [including] Laser Materials Processing Conference, Laser Microprocessing Conference, Nanomanufacturing Conference*. LIA pub 616 = 106. LIA, Orlando, Fla., S.252–261.

[12] Muhammad, S., Miikka, K., Suck-Joo, N., Sang-Woo, H., and Veli, K. Effect of leading and trailing torch configuration on mixing and fluid behavior of laser-gas metal arc hybrid welding. In: Journal of Laser Applications, 4. 2017, S.42009.

[13] Zhao, L., Sugino, T., Arakane, G., and Tsukamoto, S. Influence of welding parameters on distribution of wire feeding elements in CO 2 laser GMA hybrid welding. In: Science and Technology of Welding and Joining, 5. 2009, S.457–467.

[14] Üstündağ, Ö., Bakir, N., Gumenyuk, A., and Rethmeier, M. Influence of an external applied AC magnetic field on the melt pool dynamics at high-power laser beam welding. In: IOP Conference Series: Materials Science and Engineering, 1. 2021, S.12017.

IOP Conf. Series: Materials Science and Engineering 1296 (2023) 012029 doi:10.1088/1757-899X/1296/1/012029

Influence of shielding gas on filler wire mixing at laser hybrid welding of thick high strength steels

Ömer Üstündag[1,*], Dirk Kampffmeyer[2], Michael Wolters[2], Nasim Bakir[1], Andrey Gumenyuk[1,2] and Michael Rethmeier[4,1,3]

[1] Bundesanstalt für Materialforschung und -prüfung, Unter den Eichen 87, 12205 Berlin, Germany
[2] Messer SE & Co. KGaA, Kleinewefersstraße 1, 47803 Krefeld, Germany
[3] Fraunhofer Institute for Production Systems and Design Technology, Pascalstraße 8-9, 10587 Berlin, Germany
[4] Institute for Machine Tools and Factory Management, Technische Universität Berlin, Pascalstraße 8-9, 10587 Berlin, Germany

*Corresponding author: oemer.uestuendag@bam.de

Abstract. The laser hybrid welding process offers many advantages during welding of thick-walled steels, such as the increased penetration depth and, thus, reduced number of layers, reduced heat input and decreased distortion compared to arc-based welding processes. Especially, when welding high-strength steels (HSS), the reduced heat input plays an essential role. However, a major challenge when laser hybrid welding of thick-walled steels is the limited filler wire mixing over the entire seam thickness, which can lead to changed mechanical properties over the depth. To overcome this issue, the add of oxygen into the shielding gas and its influence on the filler wire mixing and finally to the mechanical properties were investigated within this work. Therefore, 20 mm thick S690QL steels were laser hybrid welded in a single-pass. A contactless electromagnetic backing was used to avoid sagging. The admixture of oxygen was performed by a gas mixer, where the oxygen content was varied between 0 % and 7.2 %. The experiments were also accompanied by laser beam welding tests in steel/glass configuration, where the melt pool geometry as well as the melt flow characteristics were captured by a high-speed camera. It can be concluded, that adding of approx. 4 % oxygen into the shielding gas had a positive effect on the filler wire mixing, were up to a depth of 18 mm elements of the filler wire could be observed.

Introduction

The laser hybrid welding process was developed at the end of the 1970s as arc augmented laser welding by coupling tungsten inert gas (TIG) welding and laser beam welding (LBW) in a common interaction zone [1]. Since then, the process has been further developed and has already achieved an industrial breakthrough in some industries, such as shipbuilding and crane construction. The goal of coupling these welding processes is to remove their limitations and capitalize on the individual processes' benefits. Due to the high penetration depth and high productivity, the laser hybrid welding process offers an alternative to conventional arc welding processes, especially for thick plate welding. But some process-specific challenges limit the industrial use of the laser hybrid welding process in thick-plate welding, where plate thicknesses exceed 15 mm.

One of the limiting factors is the restricted filler wire mixing, where a homogeneous distribution of the added filler wire supplied from above cannot be provided over the entire seam thickness. This can lead to different microstructures and finally to different mechanical properties over the seam thickness. Basically, a laser hybrid weld is characterized by a wine-cup shape with a wide seam and HAZ on the upper surface, which is mainly influenced by the arc welding process and a narrow seam with nearly parallel seam flanks in the lower part, which is typical for LBW seam. Therefore, the terminology of arc-dominated and laser-dominated zone has been established and accepted by many researchers, when reporting on the mechanical properties of laser hybrid welded thick-walled steels, due to different cooling conditions and microstructure in the respective parts [2]. In [3] it was reported that the Charpy impact toughness of a single-pass hybrid laser-arc welded 25 mm thick structural steel plate is decreased up to 60 % in the root part (laser-dominated zone) compared to the top part (arc-dominated zone), where the decrease in cooling time over depth, the resulting microstructure and grain size, together with the inhomogeneous filler wire mixing were the reasons for the inhomogeneous course of the mechanical properties over the seam depth.
The uneven filler wire mixing over the entire seam depth is a well-known challenge when laser hybrid welding of thick-walled steels, where elements of the filler wire could be measured up to a depth of 7 mm [4,5]. After this penetration depth no elements of the filler wire was detectable. An element mapping on the cross-section of a laser hybrid welded 10 mm thick steel is shown in Figure 1, where a Ni-based filler wire was used as a tracing element. It serves as an illustration of the inhomogeneous filler wire mixing.

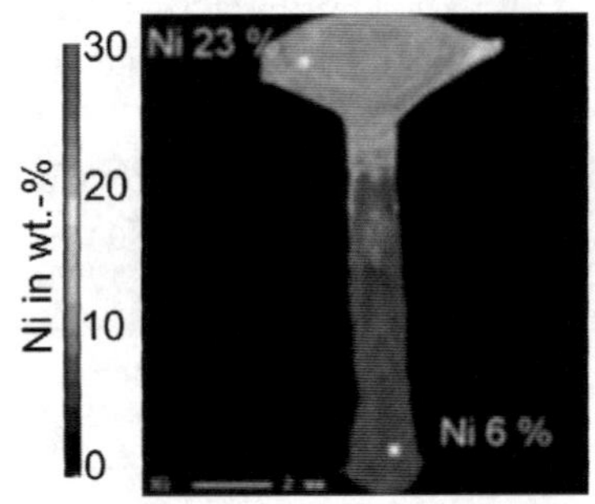

Figure 1. Element mapping (Ni-distribution) on a single-pass laser hybrid welded 10 mm thick steel plate [5]

Therefore, several studies dealing with the filler wire mixing were conducted in the past. Thereby, the influence of the welding parameters on the filler wire mixing were mainly studied. The influence of the arc mode showed a great influence on the filler wire mixing in contrast to the arc power when laser hybrid welding of 20 mm thick steels with a laser power of 16 kW as reported in [5]. A modified spray arc rather than a pulsed GMAW was recommended because it provided a deeper penetration and formation of a deeper arc-dominated zone due to its more powerful dynamic effect on the weld pool. Similar findings were reported in [6] where the proportion of the deposited filler wire to the volume of the fusion zone was a critical aspect impacting the filler wire mixing. Another parameter which influenced the filler wire mixing was the interdistance between the wire tip extension and the laser beam. A short distance of 2 mm to 6 mm was recommended when laser hybrid welding of 10 mm thick steels using 8 kW laser power [6]. A further parameter influencing the filler wire mixing was the process order. A laser leading orientation was recommended when welding 10 mm thick steels with a laser power of 10 kW [7]. Differences in the flow pattern and temperature contours could be observed depending on the process order numerically. The transport mechanism for transferring the filler material during laser hybrid welding was changed during laser leading process order [8,9]. Two recirculating flows, one on the upper part and another on the root surface could be detected. In contrast to the laser leading process order, the solidification front revealed that the molten flow between the upper and root part was separated in the arc leading process order [9]. A formation of an inward flow could be detected also at

a laser leading process order, where this inward flow was formed when the fluid flowed from the rear end to the keyhole on the pool surface before flowing to the bottom, which also improved the filler wire mixing [10]. Another effective method to improve the filler wire mixing when thick-plate welding to set an air gap of 0.3 mm to 0.4 mm between the joining parts as reported in [7,11]. An air gap of 0.4 mm was recommended when laser hybrid welding of 45 mm thick steels in counter-layer technique with a laser power of 15 kW at welding speeds of < 1 m/min. At faster welding speeds, the stability of the weld pool and filler metal transportation efficiency decreased [11]. The shielding gas can also have an influence the melt flow characteristics and finally the filler wire mixing. A shielding gas containing more than 2 % O_2 resulted in a more homogenous distribution of alloying elements [10].
Another approach to influence the melt flow and the filler wire mixing was performed with the aid of externally applied oscillating magnetic fields. First, a magnetic field with a low frequency of 10 Hz to 50 Hz was applied for wire-feed LBW of 3 mm thick aluminium plates [12-14]. The filler wire mixing was improved by oscillating magnetic fields and its effect on the flow dynamics. It was also reported that the oscillating frequency of the magnetic field and the magnetic flux density had an influence on the filler wire mixing. Best results were obtained with a coaxially inhomogeneous magnetic field. In [15] an external magnetic field was applied for wire-feed LBW of 10 mm thick austenitic stainless steel to improve the mixing behaviour. It was found that a bulging region is formed, which narrows the metal transfer channel from the top to the bottom region. A change of the magnetic field orientation of 10° to 20° to the welding direction was recommended numerically as well as experimentally to eliminate the bulging phenomenon, thus providing a downward transfer channel for the melt flow. In these studies, the magnet was operated on the top side during wire-feed LBW in partial penetration mode.
With a magnet operating at the bottom side to prevent sagging as well as to improve the filler wire mixing at single-pass HLAW of steel plates with a wall thickness of 20 mm and above was demonstrated in previous studies [16,17]. Therefore, an AC magnet was positioned under the workpiece and operated at a magnetic flux density of approx. 180 mT. It could be shown that a turn of the magnetic field orientation to the welding direction and the oscillating frequency had a major influence on the filler wire mixing. With the turn of the magnetic field orientation a rotational component of the generated Lorentz force could be formed, which had a positive effect on the filler wire mixing. The oscillating frequency of the magnetic field influenced the skin layer depth and thus the penetration depth of the magnetic field and the induced eddy currents into the material, which also influenced the filler wire mixing. During observations of the melt flow geometry during LBW in steel/glass configuration it could be determined that a vortex was formed at the seam root in the opposite direction to the flow on the upper part. A restricted area was found at the mid-thickness which favours the limited transport of the filler wire from the top to the bottom of the seam. The shape of the molten pool, the flow direction and their velocities during LBW with the mentioned narrowing effect which probably inhibits the filler wire mixing in the mid-thickness is shown in Figure 2.

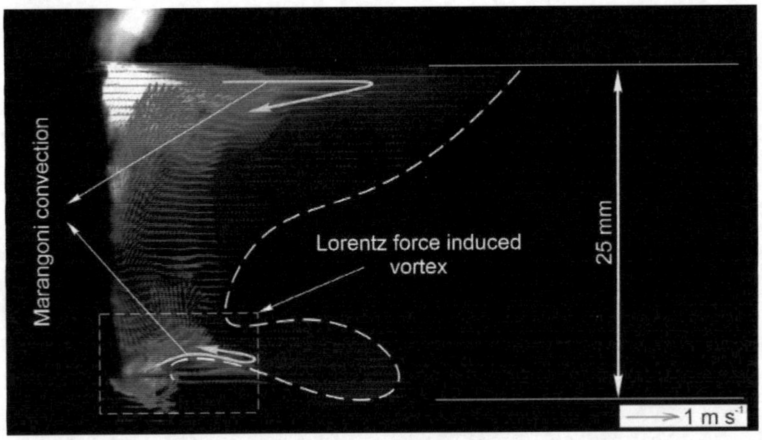

Figure 2. Observations of the melt flow geometry during LBW in steel/glass configuration [17]

The aim of this study is to investigate the filler wire mixing during laser hybrid welding of 20 mm thick steels of S690QL with externally applied oscillating magnetic fields and the addition of oxygen into the shielding gas in the range of 2 % to 7.2 %.

Experimental setup

LBW experiments in steel-glass configuration

The high-power fibre laser IPG YLR-20000 with a maximum output power of 20 kW was used as the laser beam source. The emission wavelength and beam parameter product were 1070 nm and 11 mm x mrad, respectively. The laser radiation was transmitted through an optical fibre with a core diameter of 200 µm. A laser-processing head BIMO HP has been selected, which provides a magnification of 2.8 so that the laser beam can be focused into a spot with a diameter of 560 µm. To detect the keyhole during high-power laser beam welding a special setup was necessary. Welding trials in butt joint configuration of 20 mm thick structural steel plate (S690QL) and quartz glass were conducted. Side views of the molten pool were taken with help of a highspeed camera Fastcam 1024PCI and interference band-pass filter at 808 nm and band width of 20 nm. The frame rate and the frame size were 2000 fps and 1200 pixels to 1200 pixels, respectively. An oscillating magnetic field generated by an AC electromagnet was applied to the root side of the weld specimen, where the magnetic field was turned 15° perpendicular to the welding direction. The schematic representation of the experimental setup during LBW in steel/glass configuration is shown in Figure 3. The LBW experiments on steel-glass configuration were performed using a laser beam power of 17.7 kW at a welding speed of 1.2 m/min and a focal position of -5 mm. The AC magnet was operated at an oscillation frequency of 780 Hz and a magnet power of 1.7 kW (corresponds to approx. 180 mT).

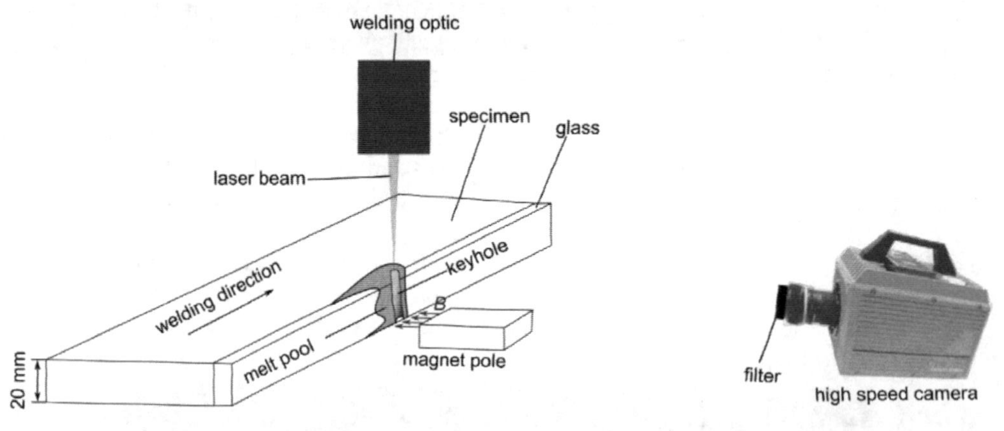

Figure 3. Experimental setup for LBW in steel-glass configuration.

Laser hybrid welding experiments

The HLAW experiments were performed using the same laser source as described. Additionally, a welding machine Cloos Quineo with a maximum current of 600 A was used as an arc power source. The laser optics and GMAW torch were mounted on the robot arm, where the laser axis was positioned 90° to the weld specimen surface and the GMA torch was tilted 25° relative to the laser axis. The processing head and the magnet remained in a fixed position during the welding, where the specimens were moved by an external axis. Figure 4 shows the setup for the HLAW experiments.

Figure 4. Experimental setup for HLAW of 20 mm thick S690QL in butt joint configuration with 15°-turn of the magnetic field orientation.

The experiments were carried out on 20 mm thick steel plates (S690QL) in butt joint configuration with an arc leading process order and a distance of 4 mm between the two heat sources and under the following welding parameters: laser beam power of 17.7 kW; welding speed of 1.2 m/min; focal position of the laser beam of -5 mm; wire feed speed of 13 m/min and stick-out of 16 mm. For the experiments a gas mixer with four connections for different welding gases was used, where three different welding gases were connected to generate the desired compositions: one bottle each 100% Ar, M21 (18 % CO_2 in Ar) and M22 (8 % O_2 in Ar). The desired compositions are summarized in Table 1. The volume flow rate was set to 20 l/min. A solid wire Thermanit316 with a diameter of 1.2 mm was used as a filler wire. The magnet parameter remained unchanged.

Table 1. Gas compositions during laser hybrid welding experiments

Gas/percentage	Ar (100 % Ar)	M21 (18 % CO_2 in Ar)	M22 (8 % O_2 in Ar)	Total gas composition
Gas 1 (Reference)	10 %	90 %	-	83.8 % Ar; 16.2 % CO_2
Gas 2	10 %	65 %	25 %	86.3 % Ar; 11.7 % CO_2; 2 % O_2
Gas 3	10 %	40 %	50 %	88.8 % Ar; 7.2 % CO_2; 4 % O_2
Gas 4	10 %	15 %	75 %	91.3 % Ar; 2.7 % CO_2; 6 % O_2
Gas 5	10 %	-	90 %	92.8 % Ar; 7.2 % O_2

For metallographic inspection, longitudinal and cross-sections of the laser hybrid welded samples were cut. After polishing and etching the samples using 2 % nital solution, the mixing of the filler wire over the thickness was evident.

Welding materials

For the experiments structural steel plates of S690QL with a thickness of 20 mm were used. The filler wire diameter was 1.2 mm. The chemical composition of the materials used are listed in Table 2.

Table 2. Chemical composition of materials used, shown in wt-%.

Material/Element	C	Mn	Si	P	S	Cr	Ni	Mo	Cu	Nb	Fe
S690QL	0.2	1.7	0.8	0.02	0.01	1.5	2	0.7	0.5	0.06	bal.
Thermanit 316	0.03	1.5	0.7	-	-	19	12	2.7	-	-	bal.

Results and discussion

LBW experiments in steel-glass configuration

Melt pool geometry recorded by high-speed imaging during LBW in a steel/glass configuration shows differences when oxygen is added to the shielding gas. For the reference case without oxygen (gas 1), the geometry of the melt pool shows that the length of the melt pool is about 23.6 mm at the top, 12 mm in the middle part and 19.7 mm at the bottom. Two main flows can be identified from the high-speed images. The flows that may be in the beam path of the laser, e.g. on the wall of the keyhole, have two directions: downwards and upwards. The upward and downward flows are mainly driven by the vapour velocity within the keyhole. The Marangoni flow was also observed on the surface of the melt pool. It is obvious that two different vortices form on the top surface or at the weld root. In addition, a narrowed or constricting region with a low melt flow rate forms between the two vortices. This phenomenon can lead to a separation of the material transport from the top to the weld root, as opposite vortices are formed in the top and the root part, separated by a constricting region. Similar results were also obtained by Meng et al. [15] and Sohail et al. [9] and identified as the cause of the limited filler wire mixing. With the addition of oxygen to the shielding gas, the width of the constricted area is increased. At an O_2 content of 4 %, the result is a melt pool geometry that is almost free of necking, as can be seen in Figure 5.

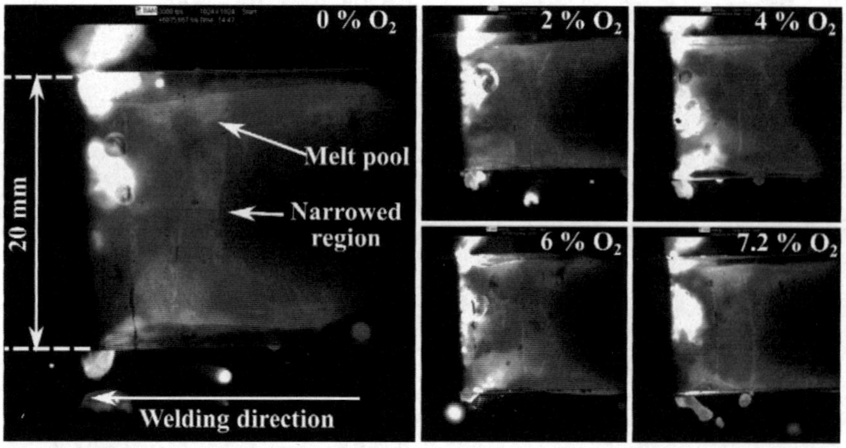

Figure 5. Records of the melt pool geometry during LBW in steel/glass configuration with varying oxygen contents.

Figure 6 shows the estimated melt pool geometries during LBW I steel/glass configuration with varying shielding gas compositions. As it can be noted, the melt pool length is increased on the free surfaces due to the Marangoni flow.

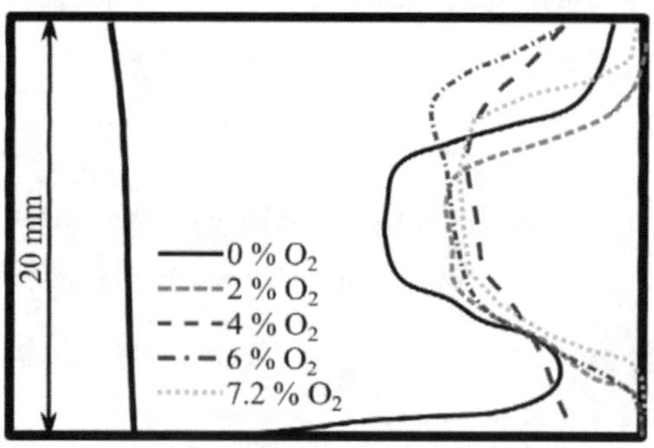

Figure 6. Estimated melt pool geometry during LBW in steel/glass configuration for varying shielding gas compositions.

The length of the melt pool on the free surfaces and the necking area for the different LBW with varying shielding gas contents are summarized in Table 3.

Table 3. Gas compositions during laser hybrid welding experiments

Gas/melt pool length	Melt pool length top in mm	Melt pool length necking area in mm	Melt pool length root in mm
Gas 1 (Reference)	23.6	12	19.7
Gas 2	25.4	15.1	23.6
Gas 3	21.3	16.8	20.5
Gas 4	21.4	15.2	24
Gas 5	24.8	15.9	24.2

Laser hybrid welding experiments

Figure 7 shows a laser hybrid welded specimen (gas 1 - reference) with a length of 300 mm as an example. The root area is effectively compensated by the contactless electromagnetic backing system, as shown. Visual inspection of the weld did not reveal any defects such as incomplete fusion or lack of penetration. The quality of the root corresponds to quality level B according to ISO 12932 and ensures a uniform weld penetration over the entire length of the seam.

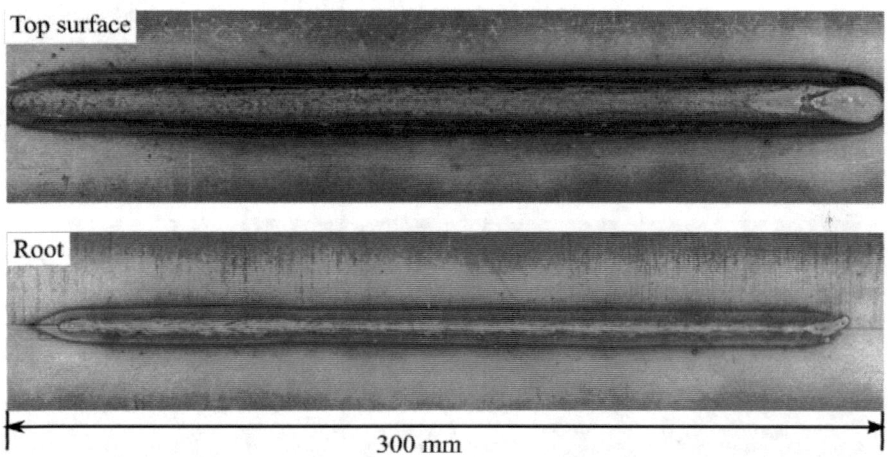

Figure 7. Outer appearance of the top surface and the root of the laser hybrid welded joint with contactless electromagnetic backing system; Gas 1 – Reference.

Figure 8 and Figure 9 show the cross and longitudinal sections of the laser hybrid welded samples with different shielding gas compositions, as shown in Table 1, respectively. The element distribution over the entire seam thickness is evident. In all longitudinal sections three regions are formed regarding the filler wire mixing behaviour: well-mixed region on the top side, a transition region, and a poor-mixed region in the bottom. For the reference weld (Gas 1) the depth of well-mixed region reaches up to approx. 9 mm which is an improvement regarding to the results shown in [4,5]. The electromagnetic fields applied to the root side alone create a vortex in the root area, as shown in Figure 2, which favours the filler wire mixing. The vortex is formed by the generated Lorentz forces, especially by forming a rotatory component of the Lorentz force due to the inhomogeneous course of the eddy currents, as also described in [16]. Gatzen et al. [12-14] and Meng et al. [15] also showed similar results. With the addition of oxygen to the shielding gas, especially 4 % O_2 (Gas 3) the depth of the well-mixed region could be increased up to 18 mm, which corresponds to an almost homogeneous distribution of the filler wire over the entire seam thickness. It is probably due to the melt pool geometry and the low constriction and the resulting separation of the flows in the melt pool. The oxygen content leads to a change of the surface tension, so that a different melt pool geometry is formed. Similar results were also obtained by Zhao et al. [10] at an oxygen level of 2 %. At higher oxygen levels, a decrease in the filler wire mixing behaviour can be observed.

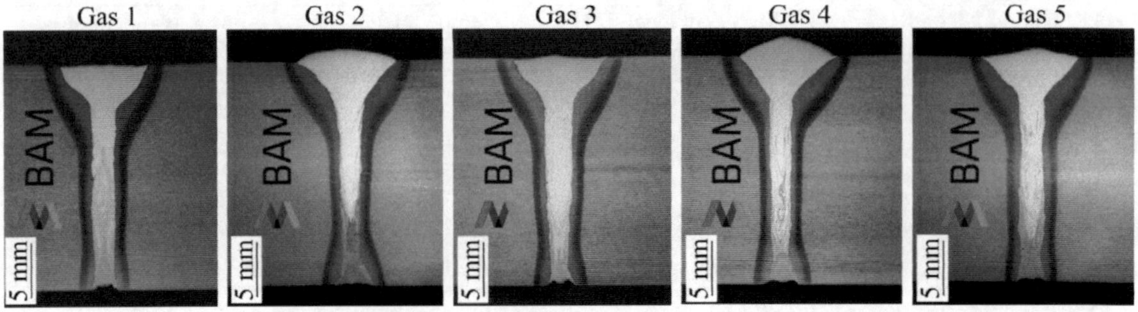

Figure 8. Cross sections of single-pass laser hybrid welded 20 mm thick steels with varying shielding gas compositions.

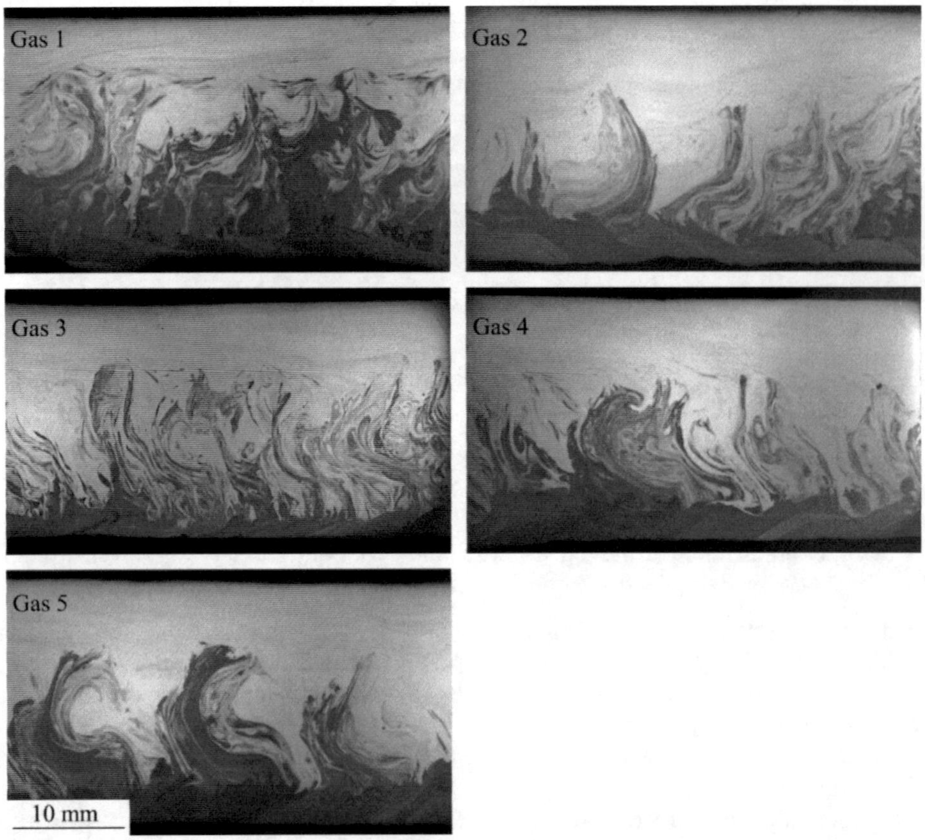

Figure 9. Longitudinal sections of single-pass laser hybrid welded 20 mm thick steels with varying shielding gas compositions.

Conclusion

Single-pass LBW in steel/glass configuration and laser hybrid welding of 20 mm thick high-strength steels with contactless electromagnetic backing were performed during this work. The aim of this study was to investigate the filler wire mixing. Therefore, different shielding gas compositions with varying oxygen contents were added to the process. The gases were prepared by a gas mixer. During the LBW experiments in steel/glass configuration, the melt pool geometry captured by a high-speed camera shows changes. The melt pool length varies for different gases. In all experiments a necking area was found which can inhibit the material flow from the top surface to the root part. At an oxygen level of 4 % an almost necking-free melt pool geometry was detected. It could be confirmed during the laser hybrid welding experiments that the filler wire mixing was improved by adding 4 % oxygen to the shielding gas in combination with the formation of a vortex on the weld root by external applied oscillating magnetic fields. Thus, an almost homogeneous filler wire mixing over the entire seam thickness could be achieved.

Acknowledgments

The research project was carried out in the framework of the Industrial Collective Research programme (IGF No.20.827N). It was supported by the Federal Ministry for Economic Affairs and Climate Action (BMWK) through the AiF (German Federation of Industrial Research Associations e.V.) based on a decision taken by the German Bundestag. Financial funding is gratefully acknowledged.

References

[1] M. Eboo, W.M. Steen, and J. Clarck. "Arc augmented laser processing of materials". In: Proceedings of the International Conference of Advances in Welding Processes, Harrogate, UK, 9-11 May 1978.
[2] S. Olschok. "Laserstrahl-Lichtbogen-Hybridschweißen von Stahl im Dickblechbereich". PhD-Thesis, Shaker-Verlag, 2008.
[3] Ö. Üstündag, S. Gook, A. Gumenyuk, and M. Rethmeier. "Mechanical properties of single-pass hybrid laser arc welded 25 mm thick-walled structures made of fine-grained structural steel". In: Procedia Manufacturing 36 (2019), pp. 112-120.
[4] U. Dilthey, K. Woeste, S. Olschok, and P. Sha. "Hydrodynamische Prozessanalyse beim Laserstrahl-MSG-Hybridschweissverfahren–Teil 2: Aufmischung". In: Schweißen und Schneiden 9.9 (2007), pp. 492–498.
[5] S. Gook, A. Gumenyuk, and M. Rethmeier. "Hybrid laser arc welding of X80 and X120 steel grade". In: Science and Technology of Welding and Joining 19.1 (2014), pp. 15–24.
[6] W. Suder, S. Ganguly, S. Williams, and B. Y. Yudodibroto. "Penetration and mixing of filler wire in hybrid laser welding". In: Journal of Materials Processing Technology 291.1 (2021), pp. 1–11 (article number 117040).
[7] M. Karhu, V. Kujanpää, H. Eskelinen, and A. Salminen. "Filler metal mixing behaviour of 10 mm thick stainless steel butt-joint welds produced with laser-arc hybrid and laser cold-wire processes". In: Applied Sciences 9.8 (2019), pp. 1–20 (article number 01685).
[8] U. Dilthey, A. Gumenyuk, and S. Olschok. "Hydrodynamische Prozessanalyse beim Laserstrahl-MSG-Hybridschweissverfahren–Teil 3: Prozesssimulation und Vergleich mit Schweissversuchen". In: Schweißen und Schneiden 60.3 (2008), pp. 151–155.
[9] M. Sohail, M. Karhu, S. J. Na, S. W. Han, and V. Kujanpaa. "Effect of leading and trailing torch configuration on mixing and fluid behavior of laser-gas metal arc hybrid welding". In: Journal of Laser Applications 29.4 (2017), pp. 1–14 (article number 042009).
[10] L. Zhao, T. Sugino, G. Arakane, and S. Tsukamoto. "Influence of welding parameters on distribution of wire feeding elements in CO2 laser GMA hybrid welding". In: Science and Technology of Welding and Joining 14.5 (2009), pp. 457–467.
[11] I. Bunaziv, S. Wenner, X. Ren, J. Frostevarg, A. F. Kaplan, and O. M. Akselsen. "Filler metal distribution and processing stability in laser-arc hybrid welding of thick HSLA steel". In: Journal of Manufacturing Processes 54.1 (2020), pp. 228–239.
[12] M. Gatzen. "Influence of low-frequency magnetic fields during laser beam welding of aluminium with filler wire". In: Physics Procedia 39 (2012), pp. 59–66.

[13] M. Gatzen, Z. Tang, and F. Vollertsen. "Effect of electromagnetic stirring on the element distribution in laser beam welding of aluminium with filler wire". In: Physics Procedia 12 (2011), pp. 56–65.

[14] M. Gatzen and Z. Tang. "CFD-based model for melt flow in laser beam welding of aluminium with coaxial magnetic field". In: Physics Procedia 5 (2010), pp. 317–326.

[15] X. Meng, M. Bachmann, A. Artinov, and M. Rethmeier. "The influence of magnetic field orientation on metal mixing in electromagnetic stirring enhanced wire feed laser beam welding". In: Journal of Materials Processing Technology 294.1 (2021), pp. 1–12 (article number 117135).

[16] Ö. Üstündag, V. Avilov, A. Gumenyuk, and M. Rethmeier. "Improvement of filler wire dilution using external oscillating magnetic field at full penetration hybrid laser-arc welding of thick materials". In: Metals 9.5 (2019), pp. 1-10 (article number 00594).

[17] Ö. Üstündag, N. Bakir, A. Gumenyuk, and M. Rethmeier. "Influence of an external applied AC magnetic field on the melt pool dynamics at high-power laser beam welding". In: IOP Conference Series: Materials Science and Engineering 1135.1 (2021), pp. 1-10 (article number 012017).

Laser welding of additively manufactured parts - A review

Saeid Parchegani [1], Heidi Piili [1], Ashish Ganvir [1] and Antti Salminen [1]
[1] Department of Mechanical and Materials Engineering, University of Turku (UTU), 20500Turku, Finland

Saeid.parcheganichozaki@utu.fi

Abstract. Additive manufacturing (AM) is a process in which parts are built up layer by layer, introducing novel approaches to how parts can be manufactured with less material waste, shorter lead times and lower costs than traditional manufacturing. One of the key advantages of AM over conventional manufacturing is its design flexibility, which enables for manufacturing of parts with highly detailed geometries in one go, leaving out the need for molding, casting, etc. However, due to the chamber size of the machines, the size of AM parts is limited. To overcome this limitation, joining AM parts together or to wrought or cast material has been proposed. Among the various welding technologies, laser welding is considered a suitable candidate for joining AM parts because of its low heat input, resulting in low deformation, high welding speed, and full automation capability. This study will provide a fundamental understanding of laser welding of AM parts by reviewing current research in the field. The possibility of joining most commonly used AM parts such as AlSi10Mg, AISI 316L, Ti6Al4V and Nickel alloy 718 by laser welding are investigated. Furthermore, the effect of laser welding parameters on mechanical and microstructural properties of joined AM parts are discussed.
Keywords: Additive manufacturing, laser welding, AISI 316L, Manufacturing large parts

1. Introduction

Additive manufacturing (AM), known also as 3D printing, has transformed the manufacturing process in such a way that producing parts with highly complicated geometries is now possible due to its freedom in design using a computer-aided design (CAD) model. AM parts can be manufactured layer by layer directly from a CAD model, eliminating the need for costly machining processes compared to traditional manufacturing [1]. Additionally, minimizing waste material is another advantage of additive manufacturing, making it more cost-effective to produce highly complex parts [2].

AM technology can be categorized in seven different methods based on different materials and applications. These methods are laser powder bed fusion (PBF-LB), material jetting (MJT), binder jetting (BJT), material extrusion (MEX), vat photopolymerization (VPP), direct energy deposition (DED) and sheet lamination (SHL) [3]. AM can process a wide range of materials, including ceramics, polymers, and metal alloys [4]. AM has gained particular attention in printing metal alloys due to its near-net-shape capability in manufacturing dens metallic parts with mechanical performances comparing to their conventionally manufactured counterparts. PBF-LB and DED are the most commonly used AM methods for manufacturing metallic parts [5]. Both processes use a laser beam as a heat source to interact with feedstock to build a final part layer-by-layer. The LPBF process is the most widely used and versatile method for producing a wide variety of materials. Consequently, PBF-LB has emerged as a critical manufacturing technology in high-level industries such as aerospace, automotive, medical, and energy [6]. However, the PBF-LB process has some disadvantages, such as dimensional limitation, due to the chamber size of printers, which is why the PBF-LB process cannot be used to produce large-scale parts [7]. To address this limitation, small parts manufactured with PBF-LB can be joined together to form a larger part using different welding technologies [8]. Some of the most popular

welding technologies used in AM are gas tungsten arc welding (GTAW/TIG), plasma arc welding (PAW), friction stir welding (FSW), electron beam welding (EBW) and laser beam welding [9]. Laser welding has been demonstrated to be a suitable joining method for AM parts due to its low heat input and small spot size [10]. This is mostly due to the importance of precision and narrow welds in joining highly complex and thin-walled PBF-LB parts. In other words, laser welding will result in low distortion and a narrow heat affected zone in PBF-LB parts, offering high-quality joints.
The aim of this study is to review the most recent research on laser welding of parts manufactured with PBF-LB. The effect of welding parameters on weld bead geometry of the most commonly used metal materials manufactured with PBF-LB are investigated. In addition, the mechanical and microstructural properties of the laser-welded joint are discussed. This study is particularly important in providing more insights into the fundamental knowledge of the topic because there have been very few studies and research in this area, and thereby paving the way for further research in the future.

2. Additive manufacturing

2.1. Laser powder bed fusion (PBF-LB)

PBF-LB is a powder-based process that melts and fuses powders together layer-by-layer using a laser beam as a heat source. The process is divided into several steps, the first and most important of which is modeling and designing the part with CAD software [11]. This step determines the laser scan path on the powder bed, where the laser selectively exposes only the CAD model-defining powder. The CAD model is divided into layers based on the layer thickness desired, and each layer is built on top of the previous layer until the final part is created. A typical PBF-LB machine is shown in Figure 1. It includes several different components such as a laser source, a build platform, a recoater (used to spread the powders), etc. [12].

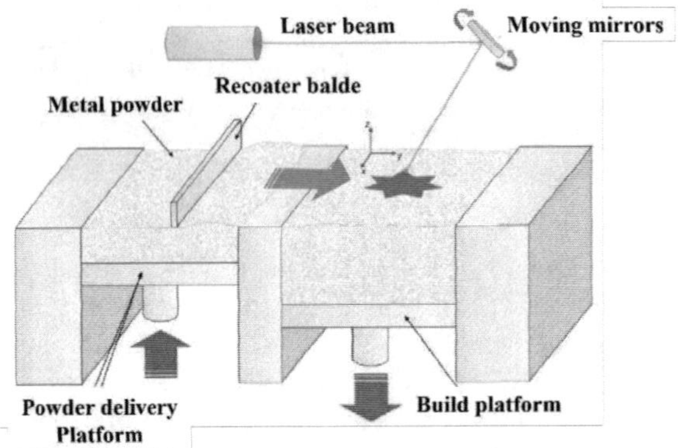

Figure 1. A schematic of PBF-LB machine [12].

The particle size used in PBF-LB is typically between 10 and 70 μm, which allows parts to be printed with a thinner layer thickness, resulting in a dimensionally precise final part [13]. Parts manufactured with PBF-LB have a higher density than parts made with other AM technologies with a superior mechanical property [14]. Hence, PBF-LB is a suitable candidate for manufacturing highly complex parts with excellent surface quality and precise final dimensions for use in medical, aerospace, automotive, etc. Figure 2 displays some of the parts manufactured with PBF-LB [15]. It is important to note that the quality of final parts manufactured with PBF-

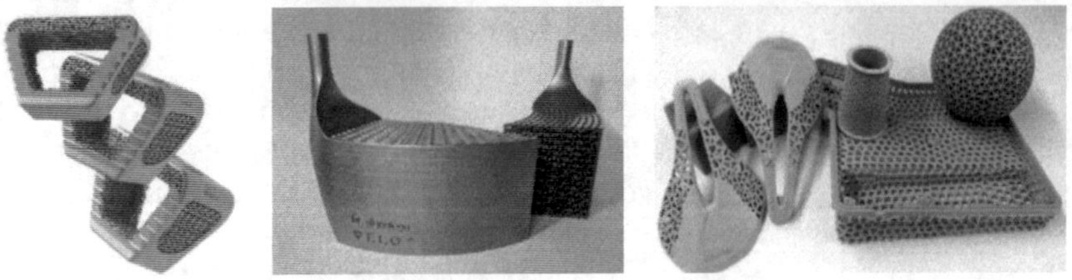

Figure 2. Parts manufactured with PBF-LB [15]

LB is primarily determined by process parameters, as some defects, such as pores and lack of fusion, have been reported to degrade part mechanical performance. Developing optimized process parameters is thus a critical step in discovering the full potential of PBF-LB parts [16].

3. Welding technologies in AM

3.1. Laser welding

Laser welding is a method of joining materials in which a high-intensity energy source is focused into a small spot size. The typical spot size of the laser beam is in range between 0.2-0.7 mm for most common welding applications [17]. Because of its heat is concentrated in small spot size compared to other welding technologies, laser welding results in a narrower heat-affected zone and less distortion, which are crutial for high quality joints in most applications, such as the automotive industry, energy sector. Furthermore, recent advancements in fiber lasers have resulted in improved laser source delivery systems for various welding conditions, with higher laser powers reaching several kilowatts [18]. Also, the ability to be fully automated is another advantage of laser welding, which produces high-quality welds at a faster rate than other welding technologies.

Laser welding can be classified into two modes based on the power density, as shown in Figure 3 [19]. In Conduction laser welding, laser power density is lower than 0.5 MW/cm^2 in which a shallow and wide weld seams are the characteristic of this mode. This is because heat is conducted from the centre to the edges of the weld pool during conduction laser welding (Figure 3a). The second laser welding mode is known as keyhole welding in which laser power density is beyond 1.5 MW/cm^2 [20]. Deep narrow welds distinguish the keyhole mode (Figure 3b). Keyhole welding is especially useful for welding of thick metals where high speed welding with fully penetrate welds is required; however, laser welding parameters such as laser power, speed, and focal length need be optimized for each application because keyhole welding can cause defects such as pores due to keyhole instability [21].

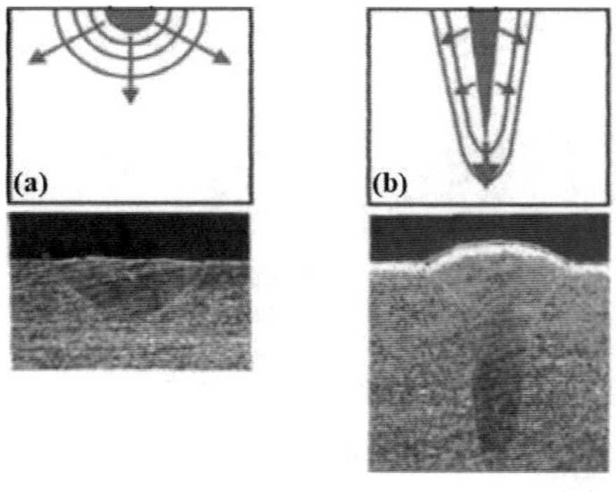

Figure 3. Different laser welding modes a) conduction b) keyhole [19]

4. Laser welding of PBF-LB components

4.1. Stainless steel

Stainless steel is a low-carbon iron alloy used in many different industrial applications due to its high strength and good mechanical performance, as well as good corrosion resistance. The most commonly used stainless steel is AISI 316L, which has superior corrosion resistance to other grades of stainless steel due to the added chromium and molybdenum content. AISI 316L is a good choice for harsh chemical environments such as chemical plants, marine industries, etc. [22]. Welding of 316L has been extensively studied over the last few decades [23-24] whereas data on laser welding of PBF-LB/M/316L is limited.

Matilainen et al. [25] investigated the effect of laser heat input on the weldability of PBF-LB/M/316L and cold-rolled 316L sheet plates. The cross-section of welded samples using bead-on-plate laser welding revealed that full penetration joint in PBF-LB/M/316L sample can be accomplished with lower energy input than in cold-rolled 316L sample. The better absorptivity of PBF-LB/M/316L is attributed to the easier formation of keyhole. In other words, the higher surface roughness of PBF-LB/M/316L leads to higher absorptivity of laser light into the material, hence keyhole is formed easier in PBF-LB/M/316L even at lower laser energies. Furthermore, despite having different weld geometry, both PBF-LB and cold-rolled 316L joints showed the same melting efficiency. The correlation between microstructure and mechanical properties of laser welded PBF-LB/M/316L was studied by Zhang et al. [26]. The average grain size in the fusion zone was found to be five times that of the base metal, resulting in weaker joint strength in the weld zone due to the coarser grain size. The study found that build orientation had no significant impact on joint strength and concluded that laser welding can be used to successfully join PBF-LB/M/316L components; however, the lower mechanical properties of the joints in comparison to the base metal must be considered in the final application. The effect of heat treatment on the joints of PBF-LB/M/316L to sheet metal 316L was examined by Rautio et al. [27]. The stress-relieving heat treatment (600ºC followed by 2 h annealing) performed after and before welding resulted in no significant changes in the hardness of the joint between PBF-LB/M/316L and sheet metal 316L. Despite the fact that the tensile test revealed that the PBF-LB/M/316L joint had the highest yield strength compared to the 316L to 316L and PBF-LB/M/316L to 316L joints, ductility decreased dramatically. The heat treatment also revealed no difference in yield strength and ultimate tensile strength of dissimilar joints of PBF-LB/M/316L to 316L, which is consistent with the hardness result. Yang et al. [28] investigated the effect of laser welding on microstructure and mechanical properties as well as corrosion resistance of the PBF-LB/M/304 and wrought 304. It was reported that the joint of PBF-LB/M/304 despite having an irregular grain orientation along with coarser grain size in the weld zone (Figure 4) which results in a weaker weld joint compared to the base metal, the corrosion resistance of the joint samples was comparable to the wrought counterpart and better than the non-welded PBF-LB/M/304. This was contributed to the fact that the coarser grain size in the weld zone results in lesser grain boundaries and thus superior corrosion resistance compared to the base PBF-LB/M/304 metal.

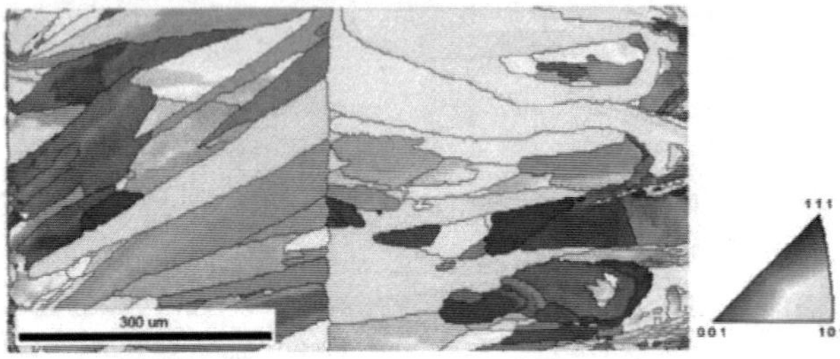

Figure 4. EBSD inverse pole figure mapping of grain size distribution of laser-welded joint of PBF-LB/M/304 [28]

Abroug et al. [29] investigated the effect of laser welding on the fatigue strength of cold-rolled 316L sheet plates and PBF-LB/M/316L. The fatigue strength of cold-rolled 316L sheet samples without welding was the highest, while it was reduced by 40% in non-welded PBF-LB/M/316L. This is because of the high surface roughness and lack of fusion in the PBF-LB process which facilitate the crack initiation. Cold-rolled 316L sheet plates had a 35% reduction in fatigue strength after welding, but PBF-LB/M/316L had the same fatigue strength as non-welded PBF-LB/M/316L samples. The joint of cold-rolled 316L broke at the weld zone while the joint of PBF-LB/M/316L broke at the base metal, demonstrating the possibility of achieving a high strength weld in PBF-LB/M/316L. However, more research is needed to validate this finding due to the scarcity of data on the fatigue life of laser welded PBF-LB/M/316L.

4.2. Aluminium alloy

Aluminium alloys are attractive materials for various industrial applications such as aerospace, automotive, marine, etc. due to their high strength and light weight properties along with excellent corrosion resistance [30]. Due to their low weight, these alloys have been widely used in the body structure of satellites and automobiles. AlSi10Mg is the most used aluminium alloy in PBF-LB process due to its high mechanical properties and low recycling cost compared to the cast and wrought counterparts. Despite having some appealing properties for many applications, PBF-LB/M/AlSi10Mg suffers from pore formation during the PBF-LB and welding processes, which degrades the mechanical performance of the welded parts [31]. High hydrogen content in PBF-LB/M/AlSi10Mg and small pore formation during the PBF-LB process that could be pushed to the weld zone during welding are two of the main reasons for high pores susceptibility of welded PBF-LB/M/AlSi10Mg [32]. Therefore, a better understanding of microstructural and chemical change in PBF-LB/M/AlSi10Mg during welding is vital to improve the weldability of this important alloy.

Biffi et al. [33] examined the microstructure and mechanical properties of laser welded PBF-LB/M/AlSi10Mg. It was found that the full penetration welding is feasible utilizing laser welding, however, porosities both from gas trapped and high hydrogen content were inevitable in the weld zone. Thus, the ultimate tensile strength (UTS) and ductility were drastically reduced in welded samples compared to as-built PBF-LB/M/AlSi10Mg. In addition, the standard heat treatment (320 °C for 2 h with air cooling) of as-built PBF-LB/M/AlSi10Mg after welding, resulted in further reduction in ductility and UTS due to the stress relieved nature of the heat treatment. Gui et al. [34] introduced a new way of welding PBF-LB/M/AlSi10Mg using layer deposition welding. The cross-section of the joint of PBF-LB/M/AlSi10Mg (Figure 5a), cast AlSi10Mg (Figure 5b) and layer deposition welding of PBF-LB/M/AlSi10Mg (Figure 5c) are shown in figure 5. The layer deposition welding was done on 45° V-groove configuration with root gap where AlSi10Mg powders were deposited in three different layers with different laser parameters. Figure 5 Shows that the layer deposition welding has less porosity compared to the joint of PBF-LB/M/AlSi10Mg, while the joint of cast AlSi10Mg is almost porosity free. This is due to the slower solidification rate of the laser welding compared to the layer deposition welding which means that the weld zone microstructure of layer deposition welding is finer than the weld zone of the PBF-LB/M/AlSi10Mg. Hence, it was reported that the tensile properties of the layer deposition welding were superior to laser welded PBF-LB/M/AlSi10Mg.

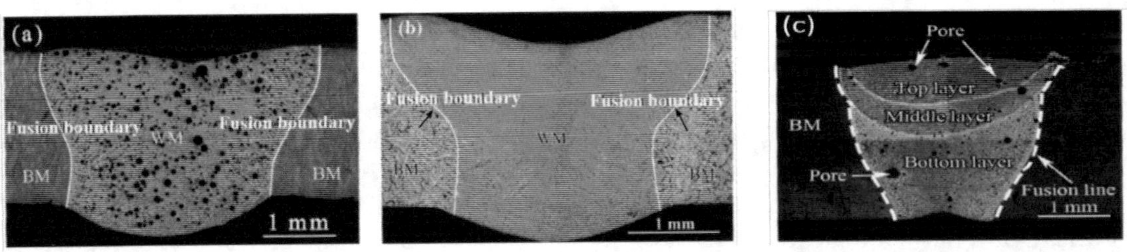

Figure 5. Cross-section of joints (a) laser welded PBF-LB/M/AlSi10Mg (b) laser welded cast AlSi10Mg (c) layer deposition welding of PBF-LB/M/AlSi10Mg [34]

Peng et al. [35] studied the effect of Zr and Er addition in to the filler metal during laser welding of PBF-LB/M/AlSi10Mg. The size of pores and also total porosity of the welded samples were reduced with Zr and Er addition as shown in Figure 6. The pore morphology revealed that the Si-rich phases at inner walls of the pores has transformed from columnar dendrite without Er and Zr addition to the equiaxed dendrite with Er and

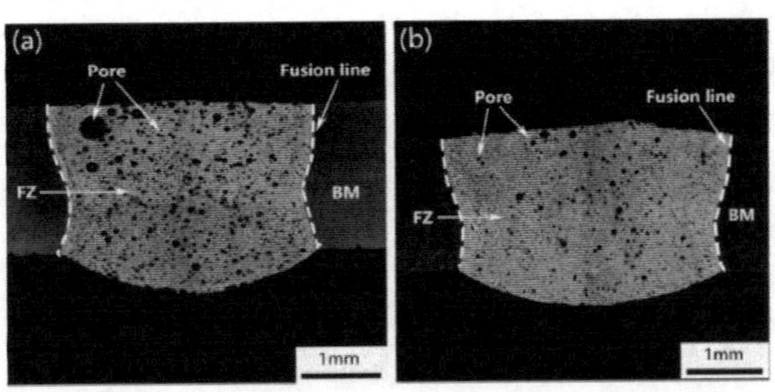

Figure 6. Laser welding of PBF-LB/M/AlSi10Mg (a) without Zr and Er addtion (b) with Zr and Er addition [35]

Zr addition as shown in Figure 7. On the other hand, Si-rich eutectic network has significantly enhanced with Er and Zr addition, resulting in strengthening the welded samples as both hardness and UTS was reported to be higher than the weld samples without Er and Zr addition.

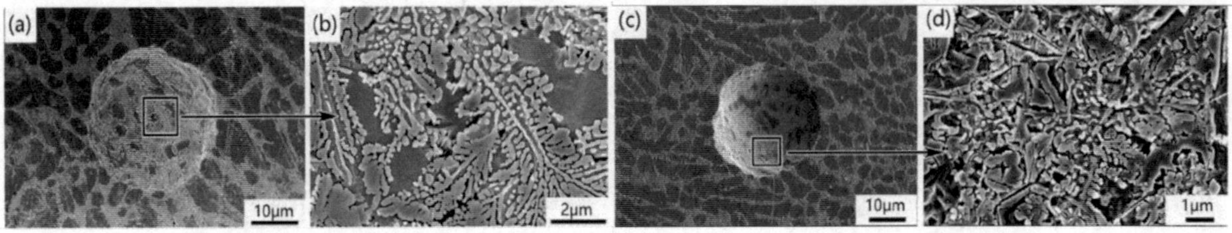

Figure 7. Pore morphology of laser welded PBF-LB/M/AlSi10Mg (a)(b) without Zr and Er addition (c)(d) without Zr and Er addition [35]

The hydrogen pore formation during welding seems to be still the main challenge regarding the laser welding of PBF-LB/M/AlSi10Mg. This is mainly due to the high hydrogen content in the powder itself. The powder has a larger surface area than cast or conventional parts, making it more susceptible to moisture absorption in the environment. Moreover, the powder is exposed to an open air during the powder handling steps which causes more hydrogen absorption from the environment. Therefore, more research is needed to either reduce the hydrogen content of AlSi10Mg powders or keep the powder dry during the various powder processing steps in PBF-LB.

4.3. Titanium alloys

Titanium alloys first introduced to aerospace industries due to their specific strength. However, the application of titanium alloys rapidly extended to some other industries due to other important features such as excellent corrosion resistance, good mechanical strength at high temperature, good creep resistance. In addition, titanium alloys have been vastly exploited in biomedical application due to their unique biocompatibility, elastic modules comparable to human bone and high fatigue strength [36]. Among titanium alloys, Ti6Al4V has been the workhorse alloy used extensively in number of industries such as, aerospace, biomedical, gas turbine engine, etc [37]. Thus, the joining of PBF-LB/M/Ti6Al4V to produce larger pats has always been in high demand. Yu et al. [38] Investigated the weldability of PBF-LB/M/Ti6Al4V using laser welding. Three different welding scenarios were tested in this study as follows, joint of PBF-LB/M/Ti6Al4V followed by stress-relieving (500°C for 2 h followed by furnace cooling), joint of PBF-LB/M/Ti6Al4V to wrought followed by stress-relieving, joint of PBF-LB/M/Ti6Al4V followed by heat treatment (850°C for 2 h followed by furnace cooling) and stress-relieved as-built PBF-LB/M/Ti6Al4V base material (BM) was used to compare the mechanical properties of the joints. In terms of microstructure, it was noticed that the weld zone microstructure of the joint of PBF-LB/M/Ti6Al4V and PBF-LB/M/Ti6Al4V to wrought are similar to as-built PBF-LB/M/Ti6Al4V, however, heat treated joint of PBF-LB/M/Ti6Al4V resulted in coarser acicular structure as shown in Figure 8. The coarser weld zone grains of heat-treated weld samples resulted in inferior microhardness and UTS and YS compared to other welded samples while PBF-LB/M/Ti6Al4V and PBF-

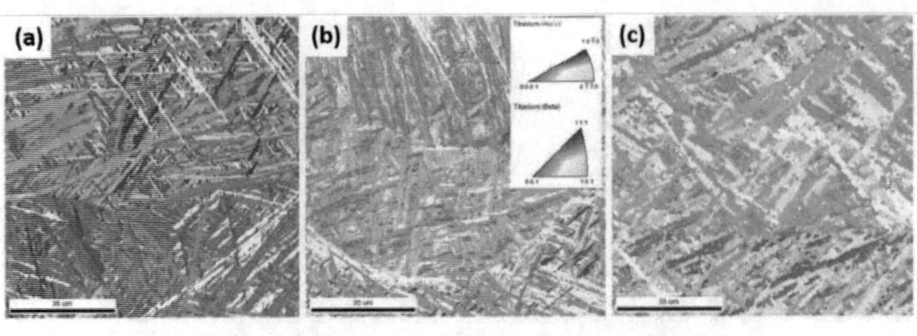

Figure 8. EBSD orientation maps of fusion zone of (a) PBF-LB/M/Ti6Al4V (b) PBF-LB/M/Ti6Al4V to Wrought Ti-6Al-4V (c) heat-treated PBF-LB/M/Ti6Al4V [38]

LB/M/Ti6Al4V to wrought Ti6Al4V and as-built PBF-LB/M/Ti6Al4V all showed similar microhardness and UTS and YS which are much higher than the wrought Ti6Al4V base metal. Furthermore, the fatigue life and fatigue crack growth rate (FCGR) of welded and as-built PBF-LB/M/Ti6Al4V samples revealed that the joint of PBF-LB/M/Ti6Al4V and PBF-LB/M/Ti6Al4V to wrought and heat-treated welds have similar fatigue life despite having different base metals and heat treatment. This means that the microstructure of the weld zone has no effect on fatigue life, and the pore in the weld zone is the most important factor influencing fatigue life. All the welded sample broke at weld zone where cracks initiated from pores and for the as- built PBF-LB/M/Ti6Al4V the crack initiated from pores on the surface. The BM wrought possess the highest fatigue life, owing to its lower porosity, while having the inferior mechanical properties compared with other welded samples. Mechanical properties of laser welded PBF-LB/M/Ti6Al4V in relation to the layer thickness examined by Rautio et al. [39]. The layer thickness of the PBF-LB/M/Ti6Al4V impacted the UTS and YS and elongation, as decreasing the layer thickness from 50 μm to 30μm led to reduction of UTS and YS by 6 % however elongation increased by 40 %. It is worth noting that the UTS, YS, and elongation of as-built and welded samples were comparable, indicating that laser welding can be used to join PBF-LB/M/Ti6Al4V without losing elongation which in contrast with other studies [40].

4.4. Nickel alloys

Nickel alloys are well-known for their high strength at high temperatures and excellent corrosion resistance in harsh environments. Nickel alloys are widely used in aerospace, turbine engineering, power plants, and chemical applications [41]. Alloy 718 is one the most important nickel-based supper alloys that is used in jet engine and gas turbine manufacturing due to its high strength at high temperature. Alloy 718, on the other hand, is a very tough material for machining tools, which raises the cost of the finished product., particularly for jet engine components, which rely heavily on machining due to their complex features [42]. This is where additive manufacturing (AM) can help by eliminating the machining process and thus lowering the cost of final components. It is critical to investigate the possibility of joining PBF-LB/M/IN718 in order to broaden their use and application in other areas. However, Alloy 718 is highly prone to solidification and hot cracking in weld zone and heat affected zone during welding [43].

Schwer et al. [44] examined the effect of low and high energy input on the weld geometry of PBF-LB/M/IN718 and wrought 718 utilizing bead-on-plate laser welding. PBF-LB/M/IN718 samples were produced horizontally, vertically and at 45°. All PBF-LB/M/IN718 joint samples showed high porosity and liquidation cracking in the weld zone. At lower energy input (0.6 kJ/cm PBF-LB/M/IN718 built horizontally showed 31% less molten volume compared to other welded samples. At this energy input only, the partial penetration reached in all samples. At higher energy input (1.8 kJ/cm) full penetration was achieved in all samples however, PBF-LB/M/IN718 in all different built orientation showed excessive weld at the bottom of the weld compare to wrought 718. It was concluded that PBF-LB/M/IN718 joints have up to 18% more molten volume compared to wrought 718. Raitiuo et al. [45] investigated the effect of heat treatment (solution followed by double aging) before and after welding on mechanical and microstructural properties of PBF-LB/M/IN718. The as-built PBF-LB/M/IN718 had a very fine columnar dendrite morphology due to rapid cooling and heating while heat treated PBF-LB/M/IN718 exhibited irregular grains shape where columnar microstructure seemed to be disappeared. The microstructure of as-built PBF-LB/M/IN718 after welding is shown in Figure 9. The weld zone exhibits an elongated large grain morphology toward the fusion zone line with a needle-shape δ phase are precipitated in the interdendrite regions (Figure 9c) and also some microcracks have developed in the weld zone (figure 9d). The as-built PBF-LB/M/IN718 heat treated before welding exhibits MC-type carbide throughout the weld zone while the microstructural of as-built PBF-LB/M/IN718 heat treated after welding shows γ' strengthening phase in the weld zone. The microhardness of the weld zone of heat-treated PBF-LB/M/IN718 after welding was significantly decreased compared with other welded joints. The tensile results were also in agreement with microhardness test, showing that laser welding deteriorated the strength of heat treated PBF-LB/M/IN718. It was concluded that heat treatment can strengthen the weld however it significantly reduces the ductility.

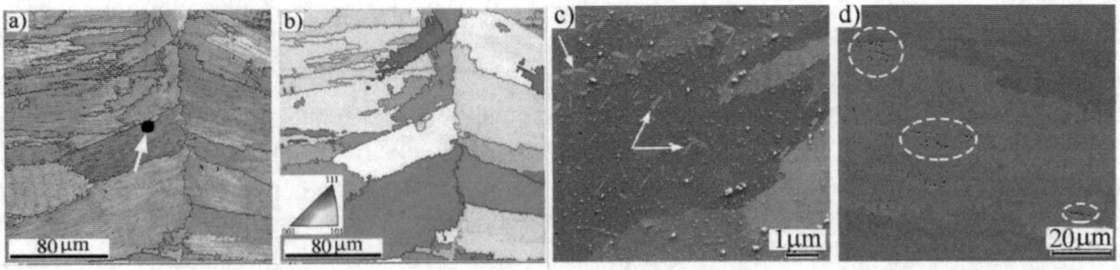

Figure 9. (a)(b) the weld zone microstructure of PBF-LB/M/IN718 (c) δ phase precipitation in the weld zone (d) microcracks [45]

Zhang et al. [46] examined the effect of laser welding on fatigue behaviour of PBF-LB/M/IN625. A low-cycle fatigue at 815 ºC were performed on the welded samples and also as-built PBF-LB/M/IN625. The results showed that fatigue behaviour of welded samples was better that the BM as all the cracks initiated at the surface of BMs. It was shown that the hardness of the weld zone increased after the test due to the precipitation of γ'' and δ resulting in better fatigue and durability of the welded samples. It was concluded that laser welding is a feasible way to join PBF-LB/M/IN625.

5. Summary

AM as an emerging technology is changing the concept of manufacturing by introducing a digitalized manufacturing based on CAD models utilizing powder materials in order to minimize waste and maximize the production rate. The concept of design-based manufacturing has enabled the possibilities of producing highly complex components that are almost impossible to be produced by any other conventional technologies. However, the full potential of AM in manufacturing larger size components has been hindered by the size limitation of the AM machines. Laser welding as a robust and versatile joining method particularly for metallic material, is introduced to overcome this limitation. This review is aimed to provide more deep insights into the current studies and research on weldability of PBF-LB components using laser welding. This review is particularly important because of the lack of data in this topic. The outcome of this review can be summarized as follow:

- Laser welding of a wide range of PBF-LB alloys has been demonstrated to be feasible in terms of weld strength, however, ductility may decrease in the weld zone.
- laser welding parameters play a crucial role in maximizing the joint integrity and strength of PBF-LB components as the optimized parameters have shown to increase the mechanical properties of the PBF-LB joints.
- Post processing (heat treatment) can be utilized depending on the final use and application of welding PBF-LB components, however, it has been concluded that in most cases, heat treatment before welding may lower the joint strength specially in PBF-LB/M/IN718. It needs to be researched more to fully understand the nature of heat treatment on weldability of LPBF parts.
- Laser welding of aluminium PBF-LB seems to be really challenging even using optimized laser parameter due to the high porosity. However, adding some other elements into the raw powder has positively lowered the pores formation.
- A complete understanding of microstructural evolution of welded PBF-LB parts will lead to better understanding of mechanical and fatigue behaviour of the joints.
- Fatigue life of welded PBF-LB and as-built PBF-LB are inferior to their wrought counterpart due to the formation of pores in both as-built and welded PBF-LB while mechanical performance such as UTS and YS might be superior.

Acknowledgement
Authors of this article acknowledge project of DREAMS (Database for Radically Enhancing Additive Manufacturing and Standardization). This project was funded by Business Finland and the duration of

project is from April 2022 to February 2024. The Participant of project were the Finnish Additive Manufacturing Ecosystem (FAME). Aim of the project was to establish standards for AM and develop a powder material database for AM. Authors would like to thank partners of this project for their contribution to this article.

Authors contribution:

Parchegani conducted the literature review and wrote the initial draft of this publication. Piili and Ganvir modified and provided comments to the literature review data base. The final draft was commented, edited and modified by all authors. Salminen reviewed, supervised, provided resources and funding acquisition.

References

[1] Frazier WE. Metal additive manufacturing: a review. J Mater Eng Perform 2014; 23:1917e28.
[2] Attaran M. The rise of 3-D printing: the advantages of additive manufacturing over traditional manufacturing. Bus Horiz 2017; 60:677e88.
[3] Sames WJ, List FA, Pannala S, Dehoff RR, Babu SS. The metallurgy and processing science of metal additive manufacturing. Int Mater Rev 2016;61(5):315–60.
[4] Wong KV, Hernandez A. A review of additive manufacturing. International Scholarly Research Network Mechanical Engineering; 2012.
[5] Herzog D, Seyda V, Wycisk E, Emmelmann C. Acta Mater 2016; 117:371e92.
[6] Thompson MK, Moroni G, Vaneker T, Fadel G, Campbell RI, Gibson I, et al. CIRP Ann – Manufacturing Technol 2016; 65:737e60.
[7] Buchanan C, Gardner L. Metal 3D printing in construction: A review of methods, research, applications, opportunities and challenges. Eng Struct. 2019; 180:332–348.
[8] Prashanth KG, Damodaram R, Scudino S, Wang Z, Rao KP, Eckert J. Friction stir welding of Al-12Si, parts produced by selective laser melting. Mater Des 2014; 57:632–7.
[9] Zhao Z, Guo Y, Bai L, Wang K, Han J. Quality monitoring in wire-arc additive manufacturing based on cooperative awareness of spectrum and vision. Optik 2019; 181:351–60.
[10] Zhang C, Bao Y, Zhu H, et al. A comparison between laser and TIG welding of selective laser meltedAlSi10Mg. Opt Laser Technol. 2019; 120:105696, doi: 10.1016/j.optlastec.2019.105696.
[11] Attaran M. The rise of 3D printing: the advantages of additive manufacturing over traditional manufacturing.Business Horizon 2017;60:677–88.
[12] Cobbinah, P.V.; Nzeukou, R.A.; Onawale, O.T.; Matizamhuka, W.R. Laser Powder Bed Fusion of Potential Superalloys: A Review. Metals 2021, 11, 58.
[13] Vock S, Kloden B, Kirchner A, Weibgarber T, Kieback B. Powders for powder bed fusion: a review Prog. Addit. Manuf. (2019), pp. 1-15
[14] Shapiro AA, Borgonia JP, Chen QN, Dillon RP, McEnerney B, Polit-Casillas R, et al. Additive manufacturing for aerospace flight applications. J Spacecraft Rockets 2016;53(5):952e9.
[15] https://www.ntop.com/resources/blog/
[16] Leuders S, Thone M, Riemer A, Niendorf T, Troster T, Richard HA, et al. Int J Fatigue 2013;48:300.7.
[17] Katayama S. *Laser welding, joining, or brazing of dissimilar materials Topics in mining, metallurgy and materials engineering* (2020), pp. 167-184
[18] Katayama S. *Fundamentals and Details of Laser Welding Topics in mining, metallurgy and materials engineering* (2020), pp. 167-184 ISBN: 978-981-15-7932-5
[19] https://amadaweldtech.com/blog/laser-welding-modes-conduction-transition-keyhole-welding
[20] Xiao R, Zhang X. Problems and issues in laser beam welding of aluminum–lithium alloys, Journal of Manufacturing Processes, Volume 16, Issue 2, 2014, Pages 166-175, ISSN 1526-6125
[21] Zhang D, Wang M, Shu C, Zhang Y, Wu D, Ye Y, Optics & Laser Technology, Volume 114, 2019, Pages 1-9, ISSN 0030-3992.

[22] Trelewicz J, Halada G, Donaldson O, Manogharan G, Microstructure and corrosion resistance of laser additively manufactured 316L stainless steel, JOM 68 (3) (2016) 850–859.
[23] Korinko P, Malene S, Considerations for the weldability of types 304L and 316L stainless steel, Practical Failure Analysis 1, (2001) 61–68
[24] Feng Y, Luo Z, Liu Z, Li Y, Luo Y, Huang Y, Keyhole gas tungsten arc welding of AISI 316L stainless steel, Mater. Des. 85 (2015) 24–31.
[25] Matilainen V, Pekkarinen J, Salminen A (2016) Weldability of additive manufactured stainless steel. Phys Procedia 83:808–817
[26] Zhang R, Buchanan C, Matilainen V, Daskalaki-Mountanou D, Ben Britton T, Piili H, Salminen A, Gardner L, Volume 208, 2021,109921, ISSN 0264-1275.
[27] Rautio T, Mäkikangas J, Jaskari M, Keskitalo M, Järvenpää A. Microstructure and Mechanical Properties of Laser Welded 316L SLM Parts. Key Eng. Mater. 2020, 841, 306–311.
[28] Yang J, Wang Y, Li F, Huang W, Jing G, Wang Z, Zeng X, Journal of Materials Science & Technology, Volume 35, Issue 9, 2019, Pages 1817-1824, ISSN 1005-0302.
[29] Abroug F, Monnier A, Arnaud L, Balcaen Y, Dalverny O, Engineering Fracture Mechanics, Volume 275, 2022, 108865, ISSN 0013-7944.
[30] Ye H. An overview of the development of Al-Si-alloy based material for engine applications. J Mater Eng Perform. 2003; 12:288–297. doi:10.1361/10599490377 0343132.
[31] Ludwig W, Buffi J, Savelli S, Cloerens P. Study of the interaction of a short fatigue crack with grain boundaries in a cast Al alloy using X-ray microtomography, Acta Mater. 51 (2003) 585598.
[32] Emmelmanna P, Beckmanna D. Optimization of laser welding process for laser additive manufactured aluminum parts by means of beam oscillation and process-oriented component design, LIM 2017
[33] Biffi C, Fiocchi J, Tuissi A. Laser weldability of AlSi10Mg alloy produced by selective laser melting: microstructure and mechanical behavior, J. Mater. Eng. Perform. 28 (2019) 6714–6719.
[34] Cui L, Peng Z, Chang Y, He D, Cao Q, Guo X, Zeng Y, Porosity, microstructure and mechanical property of welded joints produced by different laser welding processes in selective laser melting AlSi10Mg alloys, Optics & Laser Technology, Volume 150, 2022, 107952, ISSN 0030-3992.
[35] Peng Z, Cui L, He D, Guo X, Zeng Y, Cao Q, Huang H, Effect of Er and Zr addition on laser weldability of AlSi10Mg alloys fabricated by selective laser melting, Materials Characterization, Volume 190, 2022,112070, ISSN 1044-5803
[36] Omoniyi, P., Mahamood, M., Jen, T. & Akinlabi, E. TIG welding of Ti6Al4V alloy: Microstructure, fractography, tensile and microhardness data. *Data Br.* **38**, 107274 (2021).
[37] Cheng, K., Xi, M., Chen, S., Cui, G. & Zhou, H. Microstructures and mechanical properties of Ti6Al4V alloy repaired by the technology laser repairing. Opt. Laser Technol. 144, 107410 (2021).
[38] Hanchen Yu, Fangzhi Li, Jingjing Yang, Jianjun Shao, Zemin Wang, Investigation on laser welding of selective laser melted Ti-6Al-4V parts: Weldability, microstructure and mechanical properties, Materials Science and Engineering: A, Volume 712, 2018, Pages 20-27, ISSN 0921-5093.
[39] Rautio T, Hamada A, Mäkikangas J, Jaskari M, Järvenpää A, Laser welding of selective laser melted Ti6Al4V: Microstructure and mechanical properties, Materials Today: Proceedings, Volume 28, Part 2, 2020, Pages 907-911, ISSN 2214-7853.
[40] Gao X, Zhang L, Liu J, Zhang J, Mater. Sci. Eng. A 559 (2013) 14–21.
[41] Popovich V, Borisov E, Popovich A, Sufiiarov V, Masaylo D, Alzina L, "Impact of heat treatment on mechanical behaviour of inconel 718 processed with tailored microstructure by selective laser melting," Materials & Design, vol. 131, pp. 12–22, oct 2017.
[42] Rakesh M, Datta S. Machining of Inconel 718 Using Coated WC Tool: Effects of Cutting Speed on Chip Morphology and Mechanisms of Tool Wear. *Arab J Sci Eng* **45,** 797–816 (2020).
[43] Shinozaki, K., Kuroki, H., Luo, X., Ariyoshi, H., and Shirai, M. Effects of welding parameters on laser weldability of Inconel 718. Study of laser weldability of Ni-base, heat-resistant superalloys (1st Report). In: Welding International, 12. 1999, S.945–951.

[44] Brunner-Schwer C, Simón-Muzás J, Biegler M, Hilgenberg K, Rethmeier M, Laser welding of L-PBF AM components out of IN718, Procedia CIRP, Volume 111, 2022, Pages 92-96, ISSN 2212 8271.

[45] Rautio T, Mäkikangas J, Kumpula J, Järvenpää A, Hamada A. Laser Welding of LPBF Manufactured IN718: Microstructure and Mechanical Properties, Engineering Materials (2021), pp. S.234-S.241.

[46] Zhang, Y., Hu, X. & Jiang, Y. Study on the Microstructure and Fatigue Behavior of a Laser-Welded Ni-Based Alloy. *J. of Materi Eng and Perform* **29**, 2957–2968 (2020).

Effect of heat treatment in dissimilar welds of additively manufactured laser welded 316L-Inconel 718 joints

M Lindqvist[1], E Kivirasi[1], K Lipiäinen[2], H Piili[3] and A Salminen[3]

[1]Delva Oy, Hämeenlinna 13430, Finland
[2]Laboratory of Steel Structures, LUT University, Yliopistonkatu 34, Lappeenranta 53850, Finland
[3]Department of Mechanical Engineering Faculty of Technology, University of Turku, Turku 20520, Finland

markku.lindqvist@delva.fi

Abstract. Laser powder bed fusion for metals is a method of producing end use components for industrial use. Powder bed fusion machines are relatively small and are usually used to create only the critical part of the larger assembly. Therefore, L-PBF manufactured parts must be attached to each other for example by welding. The industrial world needs to be able to join the printed superalloy components to the traditionally manufactured components to reach better corrosion, wear and/or heat resistance. The problem is that there is limited amount of information about the suitable welding parameter values for these applications. This study examines how the standard heat treatment cycles affect to the quality of the weld, and if the post-heat treatment is reducing undesired phases in the heat affected zone. It is also examined how post weld heat treatment for Inconel 718 affects the base material in IN718-316L dissimilar welds. Tests show that post-weld heat treatment has a positive effect on weld hardness on IN718 side. After welding, the hardness drops to 50% of the nominal hardness. When the welds are heat treated according to AMS5662, the hardness values are restored. Tests also show that the heat treatment does not have significant effect on tensile or hardness properties of 316L.

1. Introduction

Dissimilar joints, for example of stainless steel 316L and Inconel super alloys, are used for example in aerospace and energy industries**[1–4]**. The research done so far of welded AM components is still sparse and must be continued further to provide enough information for the industry. In the Finnish industry, there is generally a need of large component manufacturing**[5]**. Because as of now it is not feasible to produce those fully with laser-based powder bed fusion (PBF-LB), it is important to know how the AM components and materials behave during the welding process. Welding of the Inconel 718 is a commonly used process in certain industries as it has good weldability**[6–8]**. Process parameter values need to be carefully considered during laser welding. It is easy to achieve welds with undesired microstructures due to high heat input during the welding process. Undesirable niobium rich Laves phase in Inconel 718 can reduce the material properties **[8–14]**. Laves phase is mainly generated due to the high heat input**[15]** and slow cooling rates after the welding process. Due to lower heat input and faster cooling rates, laser welding can be more effective. Higher cooling rates greatly reduce the formation of Laves phase in the part.**[16,17]** Laser welding and fast cooling rates cannot however be achieved in every case and Laves phase formation can also be controlled with different filler metals.

When using low-niobium fillers, formation of the Laves phase can be controlled**[2]**. Due to the AM microstructure, the phase formation differs from traditionally manufactured Inconel 718. Post-weld heat treatments can be used, and the research shows that the standard 980°C heat treatments, according to standard AMS 5662**[9]**, can reduce the Laves phase and improve the properties of the weld**[8,17,18]**. Higher temperatures in the solution step of the heat treatment can dissolve Laves phase more efficiently**[13,19]**. The heat treatment can be used to increase the strength and toughness considerably compared with dissimilar materials in welded condition. When combining benefits from AM and traditional methods, welding of the smaller AM components to a larger structure can often be needed. Especially in cases of large structures, AM (PBF-LB) is not feasible as the only manufacturing method.**[20]** In these structures, it is not always feasible to do the heat treatments, which can be problematic if undesired phase structure is formed during the welding operations. Therefore, it is important to study the effects of welding to AM Inconel 718 as well as dissimilar welds to achieve acceptable properties to end use structures.

2. Aim and purpose of the study

Authors have executed a study of laser welding and effect of post-weld heat treatment of additively manufactured materials. Study consists of IN718 – 316L and IN718 – IN718 material pairs. This study focuses on one material pair and heat treatment. This is done to have a deeper analysis for different cases. Study structure is shown in Figure 1.

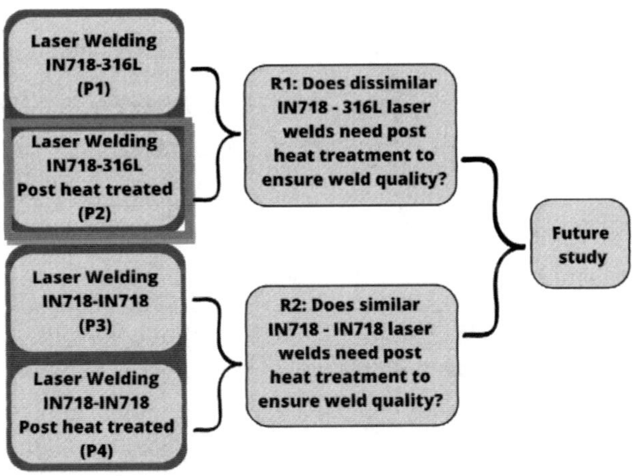

Figure 1. Structure of whole study of laser welding of IN718-316L and IN718-IN718. The highlighted (P2) part was the focus of this study.

Aim of this study is to generate data on behaviour of studied materials and material pairs in laser welding and subsequent heat treatment. This data eases the use of different additively manufactured materials in industrial use.

3. Material and methods

Test samples for this test were manufactured with EOS M290 machine by Delva Oy. Sample size was 10,5 x 80 mm round bar. All the samples were printed using standard EOS parameter values. 316L_040_FlexM291_1.11 for 316L stainless steel and IN718_PerformanceM291_2.11 for Inconel 718. Samples for this test were printed vertically. Layer thickness of 40 micron was used for both materials. Powder was supplied by EOS. Chemical composition for the powders can be found in Table 1 and Table 2.

Table 1. Chemical nominal composition of EOS IN718.

Element	Min [wt.-%]	Max [wt.-%]
Ni	50	55

Cr	17.0	21.0
Nb	4.75	5.5
Mo	2.8	3.3
Ti	0.65	1.15
Al	0.20	0.80
Co	N/A	1.0
Cu	N/A	0.3
C	N/A	0.08
Si, Mn	N/A	0.35
P, S	N/A	0.015
B	N/A	0.006
Fe	N/A	Balance

Table 2. Chemical nominal composition of EOS StainlessSteel 316L.

Element	Min [wt.-%]	Max [wt.-%]
C	N/A	0.03
Si	N/A	1.00
Mn	N/A	2.00
P	N/A	0.045
S	N/A	0.015
Cr	16.50	18.50
Ni	10.00	13.00
N	N/A	0.10
Mo	2.00	2.50
Fe	N/A	Balance

After printing, all the samples were heat treated. For 316L, standards solution annealing for one hour in 900ºC was used. Inconel 718 samples were treated according to standard AMS5662 heat treatment cycle. First the samples are heated to 954ºC for 30 minutes followed by air cooling. After that follows double aging which is performed by holding the samples in 718ºC for 8 hours followed by furnace cooling to 621ºC at a rate of 50ºC/h. The samples are then held at 621ºC for 8 hours followed by cooling to room temperature.

3.1. Experimental setup

Autogenous welding tests for the samples were done at LUT University using IPG YLS-10000 fiber laser with maximum power of 10 kW. Welding was performed with Precitec YW50 welding optics by Precitec with 300 mm focal length focusing lens and 150 mm collimator lens. Used multimode fiber core diameter was 600 µm. Shielding gas was not used during the welds. Focal point of the laser was on the surface of the test samples. Welds were performed by butt welding the two circular rods together. Photograph from the test setup can be seen in (Figure 2)

Figure 2. Image of the test setup for welding.

Pretest samples are cut half from the middle of the sample to form suitable sample size for the microscopic analysis. Cut was done along the welded rods to form two semicircular samples. (Figure 3) All test samples were photographed after cutting. Samples were polished and etched with Adlers Etchant. Microscopic images were taken from all the samples, from fusion zone, HAZ, and base materials (Figure 3). SEM-imaging was used for some to get more information from the weld.

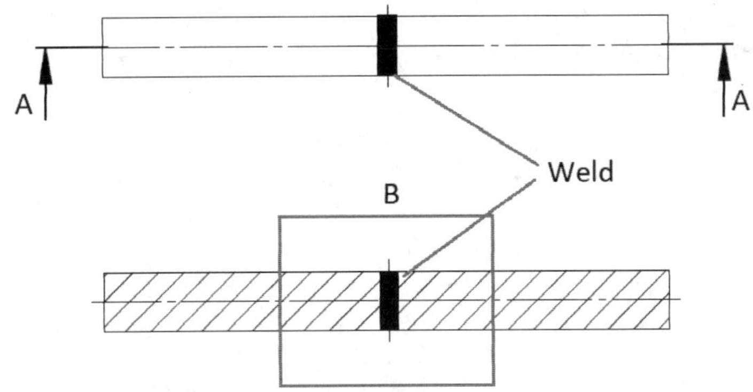

Figure 3. Cut was done after welding the rods together along axis A-A. Pictures from the sample were taken from area B.

Laser welding tests were done with constant laser power and varied welding speed. Five different welding speeds were used 1.35 (samples 15-1), 1.60 (samples 60-2), 1.85 (samples 17-3), 2.10 (samples 18-4) and 2.35 m/min (samples 19-5). Different welding parameters for samples were used to find optimal energy and heat input for the welds and to study weld quality and effects in microstructure. The change of welding speed caused naturally change in energy input also, which varied accordingly with values of 111, 94, 81, 71 and 64 J/mm. The energy input of laser welding was calculated by using equation:

E = P/v, where E is energy input (J/mm), P is the laser power (W) and v is the welding speed (mm/s)

After welding and cutting the samples, other half of the cut pieces were heat treated again. The heat treatment followed AMS5662 standard. Hardness measures were taken from the weld from both parts of the sample. Measurements were taken approximately 1 mm from the sample surface. The results from heat treated and in welded condition samples were compared.

4. Results and discussion

The energy input difference is clearly seen when comparing the samples with different heat inputs. This can be seen in Figure 4.

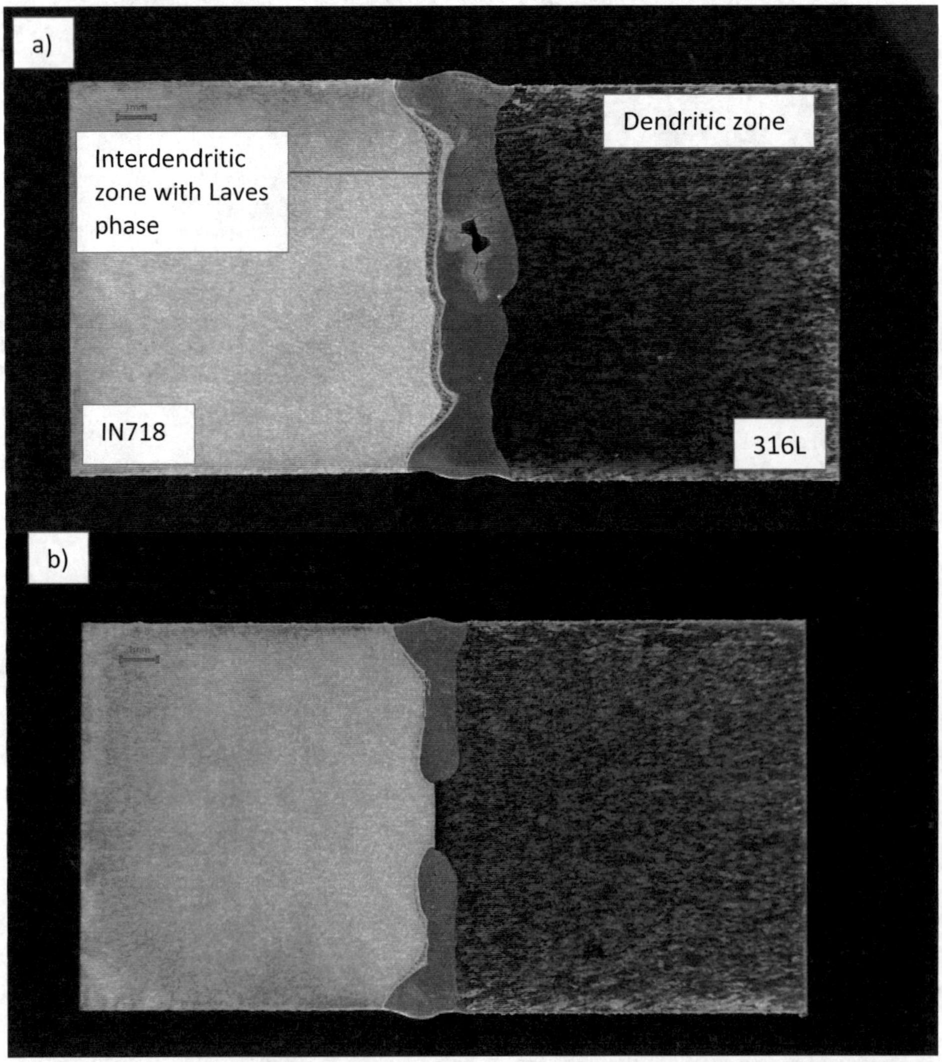

Figure 4. High energy input weld (samples 15-1) (a). Low energy input weld (samples 19-5) (b). Comparison between highest and lowest energy inputs.

Increasing the energy more significantly affects the 316L side of the material and the melt pool in 316L side is larger but the welds are generally quite symmetrical. Weld penetration increases with energy input from around 4 mm to over 6 mm. It also increases the width of the weld from 1 mm to over 2 mm. Some defects, mainly porosity and cracking (ISO 6520-1 reference 100 and 200 as stated in standard ISO/DIS 13919-9) can also be seen when using high energy inputs. These defects are introduced as the weld roots cross on the bottom as the penetration depth is more that the sample radius.. Pre-heating can reduce cracking as the cooling rate is more uniform, but it also introduces more Laves phase formation as the cooling rate decreases. Pre-heating was not used in these tests. Laves phase rich area is also larger when using higher energy inputs. This happens as the cooling rates are lower.

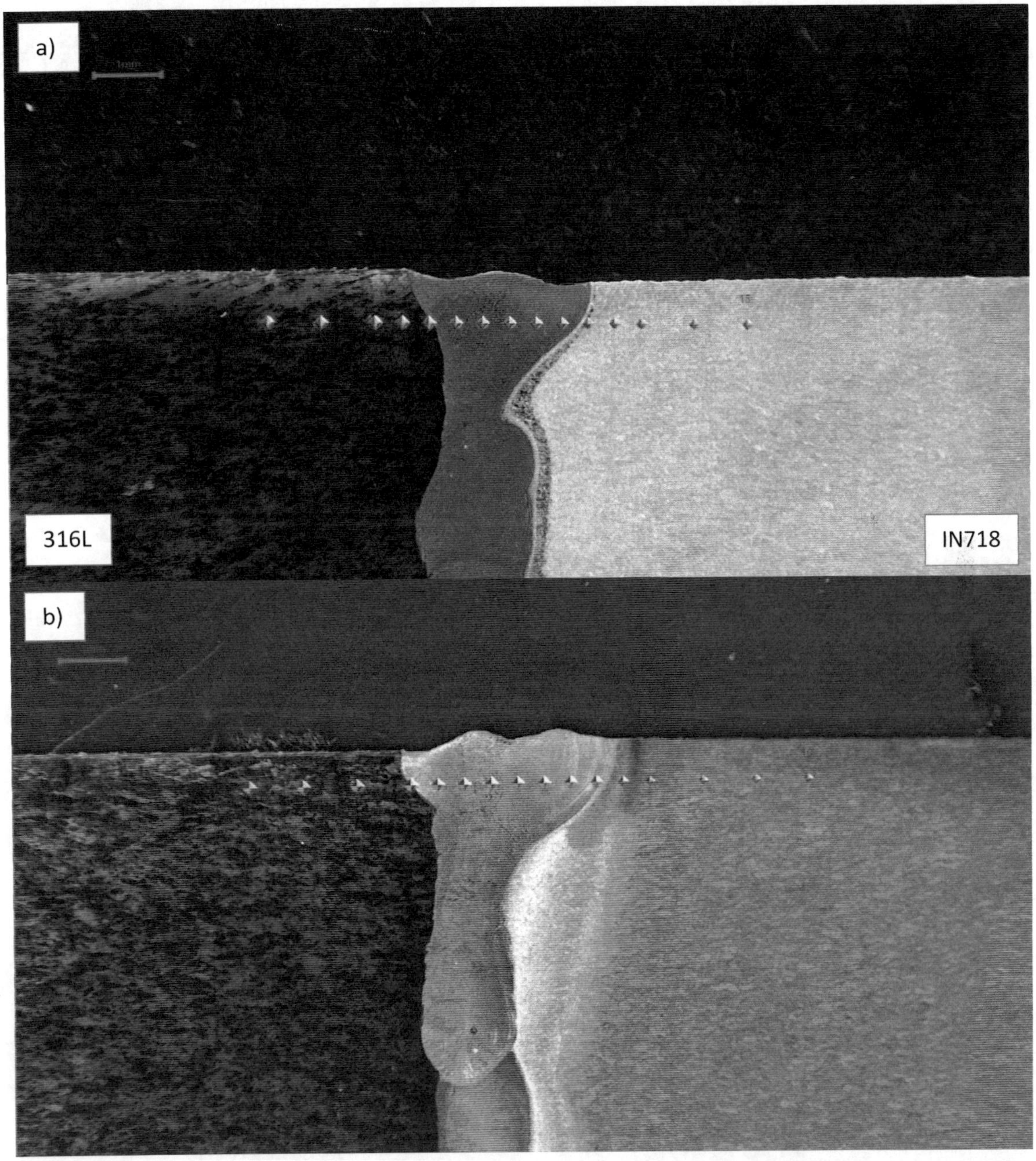

Figure 5. IN718 HAZ differences can be seen in as welded (a) and post-weld heat treated parts (b). Welding test with sample pair 15-1, high energy input.

Fine dendritic structures are formed in the weld. Laves phase rich dark zone is seen in IN718 side of the HAZ and its visibility is reduced with heat treatment (Figure 5). The material microstructure becomes more uniform and this dark region near weld is removed as the solution + double aging treatment (AMS 5662) removes the unwanted phases.

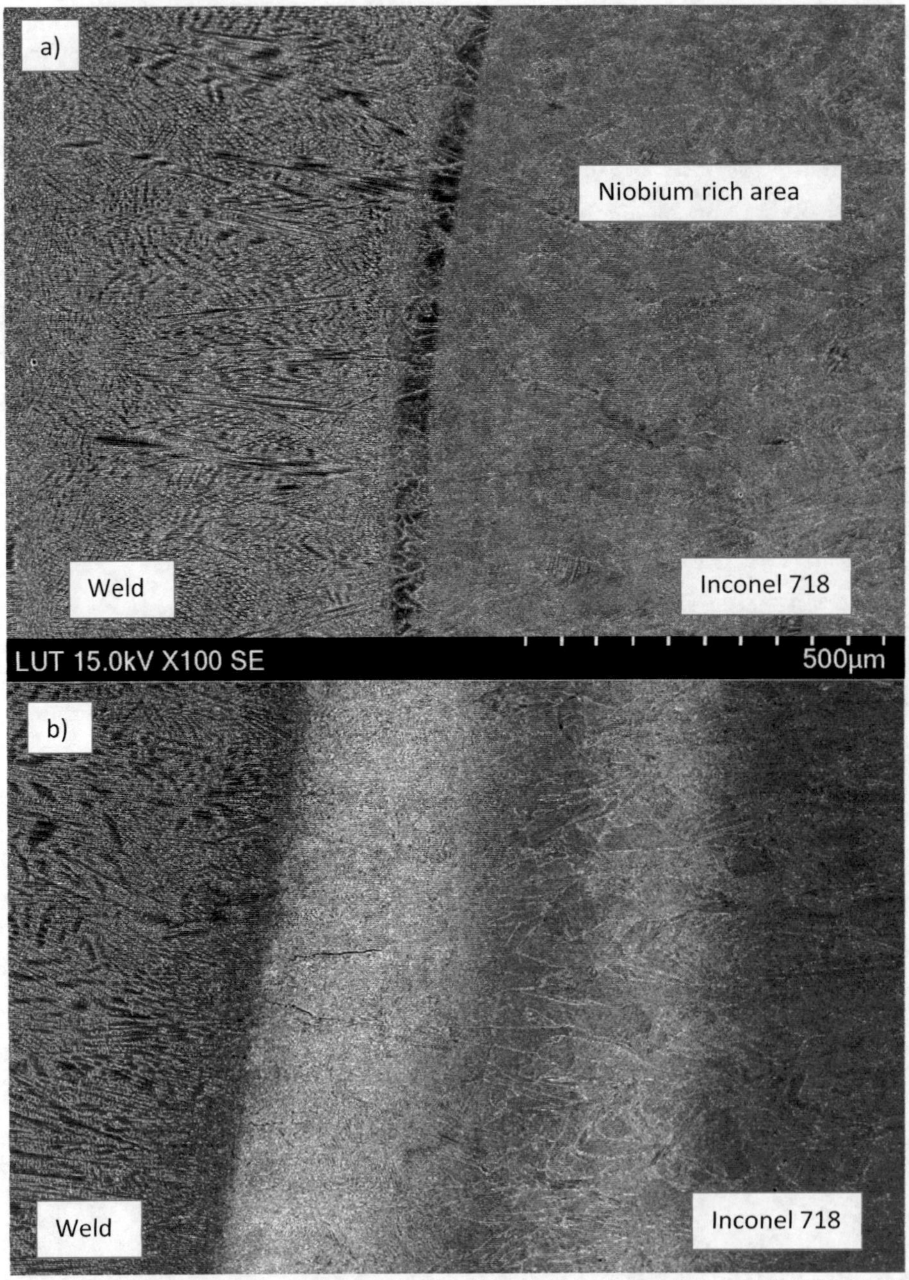

Figure 6. SEM image of the welds before heat treatment (a) and after heat treatment (b).

SEM images highlights the dark niobium rich zone between the weld and base material. SEM imaging shows Laves phase as white concentrations in the dark zone.

In addition to visual comparison, hardness tests were performed. Hardness profiles were measured across the welds in both heat treated and as welded states. Comparison of the hardness profiles can be

IOP Conf. Series: Materials Science and Engineering 1296 (2023) 012031 doi:10.1088/1757-899X/1296/1/012031

seen in Figure 3. Weld sample for this hardness test was chosen from the heat input that was midpoint of the test series. Other weld hardness's behaved similarly.

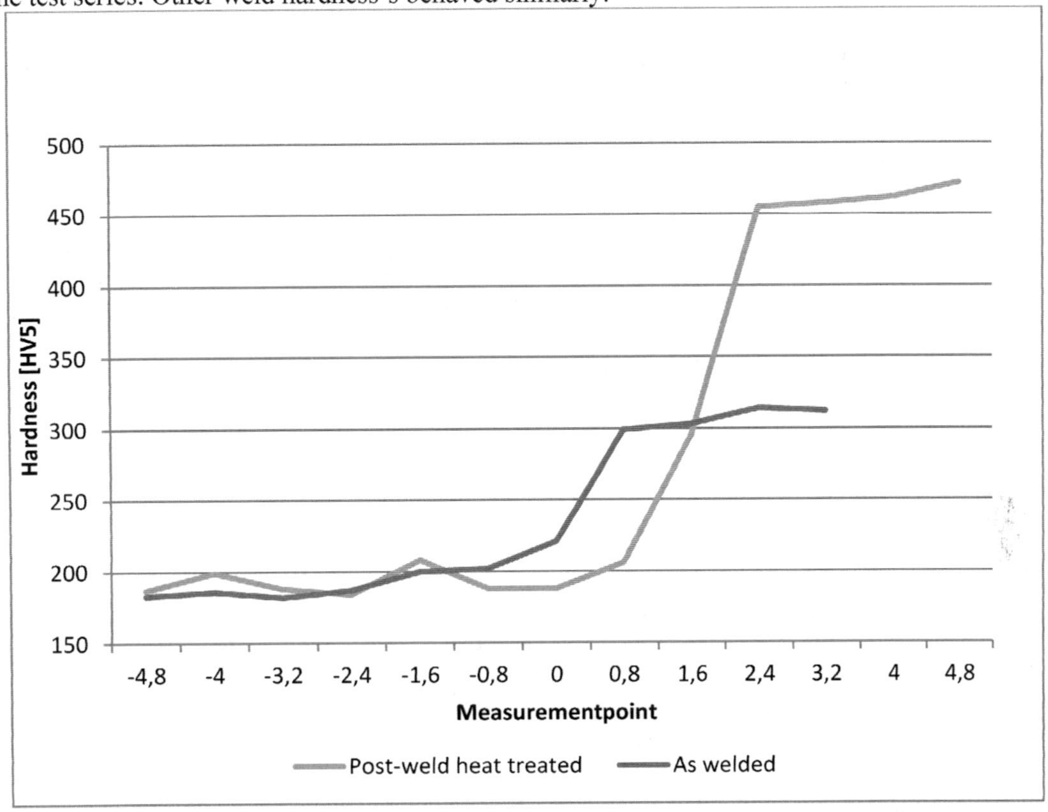

Figure 7. Hardness profile of the samples before and after post-weld heat treatment. Test sample for this hardness tests was 17-3.

In these results post-weld heat treatment greatly increases the hardness of the Inconel part of the sample. In as welded state, hardness's observed were around 300 HV. After heat treating the parts, hardness of Inconel increased to 450-470 HV. This 450-470 HV range is expected in printed and heat treated IN718 alloy. It can also be interpreted that mixing of materials is happening as the hardness value starts to increase in the weld and sharply increases as the measurements move to the IN718 side of the weld. Drop in hardness values in the weld that was observed by study done in Oulu university**[21]** was not seen in these tests. It can also be seen, that AMS5662 heat treatment for Inconel 718 has no effect on 316L hardness properties in PBF-LB 316L-IN718 dissimilar welded joints.

5. Conclusions

Dissimilar welded joints of 316L and Inconel 718 materials were studied. It seems that laser welding of these materials shows good weldability with good material properties for Inconel 718 super alloy when post-weld heat treatment is used. It can be seen that hardness values for Inconel 718 are higher than 316L even without this post-weld heat treatment. It can be expected the material to fail in the 316L side of the joint in tensile test. The tests show that softening of the IN718 material during the welding can be countered by heat treating the part again after welding using the standard AMS 5662 heat treatment. Weld parameters should be optimized to find optimal for dissimilar materials. Some cracking and porosity in the weld were observed in the welded samples. The microstructures and phases in the samples need to be examined further. Follow up Gleeble tests are planned to examine the right cooling time and adjust the welding parameters accordingly. Also, tensile tests for welds are planned.

IOP Conf. Series: Materials Science and Engineering 1296 (2023) 012031 doi:10.1088/1757-899X/1296/1/012031

Credit authorship contribution

All authors contributed to the publication by either conceptualizing the idea, by analyzing the data or providing testing possibilities. Publication was commented and modified by all authors.
All authors contributed to chapters 2. Aim and purpose of the study and 3. Material and methods. Markku Lindqvist was main responsible for chapters 1. Introduction and 4. Results and discussion. All authors participated in conclusions and final analysis.

References

[1] Sonar T, Balasubramanian V, Malarvizhi S, Venkateswaran T, Sivakumar D. An overview on welding of Inconel 718 alloy - Effect of welding processes on microstructural evolution and mechanical properties of joints. Materials Characterization 2021;174:110997. https://doi.org/10.1016/j.matchar.2021.110997.

[2] Devendranath Ramkumar K, Patel SD, Sri Praveen S, Choudhury DJ, Prabaharan P, Arivazhagan N, et al. Influence of filler metals and welding techniques on the structure–property relationships of Inconel 718 and AISI 316L dissimilar weldments. Materials & Design (1980-2015) 2014;62:175–88. https://doi.org/10.1016/j.matdes.2014.05.019.

[3] Voropaev A, Stramko M, Sorokin A, Logachev I, Kuznetsov M, Gook S. Laser welding of Inconel 718 nickel-based alloy layer-by-layer products. Materials Today: Proceedings 2020;30:473–7. https://doi.org/10.1016/j.matpr.2019.12.399.

[4] De Bartolomeis A, Newman ST, Jawahir IS, Biermann D, Shokrani A. Future research directions in the machining of Inconel 718. Journal of Materials Processing Technology 2021;297:117260. https://doi.org/10.1016/j.jmatprotec.2021.117260.

[5] Lindqvist M, Piili H, Salminen A. Benchmark Study of Industrial Needs for Additive Manufacturing in Finland. Physics Procedia 2016;83:854–63. https://doi.org/10.1016/j.phpro.2016.08.089.

[6] Jiang R, Mostafaei A, Wu Z, Choi A, Guan P-W, Chmielus M, et al. Effect of heat treatment on microstructural evolution and hardness homogeneity in laser powder bed fusion of alloy 718. Additive Manufacturing 2020;35:101282. https://doi.org/10.1016/j.addma.2020.101282.

[7] Xu L, Chai Z, Peng B, Zhou W, Chen X. Effect of heat treatment on microstructures and mechanical properties of Inconel 718 additively manufactured using gradient laser power. Materials Science and Engineering: A 2023;868:144754. https://doi.org/10.1016/j.msea.2023.144754.

[8] Huang W, Yang J, Yang H, Jing G, Wang Z, Zeng X. Heat treatment of Inconel 718 produced by selective laser melting: Microstructure and mechanical properties. Materials Science and Engineering: A 2019;750:98–107. https://doi.org/10.1016/j.msea.2019.02.046.

[9] Cao X, Rivaux B, Jahazi M, Cuddy J, Birur A. Effect of pre- and post-weld heat treatment on metallurgical and tensile properties of Inconel 718 alloy butt joints welded using 4 kW Nd:YAG laser. J Mater Sci 2009;44:4557–71. https://doi.org/10.1007/s10853-009-3691-5.

[10] Tucho WM, Cuvillier P, Sjolyst-Kverneland A, Hansen V. Microstructure and hardness studies of Inconel 718 manufactured by selective laser melting before and after solution heat treatment. Materials Science and Engineering: A 2017;689:220–32. https://doi.org/10.1016/j.msea.2017.02.062.

[11] Kumara C, Balachandramurthi AR, Goel S, Hanning F, Moverare J. Toward a better understanding of phase transformations in additive manufacturing of Alloy 718. Materialia 2020;13:100862. https://doi.org/10.1016/j.mtla.2020.100862.

[12] Górka J, Jamrozik W, Kiel-Jamrozik M. The effect of TIG welding on the structure and hardness of butt joints made of Inconel 718. Heliyon 2023;9. https://doi.org/10.1016/j.heliyon.2023.e13175.

[13] Ghaemifar S, Mirzadeh H. Precipitation kinetics of niobium carbide (NbC) during homogenization heat treatment of additively manufactured Inconel 718 superalloy. Journal of Materials Research and Technology 2023. https://doi.org/10.1016/j.jmrt.2023.06.069.

[14] Kruk A, Cempura G. Decomposition of the Laves phase in the fusion zone of the Inconel 718/ATI 718Plus® welded joint during isothermal holding at a temperature of 649 °C. Materials Characterization 2023;196:112560. https://doi.org/10.1016/j.matchar.2022.112560.
[15] Popovich VA, Borisov EV, Popovich AA, Sufiiarov VSh, Masaylo DV, Alzina L. Impact of heat treatment on mechanical behaviour of Inconel 718 processed with tailored microstructure by selective laser melting. Materials & Design 2017;131:12–22. https://doi.org/10.1016/j.matdes.2017.05.065.
[16] Anbarasan N, Gupta BK, Prakash S, Muthukumar P, Oyyaravelu R, Kumar RJF, et al. Effect of Heat Treatment on the Microstructure and Mechanical Properties of Inconel 718. Materials Today: Proceedings 2018;5:7716–24. https://doi.org/10.1016/j.matpr.2017.11.448.
[17] Janaki Ram GD, Venugopal Reddy A, Prasad Rao K, Reddy GM, Sarin Sundar JK. Microstructure and tensile properties of Inconel 718 pulsed Nd-YAG laser welds. Journal of Materials Processing Technology 2005;167:73–82. https://doi.org/10.1016/j.jmatprotec.2004.09.081.
[18] Lalam SV, Reddy GM, Mohandas T, Kamaraj M, Murty BS. Continuous drive friction welding of Inconel 718 and EN24 dissimilar metal combination. Materials Science and Technology 2009;25:851–61. https://doi.org/10.1179/174328408X369384.
[19] Hasani N, Dharmendra C, Sanjari M, Fazeli F, Amirkhiz BS, Pirgazi H, et al. Laser powder bed fused Inconel 718 in stress-relieved and solution heat-treated conditions. Materials Characterization 2021;181:111499. https://doi.org/10.1016/j.matchar.2021.111499.
[20] Zhang R, Buchanan C, Matilainen V-P, Daskalaki-Mountanou D, Britton TB, Piili H, et al. Mechanical properties and microstructure of additively manufactured stainless steel with laser welded joints. Materials & Design 2021;208:109921. https://doi.org/10.1016/j.matdes.2021.109921.
[21] Rautio T, Mäkikangas J, Kumpula J, Mustakangas A, Järvenpää A. Hybrid Products from Laser Powder Bed Fusion Manufactured Inconel 718 and 316L Utilizing Laser Welding. Materials Science Forum 2022;1053:199–204. https://doi.org/10.4028/p-doj951.

Acknowledgements

The authors would like to acknowledge all relevant personnel in Delva Oy for enabling the experiments carried out for this study and thank for their assistance and knowledge. Authors would also like to gratitude LUT University for their assistance and knowledge.

Manufacturing 4.0 (MFG4.0) project funded by the Strategic Research Council of Finland, 1.1.2018 – 31.12.2024. The Participant of project were MFG4.0 project is carried out by University of Turku, University of Jyväskylä, University of Helsinki and LUT University. The project aims to investigate on how the digital manufacturing will change fabrication in Finland.

DREAMS (Database for Radically Enhancing Additive Manufacturing and Standardization). This project was funded by Business Finland, 1.4.2022 – 28.2.2024. The Participants of project were the Finnish Additive Manufacturing Ecosystem (FAME). University of Turku, University of Oulu and LUT University. Aim of the project was to establish standards for AM and develop a powder material database for AM. Authors would like to thank partners of this project for their contribution to this article.

The effects of laser welding parameters on weldability and quality of the microstructure of additively manufactured Inconel 718-316L joints

E Kivirasi[1], M Lindqvist[1], K Lipiäinen[2], H Piili[3] and A Salminen[3]

[1]Delva Oy, Hämeenlinna 13430, Finland
[2]Laboratory of Steel Structures, LUT University, Lappeenranta 53850, Finland
[3]Department of Mechanical Engineering Faculty of Technology, University of Turku, Turku 20520, Finland

eetu.kivirasi@delva.fi

Abstract. Additive manufacturing of metals is a way of producing high-quality end-use parts. Technical alloys, for example Inconel 718, can be used to obtain a lot of benefits for example wear, corrosion, and heat resistance. Laser welding of Inconel 718 is a standard process, but there is rather limited amount of information of welding of additively manufactured nickel superalloys to the additively manufactured stainless steel. The process parameters need to be considered in laser welding. Undesired microstructure can occur due to wrong heat input during the welding process. This study examines laser welding of additively manufactured Inconel 718-316L parts and the effects of the welding parameters to the quality of the weld by analysing microstructure from the heat affected zone. This is done to achieve better part quality more cost efficiently compared to traditionally produced parts and to optimize the welding parameters. It is not feasible to manufacture the full large structure with IN718 and AM could be used to manufacture just the functional parts of the assembly. Tests have shown that welding heat input and cooling time affect to the quality. Inconel 718 hardness decrease across the fused zone because of the mixing of different elements in the molten weld pool. Laser welding highlights cuboidal shaped niobium rich carbides throughout the material to the heat affected zone grain boundaries on Inconel side.

1. Introduction

Dissimilar joints, for example of 316L and IN718, are commonly used in aerospace and power industries [1]. The research done so far of welded additive manufacturing (AM) components is still very sparse and must be continued further to provide enough information for the industry [2,3]. In the Finnish industry the most common large parts are manufactured, because it is not feasible to produce those fully with AM. That is the reason why it is so important to know how the AM components and materials behave during the welding process.

Laser welding of Inconel 718 is a standard process in oil and gas industry [4]. The possible filler metal and process parameters need to be considered in laser welding. It is easy to achieve welds with undesired microstructures due to heating during the welding process. Undesirable niobium rich laves phase in Inconel 718 can reduce the properties of the material in heat affected zone (HAZ) [5]. Laves phase is mainly generated due to the high heat input and slow cooling rates [6]. Faster cooling rates largely reduce the formation of laves phase in the part [7]. Laser welding cannot be used in all of the cases, for example burner nozzles because of the multiple internal channels. The formation of laves phase can also be controlled with different filler metals. When using low-niobium fillers, formation of laves phase can be controlled. Due to the AM microstructure, the phase formation differs from traditionally manufactured Inconel 718 [8]. Laser welding of smaller AM components to a larger traditionally manufactured structure can be beneficial when combining the advantages from additive manufacturing and traditional methods [9]. Especially in cases of large welded structures, it is not always possible to do the heat treatments, which can be problematic if a wrong microstructure is formed during the laser welding [10]. Therefore, it is important to study the effects of welding to AM Inconel 718 – 316L joints to achieve acceptable properties to end use structures.

2. Aim and purpose of the study

Authors of this paper have executed study of laser welding of IN718-316L and IN718-IN718 which is shown in Figure 1. This study focuses on different material pair and post weld heat treatment. This is done to ensure deeper analysis and conclusions for each different case.

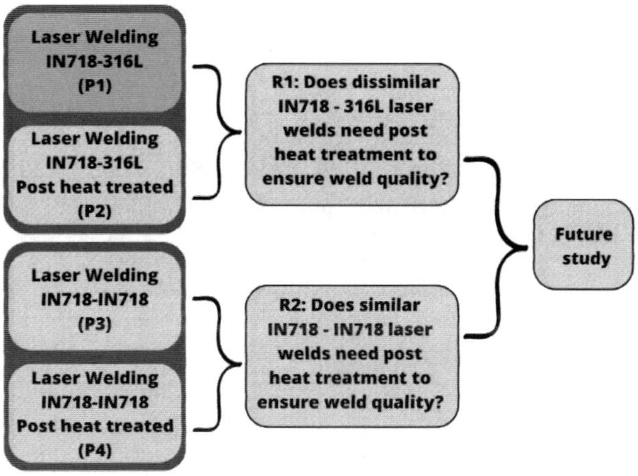

Figure 1. Study of laser welding of IN718-316L and IN718-IN718.

As it can be seen from Figure 1 as orange, this study focus on laser welding IN718-316L pair without weld post heat treatments. The aim of this study is to create initial data for the industry about joining additively manufactured material pairs by laser welding. This study consists of Inconel 718-316L laser welds. The goal is to create benchmark for laser welding AM components and to investigate possible differences in material behavior during laser welding. This research eases the use of additive manufacturing in industrial applications. Scientific relevance of this article is to produce novel information about the laser welding of additively manufactured Inconel 718-316L dissimilar joints. Industrial relevance of this article is to produce laser welding parameters which can be used to weld additively manufactured Inconel 718-316L parts and to produce information of the manufacturing methods that can be used to manufacture parts more cost efficiently.

3. Experimental set up and procedure

3.1. Materials and methods

10.5 x 80 mm Inconel 718 round rods were additively manufactured. Rods were manufactured with EOS M290 machine with standard EOS Inconel 718 powder and standard parameters. The chemical composition of the Inconel 718 powder is shown in table 1.

Table 1. Material nominal composition of EOS NickelAlloy Inconel 718 [11].

Element	Min [wt.-%]	Max [wt.-%]
Ni	50	55
Cr	17.0	21.0
Nb	4.75	5.5
Mo	2.8	3.3
Ti	0.65	1.15
Al	0.20	0.80
Co	N/A	1.0
Cu	N/A	0.3
C	N/A	0.08
Si, Mn	N/A	0.35
P, S	N/A	0.015
B	N/A	0.006
Fe	N/A	Balance

The round rods were heat treated according to EOS standard heat treatment for Inconel 718 conforming to AMS 5662 and AMS 2774 aerospace material specifications. Heat treatment consists of solution annealing and ageing phases. Solution annealing phase is performed by heating the rod to 954 °C for one hour/25 mm of material thickness followed by rapid cooling in air. After the solution annealing phase the ageing phase is performed by holding the rod at 718 °C for eight hours, cooling it to 621 °C in two hours and holding eight hours in 621 °C followed by cooling in air.
The second material used in this study was EOS StainlessSteel 316L 1.4404. 10.5 x 80 mm round rods were additively manufactured. Rods were manufactured with EOS M290 machine with standard EOS StainlessSteel 316L powder and standard parameters. [11,12] The chemical composition of the EOS StainlessSteel 316L powder is shown in table 2.

Table 2. Material nominal composition of EOS StainlessSteel 316L 1.4404 [12].

Element	Min [wt.-%]	Max [wt.-%]
C	N/A	0.03
Si	N/A	1.00
Mn	N/A	2.00
P	N/A	0.045
S	N/A	0.015
Cr	16.50	18.50
Ni	10.00	13.00
N	N/A	0.10
Mo	2.00	2.50
Fe	N/A	Balance

3.2. Experimental set-up

This study was carried out at LUT University (Lappeenranta, Finland) using IPG YLS-1000 fiber laser which has maximum power of 10 kW. Used welding head was YW50 with 300 mm focal length focusing lens and 150 mm collimator lens. Used fiber diameter was 600 µm. Shielding gas was not used during the welds. Inconel 717 – 316L. Experimental set up can be seen in figure 2.

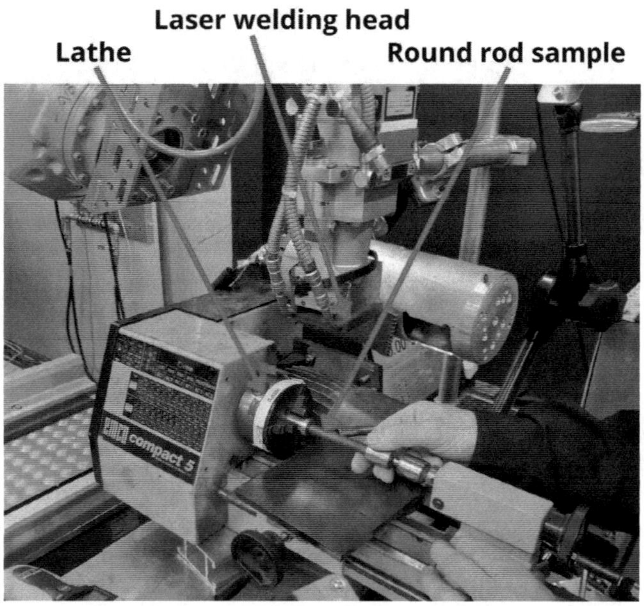

Figure 2. Experimental set up.

After the welding all the tests samples were photographed. Pretest sample rods are cut half from the middle to form suitable sample size for the microscopic analysis. Cutting was done using Struers Secotom cutting machine. Samples were polished using Struers LaboForce-100 machine and etched with Alders Etchant. Weld beads from the polished sections were macro photographed and weld widths and areas were measured. Quality of the welds was evaluated using the ISO 13919-1 Standard: electron and laser beam welded joints. Guidance on the quality levels for imperfections Part 1: Steel (ISO 13919-1 Standard).

3.3. *Experimental procedure*

The focal point position was set on top of the test rods in all welding tests. Welds were butt welds. Pre-tests were done first to find suitable laser power-welding speed combination for the test series. Six different laser powers with constant welding speeds were tested. All welds were done in room temperature (+ 22 °C). Welding parameters of pre-tests are shown in table 3.

Table 3. Pre-test welding parameters for Inconel 718-316L.

	Laser Power [W]	*Welding Speed [m/min]*	Stage Optical fiber diameter [μm]	*Energy input [J/mm]*
No. 1	2250	1.75	600	77
No. 2	2500	1.75	600	86
No. 3	2750	1.75	600	94
No. 4	3000	1.75	600	103
No. 5	3250	1.75	600	111
No. 6	3500	1.75	600	120

The energy input of laser welding was calculated by using equitation:

$$E = P/v \tag{1}$$

Where E is energy input (J/mm), P is the laser power (W) and v is the welding speed (mm/s).

IOP Conf. Series: Materials Science and Engineering 1296 (2023) 012032 doi:10.1088/1757-899X/1296/1/012032

Based on the pre-test results laser power of 2500 W was selected for the test series and welding speed was varied. Laser welding parameters of the test series are shown in table 4.

Table 4. Laser welding parameters.

	Laser Power [W]	Welding Speed Stage [mm/min]	Optical fiber diameter [µm]	Energy input [J/mm]
15-1	2500	1.35	600	111
60-2	2500	1.60	600	94
17-3	2500	1.85	600	81
18-4	2500	2.10	600	71
19-5	2500	2.35	600	64

After welding all test samples were photographed with Canon EOS 5Ds digital single-lens reflex camera (DSLR). Samples are cut half from the middle to form a suitable sample size for the microscopic analysis. Samples were polished and etched with Alders Etchant. Weld beads from the polished sections were macro photographed and weld widths and areas were measured. Deeper analysis were done using Hitachi SU3500 Scanning Electron Microscope (SEM) with energy dispersive X-ray analysis (EDS). Quality of the welds was evaluated using the ISO 13919-1 Standard: electron and laser beam welded joints. Guidance on the quality levels for imperfections Part 1: Steel (ISO 1319-1 Standard).

4. Results and discussion

The heat input affects to the weld dimension and the quality. Higher heat input causes root boiling the second time laser beam hits the weld which causes unwanted tension cracking inside of the weld. Tension cracking can be seen in figure 2 (15-1). Uneven cooling rate causes tension cracking inside the round rod . Lower heat input removes the tension cracking from the weld. Samples were not preheated before laser welding and that is one reason for the rapid cooling rate of the laser weld. Surface of the weld cools faster and solidificates before the root inside the round rod which causes tension to the bottom of the weld. This tension causes solidification cracking. This can be seen in figure 2 (60-2). Heat input degreases roughly 15 % in every step when moving left from the sample 15-1. It can be seen that weld penetration depth degreases, but also the interdendritic zone degreases as the laser weld cools down faster. Heat input correlates to the size of the interdendritic zone. Interdendritic zone is formed from unwanted phases in Inconel 718 during the laser welding. Size of the interdendritic zone can be used as quality measurement when comparing laser welding parameters on AM Inconel-316L laser welds. Based on ISO 6520-1 standard table 1 – Imperfections, samples15-1 and 60-2 have cracks in the root. Samples 17-3, 18-4 and 19-5 have incomplete penetration. Sample 17-3 has pores inside the weld. Sample 18-4 has sagging. Sample 17-3 has linear misalignment.

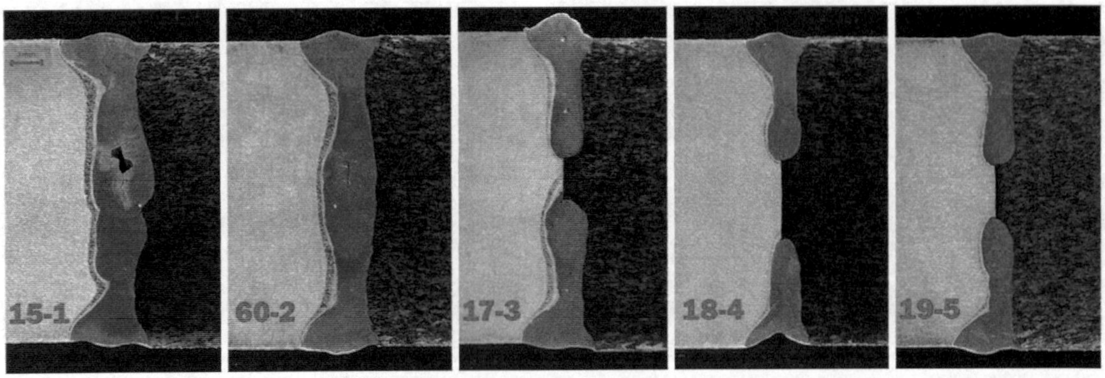

Figure 3. Heat input effect to the laser weld penetration depth on round bar.

Fine dendritic structures can be seen in interdendritic grain growth zone. Laves phases are formed because of lower cooling rate than the laser weld. Etching reveals dendritic zone as white and interdendritic zone as dark. [13] This can be seen in all samples without post weld heat treatment [14]. This can be seen in figure 3.

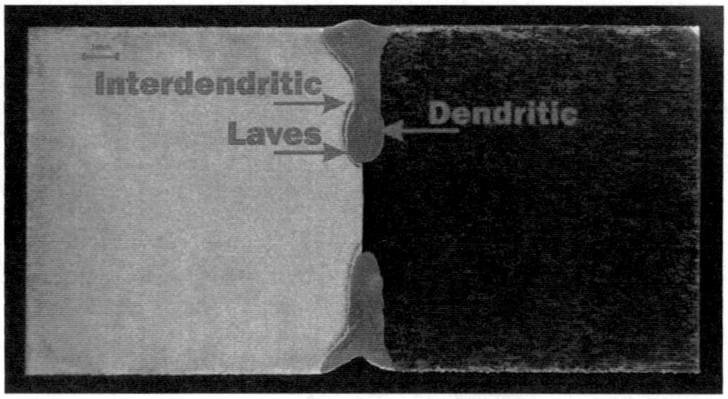

Figure 4. Different microstructures after etching in Inconel 718-316L sample.

Scanning electron microscope (SEM) images shows that carbides gather in the grain boundaries [15]. This can be seen in figure 4 a as dark spots. SEM-EDS mapping reveal that grain boundaries are niobium rich. Niobium forms hard and fragile Niobium carbide (NbC) to the grain boundaries which increase the risk of cracking, especially in oxygen rich environment.[16] This can be seen in figure 4 b as red. Figure 4 c shows elemental distribution in EDS mapping.

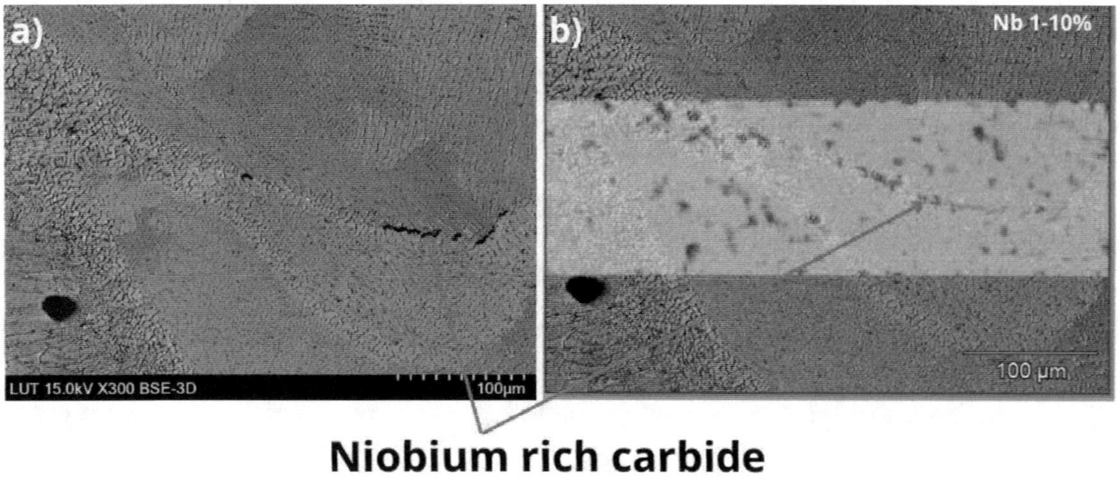

Figure 5. Grain boundaries in laser weld.

Closer images reveals that grain boundary has a lot of carbide formation which can be seen as dark spots in figure 5 a. Laves phases can be seen in figure 5 as small pure white formations [5,17]. Most of the dark carbides have niobium rich content which can be seen as blue, green, yellow and red in figure 5 b.

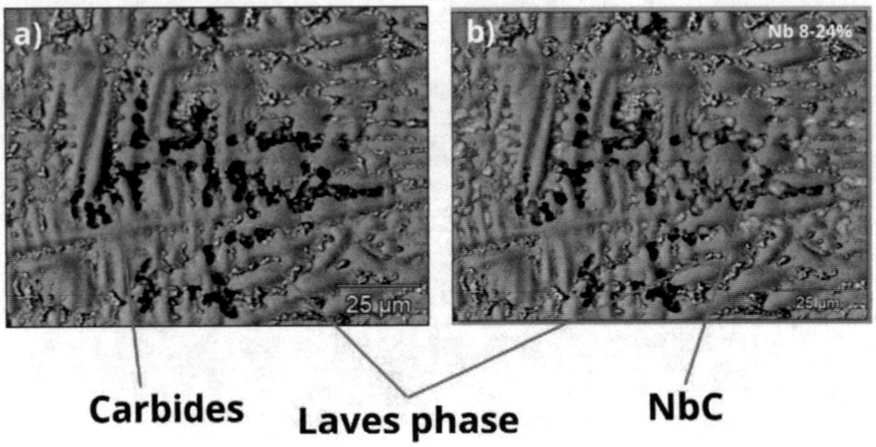

Figure 6. Niobium rich carbides and laves.

Hardness profile was measured over the HAZ and the weld. Vickers hardness testing, test method HV5 based on ISO 6507 standard was used to measure hardness. Struers DURAMIN-100 hardness testing machine was used in tests.
Hardness in the middle of the weld is 200 HV, Inconel 718 HAZ is 300 HV and 316L HAZ is 180 HV. This can be seen in figure 6. Hardness of AM Inconel 718 is 460 HV and AM 316L is 190 HV. It can be seen that laser welding does not affect to the material properties of the AM 316L. Mixing occurs inside the laser weld and the hardness from the 316L side increases from 180 HV to 200 HV. Moving closer to the Inconel 718 side hardness increases from 200 HV to the 310 HV. Heat input of the laser weld degreases hardness of the Inconel 718 in the HAZ zone 32 % compared to the Inconel 718 base material.

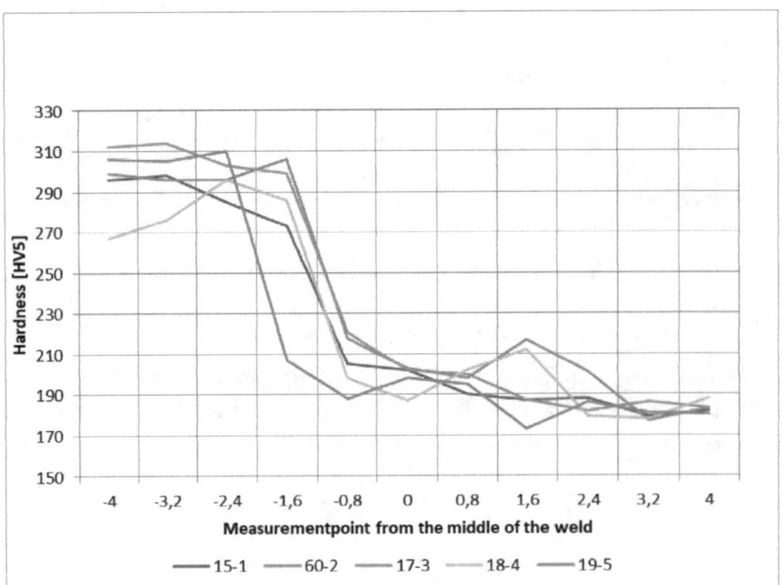

Figure 7. Hardness measurements.

5. Conclusions

Laser welding causes Inconel 718 to soften in the HAZ. The mixing of the materials is minimal, so the hardness does not increase significantly in the HAZ of the 316L compared to the base material. However, there is mixing in the weld, so the weld is harder than the base material. From the hardness curves of the samples welded with different heat inputs, it can be concluded that the optimal material properties for the weld are obtained by using a moderate heat input of 81 J/mm. Too much heat output increases the HAZ and the properties of the base material loses properties. With insufficient heat input in the weld, sufficient mixing is not achieved at a time when the weld on the 316L side is softer than the base material.

SEM images in the HAZ show niobium-rich carbides that accumulate at the grain boundaries. NiC and Nb rich areas are responsible for oxygen assisted crack growth in Inconel 718. Niobium accumulates at the grain boundaries of Inconel 718 when the heat input is excessive and the T 8/5 time is not suitable for the material. As a follow-up study, it is planned to find out more precisely the suitable T 8/5 time for Inconel 718 with the help of Gleeble tests, to plan laser welding tests based on Gleeble tests and to do SEM EDS analyzes of the niobium deposits/interdendritic zone accumulating on the grain boundaries.

The test arrangements must be developed better in the future so that welding can be done in shielding gas. Mison Ar or/and Mison H2 will be used as shielding gas.

All authors contributed to the publication by either conceptualizing the idea, by analysing the data or providing testing possibilities. Publication was commented and modified by all authors.
All authors contributed to chapters 2. aim and purpose of this study and 3. experimental set up and procedure. Eetu Kivirasi was main responsible for chapters 1. introduction and 4. results and discussion. All authors participated in conclusions and final analysis.

References

[1] Henderson MB, Arrell D, Larsson R, Heobel M, Marchant G. Nickel based superalloy welding practices for industrial gas turbine applications. Science and Technology of Welding and Joining 2004;9:13–21. https://doi.org/10.1179/136217104225017099.
[2] Effect of welding parameters on bead characteristics and mechanical properties of wire and arc additive manufactured inconel 718 - Turku University 2023. https://utuvolter.fi (accessed May 27, 2023).
[3] Liu B, Ding Y, Xu J, Gao Y, Wang X, Zhang H, et al. Outstanding strength-ductility synergy in Inconel 718 superalloy via laser powder bed fusion and thermomechanical treatment. Additive Manufacturing 2023;67:103491. https://doi.org/10.1016/j.addma.2023.103491.
[4] Sonar T, Balasubramanian V, Malarvizhi S, Venkateswaran T, Sivakumar D. An overview on welding of Inconel 718 alloy - Effect of welding processes on microstructural evolution and mechanical properties of joints. Materials Characterization 2021;174:110997. https://doi.org/10.1016/j.matchar.2021.110997.
[5] Evangeline A, Sathiya P, Arivazhagan B. Laves Phase Formation and Segregation of Nb in Ni–Cr–Mo Superalloy over 316L by Hot Wire (HW) TIG Cladding Process. Arab J Sci Eng 2020;45:9685–98. https://doi.org/10.1007/s13369-020-04873-0.
[6] Sharma SK, Biswas K, Nath AK, Manna I, Dutta Majumdar J. Microstructural change during laser welding of Inconel 718. Optik 2020;218:165029. https://doi.org/10.1016/j.ijleo.2020.165029.
[7] Popovich VA, Borisov EV, Popovich AA, Sufiiarov VSh, Masaylo DV, Alzina L. Impact of heat treatment on mechanical behaviour of Inconel 718 processed with tailored microstructure by selective laser melting. Materials & Design 2017;131:12–22. https://doi.org/10.1016/j.matdes.2017.05.065.

[8] Li L, Peng G, Wang J, Gong J, Li H. Experimental study on weld formation of Inconel 718 with fiber laser welding under reduced ambient pressure. Vacuum 2018;151:140–7. https://doi.org/10.1016/j.vacuum.2018.02.008.
[9] Zhang R, Buchanan C, Matilainen V-P, Daskalaki-Mountanou D, Britton TB, Piili H, et al. Mechanical properties and microstructure of additively manufactured stainless steel with laser welded joints. Materials & Design 2021;208:109921. https://doi.org/10.1016/j.matdes.2021.109921.
[10] Gruber K, Szymczyk-Ziółkowska P, Dziuba S, Duda S, Zielonka P, Seitl S, et al. Fatigue crack growth characterization of Inconel 718 after additive manufacturing by laser powder bed fusion and heat treatment. International Journal of Fatigue 2023;166:107287. https://doi.org/10.1016/j.ijfatigue.2022.107287.
[11] Nickel Alloys for Industrial 3D Printing | EOS n.d. https://www.eos.info/en/3d-printing-materials/metals/nickel-alloys (accessed June 25, 2023).
[12] Stainless Steel for 3D Printing| EOS n.d. https://www.eos.info/en/3d-printing-materials/metals/stainless-steel (accessed June 25, 2023).
[13] Mohd Yusuf S, Mazlan N, Musa NH, Zhao X, Chen Y, Yang S, et al. Microstructures and Hardening Mechanisms of a 316L Stainless Steel/Inconel 718 Interface Additively Manufactured by Multi-Material Selective Laser Melting. Metals 2023;13:400. https://doi.org/10.3390/met13020400.
[14] Wen Y, Gao J, Narayan RL, Wang P, Zhang L, Zhang B, et al. Microstructure-property correlations in as-built and heat-treated compositionally graded stainless steel 316L-Inconel 718 alloy fabricated by laser powder bed fusion. Materials Science and Engineering: A 2023;862:144515. https://doi.org/10.1016/j.msea.2022.144515.
[15] Gao M, Wei RP. GRAIN BOUNDARY NIOBIUM CARBIDES IN INCONEL 718. Scripta Materialia 1997;37:1843–9. https://doi.org/10.1016/S1359-6462(97)00373-4.
[16] K. DR, Sidharth D, K.v. PP, R. R, K. GM, S. N. Microstructure and properties of inconel 718 and AISI 416 laser welded joints. Journal of Materials Processing Technology 2019;266:52–62. https://doi.org/10.1016/j.jmatprotec.2018.10.039.
[17] Kumara C, Balachandramurthi AR, Goel S, Hanning F, Moverare J. Toward a better understanding of phase transformations in additive manufacturing of Alloy 718. Materialia 2020;13:100862. https://doi.org/10.1016/j.mtla.2020.100862.

Acknowledgements

The authors would like to acknowledge all relevant personnel in Delva Oy for enabling the experiments carried out for this study and thank for their assistance and knowledge. Authors would also like to gratitude LUT University for their assistance and knowledge.

Manufacturing 4.0 (MFG4.0) project funded by the Strategic Research Council of Finland, 1.1.2018 – 31.12.2024. The Participant of project were MFG4.0 project is carried out by University of Turku, University of Jyväskylä, University of Helsinki and LUT University. The project aims to investigate on how the digital manufacturing will change fabrication in Finland.

DREAMS (Database for Radically Enhancing Additive Manufacturing and Standardization). This project was funded by Business Finland, 1.4.2022 – 28.2.2024. The Participants of project were the Finnish Additive Manufacturing Ecosystem (FAME). University of Turku, University of Oulu and LUT University. Aim of the project was to establish standards for AM and develop a powder material database for AM. Authors would like to thank partners of this project for their contribution to this article.

Coaxial Wire Laser-based Additive Manufacturing of AA7075 with TiC Nanoparticles

W Meneses[1], J Tuominen[1 *], A Ylä-Autio[2], O Wiikinkoski[2], A Sabr[1] and P Peura[1]

[1]Tampere University, Materials Science and Environmental Engineering, Tampere, Finland
[2]Tampere University, Automation Technology and Mechanical Engineering, Tampere, Finland
[*] Corresponding author e-mail address: jari.tuominen@tuni.fi

Abstract. AA7075 is a heat treatable aluminium alloy widely used in aerospace and automotive applications due to its outstanding high strength-to-weight ratio. However, the implementation of this alloy in Additive Manufacturing (AM) processes has been limited due to its susceptibility to hot cracking. Moreover, selective evaporation of low boiling point elements Zn and Mg can cause gas porosity and diminish the mechanical properties of AM parts. Recent research revealed the effectiveness of nanoparticles additives to change the solidification behaviour of high-strength aluminium alloys and improve their weldability/printability. In this study, AA7075 enhanced with TiC nanoparticles was utilized as wire feedstock to create single and multi-layer samples with coaxial laser-directed energy deposition (L-DED). The response of the samples to precipitation hardening was studied, evaluating the microstructure and the microhardness before and after T6 heat treatment. Specimens were characterized using optical and electron microscopy and electron backscatter diffraction (EBSD). Crack-free and virtually porosity-free samples were fabricated, which exhibit a refined equiaxed grain structure with grain size <10μm. This confirms the ability of TiC nanoparticles to prevent columnar dendritic growth and promote heterogeneous nucleation. Microhardness values increased by 51 HV after T6 heat treatment and were uniform across the sample. Energy Dispersive Spectroscopy (EDS) analysis showed that there are evaporation losses of Zn and Mg. Considering the boiling temperatures of these elements, it is inferred that the molten pool reaches temperatures above 1090°C, and the partially melted zone temperature is between 907°C and 1090°C.

Keywords: Laser directed energy deposition, coaxial wire, AA7075, microstructure, T6 heat treatment, TiC nanoparticles.

1. Introduction

Additive manufacturing (AM) is gaining increasing interest both in industry and academia due to its possibility to build parts with intricate shapes that were difficult or even impossible to create using traditional manufacturing methods [1,2]. Furthermore, AM enables the elimination of manufacturing steps, light-weighting and consolidation of multiple components, which results in reduction of raw material and the environmental footprint of manufacturing activities [3]. There are two major AM techniques capable of processing fully dense metallic parts: powder-bed fusion (PBF), and directed energy deposition (DED) [4,5]. In PBF, localized heat is used to fuse specific portions of a powder previously spread onto a building platform. The heat source moves in a raster fashion following a pre-defined pattern according to the geometry of the desired part [6,7]. On the other hand, DED systems use a heat source of high energy density (laser, electric arc, plasma or electron beam) to melt a wire or powder feedstock material and build 3D components through a layer-by-layer deposition method [5]. PBF creates layers of about 50 to 100 μm in thickness. Whereas with DED processes, the height of each layer is around 1-2 mm resulting in much higher deposition rates [8]. Therefore, PBF is preferred when higher dimensional accuracy and lower surface roughness are needed. In contrast, DED has the capacity to produce very large parts at a high production rate and the ability to work with multi-materials [4,5]. In DED systems, feedstock can be wire or powder. Wire-fed DED processes offer several advantages over powder-fed DED, namely that feedstock is cheaper and there is more availability of different alloys, the utilization rate is 100%, the set-up cost is lower and it is easier to control the feed rate [1,7,9]. Furthermore, wires have considerably less surface area per kilogram than powders and is thus less prone to oxidize and absorb contaminants or moisture [10]. This last characteristic makes the processes fed with wire material less likely to porosity formation [5,9].

The development of AM methods for high strength aluminium alloys of 2xxx and 7xxx series has been difficult due largely to their susceptibility to hot cracking and porosity. One alloy of particular interest is AA7075, which is widely used in aerospace applications due to its outstanding high strength-to-weight ratio [11]. The excellent mechanical properties of this alloy are attributed to a fine dispersion of zinc, magnesium and copper precipitates, that are obtained after precipitation hardening heat treatment (solid solution + quenching + artificial aging) [12–14]. During this treatment, immediately after quenching, vacancy-rich clusters start to form through homogeneous nucleation in the supersaturated solid solution. These clusters then grow to coherent spherical Guinier–Preston (GP) zones with an average Zn:Mg ratio of 1:4 [15]. As aging continues for a few hours, GP zones grow to a size where intermediate phase η' can precipitate, which is semi-coherent and is the main hardening phase in AA7075 [15,16]. Furthermore, since Mg content in AA7075 is rather high, an intermediate phase T' can nucleate directly from the supersaturated solid solution, forming a semi-coherent precipitate that can contribute to hardness increase [17]. Finally, after approximately 24 hours of aging the growth and coarsening of η' results in the formation of the equilibrium phase η, with composition $Mg(Zn,Cu,Al)_2$. This phase is incoherent and therefore, as its content increases the mechanical strength of the alloy is reduced, e.g. over-aging occurs [15,17].

The two principal challenges that hinder the processability of AA7075 with DED methods are high tendency to porosity and hot cracking. There are two types of porosity: (i) gas pores, which have a spherical shape, and (ii) lack-of-fusion pores, which have an irregular morphology and form in specific regions like in the overlap areas between adjacent beads or at the interface with the substrate [18]. The presence of lack-of-fusion pores can be attributed to inadequate process parameters, for instance high wire feed rate or too fast scanning speed [18]. On the other hand, gas pores can appear due to several reasons. First, liquid aluminium can quickly dissolve high amounts of hydrogen, and because the solubility rapidly decreases during solidification, most of this hydrogen precipitates as gas and generates pores [19,20]. Second, trapped gases generated for the evaporation of low-boiling-point elements zinc and magnesium (907°C and 1090°C, respectively) [21]. Furthermore, several studies have shown that Zn and Mg evaporation in 7xxx Al alloys also results in diminished hardness and tensile strength values in both as deposited and heat-treated states [18,21–23]. Moreover, contamination of the substrate or feedstock and shielding gas entrapment can also lead to gas porosity [18]. Convection forces that exist in the molten pool can potentially expel the gas bubbles; however, porosity problems can be particularly acute in DED processes because aluminium's rapid solidification rates make it very difficult for any gases to escape [24].

Hot cracking can be divided into solidification cracking and liquation cracking [25]. AA7075 has a large solidification range of about 100°C [11]. As the molten pool cools down, the constituents with the highest solidification temperature start to solidify in the form of nuclei that grow to create dendrites. As dendrites grow bigger, they collide with adjacent ones, creating boundaries that are occupied by the constituents with lower solidification temperatures, still in liquid state. Solidification contraction produces tensile forces that can be strong enough to tear the liquid film, and if there is not enough melt phase to backfill and 'heal' these voids, solidification cracking will occur [18,19,25]. On the other hand, liquation cracking occurs because of reheating of previously solidified layer. As layers are deposited, the temperature away from the molten pool can be high enough to partially melt eutectic phases with low melting points (such as Al_2CuMg) and create a liquid film along the grain boundaries [18]. The tensile stresses produced during solidification shrinkage can break this film producing liquation cracks [25].

Several ways have been implemented to prevent or reduce cracking in AA7075 processed by laser-based AM, which can be grouped in two categories: process optimization and modification of raw materials [18]. The first category includes for example adjustments of power input and traverse speed to optimize heat input [26], preheating of the substrate to reduce the residual stresses and thus prevent cracking [27]. Furthermore, hot isostatic pressing method was successfully employed to heal liquation cracks [28]. Wang et at. [18] evaluated the effect of process parameters on microstructure and mechanical properties of lateral-fed wire L-DED of AA7075. Multi-layer crack-free samples were manufactured by the optimization of scanning strategies to control the thermal gradients and lower the cooling rate, however samples exhibited porosity and columnar grains [18]. With respect to raw materials, additives such as Si can be included, that enhances the flowability and improves the backfilling of cracks during solidification [29,30]. Likewise, studies have proved the efficiency of nanoparticles such as TiB_2 and TiC to promote heterogenous nucleation and produce a refined equiaxed grain microstructure with better resistance to cracking as grains are more capable of rotating and deforming to accommodate the strains originated from solidification shrinkage [22,31,32]. This has enabled the building of crack-free samples of AA7075 using GTAW-based DED [12], powder-based L-DED [21] and PBF [30,33]. However, to the best of the authors' knowledge no reports exist that deal with coaxial wired L-DED of nanofunctionalized AA7075, which is a very promising processes because of the aforementioned advantages of wire-based methods over powder-based ones and considering that laser as energy source in DED processes offers the possibility to have shorter interaction times and higher cooling rates than arc-based methods, which controls micro-segregation and produces fine-grained microstructures [34,35]. Furthermore, compared to lateral-fed systems, coaxial offers the advantage of omnidirectionality [36]. In the present paper, AA7075 enhanced with TiC nanoparticles was utilized as wire feedstock to create single and multi-layer samples with coaxial wire L-DED. The response of the samples to precipitation hardening was studied, evaluating the microstructure and the microhardness before and after T6 heat treatment.

2. Materials and methods

2.1. Coaxial wire L-DED

The DED equipment set-up consisted of four main components: laser, coaxial wire system, wire feeder and CNC stage. The laser system was a CORELASE 3 kW fiber laser (Coherent Inc. Santa Clara, USA) with 0.4 mm fiber diameter assembled with QBH connector to the coaxial wire head (Fraunhofer IWS, Dresden, Germany) integrated with a wire feeder system DINSE FDE100 (Dinse GmbH, Hamburg, Germany). The laser beam focal point had a diameter of 1.2mm, and was defocused -4mm, so that the spot diameter right on the substrate surface was ~3mm. Details about the coaxial wire system can be found elsewhere [37]. This whole laser assembly was mounted on a 3-axis CNC stage Aerotech ATS11010 (Pittsburgh, Pennsylvania, USA).

Samples were built with a 1.2mm diameter wire of AA7075 nanofunctionalized with TiC particles of 40–60 nm diameter (MetaLi LLC, California, USA). The chemical composition of the wire is presented in Table 1. The substrate was a plate of EN AW-5754-H111 of 8 mm thickness, which was cleaned with solvent degreaser before deposition. Preliminary testing with different powers, traverse speeds and wire feed rates were conducted to establish a convenient set of parameters. A power decay strategy was employed as layers were built on top of each other, to offset the reduction of heat conduction through the substrate plate. Built samples consisted of single-wall specimens of 70mm length and up to 6 beads (height of ~6mm). In total 22 single-layer and 6 multi-layer samples were built. Process parameters are outlined in Table 2.

Table 1. Chemical composition of AA7075 nanofunctionalized with TiC (wt%)

Al	Zn	Mg	Cu	Ti	Fe	Si	Cr	Mn
Bal.	6.59	2.91	1.98	1.41	0.13	0.03	0.03	0.03

Table 2. Coaxial wire L-DED process parameters

Parameter	Value
Power	2700 W (for 1st and 2nd bead) and 2500 W (for the rest)
Traverse speed (TS)	15 mm/s
Wire feed rate (WFR)	2 m/min
Waiting time	3 ms
Shielding gas flow (Ar)	12 l/min
Step size	1.1 mm

Interpass time	60 s

2.2. Thermal post-processing
A multi-layer sample of 6 beads on top of each other was heat-treated to T6 temper condition. The sample was solution treated in an electric furnace at 480°C. As per [15] and based on the sample thickness, 50 minutes was deemed enough time to get all the maximum practical concentration of alloying elements in solid solution. Immediately after extraction from the furnace, the sample was quenched in room temperature water, to avoid any precipitation and obtain a super saturated solid solution with high concentration of vacancies. According to the CCT diagram presented in Figure 1a, a cooling rate of at least 100°C/s is needed to avoid the formation of $MgZn_2$, which starts to precipitate at 380°C after 1 second. The sample was then artificially aged at 120°C for 21 hours, which according to the isothermal diagram in Figure 1b, is when the maximum concentration of precipitate phase η′ is achieved, which provides the maximum hardening effect to AA7075. After aging, sample was let to cool down to room temperature in air.

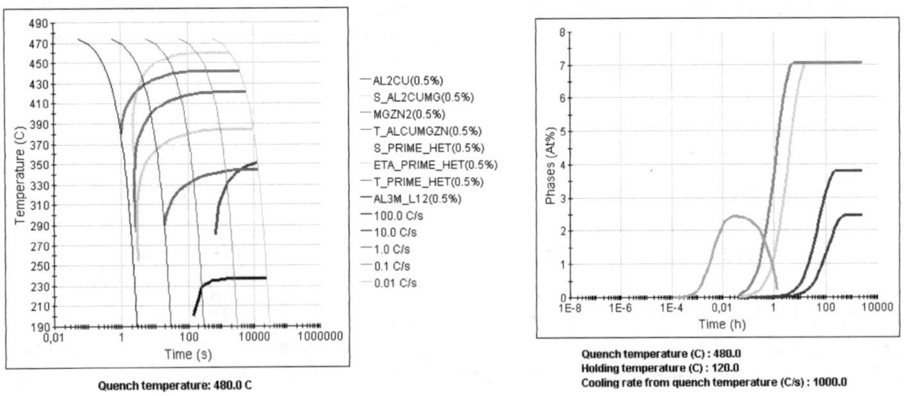

(a) **(b)**

Figure 1. (a) CCT diagram and (b) Isothermal diagram showing the kinetics of precipitation calculated for the wire composition using software JMatPro™ version 13.2.

2.3. Characterization
The deposited samples were cross sectioned closed to the centre, then were grinded with SiC papers down to 7μm grain size and polished using diamond suspensions of 3 μm and 1 μm, finalizing with colloidal silica suspension of 0.05 μm. After that, samples were cleaned in ethanol with an ultrasonic cleaner. Additionally, some samples were etched with Keller's reagent. Microstructure characterization was done with optical microscope Leica DM2500 M (Leica Microsystems GmbH, Wetzlar, Germany), Alicona InfiniteFocus G5 optical 3D measurement sensor (Alicona Imaging GmbH, Graz, Austria) and with SEM JEOL JSM-IT500 (Jeol Ltd., Tokyo, Japan) which is equipped with an EDS detector JEOL Dry SD30. Porosity was characterized by analysing optical microscopy images in ImageJ software. For EBSD analysis, Zeiss ULTRAplus SEM (Carl Zeiss AG, Oberkochen, Germany) equipped with symmetry CMOS EBSD detector (Oxford Instruments plc) was utilized, using an accelerating voltage of 20 kV, working distance of 15 mm, tilt angle of 70° and a step size of 0.7 μm for a 250 × 250 μm scanning area located in the middle of the sample. Grains were detected using a boundary disorientation angle of 10° and a minimum size of 10 pixels. Microhardness was measured using tester Matsuzawa MMT-X7 (Matsuzawa Co., Ltd., Akita, Japan), with a 200 g load applied for 10 seconds. Hardness was measured along the build direction at points separated by 0.5 mm and starting 0.1 mm from the top.

3. Results and discussion
Initially, a screening of process parameters was conducted with single-layer samples to determine the values that would result in beads with good surface morphology (no dripping or stubbing defects). Cross-section

analysis of these samples revealed that there is a clear relationship between energy input and percentage of porosity; as energy increased from 60 J/mm to 90 J/mm, porosity decreased from 2% to 0.02%. In all cases porosity can be classified as gas pores and not lack of fusion. These results confirm the correlation between porosity and energy input that has been reported by other authors [26,36], as the energy increases, there is more time for the gases in the melt pool to escape, promoting a low porosity. Even though it was not observed in the current study, it is described in the same literature that excessive increase in power can result in severe evaporation of Zn and Mg, which could lead to an increase in porosity. Using this information, power input setting was optimized to produce beads with minimum porosity. During the building of multilayer samples, a power decay strategy was employed to compensate for the change in heat dissipation conditions. Multilayer samples of up to 6 layers were built that in all cases exhibited no cracks and a porosity of less than 0.08%. A low magnification micrograph of a 6-layer sample is presented in Figure 2.

Figure 2. Optical microscope micrograph at 5X of AA7075-TiC 6-layer sample in as-built condition with 0.075% of porosity. Scale bar reads 2.5mm.

The chemical composition of different beads is presented in Table 3. Comparing these values to the wire composition (Table 1), regarding Zn content, the single-layer sample showed a decrease of 5%, whereas for a multilayer sample the composition indicates a reduction of 20% in the middle section and a reduction of 8% in the last bead. In the case of Mg content, the reduction is approximately equal for the single and multilayer cases, and corresponds to a 13-14% loss. This indicates that the content of Zn and Mg decreases during L-DED process, even for sections that do not experience any reheating (single-layer and last bead in multilayer sample). Additionally, during subsequent depositions, reheating causes further reduction of the Zn content, evidenced by the lower Zn content in the middle section of the multilayer sample. In contrast, deposition of successive layers does not result in additional losses of Mg. Considering boiling temperatures of these elements, it can be inferred that the molten pool reaches temperatures above 1090°C, and the partially melted zone temperature is between 907°C and 1090°C. Another observation that can be made from Table 3 is that the single layer sample exhibits the least decrease of Zn and Mg, which can be attributed to the quicker heat dissipation from the substrate, that acts as a heat sink. Despite evaporation losses, the content of major strengthening elements Zn, Mg and Cu is still within the composition limits for commercial AA7075. Regarding the low Cr content in the wire, the following can be mentioned. The main reason for chromium addition to AA7075 is to provide grain refinement [11]. Therefore, since TiC nanoparticles serve this purpose in nano-functionalised 7075-TiC, Cr content is reduced to values below the limit of 0.18wt%.

Table 3. Chemical composition of L-DED deposits of 7075-TiC (wt%)

Sample	Al	Zn	Mg	Cu	Ti	Fe	Si	Cr	Mn

Single-layer	Bal.	6.25	2.54	1.65	1.48	0.12	0.05	0.02	0.04
Middle section multilayer	Bal.	5.27	2.48	1.66	1.42	0.12	0.05	0.04	0.00
Last bead multilayer	Bal.	6.04	2.52	1.33	1.45	0.11	0.04	0.01	0.02
Commercial Wrought 7075 [11]	Bal.	5.1-6.1	2.1-2.9	1.2-2.0	0.20 max	0.50 max	0.40 max	0.18-0.28	0.30 max

Multilayer samples were characterized using SEM and EDS, results are presented in Figures 3, 4 and 5. The microstructure consists of a dark-grey aluminium matrix with light-grey secondary phases homogenously distributed. In the as-built condition, the secondary phase corresponds mostly to eutectic $Mg(Zn,Cu,Al)_2$, which is discontinuous and randomly oriented along the grain boundaries. This microstructure is in line with results reported in [26], where authors explained that as the primary solution phase solidifies, the alloying elements are pushed to the grain boundaries, with the eutectic phases being the last to solidify. This can be seen in the quantification maps in Figure 4, that show that the concentration of Mg, Zn and Cu is higher at the second phases. When compared to as-built condition, the heat-treated sample presents a reduced secondary phase, with much finer features. During T6, all secondary phases are dissolved and during aging new fine coherent and semi-coherent precipitates are formed, which are smaller than the phases that precipitate during the several heating-cooling cycles that the material experienced during deposition. Furthermore, quantification maps in Figure 4 and Figure 5 indicate that TiC nanoparticles are mostly concentrated around the secondary phase. This agrees with [13,14] that reported that TiC prefer to stay with the eutectic phases during solidification due to its good lattice matching of 5.8%. These images also evidence that clusters of TiC are still present after heat treatment, showing their good thermal stability. When compared to the microstructure reported in other works where multilayer samples of non-nanofunctionalized AA7075 were built using L-DED [18,36,38], the eutectic features seen in Figure 3 are smaller and discontinuous, even though the energy input was similar in all cases. These results are consistent with earlier studies [13,14] that stated that the presence of TiC nanoparticles promotes the formation discontinuous eutectic features, which is beneficial in preventing crack formation. The small and discontinuous eutectic phase along with the accumulation of TiC nanoparticles around these features match the results reported by [31], which affirms that nanoparticles are pushed by the solidification front and gather around the eutectic phase that solidifies last, interrupting its growth and ultimately modifying the morphology of the secondary phase to small discontinuous features.

The heat-treated sample in Figure 3 exhibits the presence of some elongated plate-like precipitates, which are shown in a higher magnification in Figure 5. Based on the quantification maps, these precipitates do not correspond to TiC and have a very high concentration of Cu. The shape of this precipitates resembles the Widmanstatten morphology. It is explained in [39] that the elongated shape is attributed to the high mobility of incoherent interfaces at the edges in contrast to the low mobility of semi-coherent interfaces at the broad faces. Considering its composition and morphology, this precipitate is most probably S' (Al_2CuMg), which is reported in literature as incoherent elongated laths [17]. According to the isothermal diagram in Figure 1b, during artificial aging GP zones form first, but after about 1 hour of ageing get completely replaced by η'. Over-aging of the sample would lead to precipitation of S' and θ', which are coarse incoherent precipitates that according to JMatPro simulation, should appear after 30 hours (Figure 1b). A possible explanation for the early precipitation of S' might be that the presence of TiC nanoparticles has accelerated the aging kinetics. Other studies have shown that nanoparticles can act as heterogeneous nucleation sites, interact preferentially with quenched-in vacancies, and create dislocations that facilitate diffusion of alloying elements, thus inducing a faster aging behaviour [40–42].

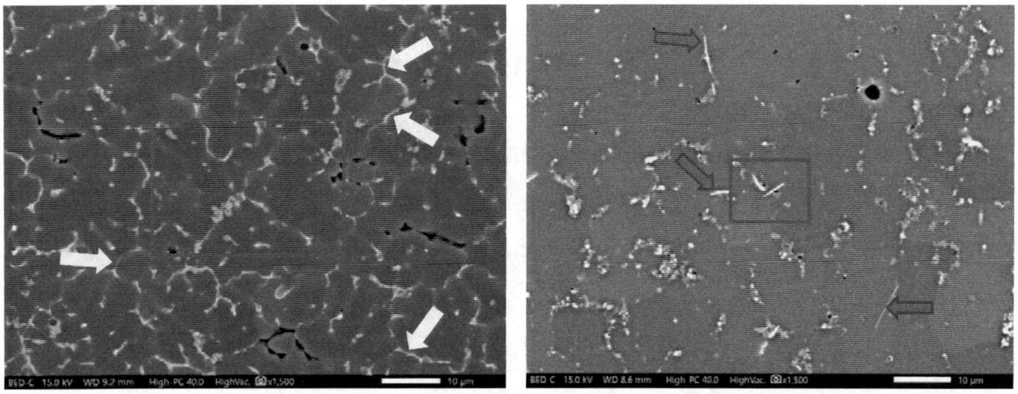

Figure 3. SEM micrographs of AA7075-TiC multilayer samples before (left) and after (right) T6 heat treatment. Yellow arrows point at grain boundaries. Blue arrows indicate incoherent S' precipitate (Al_2CuMg). Dark particles are oxide inclusions. Red squares mark areas that are further analysed in Figures 4 and 5.

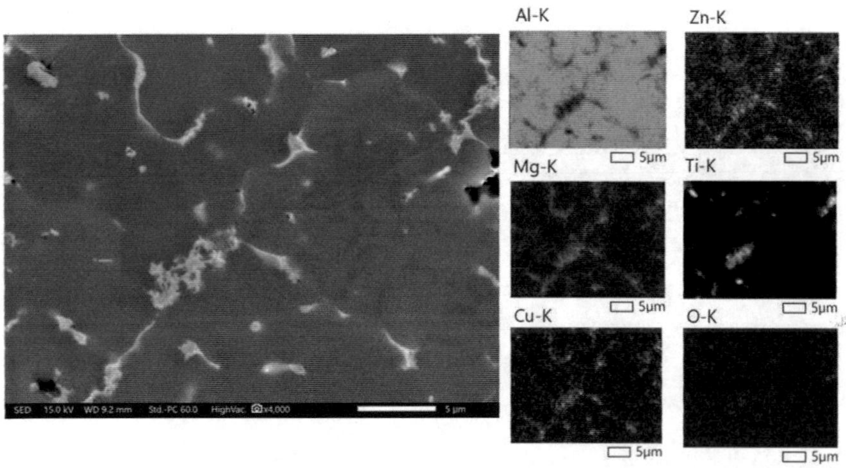

Figure 4. SEM micrograph and EDS quantification map of AA7075-TiC multilayer sample in as-built condition. TiC nanoparticles are mostly concentrated around the secondary phases.

The morphology of the eutectic phase in Figure 3 suggests that the sample displays an equiaxed grain structure. This is confirmed with the EBSD images presented in Figure 6, that show equiaxed grains for both the as-built and the heat-treated samples. DED processes are characterized by high thermal gradients along the building direction, which produces columnar grains [6,18,43]. The results in this case confirm the ability of TiC to promote nucleation and prevent the columnar structure as reported in literature [13,14,21,22,31]. The grain size in the as-built condition is 8-9μm, and after T6 it increased slightly to 9-10μm. This indicates a good thermal stability of the grains, that remain small even after solution treatment at 480°C and artificial aging at 120°C. These results are in accord with recent studies [14,40] which showed that TiC nanoparticles prevent grain growth during heating due to the pinning effect they have on grain boundaries. Figure 6 indicates that heat treatment causes the percentage of high angle grain boundaries to increase, as it changed from 67% in the as-built condition, to 94% in the heat-treated sample. This is due to recrystallization during solution treatment. Authors in [26,44] have explained that at high temperatures, diffusion results in the dissociation and merging of sub-grain boundaries with smaller orientation differences to form high angle grain boundaries. It is also mentioned in the same reports that the higher mobility of high angle boundaries allows them to absorb low angle ones.

IOP Conf. Series: Materials Science and Engineering 1296 (2023) 012033 doi:10.1088/1757-899X/1296/1/012033

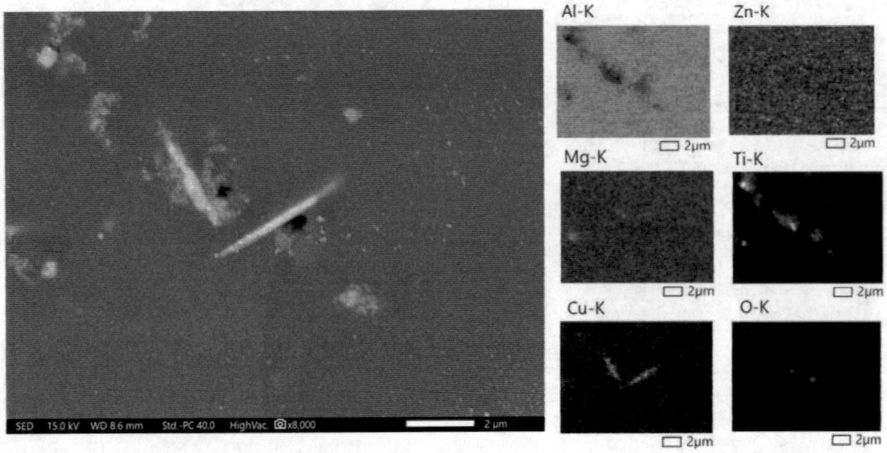

Figure 5. SEM micrograph and EDS quantification map of AA7075-TiC multilayer sample in T6 condition. Small features correspond to fine η' and T' precipitates along with TiC nanoparticles. The two big plate shape precipitates are S' (Al_2CuMg).

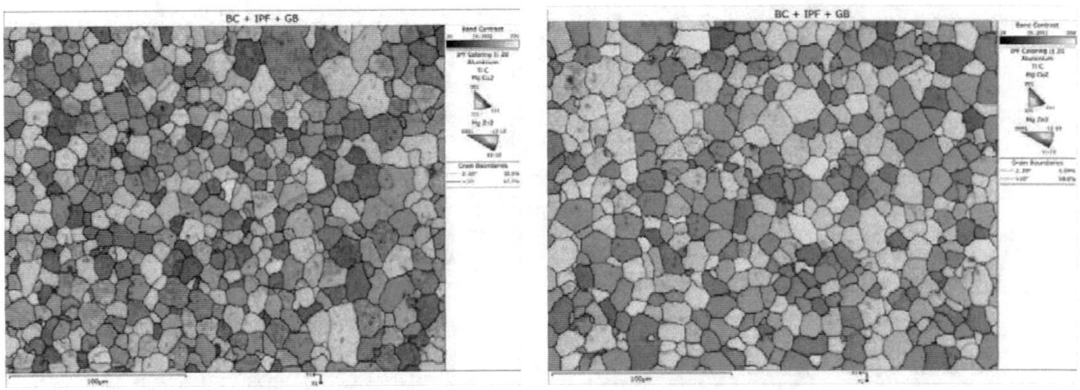

Figure 6. EBSD band contrast, inverse pole figures and grain boundary angles for AA7075-TiC multilayer coaxial wire L-DED sample in as-built (left) and T6 (right) conditions.

Microhardness results for as-built and heat-treated samples are outlined in Figure 7; in both cases, hardness is uniform along the building direction, which is in accordance with the uniform distribution of strengthening elements (Figure 4 and 5) and uniform grain microstructure (Figure 6) of the samples. The average values in the as-built and T6 condition are 121 HV and 172 HV, respectively, which represents an increment of more than 50 HV, confirming the positive response of the sample to precipitation strengthening heat treatment. Chemical compositions in Table 3 indicate that there were evaporation losses of up to 20% of Zn and 14% of Mg. Given that these alloy elements contribute to the formation of coherent and semi-coherent precipitates, if their concentration is decreased, the improvement of mechanical properties after solution treatment and quenching will be limited. Despite the loss of some Zn and Mg, composition values are still within the standard values for AA7075, and so is the hardness after heat treatment, which is indicated in literature to be around 160-175 HV for cast and wrought AA7075 after T6 [11,27]. Other studies have reported average values for L-DED processed AA7075 in T6 condition of 97 HV [27], 115 HV [21] and 145 HV [26]. These lower values were attributed to evaporation losses of Zn and Mg and to the presence of porosity.

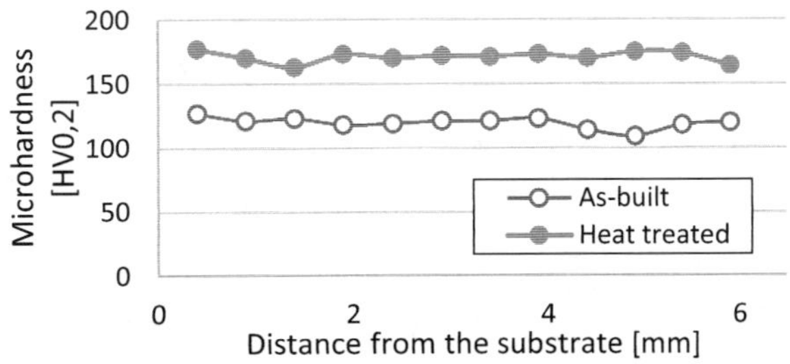

Figure 7. Microhardness profile along the building direction for AA7075-TiC multilayer coaxial wire L-DED sample in as-built and heat-treated conditions.

4. Conclusions

In this work, the processability of nanofunctionalized AA7075 with coaxial wire L-DED was investigated. Process parameters were optimized to produce crack-free and virtually porosity-free single and multi-layer specimens of AA7075-TiC. The response of the samples to precipitation hardening was studied, evaluating the microstructure and the microhardness before and after T6 heat treatment. The following conclusions are drawn:

- The percentage of porosity directly depends on the energy input. For single layer samples, the porosity was reduced from 2% to 0.02% when the power input increased from 60 J/mm to 90 J/mm. As the energy is increased, the gases in the melt pool have more time to escape through the action of convection forces before the material solidifies.
- EDS results indicate that there are evaporation losses of Zn and Mg during the deposition process of single-layer and multi-layer samples. However, during deposition of subsequent layers, reheating of adjacent layer causes further evaporation loss of Zn only, and not Mg. This indicates that the temperature in the partially melted zone is between 907°C and 1090°C. Comparing to wire composition, Zn and Mg contents show reductions of up to 20% and 14%, respectively.
- Samples exhibit a fine grain microstructure and a discontinuous and small eutectic phase. Grain size is <10µm for both as-built and T6 conditions. This confirms the ability of TiC nanoparticles to act as nucleation sites and promote an equiaxed grain structure with fine eutectic phase, which is resistant to hot cracking and present good mechanical properties. Furthermore, TiC nanoparticles have a pinning effect on boundaries, preventing grain growth during heat treatment.
- Samples exhibit a positive response to T6 heat treatment. Microhardness increased from 121HV to 172HV, with uniform hardness values along the build direction.
- The presence of TiC nanoparticles might be the cause for early precipitation of S' (Al_2CuMg) phase, which is an incoherent precipitate that can be detrimental for the mechanical properties of 7075-T6. The effect of TiC nanoparticles on aging kinetics needs to be further investigated to optimize the precipitation strengthening heat treatment for nanofunctionalized AA7075.

References

1. Ding D, Pan Z, Cuiuri D, Li H. Wire-feed additive manufacturing of metal components: technologies, developments and future interests. Int J Adv Manuf Technol. 2015 Oct;81(1–4):465–81.
2. Rodrigues TA, Duarte V, Miranda RM, Santos TG, Oliveira JP. Current Status and Perspectives on Wire and Arc Additive Manufacturing (WAAM). Materials. 2019 Apr 4;12(7):1121.
3. Blakey-Milner B, Gradl P, Snedden G, Brooks M, Pitot J, Lopez E, et al. Metal additive manufacturing in aerospace: A review. Materials & Design. 2021 Nov;209:110008.
4. Çam G. Prospects of producing aluminum parts by wire arc additive manufacturing (WAAM). Materials Today: Proceedings. 2022;62:77–85.
5. Svetlizky D, Das M, Zheng B, Vyatskikh AL, Bose S, Bandyopadhyay A, et al. Directed energy deposition (DED) additive manufacturing: Physical characteristics, defects, challenges and applications. Materials Today. 2021 Oct;49:271–95.
6. DebRoy T, Wei HL, Zuback JS, Mukherjee T, Elmer JW, Milewski JO, et al. Additive manufacturing of metallic components – Process, structure and properties. Progress in Materials Science. 2018 Mar;92:112–224.
7. Gibson I, Rosen DW, Stucker B. Additive manufacturing technologies: 3D printing, rapid prototyping and direct digital manufacturing. Second edition. New York ; London: Springer; 2015. 498 p.

8. Bourell DL, Wohlers T. Introduction to Additive Manufacturing. In: Bourell DL, Frazier W, Kuhn H, Seifi M, editors. Additive Manufacturing Processes [Internet]. ASM International; 2020 [cited 2023 Jan 20]. p. 3–10. Available from: http://dl.asminternational.org/handbooks/book/119/chapter/2264049/Introduction-to-Additive-Manufacturing
9. Kumar Sinha A, Pramanik S, Yagati KP. Research progress in arc based additive manufacturing of aluminium alloys – A review. Measurement. 2022 Aug;200:111672.
10. Toyserkani E, Sarker D, Ibhadode OO, Liravi F, Russo P, Taherkhani K. Metal additive manufacturing. Hoboken, NJ: Wiley; 2021. 1 p.
11. Anderson K, Weritz J, Kaufman JG, editors. 7075 and Alclad 7075: High-Strength Structural Alloy. In: Properties and Selection of Aluminum Alloys [Internet]. ASM International; 2019 [cited 2023 Mar 13]. p. 432–8. Available from: http://dl.asminternational.org/handbooks/book/91/chapter/2087406/7075-and-Alclad-7075High-Strength-Structural-Alloy
12. Oropeza D, Hofmann DC, Williams K, Firdosy S, Bordeenithikasem P, Sokoluk M, et al. Welding and additive manufacturing with nanoparticle-enhanced aluminum 7075 wire. Journal of Alloys and Compounds. 2020 Sep;834:154987.
13. Sokoluk M, Cao C, Pan S, Li X. Nanoparticle-enabled phase control for arc welding of unweldable aluminum alloy 7075. Nat Commun. 2019 Dec;10(1):98.
14. Zuo M, Sokoluk M, Cao C, Yuan J, Zheng S, Li X. Microstructure Control and Performance Evolution of Aluminum Alloy 7075 by Nano-Treating. Sci Rep. 2019 Dec;9(1):10671.
15. MacKenzie DS. Metallurgy of Heat Treatable Aluminum Alloys. In: Anderson K, Weritz J, Kaufman JG, editors. Aluminum Science and Technology [Internet]. ASM International; 2018 [cited 2023 Mar 1]. p. 411–37. Available from: http://dl.asminternational.org/handbooks/book/63/chapter/1397401/Metallurgy-of-Heat-Treatable-Aluminum-Alloys
16. Wang J. Physical Metallurgy of Aluminum Alloys. In: Anderson K, Weritz J, Kaufman JG, editors. Aluminum Science and Technology [Internet]. ASM International; 2018 [cited 2023 Feb 27]. p. 44–79. Available from: http://dl.asminternational.org/handbooks/book/63/chapter/1396691/Physical-Metallurgy-of-Aluminum-Alloys-1
17. Banhart J. Age Hardening of Aluminum Alloys. In: Totten GE, editor. Heat Treating of Nonferrous Alloys [Internet]. ASM International; 2016 [cited 2023 Mar 13]. p. 214–39. Available from: http://dl.asminternational.org/handbooks/book/10/chapter/115673/Age-Hardening-of-Aluminum-Alloys
18. Wang M, Ventzke V, Kashaev N. Wire-based laser directed energy deposition of AA7075: effect of process parameters on microstructure and mechanical properties. Journal of Materials Research and Technology. 2022 Nov;21:388–403.
19. Berube P. Weldability of Aluminum Alloys. In: Anderson K, Weritz J, Kaufman JG, editors. Aluminum Science and Technology [Internet]. ASM International; 2018 [cited 2023 Mar 7]. p. 655–92. Available from: http://dl.asminternational.org/handbooks/book/63/chapter/1400355/Weldability-of-Aluminum-Alloys-1
20. Mathers G. The welding of aluminium and its alloys. Repr. Boca Raton, Fla.: CRC Press; 2005. 236 p.
21. Cobian Gonzalez L, Tuominen J, Ahmed S, Patnamsetty M, Peura P. Directed energy deposition of AA7075 - Effect of TiC nanoparticles on microstructure. Results in Materials. 2022 Dec;16:100341.
22. Martin JH, Yahata BD, Hundley JM, Mayer JA, Schaedler TA, Pollock TM. 3D printing of high-strength aluminium alloys. Nature. 2017 Sep;549(7672):365–9.
23. Svetlizky D, Zheng B, Vyatskikh A, Das M, Bose S, Bandyopadhyay A, et al. Laser-based directed energy deposition (DED-LB) of advanced materials. Materials Science and Engineering: A. 2022 Apr;840:142967.
24. Miller E, Liu S. Laser Welding of Aluminum Alloys. In: Anderson K, Weritz J, Kaufman JG, editors. Aluminum Science and Technology [Internet]. ASM International; 2018 [cited 2023 Mar 6]. p. 728–42. Available from: http://dl.asminternational.org/handbooks/book/63/chapter/1401014/Laser-Welding-of-Aluminum-Alloys
25. Kou S. Welding metallurgy. Third edition. Hoboken, NJ: John Wiley & Sons, Inc; 2020.
26. Su Y, Wang Y, Shi J. Microstructure and mechanical properties of laser DED produced crack-free Al 7075 alloy: Effect of process parameters and heat treatment. Materials Science and Engineering: A. 2022 Nov;857:144075.
27. Bhagavatam A, Ramakrishnan A, Adapa V, Dinda GP. Laser Metal Deposition of Aluminum 7075 Alloy. Int J Mater Sci Res. 2018 Oct 23;2(1):50–5.
28. Stopyra W, Gruber K, Smolina I, Kurzynowski T, Kuźnicka B. Laser powder bed fusion of AA7075 alloy: Influence of process parameters on porosity and hot cracking. Additive Manufacturing. 2020 Oct;35:101270.
29. Li L, Li R, Yuan T, Chen C, Zhang Z, Li X. Microstructures and tensile properties of a selective laser melted Al–Zn–Mg–Cu (Al7075) alloy by Si and Zr microalloying. Materials Science and Engineering: A. 2020 Jun;787:139492.
30. Zhou SY, Su Y, Wang H, Enz J, Ebel T, Yan M. Selective laser melting additive manufacturing of 7xxx series Al-Zn-Mg-Cu alloy: Cracking elimination by co-incorporation of Si and TiB2. Additive Manufacturing. 2020 Dec;36:101458.
31. Sokoluk M, Yuan J, Pan S, Li X. Nanoparticles Enabled Mechanism for Hot Cracking Elimination in Aluminum Alloys. Metall Mater Trans A. 2021 Jul;52(7):3083–96.
32. Choi H, Cho W hyun, Konishi H, Kou S, Li X. Nanoparticle-Induced Superior Hot Tearing Resistance of A206 Alloy. Metall and Mat Trans A. 2013 Apr;44(4):1897–907.
33. Zheng T, Pan S, Murali N, Li B, Li X. Selective laser melting of novel 7075 aluminum powders with internally dispersed TiC nanoparticles. Materials Letters. 2022 Jul;319:132268.
34. Näkki J. Properties of alloy 625 claddings made with laser and CMT methods [Doctoral Thesis]. [Tampere]: Tampere University of Technology; 2018.
35. Tuominen J. Engineering coatings by laser cladding - the study of wear and corrosion properties [Doctoral Thesis]. [Tampere]: Tampere University of Technology; 2009.
36. Wang M, Kashaev N. Investigation of process window for AA7075 considering effects of different wire feed directions in lateral Laser Metal Deposition. Procedia CIRP. 2022;111:218–23.
37. Pajukoski H, Näkki J, Thieme S, Tuominen J, Nowotny S, Vuoristo P. High performance corrosion resistant coatings by novel coaxial cold- and hot-wire laser cladding methods. Journal of Laser Applications. 2016 Feb;28(1):012011.
38. Benoit MJ, Sun SD, Brandt M, Easton MA. Processing window for laser metal deposition of Al 7075 powder with minimized defects. Journal of Manufacturing Processes. 2021 Apr;64:1484–92.
39. Porter DA. Phase transformations in metals and alloys. 4th edition. Boca Raton: CRC Press, Taylor & Francis Group; 2022.
40. Cheng Y, Yuan W xia, Xu J hua, Yu L hua, Hu Y xuan, Huang T, et al. Ceramic particle-induced accelerated solution-aging behavior of spray-formed 7055 aluminum alloy TIG weld metal. J Mater Sci. 2023 Jan;58(2):802–17.
41. Ma Y, Addad A, Ji G, Zhang MX, Lefebvre W, Chen Z, et al. Atomic-scale investigation of the interface precipitation in a TiB2 nanoparticles reinforced Al–Zn–Mg–Cu matrix composite. Acta Materialia. 2020 Feb;185:287–99.
42. Li X, Cai Q, Zhao B, Liu B, Li W. Precipitation behaviors and properties of solution-aging Al-Zn-Mg-Cu alloy refined with TiN nanoparticles. Journal of Alloys and Compounds. 2018 May;746:462–70.
43. Martukanitz RP. Directed-Energy Deposition Processes. In: Bourell DL, Frazier W, Kuhn H, Seifi M, editors. Additive Manufacturing Processes [Internet]. ASM International; 2020 [cited 2023 Jan 21]. p. 220–38. Available from: http://dl.asminternational.org/handbooks/book/119/chapter/2348120/Directed-Energy-Deposition-Processes
44. Liu D, Wu D, Ma G, Zhong C, Niu F, Gasser A, et al. Effect of post-deposition heat treatment on laser-TIG hybrid additive manufactured Al-Cu alloy. Virtual and Physical Prototyping. 2020 Oct 1;15(4):445–59.

IOP Conf. Series: Materials Science and Engineering 1296 (2023) 012034 doi:10.1088/1757-899X/1296/1/012034

Temperature and distortion in laser melt injection of MMC wear-resistant coatings in a copper beryllium alloy

A Bohlen[1], A Langebeck[1], T Seefeld[1,2]

[1] BIAS – Bremer Institut für angewandte Strahltechnik GmbH, 28359 Bremen, Germany
[2] MAPEX Center for Materials and Processes - Universität Bremen, 28359 Bremen, Germany

bohlen@bias.de

Abstract. Wear-resistant coatings made of MMC materials (metal matrix composites) are ideal for increasing the service life of tools subjected to high abrasive stresses. Such MMC surfaces can be produced by laser melt injection during the manufacturing process of the tools or subsequently as a retrofit near-net-shape. However, during laser melt injection (LMI) thermally induced residual stresses can be introduced into the MMC surfaces. These can have a detrimental effect on the service life of these surfaces and thus reduce the positive effect of wear resistance.

In order to determine the influence hard particles injected during LMI have on residual stresses within the surface layer the distortion of the surface layer was evaluated. MMC layers consisting of a copper beryllium matrix and fused tungsten carbide hard particles were created by LMI. The MMC layers are compared to parts with remelted surfaces. Laser power and process velocity were varied to examine their influence on distortion. The process temperature was recorded on the melt pool by a two-color pyrometer. After the process the top layer of the MMC surface was cut off by wire EDM and then measured with a coordinate measuring machine. It is shown that an increase of incorporated powder mass per unit length and process temperature lead to higher distortions.

1. Introduction

Beryllium copper alloys, such as CuBe2, are of high interest due to their high electrical and thermal conductivity, as well as their high chemical resistance. The alloy is used for mechanically highly strained jaws, holders and guide rails used in flash butt welding and projection welding, as well as for injection molds, die casting pistons and for marine applications. In addition, beryllium copper can be used for non-sparking tools in the mining, gas and petrochemical industries.

Laser melt injection (LMI) is an efficient method for creating surface modifications with high abrasion resistance. A melt pool is generated on a base materials surface and hard particles are injected. Metal matrix composites (MMC) layers can be applied to tools directly with this process. These MMC layers can combine a high thermal conductivity matrix with high abrasive wear resistant hard particles to be used as protection layers against abrasive wear [1]. LMI is therefore beneficial for a variety of wear-exposed tool surfaces, such as casting tools and injection molds, which are often made of copper alloys. LMI with fused tungsten carbide (FTC) particles in the copper alloy Hovadur® CNCS led to a significant improvement in wear resistance against the hardened and gas nitrided steel 1.2343 [2].

IOP Conf. Series: Materials Science and Engineering 1296 (2023) 012034 doi:10.1088/1757-899X/1296/1/012034

Tribological tests for an aluminum bronze and FTC MMC have shown that wear can be reduced by 76% [3]. The service life of these MMC surfaces with ceramic particles are dependent on the residual stress state [4]. Surface modifications by laser processing lead to high temperature gradients [5], which result in residual macro stress. These stresses can lead to various types of out-of-plane distortions, such as angular change, buckling and longitudinal bending. In Laser Metal Deposition a relationship between melt pool temperature and distortion was found for cold-rolled thin DC01 sheets and 316L powder [6]. Plati et al found that in cladding MMC (nickel matrix with spherical tungsten carbide) onto a thick substrate has a little effect on the substrate but leads to substantial residual stress in the deposited layer [7].

This work adresses the measurement of distortion in MMC surfaces for wear protection, which are produced by LMI. The laser power, process velocity and the powder mass flow per unit length are varied to study their influence on the measured distortion. The melt pool temperature resulting from these process parameters is also considered as an influence on the resulting distortion.

2. Experimental

2.1. Laser Melt Injection

For laser melt injection a Trumpf TruDisk 12002 laser with a wavelength of 1030 nm was used. The beam was guided via a laser light cable with 200 μm core diameter to the processing head. The processing head has a collimation length of 200 mm and a focal length of 250 mm. The laser beam was defocused to a spot diameter of 3.6 mm. A discrete coaxial six-jet powder nozzle by GTV with 25 mm working distance was used to carry the hard particles into the melt pool. A Sensortherm Metis H322 two-color-pyrometer with a measuring range between 700 °C and 2300 °C and calibrated for the molten substrate material was used for the thermography. The pyrometer was moved together with the laser beam to monitor the temperature of the process zone. The measurements of the quotient channel were smoothed, and a mean value was calculated to characterize the process zone temperature.

As substrate material HOVADUR® CB2 alloy from Schmelzmetall was used, which is similar to EN standard CuBe2. The substrates are 50 mm in length, 20 mm in width and 12 mm in height. As hard particles for the MMC layer spherical fused tungsten carbide powder with a particle size of -106 μm +45 μm was used. By using an integrating sphere, the absorption for 1080 nm wavelength could be measured for the substrate material and the hard particles (see **Table 1**).

Table 1: Absorption for HOVADUR CB2 and Fused Tungsten Carbide Powder

	Absorption for 1080 nm wavelength (%)
HOVADUR CB2	16 ± 0.2
Fused Tungsten Carbide Powder	84 ± 0.5

Laser power and process velocity were varied, the powder mass flow rate was adapted to the process velocity to have different powder mass flow per unit length. All other parameters are held constant, shown in **Table 2**. Per specimen 4 tracks were welded next to each other with 2.1 mm overlap to generate a MMC surface. Specimens were weighed before and after process to calculate the incorporated powder mass per unit length. Each parameter set was welded three times.

Table 2: Process parameters for Laser Melt Injection

Process parameter	Value(s)
Laser spot diameter	3.6 mm
Laser power	2.0 kW 2.5 kW 3.0 kW
Process velocity	200 mm/min 250 mm/min 300 mm/min
Powder mass flow per unit length	10 g/m 20 g/m 40 g/m
Carrier gas (Ar)	3.5 l/min
Shielding gas (Ar)	15 l/min
Track length	40 mm
Surface condition	Glass bead blasted

2.2. Distortion Measurement and Analysis
The distortion test was performed with a Mitutoyo Crysta Apex coordinate measuring machine. In a first step, the whole specimens were measured in their initial state. After LMI, the samples were measured again. For this purpose, the MMC layer was separated from the rest of the substrate by electrical discharge machining (EDM) to a thickness of 1 mm from the rest of the substrate (see **Figure 1a**). The measurement of the distortion was performed on the back side of the specimens in order to exclude influences of the uneven MMC layer. Since the EDM process could lead to residual stresses and distortion, an unprocessed but cut specimen was also measured. Measurements were done on 17 lines along the length of the specimen and 6 lines along the width (see **Figure 1b**).
The data grid was smoothed and the distortion in longitudinal and transversal direction was calculated by the difference between highest and lowest data point. For this only the data from the center longitudinal and transversal line were used (see dotted white lines in **Figure 1c)**).

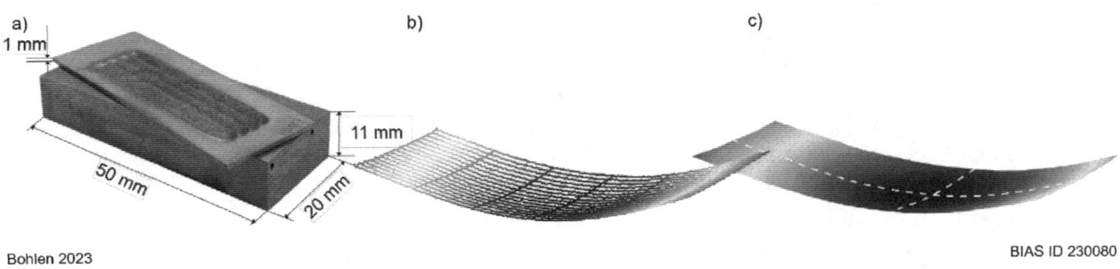

Figure 1: a) specimen with MMC surface cut by EDM, b) data points measured by coordinate measuring machine and c) data points for longitudinal and transversal distortion calculation.

3. Results
Figure 2 shows the distortion measurement for the example of 3 kW laser power and a process velocity of 300 mm/min. The longitudinal (**Figure 2 a)**) and transversal distortion (**Figure 2 b)**) show that the distortion in the welding direction is curved upwards and much higher than transversal to the welding direction. The dotted black line in the graph shows the measured distortion from the EDM cutting on an unprocessed specimen. With a grey line the distortion from remelting the specimen without hard particle addition is shown. 40 g/m powder mass flow per unit length shows the most distortion. 10 g/m and

20 g/m powder flow per unit length show distortions similar to the distortions of specimens which are remelted or just cut by EDM.

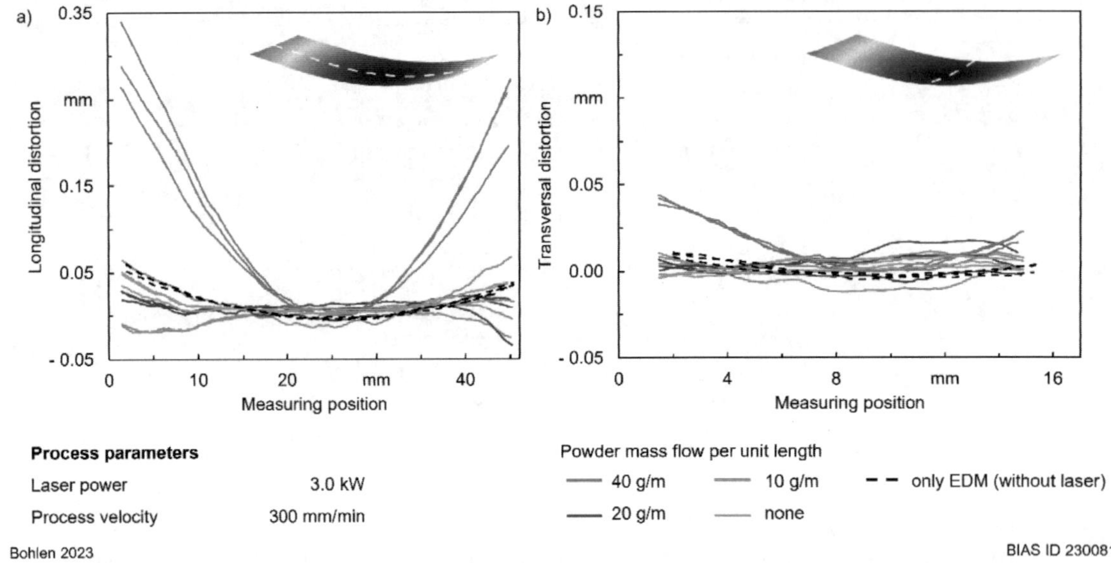

Figure 2: a) Longitudinal distortion and b) transversal distortion in z direction after LMI (red, violet, green), after laser remelting (grey) and without any laser processing (dotted black line)

The following results show the maximum distortion. Both, laser power and process velocity were varied, resulting in different laser energies per unit length. Maximum measured distortions in welding direction for different energies per unit length are shown in **Figure 3** for a constant powder mass flow rate of 40 g/m. No direct influence of energy per unit length can be seen on the measured distortion. However, a higher laser power, while having the same energy per unit length results in higher distortions.

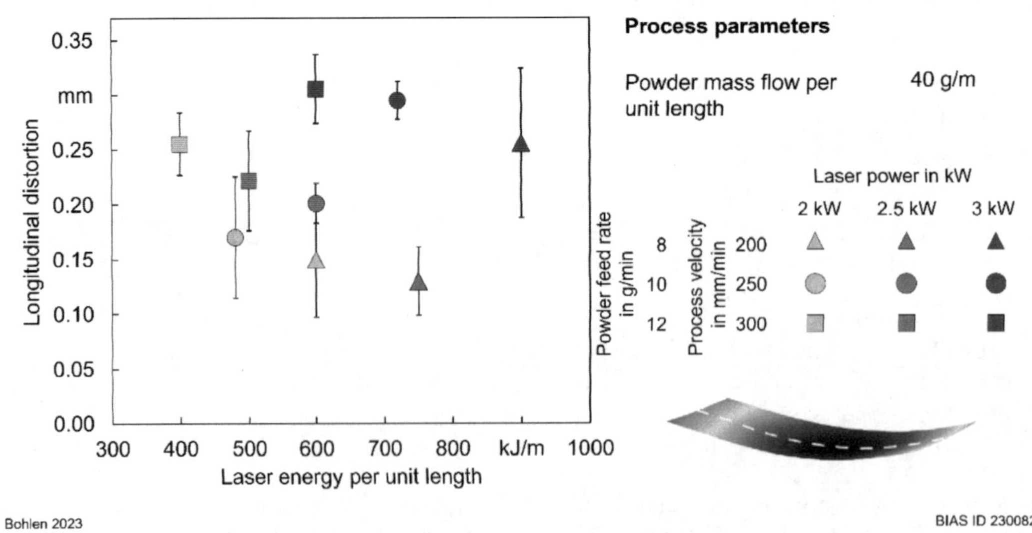

Figure 3: Maximum distortion for different energies per unit length

IOP Conf. Series: Materials Science and Engineering 1296 (2023) 012034 doi:10.1088/1757-899X/1296/1/012034

Depending on the laser power the incorporated powder mass per unit length changes (see **Figure 4**). Higher laser power leads to a higher incorporated powder mass per unit length. Contrary to expectations, for a constant laser power, it can be seen that a higher process velocity leads to an increase in incorporated powder mass per unit length.

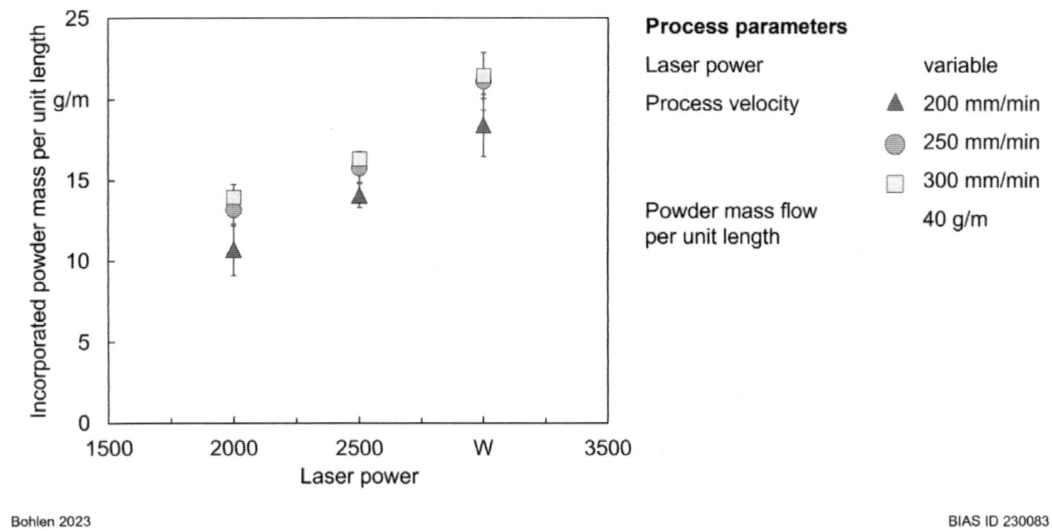

Figure 4: Incorporated powder mass per unit length for varied laser power and process velocity

The measured process temperature in relation to the incorporated powder mass per unit length is shown in **Figure 5**. An influence of the incorporated powder mass per unit length on the process zone temperature can be seen. Increase in incorporated powder mass per unit length leads to an increase of process temperature. Meanwhile the process velocity in itself also has an influence on the process temperature. Unexpectedly higher process velocities lead to higher process temperatures.

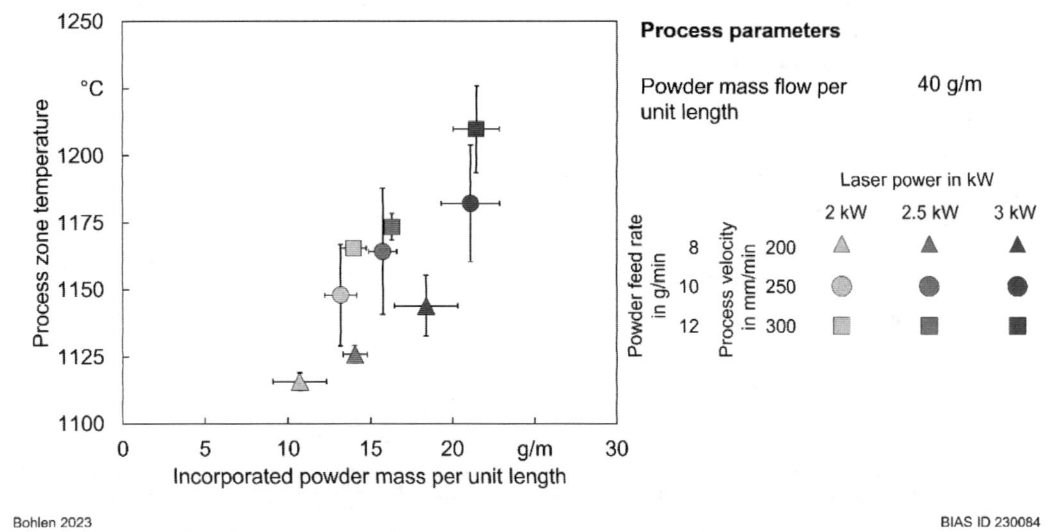

Figure 5: Influence of incorporated powder mass per unit length on process temperature.

The maximum distortion in respect to the incorporated powder mass per unit length can be seen in **Figure 6**. Low powder mass flows per unit length (10 g/m & 20 g/m) lead to small distortions, which are in the same range as distortions resulting from EDM cutting (dotted line). For a higher powder mass flow per unit length (40 g/m) the resulting incorporated powder mass per unit length varies. It can be seen that with increasing incorporated powder mass per unit length the distortion linearly increases.

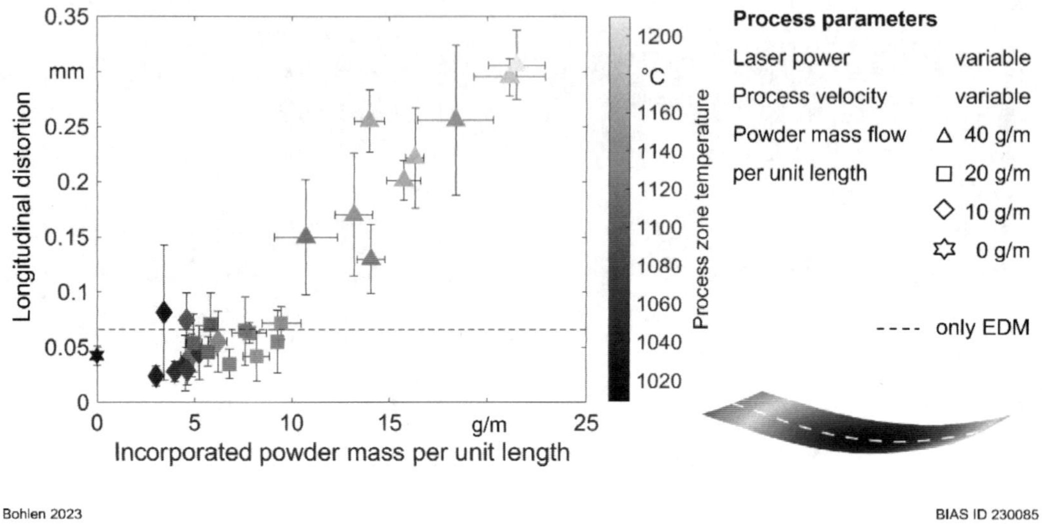

Figure 6: Maximum distortion depending on the incorporated powder mass per unit length.

4. Discussion

It has been shown that the incorporation of hard particles introduces more distortion into specimens compared to just remelting them. This can be seen in **Figure 2** where the other process parameters such as laser power and process velocity were the same. For 40 g/m powder mass flow per unit length in **Figure 3** at 3 kW laser power, an increase in distortion can be seen, while at 2 kW and 2.5 kW this effect is not clearly visible. But when the process velocity increases, the distortion increases for all laser powers, which appears counterintuitive. The powder mass flow per unit length was set to constant values, therefore at 40 g/m powder mass flow per unit length and 300 mm/min process velocity more powder was transported (12 g/min) than at lower process velocities (at 200 mm/min 8 g/min were transported). The amount of incorporated powder mass per unit length changes with the process parameters (see **Figure 4**). At higher laser powers, more particles are incorporated, which could be attributed to larger melt pools, among other factors. However, low process velocities would also result in larger melt pools, but they show a decrease in the effective amount of incorporated powder. This leads to the assumption that a higher powder mass flow leads to a better incorporation of the powder. This assumption is supported by the higher process temperatures at higher process velocities, shown in **Figure 5**. Here, as an example, with an incorporated powder mass per unit length of approx. 14 g/m, it can be seen that for 2 kW laser power and 300 mm/min process velocity, a higher process zone temperature is achieved than with 2.5 kW laser power and 200 mm/min process velocity. From the approximately equal incorporated powder mass per unit length, it can be concluded that around the same amount of powder particles entered the melt pool. Due to the higher powder mass flow at higher process velocities, the transmission of the laser beam to the substrate will be reduced, i.e., less energy should be absorbed into the substrate. The powder particles also absorb energy from the laser beam. More heating of the particles at a higher powder mass flow rate can be attributed to multiple reflection within the powder stream [8]. Which means that for the same incorporated powder mass per unit length a higher powder mass flow transports more heat into the melt pool, with less laser power. This effect is likely amplified by the absorption coefficients of the powder and substrate material (at 1080 nm wavelength 84% ± 0.5% and 16% ± 0.2%, respectively). Overall, the laser power added to the process via the powder particles seems to have a stronger impact on the process temperature than the transmissive heating of the substrate by the laser.

5. Conclusion

Distortion was measured for specimens with an MMC surface generated by LMI. It was shown that the addition of powder compared to just remelting the surfaces introduces significant distortions. If the powder mass flow was higher more laser energy was absorbed by the powder. Hereby, more heat was transported to the substrate. This way more of the particles could be incorporated. Thus, leading to higher distortions when the incorporated powder mass per unit length increases.

Acknowledgements

The IGF-Project with the IGF-No.: 21079 N / DVS-No.: 06.3341 of the "Forschungsvereinigung Schweißen und verwandte Verfahren e. V." of the German Welding Society (DVS), Aachener Str. 172, 40223 Düsseldorf was funded by the Federal Ministry for Economic Affairs and Climate Action (BMWK) via the German Federation of Industrial Research Associations (AiF) in accordance with the policy to support the Industrial Collective Research (IGF) on the basis of a decision by the German Bundestag.

Furthermore, the authors gratefully acknowledge the collaboration with the members of the project affiliated committee regarding the support of knowledge, material and equipment over the course of the research. We thank Yalin Yakar for the experimental work.

Supported by:
Federal Ministry for Economic Affairs and Climate Action
on the basis of a decision by the German Bundestag
AiF iGF
DVS FORSCHUNG

ORCID iDs

A Bohlen: https://orcid.org/0000-0002-1984-5818
A Langebeck: https://orcid.org/0000-0002-4976-4357
T Seefeld: https://orcid.org/0000-0002-1283-4761

References

[1] Zou Y, Tan C, Qiu Z, Ma W, Kuang M and Zeng D 2021 Additively manufactured SiC-reinforced stainless steel with excellent strength and wear resistance *Additive Manufacturing* **41** p 101971

[2] Warneke P, Bohlen A and Seefeld T 2022 Improving the wear resistance of copper tools for pressure die casting by laser melt injection *Prod. Eng. Res. Devel.*

[3] Freiße, H., Langebeck, A., Köhler, H., Seefeld, T. 2016 Dry strip drawing test on tool surfaces reinforced by hard particles *Dry Met. Forming OAJ FMT* pp 1–6

[4] Vreeling J Laser melt injection of ceramic particles in metals. s.n. Groningen 2001

[5] Köhler H, Jayaraman V, Brosch D, Hutter F X and Seefeld T 2013 A Novel Thermal Sensor Applied for Laser Materials Processing *Physics Procedia* **41** pp 502–8

[6] Tebaay L M, Hahn M and Tekkaya A E 2020 Distortion and Dilution Behavior for Laser Metal Deposition onto Thin Sheet Metals *Int. J. of Precis. Eng. and Manuf.-Green Tech.* **7** pp 625–34

[7] Plati A, Tan J C, Golosnoy I O, Persoons R, van Acker K and Clyne T W 2006 Residual Stress Generation during Laser Cladding of Steel with a Particulate Metal Matrix Composite *Adv. Eng. Mater.* **8** pp 619–24

[8] Partes K Hochgeschwindigkeitsbeschichten mit dem Laserstrahl. Strahltechnik **34**. BIAS Bremen 2008

IOP Conf. Series: Materials Science and Engineering 1296 (2023) 012035 doi:10.1088/1757-899X/1296/1/012035

The properties of laser-clad coatings made of NiCrBSi alloy and recycled tungsten carbides

J Tuominen[1] and J Näkki[2]

[1]Tampere University, Materials Science and Environmental Engineering, Tampere, Finland
[2]Centria University of Applied Sciences, Kokkola, Finland

jari.tuominen@tuni.fi

Abstract. Tungsten carbide hard metals are one of the most known and successful powder metallurgical products used in environments where severe wear conditions prevail. Since tungsten belongs to the list of critical raw materials for the European Union (EU), its recycling is highly desirable. In this paper, recycled tungsten carbide powders fabricated by the zinc-reclaim process where mixed with NiCrBSi matrix and deposited with one-step laser cladding using a coaxial powder feeding on mild steel substrates. Prepared clads were tested in low-stress rubber wheel abrasion tests and compared with clads produced from commercial macrocrystalline (WC) and cast and crushed eutectoid WC/W_2C reinforced clads. The results showed that recycled WC induced porosity in clads when produced with 'production' laser cladding parameters due to small carbide size resulting in poor wear performance. When laser-clad with lower heat input and productivity, pore-free clads with excellent wear performance were obtained.

1. Introduction

Tungsten carbide reinforced hard-facings produced by laser cladding, plasma-transferred arc (PTA), and metal inert/active gas (MIG/MAG) welding are widely used in such applications as mining & mineral processing, oil & gas drilling, agriculture, and steelworks [1-3]. Tungsten belongs, however, to the EU's list of critical raw materials, which makes its recycling from hard-metal scrap highly desirable [4]. The use of recycled tungsten carbides in hard-facing was already studied in Refs. [5-7]. Zikin et al. [6] and Kulu et al. [7] fabricated, for instance, Ni- and Fe-based clads reinforced with 40 vol.% of coarse (150-400 μm) recycled WC particles by PTA welding. In abrasion and impact-abrasion tests, the clad fabricated from the commercial WC/W_2C showed, however, better wear resistance than the clad made of recycled WC.

Methods to produce recycled tungsten carbide particles can be divided into mechanical and chemical methods. Mechanical methods include disintegrator milling, where hard-metal scrap such as used cutting tools are crushed [7]. Benefits of disintegrator milling include the possibility to fabricate coarse WC particles, which suit for overlay welding and hard-facing processes [7]. Such a coarse single WC particle consists of several smaller WC particles (3-5 μm), which are bound together by cobalt binder. Large or coarse WC particles tend to survive in a melt pool without dissolution and melting better than finer carbides leading to an ideal metal matrix composite, which consists of a ductile matrix reinforced with hard particulates. Long milling times lead, however, to increased content of contaminations from the grinding media (Fe) and atmosphere (oxidation), which necessitates chemical cleaning treatments afterwards [6].

In the zinc-reclaim process, hard metal scrap is exposed to molten zinc at 900-1050℃ in Ar/N_2 atmosphere [8]. Zinc forms with cobalt a compound that expands and breaks the hard metal. After the zinc is removed, the result is an easily crushed porous hard metal that can be crushed and sieved to WC-Co powder.

Common for all these recycled tungsten carbides is that they are tungsten mono-carbides (WC) and angular in shape. Tungsten mono-carbide is somewhat softer than the eutectoid WC/W_2C, which is more frequently used in hard-facing. Tungsten mono-carbide is, however, thermally more stable than eutectoid WC/W_2C, which tends to dissolve and melt in the melt pool and embrittle the matrix via the precipitation of secondary carbides.

The objective of this paper is to evaluate the suitability of recycled WC produced by the zinc-reclaim process in laser cladding and compare its wear properties to commercial macrocrystalline WC and eutectoid WC/W_2C in a low-stress three-body abrasion wear test.

2. Materials and methods

The powder used for laser cladding was NiCrBSi mixed with recycled WC, commercial macrocrystalline WC, and cast and crushed eutectoid WC/W_2C. Mechanical mixtures of powders were prepared by weighing the desired amount of powders (40/60 wt.%) and mixing them in a plastic bottle, which was placed in the mechanical mixer for 24 h. Chemical compositions, microhardness of carbides, and granulometry of powders are given in Table 1. Microhardness values were measured from the polished cross-sections of carbides mould in epoxy resin. Eutectoid WC/W_2C was clearly the hardest carbide. The morphologies of powders are illustrated in Figure 1. Tungsten carbide powders were angular in shape with nearly the same size distributions. It should be, however, noted that recycled powder also included particles where small WC particles were bound with cobalt. The substrates employed for the cladding were mild steel plates with a thickness of 20 mm, a width of 60 mm, and a length of 200 mm. Prior to cladding, they were grit-blasted with alumina to a roughness of R_a 4.4 µm.

Table 1. Chemical compositions in wt.%, microhardness values, and grain sizes of powders. Standard deviations for microhardness values are given in parentheses.

Material	B	C	Si	Cr	Fe	Co	Ni	W	$HV_{0.1}$	Size (µm)
NiCrBSi	1.7	0.25	3.5	7.5	2.5	-	Bal.	-	-	53-125
Recycled WC									1700 (260)	45-100
Macrocrystalline WC		6.1			0.15			Bal.	1628 (249)	45-90
Eutectoid WC/W_2C		3.8-4.1			≤0.3			Bal.	2397 (321)	44-74

Laser cladding was performed using a continuous wave fiber laser (Coherent) with a maximum power of 3 kW. The experimental set-up consisted of a laser-integrated beam delivery system (400 µm fiber, f160 mm collimator unit, f250 mm focus optics), coaxial powder feeding nozzle (Coax 8), rotating disc type of powder feeder (Medicoat), and a Fanuc F-200iB parallel kinematic robot. The optics and powder feeding nozzle were kept static while the base materials were moved by the robot. The laser beam was defocused to a spot size of ~4 mm (Figure 2). The optimized cladding parameters to produce fusion-bonded metal matrix composites with low dilution are summarized in Table 2. Shielding and carrier gases used in the experiments were argon with flow rates of 12 and 6 l/min, respectively. The recycled powder had to be clad with lower specific energy due to porosity issues. After the cladding had been completed, plates were water-cut to the wear specimens (50 x 20 mm^2) and coating surfaces machined to a condition where no as-clad surface remained.

The microstructure of the specimens was studied by means of an optical microscope (OM) and a scanning electron microscope (SEM). Energy-dispersive X-ray spectroscopy (EDS) was employed to

Figure 1. SEM micrographs of powders: a) gas atomized NiCrBSi, b) cast and crushed WC/W_2C, c) macrocrystalline WC, and d) recycled WC.

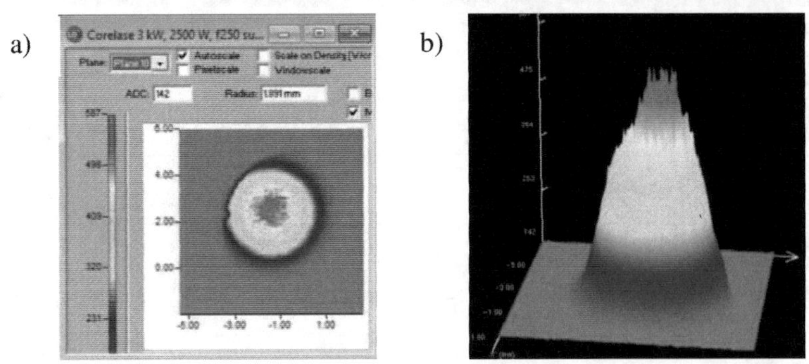

Figure 2. Intensity distribution of defocused fiber laser beam measured at working distance: a) plane and b) spatial distribution.

analyse local chemical compositions. To reveal crystal structures and phases present in studied materials, an X-ray diffractometer (Panalytic Empyrean Multipurpose Diffractometer) was used using the monochromatic Cu Kα radiation at the voltage of 45 kV and the current of 40 mA. The scanning angle 2θ covered a range of 20-80°. The Vickers microhardness measurements from the coating cross-sections were taken with a Matsuzawa MMT-X7 hardness tester using indentation loads of 1, 3, and 10 N applied for 10 s.

Three-body abrasion wear tests at room temperature were conducted with a rubber-wheel abrasion test device, a modified version of ASTM G65, where crushed dry quartz (SiO_2) sand abrasives (0.1-0.6 mm) flow between the surfaces of the rotating rubber wheel (Durometer A84-88) and the test specimen. The test device and parameters are described in more detail in Ref. [9]. The worn specimen surfaces were examined by SEM. The volume losses were calculated based on the mass loss and the theoretical density of the materials. Reference material in abrasion tests was quenched & tempered steel with a hardness of 450 HV.

Table 2. Laser cladding process parameters.

Material	Power (kW)	Traverse speed (mm/s)	Powder feed rate (g/min)	Specific energy (J/mm^2)	Inter-track advance (mm)
NiCrBSi + Recycled WC (40/60 wt.%)	0.6	10	12	20	1.25
NiCrBSi + Macrocrystalline WC (40/60 wt.%)	1.8	18.3	39	33	1.50
NiCrBSi + Eutectoid WC/W$_2$C (40/60 wt.%)	1.8	18.3	38	33	1.50

3. Results and discussion

3.1. Microstructure

The representative low-magnification cross-sectional OM images of coatings manufactured with process parameters given in Table 2 are illustrated in Figures 3a-c. Coatings are generally characterized by low dilution, a thickness of ~1.1 – 1.3 mm, homogeneous distribution of primary carbides, slight porosity, and varying amounts of vertical cracks, which stop at the fusion line and do not propagate to the base material. Some essential coating characteristics are shown in Table 3. It can be, for instance, seen that the primary carbide content is the highest and size the smallest in recycled WC coating.

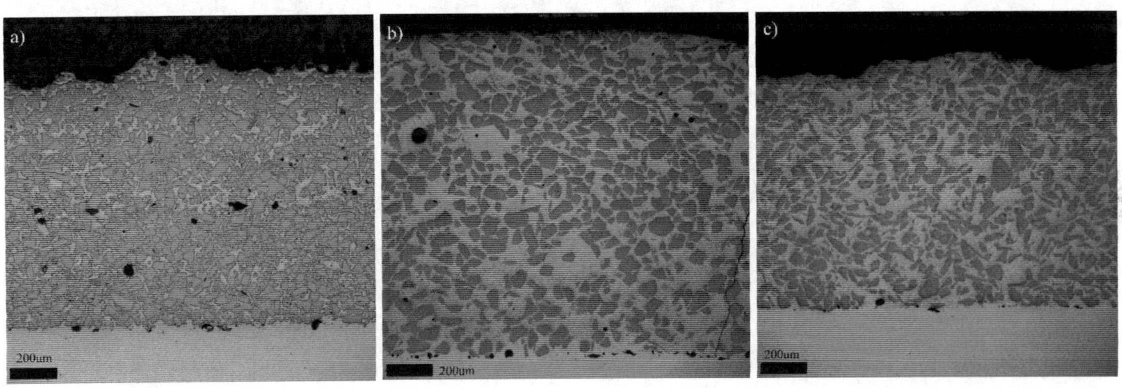

Figure 3. Low magnification OM images of a) Recycled WC, b) Macrocrystalline WC, and c) Eutectoid WC/W$_2$C coatings.

Table 3. Summary of coating and process characteristics.

Material	Primary carbide content	Average primary carbide size	Secondary carbide content	Thickness	Coverage rate
Recycled WC	56 vol.%	24 μm	5 vol.%	1.1 mm	0.02 m^2/h
Macrocrystalline WC	48 vol.%	57 μm	7 vol.%	1.3 mm	0.1 m^2/h
Eutectoid WC/W$_2$C	42 vol.%	48 μm	11 vol.%	1.1 mm	0.1 m^2/h

Low and high magnification SEM images and image analysis revealed that the amount of secondary carbides in the matrix due to the dissolution of primary carbides is the highest in the coating made of eutectoid WC/W$_2$C as displayed in Figures 4a-f, 5, and Table 3. This is attributed to the lower chemical stability of W$_2$C than WC [10]. According to Ref. [11] eutectic WC/W$_2$C decomposes into W and WC below 1250℃. Whereas monocrystalline WC is thermally stable up to 2757-2777℃. Dissolution in

IOP Conf. Series: Materials Science and Engineering 1296 (2023) 012035 doi:10.1088/1757-899X/1296/1/012035

recycled WC is less than in monocrystalline WC due to lower specific energy in the cladding process despite finer WC granulometry. Secondary carbides are mainly tungsten-based mixed carbides in Macrocrystalline WC and eutectoid WC/W_2C. Whereas the matrix of recycled WC coating includes also small black dots which are probably chromium carbides, borides, and/or silicides. Common for all the secondary carbides is that they locate in the interdendritic region of the Ni-based matrix. Light regions in the matrix are richer in chromium, iron, and tungsten compared to dark regions.

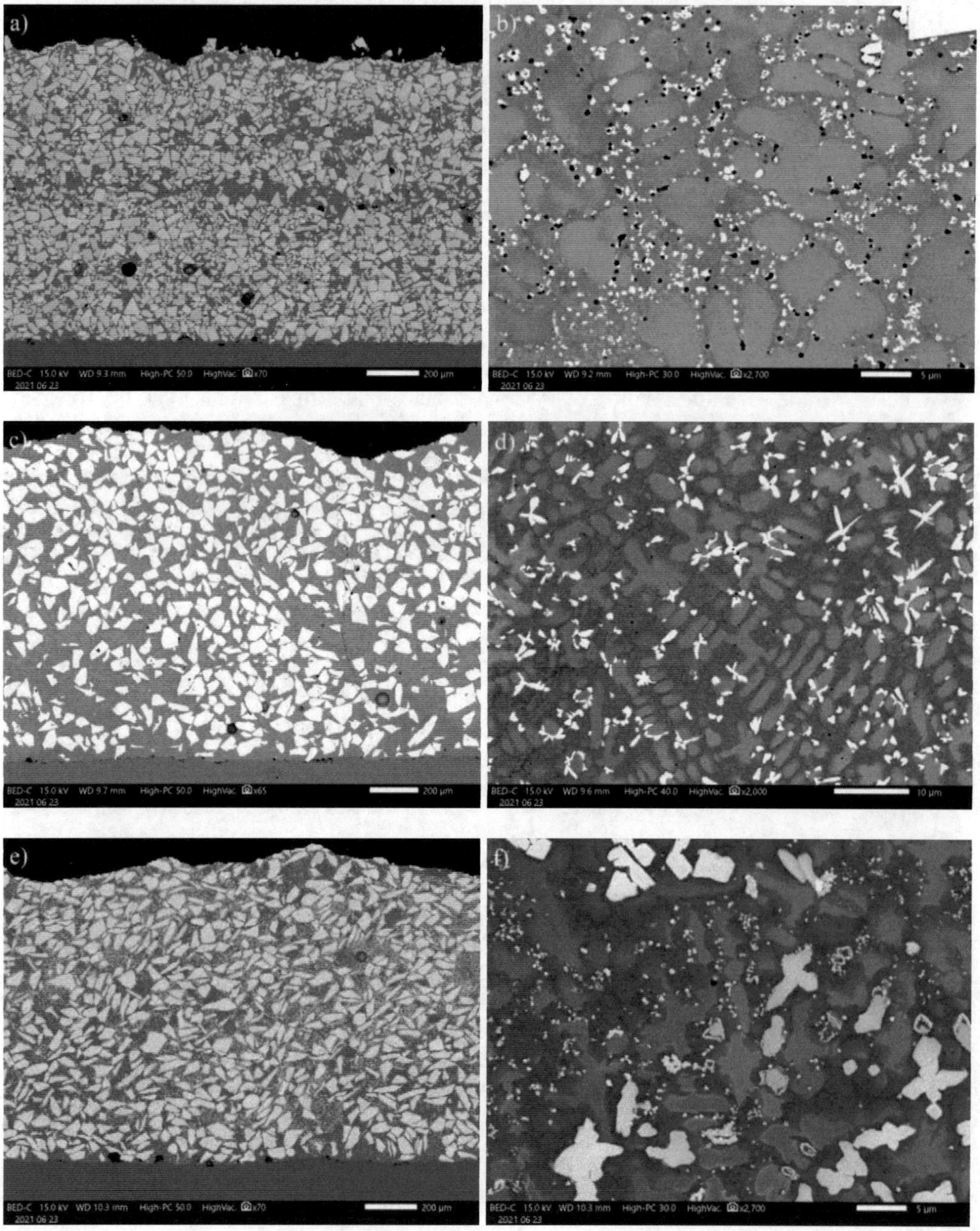

Figure 4. Low and high magnification SEM images of a,b) Recycled WC, c,d) Macrocrystalline WC, and e,f) Eutectoid WC/W_2C coatings.

According to XRD scans shown in Figure 6, the secondary carbides could be Ni_3W_3C, Ni_2W_4C, and $Ni_2W_{11}C_4$. Other phases in the matrix could be Ni_2Si, Cr_5B_3, and Ni_4B_3. As evidenced by the EDS analyses taken from the matrix, all the coatings are chemically only very little diluted (Table 4). Iron contents of the matrix are at the level of 2.0-2.8 wt.%.

Figure 5. High magnification SEM images of primary carbide and matrix interface a) Recycled WC, b) Macrocrystalline WC, and c) Eutectoid WC/W_2C.

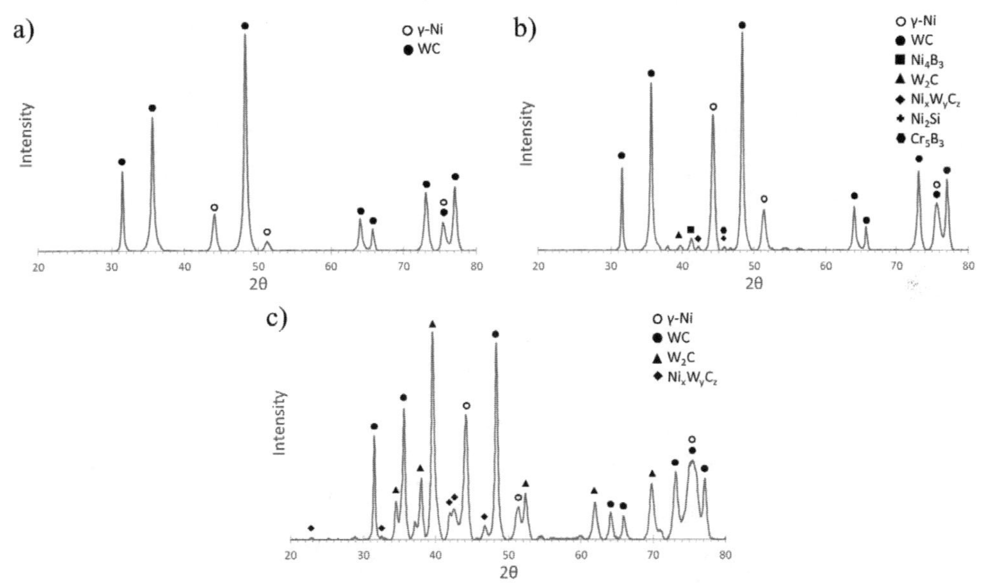

Figure 6. XRD patterns of a) Recycled WC, b) Macrocrystalline WC, and c) Eutectoid WC/W_2C.

Table 4. Average chemical composition of matrix without carbon and boron in wt.%.

Material	Ni	Co	W	Cr	Si	Fe
Recycled WC	61.8 ± 0.1	17.6 ± 0.9	11.2 ± 0.7	5.5 ± 0.1	2.0 ± 0.1	2.0 ± 0.1
Macrocrystalline WC	75.1 ± 0.7	-	12.9 ± 0.6	6.8 ± 0.0	2.4 ± 0.1	2.8 ± 0.1
Eutectoid WC/W_2C	73.1 ± 2.4	-	16.7 ± 2.8	5.5 ± 0.2	2.3 ± 0.1	2.4 ± 0.0

3.2. Microhardness

Hardness profiles measured from the coating cross-sections are illustrated in Figure 7. Average coating hardness values vary from ~730 to 950 HV_1 eutectoid WC/W_2C coating being the hardest and the macrocrystalline WC the softest. Hardness values were also measured from the primary carbides and Ni-based matrices (Table 5). Eutectoid WC/W_2C was the hardest and recycled WC was the softest primary carbide. The Ni-based matrix was the hardest in eutectoid WC/W_2C coating and softest in the recycled WC coating.

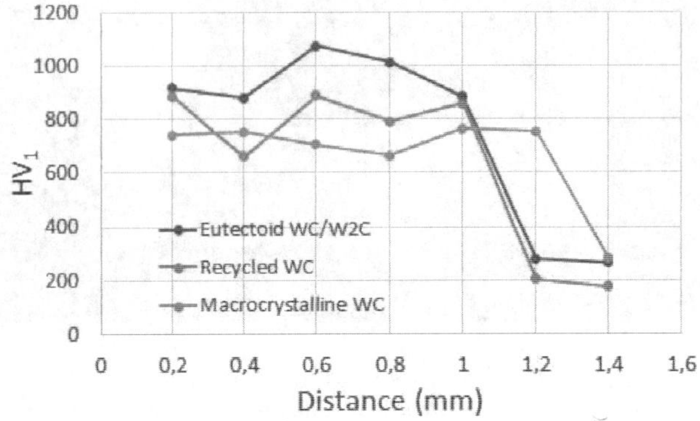

Figure 7. Microhardness values of coatings measured from cross-sections.

Table 5. Hardness values measured from the primary carbides and Ni-based matrices together with average mean free path values between primary carbides. Standard deviations are given in parentheses.

Material	Primary carbide ($HV_{0.1}$)	Matrix ($HV_{0.3}$)	Mean free path between primary carbides (μm)
Recycled WC	1478 (274)	562 (20)	~16
Macrocrystalline WC	1687 (206)	584 (15)	~45
Eutectoid WC/W_2C	2372 (164)	703 (96)	~33

3.3. Abrasion wear

The rubber wheel abrasion test (RWAT) results are presented in the form of weight and volume losses as column charts in Figures 8a and b. The results are the average of three test runs per studied material. As indicated by the results, the differences in wear resistance between different coatings are rather obvious, which suggests that the coatings can be ranked in terms of wear performance. Eutectoid WC/W_2C coating, which showed the highest primary carbide and matrix hardness, exhibited the best wear performance. Recycled WC coating was slightly better than macrocrystalline WC coating owing obviously to its higher primary carbide content (56 vs 42 vol.%) and shorter mean free path (16 vs 45 μm) between carbides given in Table 5.

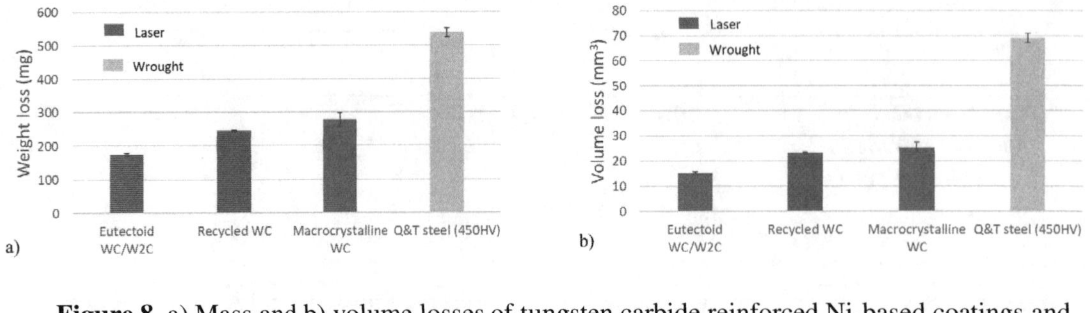

Figure 8. a) Mass and b) volume losses of tungsten carbide reinforced Ni-based coatings and reference material subjected to RWAT.

High-magnification SEM images of wear scars are displayed in Figures 9a-f. Common for all the wear scars is that matrix wore down preferentially leaving primary carbides protruding from the matrix. Obviously, abrasion conditions were hard for the matrices since the average hardness of the abrasive (~750-1200HV) was clearly higher ($H_{abrasive}/H_{material} > 1.2$) than the hardness of the matrices [12]. It was also observed that all the primary carbides remained rather intact under the used abrasion conditions. Not any differences in cracking were detected even if it is known that monocrystalline WC has lower fracture toughness than eutectoid WC/W_2C [11]. Despite only partial dissolution or modest melting of primary carbides, they were well bonded to the matrix, and carbide 'pull-outs' were not observed.

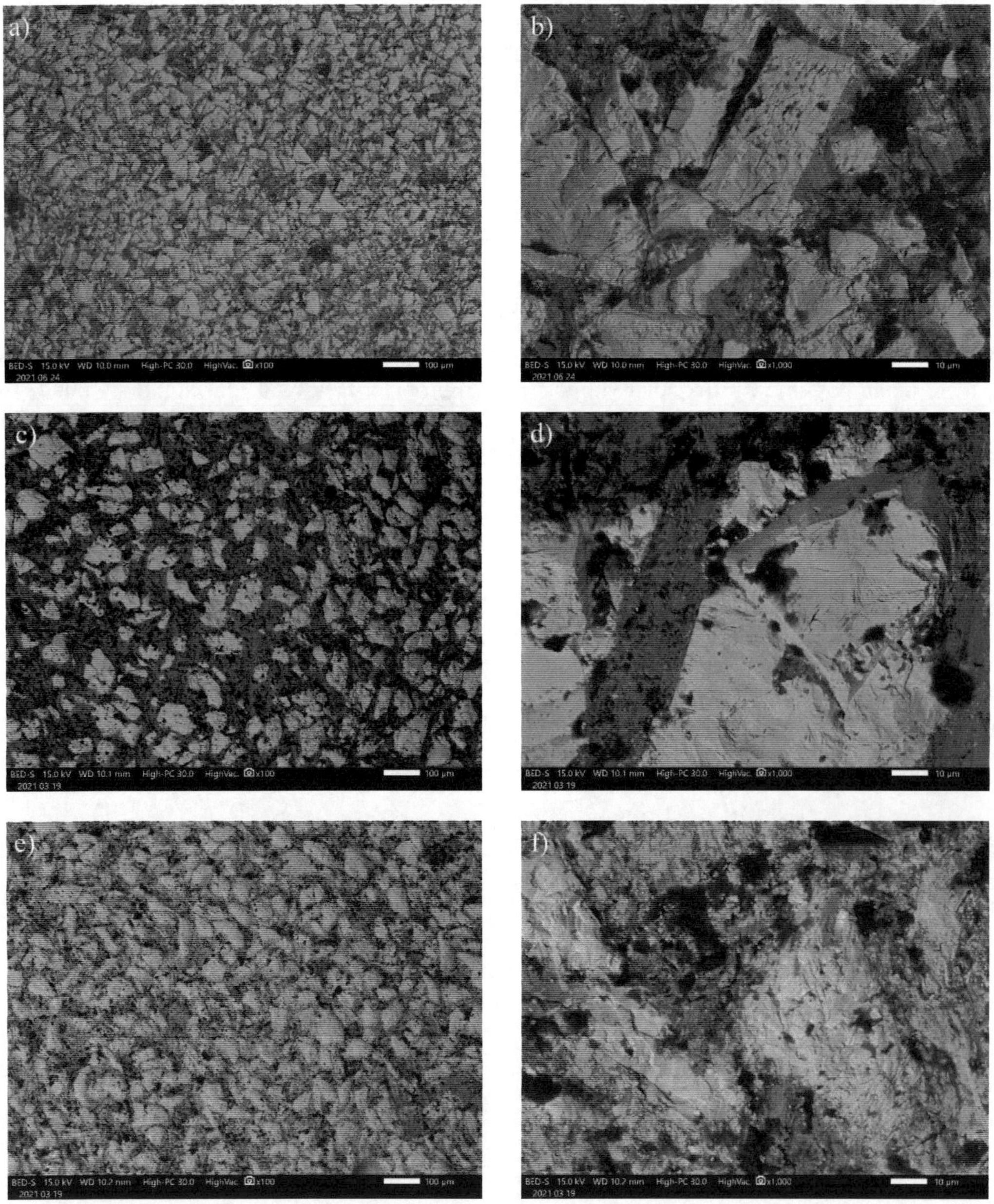

Figure 9. High magnification SEM images of wear scars of a,b) Recycled WC, c,d) Macrocrystalline WC, and e,f) Eutectoid WC/W_2C coatings.

IOP Conf. Series: Materials Science and Engineering 1296 (2023) 012035 doi:10.1088/1757-899X/1296/1/012035

4. Conclusions

The following conclusions can be drawn from the study:

1. Due to the small carbide size of recycled WC, specific energy (J/mm^2) in the laser cladding process needed to be reduced to obtain pore-free clad layers.
2. In a low-stress three-body abrasion recycled WC exhibited wear resistance equivalent to commercial macrocrystalline WC.
3. Eutectoid WC/W_2C showed, however, the best wear performance due to the highest primary carbide and matrix hardness.

The successful introduction of recycled carbide in conventional laser cladding would require a significant increase in carbide size. Potential applications could be machines and components that are subjected to impact and impact-abrasion wear, as tungsten monocarbide is less prone to embrittle the metal matrix due to diminished dissolution in the melt pool. In the future, the use of recycled carbide should be investigated in the novel extreme high-speed laser cladding (EHLA) process, which may allow a smaller carbide size due to the lower specific energy (J/mm^2) of the process.

Acknowledgements

This research was carried out as part of the 'CINEMA – Towards circular economy via eco-design and sustainable remanufacturing' project which belonged to the European V A Nord programme. The work has been made possible by financial support from European Regional Development Fund, Regional Council of Lapland, Länsstyrelsen Norbotten, Kokkolan Seudun Kehitys Ltd., Wärtsilä Finland Oy, Kokkola LCC Oy, Alfamat Oy, Tikomet Oy, Robit Plc, Somotec Oy, Carrum Oy, Castolin Eutectic, Duroc Laser Coating AB, Flowplus Oy, Gestamp Hardtech AB, InfraNord.

References

[1] Wang SH 2016 *CWA Journal* **16** 52
[2] Seger R 2013 *Effect of tungsten carbides properties of overlay welded WC/NiSiB composite coatings* (MG 202X Examensarbete inom Maskinteknik)
[3] Liu W, Bell A, Wang Z, Zhang Z and Kovacevic R 2020 *Wear* **456-457** 203364
[4] European Commission 2020 *Study on the EU's list of Critical Raw Materials*
[5] Kulu P and Zimakov S 2000 *Surf. Coat. Technol.* **130** 46
[6] Zikin A, Ilo S, Kulu P, Hussainova I, Katsich C and Badish E 2012 *Materials Science (Medziagotyra)* **18** 12
[7] Kulu P, Tarbe R, Zikin A, Sarjas H and Surzenkov A 2013 *Key Eng. Mat.* **527** 185
[8] Shemi A, Magumise A, Ndlovu S and Sacks N 2018 *Miner. Eng.* **122** 195
[9] Tuominen J 2009 *Engineering Coatings by Laser Cladding – The Study of Wear and Corrosion Properties* (Tampere University of Technology)
[10] Pierson HO 1996 *Handbook of refractory carbides and nitrides: properties, characteristics, processing, and applications* (first ed. Noyes Publications, USA)
[11] Fisher G, Wolfe T and Meszaros K 2013 *J. Therm. Spray Techn.* **22** 764
[12] Hutchings IM 1996 *Tribology: Friction and Wear of Engineering Materials* (4th ed. Arnold, London)

IOP Conf. Series: Materials Science and Engineering 1296 (2023) 012036 doi:10.1088/1757-899X/1296/1/012036

Using ultrasonic atomization to recycle aluminium bronze chips for additive laser directed energy deposition

Vinzenz Müller[1], Janek Maria Fasselt[1], Tobias Kruse[2], Christian Klötzer[2], Rafael Kleba-Ehrhardt[3], Tomasz Choma[4], Max Biegler[1], Michael Rethmeier[5,1,6]

[1] Fraunhofer Institute for Production Systems and Design Technology IPK, Pascalstraße 8-9, 10587 Berlin, Germany

[2] Mecklenburger Metallguss GmbH, Teterower Str. 1, 17192 Waren, Germany

[3] Technische Universität Berlin, Institute of Material Science and Technology, Chair of Advanced Ceramic Materials, Straße des 17. Juni 135, 10623 Berlin, Germany

[4] Division of Construction and Functional Materials, Warsaw University of Technology, Krakowskie Przedmieście 26/28, 00-927 Warszawa, Poland

[5] Institute for Machine Tools and Factory Management, Technische Universität Berlin, Pascalstraße 8-9, 10587 Berlin, Germany

[6] Bundesanstalt für Materialforschung und -prüfung (BAM), Unter den Eichen 87, 12205 Berlin, Germany

E-mail: vinzenz.mueller@ipk.fraunhofer.de

Abstract. In the post-processing of large maritime components, a considerable amount of waste in the form of milling and grinding chips is produced. At the same time, additive manufacturing technologies have shown great potential in producing high-volume parts for maritime applications, allowing novel design approaches and short lead times. In this context, this study presents a sustainable approach to recycle and use aluminium bronze waste material, generated during post-processing of large cast ship propellers, as feedstock for laser-powder directed energy deposition. The recycling technology used to produce powder batches is inductive re-melting in combination with ultrasonic atomization. The derived metal powders are characterized using digital image analysis, powder flowability tests, scanning electron microscopy as well as energy dispersive X-ray spectroscopy. Compared to conventional metal powders produced by gas atomization, the recycled material shows excellent sphericity and a powder size distribution with a higher content of finer and coarser particles. Metallographic sections of deposited additively produced specimens show an increased hardness and reduced ductility, but also competitive densities and higher yield and ultimate tensile strength compared to cast material. The process chain shows high potential for the maritime sector to enable circular and sustainable manufacturing.

1. Introduction

The majority of modern ship propellers are produced from aluminium bronze via sand casting. The alloy offers good material properties and high resistance to cavitation and corrosion in sea water which makes it especially suitable for maritime applications [1]. After casting and several machining steps, ship propellers are ground in a dry grinding process to generate the final geometry of the propeller blades. Due to the large component sizes, up to 200 kg of material is removed from the part during this process. This paper investigates an approach to recycle this aluminium bronze swarf to feedstock for additive manufacturing (AM) to be used in a powder-based directed deposition process with laser beam (DED-LB).

There are some authors who investigate the use of metal swarf as feedstock for AM processes. For example, Jackson et al. [2] produce specimens from ball milled 304L steel machining chips with good material properties of additive specimens. Razumov et al. [3] generate metal powder by melting steel machining chips with a plasma torch. They are able to produce spherical particles with acceptable particle size distribution (PSD) for a laser powder-bed fusion process. After heat treatment, the mechanical properties of produced specimens exceed the conventional material properties.

For AM feedstock, the characteristics and properties of the metal powder are very important for good processability and part quality. In order to analyze the powders, certain characterization processes can be applied. They often include measurements of PSD, morphology of particles, flowability or chemical composition. In the literature, applicable methods and procedures for characterization are investigated extensively. For example, Wimler et al. [4] and Mitterlehner et al. [5] investigate various methods to measure PSD and morphology of AM powders. Kiani et al. [6] use statistical approaches for powder characterization. The studies of Baesso et al. [7] and Spierings et al. [8] apply and compare methods of standardized flowability tests.

The raw material that was used in this study is shown in the SEM image in Figure 1. It can be stated by sight alone that these particles are not well suited for a DED-LB process where flowability is important for a stable process. In a previous study, the authors of this paper managed to deposit these grinding chips without further processing. While this was generally possible, the flowability was not sufficient and the contamination with foreign particles caused significant welding defects [9].

Therefore, this study utilizes a recycling process called ultrasonic atomization (UA) to produce metal powders from the chips. The UA approach is described by Lierke and Grießhammer [10]. More recently, Żrodowski et al. [11] investigated UA to produce AM powders. For UA, the material is melted down and dropped on a sonotrode that vibrates with frequencies in the kHz-range. As a result, fine metal droplets are ejected. Due to their surface tension they form into a spherical shape in which they solidify. The process takes place in an inert atmosphere of Argon to prevent oxidation. The particle size distribution (PSD) can be controlled by the frequency of the sonotrode.

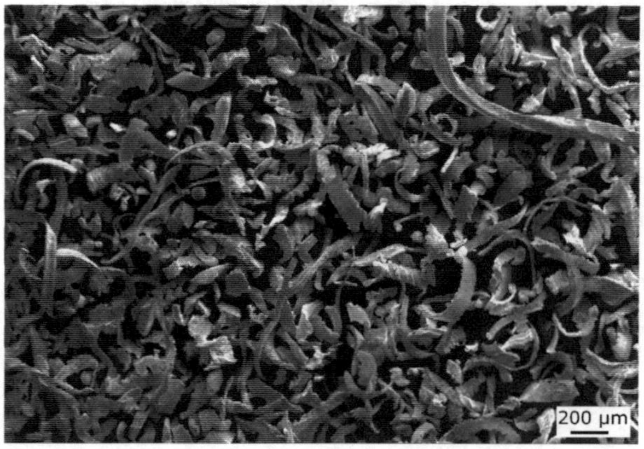

Figure 1. SEM image of grinding chips before the ultrasonic atomization process. The grinding chips consist of long and sharp-edged particles with a very inconsistent PSD and morphology.

2. Materials and Methods

2.1. Feedstock

The material used in this study was from grinding chips of aluminium bronze (CuAl10Ni5Fe5). The material was produced during post processing of cast propellers by means of grinding belts utilizing corundum as abrasive particles. The abrasive was primarily composed of aluminium oxide (Al_2O_3). No cooling lubricants were used during the process. The material was not subjected to any cleaning or purification measures before or after the ultrasonic atomization process.

To compare the properties of recycled grinding chips, commercially available gas atomized aluminium bronze powder (CuAl9.5Fe) from Oerlikon Metco was characterized as well. The nominal PSD according to the producer was 45 µm to 125 µm.

2.2. Ultrasonic atomization

The UA process was carried out using the induction crucible-based ultrasonic atomization system rePowder from Amazemet. The heating zone consisted of a graphite-based crucible and pouring system. The nozzle to feed the melt had a diameter of 1 mm to avoid clogging with possible contamination of leftover ceramic particles from the grinding process. The feedstock material was heated to T = 1300 °C and poured down through the nozzle system onto a carbon fiber sonotrode, vibrating with a frequency of f = 40 kHz. Both the melting and atomization processes were carried out in an inert gas atmosphere of Argon (grade 5.0). Powder yielded from the process was 77%. The remaining material was mainly slag consisting of organic and ceramic contaminations.

2.3. Material characterization route

To measure PSD and particle morphology, dynamic image analysis was performed on a Microtrac Retsch CamSizer X2. Scanning electron microscopy (SEM) as well as energy dispersive X-ray spectroscopy (EDX) were conducted on a Zeiss LEO 1455VP. The flowability of the material was evaluated using two characteristics: The Carney flowmeter test according to ASTM B964 and the Hausner ratio. To calculate the Hausner ratio, the bulk density was measured using a Pharma Test SV100 based on DIN ISO 3923, and tapped density was measured with a JEL STAV 2003 based on DIN ISO 3953. Furthermore, the level of oxygen content was measured on a Leco ONH836 elemental analyzer.

2.4. DED equipment

The DED-LB system consisted of a Trumpf TruLaser Cell 7020 equipped with a Yb:YAG disc laser (P_L = 2 kW, focal length of 200 mm, wavelength of 1030 nm) a Trumpf three-jet powder nozzle and a GTV PF2/2 powder disc feeder. The welding process parameters are listed in Table 1. As substrate, 10 mm thick 1.4404 steel plates were used. To determine the mass flow rate under realistic conditions, the powder feed rates for disc speeds of 1, 3, 5, 7 and 9 rpm were measured three times and averaged.

Table 1. DED-LB welding process parameters as used in this study to manufacture specimens.

Laser power (W)	Feed rate (mm/min)	Material mass flow (g/min)	Laser spot size (mm)	Argon shielding gas (l/min)	Helium carrier gas (l/min)
1300	750	11	1	10	4

IOP Conf. Series: Materials Science and Engineering 1296 (2023) 012036 doi:10.1088/1757-899X/1296/1/012036

2.5. Analyses of properties
Metallographic cross sections were prepared using multiple grinding steps, 6, 3 and 1µm diamond polishing suspensions and then etched with Adler reagent. To obtain tensile values, flat tensile specimens according to ISO 6892-1 were manufactured from the deposited cuboid in the X-Y-plane and cut out by means of wire electrical discharge machining (EDM). The tensile tests were carried out on a Zwick/Roell Z150 at room temperature.

3. Material characterization
In order to compare the recycled powder from ultrasonic atomization with the reference powder, selected material characteristics were analyzed and evaluated. The steps included dynamic image analysis to measure the PSD and morphology of particles, SEM to gather qualitative shape information and EDX to detect contaminations. Furthermore, standardized flowability tests and oxygen measurements were carried out.

3.1. Dynamic image analysis
The PSD measured in the dynamic image analysis of the recycled powder shows a distribution of $d_{10} = 45$ µm to $d_{90} = 179$ µm with $d_{50} = 96$ µm. The variable d represents the equivalent particle diameter that corresponds to the diameter x_{area} of a circle with an equivalent area. The span of the PSD is wider both on the finer and coarser end compared to the gas atomized reference powder which has a PSD of $d_{10} = 66$ µm to $d_{90} = 133$ µm ($d_{50} = 99$ µm). Both measurements are shown in Figure 2 and, together with other measurement data, are summarized in Table 2. Besides the PSD, the sphericity of particles also plays an important role in flowability. To evaluate this factor, the aspect ratio b/l as a factor of each particle's shortest chord $x_{c,min}$ and longest chord $x_{Fe,max}$ can be evaluated:

$$\frac{b}{l} = \frac{x_{c,min}}{x_{Fe,max}} \tag{1}$$

A value of $b/l = 1$ indicates a perfectly spherical particle, a deviation results in values of $b/l < 1$. The results for the powders of this study reveal that the ultrasonic powder shows a significantly higher number of spherical particles than the commercial gas atomized powder. Approx. 82% of the analyzed ultrasonic particles have an aspect ratio of $b/l \geq 0.9$. In other words, 82% of all particles are almost perfectly spherical. This is superior to the reference powder, where only 43% fall into the same sphericity range.

Table 2. Summary of the results of the different material characterization tests for sieved chips, ultrasonic atomized powder and gas atomized reference powder.

	d_{10} (µm)	d_{50} (µm)	d_{90} (µm)	b/l_{mean}	$b/l \geq 0.9$ (%)	Carney flow rate (s/150g)	Bulk density (g/cm^3)	Tap density (g/cm^3)	Hausner ratio	Oxygen content (%)
Chips[1]	81	131	188	0.51	2	44.8	1.80	2.50	1.39	0.0727
Ultrasonic	45	96	179	0.92	82	8.6	4.67	4.90	1.05	0.0125
Reference	63	91	133	0.84	43	10.5	3.97	4.44	1.12	0.0196

[1]Chips were sieved with a 125 µm mesh analysis sieve

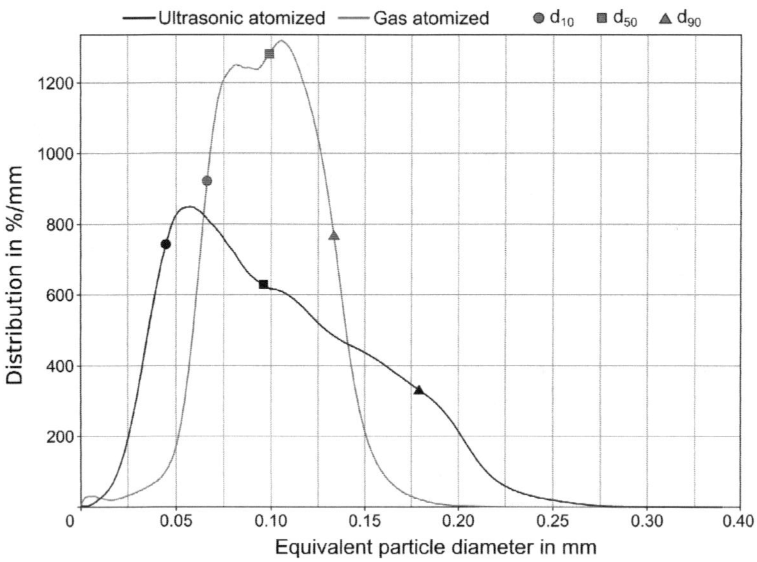

Figure 2. Particle size distribution for recycled, ultrasonic atomized powder and commercial, gas atomized powder.

4. SEM and EDX analysis

In order to obtain image-based information on the metal powders, SEM analyses were carried out. Furthermore, EDX was performed to analyze the chemical composition of the droplets and to detect possible contamination.

The EDX analyses of the ultrasonic atomized powders did not reveal any deviations in the chemical composition compared to the cast material CuAl10Ni5Fe5. While Żrodowski et al. detected a depletion of elements like Manganese in certain alloys in previous studies [11], aluminium bronze seems to be stable in ultrasonic atomization. EDX analysis of the reference material CuAl9.5Fe also shows a chemical composition of the particles that was well within the expected range.

In the SEM images depicted in Figure 3 a), the reference powder has a low number of spherical particles and contains a significant number of spattered and agglomerated particles as well as many adhesions in the form of satellites. Some elongated particles are also clearly visible. Compared to other studies, such as from Mitterlehner et al. [5], the quality of the gas atomized powder seems to be in line with industry standards where elongated particles, satellites and agglomerates can be observed quite frequently.

The images of the recycled chips (see Figure 3 b) are uniformly round in majority. Only a few agglomerated, fused particles or satellites can be seen. The findings confirm the data gathered by the dynamic image analysis in which the recycled powder showed a higher percentage of particles with a favorable aspect ratio of $^b/_l \geq 0.9$ and significantly fewer spherical particles were measured in the commercial material.

While no foreign particles could be detected in the SEM images of the reference powder, impurities were found in the recycled material. For instance, few elongated particles could be detected that seem to be larger droplets of metal, potential contamination from a chamber wall of the equipment (see Figure 3 c). The number of such particles was very low and a relevant influence on the AM process can likely be ruled out.

In the image in Figure 3 c) and d), a fiber adhering to powder particles can be observed. Such fibers of approx. 5 µm to 10 µm width were also detected on other particles. It is highly likely that these fibers originate from the vibrating sonotrode used in the ultrasonic atomization process of this study. The sonotrode was made of binder-free, woven carbon fibers. This type of material is beneficial due to its high fatigue strength and resistance to high temperatures. Apparently, however, fibers detach from the plate when exposed to the hot melt, and subsequently contaminate the powder. While these impurities are not expected to affect the flowability due to their small size, traces of carbon can have a negative influence on the welding process and the component quality, as they could cause inclusions or porosity. To avoid this, a sonotrode from a different material, ideally with a similar alloy to the chips should be used in the future.

As mentioned above, a previous study by the authors of this study [9] analyzing the as-collected grinding chips showed high contamination with organic and ceramic particles that were residues from the grinding belt used during manufacturing. In the recycled material of this study, no such impurities were found in the final powder. This is due to the complete re-melting of the swarf prior to the atomization process. During the melting, all organic and ceramic elements float to the top of the melt. This brings the advantage that nearly all foreign particles and contaminations can be removed during or after the process as a slag. This significantly improves the quality of the powder compared to other recycling approaches like milling or grinding, where additional cleaning steps must be considered.

Measurements via an elemental analyzer show that the UA powder has an average oxygen content of 0.0125%. The reference powder was measured with 0.0196% of oxygen on average. Both results are in an acceptable range for AM powders. The lower values for the recycled powder indicate a sufficient inert gas atmosphere in the UA equipment.

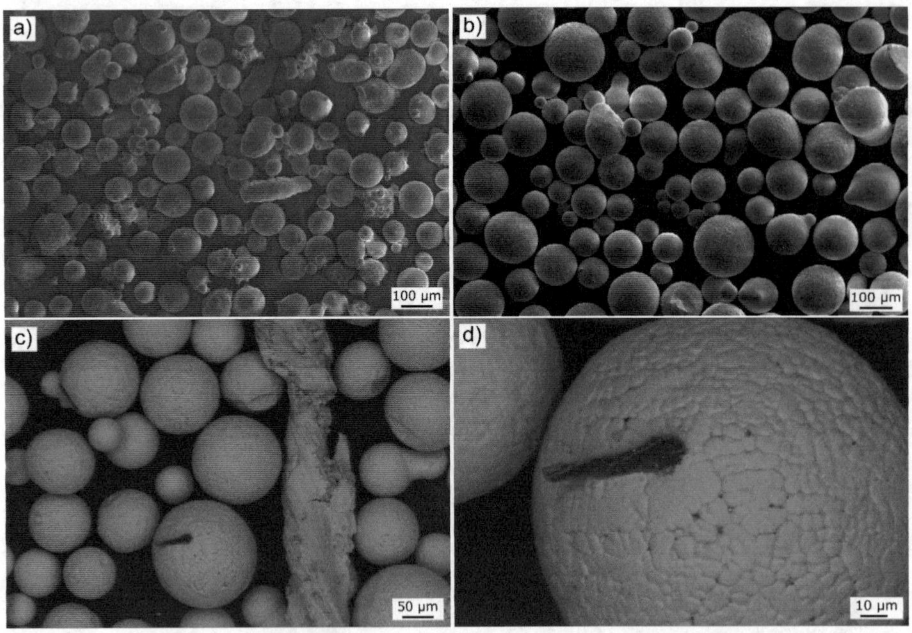

Figure 3. a) SEM images of gas atomized reference powder, b) ultrasonic atomized grinding chips, c) intact chip and carbon fiber in recycled powder and d) detail of particle with adhering fiber.

5. Standardized flowability tests

In order to quantify the powder flow characteristics, the Carney flow test was performed and the Hausner ratio was calculated as a factor of bulk and tapped density. The flowability values are summarized in Table 2. In the Carney test, 150 g of recycled materials needed 8.6 s on average to flow out of the funnel, while the gas atomized material took on average 10.5 s. The measured Hausner ratio of 1.05 for the recycled powder indicate an excellent flowability. With a Hausner ratio of 1.12, the commercial powders show a good flowability (compare Baesso et al. [7]). It can be noted that the results of the two flowability tests are consistent with the measurements from dynamic image analysis and SEM images. As pointed out by Kiani et al. [6], a wider range of particle sizes as present for the UA powder can lead to a reduced flow rate due to more contact points between the particles and tighter packing. For the UA powders, the influence of the wider PSD compared to the reference powder seems to be lower than the high sphericity and the homogenous shape of the particles, leading to better flow behavior.

In summary, the results of the material characterization show that ultrasonic atomization is well suited to produce high-quality, spherical AM powders with better flow and morphology properties compared gas atomized powders in this study. While few process-induced foreign particles (carbon fibers) can be found in the recycled powder, nearly all contaminations from the grinding process were removed during ultrasonic atomization.

6. DED-LB processing and results

6.1. Flowability and processability with disc feeder

While standardized flowability tests as performed during material characterization can give a theoretical implication to compare different powders, processability in specific DED-LB systems must be ensured as well. To verify sufficient powder flow and to determine the powder mass flow in g/min in relation to the disc rotation speed of the powder feeding systems, the powder output after 2 min was measured for increasing disc speeds. The results are shown in Figure 4. Each data point was measured three times. The average error was < 2% for all measuring points and is not displayed in the graph. Both powders show a steady and linear trend with increasing disc speeds. The powder flow from UA, however, is characterized by a mass flow that is constantly about 30% higher compared to the gas atomized material. While the relatively high difference of 30% was not expected, these results are again in line with the measurements in dynamic image analysis, SEM and standardized flow tests.

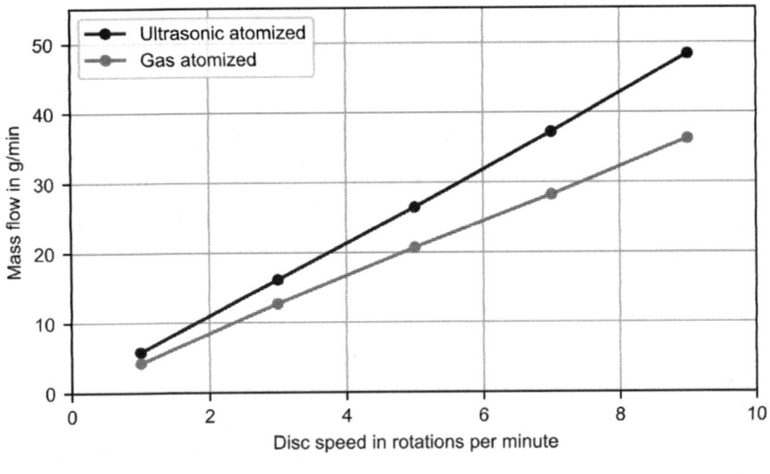

Figure 4. Results of analysis of flowability of recycled and commercial powders in a disc feed powder delivery system as used in the DED-LB system in this study.

6.2. Manufacturing of specimens

To gather mechanical properties of the recycled bronze powder, a specimen block was deposited, consisting of 35 layers with a total height of 20 mm. A meandric deposition strategy was employed with a 90° rotation of the track direction after each layer and 50% track overlap. The specimen dimensions are depicted in Figure 5 a) and were chosen to extract a series of test specimens for metallographic analysis, hardness measurements and flat tensile specimens. The manufactured cuboid after separating the specimen for microstructure analysis is shown in Figure 5 b).

It should be noted that the welding parameters used (see Table 1) were optimized for the gas atomized material and not adapted for the recycled material due to the limited amount of powder available. As can be seen in the cross-section in Figure 6, the manufactured block shows a sag at the top left edge which is due to the DED parameters used. However, the geometrical deviation did not impair the mechanical properties of the material, and the targeted four tensile specimens could still be manufactured out of the block.

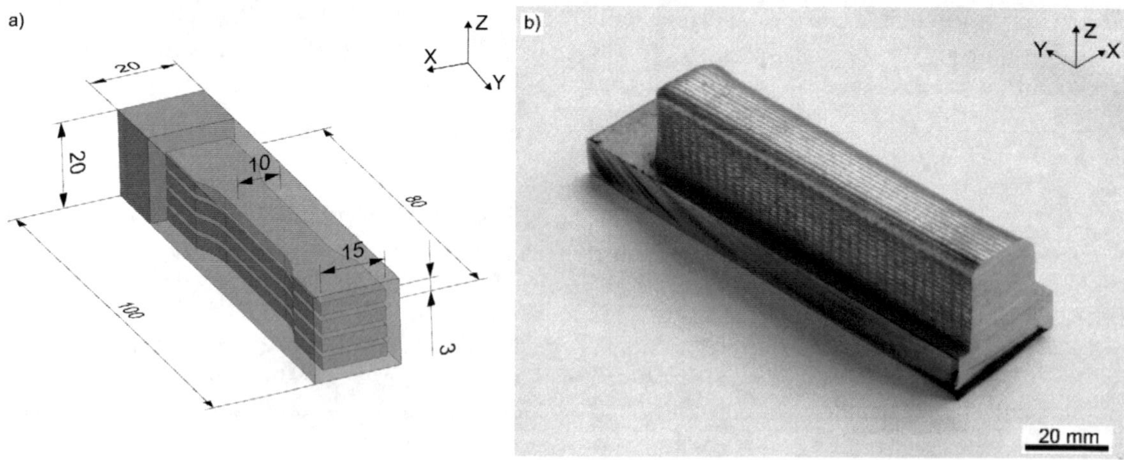

Figure 5. a) CAD drawing with nominal dimensions, the position of the micro section and flat tensile specimens, b) manufactured specimen block after sawing of substrate and micro section specimen.

6.3. Microstructure and mechanical properties

The etched metallographic cross-section in the X-Z-plane is shown in Figure 6 a) and b). No significant defects can be observed. Only very few small spherical gas pores can be detected in the cross-sections. The relative density is $\rho > 99,8\%$. The etched cross-section reveals a fine dendritic microstructure with a preferential grain orientation parallel to the build direction, which is common for AM microstructures.

The same specimen was also used to determine the hardness profile of the structure. Three parallel series of measurements were carried out with eleven measuring points, each in increasing sample height (direction of build-up). The results are plotted in Figure 7 and show a fairly constant hardness profile with an average hardness of 266 HV. According to the data sheet of the German Copper Institute for the investigated aluminium bronze [12], the hardness of cast microstructures usually is in the range of 150 HV to 180 HV. Compared to these values, the additively manufactured sample shows a hardening that exceeds the requirement for end use components.

In a casting process, usually low cooling rates prevail. For the alloy CuAl10Ni5Fe5 this leads to α- and κ-phases that precipitate and grow at the boundaries of initially formed β grains (compare Culpan and Rose [13]). However, during a DED-LB process, this precipitation process

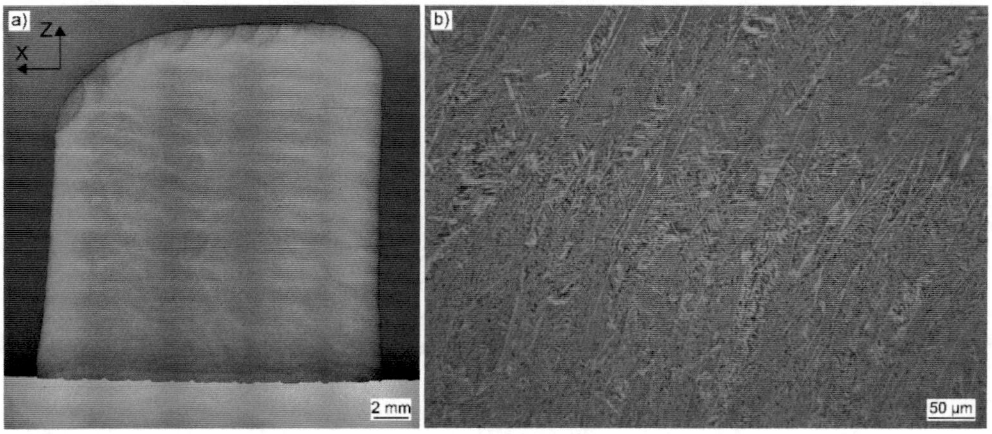

Figure 6. a) Overview image of the etched micro section, b) light microscopy close-up image of the microstructure. A grain structure with fine needles is visible.

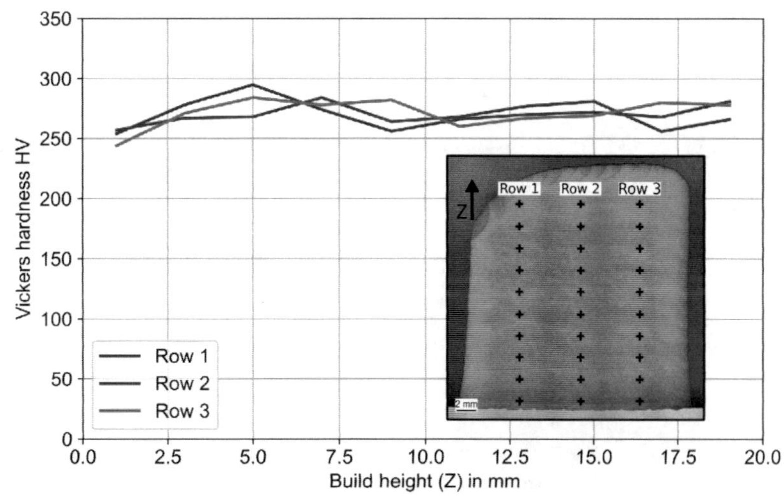

Figure 7. Lines of three hardness measurements at different positions with increasing height of the additively manufactured specimen from ultrasonic atomized powder. A uniform hardness profile can be observed.

is suppressed due to the rapid cooling. As a result, a martensitic-like structure is generated (compare Hájek et al. [14]) that mainly consists of fine needles in the β'-phase and leads to the elevated hardness of the additive specimen.

6.4. Tensile properties

In order to obtain tensile properties for the specimen from recycled powder, flat tensile specimens were machined via wire EDM and tested. The results show a relatively brittle structure with fracture elongation of around $\epsilon = 9\%$, but a high yield strength of $R_{p0.2} = 646$ MPa and tensile strength of $R_m = 911$ MPa. The deviations in the measured data between the tensile specimens are small and confirm the observation of a homogeneous deposition that was mainly free from defects or large pores. Figure 8 shows the tensile stress curves for the recycled powder. The graph

IOP Conf. Series: Materials Science and Engineering 1296 (2023) 012036 doi:10.1088/1757-899X/1296/1/012036

also includes exemplary tensile curves for casting processes commonly used for aluminium bronze. Sand casting processes are characterized by their very slow cooling rate, and the permanent mold casting process by more rapid solidification. The plot in Figure 8 clearly indicates the influence of the cooling rate for this alloy: rapid solidification leads to high tensile strength and lower ductility, while slow solidification leads to higher values in elongation at break and lower values in tensile strength.

While the tensile properties combined with the elevated hardness values do not meet the requirements for end-use parts like ship propellers, other applications are possible. The high toughness and hardness could be advantageous, e.g. for hard coatings. For additive structures, heat treatment could be applied to allow the microstructure to precipitate α- and κ-phases to reduce the hardness and increase tensile strength.

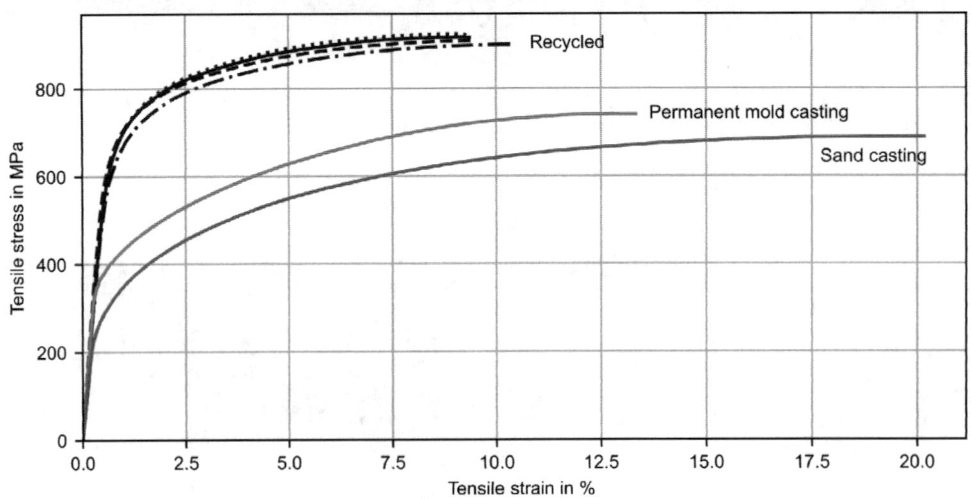

Figure 8. Results of tensile tests of DED-LB specimens from recycled powder in blue. The graph also contains tensile curves for the same alloy that were produced by sand casting in gray (very low cooling rate) and permanent mold casting in orange (faster cooling rate). The tensile test data for the casting processes are from material tests conducted by Mecklenburger Metallguss GmbH.

7. Conclusion

In this study, ultrasonic atomization was used to produce AM powder from aluminium bronze grinding chips from ship propeller manufacturing. The following summary can be drawn:

- The recycled powder was highly spherical, had excellent flowability properties, exceeding the gas atomized reference powder and was low in oxygen content. Ceramic and organic contaminants from the grinding process were successfully removed during UA.
- Metallographic cross-sections of a cuboid produced by DED-LB from the recycled powder reveal an almost defect-free microstructure with fine needle-shaped grains. Hardness measurements indicate that a martensitic-like β-phase was predominantly present which likely formed due to the high cooling rates.
- Tensile tests show a high yield strength and ultimate tensile strength but reduced fracture elongation in comparison to cast properties, caused by the relatively hard and therefore brittle microstructure of the additively manufactured specimens.

It should be noted that UA is an energy-intensive process, as all the metal must be melted in a crucible prior to atomization. This is at least partially at odds with the concept of sustainability. On the other hand, powder yields from UA are over 60% compared to typical 30-40% from gas atomization, with sustainably lower gas consumption. Therefore, mechanical recycling approaches should be investigated to reduce the energy input during recycling. However, this study shows that UA is a suitable production technology to produce high quality AM powders with excellent properties and low contamination.

Acknowledgments

This project has received funding from the European Union's Horizon 2020 research and innovation program under grant agreement No 728053-MarTERA.

References

[1] Callcut V 2002 Aluminum Bronzes: Metallurgy of Copper & Copper Alloys
[2] Jackson M A, Morrow J D, Thoma D J and Pfefferkorn F E 2020 *CIRP Annals* **69** 165–168
[3] Razumov N G, Masaylo D V, Silin A O, Borisov E V, Ozerskoy N E, Goncharov I S and Popovich A A 2021 *Journal of Manufacturing Processes* **64** 1070–1076
[4] Wimler D, Kardos S, Lindemann J, Clemens H and Mayer S 2018 *Practical Metallography* **55** 620–636
[5] Mitterlehner M, Danninger H, Gierl-Mayer C, Gschiel H, Martinez C, Tomisser M, Schatz M, Senck S, Auer J and Benigni C 2021 *Journal of Materials Engineering and Performance* **30** 7019–7034
[6] Kiani P, Scipioni Bertoli U, Dupuy A D, Ma K and Schoenung J M 2020 *Advanced Engineering Materials* **22**
[7] Baesso I, Karl D, Spitzer A, Gurlo A, Günster J and Zocca A 2021 *Additive Manufacturing* **47**
[8] Spierings A B, Voegtlin M, Bauer T and Wegener K 2016 *Progress in Additive Manufacturing* **1** 9–20
[9] Müller V, Marko A, Kruse T, Biegler M and Rethmeier M 2021 *Lasers in Manufacturing Conference 2021*
[10] Lierke E G and Grießhammer G 1967 *Ultrasonics* **5** 224–228
[11] Żrodowski Ł, Wróblewski R, Choma T, Morończyk B, Ostrysz M, Leonowicz M, Łacisz W, Błyskun P, Wróbel J S, Cieślak G, Wysocki B, Żrodowski C and Pomian K 2021 *Materials (Basel, Switzerland)* **14** 2541
[12] German Copper Institute 2005 *CuAl10Fe5Ni5 material data sheet*
[13] Culpan E A and Rose G 1978 *Journal of Materials Science* **13** 1647–1657
[14] Hájek J, Kříž A, Chocholaty O and Pakuła D 2016 *Archives of Metallurgy and Materials* **61** 1271–1276

IOP Conf. Series: Materials Science and Engineering 1296 (2023) 012037 doi:10.1088/1757-899X/1296/1/012037

High-speed laser cladding: new developments for wear and corrosion protection

J Tuominen[1], M Hallaji[1], J Kiviö[2] and J Vihinen[2]

[1]Tampere University, Materials Science and Environmental Engineering
[2]Tampere University, Automation Technology and Mechanical Engineering

jari.tuominen@tuni.fi

Abstract. Fusion-bonded and low-diluted overlay welded coatings are frequently very thick (>1mm), which results in high material consumption (kg/m^2) and expenses. High-speed laser cladding is a novel process, which can produce thin fusion-bonded and low-diluted coatings with high coverage rates and low heat input. Coating materials utilized in high-speed cladding have varied from soft Ni-based superalloys to harder martensitic stainless steels and base materials from low alloy steels to austenitic stainless steels. The ultimate objective has been to develop crack-free alternatives to environmentally hazardous hard-chrome plating. In this paper, high-speed laser cladding was used to fabricate relatively thin Ni-based coatings on various cast irons for corrosion protection and Ni-based hard-faced coatings reinforced with chromium (Cr_3C_2) and tungsten carbides (WC/W_2C), for the first time, onto low-alloy structural and quenched & tempered steels for wear applications. Obtained coatings were characterized with X-ray diffraction (XRD), optical (OM), and scanning electron microscopy (SEM). Corrosion performances were explored in long-term salt spray tests. Mechanical and wear properties were tested with Vickers microhardness measurements and three-body dry-sand rubber wheel abrasion tests (RWAT). It was shown that soft low-diluted Ni-based coatings protected the cast iron base materials in salt spray tests. Ni-based hard-faced coatings with hardness up to 1200 HV_1 exhibited high wear resistance in low-stress three-body abrasion better than hard-chrome plated coating, which was attributed to the high volume fraction of hard carbide reinforcements.

1. Introduction

Fusion-bonded and low-diluted overlay welded coatings typically range from one millimeter to several millimeters in thickness [1], which results in high material consumption (kg/m^2) and expenses. Extreme high-speed laser cladding (EHLA) and earlier introduced high-speed laser cladding are methods capable of producing thinner overlay welded coatings [2,3]. Both processes utilize high energy densities using relatively small laser beam spots 1-2 mm in diameter. Instead of plasma-transferred arc (PTA) grade powder (~50-150 μm), the EHLA process takes the advantage of smaller powder granulometry (~10-60 μm), which enables narrower powder focus. In addition, laser and powder focus locates just above the melt pool in the EHLA process, increasing the energy absorbed by the powder and decreasing the energy absorbed by the base material. Besides the ability to fabricate thin fusion-bonded coatings, the benefits of high-speed laser cladding include high coverage rates (m^2/h), high solidification rates, low heat input into the base material, and high process energy efficiency due to low heat conduction losses. Coverage rate of ~7 m^2/h was recently reached for ~100 μm thick coating using 8 kW laser power [4]. High solidification rates lead to fine grain structure, which provides hardness increase via grain boundary strengthening. Low heat input diminishes the depth of heat-affected zone (HAZ) and distortion.

Coating feedstocks utilized in high-speed laser cladding have varied from soft Ni-based superalloys to harder martensitic stainless steels [2-9]. The ultimate objective has been to develop crack-free alternatives to environmentally hazardous hard-chrome plating. Maximum hardness values reached so far have been limited to ~600 HV, which is clearly lower than those reported for hard-chrome plated coatings 800-1250 HV [10,11]. This could be the reason why wear comparison between hard-chrome and developed EHLA coatings has not been reported yet.

Hardness values similar to or higher than hard-chrome plated coatings can be obtained via metal matrix composites (MMC). They consist of ductile metal matrix and hard reinforcements. Metal matrix provides fracture toughness, transfers mechanical stresses to reinforcements, and holds the reinforcements together. Hard reinforcements provide high stiffness, hardness, and wear resistance [12]. They are typically refractory (high melting point) carbides, borides, nitrides, or oxides, which preferably do not melt but mix homogeneously with the molten matrix. Frequently used carbides in MMC hard-facings are chromium and tungsten carbides. Orthorhombic chromium carbide (Cr_3C_2, T_m ~1800°C) possesses a high degree of chemical stability and moderate hardness (~1400 HV) [13]. In overlay welding, it typically melts and precipitates as hard hexagonal close-packed (hcp) Cr_7C_3 and face-centered cubic (fcc) $Cr_{23}C_6$ secondary carbides [14,15]. Tungsten carbides exist as WC, W_2C, and WC_{1-x} (T_m ~2800°C). They are much harder (~2300 HV) than chromium carbides but less resistant to high-temperature oxidation. If melted and dissolved, they form mixed secondary carbides (M_6W_6C, M_3W_3C, M_2W_4C, M = Ni, Co, or Fe) during cooling, which are very brittle. Besides many different mechanical properties and characteristics between chromium and tungsten carbides, tungsten belongs to the EU's critical list of materials whereas chromium does not [16].

Base materials surfaced with the EHLA process have varied from mild steels [4] to tool steels [17], austenitic stainless steels [18,19], and titanium alloy Ti6Al4V [20]. Cast irons are common construction and engineering materials used in rolls, engine blocks and heads, gears, machine tool beds, etc. because of their low cost and desirable properties such as castability, excellent machinability, low friction, and good mechanical properties [21,22]. However, especially under severe service conditions the wear and corrosion performance of cast irons is not sufficient. Surfacing cast irons with overlay welding methods to improve the surface properties is challenging due to impurities (sulphur, phosphor) in cast iron, the formation of brittle chilled structures, and severe hardening of the HAZ [23]. Intermediate layers are typically needed to prevent migration of impurities to clad layer and preheating to prevent brittle cracks in the chilled zone and HAZ [23]. EHLA process is reported to be successful in surfacing brake discs made of grey cast iron but there is not any detailed study available, yet [24].

The objective of this study was to develop thin and hard (>600HV) hard-faced coatings for wear applications by high-speed laser cladding similar to the EHLA process and study its suitability for surfacing difficult-to-weld grey and ductile cast irons with Ni-based corrosion-resistant alloy.

2. Experimental

High-speed laser cladding experiments were performed using a continuous wave fiber laser (Coherent) with a maximum power of 3 kW. The experimental set-up consisted of a 400 µm fiber, 160 mm collimator unit, f250 optics, a coaxial powder feeding nozzle (Coax14V5 from Fraunhofer IWS), and a rotating disc type of powder feeder (Oerlikon Metco Twin 150). The focal point of the laser beam was above the base material resulting in a spot size of ~1.3 mm on the surface. The focal point of the powder jet was 2.5 mm above the base material. Planar wear test specimens were clamped to a processing table of Fanuc F-200iB parallel kinematic robot keeping the optics and powder feeding nozzle static. Rotationally symmetric corrosion test specimens were clad in a CNC-controlled lathe. The equipment used is illustrated in Figure 1.

2.1. *High-speed laser cladding for wear protection*

The powder feedstocks used for hard-facing were 1) mechanical mixtures of NiCrMoNb and Cr_3C_2, 2) NiBSi and spherical WC/W_2C, and 3) NiBSi and angular WC/W_2C with volume fractions of 50/50. Mechanical mixtures of powders were prepared by weighing desired amounts of powders and mixing

IOP Conf. Series: Materials Science and Engineering 1296 (2023) 012037 doi:10.1088/1757-899X/1296/1/012037

Figure 1. High-speed laser cladding equipment at Tampere University.

them in a plastic bottle, which was placed in the mechanical mixer for 24 hours. NiCrMoNb displayed in Figure 2a is a gas-atomized Ni-based alloy with a particle size of 15-45 µm composition corresponding to Alloy 625. Cr_3C_2 powder with a particle size of 5-45 µm is sintered & crushed chromium carbide rather dense and angular in morphology as shown in Figure 2b. NiBSi is a gas-atomized Ni-based self-fluxing alloy with a particle size of 10-50 µm (Figure 2c). WC/W_2C is spherical fused tungsten carbide powder with a particle size of 15-45 µm (Figure 2d). Angular WC/W_2C is fused tungsten carbide powder with a particle size of 20-53 µm (Figure 2e). The substrates employed for the cladding were mild steel plates with a thickness of 20 mm, a width of 20 mm, and a length of 50 mm. Before cladding, they were grit-blasted to a roughness of 4.4 µm. The chemical compositions of used materials are given in Table 1. The cladding parameters to produce fusion-bonded MMC coatings with low dilution are summarized in Table 2.

2.2. *High-speed laser cladding for corrosion protection*

The powder feedstock employed for corrosion protection was a gas-atomized NiCrMoNb displayed in Figure 2a with a particle size of 15-45 µm. Its composition corresponds to Alloy 625, and it was intended for thermal spraying. The base materials employed were S355 construction steel (~180 HV_1), GGG50 spherical graphite cast iron (ferritic-pearlitic), and GG25 grey cast iron (~170 HV_1) rods with a diameter of 100 mm and length of 500 mm. They were in turned condition. Their chemical compositions are shown in Table 1. Varied high-speed laser cladding process parameters are depicted in Table 3.

2.3. *Characterization*

The microstructure of the specimens was studied by applying optical microscopy (OM) and scanning electron microscopy (SEM, Jeol JSM-IT500). The polished coating specimens were etched with a 4% Nital to reveal the HAZ. Energy-dispersive X-ray spectroscopy (EDS) was employed to analyse local chemical compositions. To reveal crystal structures and phases present in the studied materials, an X-ray diffractometer (Panalytical Empyrean Multipurpose Diffractometer) was used. Phase quantifications were made with Image J software. Analysed areas were ~300 x 200 μm^2 for primary carbides and ~30 x 40 μm^2 for secondary carbides. The Vickers microhardness measurements from the coating cross-sections were taken with a Matsuzawa MMT-X7 hardness tester using indentation loads of 50 and 1000 g applied for 10 s. Microhardness measurements were also conducted separately on the primary carbides with a load of 50 g.

2.4. *Wear tests*

Three-body abrasion wear tests at room temperature were conducted for hard-faced coatings with a

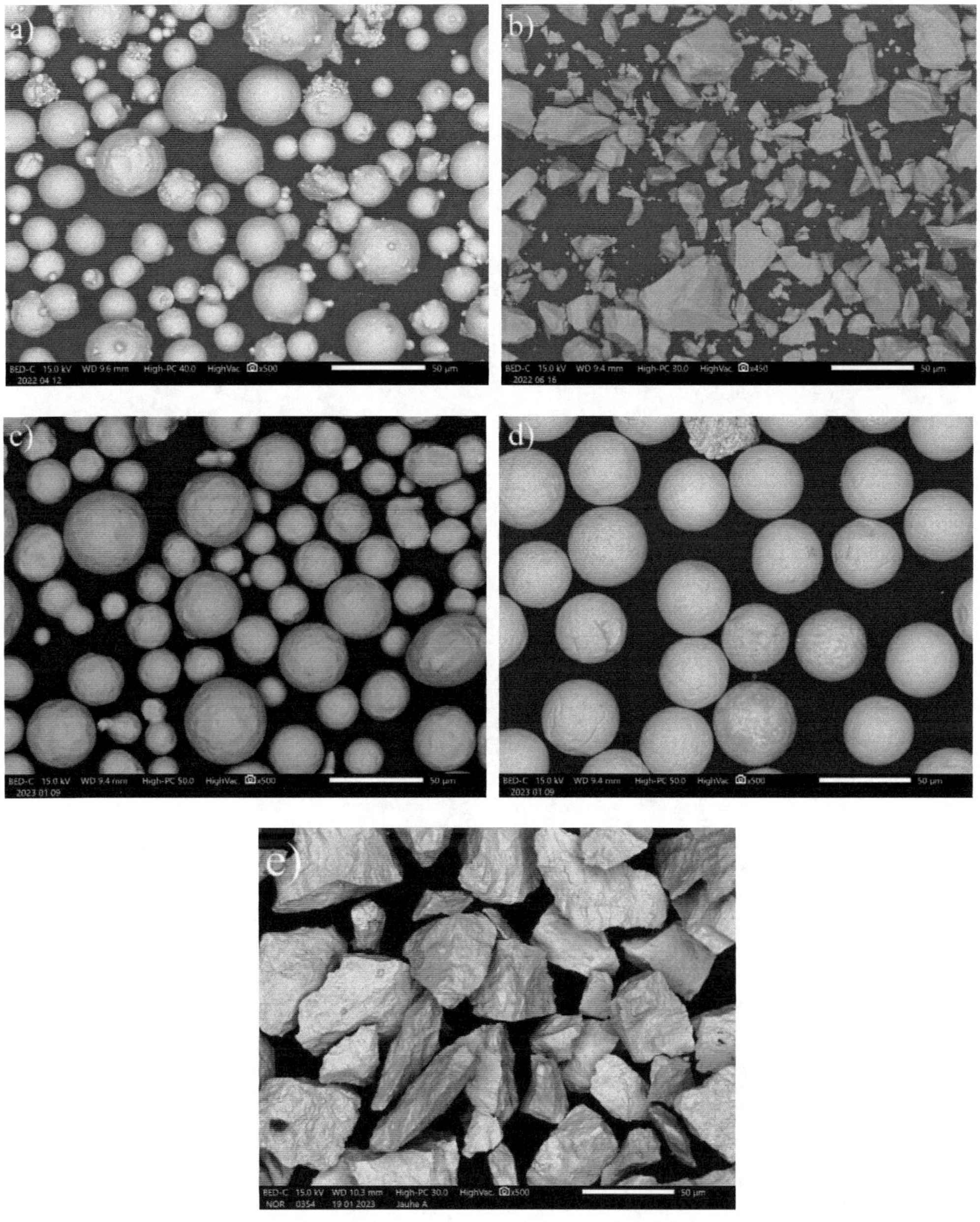

Figure 2. SEM micrographs of a) NiCrMoNb, b) Cr_3C_2, c) NiBSi, d) spherical WC/W_2C, and e) angular WC/W_2C powders.

rubber-wheel abrasion test (RWAT) device, a modified version of ASTM G65, where crushed dry quartz (SiO_2) sand abrasives (0.1-0.6 mm) flow between the surfaces of the rotating rubber wheel (Durometer A84-88) and the test specimen. The test device and parameters are described in more detail in Ref. [25]. Before wear testing, all the laser-clad surfaces were ground with #600 and #1200 SiC papers. Commercial hard-chrome plating with a surface hardness of ~950 $HV_{0.3}$ was used as a reference material. The worn specimen surfaces were examined by SEM. Wear performance was evaluated as volume loss using theoretical densities for high-speed laser-clad coatings and density values found in the literature for the hard-chrome plated coating.

Table 1. Chemical composition of powders and base materials.

Material	Ni	Cr	Mo	Nb	Fe	B	Si	W	Mn	C
NiCrMoNb	Bal.	21.0	9.0	4.5						
NiBSi	Bal.				≤0.8	1.2-1.5	2.0-2.6			≤0.05
Cr_3C_2		Bal.			0.4					13.3
WC/W_2C (spher)					≤0.3			Bal.		3.8-4.1
WC/W_2C (ang)					0.2			Bal.		3.9
S355					Bal.		0.55		1.6	0.2
GGG50					Bal.		2.4-3.0		0.1-0.3	3.3-3.7
GG25					Bal.		1.2-1.7		0.8-1.2	2.8-3.3

Table 2. High-speed laser cladding process parameters for hard-facings.

Powder	Carbide shape	Volume ratio (%)	Power (kW)	Traverse speed (m/min)	Powder feed rate (g/min)
NiCrMoNb+Cr_3C_2	Angular	50/50	1.9	25	8
NiBSi+WC/W_2C	Spherical	50/50	1.5	25	19
NiBSi+WC/W_2C	Angular	50/50	1.5	25	19

Table 3. High-speed laser cladding process parameters for corrosion-resistant coatings.

Powder	Base material	Power (kW)	Traverse speed (m/min)	Powder feed rate (g/min)
NiCrMoNb	S355	1.5-2.1	20-25	11
NiCrMoNb	GGG50	1.425-1.95	20-25	11
NiCrMoNb	GG25	1.425-1.95	20-25	11

2.5. Corrosion tests

After post-machining (turning & grinding), all the laser-clad rods on S355, GGG50, and GG25 base materials were subjected to neutral salt spray (NSS) in 5% NaCl (pH~3) for 750 h at 35°C (EN ISO 9227). The salt spray test chamber was a Liebisch S 400 M-TR. To prevent base materials degradation, both ends of the rods were mould in epoxy resin.

3. Results and discussion

3.1. Hard-faced coatings

The cross-sections of high-speed laser-clad carbide-reinforced coatings are shown in Figure 3. Coatings are generally characterized by varying numbers of primary carbides and a thickness of ~0.21-0.26 mm. Primary carbide contents are ~5 vol.% for NiCrMoNb-Cr_3C_2, ~40 vol.% for NiBSi-WC/W_2C spherical, and ~33 vol.% for NiBSi-WC/W_2C angular. Hardness values measured from the coatings (HV_1), primary carbides ($HV_{0.05}$), and matrices ($HV_{0.05}$) are tabulated in Table 4. NiBSi-WC/W_2C spherical was the hardest and NiCrMoNb-Cr_3C_2 was the softest coating.

The RWAT results are presented in the form of volume losses in Figure 4. The results are the average of three test runs per studied material. The hardest high-speed laser-clad coating, NiBSi-WC/W_2C spherical, showed the best while the softest NiCrMoNb-Cr_3C_2 had the worst wear performance. The wear scar morphologies shown in Figure 5 revealed that primary carbides protruded from the abraded surface with softer matrices preferentially removed. Importantly, coatings reinforced with WC/W_2C par-

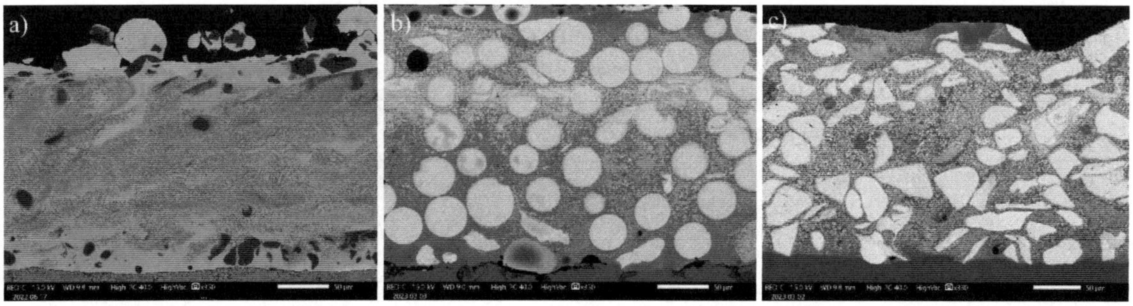

Figure 3. SEM micrographs of high-speed laser-clad a) NiCrMoNb-Cr_3C_2, b) NiBSi-WC/W_2C spherical, and c) NiBSi-WC/W_2C angular.

Table 4. Average microhardness values of carbide-reinforced hard-facings fabricated by high-speed laser cladding. Standard deviations are given in parentheses.

Powder	Carbide shape	Bulk (HV_1)	Carbide ($HV_{0.05}$)	Matrix ($HV_{0.05}$)	Mean free path between carbides (µm)
NiCrMoNb+Cr_3C_2	Angular	847 (94)	1300 (71)	791 (127)	~230
NiBSi+WC/W_2C	Spherical	1202 (244)	2805 (217)	709 (39)	~24
NiBSi+WC/W_2C	Angular	1005 (268)	2504 (190)	868 (106)	~37

ticles outperformed the hard-chrome plated coating in wear resistance suggesting that developed coatings could be used as substitutes for the hard-chrome. The best wear performance of NiBSi-WC/W_2C spherical can be attributed to the hardest carbide and shortest mean free path between the carbides shown in Table 4. The short mean free path between the carbides effectively prevented the penetration of rather large quartz sand abrasives (0.1-0.6mm), which are softer (~750-1200HV) than the carbides, to the coating.

3.2. *Corrosion-resistant coatings*

The representative examples of the cross-sections (perpendicular to cladding direction) of machined single-layer high-speed laser-clad NiCrMoNb coatings on S355, GGG50, and GG25 base materials are shown in Figure 6. The SEM micrographs and microscopic examination revealed fusion-bonded deposits without macroscopic pores. Coating thicknesses in as-clad and machined conditions were ~0.22-0.26 and 0.18-0.22 mm, respectively. High magnification SEM studies revealed compositional variation at the micro-level where Nb and Mo segregated strongly during solidification. To estimate the dilutions of the clads, chemical composition profiles throughout the coatings were obtained with EDS area analyses. The analysed areas were ~20 µm in diameter. They were taken at 40 µm intervals as

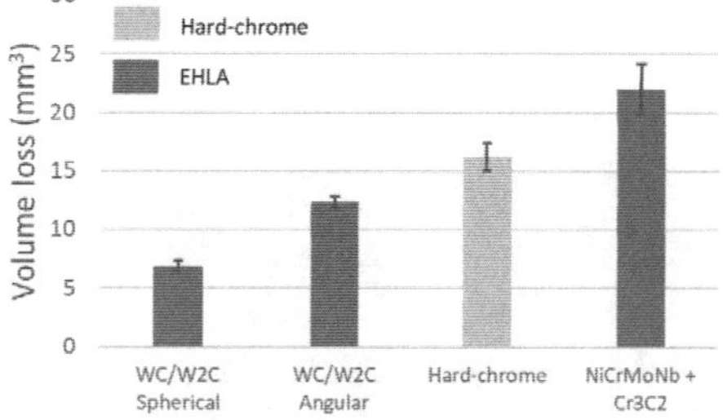

Figure 4. Volumetric wear losses of carbide reinforced hard-facings fabricated by high-speed laser cladding.

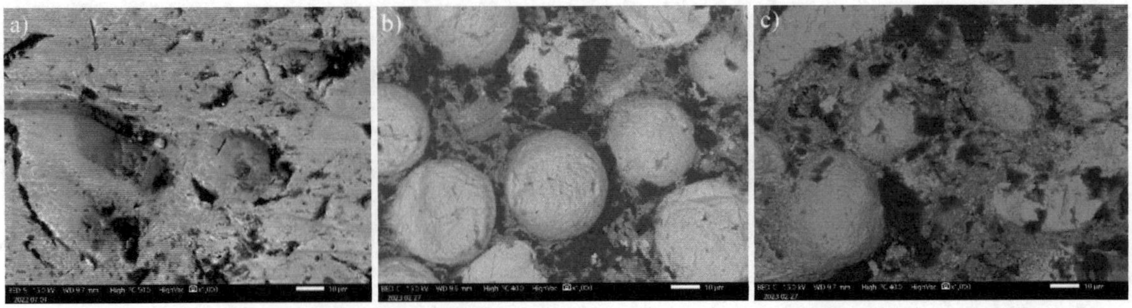

Figure 5. SEM micrographs of wear scars of a) NiCrMoNb-Cr_3C_2, b) NiBSi-WC/W_2C spherical, and c) NiBSi-WC/W_2C angular.

shown in Figure 7. It is observed that Fe concentrations in all the coatings are very low, which indicates very low chemical dilution. When neglecting the heavily diluted zone near the fusion line, Fe contents were 0.9 wt.% for NiCrMoNb on S355, 2.8 wt.% for NiCrMoNb on GGG50, and 1.7 wt.% for NiCrMoNb on GG25. Iron concentrations of coatings and regions of heavily diluted zones are somewhat larger on cast irons because they have lower melting ranges/points than S355 steel. Melting ranges of materials used are ~1370-1400°C for S355 [26], ~1165°C for cast irons [27], and ~1290-1350°C for NiCrMoNb [28].

Microhardness profiles ($HV_{0.05}$) and average hardness values (HV_1) measured from the coating cross-sections are displayed in Figure 8a-b. The average hardness of the coatings is ~300 HV_1, regardless of the base material. Hardness values are somewhat higher than reported for other low-diluted NiCrMoNb coatings fabricated by the conventional laser (234 HV_5) and CMT (220 HV_5) cladding [29]. As illustrated in Figure 8a, the maximum hardness values at HAZ were 306 for S355, 231 for GGG50, and 518 $HV_{0.05}$ for GG25 indicating moderate and severe hardening for S355 and cast irons, respectively.

To reveal possible interconnected defects or inhomogeneous distribution of intermixed iron in the coatings, single- and double-layer high-speed laser-clad NiCrMoNb coatings on various base materials were exposed to long-term salt spray tests. The results are depicted in Table 5. The results show that all the double-layer coatings survived the long-term salt spray tests without signs of corrosion. From the single-layer coatings, the coatings on 42CrMo4 and GGG50 passed the test (Figure 9a). Single-layer coatings on S355 and GG25 base materials failed probably due to interconnected hot cracks that allowed corrosive media to reach the base material and corrode it (Figure 9b). Hot cracks in Ni-based alloys are known to form mainly due to excessive amounts of boron, phosphor, and sulfur that form low melting point liquids along dendrite boundaries [29]. These impurity elements may originate from the original powder feedstock or the base material due to intermixing. The results displayed in Table 5 (low-diluted coatings on S355) suggest that the problem is in powder feedstock. The crack-sensitive boron content

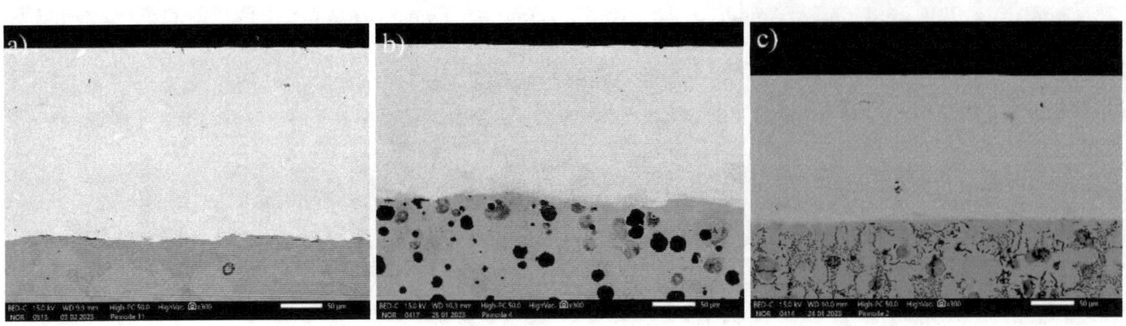

Figure 6. SEM micrographs of high-speed laser-clad NiCrMoNb alloy on a) S355, b) GGG50, and c) GG25 base materials.

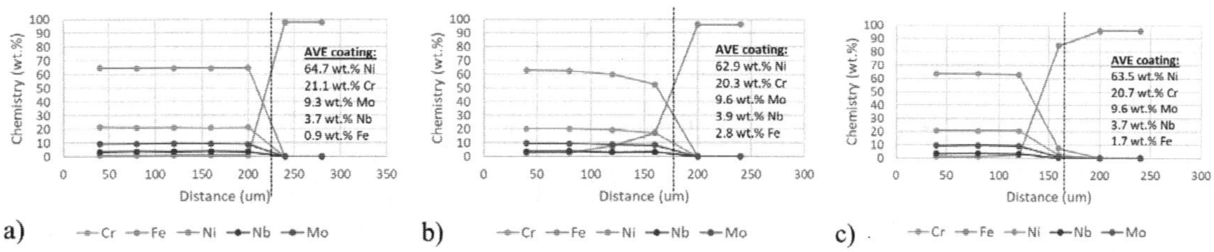

Figure 7. Chemical composition profiles of high-speed laser-clad NiCrMoNb alloy on a) S355, b) GGG50, and c) GG25 base materials.

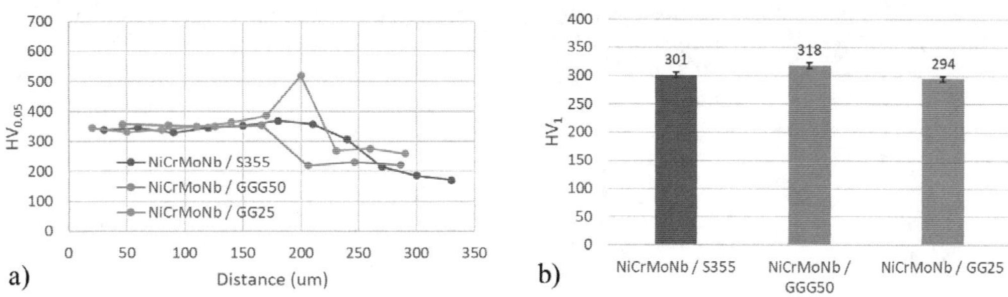

Figure 8. a) Microhardness profiles ($HV_{0.05}$) and b) average microhardness values (HV_1) of NiCrMoNb coatings on S355, GGG50, and GG25 base materials measured from the cross-sections.

of the used powder designed for thermal spraying is not known but too high boron content of the powder together with the high tensile residual stress factor due to high traverse speed may have caused the hot cracks. Therefore, important future works include the precise determination of minor alloying elements with methods given in Ref. [29]. EHLA process may need more careful control of impurity elements in feedstock than in conventional laser cladding.

Table 5. Summary of salt spray test results for NiCrMoNb coatings on various base material.

Coating	Base material	Layers	Fe content (wt.%)	Thickness as-machined (μm)	Pass/Fail	Hot cracks in cross-section
NiCrMoNb	S355	Double	0.8	~500	Pass	No
NiCrMoNb	GGG50	Double	1.2	~370	Pass	No
NiCrMoNb	GG25	Double	1.1	~460	Pass	Yes
NiCrMoNb	S355	Single	1.0	~220	Fail	No
NiCrMoNb	S355	Single	0.9	~230	Fail	No
NiCrMoNb	42CrMo4	Single	0.9	~210	Pass	No
NiCrMoNb	42CrMo4	Single	0.9	~230	Pass	No
NiCrMoNb	GGG50	Single	2.8	~180	Pass	No
NiCrMoNb	GGG50	Single	1.8	~200	Pass	No
NiCrMoNb	GGG50	Single	2.6	~230	Pass	No
NiCrMoNb	GG25	Single	1.7	~170	Fail	Yes

4. Conclusions

1. WC/W_2C reinforced Ni-based coatings fabricated by the high-speed laser cladding outperformed Cr_3C_2 reinforced NiCrMoNb and hard-chrome plated coatings in low-stress three-body

abrasion wear resistance due to a high volume fraction (~33-40%) hard carbides (~2500-2800HV).

2. Double-layer NiCrMoNb coatings produced by high-speed laser cladding protected S355, GGG50, and GG25 base materials in long-term salt spray tests due to absence of interconnected defects.
3. Occasional interconnected hot cracks originating in dendrite boundaries appeared in some single-layer NiCrMoNb coatings corroding the base material.

Figure 9. High-speed laser-clad single-layer NiCrMoNb coatings on a) GGG50 and b) GG25 cast irons after salt spray test.

Acknowledgements
This research was funded by the European Regional Development Fund, grant number A78522. The project 'Fast coating method – Benefits for the industry' is funded by the React-EU Instrument as part of the European Union's response to the Covid-19 pandemic.

References
[1] Silva CC 2013 Weld Overlay *Encyclopedia of Tribology* eds QJ Wang, RZ Smith (Springer US, Boston) pp 4094-4101
[2] Schopphoven T, Gasser A, Wissenbach K and Poprawe R 2016 *J. Laser Appl.* **28** 022501-1. https://doi.org/10.2351/1.4943910.
[3] Seefeld T, Theiler C and Sepold G 2001 *Proc. 20th Int. Congress on Applications of Lasers & Electro-Optics (ICALEO)* (Orlando: Laser Institute of America)
[4] Vogt S, Göbel M and Fu E 2022 *J. Manuf. Sci. E – T ASME* **144** 044501-1
[5] Yang J, Bai B, Ke H, Cui Z, Liu Z, Zhou Z, Xu H, Xiao J, Liu Q and Li H 2021 *Opt. Laser Technol.* **144** 107431. https://doi.org/10.1016/j.optlastec.2021.107431.
[6] Wang K, Du D, Liu G, Pu Z, Chang B and Ju J 2020 *Corros. Sci.* **176** 108922. https://doi.org/10.1016/j.corsci.2020.108922.
[7] Lampa C and Smirnov I 2019 *J. Laser Appl.* **31** 022511-1. https://doi.org/10.2351/1.5096142.
[8] Li L, Shen F, Zhou Y and Tao W 2019 *J. Laser Appl.* **31** 022511-1. https://doi.org./10.2351/1.5094378.
[9] Zhang X, Sun Y, Yu G, Chen C, Ren X and Chen L 2023 *Opt. Laser Technol.* **158** 108838. https://doi.org/10.1016/j.optlastec.2022.108838.
[10] Papatheodorou T 2005 *Sealing Technol.* **April: 5-10**. https://doi.org/10.1016/S1350-4789(05)00600-8
[11] Hutchings IM 1996 *Tribology: Friction and Wear of Engineering Materials* (fourth ed. Arnold, London)
[12] Deuis RL, Yellup JM and Subramanian C 1998 *Compos. Sci. Technol.* **58** 299

[13] Pierson HO 1996 *Handbook of refractory carbides and nitrides: properties, characteristics, processing, and applications* (first ed. Noyes Publications, USA)
[14] Tassin C, Laroudie F, Pons M and Lelait L 1995 *Surf. Coat. Technol.* **76-77** 450. https://doi.org/10.1016/0257-8972(95)02613-4.
[15] Kim TH and Kim BH 1992 *J. Mater. Sci.* **27** 2967. https://doi.org/10.1007/BF01154107.
[16] European Commission 2020 *Study on the EU's list of Critical Raw Materials*
[17] Liu M, Jiang H, Chang G, Xu Y, Ma F and Xu K 2022 *Mater. Lett.* **325** 132823. https://doi.org/10.1016/j.matlet.2022.132823.
[18] Du C, Hu L, Ren X, Li Y, Zhang F, Liu P and Li Y 2021 *Surf. Coat. Technol.* **424** 127617. https://doi.org/10.1016/j.surfcoat.2021.127617.
[19] Yong Z, Chang L, Jiang S, Xie D, Xing F, Shen H, Shen L and Tian Z 2023 *Opt. Laser Technol.* **158** 108837. https://doi.org/10.1016/j.optlastec.2022.108837.
[20] Lou LY, Zhang Y, Jia YJ, Li Y, Tian HF, Cai YJ and Li CX 2020 *Surf. Coat. Technol.* **392** 125697. https://doi.org/10.1016/j.surfcoat.2020.125697.
[21] Arabi Jeshvaghani R, Jaberzadeh M, Zohdi H and Shamanian M 2014 *Mater. Design* **54** 491. http://dx.doi.org/10.1016/j.matdes.2013.08.059.
[22] Arabi Jeshvaghani R, Shamanian M and Jaberzadeh M 2011 *Mater. Design* **32** 2028. doi:10.1016/j.matdes.2010.11.060.
[23] Li Y, Dong S, Yan, S, Li E, Liu X, He P and Xu B 2019 *Mater. Charact.* **151** 273
[24] ASM International 2021 *Brake disc coating using HICLAD,* https://www.asminternational.org/web/tss/news/-/journal_content/56/10180/43705849/NEWS
[25] Tuominen J 2009 *Engineering Coatings by Laser Cladding – The Study of Wear and Corrosion Properties* (Tampere University of Technology)
[26] https://inewsteel.com/product/S355-Alloy-Steel.html (accessed 7.4.2023)
[27] Pribulova A, Futas P and Pokusova M 2020 *Mater. Sci. Forum* **998** 42. https://doi.org/10.4028/scientific.net/MSF.998.42.
[28] https://specialmetals.com/documents/technical-bulletins/inconel/inconel-alloy-625.pdf (accessed 7.4.2023)
[29] Näkki J 2018 *Properties of alloy 625 claddings made with laser and CMT methods* (Tampere University of Technology)

IOP Conf. Series: Materials Science and Engineering 1296 (2023) 012038 doi:10.1088/1757-899X/1296/1/012038

Blue diode lasers – Understanding and influencing melt pool dynamics in copper

L. Heine[1], M. Schlett[1], S. Britten[1], A. Arndt[1], C. Spurk[2], M. Hummel[3], F. Beckmann[4], J. Moosmann[4]

[1] Laserline GmbH, Mülheim-Kärlich, Germany
[2] RWTH Aachen University, Chair for Laser Technology LLT, Aachen, Germany
[3] Fraunhofer-Institute for Laser Technology ILT, Aachen, Germany
[4] Helmholtz-Zentrum Hereon, Institute of Materials Physics, Geesthacht, Germany

luisa-marie.heine@laserline.com

Abstract. Within the last years an increasing demand of copper components was observed, especially in the fields of energy transition, e-mobility, or digitalization. For joining those components, laser welding has become more and more popular. Infrared lasers are commonly used for metal welding. They are characterized by a low absorptivity in solid copper at room temperature. In contrast, lasers in the blue wavelength spectrum are characterized by a significantly higher absorptivity. This facilitates the initial energy input into the material and makes the welding process more efficient. Further, in the liquid state, the absorptivity change in copper is much smaller for the blue wavelength. This results in a continuous heating of the material, a more stable melt pool and a lower amount of spatter. One supplier of lasers in the visible wavelength spectrum is Laserline. In 2023, they launched a 4 kW blue diode laser, the currently highest available power for this wavelength. It makes accessible a broad variety of applications. But the scientific investigations for the blue wavelength at a power of several kilowatt are still limited. It is the paper's purpose to extend them by analysing welding of copper and its alloys with a blue diode laser. One the one hand, synchrotron trials were performed allowing a look into the material during welding. A special focus was on the transition between heat conduction and deep penetration welding as well as the keyhole behaviour. On the other hand, conventional welding trials were performed analysing the impact of beam shaping on the melt pool dynamics.

1. Introduction

The energy transition, the e-mobility and the digitalization have contributed to an increasing demand for copper components which require the combination of high electrical conductivity and a certain mechanical strength. Those components pass through a broad range of production steps, among others joining processes if an electrical as well as mechanical connection is required. In this context, laser welding has become more and more popular. As contact-free joining method, e.g. in comparison to resistance welding, complex joining configurations can easily be realized. Further, the high energy density of a laser allows a well-defined heat input so that, e.g. in comparison to brazing, the warming of the whole component can be minimized. Last, laser processes are characterized by comparatively low process times enabling high production output. The basis for a fast progress in energy transition, e-mobility, and digitalization.

For metal welding, infrared lasers are commonly used. In the case of pure copper, their reflectivity measures over 90 % at room temperature [1]. Therefore, special effort is necessary to achieve a high-quality weld with sufficient penetration depth, e.g.: minimizing the spot size, power modulation or significantly increasing the laser power and process speed, as described in detail by Heider [2]. Figure 1 illustrates the absorptivity of different metals at room temperature and varying wavelengths. In the infrared region at around 1000 nm, the absorptivity of pure copper or its alloy CuSn6 measures about 10 %. Comparing this value with a visible wavelength, like e.g. blue at about 450 nm (highlighted blue in Figure 1), a significant increase to approx. 70 % can be observed. This benefit was subject of several scientific investigations within the last years. Hummel et al [3] measured the absorptivity of copper Cu-ETP during heat conduction welding for the infrared, green, and blue wavelength. They confirmed an increase in the absorption coefficient with decreasing wavelength. This results in a lower power necessary to achieve a required penetration depth under comparable boundary conditions as Punzel et al [4] determined. They compared an infrared single- as well as 2-in-1-spot setup with a green single-spot setup for copper Cu-OF. The better absorptivity of the visible wavelength results also in a lower threshold for deep penetration welding and a lower sensitivity for surface oxidation. This was observed by Engler et al [5]. They compared the infrared with the green wavelength and identified differences in the weld's cross-sectional geometry indicating a varying keyhole behaviour.

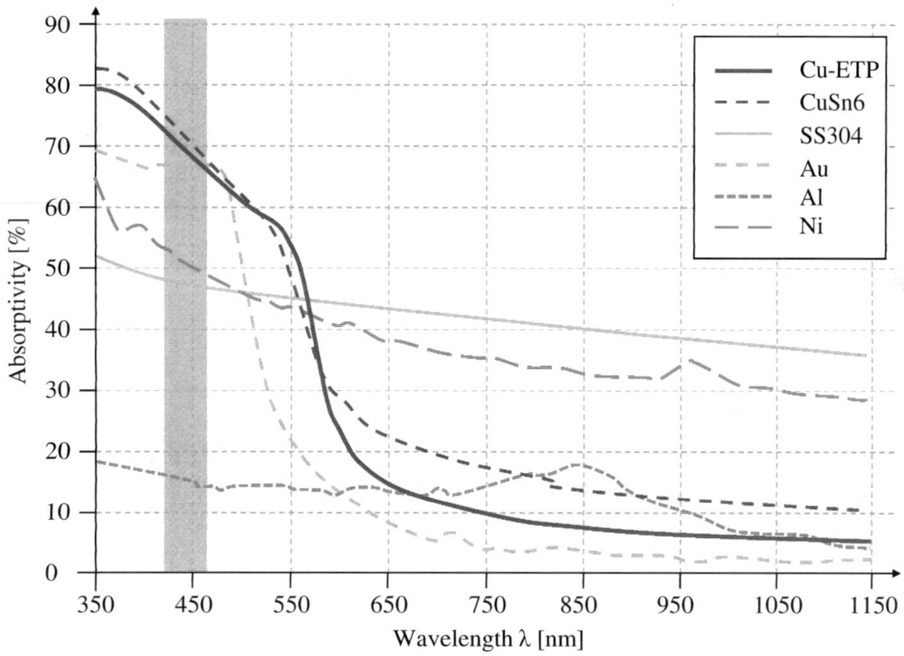

Figure 1: Absorptivity of different metals at room temperature [3]

Due to the visible wavelength's benefits when welding copper, Laserline started to develop a blue diode laser emitting at a wavelength of 445 ± 20 nm. In 2018, they launched the first system with a power of 0.5 kW after fibre. As copper is one of the best heat-conducting metals, a steady increase of the available laser power was an important goal: 1.5 kW were reached in 2019, 2 kW were launched in 2020. The corresponding beam quality was improved, too. In 2022, two new systems were introduced with a power after fibre of 1.5 respectively 3 kW and a corresponding beam quality of 20 respectively 30 mm·mrad. In 2023, a 4 kW system was launched for the latter beam quality. This power level makes accessible a broad variety of applications, reaching from brazing over additive manufacturing to heat conduction and deep penetration welding, as described in detail by Heine et al [6]. However, the scientific investigations for the blue wavelength at a power of several kilowatts are still limited and the corresponding material behaviour is not completely understood, yet.

Within the scope of a public funded research project the material behaviour beneath the surface during welding a copper alloy was investigated for the blue wavelength by synchrotron trials at the end of 2022. A special focus was on the transition between heat conduction and deep penetration welding as well as the keyhole behaviour. These insights are the basis for an optimisation of the welding process, especially as the heat conductivity of copper decreases approx. by a factor of two when reaching the melting point [7]. Another focus of investigation was the melt pool dynamic increasing with increasing penetration depth and resulting in a rougher surface appearance. In case of the infrared wavelength and for copper [4] or steel/aluminium [8] it was shown that beam shaping can improve the welding behaviour and the corresponding result. There is the question if beam shaping can also positively influence the melt pool dynamics in case of the blue wavelength. It is the paper's purpose to share these latest insights into welding of copper and its alloys with a blue diode laser and to make the benefit of the blue wavelength's absorptivity for highly reflective metals better accessible.

2. Experimental setup

All investigations were performed with blue diode lasers from Laserline on copper and its alloys. First, the synchrotron experiments, performed by LLT and ILT from Aachen together with the Helmholtz Zentrum HEREON - Institute of Materials Physics from Geesthacht, will be explained. Second, the conventional welding trials with beam shaping, performed at Laserline, as well as the corresponding metallographic analysis will be described.

2.1. Synchrotron experiments

The experimental setup for this investigation is based on the usage of the HEREON high energy beamline P07 (EH4) of Petra 3 at Deutsches Elektronen Synchrotron DESY in Hamburg, Germany. To observe the welding process, the X-ray radiation is transmitting through the metal sample and is converted into visible light by a GaGG:Ce (Gadolinium Aluminium Gallium Garnet(Ce)) scintillator. Using an ixCamera iSpeed 727, the intrinsic glow of the scintillator screen stimulated by the X-rays is recorded. The pictures were taken with 1600 x 900 pixels, with a frame rate of 5 kHz and 1 ms exposure time. The corresponding parameters are given in Table 1.

Table 1: Parameters and properties of the synchrotron imaging setup

Parameter	Unit	P07 Petra 3
Operation mode	[-]	High Beta
Photon energy	[keV]	89
Synchrotron beam area	[mm2]	2x2
Scintillator material	[-]	GaGG:Ce
Scintillator size	[mm2]	11x11
Scintillator thickness	[μm]	1200

The setup includes a high-speed axis to apply the welding speed, a scanning system with optics and the laser beam sources as shown in Figure 2. The trials in the blue wavelength spectrum were performed with the diode laser LDMblue 1800-30 and with a spot size of 0.36 mm in diameter. The samples were made of the copper alloy CuSn6. To analyse the transition between heat conduction and deep penetration welding the first experiments were performed with a constant welding speed of 100 mm/s while increasing the power from 0.5 to 1.75 kW. To investigate the keyhole behaviour during deep penetration welding the second trials were performed at a constant welding speed of 250 mm/s as well as a constant power of 1.75 kW. Table 2 summarizes the experimental boundary conditions.

Table 2: Welding parameters of the synchrotron trials

Trial	Laser source	Spot diameter	Power	Welding speed	Material
1	LDMblue	0.36 mm	0.5 - 1.75 kW	100 mm/s	CuSn6
2	1800-30		1.75 kW	250 mm/s	

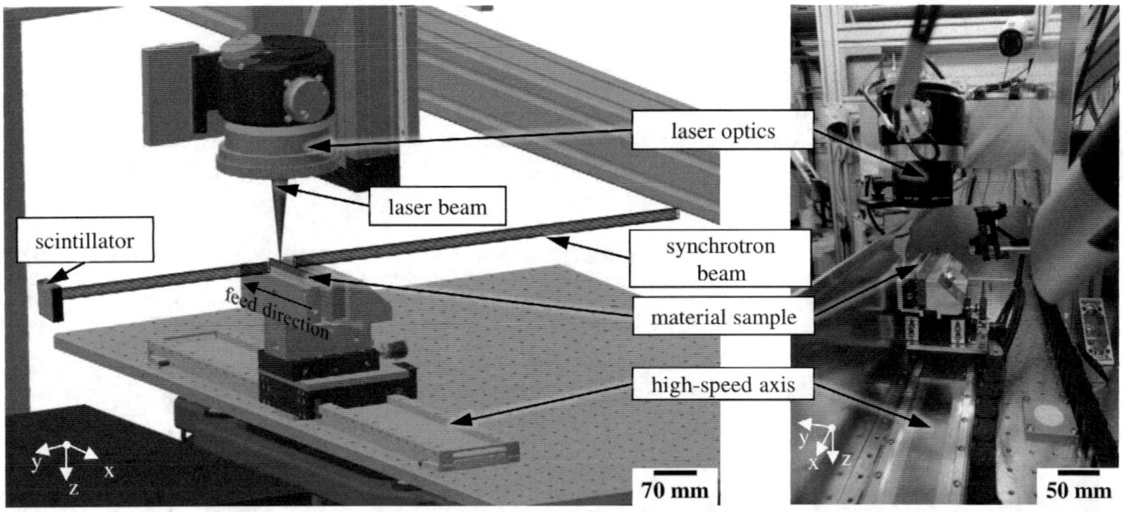

Figure 2: Experimental setup during synchrotron trials [9]

2.2. Welding trials with beam shaping

The welding experiments with beam shaping were performed at Laserline in Mülheim-Kärlich. The blue diode laser LDFblue 3000-30 was used with a circular spot size of 0.6 mm in diameter. In case of beam shaping, this centre spot is superimposed by a variable rectangular spot. This was done by using a so-called spot-in-spot module enabling an adjustable intensity distribution between the inner and outer spot. The trials were performed on 2 mm thick copper Cu-ETP and with a constant power at the sample of 2.9 kW. The welding speed was varied systematically from 33 up to 200 mm/s. The intensity distribution between the inner and outer spot was changed systematically, too. To analyse its impact on the melt pool dynamics, the welding process was documented with a high-speed camera. Table 3 summarizes the experimental boundary conditions.

Table 3: Welding parameters of trials with beam shaping

Laser source	Centre spot diameter	Power	Welding speed	Material
LDFblue 3000-30	0.6 mm	2.9 kW	33 - 200 mm/s	Cu-ETP

2.3. Metallographic analysis

The welds surfaces were documented with a Axio Zoom.V16 light microscope from Zeiss. Cross-sections were prepared by cutting and grinding. They were etched with Adler acid to make the welds visible. The penetration depth as well as the weld's width at the top were determined by means of the already mentioned microscope.

3. Results

First, the results of the synchrotron experiments will be shown with a focus on the transition between heat conduction and deep penetration welding as well as the keyhole behaviour during the latter one. Second, the welding trials' results with beam shaping will be presented.

3.1. Synchrotron experiments

Figure 3 shows extracts of a high-speed video documenting the transition between heat conduction and deep penetration welding (refers to trial 1 in Table 2). At a constant welding speed of 100 mm/s, the power was increased linearly from 0.5 to 1.75 kW. Approx. after 56 ms (Figure 3b), corresponding to an intensity of $6.3 \cdot 10^5$ W/cm², a small vapor capillary is built marking the transition from heat conduction to deep penetration welding. After approx. 88 ms (Figure 3c), corresponding to an intensity of $7.1 \cdot 10^5$ W/cm², a keyhole of about 0.3 mm is clearly visible. This keyhole increases with increasing power reaching about 1.2 mm at the maximum power of 1.75 kW (Figure 3e). In contrast to the lens-shaped geometry at the beginning of deep penetration welding (Figure 3b), the keyhole becomes steadily cone-shaped with a static front keyhole surface. However, turbulences within the subsequent melt pool increase and the formation of single pores within the weld (Figure 3d) is observed. After 503.8 ms, when the power is withdrawn, the vapor capillary collapses leaving most probably a larger pore in the solidifying material (Figure 3f).

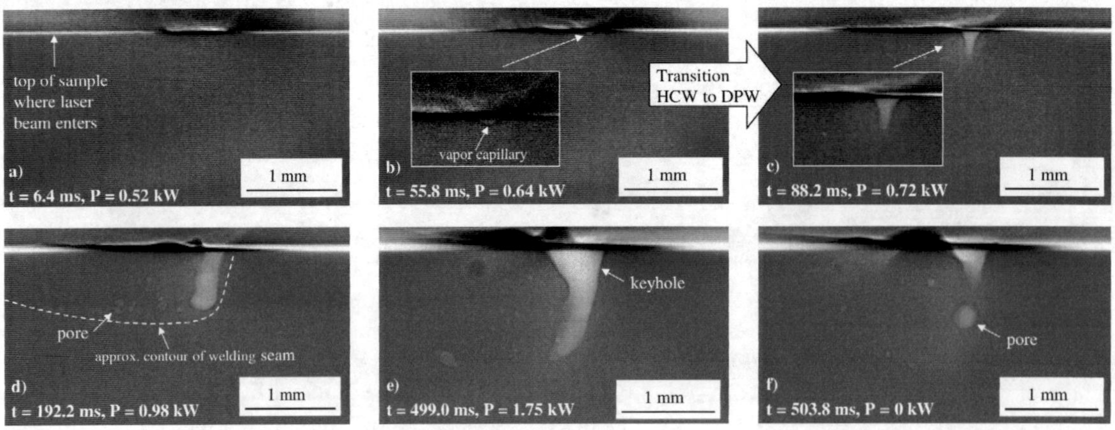

Figure 3: Extracts of high-speed video during synchrotron trials illustrating the transition between heat conduction welding (HCW) and deep penetration welding (DPW). Time *t* and power *P* are given at the left bottom of each picture. The welding speed was constant at 100 mm/s.

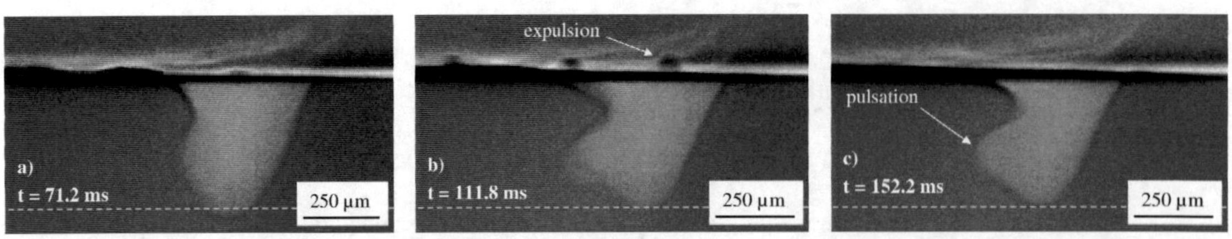

Figure 4: Extracts of high-speed video during synchrotron trials illustrating the keyhole behaviour during deep penetration welding. The welding speed was constant at 250 mm/s. The power measured 1.75 kW.

Figure 4 shows extracts of a high-speed video illustrating deep penetration welding at a constant welding speed of 250 mm/s and a power of 1.75 kW (refers to trial 2 in Table 2). The corresponding intensity measures $5.1 \cdot 10^6$ W/cm². In comparison to Figure 3 (e), the penetration depth is less due to the higher welding speed at identical power level. The keyhole is again characterized by a static front keyhole surface. Its geometrical shape is more stable and less pores are observed. The pictures shown are separated by approx. 40 ms within the welding process. The penetration depth is almost identical as indicated by the white dotted line. It is slightly influenced by pulsations observed at the vapor capillary's bottom (Figure 4c). A reduction of those is most often followed by a small expulsion at the weld's top, as illustrated in the central picture (b) of Figure 4.

3.2. Welding trials with beam shaping

By means of a so-called spot-in-spot module, beam shaping within the optics was realized by superimposing a circular inner and a rectangular outer spot. The module allows an adjustable intensity distribution between both spots. Examples, measured with the setup used during the trials, are given in Figure 5. The percentage values (here and in the following) refer to the relative power of the inner spot highlighted by a black dotted line in the pictures. The remaining relative power (e.g. 25 % in case of 75 % power in inner spot) belongs consequently to the outer spot. When increasing this value respectively reducing the power of the inner spot, a surrounding rectangular becomes more and more pronounced, as illustrated in the central and third picture of Figure 5. These and other intensity distributions were used during the trials whose results are given in the following.

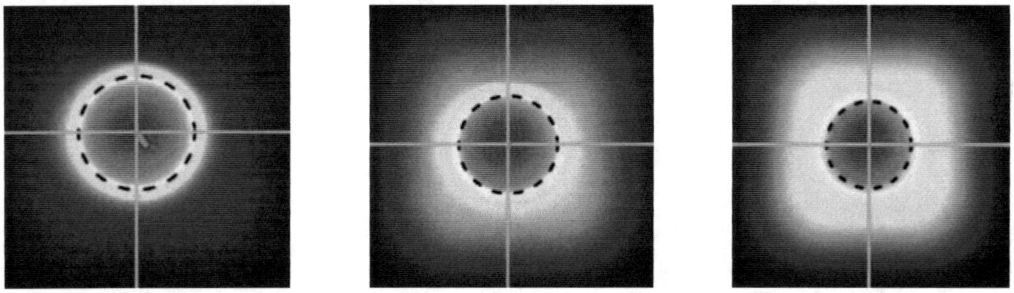

75 %   60 %   45 %

Figure 5: Schematic illustration of different intensity distributions of the applied blue laser spot (percentage values refer to the relative power in the inner spot)

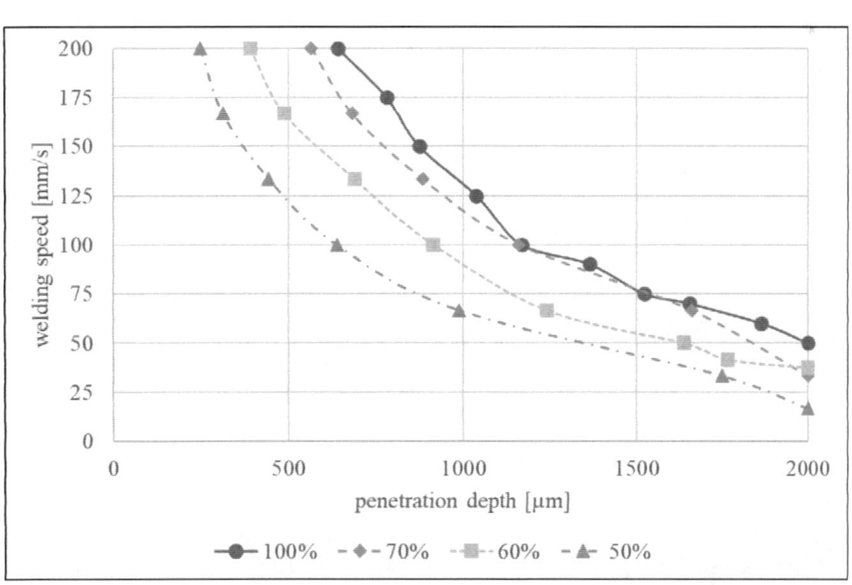

Figure 6: Impact of the relative power in the inner spot (percentage values) on the penetration depth

IOP Conf. Series: Materials Science and Engineering 1296 (2023) 012038 doi:10.1088/1757-899X/1296/1/012038

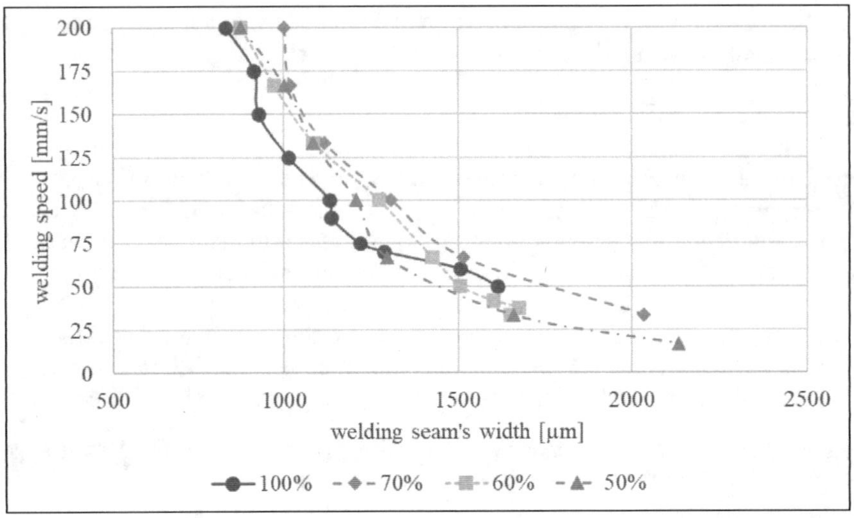

Figure 7: Impact of the relative power in the inner spot (percentage values) on the weld's width

Figure 6 and Figure 7 show the relative penetration depth respectively weld's width for varying intensity distributions and welding speeds. The power was constant at 2.9 kW. The percentage values given refer to the relative power in the inner spot. With decreasing welding speed, the penetration depth as well as the weld's width increase. In contrast, a reduction of the power in the inner spot causes a loss in penetration depth, being most pronounced for a power less than 70 %. The weld's width increases slightly with increasing power in the outer spot. This change is not as pronounced as the one of the penetration depth. It seems to disappear within the scatter below a welding speed of 75 mm/s.

Although the impact of beam shaping is almost not visible in the weld's width, it is clearly visible when looking at the qualitive surface appearance. Figure 8 juxtaposes pictures of the weld's surface for varying welding speeds and intensity distributions. At the highest welding speed of 167 mm/s, all beam shaping options show a smooth welding result. For the lower welding speeds of 100 respectively 33 mm/s the welds are characterised by small spatter, a rougher surface and more irregularities in case of 80 % power in the inner spot. These imperfections become less when reducing the power of the inner spot to 70 %, and they disappear when reducing it to 60 %.

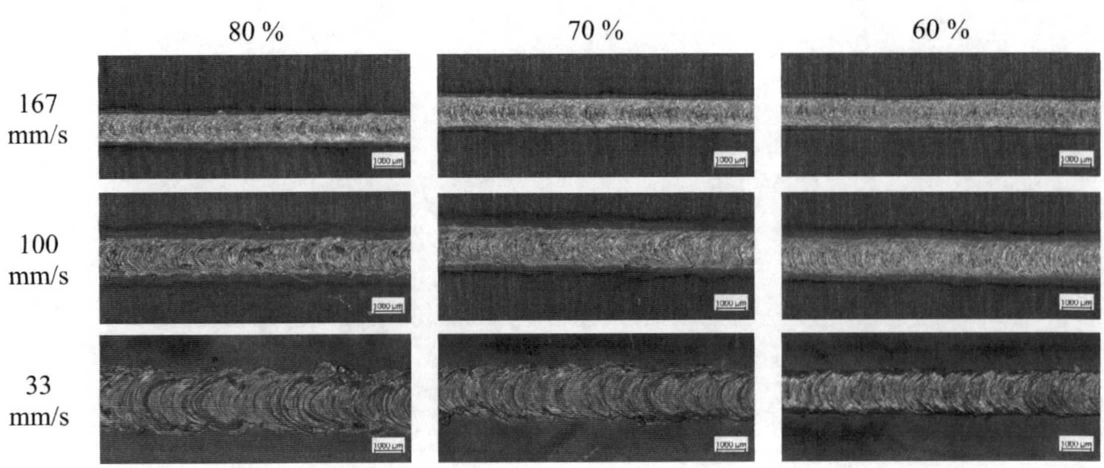

Figure 8: Weld's surface at varying welding speed and intensity distribution (percentage values at the top refer to the relative power in the inner spot)

IOP Conf. Series: Materials Science and Engineering 1296 (2023) 012038 doi:10.1088/1757-899X/1296/1/012038

Figure 9 shows extracts of the corresponding high-speed videos taken at a welding speed of 33 mm/s. In accordance with the qualitive surface appearance the process shows a turbulent melt pool with a dancing keyhole and spatter in case of 80 % power in the inner spot. The dynamics reduce and the keyhole becomes more and more stable when decreasing the power in the inner spot. In case of 60 % a stable and almost spatter-free process can be observed.

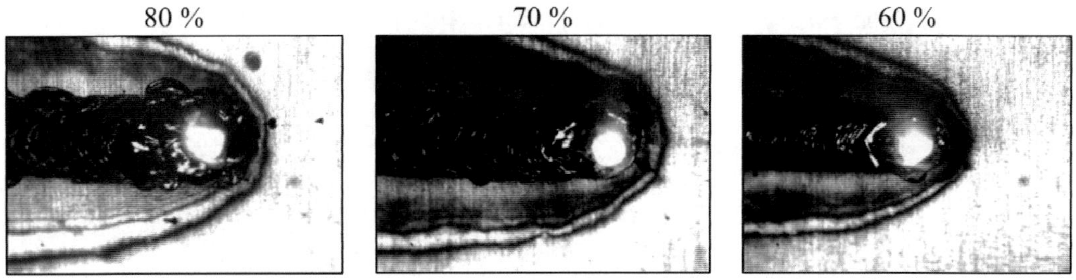

Figure 9: Extracts of high-speed videos at a varying intensity distribution (percentage values at the top refer to the relative power in the inner spot) and at a constant power of 2.9 kW as well as a constant feed speed of 33 mm/s

Figure 10 illustrates the corresponding cross-sections for a power of 80 respectively 60 % in the inner spot and a varying welding speed. As already visible in Figure 6, the penetration depth increases with decreasing welding speed and increasing power in the inner spot. Comparing the penetration depth for a constant welding speed, the difference between the two intensity levels becomes smaller with lower welding speed:

- 167 mm/s: difference of ~ 50 %,
- 100 mm/s: difference of ~ 30 %,
- 33 mm/s: difference of ~ 15 %.

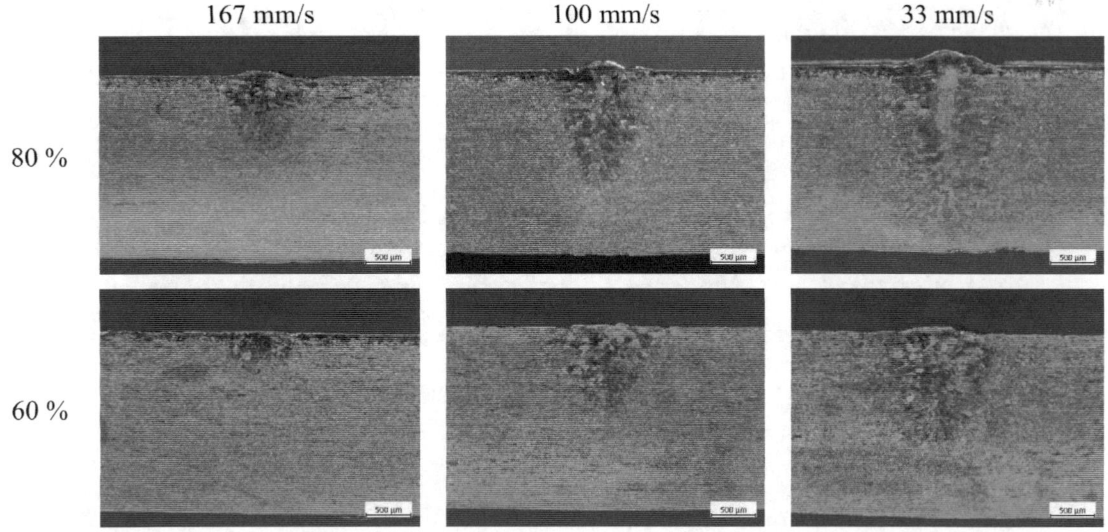

Figure 10: Welds' cross-sections at a varying welding speed and intensity distribution (percentage values at the left refer to the relative power in the inner spot)

IOP Conf. Series: Materials Science and Engineering 1296 (2023) 012038 doi:10.1088/1757-899X/1296/1/012038

4. Discussion

By means of the synchrotron experiments, the transition from heat conduction to deep penetration welding could be observed. For the welding speed of 100 mm/s, it took place at an intensity in the range of $6\ldots7\cdot10^5$ W/cm². Increasing the intensity on basis of the laser power led to an increase of the vapor capillary, corresponding turbulences, and pores. The latter ones are more frequently observed at the bottom of the capillary where a slight pulsation is visible, too. It seems that part of the capillary separates at the bottom during these pulsations. If the vapor cannot reach the weld's top surface before complete solidification it remains as pores within the weld.

Less pores are observed during the synchrotron trials with higher welding speed (250 mm/s) or during the trials with beam shaping. In the first case, the aspect ratio of the vapor capillary is lower. This seems to be beneficial for the vapor evaporation as small expulsions, following pulsations, leave the keyhole at the top instead of building pores within the material. Similar could be valid for the second case, the trials with beam shaping. The corresponding cross-sections are almost pore free even when welding with a high power in the inner spot or when turbulences of the melt pool are observed. However, during those trials, the spot size (of the inner spot) was almost twice the size of the synchrotron trials. The keyhole at the top is significantly larger and vapor can probably evaporate more easily at the top instead of searching for a way through the material.

During the trials with beam shaping, it was shown that a reduction of the inner spot's power, when using a spot-in-spot module, to approx. 60…70 % can calm down the melt pool dynamics. Although the weld's width is not affected significantly, the keyhole is stabilised within the melt pool. Comparing the extracts from the high-speed videos, it seems that especially the melt in front of the keyhole is slightly larger and more stable when shifting more power in the outer spot. In those videos, spatter formation most often takes place at the front of the weld. Stabilising this area seems to have a positive impact on spatter formation and the resulting surface appearance of the weld. The impact seems such pronounced that the loss in penetration depth, when shifting more power in the outer spot, can be compensated by reducing the welding speed without the risk of worsening again the welding process or the weld's surface appearance. This is illustrated in Figure 11 showing a weld with full penetration. The welding speed was reduced to 17 mm/s and the power in the inner spot measured 50 %.

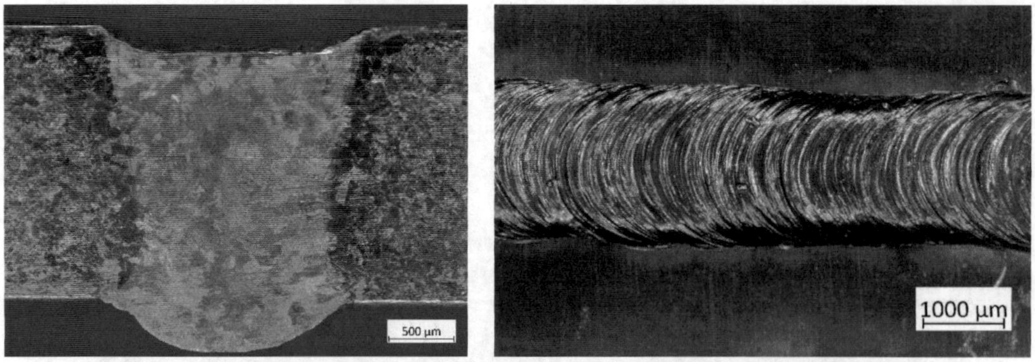

Figure 11: Bead on plate welding (2 mm Cu-ETP) with full penetration and high quality sufarce appearance using beam shaping

5. Summary

It was the paper's purpose to investigate welding of copper and its alloys with a blue diode laser. Synchrotron trials were performed allowing a look into the material during welding. A special focus was on the transition between heat conduction and deep penetration welding as well as the vapor capillary's behaviour. Further, the impact of beam shaping on the melt pool dynamics and on the weld's surface appearance was investigated.

It was observed that there exists a very smooth transition between heat conduction and deep penetration welding when using the blue wavelength. However, the turbulences increase with increasing intensity respectively penetration depth. Beam shaping can help to improve the melt pool dynamics, to minimize spatter and to optimize the surface appearance. Further, it seems beneficial to limit the aspect ratio of the vapor capillary to facilitate vapor evaporation and to reduce pore formation.

The synchrotron respectively beam shaping trials were performed with different spot sizes and at different power levels. For better comparison, it would be worth in future to complement those investigations under comparable conditions. Further, it could be of special interest to repeat the beam shaping trials in a synchrotron environment to achieve a more comprehensive understanding of the events inside the material.

Acknowledgement
The presented synchrotron investigations were carried out in cooperation with RWTH Aachen University within the framework of the Collaborative Research Centre SFB1120-236616214 "Bauteilpräzision durch Beherrschung von Schmelze und Erstarrung in Produktionsprozessen" and funded by the Deutsche Forschungsgemeinschaft e.V. (DFG, German Research Foundation). The presented investigations were carried out in cooperation with DESY in Hamburg at PETRA III, and we would like to thank F. Beckmann and J. Moosmann for their assistance in using P07 EH4. We would like to thank all people involved for their support.

References
[1] Spisz E W, Weigand A J, Bowman R L and Jack J R 1969 Solar absorptances and spectral reflectances of 12 metals for temperatures ranging from 300 to 500 K *Nasa Technical Note* **D-5353** 1–21
[2] Heider A *Erweitern der Prozessgrenzen beim Laserstrahlschweißen von Kupfer mit Einschweißtiefen zwischen 1 mm und 10 mm* (München: Herbert Utz Verlag GmbH) pp 132–1
[3] Hummel M, Schüler C, Häusler A, Gillner A and Poprawe R 2020 *J. of Advanced Joining Processes* **1** 100012
[4] Punzel E, Hugger F, Dörringer R, Dinkelbach T L and Bürger A 2020 *Procedia CIRP* **94** 587–97
[5] Engler S, Ramsayer R and Poprawe R 2011 *Physics Procedia* **12** 339–46
[6] Heine L M, Schlett M, Ocylok S and Rütering M 2022 *Cooper Alloys Conference Proceedings* (Düsseldorf: Deutsches Kupferinstitut, SWERIM, RISE) pp 24–7
[7] Blom A, Dunias P, Engen P, Hoving W and Kramer J 2003 *Proceedings of SPIE* 4977 (United States, San Jose: High-Power Lasers and Applications) pp 493–507
[8] Schäfer J, Luft A, Ullmann C and Krause V 2019 *Fügen und Konstruieren im Schienenfahrzeugbau* (Halle: 13. Fachtagung DVS) pp 66–70
[9] Hummel M, Meier C, Olowinsky A, Gillner A, Beckmann F, Moosmann J and Häfner C 2023 *Proceedings of SPIE 12414* (United States, San Francisco: High-Power Laser Materials Processing: Applications, Diagnostics, and Systems XII)

Fatigue performance of fiber laser fusion cut edges on thick plates

Kalle Lipiäinen[1], Antti Ahola[1], Shahriar Afkhami[1] and Tuomas Skriko[2]

kalle.lipiainen@lut.fi

[1] Laboratory of Steel Structures, Lappeenranta-Lahti University of Technology LUT, P.O. Box 20, FI-53851 Lappeenranta, Finland
[2] Laboratory of Welding Technology, Lappeenranta-Lahti University of Technology LUT, P.O. Box 20, FI-53851 Lappeenranta, Finland

Abstract. Fatigue performance of cut edges is an important factor in structural design, especially when geometrical stress concentration factors (SCFs) are present. High power laser cutting systems have enabled efficient and fast fusion cutting with nitrogen as an assistant gas. The fusion cutting induces recast layer, which features crack-like defects close to cut edge. In this study, plates with thickness $16 \leq t \leq 30$ mm have been fatigue-tested. A 10 kW fiber laser cutting system was used to cut plates with nitrogen as an assistant gas. High strength steel (HSS) S690 specimens included unnotched coupon specimens and welded T-profile with longitudinal gusset in post-weld treated conditions, where laser cut edge was found critical. Fatigue performance of S690 cut edges were found close to 170 MPa m = 3 when SCFs of welded specimens were included on analysis. Local properties at the cut edge including hardness measurements and scanning electron microscope (SEM) fractography. Fractography obtained local quality use with theory of critical distances -based fatigue strength assessment was demonstrated.

1. Introduction

Cut edges are important part of welded structures and the edges may become fatigue critical when the welds are located on low stressed areas or post-weld treatments are applied [1]. Longitudinal welds and post-weld treated (PWT) transverse welds and even longitudinal gussets may have higher fatigue performance than unnotched cut edges. The fatigue performance of cut edges is typically influenced by stress concentration factors (SCFs), which may be mild cut geometries or more significant holes or cut outs.

Laser cutting has been developing fast due to decreasing cost and increasing power of laser sources. Figure 1 illustrates cutting speed range with different cutting methods and laser power. First, nitrogen became popular assistant gas at medium thickness 6 mm $\leq t \leq$12 mm, which is reasonable with around 10 kW laser systems, then usable power rapidly increased up to 30 kW, which made increased efficient usability range up to t = 40 mm plates.

Levichev et al. [2] analyzed fiber laser cutting with oxygen and nitrogen by using coaxial camera and were able to monitor cutting quality during the cutting. Understanding fatigue performance and influencing factors enables to develop and optimize cutting systems and procedures. Lillemäe-Avi et al. [3] studied up to 17 mm cyclically loaded cut edges in shipyard applications. Fatigue performance of thick t = 50 mm oxy-fuel cut edges was studied by von Bock und Polach [4]. The research $t \geq 8$ mm thickness plate cut edges features, especially concerning fatigue performance of laser cut edges, is limited.

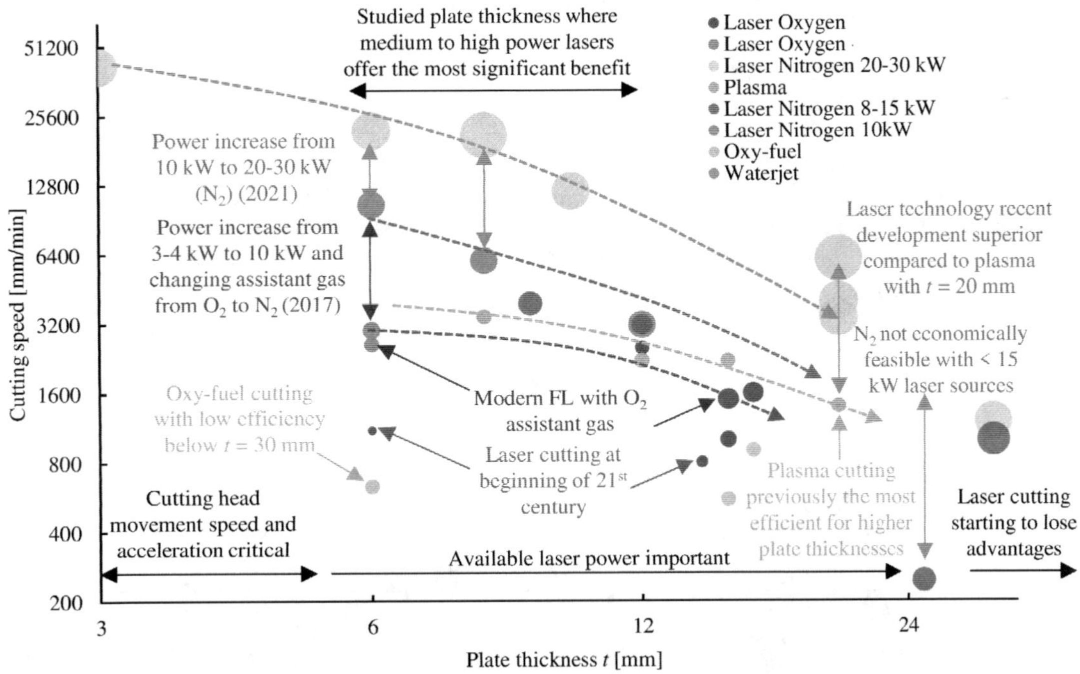

Figure 1. Insight on cutting speed and plate thickness on various cutting methods. Modified from [5].

Fatigue design for cut edges is covered e.g., by IIW recommendations [6], Eurocode 3 [7] and EN-ISO 13001 [8]. Fatigue performance is typically expressed as FAT values, that represent stress at $2 \cdot 10^6$ stress cycles. In experimental testing FAT value represent 50% survival probability whereas characteristic design values are usually estimated with 97.7% survival probability.

$$N_f = \left(\frac{FAT}{\Delta\sigma}\right)^m \cdot 2 \cdot 10^6 \qquad (1),$$

where $\Delta\sigma_{nom}$ is stress range [MPa], N_f is fatigue life [cycles], *FAT* is Allowable nominal stress range at $2 \cdot 10^6$ cycles [MPa] and m is slope parameter of SN-curve [-]

The IIW guideline recommends FAT 125 MPa $m = 3$ for cut edges in edges chamfered conditions, whereas EN-ISO 13001 features material strength and surface quality classes -based FAT classes from 140 MPa $m = 5$ to FAT 280 MPa $m = 5$. For $t = 16$ mm fiber laser fusion cut edges with typical surface quality (surface quality class 3) if Rz defined surface quality is below 100 µm, FAT 180 MPa $m = 5$ for surface quality class 3 is recommended. However, surface quality class 4 is not included and removing the Rz value of 100 µm requirement results FAT 140 MPa $m = 5$. When cutting thick plates with fiber laser (FL) N_2 method, low surface quality and possible crack-like defects below re-cast layer can be expected, and the influence on fatigue performance should be studied.

2. Specimens and testing arrangements

A 10 kW fiber laser cutting system was used to manufacture the specimens. A high-strength S690 steel grade t = 16 mm and t = 20 mm plates were used for fatigue test specimens with FL N_2 cut edges. 30 mm thick S355 plates were used to represent the maximum capacity of 10 kW fiber laser cutting system. The t = 16 mm plates were cut according to common industrial practice and t = 30 mm plates without quality requirements. The t = 30 mm plates were beyond the commercial services as the cutting speed and quality decreases significantly from t = 20 mm to t = 30 mm plates.

2.1. Experimental testing

The fatigue tests for coupon specimens were carried out with resonance-based Rumul Vibroforte 700 device with a test frequency of approximately 100 Hz. Coupon specimens with a width of 30 mm were used. SCF in welded specimens were obtained with finite element analysis (FEA). SCF value of 1.35 was obtained and used in fatigue performance analysis. The T-profile specimens were fatigue tested with servo-hydraulic 750 kN fatigue test rig with 1 Hz frequency and the detailed results are presented in [9]. Insight on test arrangement and specimens are given in Figure 2. The specimens were tested in after burr removal and some specimens with edges chamfered conditions. According to [10], the sharp burr after fiber laser fusion cutting will significantly reduce the fatigue performance.

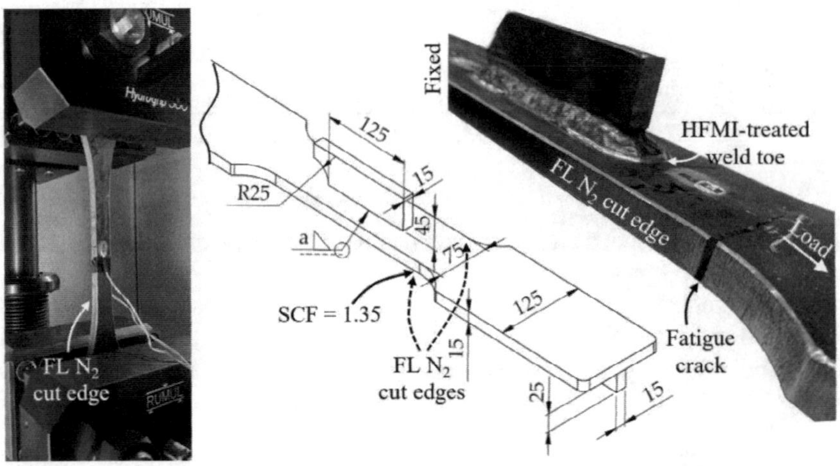

Figure 2. Unnotched specimen on test rig, welded T-profile with gusset and example of cut edge failure.

2.2. Surface roughness and integrity

Surface quality was measured with non-contact Keyence VR-3200 machine. EN 9013 surface quality ranges for t = 16 mm and t = 30 mm are 89 μm and 106 μm for range 3 to range 4 transition, respectively. The surface roughness values (RZ_5) at middle of the cut surface are given in Fig. 3 below corresponding cut edge overall image and supplemented with RZ_5 and RZ_{max} values of t = 30 mm cut edge. Hardness of S690 t = 16 mm base material was 280 HV (HV3 measurement), whereas HAZ was measured relatively constant 475 HV (HV0.2 measurement). Hardness lines and base material harnesses are given in Fig. 3.

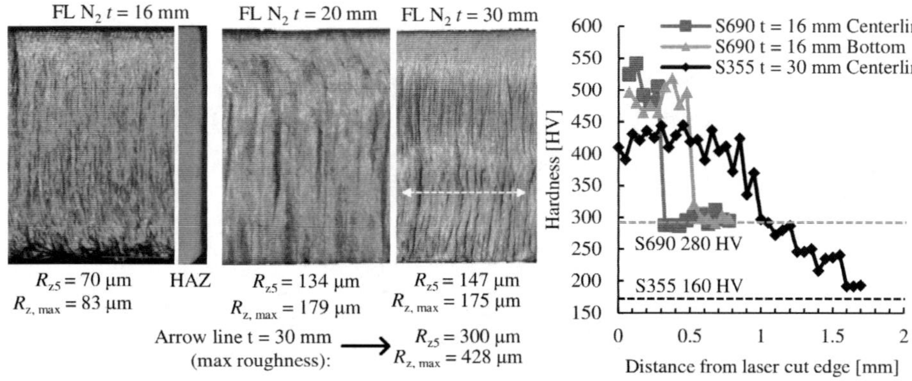

Figure 3. Surface quality of test specimens from middle of the cut edge (Images not in scale.) and hardness measurements results from t = 16 mm S690 (middle section and 2 mm from bottom surface) and t = 30 mm S355 (middle section).

Polished sections of t = 16 mm (Fig. 4a & 4b) and t = 30 mm (Fig. 4c-e) were studied to characterize cut edge features concerning geometrical aspect and discontinuity points, which are major contributors on the fatigue performance. Crack-like features, which are formed in solidification of molten material, were found on both plate thicknesses. Thick plates feature also quite sharp geometrical discontinuity point and high surface roughness points, which might cause fatigue failure.

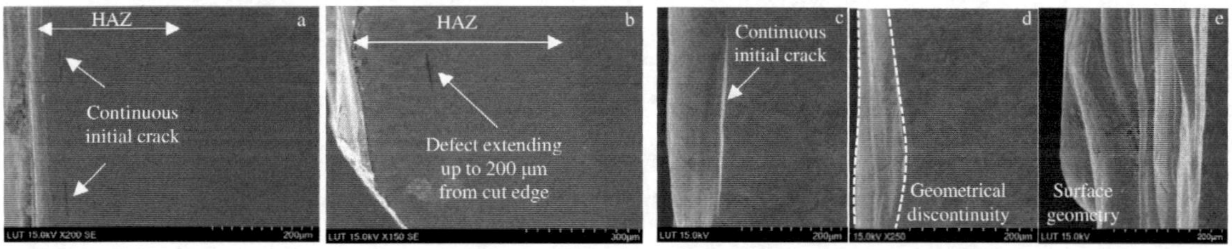

Figure 4. Polished section details from S690 t = 16 mm (a & b) and t =30 mm (c-e) edges.

3. Results

Figure 5 shows t = 16, 20 and 30 mm fatigue test results with red colored cross symbols, whereas blue and black markers illustrate fatigue test results for medium thickness ($6 \leq t \leq 12$ mm) FL N_2 cut edges studied in [10,11]. The experimental fatigue tests included only single t = 20 mm and t = 30 mm specimens. However, the specimens with continuous cut surfaces feature long critical length provide good estimation on fatigue performance and the critical defect causing fatigue failure. The results suggest that FAT 200 MPa m = 5 curve is conservative below $2 \cdot 10^5$ cycles. For low quality cut edges that include initial cracks, the slope parameter m = 3 is more suitable. The t = 16 mm S690 specimens follow FAT 170 MPa m = 3 curve.

A design curve of FAT 140 MPa m = 3 could be used for FL N_2 cut edges despite surface quality up to t = 30 mm plate thickness. Based on the test results, utilization of FAT 150 MPa m = 3 suggested in [11] could be extended for $t \geq 16$ mm FL N_2 cut edges when surface quality is controlled. Design recommendation based this study is summarized in Table 2 (Section 5).

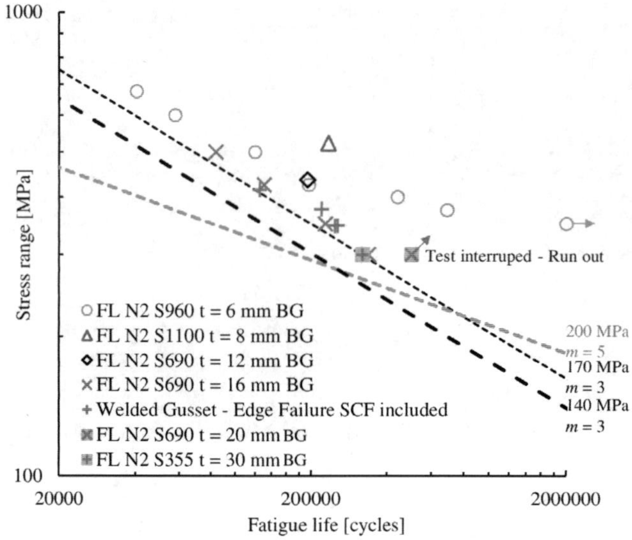

Figure 5. Fatigue test results.

Scanning electron microscope (SEM) is needed to characterize the local features influencing fatigue performance since cut edges have typically more significant differences in local features than welded joints. Fractography analyses revealed crack-like features on cut edges in a similar way than polished sections. Interestingly, a failure mode with major factor from surface quality (Fig. 6c and 6d) was found from t = 30 mm FL N_2 cut edge. In the authors' previous studies, fatigue crack has propagated from crack-like features formed under the recast layer.

It should be noted that single crack-like defect parallel to loading has limited influence on fatigue performance. The crack should be interacting with cut surface and transverse to loading to be fully effective. Consequently, crack-like features up to 200 μm are not fatigue critical (Fig 3b). Fatigue critical location was found in mid-section of cut surface, where 75 μm wide recast layer and surface roughness caused fatigue failure. Fig 6a. shows 150 μm equivalent crack as the surface roughness is only partly effective when fatigue notch factor is concerned (Table 1). The plate thickness has an influence on recast layer thickness, which explains the fatigue test results. However, the plate thickness is not the only influencing factor to recast layer thickness and fatigue performance. The kerf width (as a function of power, focal point, beam shape and cutting sped) and assistant gas flow are also significant parameters.

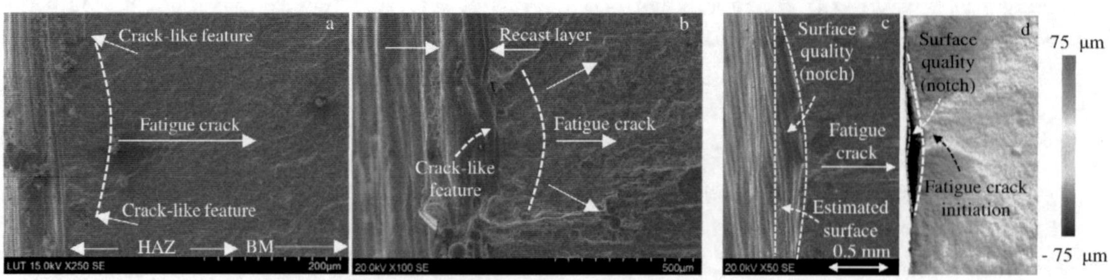

Figure 6. Fractography analyses revealing crack-crack like features due to solidification of molten material in (a) t = 16 mm S690 and (b) t = 30 mm S355. Surface quality feature (notch) of t = 30 mm S355 and surface quality connected notch causing fatigue crack initiation (c and d).

4. TCD and equivalent crack size -based fatigue strength assessment

Equivalent crack concept has been found good method to include quality parameter on fatigue strength assessment. Crack length to FEA can be extracted from SEM fractography analyses considering internal crack-like feature, which is transverse to loading and partially surface quality in some cases. The fatigue strength assessment procedure has been explained more detailed in [12]. The cut edge features in FL N_2 cutting are influenced by e.g., cutting system, nozzle type, gas flow, kerf width focal point position and chemical composition of the plate material. This explains why fatigue performance and features in fracture surfaces used as a base for modelling are not fully correlated with plate thickness. S1100 t = 8 mm was cut with relatively low cutting speed and material removal from the kerf was successful, whereas S960 t = 6 mm plate cutting was optimized to achieve high efficiency in cutting.

When equivalent crack is modelled with 2D FEA, inclusion of external geometries is easy to implement on fatigue strength assessment. Theory of critical distances (TCD) and equivalent crack is used to obtain the local stress. Major principal stresses at the distance of 0.04 mm from crack tip were extracted from FEA as illustrated in Fig. 7 with Siemens Femap performing linear analysis. Furthermore, 4R method can be used to correct the local mean stress level and include residual stresses in analysis like conducted in [13]. Due to high hardness and relatively low notch factor ($K_t \leq 2$), mean stress correction was neglected and stresses with plain TCD method is shown in Figure 7.

Table 1. Input parameters for TCD-based fatigue strength assessment method.

Specimen	Crack length [μm]	Hardness[HV]
S1100 t = 8 mm	30	550
S960 t = 6 mm	75	450
S690 t = 12 mm	75	475
S690 t = 16 mm	150	475
S355 t = 30 mm	200	425

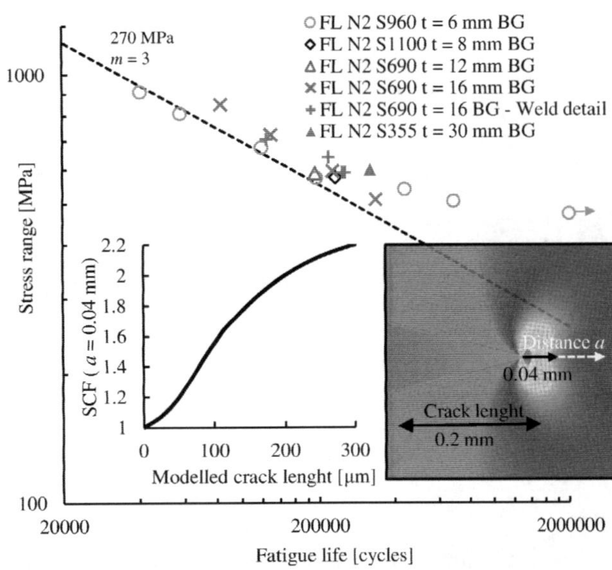

Figure 7. TCD method *S-N* curve, SCF with a = 0.04 mm in respect to crack length and an example of equivalent crack length FEA.

5. Discussion

The 10 kW fiber laser system´s cutting speed decreases significantly along plate thickness, but can be used in cutting $t = 30$ mm parts if needed. Utilizing ≥ 20 kW fiber laser setup provides reasonably fast cutting, which is competitive method compared to oxy-fuel and plasma cutting processes. Laser cutting has advantages in high dimensional accuracy and edge perpendicularity up to $t = 30$ mm steel plates, which is good for structures' assembly and welding phases.

The fatigue performance of FL N_2 cut edges was at expected level based on previous studies. Hot-dip galvanized unnotched S960 $t = 6$ mm specimens with nitrogen assisted laser cut edges were fatigue tested in [14] and FAT 144 MPa $m = 3$ was obtained with estimated initial crack length of 150 µm, which corresponded for HAZ width, due to liquid metal assisted cracking. Based on polished sections and fractography analyses, enough material should be removed (up to 0.5 mm), if fatigue performance is increasing with grinding, to remove sub-surface crack-like features. Surface quality should be evaluated in numerical values instead of thickness-based ranges to relate to fatigue performance.

More data on fatigue performance of thick steel plates is needed to increase reliability of the results. Studying the properties and performance of laser cut edges is more difficult than welded joints as high-power laser systems capable to cut $t \geq 30$ mm plates are still quite rare in industrial applications and thus, difficult to reach on fatigue testing.

Table 2. Design recommendations for laser cut edges without notches or with mild notches with SCF considered.

Cutting method	Detail	FAT class [MPa] and slope parameter *m [-]*	Note
Fiber laser (O_2 or N_2)	$t \leq 20$ mm $Rz_5 \leq 150$ µm	150 $m = 3$	Burr removed
	$t \leq 30$ mm $Rz_5 \leq 300$ µm	140 $m = 3$	
	$Rz_5 \leq 500$ µm	125 $m = 3$	

6. Conclusions

High-power fiber laser fusion cut with nitrogen as an assistant gas (FL N_2) was found well suitable method to cut thick plates. The plate thickness influenced on fatigue performance of FL N_2 cut edges due to material from the cut kerf becomes more difficult. However, cutting process and parameters have also an influence on local features and fatigue properties in addition to plate thickness.

Fatigue class FAT 125 MPa $m = 3$ can be used in fatigue performance assessment for unnotched laser cut edges practically without considering the surface quality ($Rz_5 \leq 500$ µm) when sharp burr is removed after cutting. Furthermore, it is possible to apply 150 MPa $m = 3$ design curve when $Rz_5 \leq 150$ µm is achieved even tough crack-like features would exist after fiber laser fusion cutting.

CRediT authorship contribution statement

Kalle Lipiäinen: Writing – original draft, Conceptualization, Investigation, Formal analysis. **Antti Ahola:** Writing – review & editing, Conceptualization, Software**. Shahriar Afkhami:** Writing – review & editing, Methodology and Data curation. **Tuomas Skriko:** Writing – review & editing, Conceptualization, Visualization.

Acknowledgements

The authors wish to thank Business Finland for funding via FOSSA project and SSAB Europe for providing S690 materials for experimental testing.

References

[1] Björk T, Ahola A, Skriko T, Lipiäinen K. On the backward design—an advanced concept for the design of welded structures for demanding applications. Welding in the World 2023. https://doi.org/10.1007/s40194-023-01483-w.
[2] Levichev N, Costa Rodrigues G, Vorkov V, Duflou JR. Coaxial camera-based monitoring of fiber laser cutting of thick plates. Opt Laser Technol 2021;136. https://doi.org/10.1016/j.optlastec.2020.106743.
[3] Lillemäe-Avi I, Liinalampi S, Lehtimäki E, Remes H, Lehto P, Romanoff J, et al. Fatigue strength of high-strength steel after shipyard production process of plasma cutting, grinding, and sandblasting. Welding in the World 2018;62:1273–84. https://doi.org/10.1007/s40194-018-0638-y.
[4] von Bock und Polach F, Kahl A, Braun M, Sperle J-O, von Selle H, Ehlers S. Analysis of the scatter in fatigue life testing of thick thermal cut plate edges. Ships and Offshore Structures 2022:1–14. https://doi.org/10.1080/17445302.2022.2035562.
[5] Lipiäinen K. Fatigue performance and the effect of manufacturing quality on the uncoated and hot-dip galvanized ultra-high-strength steel laser cut edges. 2022.
[6] Hobbacher AF. Recommendations for Fatigue Design of Welded Joints and Components. 2016. https://doi.org/10.1007/978-3-319-23757-2_8.
[7] Eurocode 3 — Design of steel structures — Part 1.9: Fatigue. Final Draft 2021.
[8] European Committee for Standardization. EN 13001-3-1: 2012 + a2: 2018 2018.
[9] Ahola A, Lipiäinen K, Lindroos J, Koskimäki M, Laukia K, Björk T. On the Fatigue Strength of Welded High-Strength Steel Joints in the As-Welded, Post-Weld-Treated and Repaired Conditions in a Typical Ship Structural Detail. J Mar Sci Eng 2023;11:644. https://doi.org/10.3390/jmse11030644.
[10] Lipiäinen K, Ahola A, Skriko T, Björk T. Fatigue strength characterization of high and ultra-high-strength steel cut edges. Eng Struct 2020. https://doi.org/10.1016/j.engstruct.2020.111544.
[11] Lipiäinen K, Kaijalainen A, Ahola A, Björk T. Fatigue strength assessment of cut edges considering material strength and cutting quality. Int J Fatigue 2021;149:106263. https://doi.org/10.1016/j.ijfatigue.2021.106263.
[12] Lipiäinen K, Ahola A, Kaijalainen A, Björk T. Fatigue performance of notched and hot-dip galvanized laser and mechanically cut S960 steel components considering local defects with the theory of critical distances. Int J Fatigue 2022;164:107127. https://doi.org/10.1016/j.ijfatigue.2022.107127.
[13] Lipiäinen K, Ahola A, Björk T. Fatigue performance of ultra-high-strength steel laser cut notches under variable amplitude loading. Welding in the World 2023. https://doi.org/10.1007/s40194-023-01544-0.
[14] Lipiäinen K, Ahola A, Virolainen E, Hirvi A, Björk T. Fatigue strength of hot-dip galvanized S960 cut edges and longitudinal welds. J Constr Steel Res 2022;189:107083. https://doi.org/10.1016/j.jcsr.2021.107083.

IOP Conf. Series: Materials Science and Engineering 1296 (2023) 012040 doi:10.1088/1757-899X/1296/1/012040

Effects of machining parameters on Ni-Mn-Ga-based alloys for fabrication of multifunctional micro devices using femtosecond pulse width laser

Aditya Kumthekar[1], Ville Laitinen[1], Kari Ullakko[1]

[1]Material Physics Laboratory, Lappeenranta-Lahti University of Technology LUT, Lappeenranta, 53850, Finland

E-mail: aditya.kumthekar@lut.fi

Abstract. Ni-Mn-Ga-based magnetic shape memory (MSM) alloy single crystals are known for their large magnetic-field-induced strain (MFIS). This quality makes them a promising material for use in micro actuators and devices. However, the manufacturing of single-crystal-based Ni-Mn-Ga micro actuators is challenging due to their high brittleness and other material properties – numerous machining techniques that are successfully used for the deep engraving of conventional engineering materials cannot be directly applied to Ni-Mn-Ga-based alloys. Nevertheless, previous studies have shown that a femtosecond pulse width laser (FPWL) can be successfully utilized for the defect-free micromachining various materials. This work studies the effects of different engraving parameters and introduces a novel scanning-based method for the deep micromachining of Ni-Mn-Ga-based MSM alloys with maximum surface quality. Results show that a 4-layer strategy with a 0.01 mm hatch distance provides excellent machining in terms of surface quality and dimensional accuracy. This study can be utilized within design stages to estimate minimum margin based on required machined depth and avoid defects that occur in the sample preparation stage. Additionally, evolution of structures generated by FPWL machining are characterized. The results highlight how FPWL can be considered a highly capable process for the micromachining and surface structuring of Ni-Mn-Ga-based single crystals for manufacturing multifunctional MSM microdevices.

1. Introduction

Ni-Mn-Ga-based magnetic shape memory (MSM) alloys are well recognized for their large (6–12%) magnetic-field-induced strains (MFIS), appearing when the martensitic crystal structure of the alloy's martensitic phase rearranges in reaction to an external magnetic field [1–3]. Since fine-grained randomly structured Ni-Mn-Ga polycrystal contains grain boundaries that can greatly impede twin boundary motion, large MFIS is almost solely found in oriented single crystals [4–6]. Properties of MSM alloys make them attractive for use in High-speed actuators [7], microfluid pumps [8,9], mechanical dampeners [10], and energy harvesters [11].

In spite of these qualities, however, the high costs and constraints of producing and post-processing Ni-Mn-Ga-based single crystals have prevented large-scale commercial incorporation of MSM-based devices from arising. Typically, Ni-Mn-Ga-based actuators are made by cutting oriented single-crystal rods along the crystal structure's particular planes, but the extremely brittle character of Ni-Mn-Ga-based single crystals [12] makes them challenging to cut or machine using normal mechanical

IOP Conf. Series: Materials Science and Engineering 1296 (2023) 012040 doi:10.1088/1757-899X/1296/1/012040

techniques. Previously, Ni-Mn-Ga-based single crystals were machined or cut using mechanical methods such as abrasive wire saw cutting [13], electrical discharge machining (EDM) [14], and focused ion beam (FIB) milling [7], but these methods are time-consuming and relatively expensive for the industrial-scale production of Ni-Mn-Ga-based microdevices.

Laser ablation processes, on the other hand, could make the productive, flexible, and high-precision machining of complex geometries feasible on or from Ni-Mn-Ga single crystals [15]. Recent studies have examined the potential of creating micro features on polycrystalline Ni-Mn-Ga alloys using nanosecond [16,17] and millisecond [18] pulsed lasers. Notably, these pulse widths do not prevent electron-electron interaction [19], which can lead to heat conduction into the material, resulting in cracking, local compositional changes, changes in the crystal structure, and changes in phase transition temperatures [16–18]. These detrimental impacts, however, can be eliminated by employing a femtosecond pulsed width laser as the pulse duration is shorter than the electron cooling time [19]. This addition facilitates the total local ablation of the irradiated substance without excessive thermal conduction to the machined base material.

Cavities of ~100 µm depth were generated on materials like stainless steel, titanium, aluminum alloy, and silicon using a femtosecond laser with a continuously rotating polarized beam [20]. However, the study does not focus on the finding the plateau, upon which the number of passes no longer share a linear relationship with depth. Overall, scanning the geometry in a near spiral manner has provided machining with good quality and higher material removal rates [20,21]. A hatching style might have been used to overcome the plateau reached by single-track machining as, due to increased reflectivity [22], the beam no longer being incident on flat surfaces [23][23] and difficulty of debris removal [24]. In order to develop microdevices with complex three-dimensional features using FPWL, it is crucial to understand the effect of processing parameters – in particular regarding scanning strategy, the number of passes and margin (machined width described as the perpendicular distance between external and internal geometry) as shown in Figure 1(a) below, on depth, and the surface quality and sidewall angle and dimensional accuracy of the machined area. In the present study, we introduce a simple method for deep micromachining Ni-Mn-Ga alloy using FPWL, which could easily be adapted for any kind of complex geometry.

2. Materials and methods

2.1. Laser system

The FPWL micromachining setup utilized here was designed specifically for the micromachining of Ni-Mn-Ga-based MSM alloys. An Amplitude Systems' Satsuma HP femtosecond pulsed laser was used to build the system, with the remaining components made by LASEA. The laser source has a maximum average power of 10 W and a maximum peak energy of 20 µJ. The laser itself operates on the concept of chirped pulse amplification, generating a near-Gaussian beam with a 1030 nm wavelength (near-IR laser) and a beam quality factor M2 of 1.16–1.12. The laser is capable of generating 400 fs bursts at repetition rates of up to 2 MHz. Due to an incorporated second-harmonic generation (SHG) module, the device can also provide a laser beam with a 515 nm wavelength (green laser). The beam is transferred using mirrors through different components before leaving a wavelength-specific f-theta lens with a 160 mm focal length. The machined sample is held by a three-axis movement mechanized stage. The spot size on the workpiece was approximately 33 µm. Furthermore, the scanning head includes a view module that enables viewing the sample-stage region from above via the focusing lenses. KYLA software is used to operate the laser and components from outside the system. A green laser was used to perform the designed experiments as it provided defect-free machining in the martensitic phase during initial pilot experiments.

IOP Conf. Series: Materials Science and Engineering 1296 (2023) 012040 doi:10.1088/1757-899X/1296/1/012040

2.2. Materials
The polycrystalline $Ni_{49.8}Mn_{28.4}Ga_{21.9}$ disk (Ø 40mm, ~3 mm thick) cut from larger sample using a high-precision wire saw (Ø 60 µm wire) was used for machining experiments. After cutting, the sample surfaces were ground incrementally using SiC abrasives. To characterize the micromachining performance in terms of maximum achievable depth i.e., saturation experiments, a set of ~2 mm thick semi-circular disks (Ø ~20 mm) were cut from an oriented five-layered modulated (10M) martensite $Ni_{50.4}Mn_{28.1}Ga_{21.5}$ single crystal bar manufactured by AdaptaMat Ltd., also using the high-precision wire saw. The disks were ground incrementally using SiC abrasives and subsequently electropolished in a mixture (3:1 volumetric ratio) of ethanol and 60% HNO_3 at 253 K using a constant voltage of 14 V. The Curie and martensite transformation temperatures of the disks were T_C~370 K and T_M~323 K.

2.3. Machining experiment
As the goal of this study is to explore possibility of machining deep features on Ni-Mn-Ga alloy with minimum surface quality and straight sidewalls. Square samples with 2 mm external and varied internal dimensions (0.5, 0.75, and 1 mm) were machined at ambient temperature in air – without preheating the Ni-Mn-Ga polycrystalline disk – using an average laser power of 4.8 W (100%), repetition rate of 71 kHz, scanning speed 500 mm/s, 400-fs pulse width, 2x laser beam expansion, and circular laser beam polarization. Three types of scanning strategies based on the angle of hatching scan lines were used, first two being four passes of each with constant angle i.e., 0° and 45°. The third being a 4-layer scan strategy as mentioned in [25]. Table 1 summarizes the parameters and their respective levels used to machine squares on the Ni-Mn-Ga polycrystal oval disk. The values of the parameters at different levels were chosen based on pilot experiments. A Box-Behnken design (BBD) experimental setup was used for this initial study which led to machining of total 46 sample squares on polycrystal Ni-Mn-Ga alloy disk as seen in Figure 1(d) below. Prior to micromachining, the laser beam focus was set on the surface of the disk. The samples were characterized in terms of surface quality and topological features using a Keyence VR-3200 three-dimensional measurement system.

Table 1. Parameters and their respective levels.

Parameters	level 1	level 2	level 3
Passes	10	100	190
Hatch distance	0.01 mm	0.035 mm	0.06 mm
Scanning strategy	0°x4	45°x4	4-layer
Power	2.4 W	3.6 W	4.8 W
Margin	0.25 mm	0.5 mm	0.75 mm

2.4. Saturation experiments
Squares with 2 mm external dimensions and margin of 0.75 mm were machined on the single crystal disks mentioned earlier, utilizing 10, 100, 500, 1000, 1500, and 2000 passes to obtain limits of achievable machining depth with optimum parameters determined based on the results of machining experiments. Total 12 samples were machined with one replicate of each sample with different number of laser passes. Initial characterization was performed using the abovementioned three-dimensional measurement system, but for deep cavities – due to the limitations of a three-dimensional measurement system working on the principle of back reflection – a scanning electron microscope (SEM) from Hitachi SU3500 equipped with EDS detector was used to obtain cross-section images of the cavities. The cross-section was achieved by laser machining and grinding.

2.5. EDS analysis
To characterize the redeposited layer generated by FPWL ablation, EDS analysis of a sample machined by applying 1000 passes for saturation experiments using EDS system mentioned earlier was carried out. The measurement points were placed on machined and original surface of Ni-Mn-Ga semicircular disk. The measurements were taken post machining without any cleaning.

2.6. Characterization of the structures generated by FPWL ablation.
In order to characterize the structures formed by FPWL ablation, sections with different number of laser passes adjacent to each other were machined on one of the single crystal disks mentioned earlier. The parameters used were those optimal values obtained from the results of machining experiments. One section was machined with less than 500 laser passes and other one was machined with more than 500 laser passes. The structure was characterized with the help of images obtained using the SEM system mentioned earlier. Images of the samples were taken using flat and tilted stages with angle of titled stage set approximately at 30°.

3. Results and Discussion

3.1. Machining experiments
The initial micromachining experiments revealed the effects of various applied parameters on the surface quality and dimensional accuracy of machined cavities on the Ni-Mn-Ga polycrystalline alloy oval disk. The samples generated with 10 passes, 0.035- and 0.06-mm hatch distances were eliminated as we believed these samples did not fit under the category as deep machining but rather mild marking due to the small quantity of material removed. Therefore, samples with significant depth i.e., greater than 100 μm were considered for further characterization. Parameters related to these samples are presented in Table 2 below. Based on preliminary screening, samples represented in Figure 1(a-c) were found to be optimal in terms of surface quality and dimensional symmetry.

Table 2. Parameters related to samples with depth values higher than 100 μm.

Sample no.	Margin (mm)	Sdr	Sa (μm)	β (°)	Power (W)	Passes (N)	Strategy	Depth (μm)
17	0.5	0.038	10	32.33	3.6	100	4-layer	227
1	0.5	0.07	4.6	39.01	3.6	100	0x4	199
19	0.5	0.08	5	35	2.4	100	45x4	159
35	0.5	0.15	26	28.5	3.6	190	45x4	377
38	0.75	0.056	18	30	3.6	100	45x4	192
45	0.25	0.28	16	28	3.6	100	45x4	211
41	0.5	0.05	12	28.3	4.8	100	45x4	267

In this study, minimum surface roughness and flatness is considered as better surface quality. The surface quality of the machined surface was evaluated using two surface roughness measures. Sa (arithmetical mean height), which is the surface extension of Ra (arithmetical mean height of a line), and Sdr (developed interfacial area ratio), which is a hybrid measure stated as the proportion of the definition area's extra surface area supplied by the texture when compared to the planar definition area [26]. Geometrical accuracy was assessed based on radii measured on inner corners (r_1 and r_2) of the cavities and β (the angle between the vertical plane and sidewall of machined cavities) as shown in Figure 1 below. For geometrical accuracy, the samples machined with a 4-layer strategy provided the best-machined cavities, i.e. ($r_1 \sim r_2$ and low β) Figure 1(a), although Sa is lower compared to those cavities machined with 0x4 and 45x4, as seen in Figure 1(b and c) below; Sdr is lowest in cases of cavities machined with 4-layer strategy. Low Sdr implies the surface being close to completely leveled or flat, which is essential for efficient machining as the laser is most effective when incident on a flat surface.

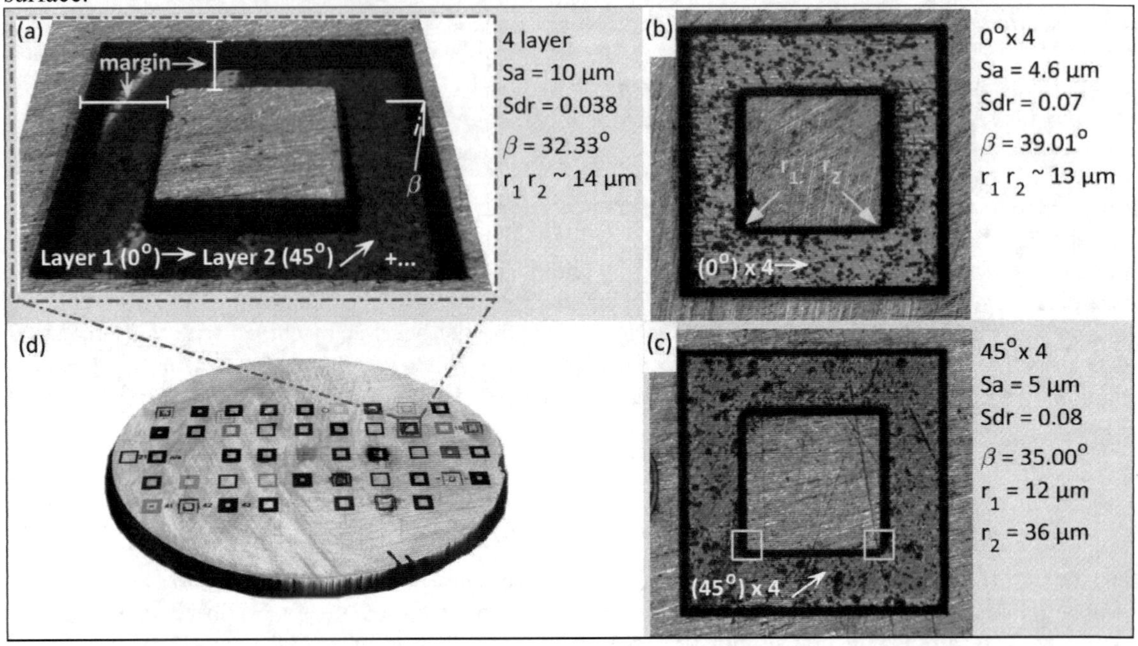

Figure 1. Characterization of FPWL machined samples on the Ni-Mn-Ga polycrystal oval disk in terms of surface quality, i.e., arithmetical mean height (Sa); developed interfacial area ratio (Sdr); dimensional quality, such as angles between the normal and sidewalls of the machined cavities (β); radii of the inner corners (r_1) and (r_2) a) three-dimensional microscopic image of cavity machined using 4-layer (sample no. 17). b) 45° x 4 (sample no. 19). c) 0° x 4 scanning strategy (sample no. 1). d) Sample squares machined on the polycrystalline Ni-Mn-Ga alloy.

3.2. Saturation experiments

Even though we can observe a trend between depth and β there was not any significant relation. The measurements of cavities were performed in as machined state it is possible that there exists a reposited layer from FPWL. To verify this and find the zone where depth is no longer increasing along number of laser passes, saturation experiments as mentioned earlier were performed. The cross-section images of the samples were observed using SEM and it was observed that there is a reposited layer. To characterize the redeposited layer generated from FPWL ablation EDS analysis was carried out and discussed in section 3.3. Therefore, the samples on single crystal disk were cleaned using acetone and light brushing before measurement of β. It is possible that the reason for no strong correlation observed between β and depth in case of previous samples machined on polycrystal Ni-Mn-Ga disk might be due to the redeposited material. We observe clearly significant relation between number of passes and β as seen in Figure 2 (a). It was observed that β is higher than 27° for cavities generated by applying less than or equal to 100 passes. For cavities generated by applying more than 100 laser passes – β is observed to be between 14–16°, as seen in Figure 2 (b–d).

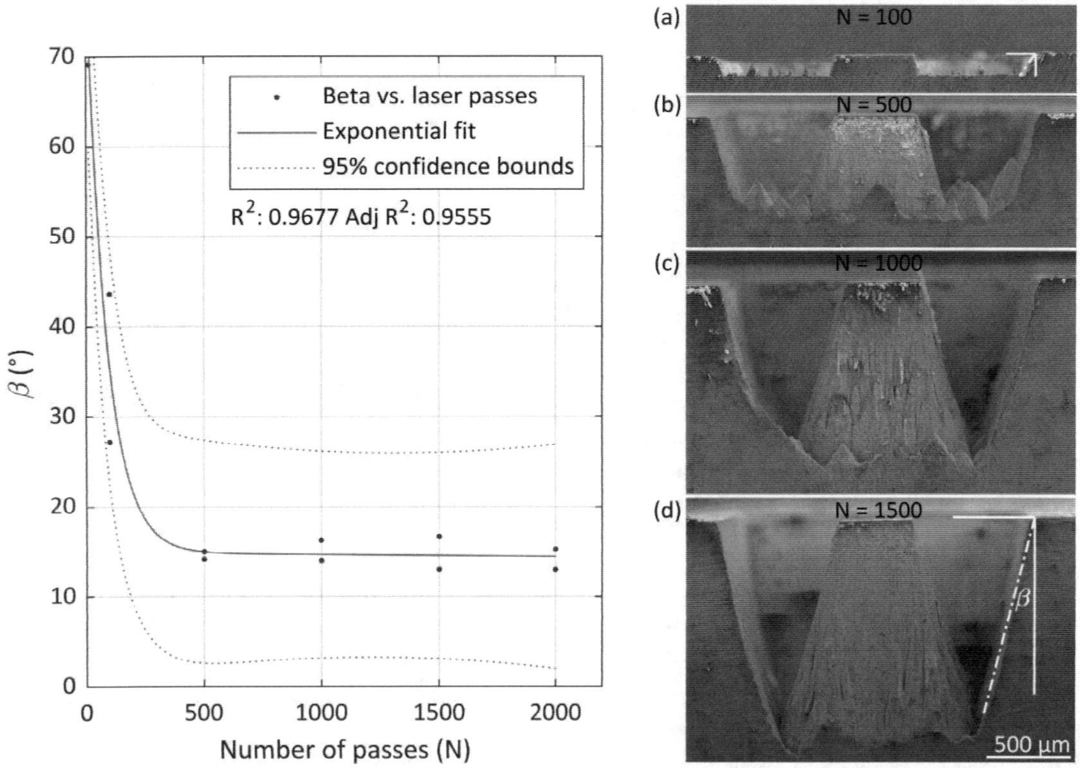

Figure 2. Variation of β with respect to the number of laser passes using a 4-layer scanning strategy and a margin of 0.75 mm until near saturation of machining/depth; a–d) secondary electron image of the cross-section of the cavities machined using a) 100, b) 500, c) 1000, and d) 1500 laser passes (near saturation).

Parameters and measured depth values of cavities machined with different number of laser passes are shown in Table 3 below. We can observe that maximum error between measurements taken on X - axis and Y-axis is 15.7 % which is the error we encountered while using three-dimensional measurement system which works on the principle of illumination and integration of images captured from different positions. The lenses are aligned in series along X- axis. For samples machined with 0.75 mm margin, we can observe the difference between depth values when compared to those of next level i.e., c values as seen in Table 3 are reducing towards higher number of passes i.e., the depth is saturating between 1000 and 1500 passes. Now as we know behavior of β being stable after 500 passes or ~600 μm, achievable depth for margin can be predicted from the results of these experiments. Future work of this study shall be generating a model for predicting minimum margin for required depth to be machined considering other factors such as material properties, thickness, temperature of disk, etc. Cleaning carried out for measurement of β revealed presence of cone shaped microstructures at the bottom of the cavity. These were characterized and the discussed in section 3.4.

IOP Conf. Series: Materials Science and Engineering 1296 (2023) 012040 doi:10.1088/1757-899X/1296/1/012040

Table 3. Parameters related to saturation experiments.

Sample number	Passes	Average depth -X axis (3 measurements) (μm)	Average depth-Y axis (3 measurements) (μm)	Error % (X -Y)	c(X)	c(Y)
1,2	10	19	22	15.7	186	211
3,4	100	205	233	13.6	462	455
5,6	500	667	688	3.1	675	799
7,8	1000	1342	1487	10.8	116	35
9,10	1500	1458	1522	4.3	145	140
11,12	2000	1603	1662	3.6		

3.3. EDS analysis

Particles (dust) deposited inside and near the cavity are caused by thermal chemical reaction of with the oxygen in the air after the material was ablated by absorbing laser pulse energy – as observed in [27,28]. The dust or particles on the edge are not visible in Figure 3 as they were blown away using pressurized gas in order to prepare them for SEM measurements, although they are still present in the cavity. Based on EDS results obtained at point 1 (original surface) and point 2 (laser machined surface), it can be stated – as per Figure 3 below – that the dust is nothing but an oxidized layer left behind while other reaction products, such as CO and CO_2, were sucked in by the suction system. An in-depth study of a thermal chemical reaction and related matters, however, remains outside the scope of this study.

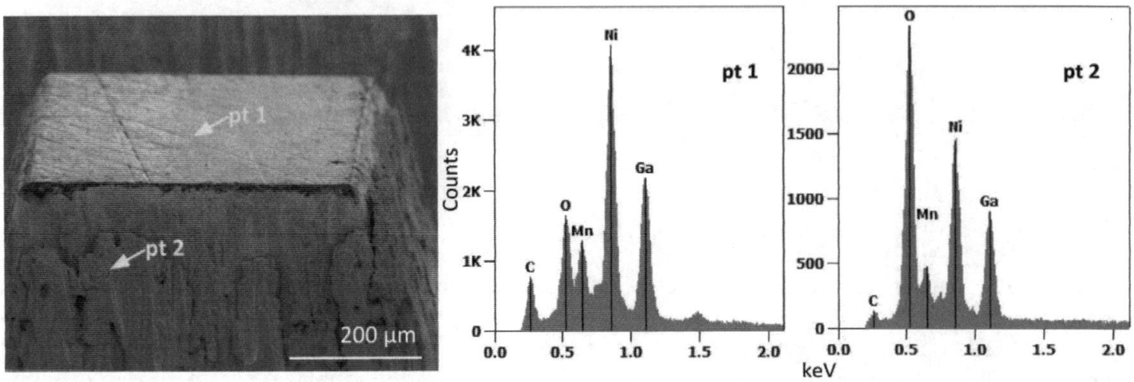

Figure 3. EDS analysis of a machined pillar.

IOP Conf. Series: Materials Science and Engineering 1296 (2023) 012040 doi:10.1088/1757-899X/1296/1/012040

3.4. Characterisation of the structures generated by FPWL ablation

One of the main interests of this study is the quality of the machined surface. Samples machined on a single crystal Ni-Mn-Ga disk with a different number of laser passes is shown in Figure 4 below, in which the evolution of these structures can be observed. The machined area can be divided into two parts; the structures observed in Figure 4 (b–d) were machined with < 500 laser passes and those observed in Figure 4 (a, e) were machined with > 500 laser passes. From here onward, these two forms will be termed 'low number of passes' (LNP) and 'high number of passes' (HNP). After cleaning the samples, it was found that structures shaped like undulating grooves – as observed in [20] on stainless steel and a few cones – are generated with LNP, as per Figure 4 (b-d). Whereas nano-sized ripples are largely formed near the inclined surface, as per Figure 4 (c–d). Ripples develop near inclined surfaces because of the dephasing between the s- and p-components. This is caused by the incidence of laser beam on tilted surface and results in the electric field distorting to become elliptical rather than circular.

A few cones covered with nano-sized particles, as described by [29], are observed on the surface machined with LNP, as per Figure 4 (d). These structures continue to grow with the number of passes, as shown in Figure 4 (a, e). Usually, they are formed in three stages: the production of early precursor sites by hydrodynamical processes, the growth of multi-scale structures, and the evolution of the structures in response to subsequent pulse illumination. Laser fluence is demonstrated to have a considerable influence on both the primary formation process and the subsequent surface structure, which is caused by a combination of favored valley ablation, fluid flow, and nanoparticle/material redeposition [30]. Notably, it was observed that grooves or defects generated from disk preparation did not disappear despite the laser machining, which persisted at least until the depths were machined with LNP, as seen in Figure 4 (e-f). This could be due to weakened material or else the grooves acting as preferential valleys that got bigger with each consecutive laser pass. This is a very important consideration from a machining point of view, as these defects can affect the final desired function of the particular device.

We have seen in Figure 1 above that at higher depths – i.e., machining saturation – these cones are machined away. This result answers the pending question in [20] concerning what happens to the structures at the bottom at the machining saturation. However, another question then subsequently arises regarding the effect of these structures on machining efficiency and the groove geometry (beta angle, surface quality of side wall, etc.). In general, the femtosecond laser initially ablates the material, and the major part is discarded in the form of gas or particles from the irradiated area, forming nanoholes and grooves along with an increasing number of laser passes, which results in these structures evolving into cones then getting completely removed.

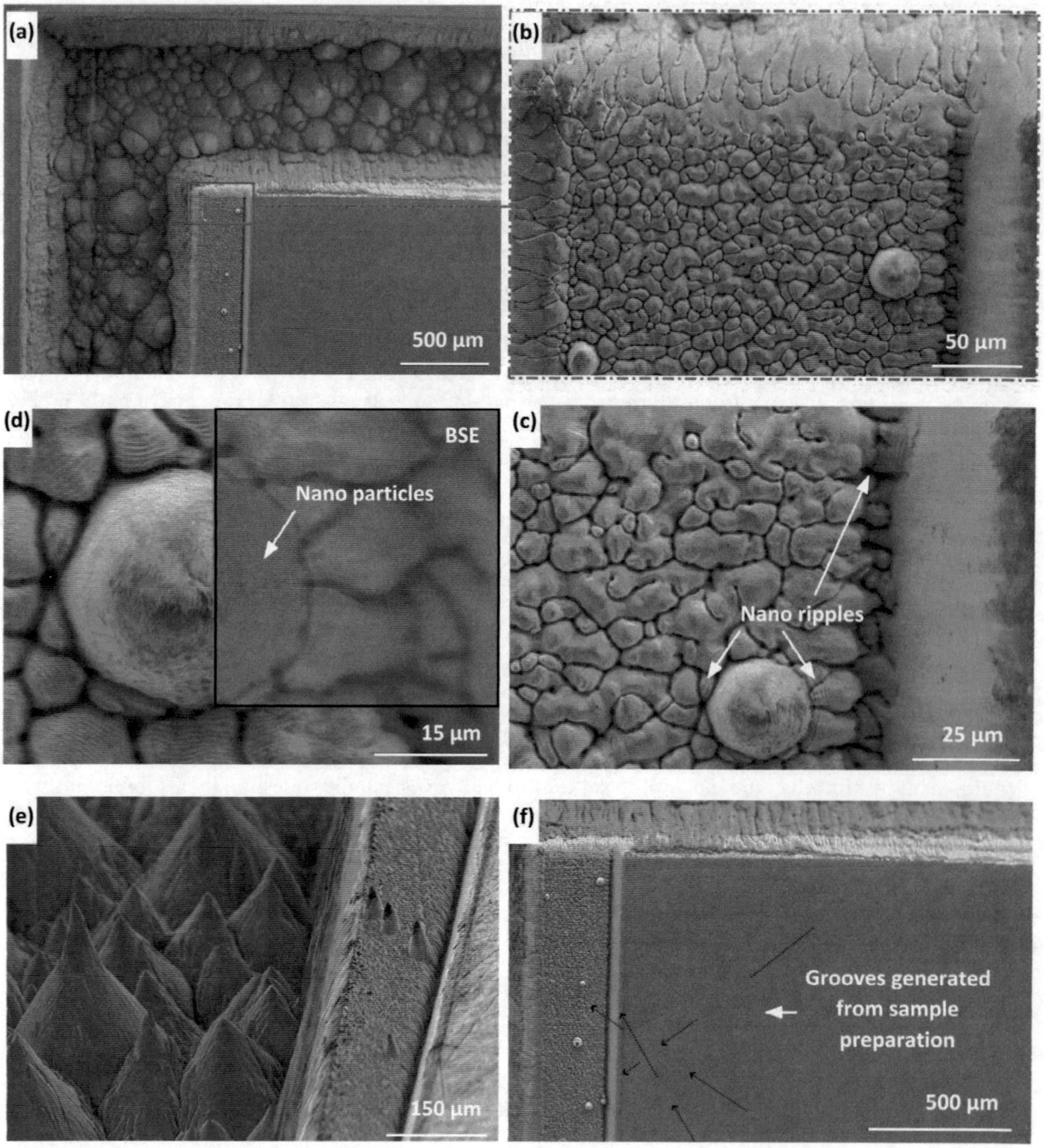

Figure 4. Characterization of structures that occurred post FPWL machining of single-crystal Ni-Mn-Ga with a different number of laser passes using SEM images: a) structures observed for different numbers of laser passes b) structures observed with a low number of laser passes are a mixture of undulating grooves and cones; c) with 2x magnification, nano ripples near inclined surfaces are observed; d) with ~4x magnification, possible nanoparticles are observed over cones using backscattered electrons; e) structures observed by tilting the disk; f) grooves formed due to mechanical machining are not removed by laser and are found to be at least a few hundred μm deep.

4. Conclusion

The effects of different engraving parameters are studied, and a novel scanning-based method is introduced for the deep micromachining of Ni-Mn-Ga-based MSM alloys. Results show that a 4-layer strategy with a 0.01 mm hatch distance provides excellent machining in terms of surface quality (Sa and Sdr), geometrical accuracy (β), and radii of inner corners (r_1, r_2). The evolution and disposal of the formed structure on the machined area are described. This study can be utilized within design stages to estimate margin based on depth, which is required to form cavities and avoid defects that occur in the disk preparation stage. Nanoparticles were found to be generated, which provides an initial base for further investigations into the benefits of using FPWL. The results do, however, highlight how FPWL is a highly capable process for micromachining Ni-Mn-Ga-based single crystals and also for creating surface modifications for multifunctional MSM microdevices. Future work will be generating a model which can predict the margin accurately based on the depth of cavity required and assess effect of microstructures generated by FPWL ablation on machining efficiency.

Acknowledgements

This work was financially supported by the Strategic Research Council of Finland (grant number 313349), the Academy of Finland (grant number 325910), and the AMBI (Analytics-based Management for Business and Manufacturing Industry) research platform of Lappeenranta-Lahti University of Technology LUT. The authors thank all participants in the projects for sharing their knowledge and for their research input.

Refrences

[1] Pagounis E, Chulist R, Szczerba M J and Laufenberg M 2014 Over 7% magnetic field-induced strain in a Ni-Mn-Ga five-layered martensite *Appl Phys Lett* **105** 1–4

[2] Murray S J, Marioni M, Allen S M, O'Handley R C and Lograsso T A 2000 6% magnetic-field-induced strain by twin-boundary motion in ferromagnetic Ni-Mn-Ga *Appl Phys Lett* **77** 886–8

[3] Sozinov A, Lanska N, Soroka A and Zou W 2013 12% magnetic field-induced strain in Ni-Mn-Ga-based non-modulated martensite *Appl Phys Lett* **102** 021902

[4] Dunand D C and Müllner P 2011 Size effects on magnetic actuation in Ni-Mn-Ga shape-memory alloys *Advanced Materials* **23** 216–32

[5] Gaitzsch U, Pötschke M, Roth S, Rellinghaus B and Schultz L 2009 A 1% magnetostrain in polycrystalline 5M Ni-Mn-Ga *Acta Mater* **57** 365–70

[6] Pötschke M, Gaitzsch U, Roth S, Rellinghaus B and Schultz L 2007 Preparation of melt textured Ni-Mn-Ga *J Magn Magn Mater* **316** 383–5

[7] Musiienko D, Straka L, Klimša L, Saren A, Sozinov A, Heczko O and Ullakko K 2018 Giant magnetic-field-induced strain in Ni-Mn-Ga micropillars *Scr Mater* **150** 173–6

[8] Nisar A, Afzulpurkar N, Mahaisavariya B and Tuantranont A 2008 MEMS-based micropumps in drug delivery and biomedical applications *Sens Actuators B Chem* **130** 917–42

[9] Guevara J D, Jones M L, Müllner P and Jankowski E 2020 Development of a Molecular Model for Understanding the Polymer-metal Interface in Solid State Pumps *The Journal of Computational Science Education* **11** 12–22

[10] Nilsén F, Aaltio I and Hannula S P 2018 Comparison of magnetic field controlled damping properties of single crystal Ni-Mn-Ga and Ni-Mn-Ga polymer hybrid composite structures *Compos Sci Technol* **160** 138–44

[11] Lindquist P, Hobza T, Patrick C and Müllner P 2018 Efficiency of Energy Harvesting in Ni–Mn–Ga Shape Memory Alloys *Shape Memory and Superelasticity* **4** 93–101

[12] Feuchtwanger J, Richard M L, Tang Y J, Berkowitz A E, O'Handley R C and Allen S M 2005 Large energy absorption in Ni-Mn-Ga/polymer composites *J Appl Phys* **97** 10M319

[13] Musiienko D, Saren A and Ullakko K 2017 Magnetic shape memory effect in single crystalline Ni-Mn-Ga foil thinned down to 1μm *Scr Mater* **139** 152–4

[14] Lindquist P G and Müllner P 2015 Working Ni–Mn–Ga Single Crystals in a Magnetic Field Against a Spring Load *Shape Memory and Superelasticity* **1** 69–77

[15] Schaeffer R D 2016 *Fundamentals of laser micromachining* (Boca Raton, Fla: CRC)

[16] Biffi C A and Tuissi A 2015 Micro-processing of NiMnGa shape memory alloy by using a nanosecond fiber laser *Opt Laser Technol* **78** 42–9

[17] Biffi C A and Tuissi A 2014 Fiber laser drilling of Ni46Mn27Ga27 ferromagnetic shape memory alloy *Opt Laser Technol* **63** 1–7

[18] Craciunescu C M, Miranda R M, Silva R J C, Assunção E and Braz Fernandes F M 2011 Laser beam interaction with NiMnGa ferromagnetic shape memory alloys *Opt Lasers Eng* **49** 1289–93

[19] Chichkov B N, Momma C, Nolte S, von Alvensleben F and Tünnermann A 1996 Femtosecond, picosecond and nanosecond laser ablation of solids *Appl Phys A Mater Sci Process* **63** 109–15

[20] Racine V, le Barh M, Killaire G and Weck A 2019 Effect of polarization on ripple formation in deep femtosecond laser machined cavities *J Mater Process Technol* **271** 162–71

[21] Liu B, Yan Y, Zhao J, Wang W and Liao K 2022 Research on hole depth in femtosecond laser deep micropore processing technology based on filament effect *Optik (Stuttg)* **249** 168307

[22] Shah L, Tawney J, Richardson M and Richardson K 2001 Femtosecond laser deep hole drilling of silicate glasses in air *Appl Surf Sci* **183** 151–64

[23] Bärsch N, Körber K, Ostendorf A and Tönshoff K H 2003 Ablation and cutting of planar silicon devices using femtosecond laser pulses *Appl Phys A Mater Sci Process* **77** 237–42

[24] Sikora A, Fares L, Adrian J, Goubier V, Delobbe A, Corbin A, Sentis M and Sarnet T 2017 Picosecond laser micromachining prior to FIB milling for electronic microscopy sample preparation *Appl Surf Sci* **418** 607–15

[25] Umer U, Mohammed M K and Al-Ahmari A 2017 Multi-response optimization of machining parameters in micro milling of alumina ceramics using Nd:YAG laser *Measurement (Lond)* **95** 181–92

[26] Townsend A, Pagani L, Scott P J and Blunt L 2019 Introduction of a Surface Characterization Parameter Sdr prime for Analysis of Re-entrant Features *J Nondestr Eval* **38**

[27] Huang J, Wei S, Zhang L, Yang Y, Yang S and Shen Z 2019 Fabricating the superhydrophobic nickel and improving its antifriction performance by the laser surface texturing *Materials* **12** 1155

[28] Li G, Li J, Zhang C, Hu Y, Li X, Chu J and Huang W 2015 Large-area one-step assembly of three-dimensional porous metal micro/nanocages by ethanol-assisted femtosecond laser irradiation for enhanced antireflection and hydrophobicity *ACS Appl Mater Interfaces* **7** 383–90

[29] Zuhlke C A, Anderson T P and Alexander D R 2013 Fundamentals of layered nanoparticle covered pyramidal structures formed on nickel during femtosecond laser surface interactions *Appl Surf Sci* **283** 648–53

[30] Zuhlke C A, Anderson T P and Alexander D R 2013 Formation of multiscale surface structures on nickel via above surface growth and below surface growth mechanisms using femtosecond laser pulses *Opt Express* **21** 8460

IOP Conf. Series: Materials Science and Engineering 1296 (2023) 012041 doi:10.1088/1757-899X/1296/1/012041

A Laser Forming Study on Time and Energy Efficiency

Georgi Nikolaev Nikolov, Anders Noel Thomsen, Morten Kristiansen

Department of Materials and Production, Aalborg University, Fibigerstræde 16, 9220 Aalborg, Denmark

E-mail: gnn@mp.aau.dk, ant@mp.aau.dk, morten@mp.aau.dk

Abstract. Laser forming is a relatively slow process utilizing temperature differences in the material to cause plastic deformation. The process requires dwell time between laser passes to cool down the material, which results in a slow production rate. Not fully cooling the workpiece has been shown to give a higher bend rate per unit of time. This, coupled with the cooling rate at higher temperatures being more pronounced permits examining the efficiency of laser forming while working in different temperature ranges. The presented work experimentally explores different temperature ranges for AISI 304 from 20 to 1300°C using a thermal camera. Therefore, for each temperature range, the workpiece is heated and cooled to a achieve a constant peak temperature. The purpose is to compare the bend rate per unit of time with different temperature ranges and to examine overall time efficiency. Finally, since different power levels are utilized for each temperature range, the total energy used to achieve each bend will also be investigated. The results show that the increased formability and rate of cooling at higher temperatures increase efficiency in terms of total time. While efficiency in terms of the equipment is higher when allowing the workpiece to fully cool.

1. Introduction

Laser forming is an iterative, flexible, and contactless forming process, that works by creating plastic deformation from heating and cooling cycles. There are three main thermal mechanisms that govern the laser forming process: the temperature gradient mechanism (TGM), the upsetting mechanism, and the buckling mechanism. The presented work utilizes TGM as its forming mechanism. With each laser pass, the heating causes localized thermal expansion, that lead to thermal stresses. When the thermal stresses surpass the yield stress of the material, further expansion is converted into plastic deformation. Upon cooling, the thermal expansion is replaced by thermal contraction which results in bending towards the laser beam. Each subsequent laser pass under TGM contributes with up to 3° to the overall bend development; therefore, multiple laser passes are necessary to achieve any significant bend angle [1]. The cooling time between the laser passes is known as the dwell time, which depends on the material properties, part geometry, and forming parameters. When considering a complicated 3D part or a part with multiple bends requiring upwards to hundreds of passes, the accumulated dwell time will result in a significant decrease in process productivity. To increase process productivity, research has been focused in a number of approaches: the use of different cooling mediums, the effects of dwell times, and scheduling of laser forming.

The possibility of using a cooling medium, known as active or forced cooling has been investigated by multiple authors: Cheng and Yao [2] investigated the effects of forced air cooling on microstructure, tensile strength and bend angle. Shen et al. [3] numerically examined the effects of forced water cooling and found that forced water cooling could reduce the dwell time without affecting the bend angle per pass. Lambiase et al. [4] investigated the effect of partially submerged passive water cooling on laser forming. The authors found an increased cooling rate compared with calm air. Shen et al [5] investigated the effect of underwater pulsed laser forming. Paramisavan et al. [6] simulated the effect of bottom surface cooling on the bend angle and the heat affected zone.

Other works have investigated the effect of the dwell time. Shen et al. [7] numerically investigated the effect of dwell time and found that a shorter dwell time increased the total deformation due to an increased peak temperature of the second pass. Lambiase et al. [8] investigated the productivity given by different process parameters and trained a neural network to determine the optimum processing settings. The authors found that allowing the part to fully cool does not lead to the highest productivity. The authors used constant settings for all passes and sought to find the optimum settings by that strategy.

Noticeably less work has been done on scheduling of laser forming in order to increase productivity, with Hao et al. [9] and Hao et al. [10] being the primary contributions. The review by Bachman et al [11] and the review of recent advances by Safari et al [12] can be examined for a broader perspective of the recent literature of laser forming.

The current literature have examined the process based on constant process settings. This is an obvious choice from the process perspective; however it fails to address a question from the material perspective. If constant power is used without fully cooling the part, the peak temperature will increase which will change the material load and ultimately lead to melting of the surface. Thereby an open question is left: What if a constant peak temperature is maintained by use of variable process settings?

The previous works have examined efficiency from a process perspective, i.e. what changes can be made to the process to improve efficiency. By use of advanced thermography, the perspective of the material can be explored instead. This work seeks to investigate different temperature profiles based on maintaining specific peak temperatures throughout processing. The purpose of these temperature profiles is to determine their effect on the overall efficiency in terms of bending rate over time and energy usage during forming.

2. Methodology

2.1. Experimental setup

The experimental setup can be seen in Figure 1. The laser is an IPG YLS-3000 SM fiber laser set in continuous mode. The laser has a beam quality of 1.2 M^2 and a wavelength of 1076 nm. The laser beam is directed using an ARGES Fiber Elephant 50 commercial galvanometric laser scanner system with dual galvanometric mirrors (+/- 10.5°) and a dynamic focus module at a focal length of 490 mm $\pm$25mm. The thermography measurements are performed with an InfraTec ImageIR 8380 having a measurement range of -10 to 1500°C using HDR operating at a combined 50 Hz. The detector is an inSb with a spectral range of 1.5 to 5.7 um and is cooled using a sterling cooler, with a resolution of 640 x 512 pixels.

2.2. Sample preparation

All samples are water jet cut workpieces of 100x50x1 mm from the same AISI 304 plate. To reduce variation in emissivity of the surface for the thermography measurements, the samples are laser marked in order to achieve a homogeneous and diffused surface. Note that this laser marking also greatly increases the absorption rate of the surface. The laser marking is performed with three out of six laser modules, power with average power output of 200W, speed of 25 mm/s, frequency of 5000 Hz, and a duty cycle of 15%. The galvanometric scanner system moves the

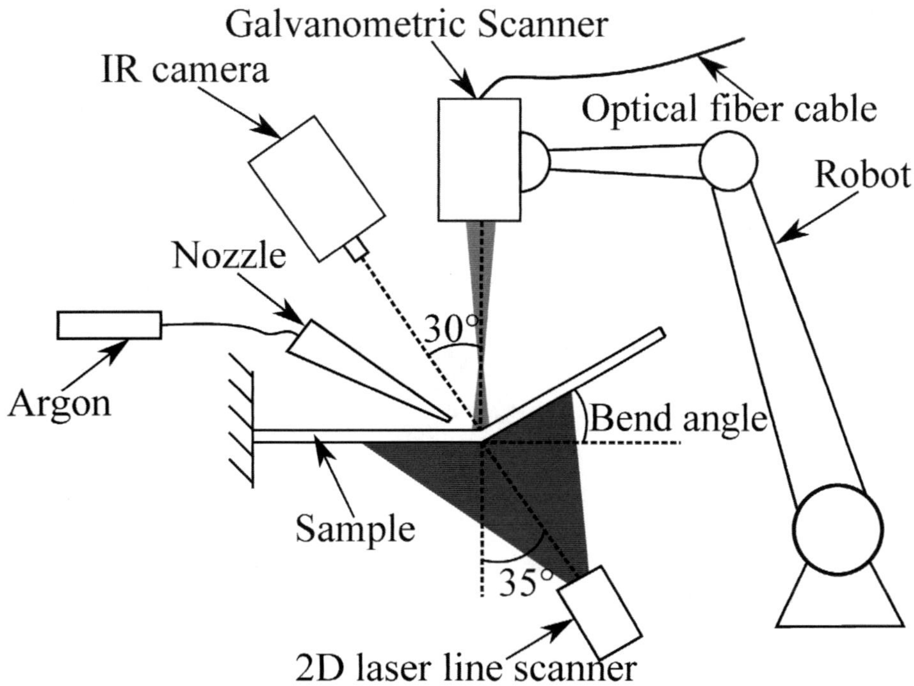

Figure 1. Schematic of experimental setup.

laser beam perpendicular to the plane of Figure 1. An overlap of 0.05 is made between the parallel passes and no height offset is applied.

2.3. Laser forming

The laser forming is performed using six out of six laser modules for the 1000 and 1300°C samples while a single laser module is used for the 700°C . The laser paths are perpendicular to the rolling direction of the samples (along the width). Argon is used as shielding gas is in order to reduce further oxidation of the laser formed surface which can affect the thermographic measurements. Furthermore, the gas flow is set to 15 L/min, which also provides limited cooling to the surface. The bend angle is measured using a Wenglor MLWL 131 2D laser line scanner operating at 20 Hz. The 2D laser line scanner is positioned to measure the bottom surface at an angle of approximately 35°. All laser forming experiments are performed using a speed of 50 mm/s with a diffused laser beam with a diameter of 3 mm. Varying laser power and dwell time is applied to achieve different temperature profiles. The settings used for each experiment can be seen in Figure 2. The experiment is run with an open loop control and the number of paths are planned based on a target bend angle of at least 30° for each sample.

2.4. Thermography measurements

The measurements are performed at an angle of approximately 30° with respect to the measured surface normal. Validation experiments using AISI 304 and a thermal oven at 900°C show a measurement error due to angle (0-45° with respect to the normal) of less than 5°C. The output data from the thermography measurements consists of 640x512 pixel frames, 50 frames per second, where every pixel contains temperature information. The temperature analysis is performed using a 16x16 pixel region of interest (ROI) of the thermal scan that is along the laser path, depicted in red seen in Figure 3. Where Figure 3 a) is a single frame showing a

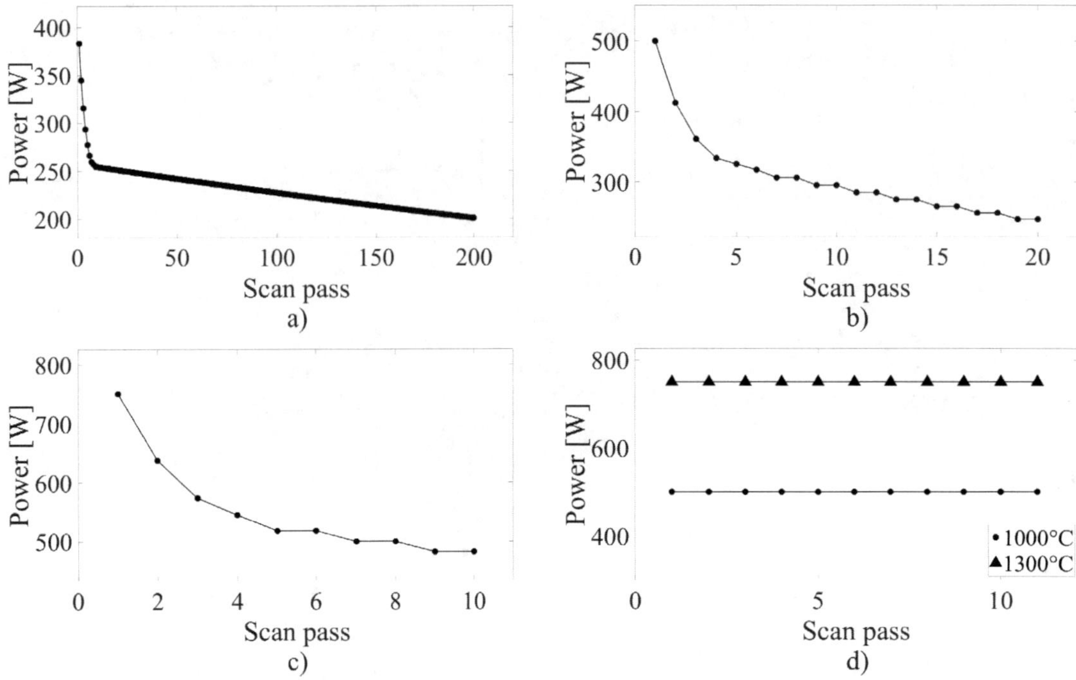

Figure 2. Power settings for temperature ranges of: a) 700°C , dwell time of 3.5 seconds. b) 1000°C , dwell time of 3.5 seconds. c) 1300°C , dwell time of 3.5 seconds. d) 1000°C and 1300°C , dwell time of 90 second.

measurement of the sample at room temperature, and Figure 3 b) shows a single frame of the same sample while its being formed. The temperature within the ROI is measured by taking the average of nine adjacent pixels. The pixels are chosen by checking how many times each individual pixel in all of the frames belonging to a single experiment have been above 600°C . The pixel that has been most often above 600°C together with it's eight adjacent pixels are chosen.

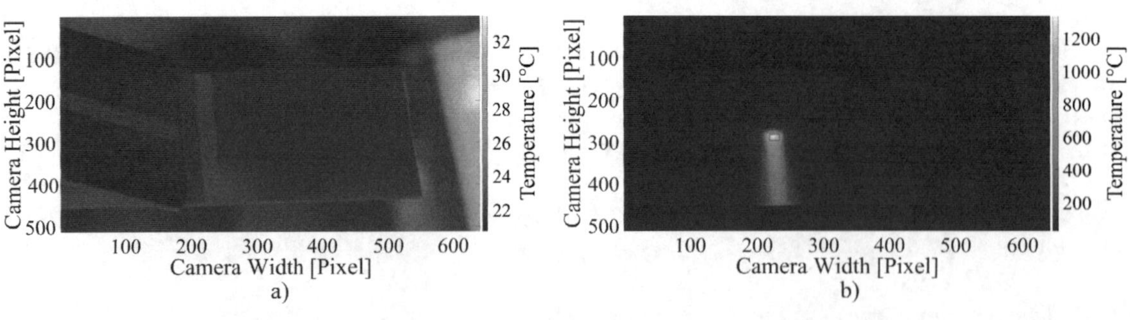

Figure 3. Thermographic measurement of a sample at: a) room temperature. b) while being formed. The highlighted region in red is the region of interest along the laser marked area.

3. Results and discussion

It should be noted that while recording the thermogrphic data, there were incomplete frames which resulted in noisy misreadings of the pixels; however the effect is limited compared to the amount of data captured and does not affect the investigated trends. The errors for each temperature profile experiment are shown in Table 1, with the total error being below 0.5% with the exception of the data for 1000°C which has an error of 0.87%. The incomplete frames are manually removed from the study.

Table 1. Number of incomplete frames in relation to the full amount of data.

Peak temperature Dwell time [s]	1000°C 90	1300°C 90	700°C 3.5	1000°C 3.5	1300°C 3.5
Total number of frames	51596	37908	52892	15150	17441
Number of incomplete frames	453	120	154	48	41
Error [%]	0.87	0.32	0.29	0.31	0.23

3.1. Bend angle

The bend development and temperature profiles for each sample can be seen in Figure 4. The results show that it was not possible to attain an angle above 25° with the current settings at 700°C . This is due to a pronounced bend rate decay, which is a result of geometrical distortion of the sample [13] and material hardening [14].

Furthermore, the results show that the fastest bend development occurs with a peak temperature of 1300°C. This is due to the elevated formability and higher gradients at higher temperatures. This is the case for both a long and short dwell time. However, it is more time efficient to form with a lower peak power than using a long dwell time.

Table 2 contains time characteristics from the processing such as total processing time, equipment time and equipment bend rate. Equipment time refers to time when the laser is on, and equipment bend rate is the ratio between the equipment time and the total developed bend angle. The results show that the fastest total processing time is with a short dwell time and a high peak temperature. This is due to the higher cooling rate at high temperatures. However the results also show that the equipment bend rate is highest when allowing a high dwell time. This is due to the higher temperature gradient. The results are notable from a scheduling perspective as it is 30-80% more efficient use of the equipment to allow the workpiece to fully cool. However, it is about three times more efficient considering the total time to allow the workpiece not to fully cool.

Table 2. Time characteristics of laser forming, where equipment time refers to work time of the laser and the equipment bend rate is the ratio between equipment time and the achieved final bend angle.

Peak Temperature Dwell time [s]	1000°C 90	1300°C 90	700°C 3.5	1000°C 3.5	1300°C 3.5
Total angle [°]	34.01	34.8	23.51	34.23	33.34
Total time [s]	988.00	727.00	879.00	242.00	217.00
Equipment time [s]	11.00	8.00	200.00	20.00	10.00
Equipment bend rate [°/s]	3.09	4.38	0.1175	1.71	3.33

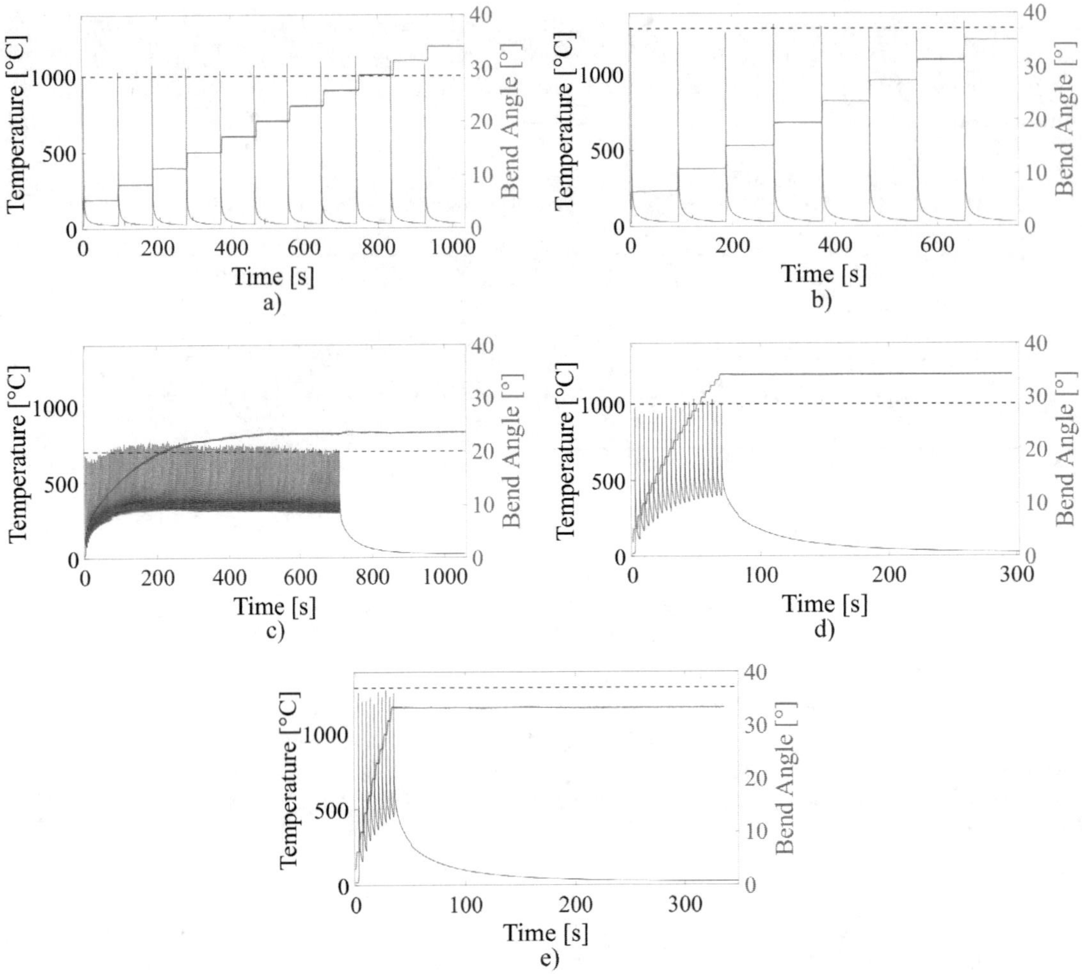

Figure 4. Temperature history for: a) 1000°C , 90 seconds dwell time. b) 1300°C , 90 seconds dwell time. c) 700°C , 3.5 seconds dwell time. d) 1000°C , 3.5 seconds dwell time. e) 1300°C , 3.5 seconds dwell time. The dashed lines indicates the target temperature.

When performing high temperature laser forming with insufficient dwell time, heat buildup in the part is bound to occur. The increased temperature affects the convection rate by reducing the thermal gradient as heat spreads through the part. The results here show that it is possible to avoid melting of the surface and maintaining a relatively constant peak temperature, but it requires careful manual work or automation with accurate prediction of the material's thermal behaviour. The advantage found here is in the order of three times faster; however, it should be noted that forced cooling will make the advantage smaller. This is because forced cooling will reduce the total time needed to bring the workpiece back to nominal temperature.

3.2. Temperature

The temperature histograms of the samples can be seen in Figure 5. It is noteworthy how little time is spent at temperatures above 800°C for all parts regardless of the settings used. The result is brief but important moments at high temperatures coupled with localized strain affecting the material kinetics. Table 3 is a summation of the time above specific temperatures

IOP Conf. Series: Materials Science and Engineering 1296 (2023) 012041 doi:10.1088/1757-899X/1296/1/012041

for the different workpieces at a single pixel along the scan line. The results show how short lived and how influential the peak temperatures are. When comparing the 1000°C with the 1300°C sample with a dwell time of 3.5 seconds, the time above 1000°C goes from 0.30 seconds to 0.42 seconds; however this difference of about 0.12 seconds seems to contribute to an almost doubling in the equipment bend rate. The data also highlights that the process, albeit slow in the total processing time, is more dynamic when the actual changes occurs. Similarly when considering the samples with 90 second dwell time, the total time at elevated temperatures is short lived as well. In contrast for the 700°C sample, when considering the bending rate it is evident to see that although some bending occurs in the range of 400-700°C it is orders of magnitude smaller than what occurs above 800°C , which again is very different from what occurs at above 1000°C . This is interesting considering the very short time the sample actually spends above 1000°C .

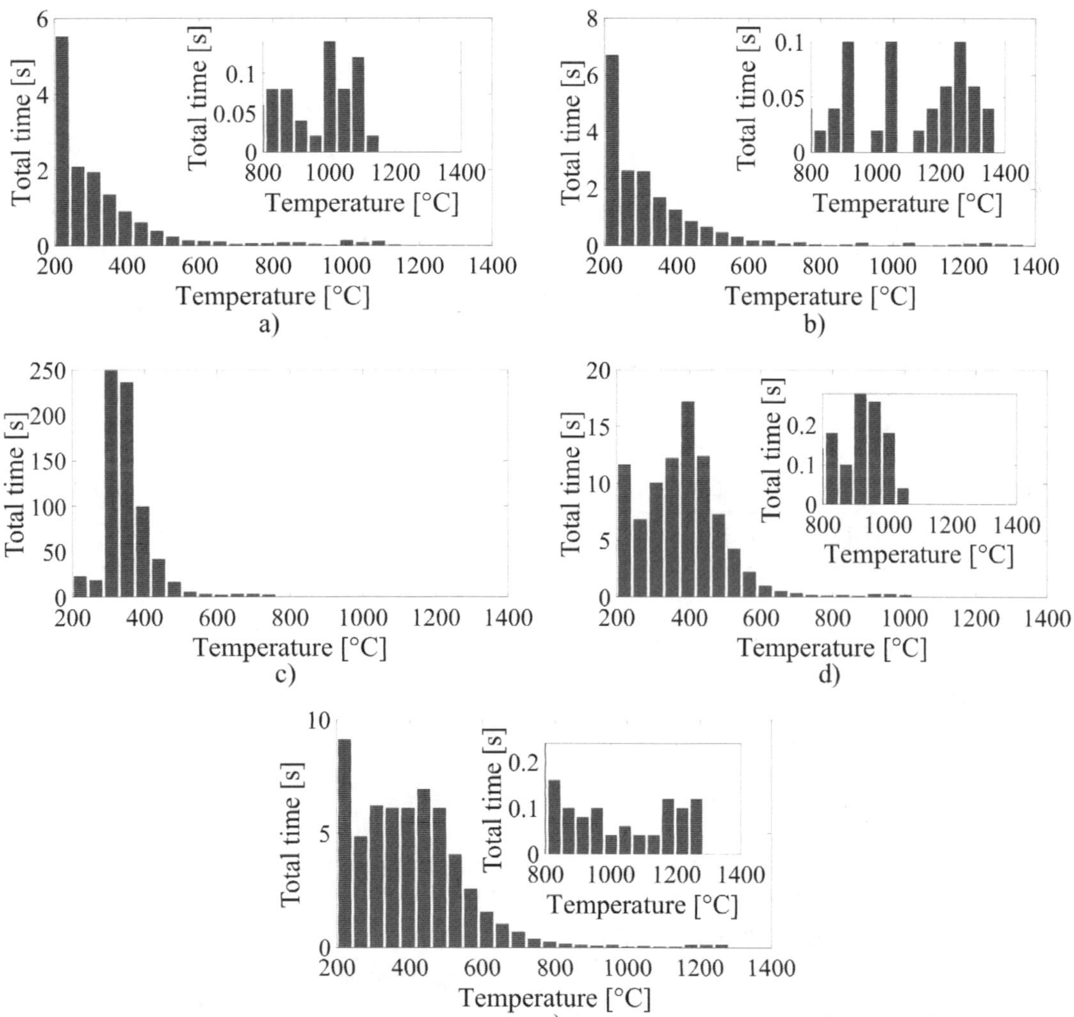

Figure 5. Total temperature history for: a) 1000°C , 90 seconds dwell time. b) 1300°C , 90 seconds dwell time. c) 700°C , 3.5 seconds dwell time. d) 1000°C , 3.5 seconds dwell time. e) 1300°C , 3.5 seconds dwell time.

IOP Conf. Series: Materials Science and Engineering 1296 (2023) 012041 doi:10.1088/1757-899X/1296/1/012041

Table 3. Time above certain temperatures for each of the samples, measured by the averaging the selected nine pixels in the ROI.

Peak Temperature Dwell time [s]	1000°C 90	1300°C 90	700°C 3.5	1000°C 3.5	1300°C 3.5
Time above 400°C	2.64	3.98	106.94	36.28	26.94
Time above 600°C	0.96	1.16	11.12	3.00	4.38
Time above 800°C	0.58	0.62	0	1.04	0.98
Time above 1000°C	0.30	0.42	0	0.12	0.50
Time above 1200°C	0	0.26	0	0	0.22

3.3. Energy

The total energy used to form each part can be seen in Table 4 along with the energy efficiency. This contains only energy output of the laser and does not account for energy usage of the system under operation. It is interesting to note that while the 700°C sample is about 10 times more inefficient compared to higher temperature forming, there is less difference between the other samples. It can appear that the higher temperatures are more efficient with a lower dwell times in comparison to longer dwell times. However, the difference is of a size that warrants more experiments before anything conclusive can be determined. Nonetheless, it is interesting to note that as it stands, it does not appear that there is a significant difference, allowing time constraints to be dominant during process planning.

Table 4. Energy consumption of the output of the laser during laser forming. Not including efficiency of laser and energy consumption of support equipment.

Peak Temperature Dwell time	1000°C 90	1300°C 90	700°C 3.5	1000°C 3.5	1300°C 3.5
Total angle	34.01	34.8	23.51	34.23	33.34
Total energy [kJ]	5.50	6.00	45.89	6.11	5.51
Efficiency [kJ/°]	0.16	0.17	1.95	0.18	0.17

4. Conclusion

Laser forming samples have been made using different peak temperatures validated using thermographic measurements. The results have been used to compare the total processing time, equipment time and energy applied.

- Bending with a long dwell time (90 seconds) creates a 30-80% more efficient use of the equipment depending on peak temperature during processing.
- Bending with a short dwell time (3.5 seconds), requires adjustments of the power used per pass but can reduce the total processing time by a factor of three.
- Bending to almost 30 degrees uses almost 10 times more energy at a peak temperature of 700°C compared with above 1000°C due to a more pronounced bend rate decay.
- There is a limited difference in the total amount of energy used when comparing long and short dwell time.

The results lastly indicate that scheduling can be used to achieve a much faster process by efficient use of the equipment either in the form of long dwell times or short dwell times depending on how many simultaneous bends can be performed.Furthermore, it would also be an interesting study to compare the metallurgical properties depending on the thermal history of the workpiece.

Acknowledgments
The authors would like to thank the Poul Due Jensen Foundation under grant with number 2020-006 entitled "Laser Forming".

References

[1] Lawrence J, Schmidt M J and Li L 2001 *International Journal of Machine Tools and Manufacture* **41** 967–977 ISSN 08906955
[2] Cheng J and Lawrence Yao Y 2001 *Journal of Manufacturing Processes* **3** 60–72 ISSN 1526-6125 URL `https://www.sciencedirect.com/science/article/pii/S1526612501700345`
[3] Shen H, Hu J and Yao Z Q 2011 *Materials Science Forum* vol 663 (Trans Tech Publ) pp 58–63
[4] Lambiase F, Ilio A and Paoletti A 2013 *International Journal of Advanced Manufacturing Technology* **64**
[5] Shen H, Ran M, Hu J and Yao Z 2014 *Optics and Lasers in Engineering* **62** 1–8
[6] Paramasivan K, Das S, Marimuthu S and Misra D 2018 *The International Journal of Advanced Manufacturing Technology* **94** 2137–2147
[7] Shen H, Shi Y and Yao Z 2006 *Proceedings of the Institution of Mechanical Engineers, Part C: Journal of Mechanical Engineering Science* **220** 507–511
[8] Lambiase F, Di Ilio A and Paoletti A 2016 *The International Journal of Advanced Manufacturing Technology* **86** 259–268
[9] Hao Y and Lien J M 2019 *Proceedings of the 3rd Annual ACM Symposium on Computational Fabrication* pp 1–11
[10] Hao Y, Guan W, Hernandez E A P and Lien J M 2021 *2021 IEEE International Conference on Robotics and Automation (ICRA)* (IEEE) pp 7788–7794
[11] Bachmann A L, Dickey M D and Lazarus N 2020 *Quantum Beam Science* **4** 44
[12] Safari M, Alves de Sousa R and Joudaki J 2020 *Metals* **10** 1472
[13] Edwardson S P, Abed E, Bartkowiak K, Dearden G and Watkins K G 2006 *Journal of Physics D: Applied Physics* **39** 382 URL `https://dx.doi.org/10.1088/0022-3727/39/2/021`
[14] Nikolov G N, Thomsen A N and Kristiansen M 2021 *IOP Conference Series: Materials Science and Engineering* **1135** 012006 URL `https://dx.doi.org/10.1088/1757-899X/1135/1/012006`

IOP Conf. Series: Materials Science and Engineering 1296 (2023) 012042 doi:10.1088/1757-899X/1296/1/012042

Laser surface texturing for superhydrophobic, icephobic and friction reduction functionalization

Antonio Ancona[1,2,3], Caterina Gaudiuso[3], Annalisa Volpe[3,4], Francesco Paolo Mezzapesa[3], Carmine Putignano[3,5], Giuseppe Carbone[3,5]

1. Università degli Studi di Bari ALDO MORO, Physics Department, via Orabona 4, 70125, Bari, Italy;
2. Univesity WEST, Department of engineering Science, 461-32 Trollhättan, Sweden;
3. CNR-IFN Institute of Photnics and Nanchnologies, Physics Department, via Amendola 173, 70126, Bari, Italy;
4. Politecnico di Bari, Physics Department, via Orabona 4, 70125, Bari, Italy;
5. Politecnico di Bari, Department of Mechanics, Mathematics and Management, via Orabona 4, 70125, Bari, Italy.

Corresponing author email: antonio.ancona@uniba.it

Abstract. Nanometre and/or micrometer scale periodic features naturally present decorating the skin of many insects, fishes or plants have been found to give them extraordinary aesthetic, adhesion, antifouling, wear resistance, superhydrophobic or drag reduction properties. With the aim of artificially conveying the same properties to products and tools, researchers have tried to reproduce such bionic topography on the surface of a huge variety of materials using different manufacturing methods. Among the different manufacturing technologies, surface texturing based on direct laser writing using ultrashort pulses offers great flexibility in terms of geometry and size of the feature to be created. This work will focus on two specific types of functionalization of metallic as well as rubber surfaces through laser surface texturing, i.e., icephobicity and friction reduction. In the first case, different texture patterns were laser written on sheet samples of Aluminum Alloy of interest for aerospace applications. Their wettability and anti-icing properties were investigated through water contact angle experiments in different environmental conditions and steam-freezing experiments were conducted in a climate chamber. The role of the laser textured pattern geometry on the frictional performances of a rubber surface under lubricated non-conformal as well as conformal point contact has been also widely investigated. In the first case, no improvements were obtained in terms of frictional performances, ascribable to the reduced number of dimples involved in the contact area. Conversely, remarkable friction reduction was achieved in mixed and in the transition between the mixed and the elasto-hydrodinamic lubrication regime, thanks to the concurrence of several effects as micro-cavitation, wear debris trapping, and lubricant entrapping within the texture.

1. Introduction

Surface science and tribology research is more and more focused on the challenging objective of designing and modifying contact interface topography at the micro and nanoscales to optimize the tribological performance of mechanical systems in terms of friction, wear, and drag reduction as well as changing the wettability of surfaces of many different materials [1]. However, engineering the topography of surfaces does not only require the designing part but also the development of a reliable manufacturing process that must be flexible enough to accomplish any kind of geometrical surface pattern and applicable to any kind of substrate material.

One promising solution that fulfils most of such requirements is Laser Surface Texturing (LST). Direct Laser ablation allows great flexibility for the creation of well-controlled microstructures on metals [2,3], semiconductors, dielectrics [4] and polymers [5]. During the ablation process, material is removed from the substrate, and the resulting features depend on laser parameters such as pulse duration, laser fluence, and repetition rate. The achievable precision and reproducibility of the laser micro-machined features is adversely affected by the duration of the pulses. The longer the pulse duration, the higher is the thermal damage to the substrate and, as a consequence, the lower the quality and the precision of the texture. Thanks to the much shorter laser-matter interaction time, laser ablation using picosecond or femtosecond pulses allows to significantly reduce the thermal load onto the material improving precision and reproducibility.

As regards the functionalities that a laser texturing process can confer on a surface, there are many examples in literature such as aesthetic, antifouling, antibacterial properties, biocompatibility, reduction of friction and wear [6], hydrophilicity [7], hydrophobicity [8], self-cleaning [9], and many other properties are under development.

Two different types of femtosecond laser texturing protocols will be examined in this work on two different kinds of materials to convey them different functionalizations, i.e. reducing lubricated friction on a fluoroelastomer layer and making an aluminum alloy substrate superhydrophobic and icephobic. The first case study has been selected because, while it is well known in literature that in metallic materials a texture consisting of an array of dimples enables friction reduction in a wide range of lubrication regimes thanks to local cavitation, reduction of viscous losses and entrapment of wear depris at the dimples locations which also serve as micro-reservoirs for lubricant [10], how can this effects be effectively replicated in soft lubrication has been scarcely investigated. This is particularly relevant if we consider that a significant shift is occurring in the industry, transitioning from metals to polymers due to their lower cost, lighter weight, and environmental compatibility. For the second case study an aluminum alloy that is widely used for the fabrication of aircrafts has been selected as substrate material and its icephobicity after the laser texturing process has been investigated. This is relevant because aircraft wings and fuselage are susceptible to ice accretion during their operation.

Previous works have shown that superhydrophobic surfaces [11], possess exceptional self-cleaning properties and notable anti-icing capabilities [12]. The fabrication of superhydrophobic surfaces using ultrashort lasers has been demonstrated on various metallic substrates, such as AA7075 [13] and stainless steel [14]. Additionally, the fabrication of metallic superhydrophobic surfaces through direct laser writing and direct laser interference patterning has also been reported [3,15]. However, only a few research papers in the literature explore the anti-icing properties of directly fabricated ultrashort laser-induced superhydrophobic metallic surfaces. Here, the wetting properties of laser textured AA2024 surfaces is investigated and their ability to hinder ice formation at sub-zero temperatures.

2. Materials and methods

2.1. Preparation and characterization of friction reduced rubber surfaces

A Yb:KGW laser from Light Conversions (mod. Pharos) was used to texture fluroelastomer sheets. This laser source provides pulses with a pulse duration of 200 fs, at central wavelength of 1030 nm, and a maximum pulse energy of 1.5 mJ. The repetition rate can be varied between 1 kHz and 1 MHz with maximum average power of 6 W. The linear polarization of the pulses exiting the laser source is converted into circular with a quarter waveplate, before guiding the laser beam into a galvoscanner (IntelliSCANse 14, SCAN-LAB, Puchheim, Germany) equipped with a 163 mm F-Theta lens which focuses the beam to a spot of 70 μm diameter. The repetition rate and the applied laser fluence exploited for the texturing were set at 1 kHz and 0.2 J/cm^2, respectively, to inhibit heat accumulation which would be detrimental to the precision of the texturing process.

Two texture patterns were fabricated on a fluoroelastomer characterized by a rubbery modulus of about 10 MPa and a reduced viscoelasticity at room temperature. Such microtextured patterns consisted of uniform squared matrices of circular dimples differing for the diameter and the mutual center-to-center distance. In particular, the Texture A was characterized by bigger (diameter = 100 μm) and deeper (depth = 50 μm) dimples, placed at a distance of 150 μm, designed to store and provide a high quantity of lubricant. The Texture B was instead produced with dimples with a diameter of 50 μm, a depth of 10 μm and a mutual distance of 100 μm. This latter pattern was designed to minimize the formation of eddies inside the dimples, though still fulfilling the role of lubricant storage. To micromachine these two different patterns, the laser beam was moved onto the sample surface along parallel lines having a hatch distance of 2 μm with a speed of 3 mm/s (Texture A) and 16 mm/s (Texture B), respectively.

The tribological performances of the fabricated textures in terms of their coefficient of friction (COF) were tested by using a CSM High Temperature pin-on-disk Tribometer (THT), in two different configurations, In the first case, a ball-on-disk configuration was used with a 100Cr6 steel ball, with a diameter of 10 mm, in a non-conformal contact with the textured rubber samples. Conversely, conformal contact was tested by using a spherical cap on the polymer patterned samples: in this case, the contact area was a 5mm-diameter disk. In both cases, the finished metal counterpart had a root-mean-square roughness of around 20 nm, while that of the rubber layer was approximately 500 nm.

Different lubrication regimes were investigated by exploiting three different lubricants: 1) pure water, with a viscosity η equal to 0.001 Pa·s; 2) a water-glycerol mixture with $\eta = 0.0155$ Pa·s; 3) pure glycerol with $\eta = 0.644$ Pa·s. Moreover, the linear speed tested on the tribometer was varied between 0.002 m/s to 0.4 m/s, thus exploring a wide viscosity-speed η U range. This procedure is widely accepted when the contribution to friction is only ascribable to lubricant viscous losses [16-19].

In addition, static water contact angle (WCA) measurements were performed to test the influence of the wettability on the tribological properties of the textured samples in the boundary regime. To this aim, a digital goniometer consisting of a Dino-lite portable microscope backlighted by a cold light lamp was used. To this aim, 5 μl water microdroplets were deposited on the samples, under static mode.

2.2. Preparation and characterization of icephobic metallic surfaces

Flat Aluminum Alloy 2024 sheets of thickness 1 mm were used. This kind of alloy has copper as primary alloying element and, thanks to its high strength to weight ratio, as well as good fatigue resistance, it finds wide use in the fabrication of aircraft wings and fuselage. The samples were washed with isopropyl alcohol in an ultrasound bath for 10 minutes and then processed with an ultrafast solid state laser system (mod. TruMicro Femto Ed. From Trumpf GmbH). The laser source emits at a wavelength of 1030 nm and provides an almost diffraction limited ($M^2 < 1.3$) linearly polarized beam, with pulse duration of 900 fs.

The beam was moved over the sample thorough a galvo-scan head (IntelliSCANse 14, SCAN-LAB, Puchheim, Germany), equipped with a 100-mm telecentric lens which focalized the beam with an apparent spot size with diameter of about 25 μm. A crosshatch texture, as previously optimized in [20], was scribed on the sample surface exploiting the parameters in Tab.1. Each laser texturing has been carried out in air atmosphere on target area of 2-cm × 2-cm.

Table 1. Process parameters for the microstructuring of the Al2024 via laser ablation.

Pulse energy E_p(μJ)	12.2
Repetition rate f (kHz)	100
Scan seed v (mm/s)	50
Hatch distance h (μm)	50

After the laser treatment, the samples were thermal aged in a climate chamber (Temperature Test Chambers-ACS 16) at 70 °C for 15 h. The climate chamber temperature can range from -30°C to 130°C, so the same chamber was used for the icing tests conducted at sub-zero temperatures. The topography of the surface structures was characterized using a scanning electron microscope (SEM, Carl Zeiss mod. Σigma). The wettability tests were performed with the apparatus described in [21]. For the icing tests, the sample was fixed on a holder inclined at 10° and put in a climate chamber at -20°C, as presented schematically in Fig. 1. The experiment was recorded by Dino-PLUS camera. The 10 μL deionized water droplets were continuously and uniformly dropped thanks to a peristaltic pump Minipuls 3 (Gilson, Middleton, WI, USA) at a rate of 2 droplets/s from a height of about 5 cm above the sample.

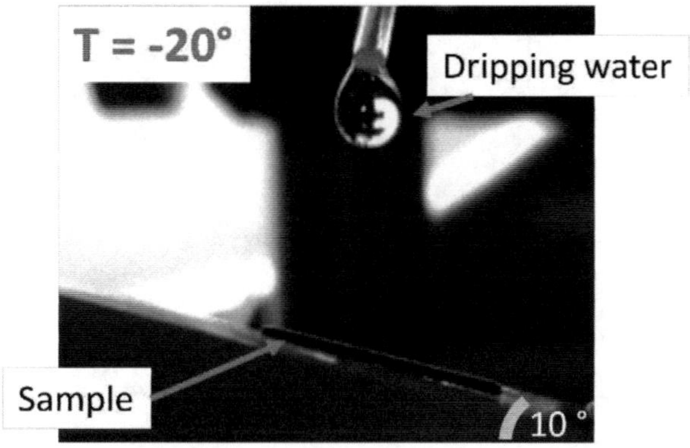

Figure 1. Dripping test setup. The sample was fixed on a holder inclined at 10°

3. Results and discussion

3.1. Friction reduction

In Figure 2, the frictional properties of the two laser fabricated textures under non-conformal contact are shown. Firstly, it can be seen that the textured surfaces were tested in a wide range of lubrication regimes, i.e., the boundary, mixed and elasto-hydrodinamic lubrication, thus it was possible to reconstruct the entire Stribeck curve for each sample. This was possible, thanks to the possibility of varying not only the sliding speed of the tribometer but also the viscosity of the lubricant by mixing with different percentages water and glycerol (only water, mixture between water and glycerol, only glycerol). Results show that the frictional performances of both laser textured rubber surfaces were not improved in comparison with the untextured reference, rather a significant worsening was revealed with an increase of the COF along the entire Stribeck curve. This could be ascribed to the reduced number of dimples involved in the contact. In fact, using Herzian relations and taking into account the applied normal load of 1 N, the number of dimples participating in the contact was estimated to be approximately a few tens for the Texture B and even fewer for Texture A, where the distance between the dimples is larger. This causes the contribution expected from cavitation to reduce friction to be negligible, thus resulting in a worsening of the frictional behavior. However, no improvements were observed in the boundary lubrication regime, too. Here, a reduction of friction was expected due to the fact that the dimples act as microtraps for wear debris, as it was observed for laser textured metal surfaces [22]. This mechanism was not observed in laser textured rubber surfaces probably due to the fact that the polymeric particles generated by the wear of the rubber samples are bigger in size and cannot be entrapped by the dimples [23].

IOP Conf. Series: Materials Science and Engineering 1296 (2023) 012042 doi:10.1088/1757-899X/1296/1/012042

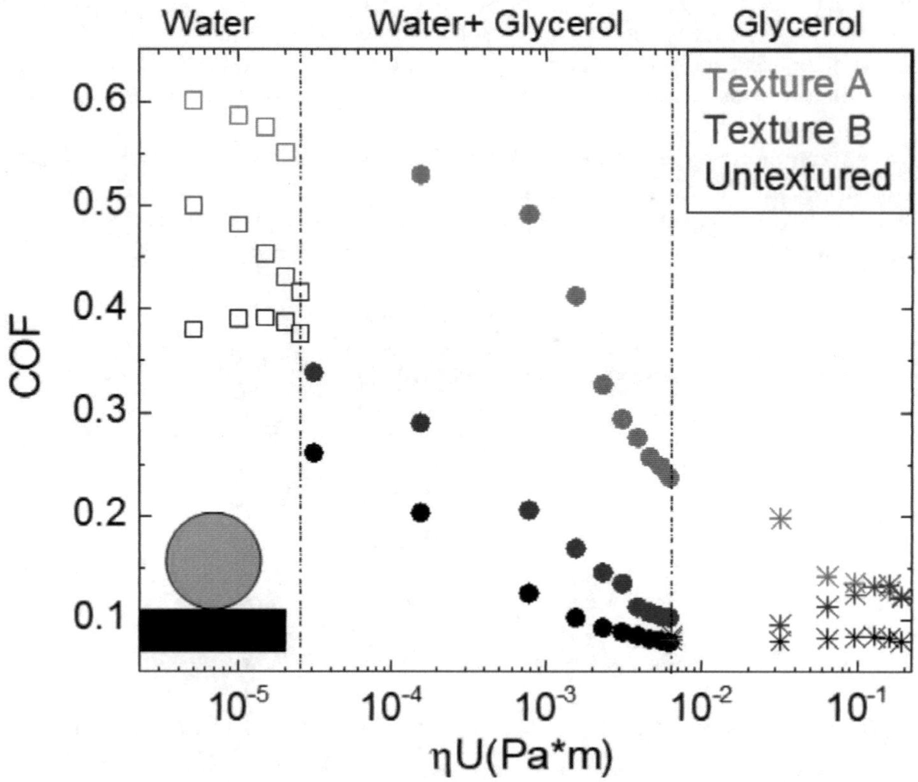

Figure 2. Coefficient of friction (COF) versus ηU in the non-conformal ball-on-disk configuration for the two laser textured samples and the untextured one. Open squares, solid circles and asterisks refer to tests performed with water, water and glycerol and pure glycerol, respectively. The normal load P is equal to P = 1 N and measurement scatter is always below 10%.

Moreover, it is known from literature that a laser surface texturing causes a significant change of the wettability of many different materials, making them hydrophobic or super-hydrophobic [8,21]. If this is the case also for rubber, a change to hydrophobic behavior would hinder the formation of a uniform lubricant film on the surface thus causing an increase of the COF. Water contact angle measurements performed on the pristine rubber surfaces and the laser textured ones are reported in Tab. 2, which confirm that the laser texturing produced an increase of the WCA from 70° up to 132° for texture A.

Table 2. Water contact angle (WCA) of the untextured and textured surfaces.

Surface Type	WCA (°)
Untextured	70 ± 5
Texture A	132 ± 2
Texture B	114 ± 2

Under conformal contact the spherical cap determined a significantly larger contact area compared to the non-conformal case and, as a consequence, a notable improvements of the frictional performances were found with the Texture B, especially in mixed and in elasto-hydrodinamic lubrications, where a reduction of up to 60% of the COF was measured compared to the untextured (see Figure 3).

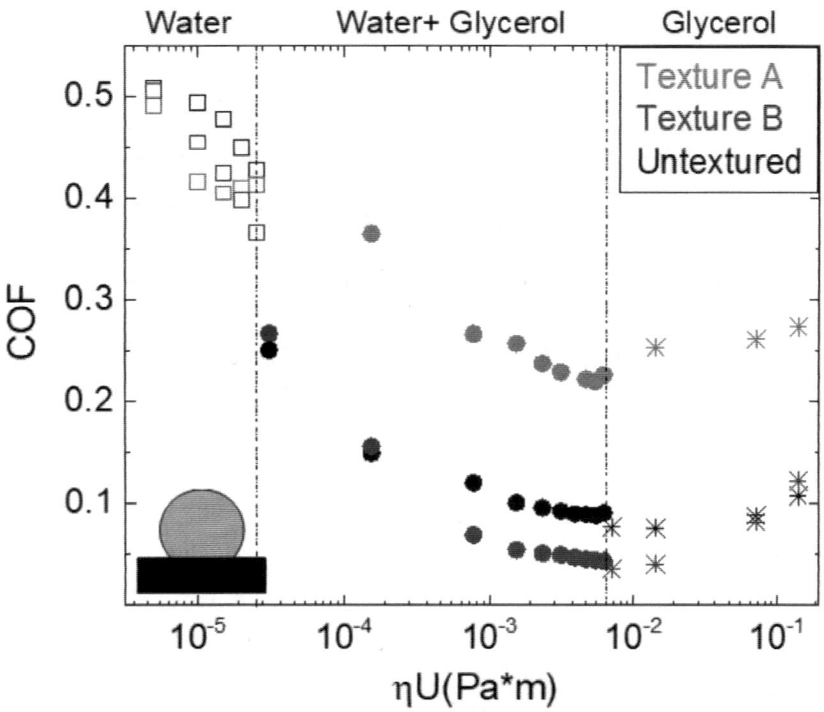

Figure 3. Coefficient of friction (COF) versus ηU in the conformal cap-on-disk configuration for the two laser textured samples and the untextured one. Open squares, solid circles and asterisk refer to tests performed with water, water and glycerol and pure glycerol, respectively. The normal load P is equal to P = 1 N and measurement scatter is always below 10%.

Different tribo-contact mechanisms can explain this behavior. Firstly, the dimples act as lubricant micro-reservoir, contributing to keep the two rubber interfaces separated, thus reducing friction. On the other hand, the geometry of the dimples determines two important effects which have two opposites consequences on the COF: cavitation and the fluid turbulence. While cavitation is expected to increase the load-carrying capacity of the lubricants and then contributes to friction reduction, the arousing of turbulences within the lubricant film causes fluid losses and, consequently, an increase of the COF [6,24]. The different behaviors of textures A and B observed in Figure 3 can thus be explained in terms of the interplay between these two competing mechanism. The different geometrical characteristics of the two textures make one effect prevail to the other. In fact, the shallow dimples of Texture B promoted cavitation and hindered the formation of eddies. Instead, the higher depth of the dimples of the Texture A fostered the formation of eddies which caused a notable deterioration of the frictional performance [6].

A minor reduction of the COF (up to 20%) was measured in the boundary lubrication regime, where the dimples acted as lubricant micro-reservoirs, which also explains the better performances obtained with Texture A characterized by deeper dimples. At the same time, in this lubrication regime the sliding speed was too slow to trigger cavitation and eddies formation.

3.2. Icephobic

In Figure 5 the pristine aluminum sample (P) is compared with the laser textured one (T). Sample T was obtained exploiting the laser texturing parameters summarized in Table 1.

IOP Conf. Series: Materials Science and Engineering 1296 (2023) 012042 doi:10.1088/1757-899X/1296/1/012042

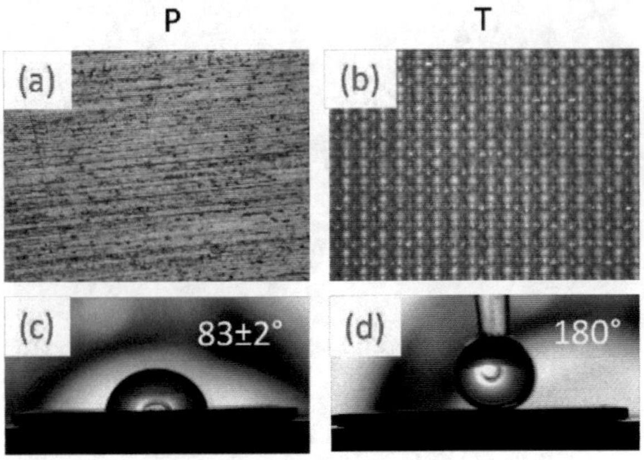

Figure 4. Optical microscope image of an untextured (a) and a laser textured sample (process parameters: hatch 50 um @1W, 5mm/s, 100kHz) (b). Water contact angle measurement (with a droplet of 10 μL) on an untextured (c) and laser textured (d) sample.

The pristine sample exhibited hydrophilicity with a measured contact angle slightly above 80° (Fig. 1c). Immediately after the laser treatment, the T surface was highly hydrophilic. In fact, as the water was dispensed on the textured surface it instantaneously spread. Thus, it was not possible to measure a CA. This behavior is known from literature and has been ascribed to the formation of hydrophilic metal oxides compounds during or immediately after the laser patterning process [25,26].

On the contrary, after that the T surface was exposed to a heat treatment with the purpose to accelerate the natural aging process, when it was attempted to gently deposit the droplet on the surface it remained attached to the micropipette (Figure 1 (d)), still maintaining its original shape even if pressed. When released, the droplet bounced off. When such behavior occurs, it is commonly assumed that the CA is equal to 180° [27]. The bouncing of the water droplet is attributed to the presence of air between the droplet and the solid surface. The air, trapped within the cross-hatched structures, acts as a cushion and reduces the direct solid–liquid interaction, as predicted by the Cassie–Baxter model [28].

The dynamic anti-icing behavior of the two samples is showed in Figure 5. The water droplets were continuously and uniformly dropped from a height of about 1.5 cm above the samples exposed in the climate chamber to a temperature of −20 °C. After 5 minutes, the untreated surface became covered with ice, while no frozen spot could be observed on the textured surface. This phenomenon can be ascribed to the different wettability proprieties of the two surfaces. When the droplet fell onto the pristine AA2024 sample, it remained and spread on the surface due to the intrinsic hydrophilic nature of the material, thus freezing after a while (Fig. 5a). Conversely, thanks to the water repellent behavior of the textured surfaces even in a sub-zero environment, the drop bounced off instantly, thus avoiding to freeze and start generating an ice layer (Fig. 5b).

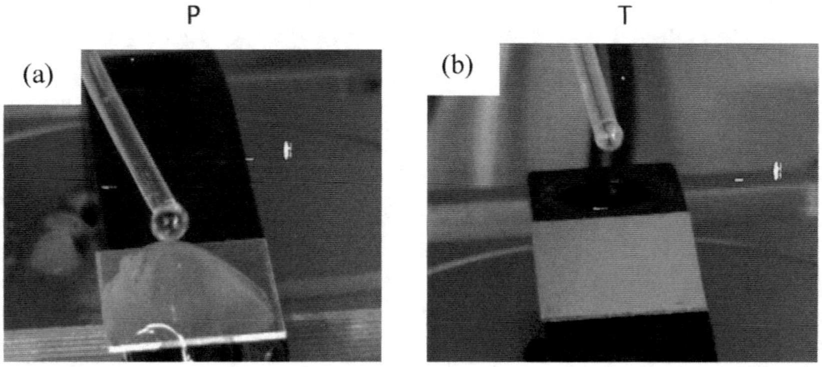

Figure 5. Ice accumulation on: (a) a pristine (P) and (b) a textured (T) surface after exposure at – 20°C for 5 minutes.

4. Conclusions

In this study, we demonstrate how laser surface texturing using ultrafast laser pulses can be successfully utilized to functionalize surfaces of very different substrate materials as metals or rubber to confer them extraordinary tribological properties that have potential applications in quite different industrial fields as automotive and aerospace. Specifically, we produced by laser ablation a texture pattern consisting of an array of microdimples on a fluoroelastomer layer with the aim to reduce friction under lubricated contact. It was found that results vary greatly depending on the type of contact and on the geometrical characteristics of the texture pattern. In non-conformal contact, the texture does not produce any friction reduction due to the limited number of dimples engaged in the contact region, resulting in limited cavitation or entrapment of polymer debris at the bottom of the dimples. This is also a consequence of the lower wettability observed in the textured samples. The outcome is quite different under conformal contact, where the contact region is larger. In the regime of elastohydrodynamic lubrication, two competing mechanisms come into play. On one hand, the micro-holes promote the formation of cavitated regions in the fluid which reduce friction, while on the other hand, turbulent eddies are likely to be generated which would increase friction. Here, the geometrical characteristics of the texture play a remarkable role. For deep micro-dimples the turbulence effects prevail and the frictional performance is deteriorated. For smaller dimples, up to a 60% reduction of friction was measured in mixed lubrication. In boundary lubrication, despite the lower wettability associated with the micro-textured surface, both textures demonstrate a modest friction reduction. Therefore, it can be concluded that reducing the coefficient of friction of rubber samples is possible as long as the texture pattern is properly designed, aiming to minimize the turbulence occurring inside the dimples while optimizing cavitation [19].

In the second case study, fs-laser laser texturing was used to produce surfaces with strong water-repellent properties and resistance against icing on AA2024 substrates. Here, a microgrid pattern was created on the surface by laser ablation. After a heat treatment to stabilize the surface chemistry, the textured samples exhibited superhydrophobic wetting behavior which was preserved also at subzero temperatures. In contrast, the untextured samples showed a natural hydrophilic wetting behavior. This led to uncontrolled ice formation at subzero temperatures for the untextured samples, while the textured ones demonstrated dynamic anti-icing performance even in extremely cold conditions (-20 °C).

Further measurements will be devoted to investigate durability of the laser texture patterns and of their tribological performance on the long-term. If such results are confirmed, this technology promises to have a tremendous impact in the functionalization of many automotive as well as aerospace components.

Acknowledgements
The authors gratefully acknowledge Pietro Paolo Calabrese for his technical support. This research was funded by the Italian Ministry of Education, University and Research (MIUR) within the projects TITAN [PON ARS01_00906] and EXTREME [PON ARS01_00849]; and the Apulian Region within the "Research in Innovation" program [Project FEB1B50F], in the framework of POR PUGLIA FESR—FSE 2014/2020.

References
[1] Holmberg K and Erdemir A 2017 *Friction* **5** 263
[2] Ancona A, Carbone G, De Filippis MD, Volpe A and Lugarà PM 2014 *Adv. Opt. Technol.* **3** 539
[3] Cardoso JT, Aguilar-Morales AI, Alamri S, Huerta-Murillo D, Cordovilla F, Lasagni AF and Ocaña JL 2018 Opt. *Lasers* Eng. **111** 193
[4] Gaudiuso C, Volpe A and Ancona A 2020 *Micromachines* **1** 327
[5] Riveiro A, Maçon ALB, del Val J, Comesaña R and Pou J 2018 *Frontiers in Physics* **6** 16
[6] Scaraggi M, Mezzapesa FP, Carbone G, Ancona A and Tricarico L 2012 *Tribol Lett* **49** 117
[7] Pou P, del Val J, Riveiro A, Comesaña R, Arias-González F, Lusquiños F, Bountinguiza M, Quintero F and Pou J 2019 *Appl. Sur. Sci.* **475** 896
[8] Giannuzzi G, Gaudiuso C, Mundo RD, Mirenghi L, Fraggelakis F, Kling R, Lugarà MP and Ancona A 2019 *Appl. Surf. Sci.* **494** 1055
[9] Jagdheesh R, Diaz M and Oca JL 2016 *RSC Adv.* **6** 72933
[10] Scaraggi M, Mezzapesa FP, Carbone G, Ancona A Sorgente D and Lugarà PM 2014 *Tribol Int* **75** 123
[11] Guo Z, Liu W and Su B 2011 *J. Colloid Interface Sci.* **353** 335
[12] Jamil MI, Ali A, Haq F, Zhang Q, Zhan X and Chen F 2018 *Langmuir* **34**, 15425
[13] Jagdheesh R, Hauschwitz P, Mužík J, Brajer J, Rostohar D, Jirícek P, Kopecek J and Mocek T 2019 *Appl. Surf. Sci.* **493** 287
[14] Wang H, He M, Liu H and Guan Y 2019 *ACS Appl. Mater. Interfaces* **11** 25586
[15] Vercillo V, Tonnicchia S, Romano J, García-Girón A, Aguilar-Morales AI, Alamri S, Dimov SS, Kunze T, Lasagni AF and Bonaccurso E 2020 *Adv. Funct. Mater.* **30** 1910268
[16] Carbone G and Putignano C 2013 *J. Mech. Phys. Solids* **61** 1822
[17] Putignano C, Carbone G and Dini D 2016 *Phys Rev* **93** 043003
[18] Putignano C and Dini D. 2017 *ACS Appl Mater Interfaces* **9** 42287
[19] Putignano C, Scarati D, Gaudiuso C, Di Mundo R, Ancona A and Carbone G 2019 *Trib. Int.* **136** 82
[20] Volpe A, Gaudiuso C, Di Venere L, Licciulli F, Giordano F and Ancona A 2020 Coatings **10** 587
[21] Volpe A, Covella S, Gaudiuso C and Ancona A 2021 Coatings **11** 369
[22] Joshi GS, Putignano C, Gaudiuso C, Stark T, Kiedrowski T, Ancona A and Carbone G. 2018 *Tribol Int* **127** 296
[23] Persson BNJ 2009 *J Phys: Condens Matter* **21** 485001
[24] Vladescu SC, Medina S, Olver AV, Pegg IG and Reddyhoff T 2016 *Tribol Lett* **62** 19
[25] Quére D, Lafuma A and Bico J 2003 *Nanotechnology* **14**, 1109
[26] Cardoso JT, Garcia-Giron A, Romano JM, Huerta-Murillo D, Jagdheesh R, Walker M, Dimov SS and Ocana JL 2017 *RSC Adv.* **7** 39617
[27] Zhang D, Chen F, Yang Q, Yong J, Bian H, Ou Y, Si J, Meng X and Hou X 2012 *ACS Appl. Mater. Interfaces* **4** 4905
[28] Cassie ABD and Baxter S 1944 *Trans. Faraday Soc.* **40** 546

IOP Conf. Series: Materials Science and Engineering 1296 (2023) 012043 doi:10.1088/1757-899X/1296/1/012043

Laser-induced reduction of iron ore: Trends of the chemical distribution across the domains

A F H Kaplan[1], T Fedina[1], F Brueckner[1,2] and H Naesstroem[1]

[1] Luleå University of Technology, Department of Engineering Sciences and Mathematics, 97 451 Luleå, Sweden
[2] Fraunhofer IWS, Winterbergstrasse 28, 01277, Dresden, Germany

alexander.kaplan@ltu.se

Abstract. The reduction of iron ore powder in a laser-induced thermal cycle using several reducing agents was studied. The laser-assisted reduction process resulted in the formation of iron-rich domains, irregularly embedded in a slag matrix, and transitional phases. The appearance of these various chemical phases was categorized and geometrically evaluated with respect to representative dimensions using scanning electron microscopy. The statistical trends of the morphology are presented, in context with trends of the chemical composition across the domains, to gain a better understanding of the mechanisms behind the reduction process. Iron domains were predominately observed in the vicinity of the Si-rich zones, indicating the occurrence of the reduction reaction as a result of the Fe_2O_3-Si interaction. Furthermore, different appearances of the Fe-rich domains and other phases in the close proximity to iron were analyzed and discussed based on diffusion and coalescence phenomena. The obtained results show that the reduction occurs, but the process is still uncontrolled and only partially understood. Further analysis and experiments are, therefore, needed to investigate the prospects of the proposed method.

1. Introduction

A method was recently developed [1-3] to achieve reduction of iron ore powder to iron rich domains, by adding a suitable reducing agent powder and laser-inducing a melting cycle to both powder components. The iron-rich zones have been evidence for a successful chemical reaction. The method turned out to be suitable to study ore reduction at small scale (~0.5…5 mm) and with fast thermal cycles (~0.1…1 s). Normally reduction processes take place during minutes [4], or even longer. The findings can be of interest for steelmaking, either for better understanding of the materials behaviour during the established steelmaking techniques or as a new route to enable ore reduction and steelmaking in a 3D-printer, additive manufacturing, which is a disruptive challenge, with pros and cons.

Established steelmaking is based on reduction of iron ore by CO via coal, generating pig iron and CO_2. The most common method of reducing iron ore to iron is still by blast furnace, and traditional iron ore reduction has concentrated on the use of coal, coke, oil, methane and natural gas [5]. However, more recent work has looked into the utilisation of electricity, biomass and hydrogen, e.g. Warner [6]. For climate change mitigation, less harmful techniques would be favourable.

One future route is reduction by hydrogen, substituting coal, or CO. When studying the reduction by hydrogen, examples for enhanced reduction rates were achieved by suitably applying a magnetic field [7], longer reaction times of the order of 30 minutes [8] or for (calculated) ionization to high H-plasma temperatures [9].

IOP Conf. Series: Materials Science and Engineering 1296 (2023) 012043 doi:10.1088/1757-899X/1296/1/012043

Reduction of iron ore by a reducing agent is only partially understood, with respect to different reducing agents, technique approaches and thermal cycles, as one motivation for the here presented approach. A published example [10] for another reducing agent is K, where the reduction rate from Fe_2O_3 to Fe_3O_4 was studied for different catalysts (V and Cr), showing opposite trends, at a temperature of 630 °C under an S/H_2-atmosphere.

A single publication of reduction of ore by a laser-induced process can be found [11], for MgO, however during the gaseous phase, as a kind of the laser chemical vapour deposition technique, LCVD. The reducing agents C, Zr and Si were compared, with respect to mass and energy efficiency; all three materials have led to successful reduction, where the generated pure Mg was condensing (in vacuum) on a target. This technique was estimated for large scale reduction facilities, estimating the use of many high power lasers.

In the present study, of reducing iron ore powder by inducing a laser melting cycle, suitable chemical elements as reducing agents were systematically identified. Silicon was ranked highest, followed by Al. Si is commercially available as pure powder and the different material combinations have similar melting and boiling temperatures, to facilitate a calm process. Aluminium in turn is well known for its thermite reduction reaction with iron ore, though with the mismatch of a much lower melting temperature.

For silicon as the additional substance, manifold supporting knowledge can be found. In particular, calculation of the free Gibbs energy for the ternary system Fe-O-Si [12] predicts a distinct regime where iron ore would reduce to Fe, plus SiO_2. It is likely that the laser-induced remelting process causes thermochemical conditions where this reduction takes places. Nevertheless, during the first studies [3] no explicit SiO_2 was found, only as component of Fayalite, Fe_2SiO_4. Further studies on silicon [13] provide knowledge on the calculated of SiO_2-saturation in an iron ore melt, as a function of the Si-content and the liquidus temperature. Relevant knowledge for the mechanisms in the melt can also be extracted from numerical simulations, even for reactions in the other direction, for oxidation from FeO to Fe_2O_3 when mixed with SiO_2, where the diffusion mechanisms provide relevant knowledge [14]. A kinetic model calculating the mechanisms at the interface between an SiO_2-melt and a steel melt describes the different atomic movements [15,16], like oxidation causing oxygen depletion by Si- and O-diffusion.

Aluminium as a reducing agent candidate was studied for different purposes, including railway welding [17], pyrotechnics or metal matrix composite synthesis [18,19]. The well explored thermite reaction between Al and iron ore [20-23] is initiated in a temperature window of 1300-1360 °C. X-ray high speed, high resolution imaging of a thermite combustion process enabled to measure the spatial and temporal evolution of liquid domains during the process kinetics [17].

Despite much experience on iron ore reduction in steelmaking and manifold relevant studies, the complex mechanisms of iron ore reduction in a melt have only been studied under specific circumstances and only partially understood. Therefore a new approach, rapid laser melting of powder of both components, was initiated, to study the mechanisms and conditions under different circumstances that enable different analysis, exploration and optimization. Here the recent successful studies on iron ore reduction by Si and Al as reducing agents are investigated further.

So far only qualitative analysis was carried out [1-3] while in the present analysis the geometrical conditions are quantitatively expressed and compared.

2. Methodology

2.1. Experiments

Laser-induced reduction of iron ore by powder melting, including a reducing agent powder, was studied. Highly iron rich domains were found, as evidence that substantial reduction took place. Because of the complex mechanisms in the melt, the mechanisms are only partially understood. Therefore indications from the irregular distribution of the domains and chemistry are supporting knowledge for interpretation of the phenomena.

Two reducing agent powders were studied, Si and Al, which have led to very different appearances, as recently reported [1-3]. For the Si-experiments, Fe_2O_3-based iron ore powder (provided by LKAB) was sieved to grain sizes <40 µm, corresponding to a mean size D_{505}=28.0 µm. The Si-powder (purchased from Elkem Bremanger) was sieved < 45 µm, with D_{50}=23.8 µm. For the experiments with Al as reducing agent, coarser powder was utilized; iron ore powder (LKAB) with a size range of 63-150 µm, D_{50}=95.4 µm and Al-powder (99.5 % Al, m-tec powder GmbH), produced by atomization, with a size range of 63-200 µm, D_{50}=103.2 µm.

Figure 1(a) shows the principle placement of two different powder beds beside each other. The laser beam travels along the interface of the two beds, optionally with a slight lateral offset to one side. It can be noted that also other arrangements were studied, while the here presented results origin from set-ups as in Fig. 1(a). The experimental set-up for the laser melting experiments is shown in Fig. 1(b).

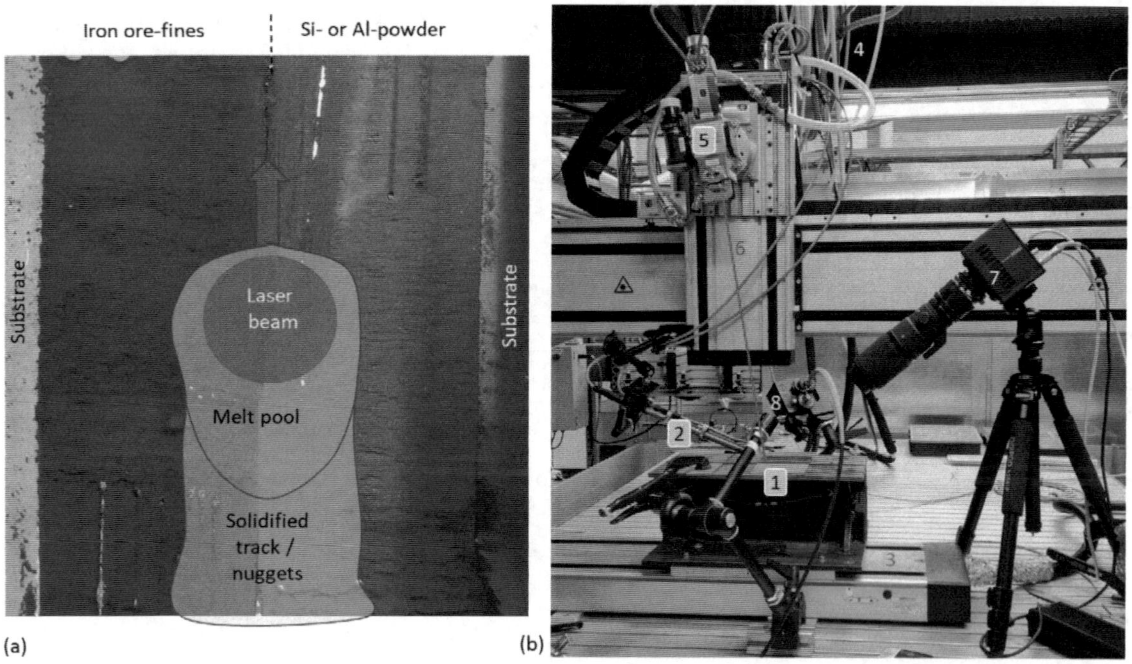

Figure 1. (a) Top view of the placement of two different powder beds beside each other; the laser beam then travels along the interface; (b) Experimental set-up (1: workpiece with powder, 2: shielding gas nozzle, 3: linear axis, 4: laser fibre, 5: processing head, 6: laser beam (illustrated), 7: high speed camera, 8: illumination laser).

The main process parameters like laser beam power, beam diameter, offset, travel speed, shielding gas flow and bed thickness were optimized to generate a smooth process. The applied process parameters are shown in Table 1.

Table 1. Main experimental process parameters.

Parameter	Si-cases	Al-cases
Laser beam power	300 W	300 W
Beam spot diameter	2.2 mm	1.1 mm
Lateral beam offset	Spot ~75% at ore-side	0, symmetric
Travel speed	0.5 m/min	0.5 or 1 m/min
Shielding gas	Ar	Ar

A continuous wave Yb:fibre laser (wavelength 1070 nm) with a maximum power of 300 W was used, model YLR-300-MM-AC, manufactured by IPG Photonics. The laser beam was delivered by an output fibre with a diameter of 50 µm, followed by a collimator lens with a focal length of 100 mm and a focusing lens with a focal length of 500 mm. The size of the defocused laser beam on the powder bed was 2.2 mm when defocusing +80 mm (above) and 1.1 mm when defocusing +40 mm. The nominal beam parameter product is 1.5 mm·mrad. The beam was inclined 8° in-line, to prevent backreflection issues. Ar was supplied as shielding gas (flow rate 10 L/min) from a tubular copper nozzle (diameter 20 mm) positioned sidewards, as can be seen in the experimental set-up in Fig. 1(b). An optimized thickness of the powder bed was 1 mm.

The solidification did not lead to continuous tracks but instead to nuggets wetted to the surface but easy to detach, see Fig. 2, which turned out to be very suitable. The nuggets were molded in epoxy and polished (but not etched), preferably at the flat bottom, for SEM- and EDS-analysis. From the many SEM-images, two representative Si-cases and four Al-cases were selected for the present study, as denominated in Table 2. They originated from three experiments and nuggets, with the shown travelling speed. Apart from the powder, all other parameters were the same, as described above and in Table 1.

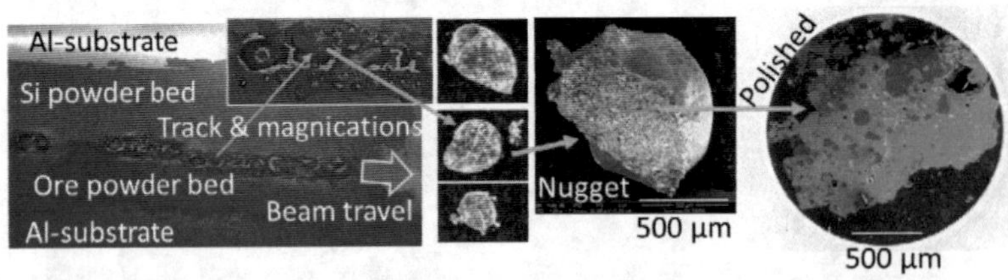

Figure 2. Resulting track after laser melting, separating into nuggets; nuggets, magnified; polished nugget surface.

Table 2. Nomenclature of the cases studied for image analysis, distinguishing Si and Al as reducing agent powder; the chosen travel speed is also listed.

Case	Speed [m/min]
Si-A	0.5
Si-B	0.5
Al-A	1.0
Al-B	1.0
Al-C	1.0
Al-D	0.5

2.2. SEM image analysis

The SEM-images and EDS-results contain important information about the mechanisms that have taken place and the resulting chemical distribution. While previous publications have qualitatively reported on the results, and quantitatively about the chemical element distributions, here quantitative evaluation of the domains is studied, as additional contribution to understand the mechanisms and formation trends. Mainly the image analysis software ImageJ (Fiji), version 1.52i has been applied. The SEM-images and EDS-images (SEM-images for a specific chemical element) have been converted to 8-bit greyscale in jpeg format. Individual threshold identification was necessary to distinguish sharp domains, which was only unambiguous for certain images, depending on the surrounding domain appearances. From particle analysis, certain properties were derived. In particular, the properties aspect ratio, roundness, solidity and area were suitable for the characterization. In contrast, circularity was not suitable, because it is very sensitive to detailed fluctuations of the domain contour. Slight roughness in the contour rapidly increases the perimeter and leads to non-representative, much lower circularity values. While the resulting SEM-images already suggest certain trends, the derived values provide quantitative comparisons, as additional knowledge.

3. Results and Discussion

From the laser melting experiments, selected nuggets were studied by SEM- and EDS-analysis, as described above. The SEM-images of the six cases are shown in Fig. 3. Note the much smaller scale for the Si-cases because the Fe-domains remained small, while Al generated significantly larger Fe-domains. Essential is also the different appearance, including the shapes and the arrangement of the Fe-domains, in the phases compound. The qualitative appearances were already discussed in the initial publications [1-3], as well as the chemical compositions from EDS-analysis.

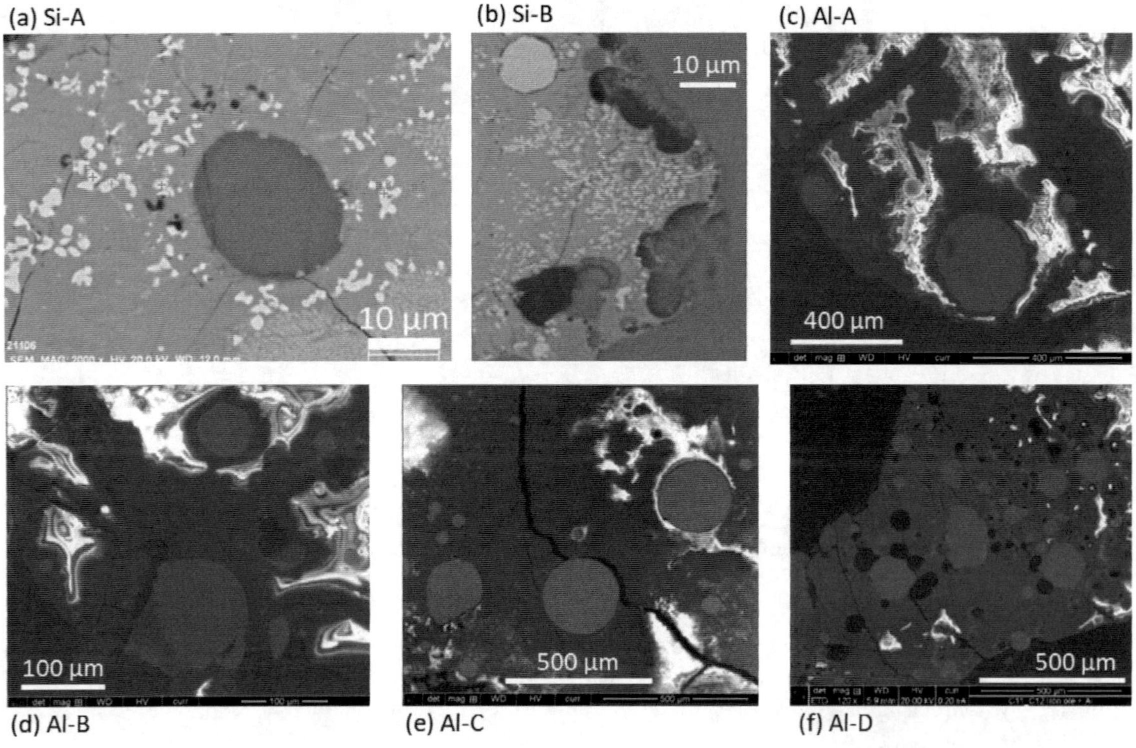

Figure 3. SEM-images: (a) case Si-A, (b) Si-B, (c) Al-A, (d) Al-B, (e) Al-C, (f) Al-D.

An example of EDS-measurements is given in Fig. 4. A large Fe-domain can be clearly distinguished from another large domain that is, despite a similar grey shade appearance, an FeAl intermetallic phase instead, all embedded in an Al_2O_3-matrix. The highly Fe-rich domains are evidence that significant reduction of iron ore to iron has taken place.

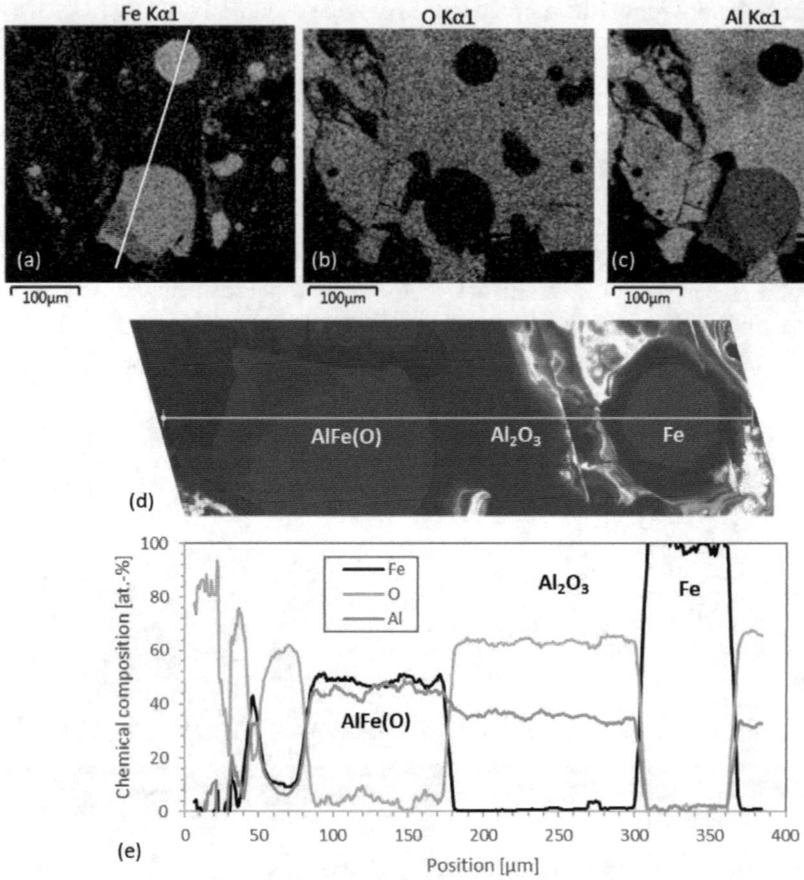

Figure 4. Example of the chemical distribution across the samples, case Al-B: (a)-(c) EDS-map scans for Fe, O and Al, respectively, (d) location for a linescan (see also (a)), (e) chemical composition Fe-O-Al along the EDS-linescan.

The largest iron-rich domains in the images were evaluated with respect to area, aspect ratio, roundness and solidity. Figure 5 shows examples how the SEM-images appear after applying a greyscale threshold, followed by numbering and evaluating separate domains, by the image analysis software.

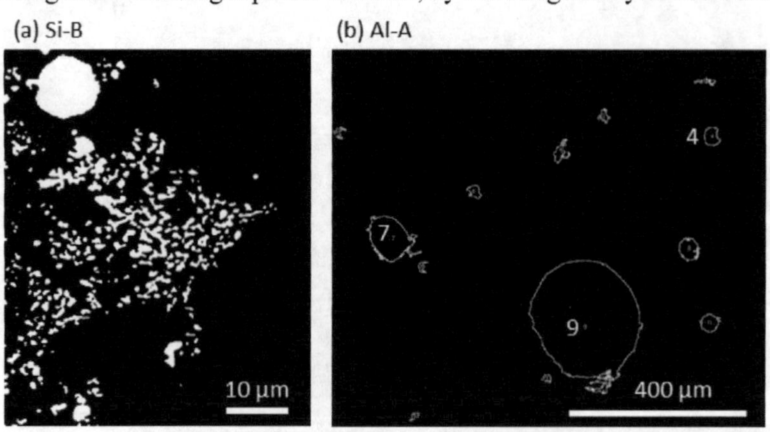

Figure 5. Examples displaying the Fe-domains after image greyscale threshold processing: (a) case Si-B, filled areas (step 1), (b) case Al-A, separated and numbered circumference loops (step 2).

From the varying data for the six cases, the respective mean values and the (absolute) standard deviations are given in Table 3. The cases Al-A,B,C are grouped together because being similar and providing less data. The histograms in Figs. 6 and 7 display the variations in more detail.

For Si, the bright domains are the iron rich domains (surrounded by a slag matrix) while for Al the grey, roundish domains are highly iron-rich (surrounded by Al_2O_3). Figure 3(b) shows a darker Si-rich domain in the centre, apparently attracting Fe-domains. In the Al-cases Fig. 3(c)-(f), bright, shiny Al-domains can be seen. The Al-case Al-D in Fig. 3(f) also displays dark domains of porosity. In contrast, the dark domains in case Si-B originate more likely from particles that have dropped out during polishing, since the scale is much smaller.

Table 3. Measured shape properties (mean value and absolute standard deviation) for the different SEM-image cases studied (Al-A,B,C grouped, since very similar and less data).

Case	Area [µm^2]	Aspect ratio [1]	Roundness [1]	Solidity [1]
Si-A	2.47±1.62	2.04±0.671	0.538±0.164	0.760±0.129
Si-B	5.75±8.21	2.22±0.905	0.504±0.159	0.721±0.119
Al-A,B,C	7886±17 208	1.22±0.403	0.857±0.139	0.883±0.107
Al-D	9269±9672	1.17±0.128	0.866±0.0868	0.887±0.0472

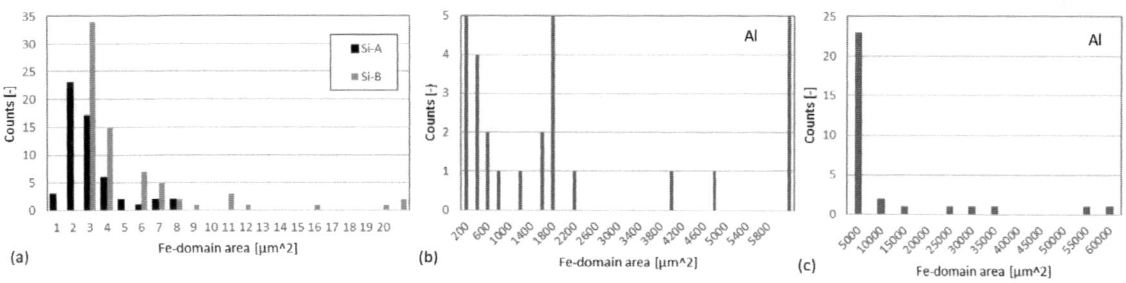

Figure 6. Histograms of the Fe-domain areas, divided into three scale ranges, for (a) Si, (b),(c) Al.

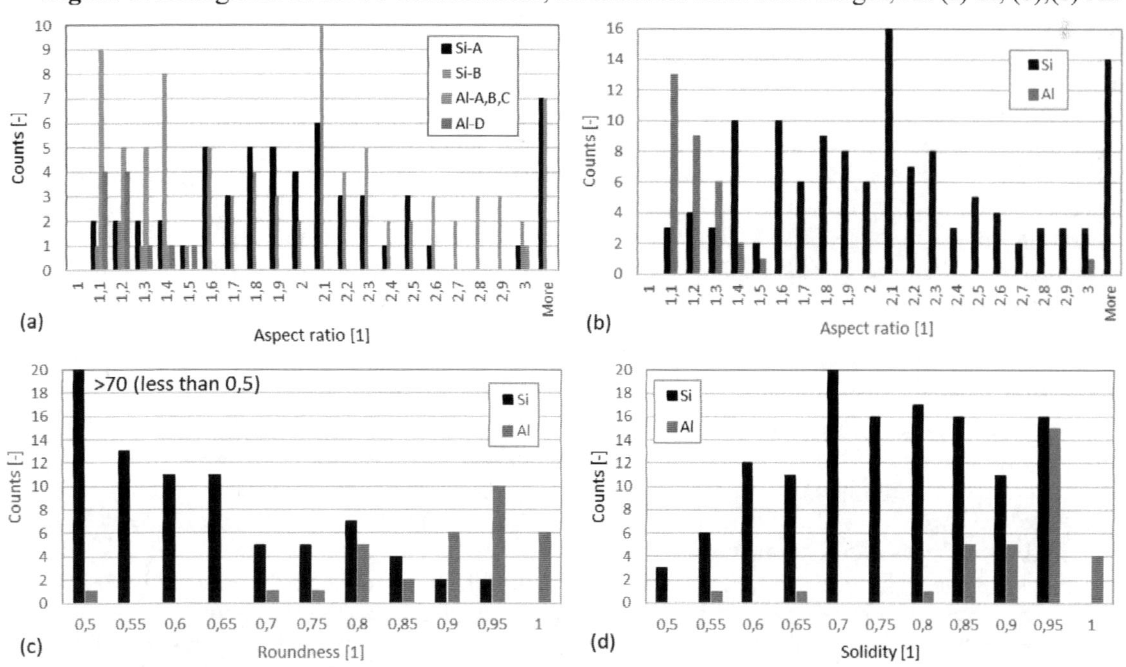

Figure 7. Histograms: (a) aspect ratio, divided into two Si- and two Al-case groups, (b) aspect ratio, all Si vs Al, (c) roundness Si vs. Al, (d) solidity Si vs. Al.

While the above evaluations enabled to quantify certain aspects of the domain shapes, another interesting aspect is the locations of and distances between domains. This is somewhat more difficult to quantitatively evaluate. Figure 8 shows a so-termed "geometry-to-distance-map" (a function provided by the ImageJ software), applied for the case Si-A. For this case many Fe-domains are located close to a larger Si-rich domain. The maps further emphasize the trend, while further quantitative evaluation of this correlation is desirable. Apparently, the Si-domain has a catalyst role to promote ore reduction and Fe-domain formation in its vicinity, though keeping some distance, while other regions do not show Fe-domains at all. These correlations support the understanding of the mechanisms that have taken place, and the corresponding conditions, see also the theoretical discussions in [3].

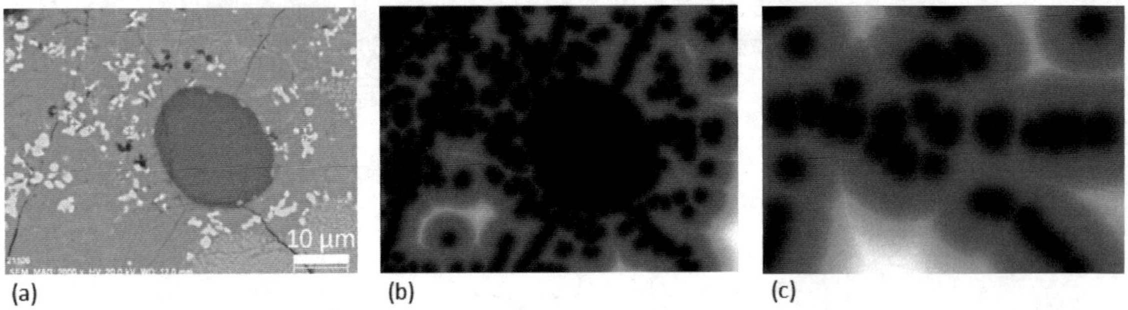

Figure 8. Case Si-A: (a) SEM-image, (b) geometry-to-distance map for threshold 150 (for the greyscale 0-255), (c) geometry-to-distance map for threshold 100.

Summarizing, the apparent differences of the Fe-domain shapes between Si and Al as reducing agents can here be quantitatively expressed, from image analysis. Al as reducing agent has easily led to much larger Fe-domains than Si, which can be straightforward expressed by the domain areas, up to three orders of magnitude larger.

For Al, the trend to circular smooth domains is reflected by all three properties, aspect ratio, roundness and solidity, all approaching to the ideal value 1. Although Si occasionally generates circular domains, too, mainly less regular shapes can be found. One possible explanation is that the domains initially grow spherically but the melt conditions enable contact between liquid Fe-domains and coalescence, just before solidification, hence without smoothening of the new larger body. The aspect ratio of these irregular shapes is significantly larger (mean value 2.1 for Si compared to 1.2 for Al), with more widespread variation (standard deviation ±0.75 for Si compared to ±0.25 for Al). For Al, different conditions in the melt apparently do not lead to these irregularities from coalescence along with solidification. Aluminium has a much lower melting temperature than iron oxide and aluminium oxide, which means that the latter two solidify much earlier, then surrounded by liquid aluminium.

While the aspect ratio describes the spatial elongation of the domains, the solidity can indicate concave-like geometry details. Roundness much lower than 1 would indicate sharper, corner-like features. For a more extensive statistical analysis of domains, these different properties can be useful to characterize the types of irregular geometry features, which can provide information about the process mechanisms, particularly smoothening by interface tension, growing of reduction zones and solidification of the multiple phases.

While the cases Al-A,B,C tend to smooth circles for the Fe-domains, for Al-D a rougher circumference can be seen, though yet very circular main geometries. In this case also significant porosity is visible. In high speed imaging strong explosion-like melt movement events along with spatter ejection can occasionally be observed for Al, probably generated by the strong thermite reaction for Al with iron ore. Intense boiling during these events can cause the porosity, maybe accompanied by rougher Fe-domains.

4. Conclusion

By quantitative analysis of the SEM-images from laser-induced reduction of iron ore powder, the following conclusions can be drawn:

(i) Al as the reducing agent powder has led to significantly larger Fe-domains than Si-powder, up to approximately three orders of magnitude larger. One explanation can be the apparently strong chemical thermite reaction of Al with iron ore, which was not observed for Si.

(ii) For Al, the Fe-domains are more circular than for Si (e.g. average aspect ratio 1.2 compared to 2.1), with a smooth contour; for Si, likely coalescence of drops shortly before solidification, at similar melting temperatures, can lead to irregular shapes, while Al might provide longer equilibrium periods for sphericity; however, occasionally Al as reducing agent also formed pores, then associated with less smooth though spherical Fe-domains.

(iii) In addition to domain area, the image analysis properties aspect ratio, roundness and solidity proved suitable to characterize different domain appearances, particularly circularity, among other features. These properties enable comprehensive statistical evaluation and characterization. In contrast, the perimeter for direct circularity derivation was too unreliable and not robust for evaluation if the image resolution is limited.

(iv) For Si, Fe-rich domains preferably formed close to Si-rich domains, while such distinct catalyst role was not visible for Al. Statistical distance quantification between domains is more complex than domain shape quantification.

Acknowledgements

The authors would like to express their gratitude to *Energimyndigheten*, the *Swedish Energy Agency*, for funding the projects AM-ORE (no. 51021-1) and AMORF (no. P2022-00202) in frame of the programme *Industriklivet*, and to the experts from the company LKAB who supported this research.

References

[1] Fedina T, Brueckner F, Kaplan A F H and Wilsnack Ch 2023 *J. Laser App.* **35** 022007
[2] Fedina T 2023 *Towards sustainability in additive manufacturing: material and process aspects*, Doctoral thesis (Luleå: Luleå University of Technology)
[3] Kaplan A F H, Fedina T and Brueckner F 2022 *Proc. CIRP* **111** 377
[4] Chung Y and Cramb A W 2000 *Metall. Mater. Trans. B* **31B** 957
[5] Fischedick M, Marzinkowski J, Winzer P and Weigel M 2014 *J. Clean. Prod.* **84** 563
[6] Warner N A 2014 *Metall. Mater. Trans. B* **45B** 2080
[7] Kim K-Y, Jang E-J, Kim D-Y and Jung S-M 2017 *Ironm. Steelm.* **44** 6
[8] Souza Filho I R, Ma Y, Kulse M, Ponge D, Gault B, Springer H and Raabe D 2021 *Acta Mat.* **213** 116971
[9] Gavilán H, Posth O, Bogart L K, Steinhoff U, Gutiérrez L and Puerto M 2017 *Acta Mat.* **125** 416
[10] Ndlela S C and Shanks B H 2006 *Ind. Eng. Chem. Res.* **45** 7427
[11] Mohamed M S, Yabe T, Baasandash C, Sato Y, Mori Y, Shi-Hua L, Sato H and Uchida S 2008 *J. Appl. Phys.* **104** 113110
[12] Hidayat T, Shishin D, Decterov S A and Jak E 2017 *J. Phase Equilib. Diffus.* **38** 477
[13] Jung I-H, Decterov S and Pelton A D 2004 *Metall. Mater. Trans. B* **35B** 493
[14] Heulens J, Blanpain B and Moelans N 2011 *Chem. Geology* **290** 156
[15] Park J H and Zhang L 2020 *Metall. Mater. Trans. B* **51B** 2453
[16] Ni P, Tanaka T, Suzuki M, Nakamoto M and Jönsson P G 2019 *ISIJ Int.* **59** 737
[17] Durães L, Costa B F O, Santos R, Correia A, Campos J and Portugal A 2007 *Mater. Sci. Eng. A.* **465** 199
[18] Travitzky N, Kumar P, Sandhage K H, Janssen R and Claussen N 2003 *Mater. Sci. Eng. A.* **344** 245
[19] Kawasaki T, Adachi T, Ohfuji H and Osanai Y 2019 *J. Mineral. Petrol. Sci.* **114** 238
[20] Fan R H, Lü H L, Sun K N, Wang W X and Yi X B 2006 *Thermochim. Acta* **440** 129

[21] Mei J, Halldearn R D and Xiao P 1999 *Scr. Mater.* **41** 541
[22] Cheng J 2012 *Reaction kinetics study of Al-Fe_2O_3 nanoenergetics*, Doctoral thesis (Singapore: Nanyang Technological University)
[23] Wang L L, Munir Z A and Maximov Y M 1993 *J. Mat. Sci.,* **28** 3693

Correlation of signal feature importance and process parameters in water jet-guided laser cutting

Roland Axel Richter[1],*, Falk Braunmüller[2], Vigneashwara Pandiyan[1], Sergey Shevchik[1], Toni Ivas[1], Amédée Zryd[2]

[1] Laboratory for Advanced Materials Processing (LAMP), Swiss Federal Laboratories for Material Science and Technology (EMPA), Thun, CH-3602, Switzerland
[2] Synova SA, CH-1266 Duillier, Switzerland

E-mail: roland.richter@empa.ch

Abstract. Laser-based cutting can deliver high accuracy and speed during the process. However, there are still some technical challenges to be solved. First, the process itself requires precise focusing of the laser beam. When using continuous wave (CW)- or pulses in the µs- to ns-range for efficient materials removal, not only the targeted material volume is heated, but also the surrounding material due to heat diffusion. This decreases the quality of the cut. Coupling the light into a water-jet solves both issues: It ensures the delivery of focused light to the target surface, while cooling the affected material during processing. This work proposes a real-time monitoring system of the cutting process by exploiting unique physical phenomena related to the process. The water jet allows both the propagation of laser light towards the target while keeping it focused and the back-propagation of process light through the focusing system. This can be captured using suitable sensors, and saved using a data acquisition system. The acquired data can be used for process characterization using signal processing. In this work we demonstrate the direct correlation between different process parameters and statistical signal features. We show that reducing the signal to a few statistical features both in time and frequency by feature extraction does not reduce the information content, but instead makes it more robust to mis-classification while decreasing the classification time. It also opens up a wide range of future applications not only to process data, but also to control the process more precise.

1. Introduction & state of the art

Laser cutting of materials is a wide-spread technology employed as an essential part of production lines for automotive industry, electronics, aviation, and others [1, 2]. However, even though it has numerous advantages, it also bears several drawbacks. Depending on laser- and system parameters not only the process zone will be affected by heat, but also material surrounding the process zone. This can influence both material parameters outside the process zone and the behavior of the process zone itself in an uncontrollable way. Molten material from the process zone can be expelled out of the cut and deposited next to the cut, resulting in surface contamination. Finally, the quality of the cut is limited by variations of the focusing of the optics. These drawbacks constrain the material type and -geometry which can be cut [3–5], and require precise adjustments throughout the cutting process.

IOP Conf. Series: Materials Science and Engineering 1296 (2023) 012044 doi:10.1088/1757-899X/1296/1/012044

To solve those issues the water jet-guided laser cutting process (WJGL) was proposed [6], where a laser beam is focused into and thereby coupled into a high-pressure water jet. This technology addresses the issues of laser cutting mentioned above. Instead of the numerical aperture of the optics being the limiting factor for the aspect ratio, the water jet acting as a multi-mode fiber defines the geometry of the cut. This allows reaching significantly higher aspect ratios [7, 8].

The water added to the process also provides active cooling, which reduces the heat-affected zone [3, 7, 9] and removes accumulated waste from the groove [10]. In addition, it has been proven that this additional cooling can reduce the amount of cracks and decrease chipping [5, 8].

These benefits allow the water jet-guided laser cutting process to be used for a wide range of applications, such as precise micro-machining [11–13], high-efficiency precision cutting of metals and composites [8, 11, 13–15], ceramics [5], semiconductors [4, 10] and other exotic materials [9].

The complexity added by the water jet, however, adds new physical phenomena in the form of noise and perturbations. These can influence the resulting quality and speed. The water jet breaks up as soon as it leaves the laminar flow regime, which limits the maximum working distance [12, 16]. Decreasing the nozzle quality will negatively influence the water jet, which in turn also reduces the working distance or makes cutting completely impossible. During material processing the plasma plume resulting from the interaction of the laser pulse with material leads to scattering, which decreases the process efficiency with up to 40 % to 50 % [10, 17]. These plasma plumes are not confined to only the bottom of the cut (when originating from the interaction between target material and laser). They can also form higher within the water jet depending on process parameters due to the non-linear interaction between laser pulse and water, which decreases the amount of light for useful processing [18].

This process is in use in industrial applications already. However, by adding constant monitoring covering a broad range of emissions it is possible to increase process stability even further, optimize the machining time and allow the integration of predictive maintenance. To the best of our knowledge no such method was reported in literature up to now. In this study, we therefore report for the first time monitoring of a water jet-guided laser cutting process, using multiple sensors to monitor both acoustic and optical emissions simultaneously. This allows a multi-faceted monitoring approach, which increases both the range of events which can be monitored and the accuracy of event identification. It also allows an improved understanding of the complex process due to the multi-faceted approach.

As the amount of collected data typically is large, one usually can not extract relevant points which can be correlated to events of interest from the raw data. Thus, it is important to reduce the acquired data to only the most relevant parts, resp. features, such that a correlation between observed events and acquired data can be made. These features are then a sparse representation of the data either in time, in frequency, or both. For this, it is necessary to know which features are relevant to be extracted, to avoid to either increase computational complexity or missing the features relevant for the task on hand.

In this manuscript we demonstrate that it is possible not only to reduce acquired data from multiple channels into a sparse representation, i.e. selected features [19], but also to deduce the importance of different data acquisition channels for the water jet-guided laser cutting process. We also show the feasibility of transferring acquired knowledge about signal behavior based on extracted features from one application to a different application, which reduces complexity and time necessary for data generation and -analysis. This in turn can improve productivity and efficiency of the process.

2. Experimental setup

For acquiring data during the process Synova SA provided a water-jet laser cut machine, with a pulsed-mode laser operating at 532 nm. This machine was sensorized with both optical and acoustic sensors. The sensor for acoustic emission (AE) was mounted directly on the coupling unit for coupling the laser beam into the water jet, as shown in Figure 1. The optical sensors are coupled into the optical path of the back-scattered process light propagating from the work piece through the water jet into the coupling unit.

Figure 1: Acoustic emission sensor (marked in bright green) mounted on the coupling unit (marked in red), which is connected to the water jet nozzle (marked in orange)

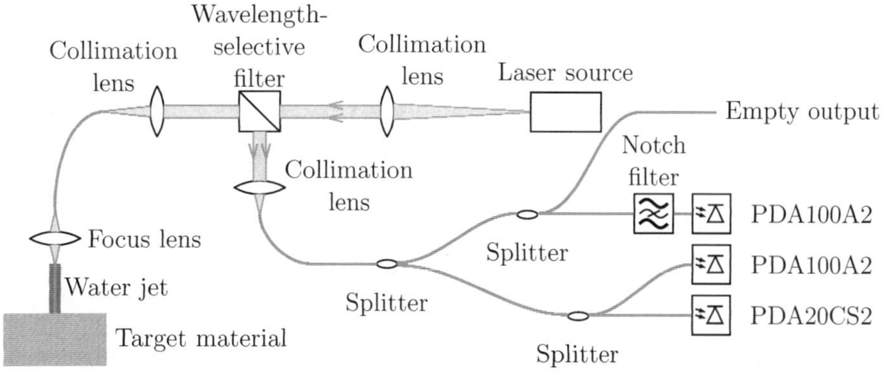

Figure 2: Schematic of setup for detecting optical emissions. The back-scattered process light is guided into the detection system by the wavelength-selective filter. Each splitter is splitting the incoming light by 50 %/50 %, before guiding it into the different optical sensors

The direct mount of the AE sensor on the water-jet nozzle allows the acquisition of structure-borne acoustic emissions from both the machine itself and the process zone. The process emissions can propagate along the water jet upwards towards the nozzle, where the sensor can detect them. This has the benefit of a material-/process-independent sensor placement with a near-constant distance between sensor and process zone, regardless of position of head and work piece relative to each other.

A similar benefit results from the placement of the optical sensors. As they operate purely on light propagating backwards through the water-jet and the focusing lens, the process zone will always be in focus without having to rely on additional external tracking. Optical emissions

IOP Conf. Series: Materials Science and Engineering 1296 (2023) 012044 doi:10.1088/1757-899X/1296/1/012044

occurring outside the process zone will be ignored, which is an additional benefit. Acoustic emissions occurring outside the process zone will be detected, but strongly attenuated.

The light generated by the process itself and back-scattered laser light is initially frequency-filtered by the focusing/combining optics inside the laser head. This removes a significant amount of the emitted spectrum from propagating to the optical sensors, which makes spectroscopic measurements through the optical head impossible. Fiber splitters with a splitting ratio of 50 %/50 % are used to split the process light into four sub-channels. These are separated into three sensors (Thorlabs PDA100A2 at 40 dB with $\lambda =$ [320 nm:1100 nm], Thorlabs PDA100A2 with notch filter at 70 dB with $\lambda =$ [320 nm:1100 nm], except the main laser line and the main Raman line originating from the interaction between water and laser, and Thorlabs PDA20CS2 at 70 dB with $\lambda =$ [800 nm:1700 nm]), and one open channel, as displayed in Figure 2.

The data from both optical and acoustic channels was digitized using an Elsys TPCEE-2016-8S DAQ module, operating at 1.17 MSample/s per channel, and saved using a custom LabVIEW-software. In this manuscript the acoustic channel corresponds to channel 0, the unfiltered visible light (320 nm to 1100 nm) channel corresponds to channel 1, the infrared light (800 nm to 1700 nm) corresponds to channel 2 and the sensor operating in the visible range together with a filter to remove the laser line corresponds to channel 3.

3. Data acquisition & treatment

To acquire data for processing, two different materials (silicon and steel) were cut at pre-defined process parameter sets, listed in Table 1. The parameters were chosen as they are known for efficient cutting in the selected materials.

Table 1: Experimental parameters used for data acquisition using the setup described in Section 2

Material	Experiment	Material thickness	Repetition rate	Power	Intensity	Gas flow
Silicon	1	3 mm	4 kHz	38 W	1.3 GW/cm^2	1.5 L/min
	2	3 mm	4 kHz	38 W	1.3 GW/cm^2	2.5 L/min
Steel	3	1 mm	4 kHz	38 W	1.3 GW/cm^2	1.5 L/min

For each experiment the cutting process was continued until the material was cut completely. This required multiple passes for each experiment, as a single pass was not sufficient to cut through the entire sample thickness at once. An example of the acquired signal for experiment 1 is shown in Figure 3.

After finishing data acquisition both an in-house custom library and the library tsfresh (Time Series FeatuRe Extraction on basis of Scalable Hypothesis tests [20]) were used for time series feature extraction and analysis of the acquired data, followed by an importance analysis based on a Random Forest Model. The Random Forest model was imported from scikit-learn [21].

The importance analysis of the different features was based on two label options:

- Difference in pass number for same material (c.f. Section 4.1)
- Difference of material (c.f. Section 4.2)

Using the tsfresh-library 196 different features were extracted from the signal. To investigate their relevance and their impact on the accuracy of the classifier, the features were ranked following their importance given by a trained RandomForestClassifier, the accuracy of the trained classifier using a test data set (distinct from the training set) was determined, and afterwards the least-relevant feature was removed from the entire database. This procedure

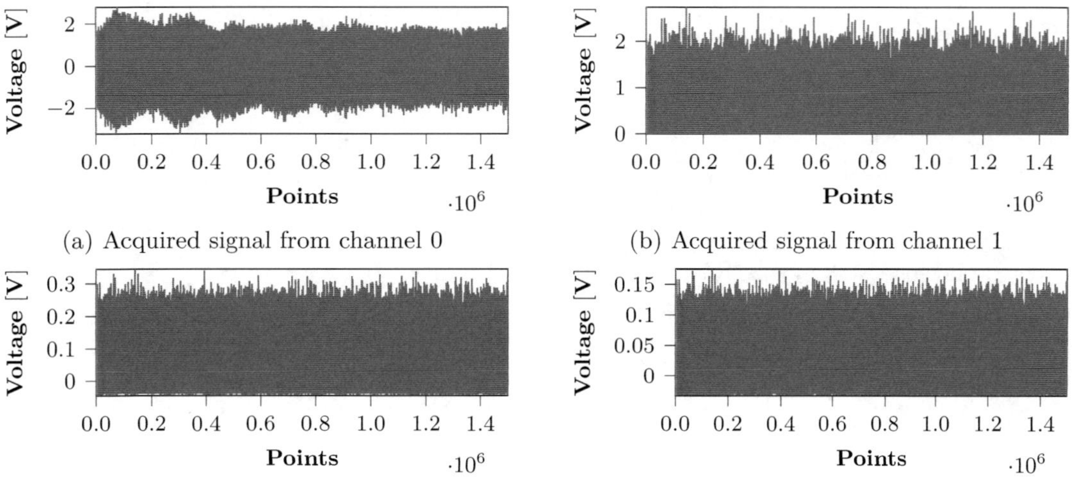

(a) Acquired signal from channel 0

(b) Acquired signal from channel 1

(c) Acquired signal from channel 2

(d) Acquired signal from channel 3

Figure 3: Example data for experiment 1, corresponding to the sensors listed in Section 2

was repeated until all features were removed from the database. To increase accuracy this procedure was repeated ten times, with changing initial values for the randomizer in both RandomForestClassifier and the splitting-function to generate training and testing set. The obtained results for the relation between used number of features and accuracy were averaged, and three thresholds were defined for determining the necessary amount of features, depending on the achieved maximum accuracy: A classification accuracy larger or equal than 70 % (labeled as **ACR70** within this study), larger or equal than 80 % (labeled as **ACR80**) and larger or equal than 90 % (labeled as **ACR90**).

4. Results

The observations and achieved results are sub-grouped into two parts following the objective detailed above, starting with the identification of pass number. For each sub-goal a short overview over achieved accuracy is given, together with the minimum number of features necessary for achieving **ACR70**, **ACR80** or **ACR90** (if applicable). The most relevant features are described together with their distributions, showing the change of the corresponding feature for different target labels. The importance of emission type (i.e. optical or acoustic) is analyzed. This can be used to optimize the number of sensors for future applications and to investigate their relevance for improving the classification accuracy.

4.1. Identification of pass number

Water jet-guided laser cutting requires multiple passes for cutting entirely through thicker materials. This makes it important to know when a certain section of the cutting path already has been cut through, to avoid spending time on retracing this segment multiple times. Knowing the current cutting depth also helps to correlate the process parameters with the efficiency of the process. If multiple passes over a target section do not sufficiently increase the cutting depth, the process parameters at that depth can be labeled as "inefficient", and must be adjusted.

For the material thickness used in the experiments in this study a direct correlation can be found between the pass number and the thickness. Therefore, it is important to be able to identify the pass number purely based on the signal, and to identify the corresponding features which can be used for separating the different pass numbers from each other.

IOP Conf. Series: Materials Science and Engineering 1296 (2023) 012044 doi:10.1088/1757-899X/1296/1/012044

For both experiment 1 and experiment 2 in Table 1 16 passes each are necessary for achieving full breakthrough. The relevance of features to differentiate between different passes is determined using a Random Forest classifier with 100 estimators. For training the generated database of all features was split into 75 %/25 % training/testing, resulting in 17925 entries in the database. With the used features and a random forest classifier it was not possible to achieve **ACR80**, while at least 21 features were necessary to achieve **ACR70** for both experiment 1 and 2. The correlation between number of features and achieved accuracy is shown in Figure 4a for experiment 1 and in Figure 4b for experiment 2, together with an indicator demonstrating the threshold for reaching **ACR70**.

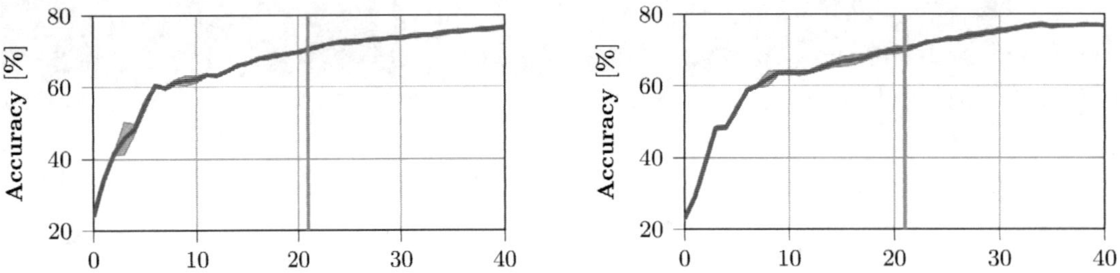

(a) Correlation for experiment 1

(b) Correlation for experiment 2

Figure 4: Correlation between number of features (along the x-axis for both figures) and achieved accuracy for a random forest classifier for experiment 1 and 2, following Table 1. The orange line determines the threshold for **ACR70** (21 features for both experiment 1 and 2). **ACR80** is not achieved. As the achieved accuracy does not increase significantly above the percentage value achieved at 40 features, the remaining 156 features are not shown here to improve readability

For both experiments it is possible to identify the features contributing most to the achieved accuracy. As those features are calculated separately for each channel, one can correlate the importance of a single channel for the correct identification of the current process stage.

For both experiments the first five most relevant features were identified, and are listed in Table 2. The randomizing process lead to a variation of which features are most relevant for identifying the process stage, resulting in more than five different features in both tables. This was taken into account by averaging the corresponding contribution.

Table 2: Most relevant features for identifying the process stage for experiment 1 on the left and for experiment 2 on the right, with the average contribution to classification accuracy (ACC) for each feature

Feature name	Channel	ACC	Feature name	Channel	ACC
Change of absolute mean	0	55 %	Change of absolute mean	0	55 %
Variance	0	42 %	Absolute energy value	0	35 %
Kurtosis	0	31 %	Skewness	0	30 %
Absolute energy value	0	30 %	Variance	0	29 %
Root mean square	0	22 %	Root mean square	0	27 %
Standard deviation	0	22 %	Benford correlation	2	23 %
Change of mean	1	18 %	Standard deviation	0	22 %
Skewness	0	12 %	Kurtosis	0	14 %
Median	2	6 %	Change of mean	1	2 %

It is possible to correlate the most relevant features with the corresponding sensor type. The channel contributing most to relevant features is channel 0, corresponding to the acoustic structure-borne sensor. For both experiments the acoustic sensor contributed with 78 % to the total amount of relevant features, indicating its importance.

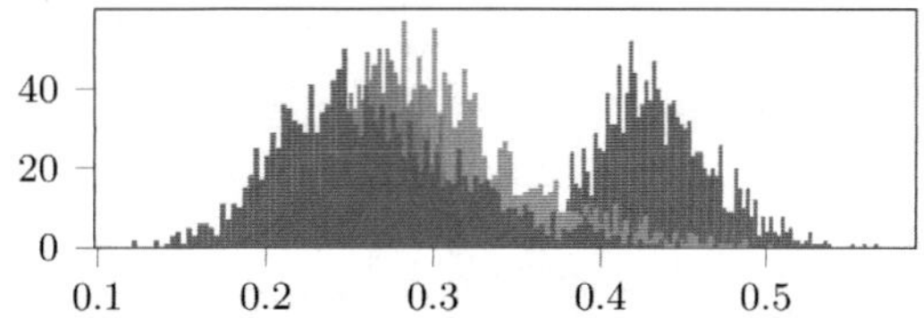

(a) Mean of absolute changes distribution for experiment 1

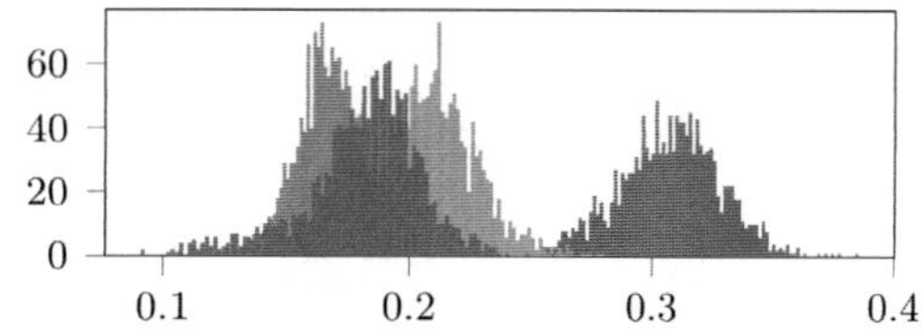

(b) Mean of absolute changes distribution for experiment 2

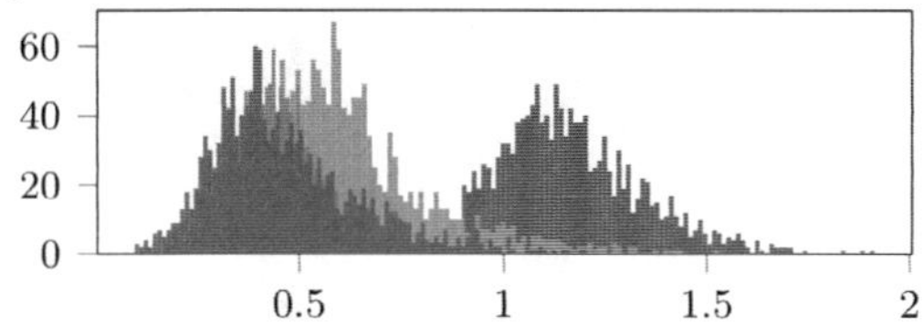

(c) Variance distribution for experiment 1

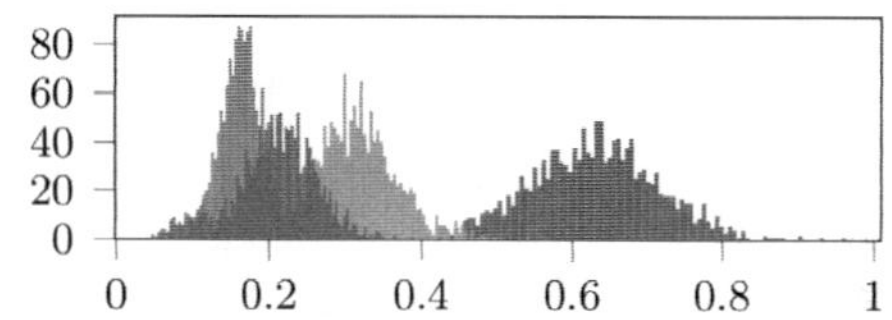

(d) Variance distribution for experiment 2

Figure 5: Distribution of the mean of absolute changes and variance for experiment 1 (in Figures 5a and 5c) and experiment 2 (in Figures 5b and 5d) across the first four target classes. The first target class **Line I** is given in blue, the second target class **Line II** in orange, the third target class **Line III** is given in green, and the last target class **Line IV** in red.

For each of the features mentioned in Table 2 it is possible to also generate the distribution of each feature, separated into the labels the features are divided in. These are shown in Figures 5a and 5c for experiment 1 and in Figures 5b and 5d for experiment 2.

By eye it is possible to differentiate the different classes for both experiment 1 and 2 in Figure 5, and to identify a trend towards smaller values for both the mean of the absolute change and the variance. This correlates closely with the water acting as damping mechanism, which thereby leads to a reduced acoustic emission when increasing the depth of the cut.

This effect is visible already during the first few passes, and thereby sufficient to at least give an initial idea of the current process stage. In comparison, as demonstrated above in Table 2, the optical back-reflection does not change sufficiently to be able to distinguish the pass numbers already during the first passes.

4.2. Material identification

To examine the changes in the signal when moving from one material to a different material, experiment 1 was compared with experiment 3, with the main difference between them being the change of workpiece material from silicon to steel. Following the approach described in Section 3 the correlation between used features and achieved accuracy could be determined, with the corresponding curve being shown in Figure 6. In addition to the average value of the accuracy, the standard deviation is given.

The features relevant for achieving **ACR90** to differentiate between silicon and steel in experiment one and three are listed in Table 3. As the calculation was repeated several times with different initial values for the random shuffle, the average contribution to accuracy of each feature is given as well.

As not every feature was within the first five relevant features for each iteration, the contribution percentages can not be summed up to 100 % within Table 3. The significant

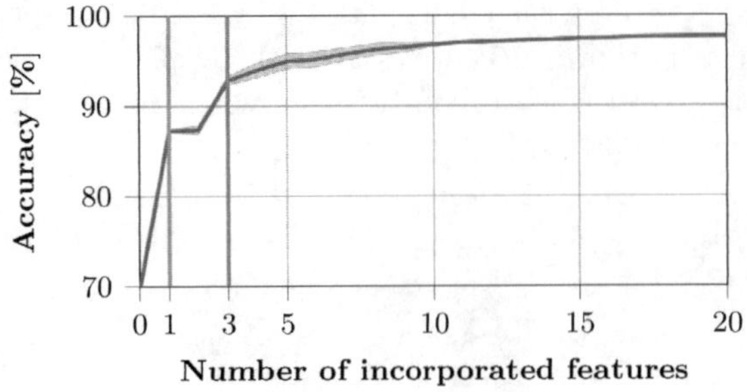

Figure 6: Correlation between number of features and achieved accuracy for distinguishing between experiment one and three. The orange line determines the threshold for **ACR80**, and the green line the threshold for **ACR90**. The x-axis is limited at 20 features, as no improvement could be detected for adding more features. Standard deviation from ten iterations is given in light orange

Table 3: Most relevant features for differentiating between silicon and steel for similar experimental parameters, sorted following the ACC

Feature name	Channel	ACC	Feature name	Channel	ACC
Root mean square	0	50 %	Variance	0	9 %
Absolute energy	0	46 %	Median	2	6 %
Kurtosis	0	44 %	Number crossing m_0	2	6 %
Standard deviation	0	13 %	Kurtosis	3	5 %
Benford correlation	2	12 %	Number crossing m_1	0	2 %
Number crossing m_{-1}	0	12 %	Mean	2	2 %

contribution of channel 0 (corresponding to the acoustic channel) is visible, compared to the optical channels.

Similar to Section 4.1 the distribution of the first four relevant features, root mean square, absolute energy, kurtosis and standard deviation (all from channel 0) can be generated, and is shown in Figure 7. The distributions for both root mean square and standard deviation show a significant difference between steel and silicon, with the distribution of both for steel being shifted towards larger values. This trend can also be observed for both absolute energy in Figure 7b and Figure 7c, with steel in general shifting the resulting distribution towards higher values, but a narrower peak.

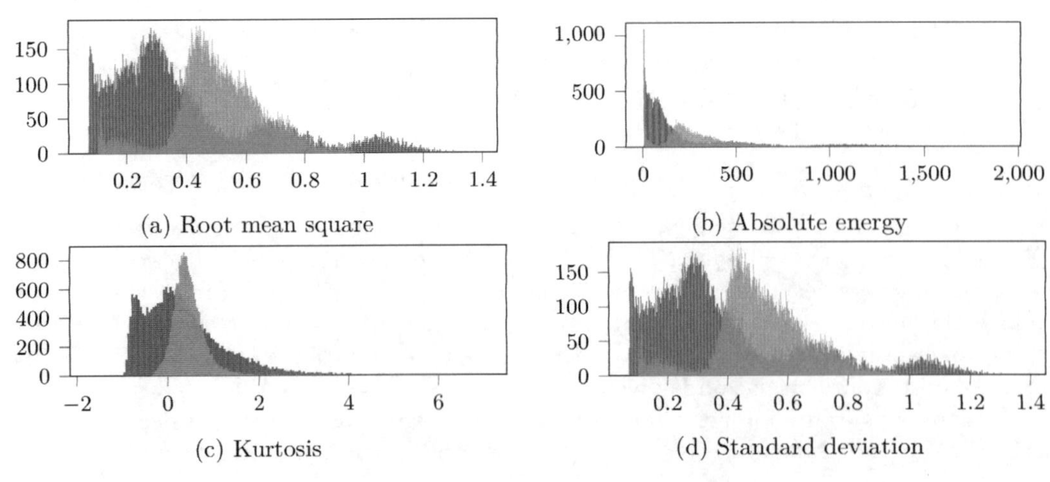

Figure 7: Distribution of the root mean square (Figure 7a), absolute energy (Figure 7b), kurtosis (Figure 7c) and standard deviation (Figure 7d) to distinguish between silicon and steel. The target class silicon is given in blue, the target class steel is given in orange

5. Conclusions & Outlook

Monitoring of water jet-guided laser processing is an important task to ensure reproducibility and stability of the process. There is, however, always a trade-off between the amount of sensors and acquired data, and the achieved result. Adding additional sensors will increase the amount of data to be analyzed and processed, while achieving only diminishing returns in accuracy improvement and increasing the computation complexity. In industrial settings, it will also increase the complexity and cost of the machine to be sold afterwards, which reduces the benefit achieved by adding active monitoring.

In this study we investigated monitoring of a water jet-guided laser process for the first time using a sensor-fusion based approach, which simultaneously incorporates both acoustic and optical emissions, and machine learning based on a simple classification model. To acquire a wider range of data, we used sensors for both optical and acoustic emissions. The data acquisition method was designed with off-the-shelf, scalable equipment. We covered two industry-relevant applications, detection of cutting status and material detection, and investigated the obtained data with respect to features relevant for classification. The relevance of the extracted features for classifying a process state was calculated using a RandomForest-model.

We could demonstrate that an acoustic sensor acquiring structure-borne acoustic data from the nozzle can acquire sufficient data for classifying both targeted applications, process-stage and material identification. Additional optical sensors operating on the back-scattered laser- and process light only provide supporting information, which is not sufficient for achieving adequate accuracy by itself when using a RandomForest-model.

However, with the used classifier it was not possible to achieve an accuracy $\geq$ 80 % for differentiating the pass number, as demonstrated in Section 4.1, even by combining all channels. This can be alleviated in a future step by changing the RandomForest-classifier to a more advanced classifier such as XGBoost [22].

Using feature-based classifiers such as RandomForest or XGBoost requires a pre-processing step, to extract features from acquired data and to reduce the amount of features to the amount necessary for accurate classification. This adds additional workload to the operator, and risks that features which are essential for identifying target classes are not taken into account during the initial pre-processing. This, in turn, can then lead to a worse performance of the used classifier. A solution for this can be a neural-network based classifier, such as ones based on PyTorch [23] or TensorFlow [24]. As those classifiers do not require a pre-processing step to extract features, but rather identify relevant features by itself, it is possible to achieve both a higher accuracy and a lower workload. Still, this also decreases the understandability of the process, as this "feature extraction" is less transparent compared to an explicit approach, as used in this study. To achieve both, high classification accuracy and an understanding of the process, a combined approach might therefore the best solution.

Acknowledgments

We would like to thank Synova SA for providing us with a WJGL-machine, which could be used for sensorization tests and data acquisition. For financial support Innosuisse - Swiss Innovation Agency with project #58389.1 IP-ICT is acknowledged.

References

[1] Genna S, Menna E, Rubino G and Tagliaferri V 2020 *Applied Sciences* **10** 4956 ISSN 2076-3417
[2] Naresh and Khatak P 2022 *Materials Today: Proceedings* **56** 2484–2489 ISSN 2214-7853
[3] Yang L J, Wang M, Wang Y, Tang J and Chen Y B 2009 *Adv. Mater. Res-switz.* **69-70** 333–337 ISSN 1662-8985
[4] Zhu H, Wang J, Li W Y and Li H Z 2014 *AMR* **1017** 193–198 ISSN 1662-8985
[5] Chang Y 2018 *IOP Conf. Ser.: Mater. Sci. Eng.* **439** 032001 ISSN 1757-899X
[6] Richerzhagen B 1999 Method and apparatus for machining material with a liquid-guided laser beam

[7] Perrottet D, Housh R, Richerzhagen B and Manley J 2005 Heat damage-free laser-microjet cutting achieves highest die fracture strength ed Fieret J, Herman P R, Okada T, Arnold C B, Bachmann F G, Hoving W, Washio K, Lu Y, Geohegan D B, Trager F and Dubowski J J (San Jose, CA) p 285
[8] Liu Y, Wei M, Zhang T, Qiao H and Li H 2021 *Optics & Laser Technology* **137** 106820 ISSN 0030-3992
[9] Sun D, Han F, Ying W and Jin C 2018 *Procedia CIRP* **71** 71–74 ISSN 2212-8271
[10] Chen X, Li X, Wu C, Ma Y, Zhang Y, Huang L and Liu W 2019 *Applied Sciences* **9** 1882 ISSN 2076-3417
[11] Porter J A, Louhisalmi Y A, Karjalainen J A and Füger S 2007 *Int J Adv Manuf Technol* **33** 961–967 ISSN 0268-3768, 1433-3015
[12] Ng E Y K and Guannan D 2015 *Int J Adv Manuf Technol* **78** 939–946 ISSN 0268-3768, 1433-3015
[13] Adelmann B, Ngo C and Hellmann R 2015 *Int J Adv Manuf Technol* **80** 2053–2060 ISSN 0268-3768, 1433-3015
[14] Li C Q, Yang L J and Wang Y 2012 *Material Science Forum* **723** 476–480 ISSN 1662-9752
[15] Chen X, Xu X, Liu W, Huang L, Li H, Wu C, Mu W and Li X 2020 *Materials* **13** 3206 ISSN 1996-1944
[16] Couty P, Spiegel A, Vago N, Ugurtas B I and Hoffmann P 2004 *Experiments in Fluids* **36**(6) 919–927 ISSN 07234864
[17] Mullick S, Madhukar Y K, Roy S and Nath A K 2015 *International Journal of Machine Tools and Manufacture* **91** 62–75 ISSN 0890-6955
[18] Feng S, Huang C, Wang J, Zhu H, Yao P and Liu Z 2017 *Precision Engineering* **47** 33–45 ISSN 0141-6359
[19] Pandiyan V, Drissi-Daoudi R, Shevchik S, Masinelli G, Logé R and Wasmer K 2020 *Procedia CIRP* **94** 392–397 ISSN 22128271
[20] Christ M, Braun N, Neuffer J and Kempa-Liehr A W 2018 *Neurocomputing* **307** 72–77 ISSN 1872-8286
[21] Pedregosa F, Varoquaux G, Gramfort A, Michel V, Thirion B, Grisel O, Blondel M, Prettenhofer P, Weiss R, Dubourg V, Vanderplas J, Passos A, Cournapeau D, Brucher M, Perrot M and Duchesnay E 2011 *Journal of Machine Learning Research* **12** 2825–2830
[22] Chen T and Guestrin C 2016 XGBoost: A Scalable Tree Boosting System *Proceedings of the 22nd ACM SIGKDD International Conference on Knowledge Discovery and Data Mining* KDD '16 (New York, NY, USA: ACM) pp 785–794 ISBN 978-1-4503-4232-2
[23] Paszke A, Gross S, Massa F, Lerer A, Bradbury J, Chanan G, Killeen T, Lin Z, Gimelshein N, Antiga L, Desmaison A, Kopf A, Yang E, DeVito Z, Raison M, Tejani A, Chilamkurthy S, Steiner B, Fang L, Bai J and Chintala S 2019 PyTorch: An Imperative Style, High-Performance Deep Learning Library *Advances in Neural Information Processing Systems 32* (Curran Associates, Inc.) pp 8024–8035
[24] Abadi M, Agarwal A, Barham P, Brevdo E, Chen Z, Citro C, Corrado G S, Davis A, Dean J, Devin M, Ghemawat S, Goodfellow I, Harp A, Irving G, Isard M, Jia Y, Jozefowicz R, Kaiser L, Kudlur M, Levenberg J, Mané D, Monga R, Moore S, Murray D, Olah C, Schuster M, Shlens J, Steiner B, Sutskever I, Talwar K, Tucker P, Vanhoucke V, Vasudevan V, Viégas F, Vinyals O, Warden P, Wattenberg M, Wicke M, Yu Y and Zheng X 2015 TensorFlow: Large-Scale Machine Learning on Heterogeneous Systems software available from tensorflow.org

AUTHOR INDEX

Institute of Physics
Dirac House, Temple Back
Bristol BS1 6BE UK

ISSN: 1757-8981
ISBN 978-1-7138-8733-1

9 781713 887331

CURRAN ASSOCIATES INC.
proceedings
.com